from Quantum to Cosmos

Fundamental Physics Research in Space

from Quantum to Cosmos

Fundamental Physics Research in Space

Editor: **Slava G. Turyshev**

NASA Jet Propulsion Laboratory
California Institute of Technology
USA

World Scientific

NEW JERSEY • LONDON • SINGAPORE • BEIJING • SHANGHAI • HONG KONG • TAIPEI • CHENNAI

Published by

World Scientific Publishing Co. Pte. Ltd.
5 Toh Tuck Link, Singapore 596224
USA office: 27 Warren Street, Suite 401-402, Hackensack, NJ 07601
UK office: 57 Shelton Street, Covent Garden, London WC2H 9HE

British Library Cataloguing-in-Publication Data
A catalogue record for this book is available from the British Library.

FROM QUANTUM TO COSMOS
Fundamental Physics Research in Space

ISBN-13 978-981-4261-20-3
ISBN-10 981-4261-20-3

Typeset by Stallion Press
Email: enquiries@stallionpress.com

Printed by FuIsland Offset Printing (S) Pte Ltd, Singapore

PREFACE

Recent progress in observational astronomy, astrophysics and cosmology has raised important questions related to the fundamental composition of the Universe, its evolution and its ultimate destiny. Results from well-conceived space-based physics experiments could provide critical clues to uncover the answers to the challenging questions facing modern physics today. There are two complementary approaches to physics research in space: one can detect and study signals from remote astrophysical objects (observational physics), or one can perform carefully designed in-situ experiments (laboratory physics). Most of the attention to date has focused on observational physics studies, while the laboratory physics approach has received relatively little attention.

The space environment offers unique conditions for precision laboratory investigation which can be improved by many orders of magnitude to explore the limits of modern physics. Examples are the availability of variable gravity potentials, large distances, and high velocity and acceleration regimes. Importantly, many recent technological advances now make it possible for researchers to take full advantage of the unique space environment.

The purpose of the recent international workshop "*From Quantum to Cosmos: Fundamental Physics Research in Space*"[a], held at the Airlie Center, Warrenton, VI, USA, May 21–24, 2006, was to demonstrate how fundamental physics research in space can provide the knowledge needed to address outstanding questions at the intersection of physics and astronomy. The focus of the workshop was on laboratory physics, although a large portion was also dedicated to explore observational physics aspects.

The meeting participants addressed motivation and current status of space-based laboratory experiments in fundamental and gravitational physics. Specific research areas discussed in detail at the workshop included various tests of general relativity and alternative theories, search of physics beyond the Standard

[a]International workshop "From Quantum to Cosmos: Fundamental Physics Research in Space", Airlie Center, Warrenton, VI, USA, May 21–24, 2006, http://physics.jpl.nasa.gov/quantum-to-cosmos/. There were two recent successful workshops in the "From Quantum to Cosmos" series: "From Quantum to Cosmos: Space-Based Research in Fundamental Physics and Quantum Technologies", Bremen, Germany, June 10–13, 2007, details at http://www.zarm.uni-bremen.de/Q2C2/ and "From Quantum to Cosmos: Fundamental Physics in Space for the Next Decade", Airlie Center, Warrenton, VI, USA, July 6–10, 2008, details at http://physics.jpl.nasa.gov/Q2C3/

Model, investigations of possible violations of the Equivalence Principle, search for new hypothetical long- and short-range forces, variations of fundamental constants, tests of Lorentz invariance and attempts at unification of the fundamental interactions. The scope of the workshop also encompassed experiments aimed at the discovery of novel phenomena including dark matter candidates and studies of dark energy. The meeting featured new technologies for space experiments including atom interferometry, precisions optical clocks and their applications for space-based physics experiments and also provided a forum for scientists to discuss policy and the long-term future of space experiments.

From the enthusiasm of the over 100 participants at the workshops, it was clearly seen that space-based laboratory research in fundamental physics is an emerging research discipline that offers great discovery potential and at the same time could drive the development of technology advances that are likely to be important to scientists and technologists in many other different research fields.

It was also clearly demonstrated at the workshop that many of the fundamental physics investigations that were discussed at the meeting can be carried out in space with much higher precision than on the ground; moreover, some of these activities can only be carried out in space.

It is my great pleasure to thank all the speakers for their participation in the workshop and especially those who were willing to contribute to this volume.

Slava G. Turyshev
Jet Propulsion Laboratory
Pasadena, California,
December 2008

CONTENTS

PART 1

POLICY

SPACE-BASED RESEARCH IN FUNDAMENTAL PHYSICS AND QUANTUM TECHNOLOGIES

SLAVA G. TURYSHEV*, ULF E. ISRAELSSON, MICHAEL SHAO and NAN YU

Jet Propulsion Laboratory, California Institute of Technology,
4800 Oak Grove Drive, Pasadena, CA 91109-0899, USA
**turyshev@jpl.nasa.gov*

ALEXANDER KUSENKO and EDWARD L. WRIGHT

Department of Physics and Astronomy, University of California,
Los Angeles, CA 90095-1547, USA

C. W. FRANCIS EVERITT, MARK KASEVICH, JOHN A. LIPA and JOHN C. MESTER

Hansen Experimental Physics Laboratory,
Department of Physics, Stanford University,
Stanford, CA 94305-4085, USA

ROBERT D. REASENBERG and RONALD L. WALSWORTH

Harvard-Smithsonian Center for Astrophysics,
60 Garden Street, Cambridge, MA 02138, USA

NEIL ASHBY

Department of Physics, University of Colorado,
Boulder, CO 80309-0390, USA

HARVEY GOULD

Lawrence Berkeley National Laboratory,
One Cyclotron Road, Berkeley, CA 94720, USA

HO JUNG PAIK

Department of Physics, University of Maryland,
College Park, MD 20742-4111, USA

Space offers unique experimental conditions and a wide range of opportunities to explore the foundations of modern physics with an accuracy far beyond that of ground-based experiments. Space-based experiments today can uniquely address important questions related to the fundamental laws of Nature. In particular, high-accuracy physics experiments in space can test relativistic gravity and probe the physics beyond the Standard

Model; they can perform direct detection of gravitational waves and are naturally suited for investigations in precision cosmology and astroparticle physics. In addition, atomic physics has recently shown substantial progress in the development of optical clocks and atom interferometers. If placed in space, these instruments could turn into powerful high-resolution quantum sensors greatly benefiting fundamental physics.

We discuss the current status of space-based research in fundamental physics, its discovery potential, and its importance for modern science. We offer a set of recommendations to be considered by the upcoming National Academy of Sciences' Decadal Survey in Astronomy and Astrophysics. In our opinion, the Decadal Survey should include space-based research in fundamental physics as one of its focus areas. We recommend establishing an Astronomy and Astrophysics Advisory Committee's interagency "Fundamental Physics Task Force" to assess the status of both ground- and space-based efforts in the field, to identify the most important objectives, and to suggest the best ways to organize the work of several federal agencies involved. We also recommend establishing a new NASA-led interagency program in fundamental physics that will consolidate new technologies, prepare key instruments for future space missions, and build a strong scientific and engineering community. Our goal is to expand NASA's science objectives in space by including "laboratory research in fundamental physics" as an element in the agency's ongoing space research efforts.

Keywords: Fundamental physics in space; general and special theories of relativity; Standard Model extensions; gravitational waves; cosmology; astroparticle physics; cold atoms; quantum sensors; science policy.

1. Introduction

Today, physics stands on the threshold of major discoveries. Growing observational evidence points to the need for new physics. Efforts to discover new fundamental symmetries, investigations of the limits of established symmetries, tests of the general theory of relativity, searches for gravitational waves, and attempts to understand the nature of dark matter were among the topics that had been the focus of the scientific research at the end of the last century. These efforts have further intensified with the discovery of dark energy made in the late 1990's, which triggered many new activities aimed at answering important questions related to the most fundamental laws of Nature.[1,2]

The 2003 Report "Connecting Quarks with the Cosmos: Eleven Science Questions for the New Century",[a] issued by the National Academy of Science's Board on Physics and Astronomy, identified the most critical research areas that require support to meet the profound challenges facing physics and astronomy today. It became a blueprint for multiagency efforts at the National Aeronautics and Space Administration (NASA), the National Science Foundation (NSF), and the US Department

[a] "Connecting Quarks with the Cosmos: Eleven Science Questions for the New Century," Board of Physics and Astronomy (The National Academies Press, 2003). In particular, the report identified the following eleven questions that are shaping the modern research in astronomy, astrophysics, and cosmology: (i) What is the dark matter? (ii) What is the nature of the dark energy? (iii) How did the Universe begin? (iv) Was Einstein right about gravity? (v) How have neutrinos shaped the Universe? (vi) What are nature's most energetic particles? (vii) Are protons unstable? (viii) What are the new states of matter? (ix) Are there more space–time dimensions? (x) How were elements from iron to uranium made? (xi) Is a new theory of matter and light needed?

of Energy (DOE), aiming at meeting the evident challenges in our understanding of matter, space, time, and the Universe. Although the report provided a list of strategic recommendations allowing NASA, NSF, and DOE to select various fundamental physics projects among their top priorities, more work is needed. In addition, some of the science opportunities, including those offered by space-based laboratory research in fundamental physics,[b] were overlooked.

The nature of matter on the Earth and the laws governing it were discovered in laboratories on the Earth. To understand the nature of matter in the Universe and the laws governing it, we must move our laboratories outside the Earth. There are two approaches to physics research in space: one can detect and study signals from remote astrophysical objects (the "observatory" mode) or one can perform carefully designed experiments in space (the "laboratory" mode). The two methods are complementary and the latter, which is the focus of this paper, has the advantage of utilizing the well-understood and controlled environments of a space-based laboratory. Existing technologies allow one to take advantage of the unique environments found only in space, including variable gravity potentials, large distances, high velocity and low acceleration regimes, availability of pure geodetic trajectories, microgravity and thermally stable environments (see App. A for details). With recent advances in several applied physics disciplines, new instruments and technologies have become available. These include highly accurate atomic clocks, optical frequency combs, atom interferometers, drag-free technologies, low-thrust micropropulsion techniques, optical transponders, long-baseline optical interferometers, etc.[3] Some of these instruments are already space-qualified, thereby enabling a number of high-precision investigations in laboratory fundamental physics in space. As a result, space-based experiments are capable of reaching very high accuracies in testing the foundations of modern physics. Furthermore, because experimental physics research complements the observational disciplines of astronomy and astrophysics, it is possible that independent confirmation by space-based fundamental physics experiments may be required to fully explain any future observations of, for example, "detection" of dark matter particles or identification of the source of dark energy.

As was demonstrated at the two recent international "Quantum to Cosmos" workshops,[c] there is a growing community of researchers worldwide interested in performing carefully-thought-out laboratory physics experiments to address some of the modern challenges that physics faces today, by utilizing the benefits of a

[b]Reprioritization of space efforts initiated by NASA in 2004 led to the termination of a successful "Microgravity and Fundamental Physics" program managed by the former Office of Biological and Physics Research. As a result, no program for space-based laboratory research in fundamental physics currently exists.
[c]"From Quantum to Cosmos: Fundamental Physics Research in Space," Airlie Center, Warrenton, VI, USA, May 21–24, 2006; http://physics.jpl.nasa.gov/quantum-to-cosmos; "From Quantum to Cosmos — II: Space-Based Research in Fundamental Physics & Quantum Technologies," Bremen, Germany, June 10–13, 2007; http://www.zarm.uni-bremen.de/Q2C2

space environment. The recent report of the Committee on Atomic, Molecular, and Optical Sciences (AMO2010)[d] emphasized the significant discovery potential of future space-based experiments using new technologies and laboratory techniques, especially in their ability to probe the fundamental laws of Nature at the highest levels of accuracy. The 2006 Report from the Dark Energy Task Force also endorsed the important role of gravitational experiments as an effective means to discover new physics that might also be at play on cosmological scales and that might be responsible for the small observed acceleration of the cosmological expansion of the Universe.[e]

The 2005 Position Paper by the European Physical Society (EPS)[f] highlighted the strong discovery potential of space-based experiments in fundamental physics and argued for space flight opportunities specifically dedicated to this area of research. The EPS' recommendations further supported the efforts of the European Space Agency's (ESA) Fundamental Physics Advisory Group (FPAG)[g] — an influential group of European scientists that advises ESA on scientific direction in fundamental physics research in space. As a result, ESA's Cosmic Vision 2015–25 process[h] marks a breakthrough for fundamental physics: for the first time, a major space agency has given full emphasis in its forward planning to missions dedicated to exploring and advancing the limits of our understanding of many fundamental physics issues, including gravitation, unified theories, and quantum theory. NASA would benefit from a similar bold and visionary approach.

This paper is organized as follows. In Sec. 2 we discuss the status of space-based research in fundamental physics, present examples of experiments that could provide significant advances in the field in the near future, and emphasize scientific and societal benefits of this space science discipline. Each subsection discusses the significance of physics to be addressed, emphasizes the role of space for a particular kind of experiment, and presents a list of potential missions. In Sec. 3 we argue for coordinated multiagency support for space-based research in fundamental physics and present a set of policy recommendations which, if adopted, would re-energize the entire field of research in fundamental physics. In App. A we present the benefits

[d] "Controlling the Quantum World," Committee on Atomic, Molecular, and Optical Sciences (AMO2010), Board on Physics and Astronomy (The National Academies of Press, July 2006). An Electronic version of the report is available from the NAP website http://www.nap.edu/catalog.php?record_id=11705

[e] "Report from the Dark Energy Task Force," June 6, 2006, at [astro-ph/0609591]; an electronic version of the Report is available at http://www.nsf.gov/mps/ast/aaac/dark_energy_task_force/report/detf_final_report.pdf

[f] "The Need for Space Flight Opportunities in Fundamental Physics," a Position Paper of the European Physical Society (EPS), published on the occasion of the centenary of Albert Einstein's *annus mirabilis* (2005) and available from EPS' website: http://www.eps.org/papers_position/paper_index.html

[g] Webpage of ESA's Fundamental Physics Advisory Group (FPAG): http://sci.esa.int/science-e/www/object/ index.cfm?fobjectid=33212

[h] For details on the ESA's Cosmic Vision 2015–25 process and the recent Call for Mission Proposals, visit http://sci.esa.int/science-e/www/object/index.cfm?fobjectid=40794

of space-based deployment for precision physics experiments. In App. B we discuss the history of fundamental physics research at NASA and its current programmatic status.

2. Fundamental Physics in Space: Great Potential for Discovery

The fundamental physical laws of Nature are described by the Standard Model and Einstein's general theory of relativity. The Standard Model specifies the families of fermions (leptons and quarks) and their interactions by vector fields which transmit the strong, electromagnetic, and weak forces. General relativity is a tensor field theory of gravity with universal coupling to the particles and fields of the Standard Model.

Despite the beauty and simplicity of general relativity and the success of the Standard Model, our present understanding of the fundamental laws of physics has several shortcomings. Although recent progress in string theory[4,5] is very encouraging, the search for a realistic theory of quantum gravity remains a challenge. This continued inability to merge gravity with quantum mechanics indicates that the pure tensor gravity of general relativity needs modification or augmentation. It is now believed that new physics is needed to resolve this issue. Recent work on scalar-tensor extensions of gravity and brane-world gravitational models, and also efforts to modify gravity on large scales, motivate new searches for experimental signatures of very small deviations from general relativity on various scales, including on the spacecraft-accessible distances in the solar system.

In addition, the Higgs boson, the particle predicted by the Standard Model, has yet to be discovered. It is widely expected that the Large Hadron Collider (LHC) at CERN will be able to probe the nature of electroweak symmetry breaking and verify the prediction of the Higgs boson in the near future. In addition to this long-anticipated discovery, one hopes to find new physics beyond the Standard Model at the LHC. The new physics could explain the hierarchy of scales and resolve the naturalness problems associated with the Standard Model. Physics beyond the Standard Model is required to explain dark matter and the matter–antimatter asymmetry of the Universe. Furthermore, the Standard Model does not offer an explanation for the observed spectrum of fermion masses and their mixing angles. The exact conservation of the charge-conjugation and parity (CP) symmetries in strong interactions appears mysterious in the Standard Model, because it requires the exact cancellation of two seemingly unrelated contributions to the measurable quantity θ, which *a priori* can take any value between 0 and 2π. New physics is expected to shed light on this mystery as well.

Theoretical models of the kinds of new physics that can solve the problems above typically involve new interactions, some of which could manifest themselves as violations of the equivalence principle, variation of fundamental constants, modification of the inverse square law of gravity at short distances, Lorenz symmetry breaking, as well as large-scale gravitational phenomena. Each of these manifestations offers an opportunity for space-based experimentation and, hopefully, a major discovery.

Our objective is to emphasize the uniqueness and advantages of space as an experimental site when addressing the challenges above and, thereby, to demonstrate that the space-based laboratory research in fundamental physics is a unique area of space science that offers the best possible science investigations to be done from space.

In this section we discuss the current status of space-based research in fundamental physics, including gravitational experiments, the search for new physics beyond the Standard Model, efforts at direct detection of gravitational waves and their use as a probe of physics in the strong gravitational limit, and also cosmology, astroparticle, and atomic physics.

2.1. *Search for a new theory of gravity and cosmology with experiments in space*

The recent remarkable progress in observational cosmology has subjected the general theory of relativity to increased scrutiny by suggesting a non-Einsteinian model of the Universe's evolution. From a theoretical standpoint, the challenge is even stronger — if gravity is to be quantized, general relativity will have to be modified. Furthermore, recent advances in the scalar-tensor extensions of gravity[6–11] have motivated searches for very small deviations from Einstein's theory, at the level of three-to-five orders of magnitude below the level currently tested by experiment.[12,13] For many of the modern gravitational experiments, space is an essential laboratory that, in combination with modern technologies, offers unique conditions that are much purer than those achievable in the best ground-based laboratories.[14,15]

We discuss below a number of laboratory experiments that benefit from the space deployment.

2.1.1. *Test of Einstein's equivalence principle*

Einstein's equivalence principle (EP)[6,7,16,17] is at the foundation of Einstein's general theory of relativity; therefore, testing the principle is very important. The EP includes three hypotheses: (i) local Lorentz invariance (LLI), (ii) local position invariance (LPI), and (iii) universality of free fall (UFF). Using these three hypotheses Einstein deduced that gravity is a geometric property of space–time.[18] One can test both the validity of the EP and of the field equations that determine the geometric structure created by a mass distribution. We shall discuss below two different "flavors" of the principle: the weak and the strong forms of the EP that are currently tested in various experiments performed with laboratory testmasses and with bodies of astronomical sizes.[17]

The *weak form of the EP* (the WEP) states that the gravitational properties of strong and electroweak interactions obey the EP. In this case the relevant test-body differences are their fractional nuclear-binding differences, their neutron-to-proton ratios, their atomic charges, etc. Furthermore, the equality of gravitational and inertial masses implies that different neutral massive test bodies will have the same free fall acceleration in an external gravitational field, and therefore in freely falling inertial frames the external gravitational field appears only in the form of a tidal interaction.[19] Apart from these tidal corrections, freely falling bodies behave as if external gravity was absent.[20]

General relativity and other metric theories of gravity assume that the WEP is exact. However, extensions of the Standard Model of particle physics that contain new macroscopic range quantum fields predict quantum exchange forces that generically violate the WEP because they couple to generalized "charges" rather than to mass/energy as does gravity.[6–11]

Currently, the most accurate results in testing the WEP were reported by ground-based laboratories.[17,21] The most recent result[22,23] for the fractional differential acceleration between beryllium and titanium test bodies was given as $\Delta a/a = (1.0 \pm 1.4) \times 10^{-13}$. Significant improvements in the tests of the EP are expected from dedicated space-based experiments.

The composition independence of acceleration rates of various masses toward the Earth can be tested to many-additional-orders-of-magnitude precision in space-based laboratories, down to levels where some models of the unified theory of quantum gravity, matter, and energy suggest a possible violation of the EP.[6–11] Interestingly, in some scalar-tensor theories, the strength of EP violations and the magnitude of the fifth force mediated by the scalar can be drastically larger in space compared with that on the ground,[24–26] which further justifies a space deployment. Importantly, many of these theories predict observable violations of the EP at various levels of accuracy, ranging from 10^{-13} down to 10^{-16}. Therefore, even a confirmation of no EP violation will be exceptionally valuable, placing useful constraints on the range of possibilities in the development of a unified physical theory.

Compared with Earth-based laboratories, experiments in space can benefit from a range of conditions including free fall and significantly reduced contributions due to seismic, thermal, and many other sources of non gravitational noise (see App. A). As a result, there are many experiments proposed to test the EP in space. We present below a partial list of these missions. Furthermore, to illustrate the use of different technologies, we present only the most representative concepts.

The MicroSCOPE mission[i] is a room temperature EP experiment in space relying on electrostatic differential accelerometers.[27] It is currently under development by CNES[j] and ESA, and is scheduled for launch in 2010. The design goal is to achieve a differential acceleration accuracy of 10^{-15}. MicroSCOPE's electrostatic

[i]Micro-Satellite à traînée Compensée pour l'Observation du Principe d'Equivalence (MicroSCOPE). For more details, see http://microscope.onera.fr

[j]Centre National d'Etudes Spatiales (CNES) — the French Space Agency. See website: http://www.cnes.fr

differential accelerometers are based on flight heritage designs from the CHAMP, GRACE, and GOCE missions.[k]

The Principle of Equivalence Measurement (POEM) experiment[28] is a ground-based test of the WEP, now under development. It will be able to detect a violation of the EP with a fractional acceleration accuracy of 5 parts in 10^{14} in a short (a few days) experiment and 3–10-fold better in a longer experiment. The experiment makes use of optical distance measurement (by TFG laser gauge[29]) and will be advantageously sensitive to short-range forces with a characteristic length scale of $\lambda < 10$ km. SR-POEM, a POEM-based proposed room temperature test of the WEP during a suborbital flight on a sounding rocket, was recently also presented.[30] It is anticipated to be able to search for a violation of the EP with a single-flight accuracy of one part in 10^{16}. Extension to higher accuracy in an orbital mission is under study. Similarly, the Space Test of Universality of Free Fall (STUFF)[31] is a recent study of a space-based experiment that relies on optical metrology and proposes to reach an accuracy of one part in 10^{17} in testing the EP in space.

The Quantum Interferometer Test of the Equivalence Principle (QuITE)[32] is a proposed test of the EP with cold atoms in space. QuITE intends to measure the absolute single-axis differential acceleration with an accuracy of one part in 10^{16}, by utilizing two colocated matter wave interferometers with different atomic species.[l] It will improve the current EP limits set in similar experiments conducted in ground-based laboratory conditions[33,34,m] by nearly seven-to-nine orders of magnitude. Similarly, the ICE project,[n] supported by CNES in France, aims to develop a high-precision accelerometer based on coherent atomic sources in space,[35] with an accurate test of the EP being one of the main objectives.

The Galileo Galilei (GG) mission[36,37] is an Italian space experiment[o] proposed to test the EP at room temperature with an accuracy of one part in 10^{17}. The key instrument of GG is a differential accelerometer made up of weakly coupled coaxial, concentric test cylinders rapidly spinning around the symmetry axis and sensitive in the plane perpendicular to it. GG is included in the National Aerospace Plan of the Italian Space Agency (ASI) for implementation in the near future.

The Satellite Test of Equivalence Principle (STEP) mission[38–40] is a proposed test of the EP to be conducted from a free-falling platform in space provided by

[k]Several gravity missions were recently developed by the German National Research Center for Geosciences (GFZ). Among them are CHAllenging Minisatellite Payload (CHAMP), GRACE (Gravity Recovery and Climate Experiment Mission; together with NASA), and GOCE (Global Ocean Circulation Experiment; together with ESA and other European countries). See http://www.gfz-potsdam.de/pb1/op/index_GRAM.html

[l]Compared to the ground-based conditions, space offers a factor-of-nearly-10^3 improvement in the integration times in observation of the free-falling atoms (i.e. progressing from ms to s). The longer integration times translate into the accuracy improvements (see discussion in Sec. 2.6.3).

[m]Its ground-based analog, called the "Atomic Equivalence Principle Test (AEPT)," is being built at Stanford University. AEPT is designed to reach a sensitivity of one part in 10^{15}.

[n]Interférométrie à Source Cohérente pour Applications dans l'Espace (ICE). See http://www.ice-space.fr

[o]Galileo Galilei (GG) website: http://eotvos.dm.unipi.it/nobili

a drag-free spacecraft orbiting the Earth. STEP will test the composition independence of gravitational acceleration for cryogenically controlled test masses by searching for a violation of the EP with a fractional acceleration accuracy of one part in 10^{18}. As such, this ambitious experiment will be able to test very precisely for the presence of any new nonmetric, long-range physical interactions.

In its *strong form the EP* (SEP) is extended to cover the gravitational properties resulting from gravitational energy itself.[17] In other words, it is an assumption about the way that gravity begets gravity, i.e. about the nonlinear property of gravitation. Although general relativity assumes that the SEP is exact, alternate metric theories of gravity such as those involving scalar fields, and other extensions of gravity theory, typically violate the SEP. For the SEP case, the relevant test body differences are the fractional contributions to their masses by gravitational self-energy. Because of the extreme weakness of gravity, SEP test bodies must have astronomical sizes.

Currently, the Earth–Moon–Sun system provides the best solar system arena for testing the SEP. Lunar laser ranging (LLR) experiments involve reflecting laser beams off retroreflector arrays placed on the Moon by the Apollo astronauts and by an unmanned Soviet lander.[16,17] Recent solutions using LLR data give $(-0.8 \pm 1.3) \times 10^{-13}$ for any possible inequality in the ratios of the gravitational and inertial masses for the Earth and Moon. This result, in combination with laboratory experiments on the WEP, yields a SEP test of $(-1.8 \pm 1.9) \times 10^{-13}$ that corresponds to the value of the SEP violation parameter of $\eta = (4.0 \pm 4.3) \times 10^{-4}$, where $\eta = 4\beta - \gamma - 3$ and both β and γ are post-Newtonian parameters.[17,41–45]

With the new APOLLO[p] facility (jointly funded by NASA and NSF; see details in Refs. 46 and 47), the LLR science is going through a renaissance. APOLLO's 1-mm-range precision will translate into order-of-magnitude accuracy improvements in the test of the WEP and SEP (leading to accuracy at the level of $\Delta a/a \lesssim 1 \times 10^{-14}$ and $\eta \lesssim 2 \times 10^{-5}$ respectively), in the search for variability of Newton's gravitational constant (see Sec. 2.1.2), and in the test of the gravitational inverse square law (see Sec. 2.1.3) on scales of the Earth–Moon distance (the anticipated accuracy is 3×10^{-11}).[47]

The next step in this direction is interplanetary laser ranging,[48–53] for example to a lander on Mars. The technology is available to conduct such measurements with a-few-picoseconds timing precision which could translate into centimeter-class accuracies achieved in ranging between the Earth and Mars. The resulting Mars Laser Ranging (MLR) experiment could test the weak and strong forms of the EP with an accuracy at the 3×10^{-15} and 2×10^{-6} levels respectively, to measure the PPN parameter γ (see Sec. 2.1.4) with an accuracy below the 10^{-6} level, and to test the gravitational inverse square law at ~2 AU distances with an accuracy of 1×10^{-14}, thereby greatly improving the accuracy of the current tests.[51] MLR could

[p]The Apache Point Observatory Lunar Laser-ranging Operations (APOLLO) is the new LLR station that was recently built in New Mexico and successfully initiated operations in 2006.

also advance research in several areas of science, including remote sensing geodesic and geophysical studies of Mars.

Furthermore, with the recently demonstrated capabilities of reliable laser links over large distances (e.g. tens of millions of kilometers) in space,[48–50] there is a strong possibility of improving the accuracy of gravity experiments with precision laser ranging over interplanetary scales.[51–53] Science justification for such an experiment is strong, the required technology is space-qualified and some components have already flown in space. By building MLR, our very best laboratory for gravitational physics will be expanded to interplanetary distances, representing an upgrade in both the scale and the precision of this promising technique.

The experiments above are examples of the rich opportunities offered by the fundamental physics community to explore the validity of the EP. These experiments could potentially offer up-to five-orders-of-magnitude improvement over the accuracy of the current tests of the EP. Such experiments would dramatically enhance the range of validity for one of the most important physical principles, or they could lead to a spectacular discovery.

2.1.2. *Test of the variation of fundamental constants*

Dirac's 70-year-old idea of cosmic variation of physical constants has been revisited with the advent of models unifying the forces of Nature based on the symmetry properties of possible extra dimensions, such as the Kaluza–Klein-inspired theories, Brans–Dicke theory, and supersymmetry models. Alternative theories of gravity[18] and theories of modified gravity[54] include cosmologically evolving scalar fields that lead to variability of the fundamental constants. Furthermore, it has been hypothesized that a variation of the cosmological scale factor with the epoch could lead to temporal or spatial variation of the physical constants, specifically the gravitational constant, G, the fine-structure constant, α, and the electron–proton mass ratio ($m_{\rm e}/m_{\rm p}$).

In general, the constraints on the variation of fundamental constants can be derived from a number of gravitational observations, such as the test of the universality of free fall, the motion of the planets in the solar system, stellar and galactic evolutions. They are based on the comparison of two time scales, the first (gravitational time) dictated by gravity (ephemeris, stellar ages, etc.) and the second (atomic time) determined by a nongravitational system (e.g. atomic clocks).[55,56] For instance, planetary and spacecraft ranging, neutron-star-binary observations, paleontological and primordial nucleosynthesis data allow one to constrain the relative variation of G.[57,58] Many of the corresponding experiments could reach a much higher precision if performed in space.

A possible variation of Newton's gravitational constant G could be related to the expansion of the Universe depending on the cosmological model considered. Variability in G can be tested in space with a much greater precision than on the Earth.[16,57,58] For example, a decreasing gravitational constant, G, coupled with angular momentum conservation is expected to increase a planet's semimajor axis, a, as $\dot{a}/a = -\dot{G}/G$. The corresponding change in the orbital phase grows quadratically with time, providing strong sensitivity to the effect of $\dot{G}$.

Space-based experiments using lunar and planetary ranging measurements are currently the best means of searching for very small spatial or temporal gradients in the values of G.[16,17] Thus, recent analysis of LLR data strongly limits such variations and constrains a local ($\sim$1 AU) scale expansion of the solar system as $\dot{a}/a = -\dot{G}/G = -(5 \pm 6) \times 10^{-13}$ yr^{-1}, including that due to cosmological effects.[41,59] Interestingly, the achieved accuracy in $\dot{G}/G$ implies that, if this rate is representative of our cosmic history, then G has changed by less than 1% over the 13.4 Gyr age of the Universe.

The ever-extending LLR data set and increase in the accuracy of lunar ranging (i.e. APOLLO) could lead to significant improvements in the search for variability of Newton's gravitational constant; an accuracy at the level of $\dot{G}/G \sim 1 \times 10^{-14}$ yr^{-1} is feasible with LLR.[51] High-accuracy timing measurements of binary and double pulsars could also provide a good test of the variability of the gravitational constant.[60,61]

The current limits on the evolution of α are established by laboratory measurements, studies of the abundances of radioactive isotopes and those of fluctuations in the cosmic microwave background, as well as other cosmological constraints (for review see Refs. 57 and 58). Laboratory experiments are based on the comparison either of different atomic clocks or of atomic clocks with ultrastable oscillators. They also have the advantage of being more reliable and reproducible, thus allowing better control of the systematics and better statistics compared with other methods. Their evident drawback is their short time scales, fixed by the fractional stability of the least precise standards. These time scales usually are of the order of a month to a year, so that the obtained constraints are restricted to the instantaneous variation today. However, the shortness of the time scales is compensated by a much higher experimental sensitivity.

There is a connection between the variation of the fundamental constants and a violation of the EP; in fact, the former almost always implies the latter.[q] For example, should there be an ultralight scalar particle, its existence would lead to variability of the fundamental constants, such as α and m_e/m_p. Because masses of nucleons are α-dependent, by coupling to nucleons this particle would mediate an isotope-dependent long-range force.[8,9,57,58,62–64] The strength of the coupling is within a few of orders of magnitude from the existing experimental bounds for such forces; thus, the new force can potentially be measured in precision tests of the EP. Therefore, the existence of a new interaction mediated by a massless (or very-low-mass) time-varying scalar field would lead to both variation of the fundamental constants and violation of the WEP, ultimately resulting in observable deviations from general relativity.

[q]Note that the converse is not necessarily true: the EP may be violated without any observable variation of fundamental constants.

Following the arguments above, for macroscopic bodies, one expects that their masses depend on all the coupling constants of the four known fundamental interactions, which has profound consequences concerning the motion of a body. In particular, because the α dependence is *a priori* composition-dependent, any variation of the fundamental constants will entail a violation of the universality of free fall.[57,58] This allows one to compare the ability of two classes of experiments — clock-based and EP-testing ones — to search for variation of the parameter α in a model-independent way.[61] EP experiments have been superior performers. Thus, analysis of the frequency ratio of the 282 nm $^{199}Hg^{+}$ optical clock transition to the ground state hyperfine splitting in ^{133}Cs was recently used to place a limit on its fractional variation of $\dot{\alpha}/\alpha \leq 1.3 \times 10^{-16}$ yr^{-1}.[65] At the same time, the current accuracy of the EP tests[17] already constrains the variation as $\Delta\alpha/\alpha \leq 10^{-10}\Delta U/c^2$, where ΔU is the change in the gravity potential. Thus, for ground-based experiments (for which the variability in the gravitational potential is due to the orbital motion of the Earth) in one year the quantity $U_{\rm Sun}/c^2$ varies by 1.66×10^{-10}, and so a ground-based clock experiment must be able to measure fractional frequency shifts between clocks to a precision of a part in 10^{20} in order to compete with EP experiments on the ground.[61]

On the other hand, sending atomic clocks on a spacecraft to within a few solar radii of the Sun where the gravitational potential grows to $10^{-6}c^2$ could, be a competitive experiment if the relative frequencies of different on-board clocks could be measured to a precision better than a part in 10^{16}. Such an experiment would allow a direct measurement of any α variation, thus further motivating the development of space-qualified clocks. With their accuracy surpassing the 10^{-17} level in the near future, optical clocks may be able to provide the needed capabilities to directly test the variability of the fine-structure constant (see Sec. 2.6.1 for details).

SpaceTime is a proposed atomic clock experiment designed to search for a variation of the fine-structure constant with a detection sensitivity of $\dot{\alpha}/\alpha \sim 10^{-20}$ yr^{-1} and will be carried out on a spacecraft that flies to within six solar radii of the Sun.[58,66] The test relies on an instrument utilizing a triclock assembly that consists of three trapped-ion clocks based on mercury, cadmium, and ytterbium ions that are placed in the same vacuum, thermal, and magnetic field environment. Such a configuration allows a differential measurement of the frequency of the clocks, and the cancellation of perturbations common to the three. For alkali atoms, the sensitivity of different clocks, based on atoms of different Z, to a change in the fine-structure constant displays specific signatures. In particular, the Casimir correction factor, $F(\alpha Z)$, leads to the differential sensitivity in the alkali microwave hyperfine clock transition frequencies. As a result, different atomic systems with different Z display different frequency dependencies on a variation of α through the αZ-dependent terms. A direct test for a time variation of α can then be devised through a comparison of two clocks, based on two atomic species with different atomic number, Z. This is a key feature of the SpaceTime instrument which, in conjunction with the individual sensitivity of each atomic species to an α variation, can produce clear and unambiguous results. Observation of any frequency drift between the three pairs of clocks in response to the change in gravitational potential, as the triclock instrument approaches the Sun, would signal a variation in α.

Clearly, a solar fly-by on a highly eccentric trajectory with very accurate clocks and inertial sensors makes for a compelling relativity test. A potential use of highly accurate optical clocks (see Sec. 2.6.1) in such an experiment would likely lead to additional accuracy improvement in the tests of α and m_e/m_p, thereby providing a good justification for space deployment.[68] The resulting space-based laboratory experiment could lead to an important discovery.

2.1.3. *Search for new physics via tests of the gravitational inverse square law*

Many modern theories of gravity, including string, supersymmetry, and brane-world theories, have suggested that new physical interactions will appear at short ranges. This may happen, in particular, because at submillimeter distances new dimensions can exist, thereby changing the gravitational inverse square law[69,70] (for a review of experiments, see Ref. 71). Similar forces that act at short distances are predicted in supersymmetric theories with weak scale compactifications,[72] in some theories with very-low-energy supersymmetry breaking,[73] and also in theories of very-low-quantum gravity scale.[74,75] These multiple predictions provide strong motivation for experiments that would test for possible deviations from Newton's gravitational inverse square law at very short distances, notably on ranges from 1 mm to 1 μm.

Recent ground-based torsion balance experiments[76] tested the gravitational inverse square law at separations between 9.53 mm and 55 μm, probing distances less than the dark energy length scale $\lambda_d = \sqrt[4]{\hbar c/u_d} \approx 85\ \mu$m, with energy density $u_d \approx 3.8$ keV/cm^3. It was found that the inverse square law holds down to a length scale of 56 μm and that an extra dimension must have a size less than 44 μm (similar results were obtained by Ref. 77). These results are important, as they signify the fact that modern experiments reached the level at which dark energy physics can be tested in a laboratory setting; they also provided a new set of constraints on new forces,[78] making such experiments very relevant and competitive with particle physics research.

Sensitive experiments searching for weak forces invariably require soft suspension for the measurement degree of freedom. A promising soft suspension with low dissipation is superconducting magnetic levitation. Levitation in $1g$, however, requires a large magnetic field, which tends to couple to the measurement degree of freedom through metrology errors and coil nonlinearity, and stiffen the mode. The high magnetic field will also make suspension more dissipative. The situation improves dramatically in space. The g level is reduced by five to six orders of magnitude, so the test masses can be supported with weaker magnetic springs, permitting the realization of both the lowest resonance frequency and the lowest dissipation. The microgravity conditions also allow for an improved design of the null experiment, free from the geometric constraints of the torsion balance.

The Inverse Square Law Experiment in Space (ISLES) is a proposed experiment whose objective is to perform a highly accurate test of Newton's gravitational law in space.[79–81] ISLES combines the advantages of the microgravity environment with superconducting accelerometer technology to improve the current ground-based limits in the strength of violation[82–84] by four to six orders of magnitude in the range below 100 μm. The experiment will be sensitive enough to probe large extra dimensions down to 5 μm and also to probe the existence of the axion,[r] which, if exists, is expected to violate the inverse square law in the range accessible by ISLES.

The recent theoretical ideas concerning new particles and new dimensions have reshaped the way we think about the Universe. Thus, should the next generation of experiments detect a force violating the inverse square law, such a discovery would imply the existence of either an extra spatial dimension or a massive graviton, or the presence of a new fundamental interaction.[91]

While most attention has focused on the behavior of gravity at short distances, it is possible that tiny deviations from the inverse square law occur at much larger distances. In fact, there is a possibility that noncompact extra dimensions could produce such deviations at astronomical distances[92] (for a discussion see Sec. 2.1.4).

By far the most stringent constraints on a test of the inverse square law to date come from very precise measurements of the Moon's orbit about the Earth. Even though the Moon's orbit has a mean radius of 384,000 km, the models agree with the data at the level of 4 mm! As a result, analysis of the LLR data tests the gravitational inverse square law to 3×10^{-11} of the gravitational field strength on scales of the Earth–Moon distance.[47]

Interplanetary laser ranging could provide conditions that are needed to improve the tests of the inverse square law on the interplanetary scales.[51] MLR could be used to perform such an experiment that could reach the accuracy of 1×10^{-14} at 2 AU distances, thereby improving the current tests by several orders of magnitude.

Although most of the modern experiments do not show disagreement with Newton's law, there are puzzles that require further investigation. The radiometric tracking data received from the Pioneer 10 and 11 spacecraft at heliocentric distances between 20 and 70 AU have consistently indicated the presence of a small, anomalous Doppler drift in the spacecraft carrier frequency. The drift can be interpreted as being due to a constant Sun-ward acceleration of $a_{\rm P} = (8.74 \pm 1.33) \times 10^{-10}$ m/s^2 for each particular craft.[93–96] This apparent violation of the inverse square law has become known as the Pioneer anomaly.

[r]The axion is a hypothetical elementary particle postulated by Peccei–Quinn theory in 1977 to resolve the strong-CP problem in quantum chromodynamics (QCD); see details in Ref. 87–90.

The possibility that the anomalous behavior will continue to defy attempts at a conventional explanation has resulted in a growing discussion about the origin of the discovered effect, including suggestions for new physics mechanisms[97–106] and proposals for a dedicated deep space experiment.[107–109,s] A recently initiated investigation of the anomalous signal using the entire record of the Pioneer spacecraft telemetry files in conjunction with the analysis of much-extended Pioneer Doppler data may soon reveal the origin of the anomaly.[110,111]

Besides the Pioneer anomaly, there are other intriguing puzzles in the solar system dynamics still awaiting a proper explanation, notably the so-called fly-by anomaly,[112–114] which occurred during Earth gravity assists performed by several interplanetary spacecraft.

2.1.4. *Tests of alternative and modified gravity theories with gravitational experiments in the solar system*

Given the immense challenge posed by the unexpected discovery of the accelerated expansion of the Universe, it is important to explore every option to explain and probe the underlying physics. Theoretical efforts in this area offer a rich spectrum of new ideas (some of them are discussed below) that can be tested by experiment.

Motivated by the dark energy and dark matter problems, long-distance gravity modification is one of the radical proposals that have recently gained attention.[115–117] Theories that modify gravity at cosmological distances exhibit a strong coupling phenomenon of extra graviton polarizations.[118,119] This strong coupling phenomenon plays an important role for this class of theories in allowing them to agree with solar system constraints. In particular, the "brane-induced gravity" model[120] provides a new and interesting way of modifying gravity at large distances to produce an accelerated expansion of the Universe, without the need for a nonvanishing cosmological constant.[121,122] One of the peculiarities of this model is the way one recovers the usual gravitational interaction at small (i.e. noncosmological) distances, motivating precision tests of gravity on solar system scales.

The Eddington parameter γ, whose value in general relativity is unity, is perhaps the most fundamental parametrized-post-Newtonian (PPN) parameter,[18,42] in that $\frac{1}{2}(1-\gamma)$ is a measure, for example, of the fractional strength of the scalar gravity interaction in scalar-tensor theories of gravity.[6,7] Currently, the most precise value for this parameter, $\gamma - 1 = (2.1 \pm 2.3) \times 10^{-5}$, was obtained using radiometric tracking data received from the Cassini spacecraft[123] during a solar conjunction experiment. This accuracy approaches the region where multiple tensor-scalar gravity models, consistent with the recent cosmological observations,[124] predict a lower bound for the present value of this parameter at the level of $1-\gamma \sim 10^{-6}$–10^{-7}.[6–11,125,126] Therefore, improving the measurement of this

[s]For details, see the webpage of the Pioneer Explorer Collaboration at the International Space Science Institute (ISSI), Bern, Switzerland: http://www.issi.unibe.ch/teams/Pioneer

parameter[t] would provide crucial information to separate modern scalar-tensor theories of gravity from general relativity, probe possible ways of gravity quantization, and test modern theories of cosmological evolution.

Interplanetary laser ranging could lead to a significant improvement in the accuracy of the parameter γ. Thus, precision ranging between the Earth and a lander on Mars during solar conjunctions may offer a suitable opportunity (i.e. MLR).[u] If the lander were to be equipped with a laser transponder capable of reaching a precision of 1 cm, a measurement of γ with an accuracy of a part in 10^6 is possible. To reach accuracies beyond this level one must rely on a dedicated space experiment.[51]

The Gravitational Time Delay Mission (GTDM)[131,132] proposes to use laser ranging between two drag-free spacecraft (with spurious acceleration levels below 1.3×10^{-13} m/s$^2/\sqrt{\mathrm{Hz}}$ at 0.4 μHz) to accurately measure the Shapiro time delay for laser beams passing near the Sun. One spacecraft would be kept at the L1 Lagrange point of the Earth–Sun system, with the other one being placed on a 3:2 Earth-resonant, LATOR-type orbit (see Ref. 133–136 for details). A high-stability frequency standard ($\delta f/f \lesssim 1 \times 10^{-13}$ $1/\sqrt{\mathrm{Hz}}$ at 0.4 μHz) located on the L1 spacecraft permits accurate measurement of the time delay. If requirements on the performance of the disturbance compensation system, the timing transfer process, and high-accuracy orbit determination are successfully addressed,[132] then determination of the time delay of interplanetary signals to 0.5 ps precision in terms of the instantaneous clock frequency could lead to an accuracy of 2 parts in 10^8 in measuring the parameter γ.

The Laser Astrometric Test of Relativity (LATOR)[133–137] proposes to measure the parameter γ with an accuracy of a part in 10^9, which is a factor of 30,000 beyond the currently best Cassini 2003 result.[123] The key element of LATOR is a geometric redundancy provided by the long-baseline optical interferometry and interplanetary laser ranging. By using a combination of independent time series of gravitational deflection of light in the immediate proximity to the Sun, along with measurements of the Shapiro time delay on interplanetary scales (to a precision better than 0.01 picoradians and 3 mm, respectively), LATOR will significantly improve our knowledge of relativistic gravity and cosmology. LATOR's primary measurement, precise observation of the non-Euclidean geometry of a light triangle that surrounds the Sun, pushes to unprecedented accuracy the search for cosmologically relevant scalar-tensor theories of gravity by looking for a remnant scalar field in today's solar system. LATOR could lead to very robust advances in the tests of fundamental physics — it could discover a violation or extension of general relativity or reveal the presence of an additional long-range interaction.

[t] In addition, any experiment pushing the present upper bounds on another Eddington parameter β, i.e. $\beta - 1 = (0.9 \pm 1.1) \times 10^{-4}$ from Refs. 16 and 17, will also be of interest.

[u] In addition to Mars, a Mercury lander[127,128] equipped with a laser ranging transponder would be very interesting as it would probe a stronger gravity regime while providing measurements that will not be affected by the dynamical noise from the asteroids.[129,130]

If implemented, the missions discussed above could significantly advance research in fundamental physics; however, due to lack of dedicated support none of these experiments could currently be performed in the US. The recent trend of allowing "laboratory" fundamental physics to compete for funding with "observational" disciplines of space science may change the situation and could help NASA to enhance its ongoing space research efforts, so that the space agency will be doing the best possible science in space — a truly noble objective.

2.2. *Detection and study of gravitational waves*

Gravitational waves, a key prediction of Einstein's general theory of relativity had not yet been directly detected. The only indirect evidence for the existence of gravitational waves came from binary pulsar investigation, the discovery that led to the 1993 Nobel Prize in physics.

Gravitational wave observatories in space will provide insight into the structure and dynamics of space and time. They will also be able to detect the signatures of cosmic superstrings or phase transitions in the early Universe and contribute to the study of the dark energy that dominates the evolution of the Universe. Space-based gravitational wave observatories, such as the planned Laser Interferometer Space Antenna (LISA),[v] will offer access to a range of the gravitational wave frequency spectrum that is not accessible on the Earth. It promises to open a completely new window into the heart of the most energetic processes in the Universe, with consequences fundamental to both physics and astronomy. LISA is a joint NASA-ISA mission that expects to detect gravitational waves from the merger of massive black holes in the centers of galaxies or stellar clusters at cosmological distances, and from stellar mass compact objects as they orbit and fall into massive black holes. LISA will measure the signals from close binaries of white dwarfs, neutron stars, or stellar mass black holes in the Milky Way and nearby galaxies.

LISA will consist of an array of three spacecraft orbiting the Sun, each separated from its neighbor by about 5 million km. Laser beams will be used to measure the minute changes in distance between the spacecraft induced by passing gravitational waves. For this purpose, the spacecraft have to be drag-free, a requirement common for many fundamental physics missions. The preparatory LISA Pathfinder[w] mission is aimed at demonstrating the ability to achieve free-fall conditions at the required levels of accuracy.

Studies of gravitational waves can provide potentially powerful insight into large-distance modified-gravity theories. The main reason is that in all such theories the graviton carries three extra polarizations (for details, see Ref. 138), which follows

[v]The Laser Interferometric Space Antenna (LISA) is an international space mission whose development is currently funded through a collaborative agreement between NASA and ESA. See the websites http://lisa.nasa.gov/ and http://www.lisa-science.org, and resources therein.
[w]LISA Pathfinder is a technology demonstration for the future LISA mission. See http://sci.esa.int/science-e/www/area/index.cfm?fareaid=40

from similar properties,[139,140] of the massive graviton.[141] This is in contrast to the normal massless graviton in general relativity, which carries only two helicities. In addition, in such theories the dispersion relation of the graviton is no longer that of a massless spin-2 particle, but rather acquires nontrivial frequency dependence.[138] As a result, both emission and propagation properties of the gravitational waves are altered in modified-gravity theories. LISA and future gravitational wave missions will be able to address these important questions and provide the insight needed to explore these possibilities.

Although of extreme importance, research in gravitational waves does not enjoy stable funding. NASA support for research in this area is conducted via the Beyond Einstein Program[x] which was approved by the US Congress in 2004 and is managed by the Astrophysics Division of NASA's Science Mission Directorate. The program was recently assessed by the Beyond Einstein Program Assessment Committee (BEPAC), formed by the Space Studies Board and the Board on Physics and Astronomy of the National Academy of Sciences (NAS) with the purpose of assessing the five proposed Beyond Einstein missions [Constellation-X, LISA, Joint Dark Energy Mission (JDEM), Inflation Probe, and Black Hole Finder probe] and of recommending which of these five missions should be developed and launched first.[y] In its recently released Report,[z] BEPAC recommended NASA to immediately proceed with the development of the JDEM mission, while also investing additional funds in LISA technology development and risk reduction.

Although the NAC's assessment effectively postpones LISA's launch toward the end of the next decade, it urges NASA to develop LISA-enabling technologies, many of which are common to other fundamental physics missions. As such, these BEPAC recommendations will have a major impact on the entire field of space-based research in fundamental physics in the next decade.

2.3. *Precision research in cosmology*

The current model of the Universe includes critical assumptions, such as an inflationary epoch in primordial times, and peculiar settings, such as the fine-tuning in the hierarchy problem, which call for a deeper theoretical framework. In addition, the very serious vacuum and/or dark energy problems and the related cosmological phase transitions lead researchers to areas beyond general relativity and standard quantum field theory. Observations of the early Universe are an important tool for constraining physics beyond the Standard Model and quantum gravity. This work led to the discovery of the fluctuations in the cosmic microwave background (CMB) made by the NASA COBE mission — an experiment that revolutionized the entire field of cosmology and led to the 2006 Nobel Prize in physics.

[x] NASA Beyond Einstein Program: http://universe.nasa.gov
[y] For details see http://www7.nationalacademies.org/ssb/BeyondEinsteinPublic.html
[z] "NASA's Beyond Einstein Program: An Architecture for Implementation," Committee on NASA's Einstein Program: An Architecture for Implementation, National Research Council. For details see http://www.nap.edu/catalog/12006.html

The 2005 Report by the CMB Task Force[aa] identified two types of observations that are critical for cosmological research: (1) study of the polarization of the CMB anisotropies and (2) direct detection of primordial gravitational waves with second- or third-generation missions. Some of the relevant observations can be made by space missions inspired by the astronomy community. Other observations will be made by space missions inspired by the fundamental physics community. It seems to be quite natural that the former missions should be under the purview of astronomy, while the latter missions should fall under the purview of fundamental physics.

The recently discovered baryon acoustic oscillations[142] together with the CMB and supernovae data provide additional, very important constraints for possible models and scenarios. The 2006 Report from the Dark Energy Task Force (DETF)[bb] explicitly mentions the significance of various tests of general relativity, especially as they relate to dark energy.[143] The report also highlights the synergy between observational and experimental methods to benefit modern research in cosmology. Just as dark energy science has far-reaching implications for other fields of physics, advances and discoveries made in laboratory fundamental physics may point the way toward understanding the nature of dark energy. For instance, such a pointer could come from observing any evidence of a failure of general relativity.

The strong coupling phenomenon (discussed in Sec. 2.1) makes modified-gravity theories predictive and potentially testable at scales that are much shorter than the current cosmological horizon. Because of the key role that nonlinearities play in relativistic cosmology, namely those of its scalar sector, their presence leads to potentially observable effects in gravitational studies within our solar system. Thus, it is possible to test some features of cosmological theories in space-based experiments performed at spacecraft-accessible distances.[92]

There is a profound connection between cosmology and possible Lorentz symmetry violation[144–146] (see also Sec. 2.5). Spontaneous breaking of the Lorentz symmetry implies that there exists an order parameter with a nonzero expectation value that is responsible for the effect. For spontaneous Lorentz symmetry breaking one usually assumes that sources other than the familiar matter density are responsible for such a violation. However, if Lorentz symmetry is broken by an extra source, the latter must also affect the cosmological background. Therefore, in order to identify the mechanism of such a violation, one has to look for traces of similar symmetry breaking in cosmology, for instance in the CMB data.[cc] In other words, should a violation of the Lorentz symmetry be discovered in experiments but not supported by observational cosmology data, such a discrepancy would indicate

[aa]See the webpage of the Task Force of CMB at the NSF, http://www.nsf.gov/mps/ast/tfcr.jsp, and also "Report from the Task Force on CMB Research (TFCR)," July 11, 2005. http://www.nsf.gov/mps/ast/tfcr_final_report.pdf

[bb]Dark Energy Task Force (DETF) website at the NSF: http://www.nsf.gov/mps/ast/detf.jsp

[cc]Analyses of the CMB for Lorentz violation have already begun.[147] This provides a systematic classification of all operators for Lorentz violation and uses polarimetric observations of the CMB to search for associated effects. Lorentz symmetry violation can also have important implication for cosmology via CPT violation and baryogenesis.[148,149]

the existence of a novel source of symmetry breaking. This source would affect the dispersion relation of particles and the performance of the local clocks, but leave no imprint on the cosmological metric. Such a possibility emphasizes the importance of a comprehensive program for investigating all possible mechanisms of breaking of the Lorentz symmetry, including those accessible by experiments conducted in space-based laboratories.

Because of the recent important discoveries, the area of observational cosmology is receiving some limited multiagency support from NASA, DOE, and NSF. However, no support is available for laboratory experiments in this discipline. NASA's support for research in cosmology comes through the Beyond Einstein Program, but it is limited to observational aspects, providing essentially no support to relevant solar system laboratory experiments.

2.4. *Space-based efforts in astroparticle physics*

Astroparticle physics touches the foundation of our understanding of the matter content of the Universe. One can use cosmic rays, high-energy photons, and neutrinos to test the fundamental laws of Nature at energies well beyond the reach of terrestrial experiments. The results can play an important role in the future development of the fundamental theory of elementary particles. Many observations require going to space because the atmosphere stops the cosmic messengers, such as X-rays. In other cases, for example that of ultrahigh-energy cosmic rays, the advantage the space missions offer is in observing a large segment of the Earth's atmosphere, which can be used as part of the detector. We will discuss below some examples of how the astroparticle physics can benefit from going to space.

2.4.1. *Detection of ultrahigh-energy cosmic rays and neutrinos from space*

Ultrahigh-energy cosmic rays (UHECR's), with energies in excess of 10^{20} eV and beyond, have been observed by the AGASA,[dd] HiRes,[ee] and Pierre Auger[ff] experiments. Understanding the origin and propagation of these cosmic rays may provide a key to fundamental laws of physics at the highest energy scales, several orders of magnitude beyond the reach of particle accelerators. The propagation of cosmic rays through space will test both the fundamental symmetry of space–time and the Lorentz invariance.

Detection of UHE neutrinos will mark the beginning of a new era in astronomy and will allow mapping of the most extreme objects in the Universe, such as supermassive black holes, active galactic nuclei, and, possibly, cosmic strings and other topological defects.

[dd]The Akeno Giant Air Shower Array (AGASA) experiment. See http://www-akeno.icrr.u-tokyo.ac.jp/AGASA
[ee]The High Resolution Fly's Eye (HiRes) experiment. See http://hires.phys.columbia.edu
[ff]The Pierre Auger Cosmic Ray Observatory. See http://www.auger.org

By comparing the rates of upgoing and downgoing neutrino-initiated air showers, one can measure the neutrino-nucleon cross section at a center-of-mass energy as high as 10^6 GeV, orders of magnitude beyond the reach of collider experiments.[85,86] There are differing theoretical predictions for this cross section; its measurement can probe fundamental physics at the highest scales.[85,86]

Space-based instruments, such as the proposed EUSO[gg] and OWL,[hh] can use a large segment of the Earth's atmosphere as a medium for detecting cosmic rays and neutrinos. These observations can test the fundamental laws of physics at the highest energy frontier. They also provide information about the most extreme objects in the Universe, such as supermassive black holes.

2.4.2. *Identifying the dark matter particles by their properties*

There is now overwhelming evidence that most of the matter in the Universe is not made of ordinary atoms, but, rather, of new, yet-undiscovered particles (see Ref. 153 for a review). The evidence for dark matter is based on several independent observations, including the anisotropies of the CMB radiation, gravitational lensing, optical observations of the galactic rotation curves, and X-ray observations of clusters. None of the Standard Model particles can be dark matter. Hence, the identification of dark matter will be a discovery of new physics beyond the Standard Model.

One of the most popular theories for physics beyond the Standard Model is supersymmetry (SUSY). A class of supersymmetric extensions of the Standard Model predicts dark matter in the form of either the lightest supersymmetric particles (LSP's) (see Ref. 154 for a review) or SUSY Q balls.[155–159] Another theoretically appealing possibility is dark matter in the form of axions.[87–90] An axion is a very weakly interacting field that accompanies the Peccei–Quinn solution to the strong CP problem.[ii] There are several other dark matter candidates that are well motivated by theoretical reasoning.

The right-handed or sterile neutrinos can be the cosmological dark matter.[160–169] The existence of such right-handed neutrino states is implied by the discovery of the active neutrino masses. Although it is not impossible to explain the neutrino masses otherwise, most models introduce gauge singlet fermions that give the neutrinos their masses via mixing. If one of these right-handed states has a mass in the range of $\sim$1–50 keV, it can be the dark matter.[jj] Several indirect

[gg]The Extreme Universe Space Observatory (EUSO). See http://www.euso-mission.org

[hh]The Orbiting Wide-angle Light collectors (OWL). See http://owl.gsfc.nasa.gov

[ii]The existence of the axion is suggested by models attempting to solve symmetry problems of the Standard Model.[87–90] The axion would violate the $1/r^2$ law of gravity at short distances and would thus be detectable experimentally.

[jj]The LSND experiment claimed[170–173] to observe a sterile neutrino with a much smaller mass ($m \sim$ eV) and a much larger mixing angle ($\sin^2\theta \sim 10^{-1}$) than those needed for dark matter ($m \sim$ keV, $\sin^2\theta \sim 10^{-9}$). The LSND neutrino would pose some serious problems for cosmology. Recent results from the MiniBooNE experiment[174] have refuted the LSND claim.

astrophysical clues support this hypothesis. Indeed, if sterile neutrinos exist, they can explain the long-standing puzzle of pulsar velocities.[175–180] In addition, the X-rays produced in decays of the relic neutrinos could increase the ionization of the primordial gas and can catalyze the formation of molecular hydrogen at redshifts as high as 100. Since molecular hydrogen is an important cooling agent, its increased abundance could play an important role in the formation of the first stars.[181–184] Sterile neutrinos can also help the formation of supermassive black holes in the early Universe, as well as explain the matter–antimatter asymmetry.[185,186] The consensus of these indirect observational hints makes a stronger case for the sterile dark matter.[187]

Depending on the properties of dark matter particles, they can be identified by one of several techniques. In the case of LSPs, their annihilations in the center of our galaxy can produce gamma rays. Thus, the search for gamma rays from the galactic center with instruments such as GLAST[kk] may lead to the discovery of this form of dark matter.[188] The spectrum, the flux, and the distribution of these gamma rays can be used to distinguish the SUSY signal from the alternatives. The LSP annihilations are also expected to produce an identifiable flux of antiprotons and antideuterons.[189] Superheavy dark matter can be discovered by the proposed EUSO or OWL if the heavy particles decay producing the UHECR's, as suggested by some theories.

If dark matter is made up of sterile neutrinos with mass in the keV range,[160–169] their decays into the lighter left-handed neutrinos and X-rays offer an opportunity to discover dark matter using a high-resolution X-ray spectrometer. Since this decay is a two-body process, the decay photons produce a narrow spectral line, Doppler-broadened due to the motion of dark matter particles. For the most plausible masses, one expects a line between 1 and 50 keV, with a width of about 1 eV. To detect this line and distinguish it from the gaseous lines, one needs an instrument with a good energy resolution. Current limits are based on the observations of Chandra and XMM-Newton.[190–200] A dedicated search by the Suzaku telescope is under way. Further ideas for space-based experiments searching for sterile neutrinos are being investigated. In particular, the recently proposed X-ray telescope in space, EDGE,[ll] would be able to search for dark matter in the form of sterile neutrinos.

Ongoing efforts in the search for dark matter particles include a number of underground detectors, as well as GLAST. The search for dark matter must cover a number of avenues because different candidate particles have different interactions. NASA has a unique opportunity to initiate and lead the work in this very important area of space-based research, by funding missions such as EDGE, which could potentially result in a major discovery. Similar opportunities for a significant discovery exist in other areas of astroparticle physics.

[kk]The Gamma-ray Large Area Space Telescope (GLAST). See http://glast.gsfc.nasa.gov
[ll]The Explorer of Diffuse emission and Gamma-ray burst Explosions (EDGE). See http://projects.iasf-roma.inaf.it/edge/EdgeOverview.htm

2.5. *Search for physics beyond the standard model with space-based experiments*

The Standard Model coupled to general relativity is thought to be the effective low-energy limit of an underlying fundamental theory that unifies gravity and the matter sector at the Planck scale. This underlying theory may well include Lorentz violation[201–204] which could be detectable in space-based experiments.[205] If one takes the Standard Model and adds appropriate terms that involve operators for Lorentz invariance violation,[206] the result is the Standard Model extension (SME), which has provided a phenomenological framework for testing Lorentz invariance,[207–216] and also suggested a number of new tests of relativistic gravity in the solar system.[217] Compared with their ground-based analogs, space-based experiments in this area can provide improvements by as much as six orders of magnitude. Several general reviews of the SME and corresponding efforts are available (for a review, see Refs. 218–227). Recent studies of the "aether theories"[228–230] have shown that these models are naturally compatible with general relativity,[18] but predict several nonvanishing Lorentz-violation parameters that could be measured in experiment.

A discovery of an electron electric dipole moment (e-EDM) would be unequivocal proof of new physics beyond the Standard Model. An EDM in an eigenstate of angular momentum is possible only if both parity (P) and time reversal (T) are violated, where T violation is, by the CPT theorem, the equivalent of CP violation. No EDM of any particle or system has yet been observed: all known CP violations (in the decays of the B and K^0 systems) are consistent with the Standard Model's Cabbibo–Kobayashi–Maskawa (CKM) mechanism. The CKM mechanism directly affects only the quark sector, and the CKM-generated e-EDM is extremely small. It is estimated[231,232] to be about 10^{-10}–10^{-5} (depending upon assumptions about the number of neutrino generations and their masses) of the current e-EDM experimental limit of 2.6×10^{-48} C-m (1.6×10^{-27} e-cm).[233–236] By improving the present e-EDM limit, constraints would be placed on many SME's and possibly on current models of neutrino physics.[237]

We will discuss below examples of space-based experiments testing for physics beyond the Standard Model.

2.5.1. *Probing the special theory of relativity in space-based clock comparison experiments*

Searches for extensions of special relativity on a free-flying spacecraft or on the International Space Station (ISS) are known as "clock comparison" experiments.[mm]

[mm]The clocks referred to here take several forms, including atomic and optical clocks, masers, and electromagnetic cavity oscillators. Some clocks may be cesium and rubidium atomic clocks, enhanced to exploit fully the low-gravity environment of space. Each produces an exceptionally stable oscillating signal from energy-level transitions in alkali atoms. Other ISS-based clocks may include masers generating stimulated microwave signals, and microwave cavities creating resonant radiation in small superconducting cavities.

The basic idea is to operate two or more high-precision clocks simultaneously and to compare their rates correlated with orbit parameters such as velocity relative to the microwave background and position in a gravitational environment. The SME allows for the possibility that comparisons of the signals from different clocks will yield very small differences that can be detected in experiment.

Tests of special relativity and the SME were proposed by the Superconducting Microwave Oscillator (SUMO) group, those from the Primary Atomic Reference Clock in SPACE (PARCS),[238–240] and the Rubidium Atomic Clock Experiment (RACE),[241] originally slated for operation on the ISS in the 2005–07 time frame. SUMO, a cryogenic cavity experiment,[242] was to be linked with PARCS to provide differential redshift and Kennedy–Thorndike measurements and improved local oscillator capability.[240] Unfortunately, for programmatic reasons the development of these experiments was canceled by NASA in 2004.[nn] Currently, an experiment called the Atomic Clock Ensemble in Space (ACES) is aiming to do important tests of the SME. ACES is a European mission[243–245] in fundamental physics that will operate atomic clocks in the microgravity environment of the ISS with fractional frequency stability and accuracy of a few parts in 10^{16}. It is jointly funded by ESA and CNES and is being prepared for a flight to the ISS in 2013–14[246] for a planned mission duration of 18 months.

Optical clocks (see Sec. 2.6) offer an improved possibility of testing the time variations of fundamental constants at a high-accuracy level.[247–251] Such measurements interestingly complement the tests of local Lorentz invariance (LLI)[252] and of the universality of free fall to experimentally establish the validity of the EP. The universality of the gravitational redshift can be tested at the same accuracy level by two optical clocks in free flight in a varying gravitational potential. Constancy and isotropy of the speed of light can be tested by continuously comparing a space clock with a ground clock. Optical clocks orbiting the Earth combined with a sufficiently accurate time and frequency transfer link can improve present results by more than three orders of magnitude.

In general relativity, continuous space–time symmetries, such as the Lorentz symmetry, are part of gauge invariance, and as such their violation can only be understood as the low-energy limit of some underlying spontaneous breaking.[144–146] Any order parameter that spontaneously breaks Lorentz invariance will also couple to gravity, ultimately distorting the geometry of space–time, thereby severely constraining the value of that parameter. A consistent Lorentz-breaking inevitably requires either going beyond the usual order parameters (which are different from ordinary cosmic fluids) or going beyond general relativity at large distances. Such

[nn]See the 2003 Report "Factors Affecting the Utilization of the International Space Station for Research in the Biological and Physical Sciences," submitted by the NRC Space Studies Board's Task Group on Research on the ISS (The National Academies Press, 2003), and especially the list of the fundamental physics experiments to be flown on the ISS in the period of 2002–2008, on p. 75 of http://books.nap.edu/books/NI000492/html

a connection suggests that there must be a strong correlation between detection of Lorentz violation and results of relevant cosmological observations. If no such correlation is found, then any discovery of Lorentz violation would indicate the existence of new physical sources or new gravitational dynamics at large distances. Therefore, a consistent breaking of Lorentz-invariance is rather a profound effect that, if detected by an experiment, must be studied together with cosmological observations to determine its nature (see Sec. 2.3 for additional details).

2.5.2. *Search for the electron's electric dipole moment*

Electron EDM experiments are a sensitive test for non–Standard Model sources of CP violation.[253] New, non-CKM sources of CP violation, which directly affect leptons and which can give rise to a large e-EDM, are predicted by the SME and are within the reach of modern experiments. A non-CKM source of CP violation is thought to be necessary for generating the observed matter–antimatter asymmetry in the Universe.[254,255] Within the range of future experiments, e-EDMs are predicted to arise from couplings to new particles and with nonstandard sources of CP violation. These particles are, in some models, candidates for dark matter, or part of the mechanism for generating the observed excess of matter over antimatter, or part of the mechanism for generating neutrino mass. In fact, potentially observable e-EDMs[231,232,256] are predicted by supersymmetry,[257] multi-Higgs models, left–right-symmetric models, lepton flavor-changing models, technicolor models,[258] and TeV scale quantum gravity theories.[69] Split supersymmetry[259–261] predicts an e-EDM within a few orders of magnitude of the present experimental limit and up to the present experimental limit. Merely improving the present e-EDM limit would place constraints on these SME's and possibly on current models of neutrino physics.[237]

Cold-atom-based experiments may be used to search for an e-EDM. As a first step a ground-based demonstration of a Cs fountain e-EDM experiment has been carried out at LBNL.[262] Similar to an atomic clock, a cold-atom-based e-EDM experiment is more sensitive in the microgravity environment of space than on the ground. Such an experiment may improve the current sensitivity to the e-EDM by several orders of magnitude.

Direct measurement of any SME effects would herald a new era, fundamentally changing the perspective on the fabric of special relativity. The effect of such a discovery would include permanent changes in cosmology, high-energy physics, and other fields.

The current research efforts in this area are limited to ground-based laboratory work supported by the NSF. All NASA funding previously available to the space-based efforts described above was terminated after the 2004 cancellation of the "Microgravity and Fundamental Physics" program (see App. B for details); no NASA support for research in this area is currently available.

2.6. *Cold atom physics, new frequency standards and quantum technologies*

Studies of the physics of cold atoms and molecules have recently produced sensational scientific results and important inventions. The field of atomic, molecular, and optical physics has had an incredibly productive decade, marked by Nobel Prize awarded for discoveries in laser cooling (1997), Bose–Einstein condensation and atom lasers (2001), laser-based precision spectroscopy, and the optical frequency comb technique (2005).

Quantum principles are at the core of many advanced technologies used in high-accuracy experiments in fundamental physics. One of the areas where the progress will certainly bring societal and technological benefits is the area of condensed quantum matter. The new phenomenon of interest is seen only at extremely low temperatures, or at very high densities, when particles such as electrons in some metals, helium atoms, or alkali atoms in atomic traps, form a pattern not in "real" space, but rather in momentum space. This "condensation" into momentum space is known as Bose–Einstein condensation (BEC), which in strongly interacting systems, such as liquid helium, results in a complex phenomenon known as superfluidity. BEC in the "paired" electrons in some metals results in superconductivity, which has already revolutionized many technologies to date and is poised to produce many new applications in the future.

While gravitational and relativistic physics examine the most fundamental laws describing the Universe on the large scale, it is equally important to look at the tiny building blocks of matter and how they manifest the same fundamental laws. New techniques allow us to use laser light to cool and probe the properties of individual atoms as a starting point for exploration. Working with individual atoms as test laboratories, researchers stand at the bridge between the smallest pieces of matter and the complex behavior of large systems.

Furthermore, conducting these experiments in space allows one to remove the influence of gravity and manipulate matter freely, without having to counteract specimens "falling" within the instruments. It is possible to study clouds of atoms, cooled by laser light to very near absolute zero yet freely floating without the forces that would be needed to contain them on the Earth. This unique realization of an atom nearly at rest in free space allows longer observation times and enables measurements with higher precision (see the discussion of the experiments relying on cold-atom-based technologies in sections above).

We present below several impressive examples of the new generation of quantum technologies and discuss their role for space-based experimental research in fundamental physics.

2.6.1. *Highly accurate optical clocks*

Precision physics, particularly precision frequency measurements, has recently shown substantial progress with the introduction of new types of atomic clocks.

Recent advances in accuracy that have turned the clocks into powerful tools in the development of applications in time and frequency metrology, universal time scales, global positioning and navigation, geodesy, gravimetry, and others.

Present-day trapped-ion clocks are capable of a stability at the 10^{-15} level over a few hours of integration time.[263] Such performance allows simultaneous one-way (down-) and two-way (up- and down-) links between space and ground, with a greatly reduced tropospheric noise contribution. These clocks are lightweight, have no lasers, cryogenics, or microwave cavities and are similar to the traveling wave tube that is already on board many spacecraft.

Microwave atomic clocks with similar performance have also been space-qualified[264] and are being prepared for a flight on the ISS.[243–245] In fact, the current accuracy of fountain clocks is already at the 5×10^{-16} level and is expected to improve in the near future.

Optical clocks have already demonstrated fractional frequency stability of a few parts in 10^{17} at 1 s of integration time. The performance of these clocks has strongly progressed in recent years, and accuracies below one part in 10^{17} are expected in the near future.[265] Access to space provides conditions for further improving the clock performance, potentially reaching the 10^{-18}–10^{-19} stability and accuracy level.[266–269] In addition, the operation of optical clocks in space provides new scientific and technological opportunities,[3,68] with far-reaching societal benefits.

High-precision optical clocks can measure the gravitational redshift with a relative frequency uncertainty of a few parts in 10^{18}, demonstrating a new, efficient way of mapping the Earths gravity field at the centimeter level. The universality of high-accuracy time transfer is another important aspect of an optical clock. In particular, it is feasible that a few space-based high-precision clocks will provide a universal high precision time reference for users both ground-based and in space — a highly accurate clock synchronization and time transfer that cannot be achieved from the ground. There will be a strong impact on various disciplines of Earth sciences that will greatly benefit from the use of clocks, including studies of relativistic geodesy, Earth rotation, climate research, ocean research, earthquakes, tsunamis, and many others.

There are several benefits of space deployment for the latest generation of atomic clocks. Thus, it is known that thermal noise and vibration sensitivity are the two factors limiting the performance of the optical local oscillator.[270,271] By operating the clock in a quiescent space environment one can use longer cavity spacers to reach better short term stability for the local oscillator. That will have a direct impact on spectral resolution as better spectral resolution will lead to smaller systematic errors.

2.6.2. *Femtosecond optical frequency combs in space*

The optical frequency comb (OFC) is a recently discovered method that enables comparison of the frequency stability of optical clocks at the level of

$\delta f/f \sim 10^{-19}$.[272] In the optical domain, using a frequency comb a group at NIST observed a beat note frequency stability at the 3×10^{-17} level between Al^+ and Hg^+ ions.[265] Where the stability of the microwave-to-optical link is concerned, with the help of a mode-locked femtosecond laser emitting a series of very short laser pulses with a well-defined repetition rate, an OFC can be generated that makes it possible to compare the microwave frequency of atomic clocks (about 10^{10} Hz) with optical frequencies (about 10^{15} Hz) with an accuracy of one part in 10^{15}.[65,273]

The OFC technique will lead to further improvements of Kennedy–Thorndike tests[274] and also tests of the universality of the gravitational redshift.[275] The OFC has aided the development of better optical clocks. With reduced vibration-related problems in the space environment, further improvements can be made in the traditional optical-cavity-based speed-of-light tests (see discussion in Sec. 2). Instead of using two orthogonally oriented cavities to test the angle-dependent (Michelson–Morely) and angle-independent terms (Kennedy–Thorndike, velocity-dependent terms), using an optical clock one can do these tests with one cavity, while using the clock itself as the absolute reference. The fact that this would be an all-optical test is an additional benefit of the method.

As an example of a possible experiment, one could use a rotating wide-bandwidth optical cavity. As the cavity rotates in space, with its axis pointing to different gravitational sources, one can measure possible frequency shifts in blue, green, and red, or even all the way to microwave (group delay) frequencies, all in one setup. These measurements could give an extra constraint with respect to the optical carrier frequency and improve the experimental sensitivity.[277] There are other promising nonfrequency comb proposals, for instance those based on high finesse cavities.[278–283]

OFC's offer a unique opportunity of simultaneous time-keeping and distance ranging. Traditionally laser ranging involves measuring the delay of pulses reflected or transponded from the ranging target.[oo] Use of the highly coherent light associated with modern femtosecond combs would allow interferometry, supplemented by pulse delay ranging.[284] The idea of absolute distance measurement within an optical fringe based on a comb can be easily implemented in space due to a significantly lower contribution from any dispersive medium there. Substantial improvement in laser ranging would aid studies of relativistic gravity in the solar system as well as astrometry, geodesy, geophysics, and planetology.[3]

2.6.3. *Atomic quantum sensors*

Atomic quantum sensors based on matter-wave interferometry are capable of detecting very small accelerations and rotations (see discussion in Refs. 285–288). The present-day sensitivity of atom interferometers used as accelerometers

[oo]Lunar ranging from a reflector placed by lunar astronauts has, since the early 1970's, improved from several tens of centimeters to a level now of about a millimeter; see Refs. 16, 47 and 46.

is $\delta a \sim 10^{-9}$ m/s$^2/\sqrt{\text{Hz}}$, and as gyroscopes, $\delta\omega \sim 6 \times 10^{-10}$ rad/s$/\sqrt{\text{Hz}}$.[289] These instruments reach their ultimate performance in space, where the long interaction times achievable in a freely–falling laboratory improve their sensitivity by at least two orders of magnitude (for a review, see Ref. 290). Possibilities of improvements in the measurement techniques with differential atom interferometry, including those beyond the standard quantum limit, are also investigated.[291]

Cold atom sensors in space may enable new classes of experiments, such as testing the gravitational inverse square law at distances of a few microns, the universality of free fall, and others. Matter-wave interferometer techniques may lead to radically new tests of the EP using atoms as nearly perfect test masses,[32] measurements of the relativistic frame-dragging precession,[292] the value of G,[293,294] and other tests of general relativity.[295] Furthermore, cold atom quantum sensors have excellent sensitivity for absolute measurement of gravity, gravity gradients and magnetic fields, as well as Earth rotation, and therefore find application in Earth sciences and in Earth-observing facilities.[296] Miniaturized cold atom gyroscopes and accelerometers may lead to the development of autonomous navigation systems not relying on satellite tracking.[297–301]

Today, a new generation of high performance quantum sensors (ultrastable atomic clocks, accelerometers, gyroscopes, gravimeters, gravity gradiometers, etc.) is surpassing previous state-of-the-art instruments, demonstrating the high potential of these techniques based on the engineering and manipulation of atomic systems. Atomic clocks and inertial quantum sensors represent a key technology for accurate frequency measurements and ultraprecise monitoring of accelerations and rotations.[3] New quantum devices based on ultracold atoms will enable fundamental physics experiments testing quantum physics, physics beyond the Standard Model of fundamental particles and interactions, special relativity, gravitation, and general relativity.[303] Because of the anticipated strong impact of these new devices on the entire area of precision measurements, the development of quantum technologies for space applications has also seen increased activity.[pp]

In addition, studies of ultracold atoms, molecules, and degenerate quantum gases (BEC, Fermi gases, and Bose–Fermi mixtures) are steadily progressing.[304–306] BEC provides gases in the sub-nanokelvin range, with extremely low velocities (i.e. at the micron-per-second level), which are ideally suited for experiments in a microgravity environment. Similarly, Fermi gases have no interactions at low temperatures, which is very important for potential tests of the EP in space-based experiments. The atom chip technology and compact interferometers now under development may provide a low-power, low-volume source for atomic quantum sensors.[307]

In summary, it is clear that in recognition of such impressive progress and also because of its uniquely strong potential for space applications, research on cold atoms

[pp]For instance, a recent European workshop on "Quantum Mechanics for Space," held at ONERA, Châtillon, France, during 30 Mar.–Apr. 2005. See http://qm-space.onera.fr and the related proceedings issue.[302]

and quantum technologies would greatly benefit from NASA (and, hopefully, multiagency) support; the current European effort in this area are an excellent example.[qq] We emphasize that coordinated multiagency support could lead to significant progress in developing quantum technologies for space applications and would benefit the entire discipline of space-based research in fundamental physics.

3. Discussion and Recommendations

The recent technological progress and the availability of the quiescent environment of space have placed fundamental physics in a unique position to address some of the pivotal questions of modern science. The opportunity to gather important new knowledge in cosmology, astronomy, and fundamental physics stems from recent discoveries suggesting that the basic properties of the Universe as a whole may be intimately related to the physics at the very smallest scale that governs elementary particles such as quarks and other constituents of atoms.

The science investigations presented in this paper are focusing on the very important and challenging questions that physics and astronomy face today. Space deployment is the common factor for these investigations; in fact, their science outcomes are more significant if the experiments are performed in space rather than on the ground.

Because of the significant discovery potential offered by the space-based laboratory research in fundamental physics, it would be beneficial to set aside some dedicated multiagency funding to stimulate the relevant research-and-development efforts. Investments in this area of fundamental physics are likely to lead to major scientific advances and to the development of new technologies and applications that will strengthen national economic competitiveness and security.[rr]

Our specific recommendations are:

3.1. *Include fundamental physics in the NAS' Decadal Survey in Astronomy and Astrophysics*

We recommend that the upcoming National Academy of Sciences' Decadal Survey in Astronomy and Astrophysics[ss] include space-based research in fundamental physics as one of its focus areas. As new scientific discoveries and novel experimental approaches cut across the traditional disciplinary boundaries, the next survey

[qq] The ESA has recently established two programs to develop atomic clocks and atom interferometers for future missions in space: ESA-AO-2004-100, "Space Optical Clocks," coordinated by S. Schiller; and ESA-AO-2004-064/082, "Space Atom Interferometers," coordinated by G. M. Tino.

[rr] Note that NASA was not included to receive a funding increase as a result of the American Competitiveness Initiative (ACI) (see http://www.ostp.gov/html/ACIBooklet.pdf). ACI provided additional authorization of US$160 million for basic science and research for FY 2007 to other agencies.

[ss] In addition, the upcoming NAS' Decadal Survey in Physics should have a strong responsibility for recommendations to NASA on space-based research in fundamental physics.

should make physics an equal partner for strategic planning of future efforts in space. The space-based research community in fundamental physics is ready to contribute to this important strategic planning process for the next decade, as evidenced by "Quantum to Cosmos" meetings.

3.2. *Establish an interagency fundamental physics task force*

For the near term, we recommend that the Astronomy and Astrophysics Advisory Committee (AAAC)[tt] establish an interagency "Fundamental Physics Task Force (FPTF)" to assess benefits of space-based efforts in the field, to identify the most important focus areas and objectives, and to suggest the best ways to organize the work of the agencies involved.[308]

The FPTF can help the agencies identify actions that will optimize near- and intermediate-term synergistic fundamental physics programs carried out at NSF, NASA, DOE, and other federal agencies,[uu] and ensure progress in the development and implementation of a concerted effort toward improvement of our understanding of the fundamental laws of physics.

Given the scientific promise of "Quantum to Cosmos" and also its interdisciplinary nature, it is the right time for the National Science and Technology Council's Committee on Science[vv] to include the new program under the jurisdiction of the Interagency Working Group (IWG) on the "Physics of the Universe" and to ask the IWG to examine the investments required to support the new area of space-based laboratory research in fundamental physics and to develop priorities for further action.

3.3. *Establish a NASA-led program dedicated to space-based efforts in fundamental physics*

In the intermediate term, we recommend that NASA establish a program dedicated to space-based efforts in fundamental physics and quantum technologies. The new program, tentatively named "Quantum to Cosmos" (Q2C), would complement NASA's "Beyond Einstein" program[x] and could also offer a "space extension" to NSF and DOE ground-based research efforts by providing unique opportunities for high-accuracy investigations in fundamental physics.

The program would focus on high-precision tests of relativistic gravity in space, searches for new physics beyond the Standard Model, direct detection and studies

[tt]The Astronomy and Astrophysics Advisory Committee (AAAC). See details at http://www.ucolick.org/~gdi/aaac and also the NFS AAAC webpage http://www.nsf.gov/mps/ast/aaac.jsp
[uu]The 2006 "Quantum to Cosmos" (Q2C) workshop (see meeting details at http://physics.jpl.nasa.gov/quantum-to-cosmos) demonstrated that, in addition to NASA, NSF, and DOE-SC, agencies like the Department of Commerce's National Institute of Standards and Technology (DOC/NIST) do benefit from laboratory physics research in space and, thus, could also sponsor the FPTF (see discussion in Ref. 308).
[vv]National Science and Technology Council (NSTC) at the Executive Office of the President of the United States. See http://www.ostp.gov/nstc

of gravitational waves, searches for dark matter, discovery research in astroparticle physics, and experiments in precision cosmology. It would also develop and utilize advanced technologies needed for space-based experiments in fundamental physics, such as laser transponders, drag-free technologies, atomic clocks, optical frequency combs and synthesizers, atom and matter-wave interferometers, and many others.

The new program would allow NASA to fully utilize the science potential of the ISS by performing carefully planned fundamental physics experiments on board the space station — which would be a direct response to the 2005 designation of the ISS as a national laboratory.[ww]

With its strong interdisciplinary research focus such a program would broadly cut across agencies, academia, and industry while also overlapping the interests of several federal agencies, namely NASA, NSF, DOE/SC, DOC/NIST, NIH, and others. We believe that this program will allow the US to regain a leadership position in fundamental physics worldwide.

3.4. *Enrich and broaden NASA's advisory structure with space-based laboratory fundamental physics*

We observe that the ESA's FPAG[g] is a good example of how to engage the fundamental physics community in space-related research activities, enrich and deepen the space enterprise, and also broaden the ESA's advocacy base. NASA would benefit from access to a similar group of science advisors. The recently formed NASA Advisory Committee (NAC)[xx] does not have representation from the fundamental physics community, nor does the Astrophysics subcommittee[yy] of the NAC. Therefore, we propose:

- To include members of the fundamental physics community in the NAC and/or NAC's Astrophysics subcommittee. Include fundamental physics in the Astrophysics Division of the SMD and provide adequate representation in the advisory structure.
- To consult with the ESA on the possibility of appointing US *ex-officio* members to the ESA's FPAG. Such participation could facilitate development of ongoing (i.e. LPF, LISA) and future fundamental physics missions.

[ww] The 2005 NASA Authorization Act designated the US segment of the ISS as a national laboratory and directed NASA to develop a plan to increase the utilization of the ISS by other federal entities, the research community, and the private sector; see http://www.nasa.gov/mission_pages/station/science/nlab. The majority of NASA's ISS research effort is focused on supporting the human exploration of space program, with only about 15% going toward other activities. There is currently no NASA support for performing fundamental physics aboard the ISS. Thus, the designation of the ISS as a national laboratory rings untrue as long as major research areas, such as fundamental physics, are excluded from participating. The recent NASA Request for Information for Earth and Space Science Payloads on the ISS could be an important first step in the right direction.

[xx] The NASA Advisory Committee (NAC). For details see webpage at http://www.hq.nasa.gov/office/oer/nac

[yy] Webpage of the NAC's Astrophysics subcommittee: http://science.hq.nasa.gov/strategy/subcomm.html

We believe that the recommendations presented above will benefit space-based research in fundamental physics and quantum technologies, the area of research with its unique science and strong technological potential.

Acknowledgments

We would like to express our gratitude to our many colleagues who have either collaborated with us on this manuscript or shared with us their wisdom. We specifically thank Eric G. Adelberger, Peter L. Bender, Curt J. Cutler, Gia Dvali, John L. Hall, Wolfgang Ketterle, V. Alan Kostelecký, Kenneth L. Nordtvedt, Douglas D. Osheroff, Craig Hogan, William D. Phillips, E. Sterl Phinney, Thomas A. Prince, Irwin I. Shapiro, Jun Ye, and Frank Wilczek, who provided us with valuable comments, encouragement, support, and stimulating discussions while this document was in preparation.

We are grateful to our European colleagues who benefited us with their insightful comments and suggestions regarding the manuscript. In particular, our gratitude goes to Orfeu Bertolami, Robert Bingham, Philippe Bouyer, Luigi Cacciapuoti, Thibault Damour, Hansjoerg Dittus, Ulrich Johann, Claus Lämmerzahl, Ekkehard Peik, Serge Reynaud, Albrecht Ruediger, Christophe Salomon, Stephan Schiller, Mikhail Shaposhnikov, Guglielmo Tino, Andreas Wicht, and Peter Wolf.

We thank our colleagues at JPL for their encouragement, support, and advice regarding this manuscript. We especially appreciate the valuable contributions from Roger A. Lee, Moshe Pniel, Michael W. Werner, and Jakob van Zyl of the Astronomy and Physics Directorate, Robert J. Cesarone of the Architecture and Strategic Planning Office of the Interplanetary Network Directorate, and also William M. Folkner and James G. Williams. Our gratitude also goes to Michael H. Salamon of NASA and Nicholas White of GSFC, who kindly provided us with many insightful comments and valuable suggestions on various aspects of the manuscript.

The work described here, was in part carried out at the Jet Propulsion Laboratory, California Institute of Technology, under a contract with the National Aeronautics and Space Administration.

Appendix A. Advantages of Carrying out Physics Research in Space

Historically, experiments in fundamental physics focused on laboratory efforts involving ground-based, underground, and more recently, balloon experiments. Scientific progress in these experiments depends on clever experimental strategy and the use of advanced technologies needed to overcome the limits imposed by the environment typically present in Earth-based laboratories. As a result, the use of sophisticated countermeasures needed to eliminate or reduce contributions from various noise sources increases the overall cost of the research. Oftentimes the conditions in the Earth-based laboratories cannot be improved to the required levels of purity.

Although very expensive, access to space offers conditions that are not available on the Earth but are pivotal for many pioneering investigations exploring the limits of modern physics. We assert that, for many fundamental physics experiments, especially those aiming at exploring gravitation, cosmology, and atomic physics, while achieving the uttermost measurement precision, and increasingly so for high energy and particle astrophysics, a space-based location is the ultimate destination.

Considering fundamental physics experiments, our solar system is a unique laboratory that offers plenty of opportunities for discovery. A carefully designed space-based experiment can take advantage of a number of factors that significantly improve its accuracy; some of these factors are listed below.

A.1. *Access to significant variations of gravitational potential and acceleration*

Some general-relativistic effects (such as the gravitationally induced frequency shift) vary with the potential. Space enables far more precise tests because the change in potential can be much greater, and at the same time less affected by noise, than in any ground-based experiments. Other effects (for example, those resulting from a violation of the EP or of local Lorentz and position invariances (see Secs. 2.5 and 2.1) vary with the magnitude or direction of the acceleration. These, and the range of frequencies at which they can be made to occur in the experiment frame, can greatly exceed the values accessible in corresponding ground-based tests.

The strongest gravity potential available in the solar system is that provided by the Sun itself. Compared with terrestrial conditions, the Sun offers a factor of $\sim$3000 increase in the strength of gravitational effects. The corresponding gravitational acceleration near the Sun is nearly a factor of 30 larger than that available in ground-based laboratories.

Placing an experimental platform in heliocentric orbit provides access to conditions that are not available on the ground. For instance, a highly eccentric solar orbit with apoapsis of 5 AU and periapsis of 10 solar radii offers more than two orders of magnitude in variation of the solar gravity potential and four orders of magnitude in variation in the corresponding gravitational acceleration, clearly not available otherwise. Smaller benefits may be achieved at a lower cost in Earth orbit.

In addition, the ability to precisely track very long arcs of trajectories of test bodies in the solar system is another great advantage of space deployment.

A.2. *Greatly reduced contribution of nongravitational sources of noise*

Compared with Earth-based laboratories, experiments in space can benefit from a range of conditions, especially those of freefall, and also with significantly reduced contributions due to seismic, thermal, and other sources of nongravitational noise.

Microgravity environments ranging from $10^{-4}\,g$ to $10^{-6}\,g$ achieved with free-falling platforms enable new laser-cooling physics experiments and high-accuracy tests of gravity with colocated clocks. A long duration in a controlled free fall environment and drag-free operations benefit many experiments. It is expected that the next-generation optical atomic clocks will reach their full potential of accuracy only in space (see Sec. 2.6.1).

Purely geodesic orbits that are achieved with drag-free spacecraft and for which effects of nongravitational forces are reduced to 10^{-10}–$10^{-14}g$ levels compared with terrestrial conditions are needed for precision tests of gravity and direct detection of gravitational waves.

A.3. *Access to large distances, velocities, and separations; availability of remote benchmarks and inertial references*

Laser retroreflectors on the Moon, radio transponders on Mars, and radio science experiments on board remote spacecraft have vastly improved the accuracy of tests of relativistic gravity. In the near future, optics- and atom-based quantum technologies could provide even higher accuracy for the next generation of interplanetary laser ranging experiments.

Formation flying technologies, with spacecraft separated by distances from hundreds of meters to millions of kilometers, enable larger apertures, more complex focal plane assemblies, and longer interferometric baselines than are possible on the Earth. For example, a gravitational wave observatory with a baseline of several million km is possible only in space, opening a window for the study of low-frequency gravitational waves.

Availability of an inertial reference frame is oftentimes one of the most critical requirements for precision tests of gravity. Modern-day precision star-trackers and spacecraft attitude control systems allow the establishment of inertial reference and, thus, the carrying out experiments in the inertial or quasi-inertial environment of the solar system.

A.4. *Access to vacuum conditions of space*

Space deployment provides for a significant reduction of atmospheric interference with the propagation of optical, radio, and X-ray signals. In fact, absence of air allows perfect optical "seeing" and avoids particle annihilation in antimatter searches. Thus, space conditions allow for small point spread functions (PSF's) (due to the absence of atmospheric blurring) and PSF stability (due to the thermally stable environment only available in space) needed for many precision measurements. As a result, the space vacuum allows the construction of instruments with unique architectures enabled by highly accurate optical metrology, including image-forming instruments with large apertures and long-baseline interferometers.

A.5. *Availability of critical technologies*

Development of technologies needed for many fundamental physics experiments is a very challenging task; however, recent years have seen the maturation of a significant number of key technologies that were developed to take advantage of the unique conditions available in space. Among them are high-precision accelerometers, drag-free control using He-proportional thrusters or small ion thrusters as actuators, ultrastable lasers in space, He dewars, cryocoolers, superconducting detectors, high-precision displacement sensors, magnetic spectrometers, small trapped-ion clocks, lightweight H-maser clocks and atomic clocks using laser-cooled atoms. Many of these technologies have been space-qualified and some have already been flown in space, thereby paving the way for the development of many fundamental physics experiments.

Many of the advanced space technologies developed for gravitational experiments can be directly applied to other space sciences, including astrophysics, cosmology, astroparticle and atomic physics. Such technological cross-pollination allows other science disciplines to take advantage of space deployment opportunities, thereby stimulating the progress in many areas of space research.

Appendix B. Fundamental Physics in Space: Lessons from the Past and Prospects for the Future

In the past, NASA had recognized the potential and importance of fundamental physics research conducted in space. In particular, the program on "Microgravity and Fundamental Physics" that was established in 1996 was focused primarily on research to be conducted on the ISS, thereby contributing to the science justification for the space station.[zz] After the publication of *Roadmap for Fundamental Physics in Space* in 1999,[aaa] the NAC[xx] endorsed the Roadmap and recommended broader support dedicated to fundamental physics.

Following the initial success of NASA's "Microgravity and Fundamental Physics" program and in recognition of the emergence of fundamental physics as a space discipline not served by other commissions, the Committee on Space Research (COSPAR) established Commission H: Fundamental Physics in Space in 1996.[bbb] Since the time of its inception, the contributions from Commission H at the biannual COSPAR meetings have been growing steadily.

As for NASA efforts in the field, a suite of missions and space-based experiments representing many areas of physics were developed and planned for flight in the period of 2002–08.[nn] However, in 2004, following the NASA reorganization in a response to the US New Space Exploration Initiative, Code U became part of the new NASA Exploration Systems Mission Directorate (ESMD). With no clear alignment with the

[zz] Among the most successful outcomes of this program that existed during 1996–2004 were the Lambda Point Experiment (LPE; 1992), the Confined Helium eXperiment (CHeX, 1997), Critical Fluid Light Scattering (ZENO; 1991, 1994), and Critical Viscosity Xenon (CVX; 1997). See details at http://funphysics.jpl.nasa.gov and http://funphysics.jpl.nasa.gov/technical/ltcmp/zeno.html
[aaa] An electronic version of *Roadmap for Fundamental Physics in Space* is available at http://funphysics.jpl.nasa.gov/technical/library/roadmap.html
[bbb] See COSPAR's webpage, http://www.cosparhq.org, for information about Commission H on Fundamental Physics in Space.

ESMD's objectives, Code U's fundamental physics budget was decimated within four months and the entire program was terminated in September 2006.

In contrast, the ESA's science program, through persistent efforts by the fundamental physics community,[309] has an established "Fundamental Physics Program" highlighted by development of several missions which are nearing flight-ready status.[ccc] The ESA's recently initiated Cosmic Vision 2015–25 program seeks to respond to the most important and exciting scientific questions that European scientists want to address by space missions in the time frame 2015–25.[h] In fact, Cosmic Vision marks a significant breakthrough for fundamental physics: for the first time, a major space agency has given full emphasis in its forward planning to missions dedicated to exploring and advancing the limits of our understanding of many fundamental physics issues, including gravitation, unified theories, and quantum theory.[310,ddd] NASA would benefit from a similar bold and visionary approach. Proposals submitted to the call for "Astrophysics Strategic Mission Concept Studies" of ROSES[eee] could lead to such an opportunity and potentially result in a dedicated space-based laboratory experiment in fundamental physics.

B.1. *Dedicated missions to test relativistic gravity*

There are two distinct approaches to organizing research in fundamental physics in space — either a dedicated space mission to perform an experiment or a smaller-scale-experiment as a part of a planetary exploration mission. We discuss below these approaches in some detail.

NASA has successfully launched two PI-led missions devoted to experimental tests of general relativity, namely Gravity Probe A[fff] and Gravity Probe B.

The first convincing measurement of the third of Einstein's proposed tests of general theory of relativity — the gravitational redshift — was made in 1960 by Pound and Rebka. The Pound–Rebka experiment was based on Mössbauer effect measurements between sources and detectors spanning the 22.5 m tower in the Jefferson Physical Laboratory at Harvard. In 1976, Gravity Probe A exploited the much higher "tower" enabled by space; a suborbital Scout rocket carried a hydrogen maser to an altitude of 10,273 km and a novel telemetry scheme allowed comparison with hydrogen masers on the ground. The clocks confirmed Einstein's prediction to 70 ppm. More than 30 years later, this remains the most precise measurement of the gravitational redshift.[18]

[ccc] Notably, the MicroSCOPE and ACES missions.

[ddd] In October 2007 several candidate missions were selected for consideration for launch in 2017/18. See http://sci.esa.int/science-e/www/object/index.cfm?fobjectid=41438. Although no fundamental physics mission was chosen at that time, the Cosmic Vision opportunity strongly motivated the work in the entire area of space-based experimental research.

[eee] NASA Research Opportunities in Space and Earth Sciences (ROSES-2007). See details at http://nspires.nasaprs.com.

[fff] See http://en.wikipedia.org/wiki/Gravity_Probe_A

Gravity Probe B was launched on April 20, 2004. The goal of this experiment is to test two predictions of general relativity by means of measuring the spin direction of gyroscopes in orbit about the Earth. For the 642-km-high GP-B orbit, GR predicts two gyroscope precessions, the geodetic effect with a rate of 6.606 arcsec per year and the frame-dragging effect with a rate of 39 milliarcsec per year. A polar orbit is chosen so the two effects would occur at right angles and could be independently resolved. The science instrument was housed in the largest helium dewar ever flown in space. The helium lifetime set the experiment duration; the orbital setup and science data phase lasted 17.3 months, exceeding requirements.

GP-B was the first spacecraft with six-degrees-of-freedom active control: translation, attitude, and roll. He boil-off gas proportional thrusters provided control actuation enabling the "drag-free" control system to reduce cross tract acceleration to 10^{-11} g. The requirements on charge control, magnetic shielding, and pressure were all met with margin. Flight data confirm that the GP-B gyroscope disturbance drift rates were more than a factor of a million times smaller than the best modeled navigational gyroscopes. The small size of the relativity effects under test imposed extreme requirements on the mission; their successful demonstration in space has yielded and will yield many benefits to future fundamental physics missions in space. Final results are pending the completion of the data analysis scheduled for early 2008.

B.2. *Missions of opportunity on planetary missions*

In addition to the development of dedicated missions, another practical way to conduct fundamental physics experiments is to fly advanced instruments as Missions of Opportunity (MO's) on planetary missions and the ISS.

The LLR experiment that was initiated in 1969 by the Apollo 11 astronauts placing laser retroreflectors on the lunar surface[17] is an excellent example of a successful MO. The resulting fundamental physics experiment is still active today and is the longest-running experiment in the history of space science (see Sec. 2.1.2).

One can further improve LLR-enabled science by delivering to the Moon either new sets of laser retroreflector arrays or laser transponders pointed at the Earth, or both types of instruments.[51] A geographic distribution of new instruments on the lunar surface wider than the current distribution would be a great benefit; the accuracy of the lunar science parameters would increase several times. A bright transponder source on the Moon would open LLR to dozens of satellite laser ranging stations which cannot detect the current weak signals from the Moon. This would greatly benefit LLR — the living legacy of the Apollo program — and would also enhance the science outcome of the new lunar exploration efforts.

Highly accurate measurements of the round-trip travel times of laser pulses between an observatory on the Earth and an optical transponder on Mars could lead to major advances in gravitation and cosmology, while also enhancing our knowledge of the Martian interior. The technology is available to conduct such measurements with picosecond-level timing precision, which could translate into millimeter-class accuracies achieved in ranging between the Earth and Mars. Similar to its lunar predecessor, the resulting Mars Laser Ranging experiment could become an excellent facility to advance fundamental physics.

Other examples of successful MO's are the recently conducted gravity experiment on the Cassini mission performed on its way to Jupiter during one of the solar conjunctions[123] and a similar experiment that is planned for the ESA's Bepi-Colombo mission to Mercury.[311]

In general, research in fundamental and gravitational physics will greatly benefit from an established mechanism to participate on planetary missions as MO's that will offer sorely needed space deployment opportunities.

References

1. F. Wilczek, "Fundamental Physics in the 21st Century," talk given at the workshop From Quantum to Cosmos: Fundamental Physics Research in Space (Airlie Center, Warrenton, VA, USA, May 21–24, 2006), electronic version available at http://physics.jpl.nasa.gov/quantum-to-cosmos
2. F. Wilczek, *Int. J. Mod. Phys. A* **21** (2006) 2011 [physics/0511067].
3. W. D. Phillips, *Int. J. Mod. Phys. D* **16** (2007) 1953.
4. E. Witten, String theory, in *Proc. of APS/DPF/DPB Summer Study on the Future of Particle Physics* (*Snowmass 2001*) (Snowmass, Colorado, 30 Jun.–21 Jul. 2001), p. 337
5. E. Witten, The Past and Future of String Theory, in *The future of Theoretical Physics and Cosmology*, eds. G. W. Gibbons, E. P. S. Shellard and S. J. Rankin (Cambridge University Press, 2003), p. 455.
6. T. Damour and K. Nordtvedt, *Phys. Rev. Lett.* **70** (1993) 2217.
7. T. Damour and K. Nordtvedt, *Phys. Rev. D* **48** (1993) 3436.
8. T. Damour and A. M. Polyakov, *Gen. Relativ. Gravit.* **26** (1994) 1171.
9. T. Damour and A. M. Polyakov, *Nucl. Phys. B* **423** (1994) 532.
10. T. Damour, F. Piazza and G. Veneziano, *Phys. Rev. Lett.* **89** (2002) 081601 [gr-qc/0204094].
11. T. Damour, F. Piazza and G. Veneziano, *Phys. Rev. D* **66** (2002) 046007 [hep-th/0205111].
12. S. G. Turyshev *et al.*, *Lect. Notes Phys.* **648** (2004) 311 [gr-qc/0311039].
13. R. W. Hellings, *Int. J. Mod. Phys. D* **16** (2007) 2107.
14. D. DeBra, *Class. Quant. Grav.* **14** (1997) 1549.
15. C. Lämmerzahl, gr-qc/0402122.
16. J. G. Williams, S. G. Turyshev and D. H. Boggs, *Phys. Rev. Lett.* **93** (2004) 261101 [gr-qc/0411113].
17. J. G. Williams, S. G. Turyshev and D. H. Boggs, Lunar Laser Ranging Tests of the Equivalence Principle with the Earth and Moon, in *Proc. Testing the Equivalence Principle on Ground and in Space*, eds. C. Lämmerzahl, C. W. F. Everitt and R. Ruffini (Pescara, Italy, Sep. 20–23, 2004), [gr-qc/0507083] to be published.
18. C. M. Will, *Living Rev. Rel.* **9** (2006) 3 [gr-qc/0510072].
19. J. L. Singe, *Relativity: The General Theory* (North-Holland, Amsterdam, 1960).
20. J. D. Anderson *et al.*, *Astrophys. J.* **459** (1996) 365.
21. S. Baeßler *et al.*, *Phys. Rev. Lett.* **83** (1999) 3585.
22. E. G. Adelberger, *Class. Quant. Grav.* **18** (2001) 2397.

23. S. Schlamminger *et al.*, Improved Test of the Equivalence Principle, (APS Meeting, Apr. 14–17, 2007), abstract BAPS.2007.APR.C12.1 (2007).
24. J. Khoury and A. Weltman, *Phys. Rev. Lett.* **93** (2004) 171104 [astro-ph/0309300].
25. J. Khoury and A. Weltman, *Phys. Rev. D* **69** (2004) 044026 [astro-ph/0309411].
26. P. Brax *et al.*, *Phys. Rev. D* **70** (2004) 123518 [astro-ph/0408415].
27. P. Touboul and M. Rodrigues, *Class. Quant. Grav.* **18** (2001) 2487.
28. R. D. Reasenberg and J. D. Phillips, *Int. J. Mod. Phys. D* **16** (2007) 2245.
29. J. D. Phillips and R. D. Reasenberg, *Rev. Sci. Instr.* **76** (2005) 064501.
30. R. D. Reasenberg and J. D. Phillips, "A Test of the WEP on a Sounding Rocket Based on POEM," talk at the workshop "From Quantum to Cosmos: Space-Based Research in Fundamental Physics and Quantum Technologies" (Bremen, Germany, Jun. 10–13, 2007).
31. R. Spero, private communication (2007).
32. M. A. Kasevich and L. Maleki, "Quantum Interferometer Test of Equivalence Principle," a study funded by the NASA "Microgravity and Fundamental Physics" Program in 2003, electronic version available at http://horology.jpl.nasa.gov/quantum/pub/QuITEsingleposter2.pdf
33. A. Peters, K. Y. Chung and S. Chu, *Metrologia* **38** (2001) 25.
34. S. Fray *et al.*, *Phys. Rev. Lett.* **93** (2004) 240404 [physics/0411052].
35. R. Nyman *et al.*, *App. Phys. B* **84** (2006) 673.
36. A. M. Nobili *et al.*, *"Galileo Galilei" (GG) Phase A Study Report* (ASI, Nov. 1998), 2nd edn., Jan. 2000.
37. A. M. Nobili *et al.*, *Int. J. Mod. Phys. D* **16** (2007) 2259.
38. J. Mester *et al.*, *Class. Quant. Grav.* **18** (2001) 2475.
39. P. Worden, J. Mester and R. Torii, *Class. Quant. Grav.* **18** (2001) 2543.
40. J. Kolodziejczak and J. Mester, *Int. J. Mod. Phys. D* **16** (2007) 2215.
41. J. G. Williams, private communications (2007).
42. K. Nordtvedt, *Phys. Rev.* **169** (1968) 1014.
43. K. Nordtvedt, *Phys. Rev.* **169** (1968) 1017.
44. C. M. Will and K. Nordtvedt, *Astrophys. J.* **177** (1972) 757.
45. C. M. Will, *Theory and Experiment in Gravitational Physics* (Cambridge University Press, 1993).
46. T. W. Murphy, Jr. *et al.*, *Int. J. Mod. Phys. D* **16** (2007) 2127.
47. J. G. Williams, S. G. Turyshev and T. W. Murphy, Jr., *Int. J. Mod. Phys. D* **13** (2004) 757 [gr-qc/0311021].
48. D. E. Smith *et al.*, *Science* **311** (2006) 53.
49. X. Sun, Laser ranging between the mercury laser altimeter and an earth-based laser satellite tracking station over a 24 million kilometer distance, OSA Annual Meeting Abstracts (Tucson, AZ, USA, Oct. 16–20, 2005).
50. J. J. Degnan, *Int. J. Mod. Phys. D* **16** (2007) 2137.
51. S. G. Turyshev and J. G. Williams, *Int. J. Mod. Phys. D* **16** (2007) 2165.
52. J. F. Chandler *et al.*, Solar-system dynamics and tests of general relativity with planetary laser ranging, in *Proc. 14th International Workshop on Laser Ranging*, eds. R. Noomen *et al.* (San Fernando, Spain, 2005), electronic version at http://cddis.nasa.gov/lw14/docs/papers/sci7b_jcm.pdf
53. S. M. Merkowitz *et al.*, *Int. J. Mod. Phys. D* **16** (2007) 2151.
54. O. Bertolami, J. Paramos and S. G. Turyshev, General theory of relativity: Will it survive the next decade? in *Lasers, Clocks and Drag-Free Control: Exploration of Relativistic Gravity in Space*, astrophysics and space science library **349**, eds. H. Dittus, C. Laemmerzahl, S. G. Turyshev (Springer-Verlag, 2007), p. 27 [gr-qc/0602016].

55. V. M. Canuto and I. Goldman, *Nature* **296** (1982) 709.
56. V. M. Canuto and I. Goldman, *Int. J. Theor. Phys.* **28** (1989) 1005.
57. J.-P. Uzan, *Rev. Mod. Phys.* **75** (2003) 403 [hep-ph/0205340].
58. J.-P. Uzan, astro-ph/0409424.
59. J. G. Williams, S. G. Turyshev and D. H. Boggs, *Phys. Rev. Lett.* **98** (2007) 059002 [gr-qc/0612171].
60. M. Kramer *et al.*, *Science* **314** (2006) 97.
61. K. Nordtvedt, *Int. J. Mod. Phys. A* **17** (2002) 2711 [gr-qc/0212044].
62. T. Dent, *J. Cosmol. Astropart. Phys.* **01** (2007) 013 [hep-ph/0608067].
63. T. Damour, *Astrophys. Space Sci.* **283** (2003) 445 [gr-qc/0210059].
64. G. Dvali and M. Zaldarriaga, *Phys. Rev. Lett.* **88** (2002) 091303 [hep-ph/0108217].
65. T. M. Fortier *et al.*, *Phys. Rev. Lett.* **98** (2007) 070801.
66. L. Maleki and J. Prestage, *Lect. Notes Phys.* **562** (2001) 329.
67. L. Maleki and J. Prestage, *Int. J. Mod. Phys. D* **16** (2007) to appear in Issue 12B.
68. S. Schiller *et al.*, *Nucl. Phys. B (Proc. Suppl.)* **166** (2007) 300, [gr-qc/0608081].
69. N. Arkani-Hamed, S. Dimopoulos and G. R. Dvali, *Phys. Lett. B* **429** (1998) 263 [hep-ph/9803315].
70. N. Arkani-Hamed, S. Dimopoulos and G. R. Dvali, *Phys. Rev. D* **59** (1999) 086004 [hep-ph/9807344].
71. E. G. Adelberger, B. R. Heckel and A. E. Nelson, *Ann. Rev. Nucl. Part. Sci.* **53** (2003) 77.
72. I. Antoniadis, S. Dimopoulos and G. R. Dvali, *Nucl. Phys. B* **516** (1998) 70 [hep-ph/9710204].
73. S. Dimopoulos and G. F. Giudice, *Phys. Lett. B* **379** (1996) 105 [hep-ph/9602350].
74. R. Sundrum, *J. High Energy Phys.* **9907** (1999) 001 [hep-ph/9708329].
75. G. Dvali *et al.*, *Phys. Rev. D* **65** (2002) 024031 [hep-th/0106058].
76. D. J. Kapner *et al.*, *Phys. Rev. Lett.* **98** (2007) 021101 [hep-ph/0611184].
77. L.-C. Tu *et al.*, *Phys. Rev. Lett.* **98** (2007) 201101.
78. E. G. Adelberger *et al.*, hep-ph/0611223.
79. H.-J. Paik, V. A. Prieto and M. Vol Moody, *Int. J. Mod. Phys. D* **16** (2007) 2181.
80. H.-J. Paik, V. A. Prieto and M. Vol Moody, *J. Korean Phys. Soc.* **45** (2004) S104.
81. H.-J. Paik, M. Vol Moody and D. M. Strayer, *Gen. Relativ. Gravit.* **36** (2004) 523.
82. J. Chiaverini *et al.*, *Phys. Rev. Lett.* **90** (2003) 151101.
83. J. C. Long *et al.*, *Nature* **421** (2003) 922.
84. C. D. Hoyle *et al.*, *Phys. Rev. D* **70** (2004) 042004.
85. A. Kusenko and T. J. Weiler, *Phys. Rev. Lett.* **88** (2002) 161101 [hep-ph/0106071].
86. S. Palomares-Ruiz, A. Irimia and T. J. Weiler, *Phys. Rev. D* **73** (2006) 083003 [astro-ph/0512231].
87. R. D. Peccei and H. R. Quinn, *Phys. Rev. Lett.* **38** (1977) 1440.
88. R. D. Peccei and H. R. Quinn, *Phys. Rev. D* **16** (1977) 1791.
89. S. Weinberg, *Phys. Rev. Lett.* **40** (1978) 223.
90. F. Wilczek, *Phys. Rev. Lett.* **40** (1978) 279.
91. E. Adelberger, B. Heckel and C. D. Hoyle, Testing the Gravitational Inverse Square Law, *Phys. World*, Apr. 3, 2005, online version available at http://physicsworld.com/cws/article/print/21822
92. G. Dvali, A. Gruzinov and M. Zaldarriaga, *Phys. Rev. D* **68** (2003) 024012 [hep-ph/0212069].
93. J. D. Anderson *et al.*, *Phys. Rev. Lett.* **81** (1998) 2858 [gr-qc/9808081].
94. J. D. Anderson *et al.*, *Phys. Rev. D* **65** (2002) 082004/1–50 [gr-qc/0104064].

95. S. G. Turyshev *et al.*, The apparent anomalous, weak, long-range acceleration of pioneer 10 and 11, in *Gravitational Waves and Experimental Gravity, Proc.* XVIIIth Workshop of the Rencontres de Moriond (Les Arcs, Savoi, France, Jan. 23–30, 1999), eds. J. Dumarchez and J. Tran Thanh Van (World Hanoi, 2000), p. 481 [gr-qc/9903024].
96. S. G. Turyshev, M. M. Nieto and J. D. Anderson, *Amer. J. Phys.* **73** (2005) 1033 [physics/0502123].
97. M. Milgrom, *Acta Phys. Pol. B* **32** (2001) 3613.
98. R. Foot and R. R. Volkas, *Phys. Lett. B* **517** (2001) 13 [hep-ph/0108051].
99. O. Bertolami and J. Páramos, *Class. Quant. Grav.* **21** (2004) 3309 [gr-qc/0310101].
100. O. Bertolami and J. Páramos, *Phys. Rev. D* **71** (2005) 023521 [astro-ph/0408216].
101. J. D. Bekenstein, *Phys. Rev. D* **70** (2004) 083509 [astro-ph/0403694].
102. M.-T. Jaekel and S. Reynaud, *Class. Quant. Grav.* **22** (2005) 2135 [gr-qc/0502007].
103. J. W. Moffat, *J. Cosmol. Astropart. Phys.* **03** (2006) 004 [gr-qc/0506021].
104. J. R. Brownstein and J. W. Moffat, *Class. Quant. Grav.* **23** (2006) 3427 [gr-qc/0511026].
105. O. Bertolami *et al.*, *Phys. Rev. D* **75** (2007) 104016 [gr-qc/0704.1733].
106. J.-P. Bruneton and G. Esposito-Farèse, gr-qc/0705.4043.
107. M. M. Nieto and S. G. Turyshev, *Class. Quant. Grav.* **21** (2004) 4005 [gr-qc/0308017].
108. S. G. Turyshev, M. M. Nieto and J. D. Anderson, *Adv. Space Res.* **39** (2007) 291 [gr-qc/0409117].
109. H. Dittus *et al.*, *ESA Publication* **SP-588** (2005) 3 [gr-qc/0506139].
110. S. G. Turyshev *et al.*, *Int. J. Mod. Phys. D* **15** (2006) 1 [gr-qc/0512121].
111. V. T. Toth and S. G. Turyshev, *Can. J. Phys.* **84** (2006) 1063 [gr-qc/0603016].
112. P. G. Antreasian and J. R. Guinn, Investigations into the Unexpected ΔV Increases During the Earth Gravity Assists of Galileo and NEAR, paper AIAA 98-4287, presented at AIAA/AAS Astrodynamics Specialist Conference and Exhibit (Boston, Aug. 10–12, 1998).
113. J. D. Anderson, J. K. Campbell and M. M. Nieto, *New Astron.* **12** (2007) 383 [astro-ph/0608087].
114. C. Lämmerzahl, O. Preuss and H. Dittus, *Int. J. Mod. Phys. D* **16** (2007) to appear in Issue 12B.
115. C. Deffayet, G. Dvali and G. Gabadadze, *Phys. Rev. D* **65** (2002) 044023 [astro-ph/0105068].
116. G. Dvali and M. Turner, astro-ph/0301510.
117. S. M. Carrol *et al.*, *Phys. Rev. D* **70** (2004) 043528 [astro-ph/0306438].
118. C. Deffayet *et al.*, *Phys. Rev. D* **65** (2002) 044026 [hep-th/0106001].
119. G. Dvali, *New J. Phys.* **8** (2006) 326 [hep-th/0610013].
120. G. Dvali, G. Gabadadze and M. Porrati, *Phys. Lett. B* **485** (2000) 208 [hep-th/0005016].
121. C. Deffayet, *Phys. Lett. B* **502** (2001) 199 [hep-th/0010186].
122. C. Deffayet, G. Dvali and G. Gabadadze, *Phys. Rev. D* **65** (2002) 044023 [astro-ph/0105068].
123. B. Bertotti, L. Iess and P. Tortora, *Nature* **425** (2003) 374.
124. D. N. Spergel *et al.*, to be published in *Astrophys. J.* (2007) [astro-ph/0603449].
125. T. Damour and G. Esposito-Farese, *Phys. Rev. D* **53** (1996) 5541 [gr-qc/9506063].
126. T. Damour and G. Esposito-Farese, *Phys. Rev. D* **54** (1996) 1474 [gr-qc/9602056].
127. J. D. Anderson *et al.*, *Planet. Space Sci.* **45** (1997) 21 [astro-ph/9510081].
128. N. Ashby, P. L. Bender and J. M. Wahr, *Phys. Rev. D* **75** (2007) 022001.

129. K. L. Nordtvedt, *Icarus* **129** (1997) 120.
130. A. S. Konopliv *et al.*, *Icarus* **182** (2006) 23.
131. N. Ashby and P. L. Bender, Measurement of the shapiro time delay between drag-free spacecraft, in *Lasers, Clocks, and Drag-Free: Technologies for Future Exploration in Space and Tests of Gravity*, eds. H. Dittus, C. Laemmerzahl and S. Turyshev (Springer-Verlag, 2006), p. 219.
132. P. L. Bender *et al.*, Requirements for measuring the gravitational time delay between drag-free spacecraft, in *Proc. Advances in Precision Tests and Experimental Gravitation in Space* (Firenze, Italy, Sep. 28–30, 2006), electronic version available at http://www.fi.infn.it/GGI-grav-space/EGS_w/pdf/bender.pdf
133. S. G. Turyshev, M. Shao and K. Nordtvedt, Science, technology and mission design for the laser astrometric test of relativity mission, in *Lasers, Clocks, and Drag-Free: Technologies for Future Exploration in Space and Tests of Gravity*, eds. H. Dittus, C. Laemmerzahl and S. Turyshev (Springer-Verlag, 2006), p. 429 [gr-qc/0601035].
134. S. G. Turyshev *et al.*, *ESA Spec. Publ.* **588** (2005) 11 [gr-qc/0506104].
135. S. G. Turyshev, M. Shao and K. Nordtvedt, *Int. J. Mod. Phys. D* **13** (2004) 2035 [gr-qc/0410044].
136. S. G. Turyshev, M. Shao and K. Nordtvedt, *Class. Quant. Grav.* **21** (2004) 2773 [gr-qc/0311020].
137. J. E. Plowman and R. W. Hellings, *Class. Quant. Grav.* **23** (2006) 309 [gr-qc/0505064].
138. G. Dvali, *New J. Phys.* **8** (2006) 326 [hep-th/0610013].
139. H. van Dam and M. J. G. Veltman, *Nucl. Phys. B* **22** (1970) 397.
140. V. I. Zakharov, *J. Exp. Theor. Phys. Lett.* **12** (1970) 312.
141. M. Fierz and W. Pauli, *Proc. Roy. Soc. Lond. A* **173** (1939) 211.
142. D. J. Eisenstein *et al.*, *Astrophys. J.* **633** (2005) 560 [astro-ph/0501171].
143. Report from the Dark Energy Task Force, Jun. 6, 2006 [astro-ph/0609591], or access it directly at http://www.nsf.gov/mps/ast/aaac/dark_energy_task_force/report/detf_final_report.pdf
144. V. A. Kostelecký, *Phys. Rev. D* **69** (2004) 105009 [hep-th/0312310].
145. G. Dvali, O. Pujolas and M. Redi, hep-th/0702117.
146. O. Bertolami and C. S. Carvalho, *Phys. Rev. D* **74** (2006) 084020 [gr-qc/0607043].
147. V. A. Kostelecký and M. Mewes, astro-ph/0702379.
148. O. Bertolami *et al.*, *Phys. Lett. B* **395** (1997) 178.
149. S. Carroll and J. Shu, *Phys. Rev. D* **73** (2006) 103515.
150. K. Greisen, *Phys. Rev. Lett.* **16** (1966) 748.
151. G. T. Zatsepin and V. A. Kuzmin, *Pisma Zh. Eksp. Teor. Fiz.* **4** (1966) 114.
152. G. T. Zatsepin and V. A. Kuzmin, *J. Exp. Theor. Phys. Lett.* **4** (1966) 78.
153. G. Bertone, D. Hooper and J. Silk, *Phys. Rep.* **405** (2005) 279.
154. G. Jungman, M. Kamionkowski and K. Griest, *Phys. Rep.* **267** (1996) 195.
155. A. Kusenko, *Phys. Lett. B* **405** (1997) 108.
156. A. Kusenko and M. E. Shaposhnikov, *Phys. Lett. B* **418** (1998) 46.
157. A. Kusenko *et al.*, *Phys. Rev. Lett.* **80** (1998) 3185.
158. K. Enqvist and A. Mazumdar, *Phys. Rep.* **380** (2003) 99.
159. M. Dine and A. Kusenko, *Rev. Mod. Phys.* **76** (2004) 1.
160. S. Dodelson and L. M. Widrow, *Phys. Rev. Lett.* **72** (1994) 17.
161. X. D. Shi and G. Fuller, *Phys. Rev. Lett.* **82** (1999) 2832.
162. K. Abazajian, G. M. Fuller and M. Patel, *Phys. Rev. D* **64** (2001) 023501.
163. A. D. Dolgov and S. H. Hansen, *Astropart. Phys.* **16** (2002) 339.
164. T. Asaka, S. Blanchet and M. Shaposhnikov, *Phys. Lett. B* **631** (2005) 151.

165. K. Abazajian, *Phys. Rev. D* **73** (2006) 063506.
166. M. Shaposhnikov and I. Tkachev, *Phys. Lett. B* **639** (2006) 414.
167. A. Kusenko, *Phys. Rev. Lett.* **97** (2006) 241301.
168. D. Boyanovsky and C. M. Ho, hep-ph/0612092.
169. T. Asaka, M. Laine and M. Shaposhnikov, *J. High Energy Phys.* **0701** (2007) 091.
170. LSND Collab. (C. Athanassopoulos *et al.*), *Phys. Rev. Lett.* **75** (1995) 2650 [nucl-ex/9504002].
171. LSND Collab. (C. Athanassopoulos *et al.*), *Phys. Rev. Lett.* **77** (1996) 3082 [nucl-ex/9605003].
172. LSND Collab. (C. Athanassopoulos *et al.*), *Phys. Rev. Lett.* **81** (1998) 1774 [nucl-ex/9709006].
173. LSND Collab. (A. Aguilar *et al.*), *Phys. Rev. D* **64** (2001) 112007 [hep-ex/0104049].
174. MiniBooNE Collab. (A. A. Aguilar-Arevalo *et al.*), hep-ex/0704.1500.
175. A. Kusenko and G. Segre, *Phys. Lett. B* **396** (1997) 197.
176. A. Kusenko and G. Segre, *Phys. Rev. D* **59** (1999) 061302.
177. G. Fuller *et al.*, *Phys. Rev. D* **68** (2003) 103002.
178. M. Barkovich, J. C. D'Olivo and R. Montemayor, *Phys. Rev. D* **70** (2004) 043005.
179. A. Kusenko, *Int. J. Mod. Phys. D* **13** (2004) 2065.
180. C. Fryer and A. Kusenko, *Astrophys. J. Suppl.* **163** (2006) 335.
181. P. L. Biermann and A. Kusenko, *Phys. Rev. Lett.* **96** (2006) 091301.
182. M. Mapelli, A. Ferrara and E. Pierpaoli, *Mon. Not. Roy. Astron. Soc.* **369** (2006) 1719.
183. J. Stasielak, P. L. Biermann and A. Kusenko, *Astrophys. J.* **654** (2007) 290.
184. E. Ripamonti, M. Mapelli and A. Ferrara, *Mon. Not. Roy. Astron. Soc.* **375** (2007) 1399.
185. E. K. Akhmedov, V. A. Rubakov and A. Y. Smirnov, *Phys. Rev. Lett.* **81** (1998) 1359.
186. T. Asaka and M. Shaposhnikov, *Phys. Lett. B* **620** (2005) 17.
187. A. Kusenko, *Int. J. Mod. Phys. D* **16** (2007) to appear in Issue 12B.
188. GLAST Collab. (A. Moriselli *et al.*), *Nucl. Phys. Proc. Suppl.* **113** (2002) 213.
189. H. Baer and S. Profumo, *J. Cosmol. Astropart. Phys.* **0512** (2005) 008 [astro-ph/0510722].
190. K. Abazajian, G. M. Fuller and W. H. Tucker, *Astrophys. J.* **562** (2001) 593.
191. A. Boyarsky *et al.*, astro-ph/0512509.
192. A. Boyarsky *et al.*, *J. Exp. Theor. Phys. Lett.* **83** (2006) 133.
193. A. Boyarsky *et al.*, astro-ph/0603368.
194. A. Boyarsky *et al.*, astro-ph/0603660.
195. S. Riemer-Sorensen, S. H. Hansen and K. Pedersen, astro-ph/0603661.
196. K. Abazajian and S. M. Koushiappas, astro-ph/0605271.
197. C. R. Watson *et al.*, astro-ph/0605424.
198. K. N. Abazajian *et al.*, astro-ph0611144.
199. S. Riemer-Sorensen *et al.*, astro-ph/0610034.
200. A. Boyarsky *et al.*, astro-ph/0612219.
201. D. Colladay and V. A. Kostelecký, *Phys. Rev. D* **55** (1997) 6760 [hep-ph/9703464].
202. D. Colladay and V. A. Kostelecký, *Phys. Rev. D* **58** (1998) 116002 [hep-ph/9809521].
203. S. R. Coleman and S. L. Glashow, *Phys. Lett. B* **405** (1997) 249 [hep-ph/9703240].
204. S. R. Coleman and S. L. Glashow, *Phys. Rev. D* **59** (1999) 116008 [hep-ph/9812418].
205. V. A. Kostelecký and R. Potting, *Phys. Rev. D* **51** (1995) 3923 [hep-ph/9501341].
206. F. W. Stecker and S. L. Glashow, *Astropart. Phys.* **16** (2001) 97 [astro-ph/0102226].
207. V. A. Kostelecký and S. Samuel, *Phys. Rev. Lett.* **66** (1991) 1811.

208. V. A. Kostelecký and R. Potting, *Phys. Lett. B* **381** (1996) 89 [hep-th/9605088].
209. V. A. Kostelecký and R. Potting, *Phys. Rev. D* **63** (2001) 046007 [hep-th/0008252].
210. V. A. Kostelecký, M. Perry and R. Potting, *Phys. Rev. Lett.* **84** (2000) 4541 [hep-th/9912243].
211. V. A. Kostelecký and S. Samuel, *Phys. Rev. D* **39** (1989) 683.
212. V. A. Kostelecký and R. Potting, *Nucl. Phys. B* **359** (1991) 545.
213. B. Altschul and V. A. Kostelecký, *Phys. Lett. B* **628** (2005) 106 [hep-th/0509068].
214. V. A. Kostelecký and S. Samuel, *Phys. Rev. Lett.* **63** (1989) 224.
215. V. A. Kostelecký and S. Samuel, *Phys. Rev. D* **40** (1989) 1886.
216. O. Bertolami and C. S. Carvalho, *Phys. Rev. D* **61** (2000) 103002 [gr-qc/9912117].
217. Q. G. Bailey and V. A. Kostelecký, *Phys. Rev. D* **74** (2006) 045001 [gr-qc/0603030].
218. V. A. Kostelecký, ed., *CPT and Lorentz Symmetry* (World Scientific, Singapore, 1999).
219. V. A. Kostelecký, *CPT and Lorentz Symmetry II* (World Scientific, Singapore, 2002).
220. V. A. Kostelecký, *CPT and Lorentz Symmetry III* (World Scientific, Singapore, 2005).
221. R. Bluhm, hep-ph/0506054.
222. D. Mattingly, *Living Rev. Rel.* **8** (2005) 5 [gr-qc/0502097].
223. G. Amelino-Camelia *et al.*, *AIP Conf. Proc.* **758** (2005) 30 [gr-qc/0501053].
224. H. Vucetich, gr-qc/0502093.
225. N. Russell, *Phys. Scripta* **72** (2005) C38 [hep-ph/0501127].
226. N. Russel, hep-ph/0608083.
227. O. Bertolami, *Lect. Notes Phys.* **633** (2003) 96 [hep-ph/0301191].
228. T. Jacobson and D. Mattingly, *Phys. Rev. D* **64** (2001) 024028 [gr-qc/0007031].
229. T. Jacobson and D. Mattingly, *Phys. Rev. D* **70** (2004) 024003 [gr-qc/0402005].
230. B. Z. Foster and T. Jacobson, *Phys. Rev. D* **73** (2006) 064015 [gr-qc/0509083].
231. W. Bernreuther and M. Suzuki, *Rev. Mod. Phys.* **63** (1991) 313, [Errata **66** (1992) 633].
232. M. Pospelov and A. Ritz, *Ann. Phys.* **318** (2005) 119.
233. B. C. Regan *et al.*, *Phys. Rev. Lett.* **88** (2002) 071805.
234. J. J. Hudson *et al.*, *Phys. Rev. Lett.* **89** (2002) 023003.
235. K. Abdullah *et al.*, *Phys. Rev. Lett.* **65** (1990) 2347.
236. S. A. Murthy *et al.*, *Phys. Rev. Lett.* **63** (1989) 965.
237. R. N. Mohaparta *et al.*, hep-ph/0510213.
238. T. P. Heavner *et al.*, PARCS: A laser-cooled atomic clock in space, in *Proc. 2001 Freq. Stand. Metrology Symp.* (2001), p. 253.
239. N. Ashby, PARCS: Primary atomic reference clock in space, in *Proc. Second Meeting on CPT and Lorentz Symmetry* (Bloomington, IN, USA, Aug. 15–18, 2001), ed. A. Kostelecký (World Scientific, 2002), p. 26.
240. D. B. Sullivan *et al.*, *Adv. Space Res.* **36** (2005) 107.
241. C. Fertig *et al.*, RACE: Laser-cooled Rb microgravity clock, in *Proc. 2000 IEEE/EIA Freq. Contr. Symp.* (2000), p. 676.
242. J. A. Lipa *et al.*, *Adv. Space Res.* **35** (2005) 82.
243. C. Salomon *et al.*, *C. R. Acad. Sci. Paris t.2 Série* **4** (2001) 1313.
244. L. Cacciapuoti *et al.*, in *Proc. 1st ESA International Workshop on Optical Clocks*, Jun 8–10, 2005, Noordwijk, The Netherland (ESA Publication, 2005), p. 45.
245. L. Cacciapuoti *et al.*, *Nucl. Phys. B* **166** (2007) 303.
246. C. Salomon, L. Cacciapuoti and N. Dimarcq, *Int. J. Mod. Phys. D* **16** (2007) to appear in Issue 12B.
247. H. Marion *et al.*, *Phys. Rev. Lett.* **90** (2003) 150801.

248. S. Bize *et al.*, *Phys. Rev. Lett.* **90** (2003) 150802.
249. M. Fischer *et al.*, *Phys. Rev. Lett.* **92** (2004) 230802.
250. E. Peik *et al.*, *Phys. Rev. Lett.* **93** (2004) 170801.
251. T. M. Fortier *et al.*, *Phys. Rev. Lett.* **98** (2007) 070801.
252. P. Wolf and G. Petit, *Phys. Rev. A* **56** (1997) 4405.
253. L. Wolfenstein and T. G. Trippe, *Phys. Lett. B* **592** (2004) 1, http://pdg.lbl.gov
254. A. D. Sakharov, *Sm. Zh. Eksp. Teor. Fiz.* **5** (1967) 32.
255. A. D. Sakharov, *J. Exp. Theor. Phys. Lett.* **5** (1967) 24.
256. S. M. Barr, *Int. J. Mod. Phys.* **8** (1993) 209.
257. A. Abel, S. Khalil and O. Lebedev, *Nucl. Phys. B* **606** (2001) 151 and references therein.
258. T. Applequist, M. Piai and R. Shrock, *Phys. Lett. B* **593** (2004) 175.
259. N. Arkani-Hamed *et al.*, *Nucl. Phys. B* **709** (2005) 3.
260. D. Chang, W.-F. Chang and W.-Y. Keung, *Phys. Rev. D* **71** (2005) 076006.
261. G. F. Giudice and A. Romanino, *Phys. Lett. B* **634** (2006) 307.
262. J. M. Amini, C. T. Munger, Jr. and H. Gould, *Phys. Rev. A* **75** (2007) 063416.
263. J. D. Prestage, JPL, private communication (2007).
264. P. Laurent *et al.*, *Appl. Phys. B* **84** (2006) 683.
265. W. H. Oskay *et al.*, *Phys. Rev. Lett.* **97** (2006) 020801.
266. T. Rosenband *et al.*, An aluminium ion optical clock, to appear in *Proc. 20th European Frequency and Time Forum* (Braunschweig, Germany, Mar. 27–30, 2006).
267. T. Rosenband, *et al.*, An aluminum ion optical clock using quantum logic (American Physical Society, 37th Meeting of the Division of Atomic, Molecular and Optical Physics, May 16–20, 2006), abstract #K4.007.
268. D. B. Hume *et al.*, Quantum State Detection through Repetitive Mapping (APS Meeting, Mar. 5–9, 2007), abst. #N33.002.
269. T. Rosenband *et al.*, physics/0703067.
270. M. M. Boyd *et al.*, *Science* **314** (2006) 1430.
271. A. D. Ludlow *et al.*, *Opt. Lett.* **32** (2007) 641.
272. L.-S. Ma *et al.*, *Science* **303** (2004) 1843.
273. S. A. Diddams *et al.*, *Phys. Rev. Lett.* **84** (2000) 5102.
274. R. J. Kennedy and E. M. Thorndike, *Phys. Rev.* Series 2 **42** (1932) 400.
275. W. D. Phillips and J. Ye, private communication (2007).
276. S. Schiller, private communication (2007).
277. The authors acknowledge contributions from J. Ye of NIST in discussing this experiment (2007).
278. J. A. Lipa *et al.*, *Phys. Rev. Lett.* **90** (2003) 060403.
279. F. Müller *et al.*, *Appl. Phys. B* **80** (2005) 307.
280. S. Herrmann *et al.*, *Phys. Rev. Lett.* **95** (2005) 150401.
281. H. Müller, A. Peters and C. Braxmaier, *Appl. Phys. B* **84** (2006) 401 [physics/0511072].
282. H. Müller *et al.*, *Phys. Rev. Lett.* **91** (2003) 020401 [physics/0305117].
283. S. Schiller, P. Antonini and M. Okhapkin, A precision test of the isotropy of the speed of light using rotating cryogenic optical cavities, in *Special Relativity: Will it Hold Another 100 Years* (Springer, 2006), Lecture Notes in Physics, eds. J. Ehlers and C. Lämmerzahl [physics/0510169].
284. J. Ye, *Opt. Lett.* **29** (2004) 1153.
285. C. J. Bordé, *Phys. Lett. A* **140** (1989) 10.
286. C. J. Bordé, *C. R. Acad. Sci. Paris Serie IV* **2** (2001) 509.

287. C. J. Bordé, J.-C. Houard and A. Karasievich, Relativistic phase shifts for dirac particles interacting with weak gravitational fields in matter-wave interferometers, in *Gyros, Clocks, and Interferometers: Testing Relativistic Gravity in Space*, eds. C. Lämmerzahl, C. W. F. Everitt and F. W. Hehl (Springer-Verlag, Berlin, 2001), Lecture Notes in Physics **562** (2001) 403.
288. A. Peters, K. Y. Chung and S. Chu, *Nature* **400** (1999) 849.
289. T. L. Gustavson, A. Landragin and M. A. Kasevich, *Class. Quant. Grav.* **17** (2000) 2385.
290. A. Miffre *et al.*, quant-ph/0605055.
291. K. Eckert *et al.*, *Phys. Rev. A* **73** (2006) 013814 [quant-ph/0507061].
292. C. Jentsch *et al.*, *Gen. Relativ. Gravit.* **36** (2004) 2197.
293. A. Bertoldi *et al.*, *Eur. Phys. J. D* **40** (2006) 271.
294. J. Fixler *et al.*, *Science* **315** (2007) 74.
295. S. Dimopoulos *et al.*, *Phys. Rev. Lett.* **98** (2007) 111102.
296. N. Yu *et al.*, *Appl. Phys. B.* **84** (2006) 647.
297. Y. Le Coq *et al.*, *Appl. Phys. B.* **84** (2006) 627.
298. B. Canuel *et al.*, *Phys. Rev. Lett.* **97** (2006) 010402.
299. G.-B. Jo *et al.*, cond-mat/0703006.
300. G.-B. Jo *et al.*, *Phys. Rev. Lett.* **98** (2007) 030407 [cond-mat/0608585].
301. D. S. Durfee , Y. K. Shaham and M. A. Kasevich, *Phys. Rev. Lett.* **97** (2006) 240801.
302. Proc. European workshop on "Quantum Mechanics for Space," *App. Phys. B* **84** (2006).
303. G. M. Tino *et al.*, *Nucl. Phys. B* **166** (2007) 159.
304. W. Ketterle, *Int. J. Mod. Phys. D* **16** (2007) to appear in Issue 12B.
305. M. W. Zwierlein *et al.*, *Nature* **442** (2006) 54.
306. J. Ye *et al.*, Precision measurement based on ultracold atoms and cold molecules, *AIP Conf. Proc.* **869** (2006) 80.
307. P. Boyer, *Int. J. Mod. Phys. D* **16** (2007) to appear in Issue 12B.
308. J. H. Marburger, III, "Space-Based Science and the American Competitiveness Initiative," keynote address at the workshop From Quantum to Cosmos: Fundamental Physics Research in Space (Airlie Center, Warrenton, VA, USA, May 21–24, 2006) to be published (2007), electronic version is at http://physics.jpl.nasa.gov/quantum-to-cosmos
309. M. C. E. Huber, *Int. J. Mod. Phys. D* **16** (2007) 1967.
310. B. F. Schutz, Fundamental physics in ESA's cosmic vision plan, in *Proc. 9th Int. Conf. Advanced Technology and Particle Physics* (Villa Olmo, Como, Oct. 17–21 2005), electronic version at http://villaolmo.mib.infn.it/ICATPP9th_2005/Space_Experiments/Schutz.pdf
311. L. Iess and S. Asmar, *Int. J. Mod. Phys. D* **16** (2007) 2117.

SPACE-BASED SCIENCE AND THE AMERICAN COMPETITIVENESS INITIATIVE

JOHN H. MARBURGER, III

Director, Office of Science and Technology Policy,
Executive Office of the President,
Washington, DC 20502, USA
John_H._Marburger@ostp.eop.gov

I discuss the process by which science contributes to the setting of government priorities, and how these priorities get translated into programs and budgets at the federal agencies that fund scientific research. New technologies are now opening exciting scientific opportunities across the biological and physical sciences. I review the motivations and goals of President Bush's American Competitiveness Initiative (ACI), the importance of societal relevance to federal investments in basic research, and the ACI's impacts on discovery-oriented disciplines within the physical sciences.

Keywords: Science policy; fundamental physics; space-based science.

Thanks to the conference organizers for inviting me to speak this morning. One of the things I had to get used to as the President's Science Advisor is that I am never asked to speak at scientific meetings to report on recent important research results, which is probably evidence of good judgment by the program committee. At most what I can do is give some insight into areas of science relevant to government priorities, and lift the edge of the veil over the mysterious process by which those priorities get translated into programs and budgets. But I cannot resist beginning with some reflections on the science. This is dangerous, because I know less about the possibilities than you do, and it would be wiser for me to sit and listen and learn. Enticing science opportunities are obviously available to us now as a result of the rapid accumulation of new technologies.

Science, after all, progresses where technology and theory intersect. The technology part throughout most of the history of science has served to extend the range of empirical observation. Telescopes, microscopes, spectroscopy, means of achieving and measuring lower temperatures, higher pressures, better space and time resolution, greater energy densities — all these are ways to extend the range of our senses, and advances in each of them have revealed new structures in nature that have required advances in theory — that is, in extensions of our conceptual framework. And those extensions in turn usually produce new options for understanding the Universe, whose exploration requires yet further advances in technology. This stimulation of technology by discovery is a primary reason for society to support basic research.

During the closing decades of the 20th century, the technology of computing added a new dimension to this picture of discovery and invention. The implications of theory can now often be articulated so reliably and in such detail that the process of comparison with data has been transformed. Constraints on experimental design have relaxed because we can accommodate far more degrees of freedom in the systems under investigation. Laboratory experiments have become more like uncontrolled nature, and the natural phenomena we can analyze are much more complex. Physicists can extract useful data from high energy collisions that spew out enormous numbers of particles. Astronomers can trace events in the earliest stages of the Universe from observations of present-day large scale structures encompassing a huge multitude of galaxies. ("Enormous" is calibrated differently for physicists and astronomers, but they are converging.) The extraction of signals deeply embedded in noise, and the management of observational parameters in real time to set "event triggers" or take advantage of serendipitous events, are possible to an unprecedented degree. The information we have been able to glean from analyses of subtle properties of the cosmic-microwave-background radiation is astonishing. Instrumentation with the precision of the LIGO apparatus is almost incredible — 10^{-18} m over the 4 km interferometer arms — and it is to some extent traceable to powerful information processing. It is no accident that the agencies that support investigations into the most fundamental processes of nature have made significant investments in high-end computing, and these fields have benefited greatly from investments in computing made by other agencies for other missions.

The same information technology has become indispensable for harnessing the knowledge we already have about fundamental laws. Here the issue is not how things move or what they are made of — the issue is how they are made and how structure is related to function in highly complex objects. Astrophysics has its share of complexity, but it is difficult to match the complexity of living systems and their components. So it is amazing that we are able to simulate some important features of organic systems "from scratch" in computer studies. Similar "computer experiments" use our knowledge of fundamental forces at low energy to discover and interpret the behavior of complex molecules and materials that may have technological importance. However, we think of strategy in science or science policy, information technology has to be elevated to a strategic level in any discussion of work at the frontier.

As I understand it, this conference is predicated on the idea that "space" — however defined — should be regarded as part of the dynamic technology infrastructure that enables new science, and I think this idea has much merit. The possibility of placing scientific apparatus in free-fall outside Earth's atmosphere has created new opportunities for observational astronomy, high precision measurements, and materials studies. Even before Alan Guth linked particle physics with cosmology in 1979, we knew the Big Bang mechanism turns the entire Universe into a high-energy-physics experiment. Looking out into space is equivalent to observing nature at ever-higher energy densities and temperatures. The Big Bang means that telescopes — photon detectors — can perform the same function as the huge detectors at the world's great particle colliders–accelerators. The Universe itself is surely the grandest technology there is. At some point we are going to have to give up on Earth-based accelerators and turn to that great machinery in the sky to continue our search for the basic stuff of matter.

Meanwhile the saga of the great accelerators continues. The world physics community is grappling with the question of how to fund the next one, currently called the International Linear Collider. This is an important machine, much better suited to unraveling the symmetries likely to be involved in extensions to the Standard Model than the Large Hadron Collider currently under construction at the European accelerator center at CERN. The LHC is needed to give assurance that the current theory is on the right track, and to justify the expense of yet another huge accelerator (the ILC requires two opposing 20 km superconducting linear accelerators). A Japanese study concluded the cost of such a machine would be about $5 billion (certainly a low estimate). We should keep in mind that this is the same order of magnitude as the currently estimated cost of the James Webb Space Telescope. I jokingly referred to the difference in the definition of "enormous" between physicists and astronomers. There is a similar difference in the perceptions of what constitutes a very expensive project. For the cost of one large space project you can build apparatus for particle physics that will occupy several generations of physicists.

I will come back to issues of expense and priorities in a moment, but let me stress here that the convergence of particle physics, astronomy and cosmology is not only important for science, but for science policy and for the organization of science within the federal government. Already the Office of Management and Budget and Congress have mandated a joint advisory committee for NSF, DOE, and NASA — the Astronomy and Astrophysics Advisory Committee (AAAC) — that will be taken seriously by OMB and OSTP, and it will need to be taken seriously by the agencies as well if they expect support for their plans at the White House level. Garth Illingworth is providing outstanding leadership of this committee and I commend its recent Annual Report to this audience.

It is not only in astronomy and particle physics that space science and space-based science are playing important roles. This conference provides an important opportunity to review the entire spectrum of space-based activities that either exploit or enhance our understanding of physical science.

Everyone here is surely aware that President Bush launched two initiatives bearing on physical science in his State of the Union address in January — the American Competitiveness Initiative (ACI) and the Advanced Energy Initiative (AEI). Since then I have been speaking about these in many different forums, and I will devote most of the rest of my time this morning to the ACI. In March I spoke to HEPAP and subsequently to NASA's annual Goddard Memorial Symposium on these initiatives, and addressed particularly the fact (which was brought distinctly to my attention) that high energy and nuclear physics did not seem to be stressed in the ACI, and NASA was not included at all among the ACI's "prioritized agencies" scheduled for significant budget increases during the next ten years.

The ACI appeared following a year of high visibility advocacy from a variety of groups, culminating in a report by a National Academy of Sciences panel chaired by former Lockheed-Martin chairman Norm Augustine. It is not correct to think of the ACI as a response to the Augustine report, but the recommendations of the latter do significantly overlap the ACI and the AEI. Many other reports have appeared in recent years that make similar recommendations. They provide a policy context for understanding the significance of the presidential initiatives. My remarks on the policy context will appear in an article in the June issue of *Physics Today*, based on a speech I gave earlier this month at the 75th Anniversary Symposium of the American Institute of Physics. Most of the rest of my talk this morning will summarize these remarks.

The ACI differs from the recommendations of the Augustine report in a number of important respects. Its components include: expanded federal funding for selected agencies with physical science missions; improved tax incentives for industrial investment in research; improved immigration policies favorable to high tech talent from other countries; and a cluster of education and training initiatives designed to enhance math and science education, particularly at the K-12 level. A brochure is available on the OSTP website that goes into more detail (http://ostp.gov). A total of $910 million is slated for the FY07 budgets of three designated "physical science" agencies. This is a 9.3% increase for the selected agencies, and the plan is to double their collective budgets over 10 years, a cumulative cost of $50 billion. The three agencies are the DOE Office of Science, NSF, and what is called the NIST "core budget," which supports research as opposed to technology transfer programs.

As this audience knows, federal physical science funding has been flat in constant dollars for more than a decade. The reasons for this are well understood, but involve multiple factors. Most dramatic was the abrupt change in Department of Defense research starting in 1991, the year historians cite as the end of the Cold War. The Department of Energy too began a re-examination of the roles of its laboratories in the post–Cold War period. Recall that there was a recession during 1990–91, and Congress was looking for a "peace dividend" following the dissolution of the Soviet Union. Congress terminated the SSC project in 1992, and House Science Committee chairman George Brown exhorted scientists to rethink their case for continued funding, especially in physical science. Toward the end of the decade a new case did emerge in a document that ought to be better known. Congressman Vern Ehlers produced a report whose short title is "Unlocking the Future" which clearly stated the conclusion that the rationale for funding science was to ensure future economic competitiveness. While not emphasizing physical science, the report did stress that "It is important that the federal government fund basic research in a broad spectrum of scientific disciplines, including the physical, computational, life and social sciences, as well as mathematics and engineering, and resist overemphasis in a particular area or areas relative to others."

At the turn of the 21st century, science policy makers began to worry about a growing imbalance between support for biomedical science and support for physical science. Early in the new Bush Administration the President's Council of Advisors on Science and Technology (PCAST) released a report called "Assessing the US R&D Investment" which said: "All evidence points to a need to improve funding levels for physical sciences and engineering." At the time, the country was still suffering the economic consequences of the burst dot-com bubble, and was realigning budget priorities in response to the terrorist attacks of September 2001. Completing the commitment to double the NIH budget was the highest science priority, next to establishing an entirely new science-and-technology initiative for homeland security. Nevertheless the Administration continued to expand funding for targeted areas of physical science, including the recently introduced National Nanotechnology Initiative, and maintained funding for the Networking and Information Technology R&D program. The NSF budget continued to increase at a rate above inflation. In the first term of the Bush Administration, combined federal R&D funding soared at a rate unmatched since the early years of the Apollo program, a jump of 45% in constant dollars over four years.

The ACI improves conditions for many if not all areas of physical science, but emphasizes fields likely to produce economically important technologies in the future. These are not difficult to identify, and all developed countries recognize their importance. Chief among them is the continued exploitation of our recent ability to image, analyze, and manipulate matter at the atomic scale. New technologies can be expected to spring from improved atomic-level understanding of materials and their functional properties in organic as well as inorganic systems. This includes much of what we would call low energy physics, including atomic, molecular, and optical physics, and large parts of chemistry and biotechnology.

Opportunities exist in particle physics and space science and exploration as well, but these are not emphasized in the Competitiveness Initiative. Not that the US is withdrawing from these fields. Some of the increased budgets in NSF and DOE will increase their vigor. The overall NASA budget is sustained in the President's FY07 budget proposal, although space science is facing flat or diminished budgets for the next few years. In my view the US is devoting a very healthy budget to space science, and with 56 space science missions currently flying it would be hard to argue that our international leadership in this area is in jeopardy. The ACI priorities signal an intention to fund the machinery of science in a way that ensures continued leadership in fields likely to have the greatest impact on future technology and innovation. In particular, although the ACI will relieve some budget pressure on DOE high energy and nuclear physics, its priority thrust is toward the cluster of facilities and programs within Basic Energy Sciences (BES). BES is certainly under-funded relative to its importance to society, just as biomedical research was under-funded in the 1980's relative to its rapidly growing significance for health care. In an era of extraordinary demands on the US domestic discretionary budget, course corrections in federal science funding entail the setting of priorities, the rationale for which must recognize national objectives of the utmost importance.

Space science and space exploration remain priorities for the United States, and relative to other investments the federal funds devoted to them are substantial. Among science agencies, only NIH has a larger budget for science. Despite current stresses on the space science budget, I expect it will experience steady but not dramatic long term growth. Conferences like this one are important for raising awareness in the communities of science as well as among policy makers of the fact that space-based science is not the same as "space science" in the usual sense, and its needs and opportunities require special attention. In particular, agencies like the Department of Defense, the Department of Energy, and the Department of Homeland Security, whose missions depend on frontier technologies, need to be aware of the opportunities that space-based research and its applications hold for solving some of their problems.

From the strictly scientific point of view, the promise of space-based experiments is vast and exciting. I am grateful to the organizers of this workshop for inviting me, and I look forward to hearing and reading more about your ideas.

Reference

1. J. H. Marburger, III, *Phys. Today*, **56** (June 2006) 38.

FUNDAMENTAL PHYSICS AT NASA: TWO CRITICAL ISSUES AND FAIRBANK'S PRINCIPLE

C. W. FRANCIS EVERITT

W. W. Hansen Experimental Physics Laboratory,
Stanford University, Stanford, CA 94305-4085, USA
francis@relgyro.stanford.edu

Space offers eight distinct paths to controlled laboratory-style experiments in fundamental physics beyond the reach of any Earth-based laboratory. This impressive range of opportunity has already led to important physics but we are only at the beginning. Meanwhile, NASA is in crisis. This paper follows the wisdom of a great and inspiring physicist, William Fairbank, that every setback or difficulty is an opportunity. NASA needs new science; fundamental physics can provide it and with it fascinating new technologies — and training for a new generation of imaginative physicists and engineers.

Keywords: Fundamental physics in space; space policy; space-based research.

1. Space and Fundamental Physics

Fundamental physics at NASA is in crisis and faces possible collapse. Clear thinking will recognize this truth and be unflinching in analysis of causes, and if clear enough can generate hope. It is time to invoke "Fairbank's principle." Those of us who had the good fortune to know William Fairbank will recall his amazing gift for making disaster an opportunity for creative thought. This is what is called for now with fundamental physics in space.

The first step is one of definition, with a fair and grateful recognition of what NASA has already done, beginning with a distinction between observations and controlled experiments. Among observational missions, one thinks of the Cosmic Background Explorer (COBE) measurements of the 3 K background radiation, of WMAP, and of the future LISA and SNAP, but beyond them, space opens new pathways to controlled physics experiments that could never be done on Earth. Examples are Gravity Probe A (GP-A), Gravity Probe B (GP-B), the AMS search for antiprotons and a variety of condensed matter physics experiments. Then there is that loaded word, "fundamental." Any science has a core of problems thought of as fundamental, yet often a continuing debate on what makes them so. In physics with its many subdisciplines, different physicists may claim the term for different reasons. Given the theme of space-enabling physics experiments, our approach should be generous, by addition not exclusion, recognizing that the final point is the importance of an experiment within its own field. This is especially so in times of budget difficulty. Having cited Fairbank's principle, I quote besides him an eminent European physicist, Maurice Jacob, former Head of Theory and then of External Relations at CERN, and from 1994 to 1998, Chair of the ESA Fundamental Physics Advisory Group. His principle, combining fairness and worldly wisdom, is that when facing budget cuts, one should not cut programs but add them, enlarging the advocacy base.

In what follows, I propose to say a few words on what NASA has already contributed, then cover the admirable 1999 NASA *Roadmap for Fundamental Physics*, then describe — I hope with due courtesy — a very bad decision taken at NASA Headquarters (HQ) later in that same year, and then, returning to Fairbank and Jacob offer thoughts on how to move forward. The ultimate decision must come, of course, from NASA. Nevertheless, if we in the physics community formulate our thoughts with realism and rigor, we may have some opportunity for influencing NASA and there are reasons for thinking that the advice, if well-framed, will be heard.

One further essential point demands emphasis. All the missions under discussion require intensive technology development — more relative to their size than has been the case with solar system or astrophysics missions. This is a challenge to NASA and to physicists — for NASA to provide the necessary technology funding and for physicists to learn just what it takes to go from their normal laboratory world to the world engaging NASA Technology Readiness Levels, TRL 5, 6 and 7. Early collaboration between the physicist and the engineer is essential, and with it an alertness to risk management. We are into an arena of thought far from the everyday experience of most physicists, but one which, if viewed rightly, adds to rather than subtracts from the fascination of fundamental physics in space.

2. NASA Contributions to Date

There have been two areas where NASA and space have done wonders for experimentation, gravitation and relativity, and condensed matter physics, followed more recently by a third, particle physics. For reasons that were accidental but initially helpful, these were originated within two distinct branches of NASA: gravitation in the Office of Space Sciences (Code S), and condensed matter physics in the Office of Microgravity and Life Science (Code U), later renamed the Office of Biological and Physical Research (OBPR). For reasons of a somewhat pragmatic kind, the AMS particle physics experiment, jointly funded by NASA, the Department of Energy and international sources, was also assigned to OBPR.

In gravitation and relativity, the four principal missions in order of performance have been: (a) lunar laser ranging measurements (LURE) to retroreflectors on the Moon (1968+); (b) radar ranging to spacecraft and planets, especially the 1976 Viking Lander on Mars and, more recently, the Cassini spacecraft, to measure the Shapiro time-delay effect of general relativity; (c) the GP-A clock comparison or Einstein redshift experiment (1976); and (d) the GP-B orbiting gyroscope experiment, launched April 20, 2004. LURE and radar ranging may be regarded as falling somewhere between the two categories I have called observations and controlled physics experiments. GP-A and GP-B are manifestly experiments in the physicist's ordinary sense of the word, though the art of performing "controlled experiments" in a 10,000 km rocket flight (GP-A) and a 642 km polar orbit (GP-B) has unfamiliar aspects.

Figure 1 illustrates GP-A. The experiment consisted in comparing the rates of a hydrogen-maser clock in a Scout rocket launched from Wallops Island with two identical ground-based clocks. The predicted maximum redshift at 10,000 km was 4×10^{-10}; the Allan variances of the clocks in the required range were better than 10^{-14}. The result was a beautifully accurate confirmation of the expected rate difference to a part in 10^4. GP-A may be considered one of the most elegant missions NASA has ever carried out.

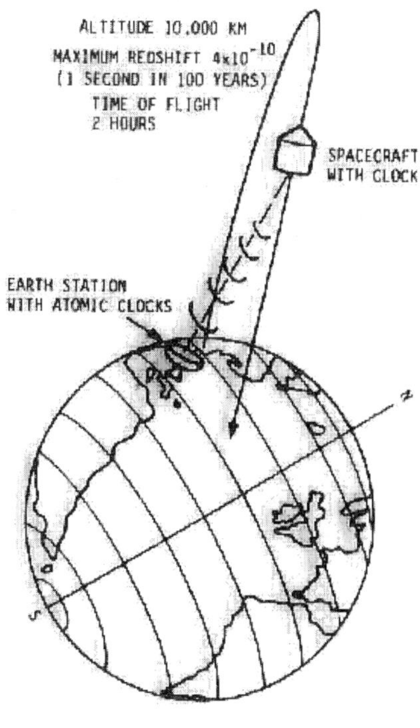

Fig. 1. Gravity Probe A.

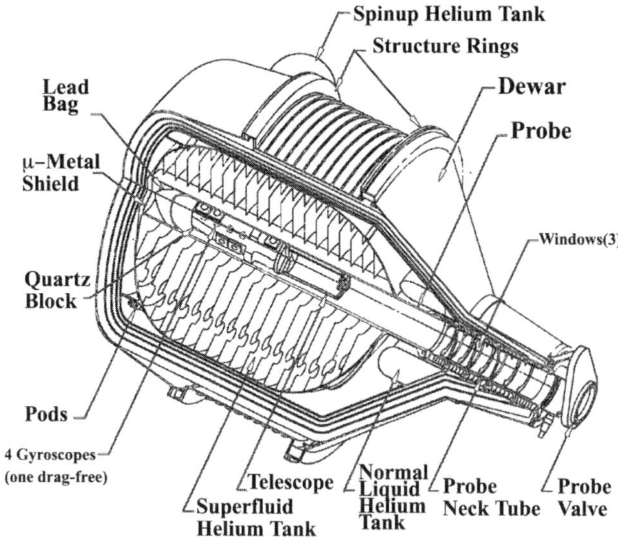

Fig. 2. Gravity Probe B payload.

GP-B measures with extreme accuracy two predicted effects of general relativity, the 6.6 arc-s/yr geodetic precession due to motion through the curved space–time around the Earth and the 40.9 marc-s/yr frame-dragging effect due to the Earth's rotation. The payload (Fig. 2) comprises four electrically suspended gyroscopes and a high precision star-tracking telescope, all maintained at cryogenic temperature in a 2420-l superfluid helium dewar. The on-orbit lifetime of the dewar was 17 months and 9 days. After initial on-orbit checkout, relativity data were gathered for 353 days, followed by a postscience calibration phase. Data analysis is now at an advanced stage; an initial science announcement is planned for April 2007.

The benefits from space deserve comment. For GP-A, it is simple but vital; the redshift depends on the difference in gravitational potential, which over a height difference of 10,000 km is far greater than in any Earth-based experiment. GP-B gains in three ways: (a) by the reduced support force on the gyroscopes, enhanced still further by making this a "drag-free" satellite with mean cross-track acceleration $\sim 10^{-11}$g; (b) by being above the atmosphere, and hence, removing otherwise horrendous "seeing" for the telescope; (c) by placing the satellite in an exact polar orbit aligned with the guide star so that the two different relativity effects are cleanly separated. It should be observed that (c) placed a strong constraint on the orbit, which had in fact a 1 s launch window beautifully met, indeed surpassed, by a combination of Boeing and luck.

Condensed-matter-physics experiments supported by NASA have included the Lambda Point Experiment (LPE, 1992) and Confined Helium eXperiment (CHeX, 1997) missions on superfluid helium, and the Critical Fluid Light Scattering

(ZENO) and Critical Viscosity Xenon (CVX) experiments on xenon near its critical temperature. All were performed on Shuttle. In each, the benefit of space comes from the reduced gravity. Ground-based tests of the λ point of liquid helium are limited by the compression of the liquid under gravity between the top and the bottom of the sample chamber; space enabled an advance of three or more orders of magnitude from a resolution of 10^{-6} K to better than 10^{-9} K. To obtain this, a new kind of thermometer had to be invented, based on measurements of a paramagnetic salt with a superconducting quantum interference device (SQUID). The resultant resolution was astonishing, $\delta T/T$ of order 10^{-11}, which in turn required the invention of a new kind of calorimeter. We have here an example of how fundamental physics in space requires and inspires new technologies. Figure 3 gives two curves, space data and ground-based data for the same instrument showing how at nK resolution pressure effects completely obscure the λ discontinuity. The LPE result has been an impressive confirmation of the logarithmic discontinuity of the specific heat of helium at the λ point and a determination of the coefficients governing it in renormalization group theory.

ZENO determined the relaxation rate of density fluctuations in xenon; CVX, an improved value for the critical exponent γ governing the relationship between viscosity and temperature near the critical point. Figure 4 compares CVX results in a ground test and in space-based measurements on STS-85 in August 1997. Microgravity conditions increased the range, and therefore resolution, of undistorted measurement, leading to a greatly improved value for γ.

One of the most baffling features of the Universe as we know it is its charge asymmetry. Some unexplained excess of matter over antimatter must have occurred very early on, after which, all of the antimatter then present must have annihilated against matter as the Universe cooled, with protons disappearing about 10 μs after

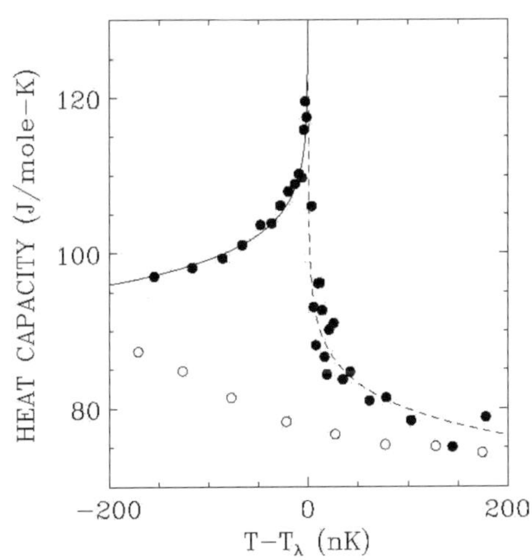

Fig. 3. The LPE results.

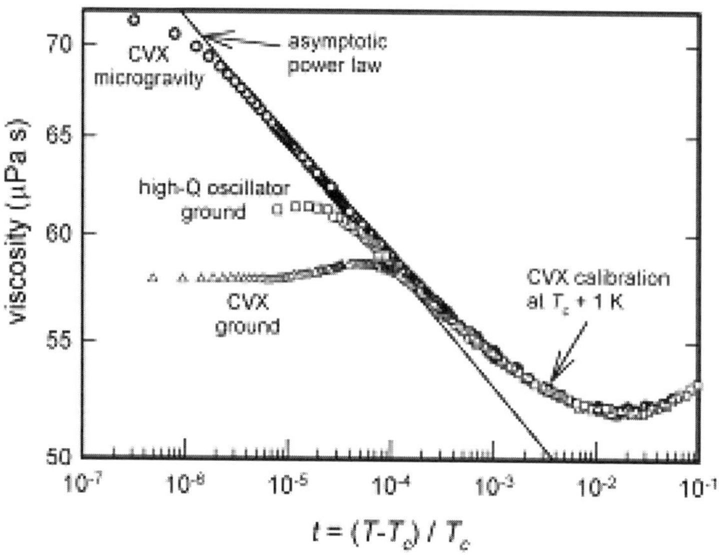

Fig. 4. The CVX results.

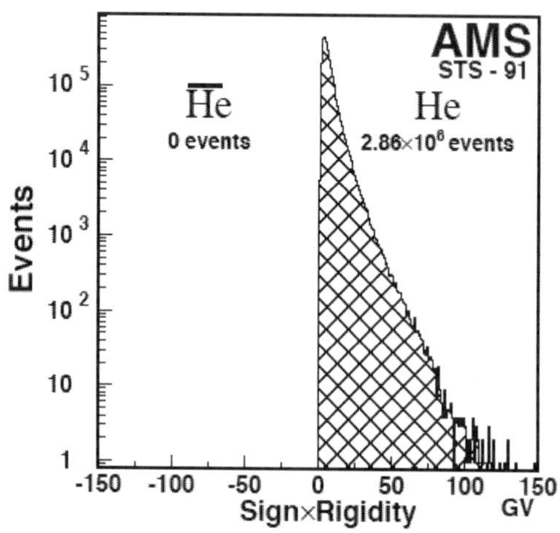

Fig. 5. Results of the AMS-1 mission.

the Big Bang. There have, however, been speculations whether this early massive annihilation was fully global, and if not, whether the lumps of antimatter might persist out of reach of the matter we know. It was to address this question that the AMS-1 instrument, comprising a 1.9 ton Nd–Fe–B ring-shaped magnet and suitable detection equipment, was flown on STS-91 in June 1998. Figure 5 shows results of a 10-day flight, with a total of 2.86×10^6 incoming He nuclei in the rigidity range 1–140 GeV but no $\overline{\text{He}}$ nuclei at any range. The result was a limit on the flux rate of $\overline{\text{He}}$ to He a factor of 3 below the 3.5×10^{-6} limit set by earlier balloon flights. The proposed AMS-2 instrument for Space Station, if flown, is expected to reach a limit of 10^{-9}.

The range revealed by the experiments just described is impressive. With the development in the 1990's of new techniques in laser cooling and atomic physics, Bose–Einstein condensation, and advanced clock technologies, many other new possibilities emerged. In March 1998, JPL, at the request of NASA HQ Code U, assembled a Working Group of over 100 persons from NASA, academia and industry to produce a *Roadmap for Fundamental Physics in Space*. This comprehensive survey, completed in the fall of 1998 and published in early 1999, is the subject of our next section

3. Roadmap for Fundamental Physics in Space

The *Roadmap* was consciously designed as a "long term framework from which to establish and advocate the National Aeronautics and Space Administration (NASA) future research and technology-development program in fundamental physics." In that definition, the words "technology-development" are key, reflecting a general recognition on the part of the community of the statement made above that fundamental physics missions require a more than usually extensive technology investment before entering the flight phase.

The framework of the *Roadmap* was "to cover groundbreaking research in fundamental physics... in the 2000–2015 timeframe" and to include "a broad-based spectrum of stakeholders — scientists, technologists and educators." Special recognition was given to the large part which undergraduate and graduate student research has played and can play in creating such missions.

After due thought, the Working Group defined two Quests and three Campaigns reflecting the different ways in which space contributes to fundamental physics, as follows:

Quest 1: *To discover and explore fundamental physical laws governing matter, space and time*, including investigations in gravitation, relativity and particle physics;

Quest 2: *To discover and understand organizing principles in nature from which structure and complexity arise*, including condensed-matter-physics experiments and future laser-cooling and atomic-physics experiments.

The three Campaigns were intersecting sets of investigations responding to the two Quests.

Campaign 1: Gravitational and relativistic physics;
Campaign 2: Laser-cooling and atomic physics;
Campaign 3: Low-temperature and condensed matter physics.

A total of 17 candidate missions were defined: 7 in gravitation, relativity and particle physics, 5 in laser-cooling and atomic physics, and 7 in low-temperature and condensed matter physics. Of these, 5 were free flyers, and 12 were candidates either for Shuttle or on the International Space Station (ISS) under a variety of conditions, but with the option in some cases of being free flyers. Critical to the condensed-matter-physics experiments was the Low Temperature Microgravity Payload Facility (LTMPF), a reusable ISS facility designed to operate two moderate-sized cryogenic payloads over periods up to six months. The LTMPF, which was close to completion in early 2004, was one of the casualties of the events to be described in the next section.

The *Roadmap* also addressed, in a constructive way, public outreach activities with all three campaigns at primary, middle and high school levels.

4. The 1999 NASA Advisory Council Recommendation I: Background

As noted earlier, historically fundamental physics at NASA has been divided between the Office of Space Sciences (Code S) and Microgravity/OBPR (Code U). The background for this divided responsibility is that whereas gravitational physics and relativity missions were (with one exception) recognized from the 1960's as belonging in Code S and its predecessors, support for condensed-matter-physics experiments and the Satellite Test of the Equivalence Principle (STEP) mission began as technology development efforts under the Physics and Chemistry Experiments in Space (PACE) program originally part of the NASA Office of Applications and Space Technology (OAST, later Code R). In the early 1980's, PACE was transferred to Microgravity and Life Science. Division of funding authority is not always a disadvantage, but as time went on, complications arose, as may be seen from the following partial list (Table 1) of reports and research opportunities from 1980 on, all of which, in one way or another, covered both interests.

In 1999, following the publication of the *Roadmap of Fundamental Physics in Space*, the NASA Advisory Council (NAC) took fundamental physics under advisement and appointed a subcommittee to review and provide recommendations, chaired by the then-Chair of the NRC Space Studies Board, Prof Claude Canizares. The matter was reviewed at the August 3–4, 1999 NAC meeting and a

Table 1. Overlapping review activities of NASA Code S and Code U for fundamental physics.

1980	Code S: "Strategy for Gravitational Physics in the 1980's" [NRC Space Science Board Report];
1988	Codes S & U: Space Science in the Twenty-First Century Imperatives for the Decades 1995 to 2015, vol. 5: Fundamental Physics & Chemistry [SSB Report];
1990	Code S *ad hoc* Committee on Gravitational Physics;
1993	Code U Research Announcement NRA 93-0SSA-12 Microgravity Sciences for Physics Missions;
1997	Code U Fundamental Physics Discipline Working Group.

Fundamental Physics

The space environment provides unique opportunities for research in fundamental physics. However, implementation of NASA programs in this area has been hampered by the absence of a clear home within the agency. NASA should review the organizational, budgetary, and interagency aspects of its fundamental physics programs, with an eye toward solving this structural problem.

Fig. 6. Text of NAC resolution transmitted to NASA Administrator on October 15, 1999.

resolution was transmitted to the NASA Administrator on October 15, 1999. The covering letter to the Administrator stated explicitly that "a theme that concerns many Council members is the lack of a 'home' in NASA for fundamental physics and we have a recommendation on the subject." Figure 6 reproduces the text of the NAC resolution.

We now come to the extremely unfortunate, and ultimately destructive, NASA response to the NAC.

5. The 1999 NAC Recommendation II: An Unfortunate NASA Response

On November 12, 1999, a memorandum "NASA's Fundamental Physics Programs" was issued with concurrence signatures from the Associate Administrators for NASA Codes U and S and the NASA Chief Scientist. Directed to the NASA Associate Administrator for Policy and Plans (Code Z), it quoted the words of the covering letter from the NAC Chair. It stated, "The same theme has been the subject of recent discussions between Code S and Code U, and with the NASA Chief Scientist," and concluded, "We have drafted a joint Code S/Code U statement that summarizes the agreement to provide a 'home room' for fundamental physics in space."

The above statement seems reassuring. Unfortunately, the accompanying Code S/Code U agreement, signed by the then-Directors of the Code S Science Program for Structure and Evolution of the Universe, and the Code U Microgravity Program, is not reassuring at all. Its essence was twofold. It divided fundamental physics into two sections: cosmic physics, the responsibility of Code S, and laboratory physics, the responsibility of Code U — exactly the same division of responsibility as before — and ended (Fig. 7) with three paragraphs of reassurance that nothing could go wrong.

The closing statement under the signatures reads, in bold letters, "We guarantee that no category of physics-in-space experiment goes homeless." This guarantee, offered in complete sincerity by serious persons, proved worthless.

6. The Demolition of Code U Fundamental Physics

The 1999 NAC resolution embodied an obvious but nonetheless profound management principle that in times of budgetary stress — and the NASA budget is always

Code S and Code U agree that there may be a limited grey area of overlap between these two areas of responsibility. We agree that no proposal should find itself homeless at NASA Headquarters. We agree that it is better for such grey area proposals have two possible avenues for funding than to have no avenue for funding.

[Examples of possible overlap areas listed here.]

We agree to steer proposals that would clearly find a more welcome and more relevant home from Code U to Code S, and from Code S to Code U, as appropriate. We agree to meet and discuss proposals that might be questionable, and if appropriate, to provide dual support.

We agree to assist our customers in the scientific community in the best possible ways to understand the distinction between Code S fields of research and Code U fields of research. Our mutual goal is that outside scientists will find a welcoming, "transparent" NASA rather than a NASA that turns proposals away.

[Original signed by Science Program Director for Structure & Evolution of the Universe, Code S and Director of Microgravity Programs, Code U.]

" We guarantee that no category of physics-in-space experiment goes homeless."

Fig. 7. Nothing can possibly go wrong.

under stress — a single, clear line of authority and a single, clear voice are essential. Difficult decisions still arise but they can be taken in a coherent way.

An example of the confusion created by not heeding the NAC came during the FY'00 Appropriations process. On April 29 through May 1, 1999, the Code U Fundamental Physics program held its annual meeting in Washington, DC. On April 29, with NASA HQ approval, a congressional breakfast was arranged at which two distinguished speakers, Prof Kip Thorne of Caltech and Dr William Phillips of NIST, gave talks respectively on gravitational physics and laser-cooling and atomic physics and the opportunities space could provide for new experiments in these areas. Copies of the *Roadmap* were also available. At the meeting, the Ranking Minority Member of the House Subcommittee, Mr Alan Mollohan, who was very impressed, asked, "What can I do to help?" It was pointed out that an increase of the Code U budget for fundamental physics from its then annual level of $35 million to $45 million would make a huge difference. A suggested distribution of funds in three areas — LTMPF, laser-cooling and atomic physics, and STEP — was transmitted to him and the Subcommittee Chair, Mr Walsh. The $10 million was duly added but through a misunderstanding to the Code S, rather than Code U, lines, none of it reached any of the congressionally mandated activities. The budget was settled after the date of the "guarantee" of Fig. 7; so much for coordination between NASA codes.

In January 2004, President Bush announced the new Moon–Mars Strategic Initiative for NASA. It is no part of my intent to question that decision; NASA did need

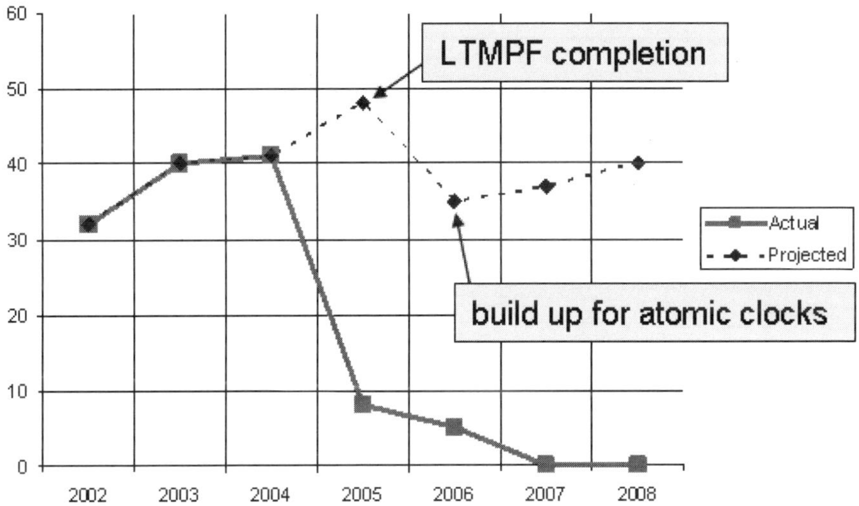

Fig. 8. Collapse of the NASA Code U fundamental physics budget.

to look beyond the ISS. But within the context of divided responsibility with no genuine home room for fundamental physics, it was a catastrophe. Code U became part of the new Exploration Directorate; management could see no contribution to exploration from the programs set out in the *Roadmap*. Within four months, the Code U fundamental physics budget was a destroyed. Figure 8 shows its collapse from a projected \$48 million, including LTMPF completion in FY '05, and \$35 million to \$40 million thereafter, to \$8 million in 2005, \$5 million in 2006 and \$0 from then on.

Let it be supposed that a Fundamental Physics Division had been established with NASA, as the NAC recommended. Its budget, too, would have come under stress. Probably several planned programs would have had to go or be slowed down, but with a single leadership, much, including technology development for the future, could have been preserved.

7. The Way Forward

That something has been irretrievably lost with the termination of the LTMPF and the closing-down of vital university research programs is evident. Nevertheless, now is the time for friends of fundamental physics in space to invoke Fairbank's principle. If we think on a time scale of a few years rather than a few months, there are grounds for hope.

The first Fairbankian observation is that NASA needs to continually renew itself in science. This elementary fact may escape us. We ask for funds and are informed that never has there been a more difficult time for NASA science than now.

Table 2. The eight fold way of fundamental physics in space.

Above the atmosphere	Optical reference, γ rays, particle physics (AMS)
Remote benchmarks	Lunar ranging, radar transponder on Mars
Large distances	LISA, ASTROD, LATOR
Reduced gravity (including drag-free)	Condensed matter, laser cooling, GP-B, LISA, STEP
Quieter seismically	Especially LISA & STEP
Varying ϕ	GP-A, SUMO
Varying g	STEP
Separation of effects	As in GP-B choice of orbit

There is the Moon–Mars Initiative, and huge programs like the James Webb Space Telescope (JWST) are overrunning; to think of any new initiative at the present time is absurd. Those of us who have had a certain amount of experience working with NASA will be aware that "this year" at NASA (whichever year it is) is always bad. We learn to take the statements of gloom with appropriate disbelief. Seen from another view, the picture undergoes a Gestalt change. Fundamental physics offers many beautiful, high-technology, relatively inexpensive experimental opportunities bringing new life to NASA. Our task is to make these opportunities visible to ourselves and to the many people within NASA who wish to build a creative future.

Earlier, in discussing GP-A, GP-B and condensed-matter-physics experiments, I remarked that space aids fundamental physics experiments in a variety of ways. Table 2 elaborates the theme by listing eight distinct kinds of opportunity available to make experiments possible that cannot be performed on Earth, with examples of missions to which they apply.

This partial list shows the richness of opportunity; the thoughtful reader will find others. Nor should we think of ourselves as without allies. The community is a growing one. Referenced above (Table 1) are just a few of the National Research Council reports and other activities in the US since 1980 which are related to fundamental physics in space. ESA, in 1994, established a long-term Fundamental Physics Advisory Group under its overall Space Science Advisory Committee. COSPAR, in 1996, established its Commission H: Fundamental Physics in Space. Interest exists at the Space Studies Board, in Congress, and in many other known and unknown locations.

What are the two critical issues? The issue for NASA is to establish for the first time an authentic "home room" for fundamental physics so that destruction of the kind observed in 2004 can be prevented. The issue for physicists, even more critical, is to recognize with full clarity the range of opportunity space brings and bend our minds to conceive experiments so intellectually compelling that they do indeed bring new life to NASA. If, along with Fairbank's principle, we can maintain the generosity of mind that is Jacob's principle, a bright future awaits us.

Acknowledgments

My work over the years has been made possible by many friends at Stanford University, NASA Marshall Center, NASA Headquarters, Lockheed Martin, Ball Aerospace, JPL and elsewhere, and supported by several NASA Grants and Contracts, most recently by NASA Contract NAS8-39225 for GP-B and NASA Cooperative Agreement NNM04AA18A for STEP.

ADDRESSING THE CRISIS IN FUNDAMENTAL PHYSICS

CHRISTOPHER W. STUBBS
Department of Physics and Department of Astronomy,
Harvard University,
17 Oxford Street, Cambridge MA 02138, USA
cstubbs@fas.harvard.edu

The observation that the expansion of the Universe is proceeding at an ever-increasing rate, i.e. the "dark energy" problem, constitutes a crisis in fundamental physics that is as profound as the one that preceded the advent of quantum mechanics. Cosmological observations currently favor a dark energy equation-of-state parameter $w = P/\rho = -1$. Awkwardly, this is the value that has the least ability to discriminate between alternatives for the physics that produces the observed accelerating expansion. If this result persists we therefore run a very real risk of stagnation in our attempt to better understand the nature of this new physics, unless we uncover another piece of the dark energy puzzle. I argue that precision fundamental measurements in space have an important role in addressing this crisis.

Keywords: Fundamental physics; dark energy; experiments.

1. Introduction: The "Standard Models" of Particle Physics and Cosmology

Over the course of one human lifetime we have developed two powerful "standard models." One pertains to particle physics while the other is cosmological. On the particle physics front we know of three families of quarks and leptons, which interact via exchange bosons. This picture is known to be incomplete: we don't know how to fit gravity into this scenario, and the dark matter continues to elude us.

On the cosmological side, driven in no small measure by the recent WMAP measurements of the structure in the cosmic microwave background,[1] we now assert that:

- Our geometrically flat Universe started in a hot big bang 13.7 billion years ago.
- The matter component of the Universe is dominated by dark matter, which is most likely outside the scope of the particles that make up the standard model of particle physics.

- Luminous matter constitutes only a few percent of the total mass of the Universe.
- The evolution of the Universe is increasingly dominated by the phenomenology of the vacuum.

This consensus cosmology is, however, ludicrous. The assertion that two regions of the vacuum experience a mutually repulsive gravitational interaction is, well, repulsive. With apologies to Kirk, Scotty, and especially Mr Spock, it's like living in a bad episode of *Star Trek*. It is worth asking, therefore, why this preposterous consensus has emerged, and on the basis of what experimental data.

The ingredients of the Universe are most conveniently expressed in terms of a cosmic sum rule, where

$$\Omega_k + \Omega_\Lambda + \Omega_m = 1,$$

with Ω_k representing the contribution from any underlying curvature in the underlying geometry of the Universe, Ω_Λ reflecting the contribution of the dark energy component, and Ω_m accounting for the matter density of the Universe (in units of the critical density).

2. The Observational Evidence for Dark Energy

The first claims for an accelerating expansion of the Universe were made in the late 1990's by two teams[2,3] that used type Ia supernovae to probe the history of cosmic expansion. Supernovae at redshifts around 0.7 were seen to have luminosities that were about 20% fainter than expected in a well-behaved Universe.

As with nearly all results in observational cosmology, the supernova measurements have the potential for unappreciated systematic error, and the type Ia Hubble diagram *alone* is (in my view) insufficient to compel us to believe in the dark energy, with $\Omega_\Lambda \sim 0.7$.

The sobering fact is that essentially all subsequent cosmological data sets drive us to the conclusion that the Universe has recently entered a time in which the scale factor is growing exponentially. The supernova data, measurements of primordial element abundances, determinations of the overall mass density of the Universe, and the structure seen in the microwave background all support this conclusion.[4]

Cognizant of the danger of making a list that appears to be comprehensive but may not be, the possibilities for the underlying physics include:

(1) A classical cosmological constant, residing in the gravitational physics sector,
(2) Vacuum energy effects, arising from quantum-mechanical fluctuations, and
(3) A modification of gravity that is manifested on the cosmological scale.

The good news aspect of this situation is that we have clear evidence of new physics. The bad news is that we presently have no idea what it means.

2.1. *Dark energy constitutes a crisis in fundamental physics*

In my opinion the dark energy situation constitutes a crisis in fundamental physics that is every bit as profound as that which preceded the advent of quantum mechanics.

One might wonder if perhaps we are somehow being misled by the data, or that the data are being misinterpreted. The convergence of results from such a wide array of measurement techniques argues against this, but we should nevertheless retain our sense of scientific skepticism. Under such circumstances we might turn perhaps to theory for some guidance....

3. The Present Theoretical Situation

From our current perspective there are only two natural theoretical values for Ω_Λ, the apparent contribution of the vacuum energy in units of the critical density. The first of these natural values comes from integrating the effect of all known quantum-mechanical fluctuations, up to the Planck mass. The resulting value of $\Omega_\Lambda = 10^{120}$ is so preposterous that it's a nonstarter. The other natural value presupposes that some cancellation mechanism steps in and performs a precise cancellation, leaving $\Omega_\Lambda = 0$ (to 120 significant figures!).

In the context of an observed value of $\Omega_\Lambda \sim 0.7$, this presents us with a problem. Some scholars have advocated that anthropic selection effects provide an explanation for this rather unexpected value, while others retain the hope that a future theory will provide some natural explanation.

Today, in the absence of a clear theoretical framework with relevant predictive power, the dark energy problem is best characterized as a data- and discovery-driven endeavor, spiced with interesting theoretical speculation.

4. The Present Observational Situation

At the present time observational cosmologists are concentrating on measuring the equation-of-state parameter of the dark energy, $w = P/\rho$, the value of which may help us distinguish between the different possible mechanisms that might underlie the dark energy. Current and upcoming cosmological observations exploit gravitational lensing, cluster abundances, baryon acoustic oscillations, and the supernova Hubble diagram to improve our understanding of the dark energy. These all have the merit of being undertaken in a regime where the signal is nonzero, but have the shared disadvantage of being susceptible to the various sources of systematic error that afflict astronomical observations. The desire to discriminate between different dark energy models will push these techniques to their limits.

At the time of this writing (mid-2006) we are starting to see preliminary determinations of $w = P/\rho$, the equation-of-state parameter of the dark energy, and its evolution over cosmic time. One such example is from the ESSENCE supernova survey, in which I participate. When the supernova data are combined with constraints from large scale structure,[5] the data favor $w = -1$. Similar results have been reported[6] by the CFHT Legacy Survey team. Unfortunately this value has the *least* power to discriminate the origin of the physics that is driving the accelerating expansion.

4.1. *It could get grim*

Looking ahead, there is a very real possibility that cosmological measurements will show that $w = -1$ at all accessible redshifts. This would be a grim circumstance, and our attempt to discover the nature of the dark energy through cosmological observations would completely stall.

Although science has very recently made huge strides in understanding the nature of the reality we inhabit, there are certainly ample examples in the past of scientific fields going through long periods of stagnation. I fear that if we arrive at $w = -1$, we may be facing such a fate unless we find another piece in this jigsaw puzzle.

5. Precision Fundamental Physics Experiments and the Dark Energy Problem

Many of the speculative ideas that have been put forward to explain the dark energy produce observable effects in other domains. We can therefore hope that a vigorous program of precision measurements and tests of fundamental physics might produce another anomalous result, which when integrated with the cosmological acceleration will lead us to a deeper understanding. Of course we do not know where this anomaly might arise, and this motivates our undertaking a broad array of experiments, including:

- Testing our understanding of gravity on all scales, including the inverse square law, the equivalence principle, the strong gravity regime, and gravitomagnetism.
- Direct tests of the fabric of space–time, including geometrical flatness tests, precision clock experiments, time delay measurements, etc.
- Tests of fundamental symmetries and Lorentz invariance,
- Probing the nature of vacuum fluctuations and Casimir forces.

Many of these projects have a long history in the precision measurement and fundamental physics arenas. In my opinion the broader community's appreciation of their importance is likely to grow as we collectively turn to face the unexpected crisis posed by the dark energy.

6. The Role of Space-Based Projects

We should distinguish at the outset between space-based *observations* and *experiments*. On the observational side, we exploit the low IR background, diffraction-limited imaging, and predictable absence of weather to collect photons from distant sources. We then use these observations to infer the properties of the Universe we inhabit.

This is to be contrasted with space-based *experiments*, where the absence of seismic perturbations, the microgravity environment, and access to various solar system gravitational potentials are combined with precision apparatus to pose well-defined experimental questions of nature.

The ability to carry out precision measurements, in which we can explicitly test the system's susceptibility to potential sources of systematic error, gives experiments a special role to play in probing for a deeper understanding of dark energy.

Our challenge is to identify those instances where we can realize large gains by combining the tools and techniques from the precision metrology and fundamental measurement communities with the special environmental aspects of space. Once these opportunities are identified we face our next daunting challenge, namely prioritizing the different options when we (for now, at least) have no idea where the signature for new physics might next emerge. As articulated by Dr Marburger in his opening remarks at this meeting, far better that we scientists make that assessment than leaving it to lobbyists and legislators.

7. An Example — Laser Ranging in the Solar System

This meeting has numerous interesting presentations on existing and proposed fundamental physics projects, many of which might exhibit the next piece of evidence for new physics. Let me pick one example and indicate how it links to the dark energy problem.

Smith *et al.* have recently demonstrated[7] the successful exchange of laser communication signals with a spacecraft at a distance of 24 million km from the Earth. See also the talk by Degnan in these proceedings. This gives reality to the notion of "piggybacking" a precision-laser-ranging capability on laser communication links. If NASA holds to its current goals of missions to the Moon and Mars, we could imagine adding a fundamental physics aspect to these flight opportunities. (See the paper by Merkowitz in this volume, and references therein.)

The scientific merit of an aggressive and coordinated solar system ranging campaign has been outlined[8] by Nordtvedt using the various gravitational interactions between the bodies in the solar system. Performing a global fit to ranging data between the Earth, the Moon, and the other planets will allow us to perform a comprehensive test of the basic foundations of gravity.

Successfully undertaking projects of this sort will require a coordination of efforts among individuals and teams drawn from diverse communities, including:

- Precision measurement and metrology,
- Atomic–molecular–optical physics,
- Astronomy,
- Gravitational theory and numerical analysis,
- Particle physics,
- Precision engineering, and
- Astronautics and spacecraft engineering.

This meeting is a great opportunity for us to renew old partnerships, and to build new ones.

8. An Exhortation

Let me then end with an exhortation. We are in the midst of a profound crisis in fundamental physics. On the theoretical side, we know that our two primary triumphs, namely general relativity and quantum mechanics, do not work well together. On the observational side we have the challenge of the increasing evidence for the accelerating expansion of the Universe. I think there is a very real possibility that the observational cosmologists will, in the decades ahead, present us with increasingly precise measurements of $w = -1$, with no evidence for evolution over cosmic time.

This will require that we push the frontiers of fundamental physics in order to search for the next piece of the dark energy puzzle. In my view the task of this meeting is to identify and refine the concepts that stand the best chance of capitalizing on space flight opportunities to address this crisis in fundamental physics.

Acknowledgments

I would like to thank the organizers for putting together such a stimulating meeting. In addition I would like to thank both Slava Turyshev and Thomas Murphy for their invitation, and for their generous help, encouragement, and valuable comments while I was preparing this talk. I am also grateful to Harvard University, to the Department of Energy (through their grant to Harvard's Laboratory for Particle Physics and Cosmology), and to the National Science Foundation (under grant AST-0507475) for supporting my own work on the dark energy problem.

References

1. D. Spergel and the WMAP team, astro-ph/0603449 (2006).
2. A. Riess *et al.*, *Astron. J.* **116** (1998) 1009.
3. S. Perlmutter *et al.*, *Astrophys. J.* **517** (1999) 565.
4. W. Freedman and M. Turner, *Rev. Mod. Phys.* **75** (2003) 1433.
5. G. Miknaitis *et al.*, *Bull. Am. Astron. Soc.* **37** (2005) 20715203.
6. P. Astier *et al.*, *Astron. Astrophys.* **447** (2006) 31.
7. D. E. Smith *et al.*, *Science* **311** (2006) 53.
8. K. Nordtvedt, *Phys. Rev. D* **61** (2000) 122001.

LABORATORY EXPERIMENTS FOR FUNDAMENTAL PHYSICS IN SPACE

WILLIAM D. PHILLIPS

Atomic Physics Division, Physics Laboratory,
National Institute of Standards and Technology,
Gaithersburg MD 20899-8424, USA

1. Introduction

This is intended as an extended abstract providing a brief summary of the talk presented at the international workshop "Quantum to Cosmos: Fundamental Physics in Space," held at the Airlie Center, Warrenton, Virginia, USA, 21–24 May 2006. In the spirit of an abstract, I have not endeavored to provide proper literature references to the research mentioned here, but only to provide the flavor of the workshop presentation.

"Laboratory Experiments" refers generally to active spaceborne experiments as opposed to the equally important passive observation in space of signals coming from more distant sources. My purpose in this presentation is to highlight a number of opportunities for addressing fundamental physics questions through such experiments. I do this from the perspective of an atomic, molecular, and optical (AMO) physicist and metrologist. Of particular importance in this context is that many of today's most exciting opportunities result from the recent development of tools that were mere dreams only five or ten years ago.

Now, at the beginning of the 21st century, it is worth recalling that the beginning of the 20th century saw predictions of the end of discovery in physics turned upside down by a period of unprecedented, explosive growth of fundamental understanding about the physical Universe. Similar predictions of the end of physics (and even science) at the close of the 20th century have already been belied by such things as the realization that 95% of the mass-energy of the Universe is stuff about which we know nothing, and that two of the greatest advances of 20th century physics, quantum mechanics and general relativity, are fundamentally incompatible.

Space is one of the most likely places where these mysteries may be resolved. The opportunities to do that stem in large part from new developments in such Earth-bound pursuits as cold atomic gases and precision frequency metrology. Taken into the environment of space, with its extended free fall, microgravity, low noise, varying gravitational potentials, etc., these and other tools promise advances that may overturn some of the most cherished ideas of physics, like Einstein's equivalence principle. Among the questions that may be addressed by space experiments are: "Does gravity behave as Einstein predicted?", "What will be the nature of a theory of quantum gravity?", "Where and how will the Standard Model fail?", "Are the fundamental constants of nature truly constant?", "What is the nature of dark matter and dark energy?".

2. Clocks

Among the new tools now available for experiments in space are superaccurate atomic clocks using cold atomic gases. These clocks not only open the possibility of new spaceborne fundamental measurements, they also offer improvements in tracking and navigation of spacecraft that can be used to improve on past experiments or on the execution of planned experiments. The performance of atomic clocks based on both ions and neutral atoms has improved in part because laser cooling reduces motion-induced errors and provides long observation times. In addition, the advent of femtosecond optical frequency combs has allowed the phase-coherent connection of optical and microwave frequencies, pushing clock frequencies higher, and relative uncertainties lower, while providing a hybrid optical source that features both a large number of stable optical frequencies and a train of optical pulses with a stable pulse interval.

The best of today's neutral-atom, microwave-frequency, atomic-fountain clocks has a fractional uncertainty better than 5×10^{-16}. But neutral atom clocks would work even better in space, where microgravity would allow even longer observation times and lower velocities, improving on many of the most important factors limiting Earth-bound performance. Trapped ion clocks operating at microwave frequencies now achieve performance that is nearly as good, while at optical frequencies the recently reported accuracy is an astounding 3×10^{-17}.

Individual clocks of such surpassing accuracy, or better still, ensembles of clocks including microwave, optical, ion, neutral atom, and even molecular clocks could perform unprecedented fundamental tests in space. Comparing a space clock in an eccentric orbit to a ground clock would test Einstein's gravitational redshift prediction to unprecedented accuracy. An ensemble of clocks in such an orbit would test the equivalence principle as it searched for changes in the ratios of the clock frequencies with position. Changes of those ratios with time would test for time variation of the fine structure constant, while molecules using both vibrational and electronic transitions for their clockwork would test for variations of the electron/nucleon mass ratio. Missions close to the Sun would provide even greater gravitational gradients, with even greater prospects of seeing a failure of the equivalence principle.

Improved atomic clocks would also impact the performance of the proposed space-based gravity wave observatory, the Laser Interferometer Space Antenna (LISA). While LISA is not a laboratory experiment in the sense used here, it is an example of an observational experiment that would benefit from the tools developed for laboratory measurements.

The search for electric dipole moments on atoms or molecules is essentially similar to the operation of atomic or molecular clocks. Such a dipole moment in eigen states of definite parity is forbidden by time reversal symmetry. While the Standard Model includes such a symmetry breaking, dipole moments predicted by the Standard Model are too small to be observed in the foreseeable future. Therefore observation of an electric dipole moment would indicate physics beyond the Standard Model, a breakdown that might be associated with quantum gravity, or might signal the existence of heretofore-hypothetical axions, a dark matter candidate.

3. Atom Interferometers

Another of the new tools available for fundamental physics experiments in space is the atom interferometer. The deBroglie wave character of atoms, along with the coherence provided by Bose–Einstein condensates or Fermi degenerate gases, opens new possibilities for atom interferometry. Atom waves can have huge advantages over light waves in interferometers that are sensitive to inertial forces. For example, an atomic Sagnac effect interferometer can be more sensitive to rotation than an equivalent optical instrument (same wavelength, flux, size, etc.) by the ratio of the rest energy of the atom to the energy of the photon. This ratio is on the order of 10^{11}, and even though atoms will not likely achieve the flux available in optical interferometers, substantial gains over optical instruments are possible.

In microgravity, where longer observation times are possible, interferometers sensitive to acceleration (and therefore to gravity) or to rotation can perform even better. This kind of sensitivity holds promise for a number of fundamental experiments. For example, measuring the acceleration, over a long period of time, of different atomic species would test the equivalence principle in the same spirit as the legendary Galileo experiment at the tower of Pisa. A Sagnac effect gryroscope could look for the frame-dragging Lense–Thirring effect (the subject of NASA's Gravity Probe-B mission) with greater precision, digging deeper for deviations from Einsteinian gravity. Using atoms as test masses to measure the universal gravitational constant G for different species or at very small distances is among the possibilities, providing tests of equivalence or searching for deviations from inverse square behavior as suggested by some string theories. Furthermore, gravimeters and gravity gradiometers could be used for exploratory missions such as the mapping of gravitational fields on asteroids and planets. On the ground this technology has the potential to improve real-time knowledge of Earth's geoid, with applications to national defense and climate change.

4. Femptocombs and Other Applications

The creation of mode-locked femtosecond lasers has resulted in highly accurate combs of frequency markers in the optical region of the spectrum, these being the

Fourier frequency components of the regular pulse trains emitted by these lasers. Such lasers offer the possibility of revolutionizing the practice of laser ranging. Normally laser ranging has involved measuring the delay of pulses reflected or transponded from the ranging target. Lunar ranging from a reflector placed by lunar astronauts has, since the early 1970's, improved from several tens of centimeters to a bit more than a centimeter. Use of the highly coherent light associated with modern femtosecond combs would allow interferometry, supplemented by pulse delay ranging. Substantial improvement to lunar ranging would aid fundamental gravitation studies as well as astrometry, geodesy, geophysics, and lunar planetology.

The application of ranging in which both high-accuracy pulse delay measurements and interferometry can be performed with the same femtosecond optical reference also holds promise for other fundamental and practical purposes. For example, future studies of what is known as the Pioneer anomaly — an apparent deviation of the trajectories of the Pioneer 10 and 11 space probes from that expected from predictions based on known gravitational and other effects — would be significantly aided by better ranging techniques. Proposed gravitation studies using laser ranging of spacecraft to detect the deflection of light passing near the Sun would also be improved by such ranging techniques, as would deep-space navigation in general.

5. Conclusions

A number of advances in AMO physics have led to improvements in metrology tools for the measurement of time and frequency, of energy shifts in atoms and molecules, and of inertial forces through atom interferometry. A few of the opportunities for laboratory experiments in space with these tools have been highlighted here. Today we are in a golden age of metrology, where new and astoundingly accurate measurement tools have some of their most promising applications in spaceborne missions. We are also in a golden age of scientific opportunity, where some of the deepest and most fundamental questions in science are within our grasp.

FUNDAMENTAL PHYSICS ACTIVITIES IN THE HME DIRECTORATE OF THE EUROPEAN SPACE AGENCY

LUIGI CACCIAPUOTI

European Space Agency, Research and Scientific Support Department,
ESTEC, Keplerlaan 1, PO Box 299,
2200 AG Noordwijk ZH, The Netherlands
Luigi.Cacciapuoti@esa.int

OLIVIER MINSTER

European Space Agency, Research and Application Division,
ESTEC, Keplerlaan 1, PO Box 299,
2200 AG Noordwijk ZH, The Netherlands
Olivier.Minster@esa.int

The Human Spaceflight, Microgravity, and Exploration (HME) Directorate of the European Space Agency is strongly involved in fundamental physics research. One of the major activities in this field is represented by the ACES (Atomic Clock Ensemble in Space) mission. ACES will demonstrate the high performances of a new generation of atomic clocks in the microgravity environment of the International Space Station (ISS). Following ACES, a vigorous research program has been recently approved to develop a second generation of atomic quantum sensors for space applications: atomic clocks in the optical domain, aiming at fractional frequency stability and accuracy in the low 10^{-18} regime; inertial sensors based on matter-wave interferometry for the detection of tiny accelerations and rotations; a facility to study degenerate Bose gases in space. Tests of quantum physics on large distance scales represent another important issue addressed in the HME program. A quantum communication optical terminal has been proposed to perform a test of Bell's inequalities on pairs of entangled photons emitted by a source located on the ISS and detected by two ground stations. In this paper, present activities and future plans will be described and discussed.

Keywords: Atomic clock; atom interferometry; Bose–Einstein condensation.

1. Fundamental Physics Research in Space

One of the most exciting challenges that physics is facing at present is represented by the harmonization of gravitation and quantum theory and, more generally, by the unification of the four fundamental interactions of Nature (strong, electromagnetic, weak, and gravitational). General relativity (GR) and quantum mechanics (QM) are the frameworks from which the developments of all the grand unification theories start.

GR explains the behavior of space–time and matter on cosmologically large scales and of very dense and compact astrophysical objects. This theory is based on Einstein's equivalence principle (EPP), which in its purest form states that:

- *Weak equivalence principle* (WEP): The trajectories of freely falling test bodies are independent of their structure and composition;
- *Local Lorentz invariance* (LLI): In local freely falling frames, the outcome of any nongravitational test experiment is independent of the velocity of the frame;
- *Local position invariance* (LPI): In local freely falling frames, the outcome of any nongravitational test experiment is independent of where and when in the universe it is performed.

For at least half a century after its first formulation, Einstein's theory of general relativity was considered "a theorist's paradise, but an experimentalist's hell."[1] No theory was in fact more beautiful and at the same time more difficult to test. A century later the technology and the scientific progress became mature enough to challenge general relativity with tests based not only on astronomical observations but also on laboratory experiments. High stability clocks are widely used to perform precision tests of LLI and LPI. WEP tests based on macroscopic objects promise unprecedented accuracy levels in the space missions STEP and GALILEO GALILEI.

QM on one hand, accounts for the behavior of matter on small scales (Angstrom and below), and ultimately leads, together with special relativity, to the so-called Standard Model of strong and electroweak interactions accounting for all the observable known forms of matter. WEP experiments based on matter-wave interferometry can be seen as a first and nontrivial attempt at interpreting the free fall of microscopic quantum objects and, more generally, general relativity in the framework of quantum theory. QM also describes quantum many-body systems and macroscopic quantum phenomena like superconductivity, superfluidity, and Bose–Einstein condensation. Laboratory sensors based on ultracold atoms have already demonstrated excellent performances for the accurate measurement of time, of tiny rotations and accelerations, and for the detection of faint forces. These instruments have opened up new, fascinating perspectives for testing general relativity as well as alternative theories of gravitation, for studying QM and for exploring the boundaries of quantum gravity.

Space is an ideal environment for improving the performance of precision instruments and pushing to the limits the accuracy of measurements testing the fundamental laws of physics. Space can ensure

- long and unperturbed "free fall" conditions,
- long interaction times,
- quiet environmental conditions and absence of seismic noise,

- absence of atmosphere,
- huge free-propagation distances and variations in altitude,
- large velocities,
- large variations of the gravitational potential,

providing unique experimental conditions, not accessible in a ground-based laboratory.

This paper aims to provide an overview of HME activities in fundamental physics.

2. The HME Program in Fundamental Physics

The HME Directorate of the European Space Agency has developed a sound program in fundamental physics based on the exploitation of quantum technologies in space.

Atomic quantum sensors based on cold atom physics have already demonstrated their potential for precision measurements. Today, atomic clocks reach stability and accuracy of a few parts in 10^{17} in the measurement of time and frequency; on the ground, atom interferometers promise sensitivities of $10^{-10}\text{g}/\sqrt{\text{Hz}}$ for acceleration measurements and $10^{-9}\,\text{rad}/\sqrt{\text{Hz}}$ for the detection of tiny rotations; the study of coherent matter waves (BEC and atom laser) in microgravity conditions will be crucial not only for basic research but also for improving the already outstanding performance of atomic quantum sensors.

The HME Directorate is presently developing the ACES mission, which will demonstrate the high potential of clocks based on laser-cooled atoms for both fundamental physics studies and applications. ACES has a key role as pathfinder of future projects exploiting the performances of atomic quantum sensors for precision measurements in space. Based on the technology inherited from ACES, HME is starting activities for the development of a second generation of atomic quantum sensors: atomic clocks in the optical domain, atom interferometry sensors for space applications, and a facility for studying Bose–Einstein condensates in microgravity.

Quantum communication on a worldwide basis represents another important issue addressed in the program. The project will test the robustness of entanglement at very long distances, study decoherence effects, and at the same time demonstrate secure quantum key exchange and distribution based on space-to-ground quantum communication.

Space is the natural environment for exploiting the high potential of these techniques. Atomic quantum sensors particularly benefit from weightlessness and free fall conditions, which will increase the interaction times and improve the sensitivity of the instruments by several orders of magnitude. Quantum communication by itself deserves long propagation distances and worldwide access.

These pioneering projects will lead to new technologies with wide applicability, covering diverse and important topics such as fundamental physics tests, very-long-baseline interferometry (VLBI), realization of SI units and metrology, global time-keeping, deep-space navigation, secure communication, prospecting for resources, GALILEO technology, geodesy, gravimetry, environment monitoring, major Earth-science themes, and planetary exploration.

All these activities have been recently approved for implementation within the ELIPS 2 (European Life and Physical Science) Programme. Consolidation studies of the new projects will be launched to demonstrate the technology readiness of the proposed systems, followed by the development of ground-based prototypes and transportable instruments which will be used as benchmarks for the design of space-qualified hardware.

3. The ACES Mission

Atomic Clock Ensemble in Space (ACES)[2–4] is a mission in fundamental physics whose aim is to demonstrate the performances of a new generation of atomic clocks in the microgravity environment of the International Space Station (ISS).

The heart of the ACES payload is represented by two atomic clocks: the primary frequency standard PHARAO (Projet d'Horloge Atomique par Refroidissement d'Atomes en Orbit), developed by CNES and the Space Hydrogen Maser (SHM), developed at the Observatory of Neuchâtel. The ACES clock signal will therefore merge together the good short and medium term frequency stability of hydrogen masers with the long term stability and accuracy of a primary frequency standard based on laser-cooled Cs atoms. One of the main objectives of the ACES mission consists in maintaining a stable and accurate on-board time scale that can be used for space-to-ground as well as ground-to-ground comparisons of frequency standards. Stable time and frequency transfer is achieved by using a microwave link (MWL), necessary not only for characterizing the ACES clock signal with respect to ground clocks, but also for performing general relativity tests of high scientific relevance.

ACES has a planned duration of 18 months. During the first 6 months, the performances in space of SHM and PHARAO will be established. The frequency standard generated by ACES will have a target long term stability of $7\cdot 10^{-14}\cdot\tau^{-1/2}$, where τ is the integration time expressed in seconds, and an accuracy of a few parts in 10^{16}. In the second part of the mission, the on-board clocks will be compared to a number of ground-based clocks operating both in the microwave and the optical domain. The space-to-ground comparisons will be used to perform measurements of the Einstein's gravitational red shift with an uncertainty of a few parts in 10^{16}, to test constancy and isotropy of the speed of light c at the $\delta c/c \simeq 10^{-10}$ accuracy level, and to measure time variations of the fine structure constant α at the level of $\alpha^{-1}d\alpha/dt \simeq 10^{-17}$/year. Further details on the ACES mission and its status can be found in Ref. 5 and in the corresponding paper reported in this special issue.

At present, ACES is the most advanced project on cold-atom sensors for space applications. It will study the physics of cold atoms in microgravity, demonstrate the outstanding performances of the space clocks PHARAO and SHM, validate a new time and frequency transfer technique with an unprecedented level of stability, perform accurate tests of Einstein's theory of general relativity, and develop applications in different areas of research. In addition, ACES is paving the way to the development of second-generation atomic quantum sensors for space, fostering the necessary technology development and validating in space a series of tools and instruments extremely important for future missions: from complex laser systems to advanced vacuum techniques, from space clocks to links for accurate time and frequency dissemination.

4. Second Generation of Atomic Quantum Sensors for Space Applications

4.1. *Space optical clocks*

In a frequency standard in the optical domain, a laser in the visible region of the electromagnetic spectrum is used to excite atoms on the clock transition. The resonance signal is detected by measuring the population at the two clock levels and used to keep the laser frequency tuned on the atomic transition. Clock cycles are then counted by a femtosecond-laser frequency-comb generator.

Optical clocks have already demonstrated fractional frequency stability of a few parts in 10^{15} at 1 s of integration time and hold the promise of accuracy down to the 10^{-18} level.[6] In space, where the weightlessness and the extremely quiet environment ensure the ideal conditions for detecting very narrow signals, these performances can be improved even further. Clocks in space represent unique tools for testing fundamental laws of physics at an unprecedented level of accuracy and for developing applications in time and frequency metrology, universal time scales, global positioning and navigation, geodesy, and gravimetry.

Constancy and isotropy of the speed of light can be tested by continuously comparing a space clock to a ground clock. LLI tests based on this technique have already been performed in 1997 by comparing clocks on board GPS satellites to a hydrogen maser.[7] Optical clocks orbiting around the Earth combined to a time and frequency transfer link not degrading the clock performances can improve present results by more than three orders of magnitude.

Optical clocks can measure Einstein's gravitational red shift with a relative frequency uncertainty of a few parts in 10^{18}, demonstrating a new, efficient way of mapping the Earth's gravity field and defining the shape of the geoid at the cm level. The universality of the gravitational red shift can be tested at the same accuracy level by two optical clocks based on appropriately chosen transitions in free flight in a varying gravitational potential.

As a direct consequence of EEP, general relativity and other metric theories of gravitation forbid any time variation of nongravitational constants. Today, optical clocks offer the possibility of testing time variations of fundamental constants at a high accuracy level.[9,10] Interestingly, such measurements complement the tests of the local Lorentz invariance and of the universality of free fall to experimentally establish the validity of Einstein's equivalence principle (EEP).

Third generation of navigation systems will benefit from the technology development related to optical clocks. New concepts for global positioning based on a reduced set of ultrastable space clocks in orbit combined with simple transponding satellites could be studied. As already mentioned, a new kind of geodesy based on the precise measurement of Einstein's gravitational red shift could be envisaged.

The Space Optical Clocks project[11] will demonstrate the high potential of this emerging technology for both fundamental physics studies and applications. Compact prototypes of optical clocks based on strontium and ytterbium atoms will be developed, characterized, and compared, preparing the background for the development of high performance instruments for space applications.

4.2. *Atom interferometry sensors for space applications*

Atomic quantum sensors based on matter-wave interferometry represent a key technology for the detection of tiny acceleration, rotations, and faint forces. These instruments reach their ultimate performances in space, where the long interaction times achievable in a free falling laboratory can improve their sensitivity by at least two orders of magnitude. Atom interferometry sensors find important applications in fundamental physics research, navigation, geology, Earth observation, etc.

An atomic gyroscope has already been proposed to measure the Lense–Thirring effect in space.[12] Such measurements would provide direct tests of metric theories of gravitation.

However, the main interest of these sensors is in the possibility of performing tests at the boundaries of QM and gravitation. Cold atom sensors in space may be the key to challenge experiments in fundamental physics: tests of Newton's law at micrometric distances, the neutrality of atoms, and the universality of free fall. The weak equivalence principle represents one of the cornerstones of Einstein's theory of general relativity. Differently from all the other fundamental interactions, gravity appears to affect all bodies in a universal way, independently of mass and internal composition. Revisiting the Pisa gedanken experiment using quantum particles[13] requires some comments on the interplay between QM and gravitation. In fact, the classical concept of deterministic trajectory is not valid anymore for a quantum system, and the identification of the world lines of a freely falling quantum particle with preferred curves having a well-defined geometrical meaning does not hold anymore. These fundamental problems, presently under study, show the importance of performing experiments with quantum systems in free fall. To date, matter-wave tests of the equivalence principle have been performed with neutrons[14] and samples of laser-cooled atoms.[15] These experiments can be improved in space where the long interaction times possible in a freely falling laboratory promise sensitivity to accelerations in the low 10^{-12} g regime on a single measurement.

Cold-atom sensors have excellent sensitivity for the absolute measurement of gravity, gravity gradients, and magnetic fields as well as Earth rotation, and therefore find direct application in Earth sciences, or more generally in Earth observation facilities. Miniaturized cold atom gyroscopes and accelerometers will be key instruments for the next generation of navigation systems not relying on satellite tracking. Matter-wave sensors have indeed have a well-defined scale factor, do not require periodic calibration, and do not suffer from problems related to bias drifts.

Atom interferometry is also an important ingredient of research in quantum information, where preservation of coherence and precise control of the atomic phase represent a key issue.

The project Atom Interferometry Sensors for Space Applications[16] will exploit this field and, as a first step, it will demonstrate this technology with a transportable sensor that will serve as a prototype for the space qualification of the final instrument.

4.3. *BEC in space*

When cooled down to very low temperatures and constrained in trapping potentials where high densities can be reached, a gas of identical particles undergoes a phase transition from a classical to a macroscopic quantum system. Identical particles start occupying the lowest energetic states behaving as a quantum many-body system with well-defined properties depending on the bosonic or fermionic nature of the particles themselves. The concept of Bose–Einstein condensation (BEC) dates back to the first theoretical studies of Einstein who predicted this phenomenon as a direct consequence of statistical properties of a gas of identical bosons. The initial scepticism of his result is well summarized in his famous words: "From a certain temperature on the molecules condense without attractive forces, that is they accumulate at zero velocity. The theory is pretty but is there also some truth to it?" The ground state of a bosonic gas is macroscopically populated when the de Broglie wavelength λ_{dB} of the particles becomes comparable to the interparticle separation. This condition, also called quantum degeneracy, occurs when $\rho = n \cdot \lambda_{\mathrm{dB}}^3 \simeq 2.16$, where n is the density of the sample and ρ is the so-called phase-space density.

After the first experimental realization of BEC,[17,18] many studies have extensively investigated the properties of this new state of matter, including the thermodynamics of the phase transition, the collective oscillation modes of the sample, its coherence properties and superfluid behavior, BEC physics in extremely confined potentials (1D or 2D geometries) or in optical lattices. Studies of degenerate Fermi gases and quantum mixtures are rapidly progressing.

At present, laboratory experiments can routinely produce BEC's with typical temperatures down to a few tens of nK. Nevertheless, even at these ultra-low temperatures, residual kinetic energy plays a significant role, masking important effects related to the quantum nature of the system. In microgravity the thermal motion of atoms can be reduced even further, and temperatures in the low pK or even fK regime are accessible. Further, very long expansion times can be achieved in an almost perturbation-free environment, where the atomic sample can evolve unbiased by gravity and without any need for levitation. These conditions set the stage for innovative studies on the physics of degenerate quantum gases in microgravity and for the utilization of coherent sources of ultracold atoms to enhance the performances of atom interferometry sensors.[19]

5. Quantum Communication for Space Experiments

An important aspect of QM is represented by the entanglement of states. Connected to the nonlocal behavior of this theory, entanglement is responsible for phenomena like quantum teleportation, Einstein–Podolski–Rosen paradoxes, quantum computing, etc. Quantum entanglement is also crucial for understanding the role of the measurement process in QM. The postulate of the wave-function reduction describing the interaction of a physical system with the measurement apparatus seems to be contradictory when the system and the measurement apparatus are both described quantum-mechanically. Decoherence could be the solution to this contradiction.

Quantum technologies are demonstrating their robustness and potential both for fundamental physics studies and for secure communication. Tests of quantum physics on a large distance scale will provide deeper insight into modern theories. At the same time, the latest achievements in the field of quantum information processing keep promising new applications in quantum communication, such as quantum cryptography and teleportation. As a matter of fact, quantum communication is becoming the cutting edge of information technology.

Space offers large free propagation distances and the possibility of worldwide access, essential ingredients for testing entanglement of quantum states and demonstrating quantum communication concepts. The project Space QUEST, Quantum Communication in Space,[20] proposes the development of a quantum communication optical terminal orbiting around the Earth to perform experiments on entangled photons propagating on very long distances and to test quantum communication protocols on a worldwide basis.

The space terminal is composed of an EPR source of light and two optical telescopes transmitting couples of entangled photons towards two receiving ground stations. Installed on the ISS, the optical terminal will allow one to establish common view contacts with ground stations at a maximum distance of 1600 km and for typical durations of 200 s.

The experiment will provide the first tests of Bell's inequalities on pairs of entangled photons over distances not accessible on Earth. This will be the first step toward the study of quantum correlations at astronomical distances. The interaction of correlated pairs with the atmosphere will be analyzed and decoherence effects taking place during the propagation will be studied. The worldwide access ensured by the low-Earth orbiting space platform will be crucial for demonstrating secure quantum key distribution between the ISS and a given ground station, or quantum key exchange between two ground stations via the ISS. A simple quantum key distribution protocol based on the transmission and detection of entangled photons on a well-defined polarization will be established and complemented by a classical transmission channel.[21,22] These experiments will be the first step toward secure communication based on quantum cryptography.

6. Conclusion

In this paper, the HME program in fundamental physics has been presented and discussed.

Recent scientific and technology developments have produced instruments based on quantum systems that with their performances are challenging our understanding of the Universe and of the physical laws of Nature.

HME is investing in quantum technology for space applications. Ground-based prototypes will provide the first demonstration of second-generation atomic quantum sensors as well as scientific data, precursors of dedicated missions exploiting the full potential of quantum technology in space. These activities are well coordinated with the scientific and technology development fostered by other ESA Directorates.

The projects on cold-atom-based systems and quantum communication will bring about outstanding scientific results and mature space-proved technology within a plausible time frame of 6–10 years. The success of these initiatives will benefit from the strong interaction between ESA, national agencies, and scientists. From this point of view, the HME program in fundamental physics represents a unique opportunity to consolidate this kind of technology and prepare key instruments for future space missions.

Acknowledgments

The authors express their warm thanks to all the members of the scientific teams involved in these project and contributing with their expertise and knowhow to the HME program in fundamental physics.

References

1. C. W. Misner, K. S. Thorne and J. A. Wheeler, *Gravitation*, 2nd edn. (Freeman, San Francisco, 1973).
2. C. Salomon and C. Veillet, ACES: Atomic Clock Ensemble in Space, in *Proc. 1st Symposium on the Utilisation of the International Space Station* (ESA Special Publication, SP 385, 1997), p. 295.
3. C. Salomon *et al.*, PHARAO: A Cold Atom Clock in Micro-gravity, in *Proc. 1st Symposium on the Utilisation of the International Space Station* (ESA Special Publication, SP 385, 1997), p. 385.
4. C. Salomon *et al.*, *C. R. Acad. Sci. Paris* **t.2 Série 4** (2001) 1313.

5. L. Cacciapuoti *et al.*, The ACES mission, in *Proc. 1st ESA International Workshop on Optical Clocks*, 8–10 June 2005, Noordwijk, The Netherlands (ESA Publication, 2005), p. 45.
6. T. Rosenband *et al.*, An aluminium ion optical clock, to appear in *Proc. 20th European Frequency and Time Forum*, 27–30 Mar. 2006, Braunschweig, Germany.
7. P. Wolf and G. Petit, *Phys. Rev. A* **56** (1997) 4405.
8. C. Lämmerzahl *et al.*, *Gen. Relativ. Gravit.* **36** (2004) 2373.
9. H. Marion *et al.*, *Phys. Rev. Lett.* **90** (2003) 150801.
10. S. Bize *et al.*, *Phys. Rev. Lett.* **90** (2003) 150802.
11. S. Schiller *et al.*, Space Optical Clocks, proposal submitted to *ESA Announcement of Opportunity 2004*.
12. C. Jentsch *et al.*, *Gen. Relativ. Gravit.* **36** (2004) 2197.
13. L. Viola and R. Onofrio, *Phys. Rev. D* **55** (1997) 455.
14. K. C. Littrel *et al.*, *Phys. Rev. A* **56** (1997) 1767.
15. S. Fray *et al.*, *Phys. Rev. Lett.* **93** (2004) 240404.
16. G. Tino *et al.*, Atom Interferometry Sensors for Space Applications, proposal submitted to *ESA Announcement of Opportunity 2004*.
17. M. H. Anderson *et al.*, *Science* **269** (1995) 198.
18. K. B. Davis *et al.*, *Phys. Rev. Lett.* **74** (1995) 3969.
19. W. Ertmer *et al.*, Bose–Einstein Condensates in Microgravity, proposal submitted to *ESA Announcement of Opportunity 2004*.
20. A. Zeilinger *et al.*, Space QUEST: Quantum Communication for Space Experiments, proposal submitted to *ESA Announcement of Opportunity 2004*.
21. A. K. Ekert, *Phys. Rev. Lett.* **67** (1991) 661.
22. C. Kurtsiefer, *Nature* **419** (2002) 450.

LESSONS FROM INTRODUCING NEW SCIENTIFIC DISCIPLINES INTO EUROPEAN SPACE RESEARCH

MARTIN C. E. HUBER
Laboratory for Astrophysics, Paul Scherrer Institut,
CH-5232 Villigen PSI, Switzerland
Martin.Huber@psi.ch

Physics experiments in space will permit us to investigate natural phenomena that cannot be observed on the ground, such as low-frequency gravitational waves, and to reach uncharted realms of accuracy — accessible only through experiments carried out in space — where current foundations of physics can be further tested and potentially falsified. Such projects require technologies that have not been in hand for a long time but are available now.

To avoid conflict of interest, the merit of space projects in physics, from the proposal stage through development, ought to be judged by experts in physics, rather than by space scientists from other fields. It is time now to set aside some funding to let missions in fundamental physics compete fairly with the established space sciences, thereby enriching and deepening the space enterprise — and broadening its advocacy base.

We look, in the context of the European space scene, at the measures and events that resurrected the initially suppressed planetary sciences and brought solar physics to blooming after a long drought; and derive ideas on how to increase the number of flight opportunities for fundamental physics in space.

Keywords: Space research; fundamental physics; science policy.

1. Introduction

The discipline of fundamental physics in space is now in the pioneering phase, i.e. in the phase where established space-science disciplines, namely solar-system exploration and astronomy, have been about 35 years ago. Although the set of technologies available early on in the space age was rather limited, these disciplines could go ahead with space experimentation — and at a time when space science was well supported financially.

In contrast, significant advances in physics through experiments carried out in space had to await the development of additional technologies: engineers and physicists interested in such physics experiments had to postpone preparing programmed space projects, until the relevant additional key technologies had been developed and refined in the laboratory. More recently, the first orbiting fundamental-physics experiment, NASA's Gravity Probe B (GP-B),[1] which is based on the technologies in question, has convincingly demonstrated that the required experimental tools are in hand today.

In principle, this lets us proceed now with exploiting the advantages of the space environment for experiments in fundamental physics, and investigate physics phenomena that cannot be studied on the ground. Among these are low-frequency gravitational waves from astrophysical objects, and experiments in the realm of hitherto-inaccessible precision — that can be reached in space only — where current foundations of physics can be further verified or falsified, thus opening doors to more accurate and comprehensive, and hopefully unified, theories.[a]

However, funding for space science is now much harder to come by than 35 years ago and, consequently, there is no "equal opportunity" for fundamental physics in space, i.e. there is no viable number of flight opportunities for orbiting physics experiments yet. The question then is: How can we achieve a situation that is commensurate with the importance of the topic, and somehow comparable to the other, now well-established space-science disciplines?

In the following, we will examine the past and present situation of fundamental physics research in space within the framework of the European Space Agency (ESA) and its scientific predecessor organization, the European Space Research Organisation (ESRO), and then look at the actions that have been taken to introduce solar physics and planetary sciences into the ESA Science Programme. Experience from the successful efforts that resulted in bringing "new" disciplines into the Programme can provide lessons on how to proceed, not only in the European but also in the international context. In the NASA framework, major fundamental-physics experiments have been performed on its two Gravity Probes (GP-A[3] and GP-B). In Europe, several projects — enumerated in detail below — are under development and will be launched in a few years.

From our analysis, we will derive ideas that may help us in obtaining a sustainable number of flight opportunities for fundamental physics in space. This will increase the size of the space-science community, and thus also the breadth of the advocacy base for science in space.

[a]The situation of fundamental physics in space has been described in the Position Paper "The Need for Space Flight Opportunities in Fundamental Physics"[2] of the European Physical Society (EPS). The Position Paper was published on the occasion of the centenary of Albert Einstein's *annus mirabilis*, which was celebrated around the globe as World Year of Physics 2005. It has been included in the "relevant material" for the present workshop and is available on the Web via the Internet address http://www.eps.org/papers_position/paper_index.html

2. Space Technology Available Then and Now

The technologies available 35–40 years ago for scientific missions restricted space research to a few subdisciplines:

- Missions addressing space-plasma physics used two-axis stabilized, spinning spacecraft that had to provide a magnetically clean environment, and therefore often were equipped with long booms.
- Solar-physics and astronomy missions had requirements for three-axis stabilized satellites with pointing accuracy and stability in the sub-arc-min range.
- Detectors were usually single-channel devices that operated under high voltage; and scanning systems were used to generate both spectra and to map the object, or the part of the sky under investigation.

Improved pointing accuracy and stability as well as advanced imaging detectors later enabled the rise of "Earth observation from orbit."

Today's technologies also include those required for research in fundamental physics. This allow us to explore physics phenomena under the advantageous conditions that can only be created in space:

- Spacecraft can provide purely gravitational orbits by counteracting atmospheric drag, light pressure etc. down to residual perturbations at a level around 10^{-9} g or less. Such spacecraft use a proof mass and a system for fine orbit and attitude control that includes μN-propulsion systems (such as He-proportional thrusters or small ion thrusters), drag-free control software with spacecraft/experiment interrelation, and proof-mass charge control (which is necessary for counteracting the deposition of charge by cosmic radiation).
- Excellent measuring accuracy is achieved by means of precision displacement sensors, so-called superconducting quantum interference devices (SQUIDs), ultra-stable lasers in space, and lightweight H-maser clocks.
- Slosh-free He dewars have been developed as well and special orbits can be achieved, where needed.

These technologies, which have been developed predominantly with NASA funding and at Stanford University, have now been employed in the first orbiting fundamental-physics experiment — in Gravity Probe B (GP-B; Fig. 1) — and provide valuable in-orbit experience.

Moreover, during the past few decades, new physical methods, such as laser cooling of atoms and, subsequently, cold-atom physics, have come into being through laboratory experiments. These methods now enable us to:

- carry out atom-beam interferometry,
- create Bose–Einstein condensates, and
- construct atomic clocks of unprecedented accuracy.

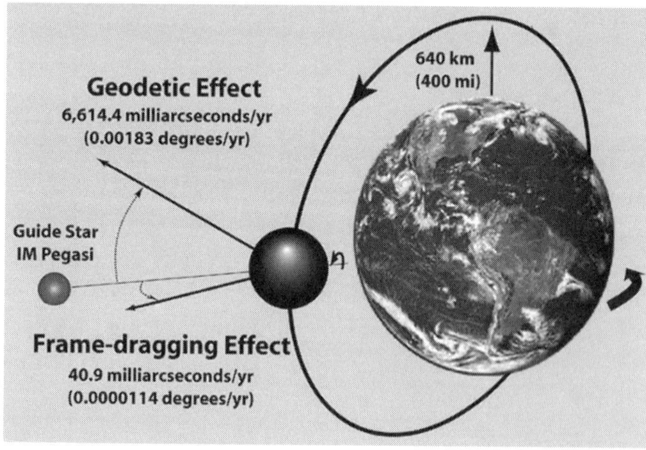

Fig. 1. The orbit and the miniscule directional changes that had to be measured by Gravity Probe B.

These three subjects herald important applications that will draw added benefits from a low-noise gravitational environment, as it can be achieved in space.

Several space experiments in fundamental physics are currently under development in Europe, in collaboration partly with NASA and partly with the French National Space Agency:

- *LISA Pathfinder*,[4] a technology mission being prepared jointly by ESA and NASA for launch in 2009, and with the aim of demonstrating, in space, critical technologies that are required for the *Large Interferometer Space Antenna* (*LISA*) — a mission devoted to the detection of low-frequency gravitational waves in the range of 10^{-4}–10^{-1} Hz.
- *MICROscope* (*MICRO-Satellite à traînée Compensée pour l'Observation du Principe d'Equivalence*),[4] a joint project of the Centre National d'Etudes Spatiales (CNES) and ESA, with a launch foreseen in 2010.[b]
- *ACES* (*Atomic Clock Ensemble in Space*),[5] consisting of a Cs clock and an H-maser, is being prepared for a launch to the International Space Station around 2010. ACES will be able to verify several principles of the general theory of relativity more precisely than ever before.[c]

[b]MICROscope will test the equivalence principle to a level of 10^{-15}, and is thus a predecessor of the Satellite Test of the Equivalence Principle (STEP), which would further advance the accuracy to 10^{-18} by the use of cryogenic techniques.

[c]The ACES experiment with its Cs clock is expected to provide time with a stability of 10^{-16}, i.e. to a level where geophysical noise begins to be felt on the ground. This might lead to global timekeeping in and distribution from orbit in the not-too-distant future. ISAS is the Japanese Institute of Space and Astronautical Science, one of the four principal sections of the Japan Aerospace Exploration Agency, JAXA.

Fig. 2. As a first "STEP," the French/ESA MICROscope satellite will verify the equivalence principle to 10^{-15}.

3. How to Insert Fundamental Physics into Space Programs

Given the inventory of the NASA GP-B and promising projects progressing toward their launch in Europe, and with one of the most forward looking, trail-blazing missions ever planned for space flight — namely LISA — in an advanced stage, albeit on hold as a project, we might look at what happened to solar physics and planetary sciences at ESA. These two subdisciplines had not, to any extent, been part of the early ESA Science Programme. Planetary sciences, in fact, had been consciously blocked out early on, mainly because of a lack of financial resources.

3.1. *Prélude: The scientific advisory structure of ESA*

To describe the events that led to solar physics becoming a strong part of the ESA Science Programme, it is necessary to briefly introduce ESA's Advisory Structure. There is a standing committee with three working groups: the Space Science Advisory Committee (SSAC), which gets discipline-specific advice from the Solar System Working Group (SSWG), the Astronomy Working Group (AWG) and the Fundamental Physics Advisory Group (FPAG). In order to develop long-term programs, such as Horizon 2000 (1985), Horizon 2000 Plus (1995) (jointly referred to as Horizons 2000), and now Cosmic Vision (2005), *ad hoc* Survey Committees were set up. These obtained input from topical teams, which were, however, not necessarily mapping the subdisciplines covered by the standing Working Groups, SSWG, AWG and FPAG. For the Horizon 2000 survey, for example, a topical team on physics

Fig. 3. The ESA Horizon 2000 Survey Committee, set up by the Director of Scientific Programmes, Roger M. Bonnet, with members of associated topical teams, during a break in their final meeting in Venice. The membership included scientists, such as Léon Van Hove, a former Director-General of CERN (Conseil Européen de Recherche Nucléaire, the European Particle Physics Laboratory in Geneva), Gian-Carlo Setti (fifth and sixth persons from right), and Ian Roxburgh (second person from left), who brought in perspectives from fields of science that had not taken advantage of the scientific potential of space at the time.

was set up, although there had been no FPAG at the time; moreover, there were scientists in the Survey Committee (see Fig. 3) whose interest went well beyond the then-practiced space-science disciplines.

3.2. *The long march to success for solar physics at ESA*

Solar physicists did not succeed in obtaining access to the ESRO and ESA programs; apart from four individual experiments that had been included in the payloads of two early ESRO satellites, on the European Retrievable Carrier (EURECA) platform, and on ESA's First Spacelab Payload. Consequently, solar physics had no dedicated observatory mission for a long time: the pursuit of studies for a Grazing Incidence Solar Telescope (GRIST), which was to be flown on Spacelab together with a US Solar Optical Telescope (SOT), was stopped in the aftermath of the cancelation of the NASA Out-of-Ecliptic probe in 1981 and, in 1983, in the competitive selection for the next ESA science project, the Infrared Space Observatory (ISO) was preferred to the Dual Irradiance and Solar Constant Observatory (DISCO).

In the course of time, however, solar physicists joined up with the communities dealing with solar-wind and space-plasma physics, and succeeded in having the SOlar and Heliospheric Observatory (SOHO) included in the program; SOHO itself included solar-wind experiments and was eventually combined with the Cluster mission in the so-called Solar–Terrestrial Science Programme (STSP) Cornerstone in the Horizon 2000 Programme. Undoubtedly, the fact that scientists from the US had participated in the studies for SOHO had also contributed to the mission being selected. The relative novelty of the helioseismology method at the time, when the studies took place, probably helped as well. At the project stage, SOHO (and Cluster) became a joint ESA/NASA mission.

The solar-physics and the heliospheric-physics communities later also obtained a Solar Orbiter Cornerstone in the Horizon 2000 Plus exercise.

Solar observations from space had been carried out starting with the very first sounding-rocket experiments, and had then been pursued from the early 1960's to the end of the century by the use of NASA and ISAS[d] satellites. Although solar/heliospheric physics is by now a mature discipline and thus not anymore an exploratory activity, this subdiscipline of astronomy remains pioneering in stellar research.[e] Specifically, some processes, like coronal heating, flares and the acceleration of solar or stellar wind, are still poorly understood and need more observational input. Likewise, further investigation of the solar interior through low-noise helioseismology with high spatial resolution, also making use of the recently developed method of local helioseismology, is bound to bring much important insight into stellar structure and dynamics. A key contribution of solar physics (by SOHO) was also to show that the so-called solar-neutrino problem was, in fact, a problem of the Standard Model of particle physics (which could be solved by introducing neutrino oscillations), rather than a solar problem.

While attempts of European solar physicists to obtain a mission dedicated to their subject were frustrated repeatedly, the scientists concerned got training by participating in NASA projects, such as the Orbiting Solar Observatories (OSO) series, the Apollo Telescope Mount (ATM) on Skylab and the Solar Maximum Mission (SMM), or by contributing to the Japanese/UK/US Yohkoh satellite. Solar-B, now called Hinode (Sunrise), a Japanese/US/UK/ESA satellite, launched in September 2006 to study the Sun, is the latest example of such a collaboration. Their participation took different forms: some solar physicists worked abroad, while others were able to work at home after persuading their national funding agencies to fly rocket experiments or to contribute experiments for flight on satellites being realized in another country.

[d]ISAS is the Japanese Institute of Space and Astronautical Science.

[e]In this sense, solar physics — although an observational rather than an experimental discipline — is somewhat similar to fundamental physics: in spite of being mature, its practice, particularly in space, is pioneering science in general, and is thus bound to bring new knowledge that will influence science beyond its own field.

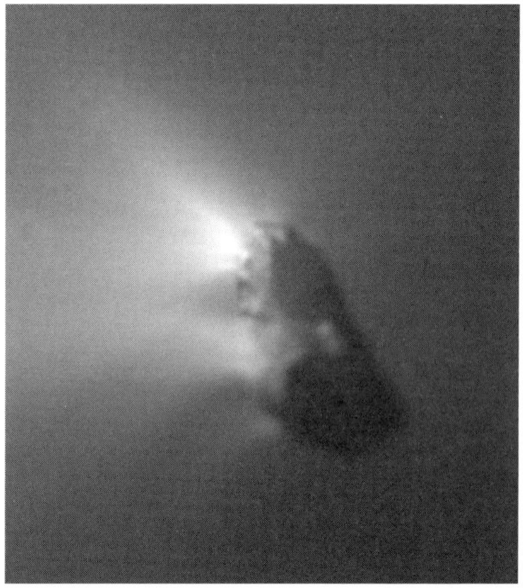

Fig. 4. Composite image taken by the Halley Multicolor Camera on the Giotto probe during the flyby of 13/14 March 1986.[6] It is often said that this Giotto picture — together with the scientifically more fundamental composition measurements made in the coma by mass spectrometers and *in-situ* measurements from other sensors — put ESA's Science Programme on the map.

3.3. *Introducing planetary sciences at ESA*

The resurrection of planetary sciences, which, as mentioned, had been excluded from the program in the 1960s, started with the Giotto probe, which flew by Comet Halley during its 1986 perihelion, and took pictures from a distance much closer — 596 km from the comet's nucleus (Fig. 4) — than that of either the Japanese or the Russian probes that had been sent to also encounter the comet at that same time.

European planetary scientists got training in space research by participating in NASA missions in the same way as some solar physicists did. Giotto's success helped the community involved to succeed in obtaining two missions in the Horizon 2000 Programme, namely the Rosetta Cornerstone (a comet rendezvous with a cometary lander) and, in addition, a so-called medium-size mission, the ESA Huygens probe, which descended on Titan, after having been delivered there by NASA's Cassini mission to Saturn.[f]

Later on, the planetary community also obtained a Cornerstone of Horizon 2000 Plus, namely a mission to Mercury, now BepiColombo. An additional, inexpensive mission to the most important planet, Mars Express, was inserted into Horizon 2000 Plus, after other missions to Mars that had been under study by ESA had not been

[f] Strictly speaking, the Huygens probe is a consequence of the deliberations of a Joint Working Group on planetary missions of the European Science Foundation and the US National Academy of Sciences.

selected in the competitive procedure for the so-called medium-sized missions. A further, low-cost Venus-Express mission that made use of the Mars-Express design was added to the list of ESA's planetary missions.

4. The Roller Coaster for Fundamental Physics at ESA

Between 1971 and 1979, ESRO, and then ESA, had a Fundamental Physics Panel[7] with a decidedly distinguished membership: Hermann Bondi was the Chairman, Ian Roxborough his deputy, and Jacques Blamont, Giuseppe Cocconi, Giuseppe (Bepi) Colombo, B. Laurent, Reimar Lüst, Giuseppe (Beppo) Occhialini, Evry Schatzman and Dennis Sciama were the members. However, in the late 1970's, the Panel came to the conclusion that the technology required for meaningful fundamental physics experiments in space was still lacking. Auspiciously, in mid-1983, when deliberations on Horizon 2000 were started, a topical team on physics was set up, and the calls for proposals for the medium-size missions of Horizon 2000 also invited proposals for missions devoted to fundamental physics. A Satellite Test of the Equivalence Principle (STEP) was subsequently proposed for the second, and reproposed for the third medium-size mission of Horizon 2000. STEP was both times selected for Feasibility and subsequently also for a Phase-A Studies. And in order to properly judge proposals as well as the results of studies in the field of fundamental physics, a Fundamental Physics Advisory Group (FPAG) was set up.[g] It was recognized at that time already that both SSWG and AWG would have a conflict of interest, if they had to evaluate the merits of a fundamental physics mission.

Although the INTEGRAL (INTErnational Gamma-RAy Laboratory) mission, and later the PLANCK mission, were preferred to STEP in the final competitive selections of the second and third medium-size ESA projects, M2 and M3, of Horizon 2000,[h] the community had established its credibility; and LISA was later designated as a Cornerstone in the Horizon 2000 Plus exercise — not least because of its profound contributions to both physics and astronomy.[i] Given the novelty of the technologies to be employed, it was decided to first test some of the LISA technologies in a Small Mission for Advanced Research in Technology (SMART) called LISA Pathfinder. As mentioned above, LISA Pathfinder is a joint ESA/NASA mission; so is LISA, albeit NASA has put it on hold.

[g] Jean-Pierre Blaser was the first chairman, and Maurice Jacob was his successor. The current Chairman is Bernard Schutz.

[h] In the selection for M2, the prime scientific goal of STEP had been augmented by a measurement of the gravitational constant, a test of the inverse-square law of gravity, an experiment to search for spin-coupling forces, a geodesy experiment, and an aeronomy experiment; in the selection for M3, the STEP mission still had retained the spin-coupling experiment.

[i] Nevertheless, there were some astronomers who were of the opinion at the time that LISA would "not contribute anything whatsoever to astronomy" — a further indication that the creation of a separate Fundamental Physics Advisory Group was crucial for the discipline.

5. The Elements of Success that Got Solar Physics into ESA

European solar physicists pursued their subjects in the absence of dedicated ESRO or ESA missions by getting training and hands-on experience through participating in non-European missions. And, once European missions were being studied, the solar-physics community united behind one given mission at a time (GRIST, DISCO, SOHO). Solar physicists also realized that trans-Atlantic collaboration was essential for reaching a viable size of the community.

With the emergence of heliospheric physics, it also had become clear that solar physicists should look toward co-operating, rather than competing, with researchers dealing with solar-wind and space-plasma physics. This recognition also led to a decisive meeting in international collaboration: Stan Shawhan's informal, international review of solar–terrestrial and space-plasma missions. Scientists related to ESA, ISAS and NASA met in Washington DC in the early summer of 1983, and reduced an excessive list of missions planned in the fields of solar, solar–terrestrial and space-plasma physics to a rational ensemble, the International Solar–Terrestrial Physics (ISTP) Program. The ISTP Program then became the main topic on the agenda of the Inter-Agency Consultative Group (IACG). This Group, which consisted of the heads of the science programs of ESA, IKI,[j] ISAS and NASA, met annually to review the progress of their programs and to consult on possibilities of collaborations. As a consequence, two satellites of the "OPEN" Program, which was originally an exclusive NASA program, were later taken on by additional players. The OPEN program was designed to thoroughly investigate the Earth's magnetosphere environment; it consisted of four satellites, namely Polar, Equator, Wind and Tail, which were named after their orbit locations. Tail became Geotail, a joint ISAS and NASA mission, and Equator became Double Star by ESA and the China National Space Administration (CNSA).[k]

That SOHO and Cluster eventually became a Cornerstone in the Horizon 2000 Programme had one of its roots in Stan Shawhan's meeting as well. There, boundaries between discipline-oriented communities had been crossed and a network of collaboration between scientists associated with ESA, ISAS and NASA — enhanced by the regular meetings of IACG — was established.

Putting SOHO[8] and Cluster[9] together in a single Solar Terrestrial Science Programme (STSP) eliminated the competition between two communities, as mentioned above. An additional bonus of inserting the STSP Cornerstone was the balance that resulted in Horizon 2000 between the disciplines represented by ESA's scientific community.

[j]IKI, the Space Research Institute of the Russian Academy of Science, is carrying out most of the Russian scientific space research.
[k]In fact, Double Star consists of two satellites. An "equatorial" spacecraft (TC-1) with an elliptical orbit of altitudes 570 km × 78,970 km and an inclination toward the equator of 28.5° can also reach the magneto tail; the other spacecraft (TC-2) is in a polar orbit of 700 km × 39,000 km.

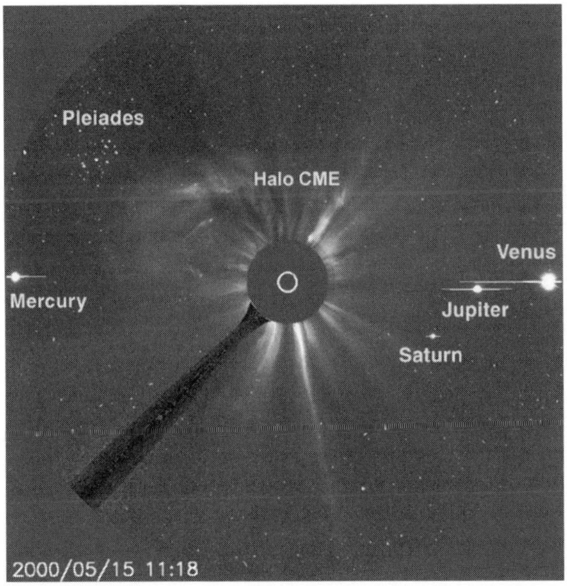

Fig. 5. SOHO's look at the heliosphere while a coronal mass-ejection (CMA) event was in progress (and at a time when four of the five innermost extraterrestrial planets appeared in the 30-solar-diameter wide field of its white-light coronagraph). This picture is the result of international collaboration, and it became reality after the boundaries between discipline-oriented communities, namely between the solar, heliospheric and space-plasma physics communities, had been crossed.

6. The Elements of Success for Planetary Sciences at ESA

Success for the planetary community — some of whose members had trained abroad, or had been funded by their national agencies to participate in missions of other countries — came through Giotto. This mission, where the space-plasma physics and mass-spectrometry communities were involved as well, established the credibility of the community.

After a Mission to Mercury was placed as a Cornerstone into the Horizon 2000 Plus Programme, but research on Mars was absent, the low-cost mission Mars-Express was initiated, because a lack of research devoted to Mars in the European Space Programme was considered a bad mistake.[101]

7. Conclusions and Suggestions

First of all, future investigators in a new space-science discipline need to be trained and to gain hands-on experience. Many European solar and planetary scientists had to get their training abroad.

[1]In hindsight, one may also ask whether concentrating on the most attractive planet, Mars, might not have brought more than the large array of planetary missions — including the very expensive BepiColombo mission (which moreover will arrive at Mercury after NASA's Messenger). The total ESA expense on planetary sciences in the years 1990–2010 will be 2.3 GEuro, i.e. the equivalent of ten Mars-Express missions.

A new discipline needs a proper advocacy group. There is no way around establishing an appropriate board or committee, such as ESA's Fundamental Physics Advisory Group, before a space agency will include and support this discipline in its program. Otherwise, independent advice, free from conflict of interest, is not granted.[m]

The community must unify itself behind one or, at most, very few key projects, and actively seek the support of administrators and politicians. In particular, a grand project, such as LISA, will succeed only if the community is united behind it. And as the mission comes closer to approval and eventual implementation, it must be stressed for persons unfamiliar with space research that a mission ready for approval and, particularly, ready for flight, is always somewhat obsolete, as far as the technologies employed are concerned.[n]

Organizing summer schools will help arouse and maintain the interest of a future generation.

When advocating a new discipline, one should also look to emerging communities and agencies, at present in Asia and the Pacific space, as well as in South America.

In this connection, one needs to ask the question whether a new, enlarged forum should not fill the gap caused by the abolition of the Inter-Agency Consultative Group. After all, we owe the co-ordination of the flights to Comet Halley and of the ISTP Program to the IACG; that Group also provided a platform for valuable co-ordination of VLBI observations both on the ground and in space.

Appendix A. What Should Fall Under the Label "Fundamental Physics"?

In due course, the LISA project will have a considerable impact on astronomy, yet for the time being it is better located under "Fundamental Physics." A historic precedent is high-energy astrophysics, which to a large extent was introduced by physicists.

In order to provide a substantial number of flight opportunities for physics experiments in space beyond LISA, a reusable, drag-free platform, such as the "Fundamental Physics Explorer" of ESA's Cosmic Vision Programme, is the right direction to go. This will help to build a robust yet geographically spread community, and this is an important resource for future, larger missions in fundamental physics in space, as well as an important aspect for the education of a future generation.

Whether cosmology belongs to observational astronomy or fundamental physics needs to be decided from case to case, depending on the prime character of the

[m] The same applies to general planning exercises, such as the Decadal Reports of the US National Academy of Sciences, and is also exemplified by the Joint Working Group of the European Science Foundation and the US National Academy of Science which led to the Huygens probe.

[n] This is a consequence of the need to have the required technologies "ready to go" once a project is approved. More advanced technologies may come into being in the time between project approval and the launch. But such technologies are generally not reliable and mature enough for space use and may lead to schedule delays and, consequently, to severe cost overruns.

mission in question. On the other hand, astroparticle physics belongs predominantly to fundamental physics. The EPS Position Paper (2005) discusses this point in the context of the ESA Science Programme.[2]

Acknowledgments

The author has learned much of what is described in this paper while editing the EPS Position Paper. The stimulating input from members of the EPS Position Paper Advisory Committee is gratefully acknowledged.

References

1. M. Keiser, Gravity Probe B, see http://einstein.stanford.edu.
2. EPS Position Paper, The Need for Space Flight Opportunities in Fundamental Physics (2005); see http://www.eps.org/about-us/position-papers
3. R. F. C. Vessot *et al.*, *Phys. Rev. Lett.* **45** (1980) 2081.
4. J. Clavel, The Fundamental Physics Programme at ESA (2006) http://funphysics.jpl.nasa.gov/quantum-to-cosmos/program.html
5. C. Salomon, *Int. J. Mod. Phys. D* **16** (2007) to appear in Issue 13.
6. H. U. Keller, The Nucleus, in *Physics and Chemistry of Comets*, ed. W. F. Huebner, (Springer-Verlag, Heidelberg, 1992).
7. R. Reinhard, Ten years of fundamental physics in ESA's space science programme, *ESA Bulletin* **98** (1999).
8. B. Fleck, V. Domingo and A. I. Poland (eds.), *The SOHO Mission, Sol. Phys.* **162** (1–2) (1995).
9. R. Schmidt and M. L. Goldstein (eds.), Cluster: Mission, Payload and Supporting Activities, ESA Special Publication **SP-371** (1993) 259.
10. L. Woltjer, *Europe's Quest for the Universe — ESO and the VLT, ESA and Other Projects* (EDP Sciences, Paris, 2006).

NATIONAL SCIENCE FOUNDATION VISION IN PARTICLE AND NUCLEAR ASTROPHYSICS

RICHARD N. BOYD

National Science Foundation,
4201 Wilson Blvd., Arlington, VA 22230, USA
rboyd@nsf.gov

The NSF has made investments in searches for dark matter, in ultrahigh energy cosmic rays and gamma rays, in neutrino physics and astrophysics, and in nuclear astrophysics. We expect the future to witness the expansion of these efforts, along with efforts to refine the measurements of the cosmic microwave background. In some of these efforts the Deep Underground Science and Engineering Laboratory is expected to play a major role.

Keywords: NSF; particle astrophysics; nuclear astrophysics.

1. Dark Matter Searches

The Universe has been shown, by the Supernova Cosmology Project, the High-z Program (both involving Type Ia supernova searches), and the Wilkinson Microwave Anisotropy Probe, to be composed of about 25% "dark matter," which acts gravitationally to affect the motions of the stars in galaxies, but otherwise has little effect on astrophysical observables, and is difficult to detect. Identification of the nature of the dark matter was recognized as one of the 11 most important questions in the National Academy of Science study "Connecting Quarks to the Cosmos," so it has been regarded as an effort of high urgency. One strong candidate for this dark matter is weakly interacting massive particles (WIMP's), which could pervade the cosmos, providing just the observed effects. Recent years have seen a number of experiments to detect these WIMP's, most notably the Cryogenic Dark Matter Search (CDMS), comprised of hockey-puck-sized detectors of Ge and Si. These are maintained at temperatures below 50 mK, and housed 2000 feet underground in the Soudan mine in northern Minnesota, to minimize noise and backgrounds. The detectors register events in which a WIMP interacts with a nucleus, imparting recoil

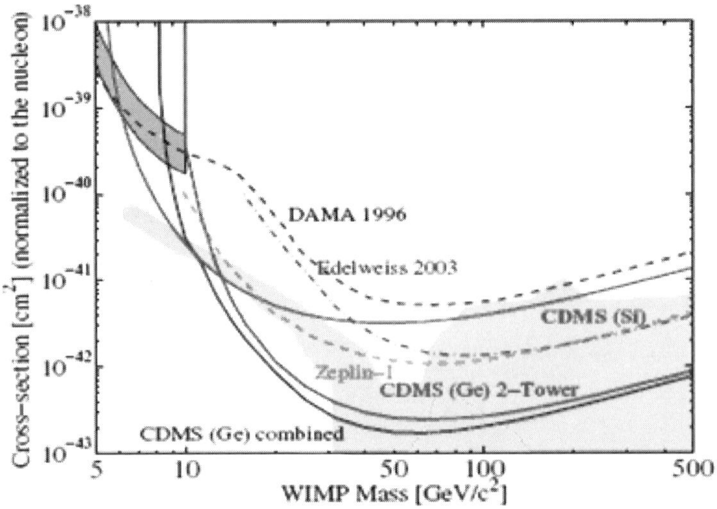

Fig. 1. CDMS limits[1] on the WIMP-nucleon cross section versus WIMP mass. They are shown as the two bottom curves (resulting from slightly different approaches to the data analysis), with results from other experiments also indicated. Predictions of supersymmetric models are shown as the shaded regions.

energy to the nucleus, which can be detected. This experiment currently holds the world's best result[1] on the nonexistence of WIMP's, at a level which is beginning to impinge on various possible theoretical candidates for their identity, e.g. supersymmetric particles. The results from the past runs are shown in Fig. 1.

The current experiment, now dubbed CDMS II, contains five "towers," each containing six of the detectors (A "tower" is several inches high!). It continued to operate in 2006–07, hopefully pushing the limit from that experiment by close to another order of magnitude, or perhaps detecting WIMP's. This project is funded jointly by the NSF and the DOE.

However, the NSF is also supporting efforts in other dark matter experiments, usually together with another funding agency. These include XENON and Zeplin II (both of which use liquid xenon as their detection medium), WARP (which uses a liquid argon detector), SuperCDMS (which will use the same basic scheme as did CDMS, but with major technical improvements), COUPP (which uses a bubble chamber detection scheme), DRIFT (which uses gas detectors, and so could indicate the directionality of WIMP events), and PICASSO (which uses superheated droplets to detect the WIMP's).

The funding levels of all the above experiments except CDMS are at the R&D or test facility level. However, the next generation version of each of these experiments will be considerably larger, and much more expensive, than the current versions. Thus the NSF and DOE have created a Dark Matter Scientific Assessment Group, (DarkSAG) to help identify the most promising technologies for pushing the state of the art beyond its present limits.

2. Ultrahigh Energy Cosmic Rays

A longstanding question in particle astrophysics is the origin of the highest energy particles that have been detected in the Universe; this is another of the 11 most important questions in "Connecting Quarks to the Cosmos." However, the first priority is to measure the actual energies of those particles. Two experiments, the Akeno Giant Air Shower Array (AGASA) and the High Resolution Fly's Eye (HiRes) have sought to do so. It has long been thought that such particles would be subject to the GZK cutoff, the energy cutoff, suggested by Greisen, Zatsepin, and Kuzmin, which would result from photon–pion interactions with high probability of any protons with energies above about 6×10^{19} eV, with the cosmic microwave background photons from the Big Bang. AGASA used surface water Cherenkov detectors to observe the energy of the air showers produced by interactions of the ultrahigh energy cosmic rays with the Earth's atmosphere by detecting their secondary muons, while HiRes detected the fluorescence produced by interactions of the shower particles with nitrogen in the Earth's atmosphere. Obviously each technique detected only part of the energy, so models had to be developed to infer the total energy of the primary cosmic ray from the radiation detected.

The results were that AGASA did not see any GZK cutoff, but HiRes did. These results[2] are indicated in Fig. 2. However, a renormalization of about 30% of the energy of one of the experiments, HiRes upward or AGASA downward, would put the results in agreement. The basic problem is that the statistics at the highest energies were very poor; such particles occur at a rate of one per km^2 per century.

Thus the Pierre Auger Observatory has been built, at a site located at the foot of the Andes mountains in Malargue, Argentina. It is a consortium of many nations and funding agencies. It consists of both surface water Cherenkov detectors and air fluorescence detectors, so that for some fraction of the events it observes it can

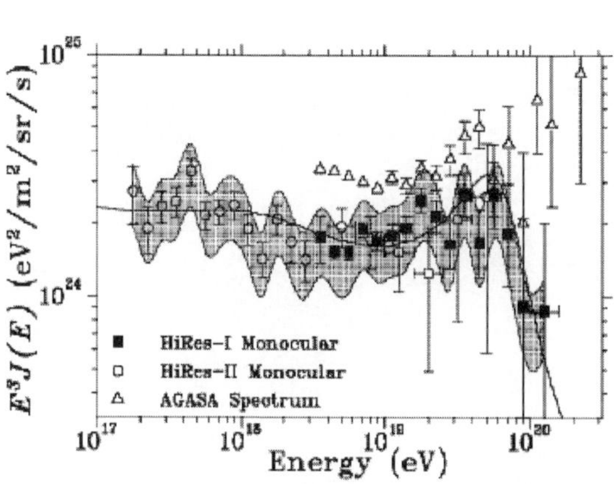

Fig. 2. Results[2] of the HiRes and AGASA experiments. The two data sets shown for HiRes were both taken in "monocular" mode, in which data from the two arrays of detectors were analyzed independently. These results are consistent with those in which the two detector arrays are used to observe the same event, i.e. "stereo" mode.

detect single events with both methods of detection, thereby seeing if there is an inconsistency between them. The Observatory is about 60% complete as of summer, 2006, but data taking has already begun. Preliminary results show that the GZK cutoff is observed in the water Cherenkov detectors if their energy is normalized to the air Cherenkov detector results. However, considerably better statistics must be obtained before any definite conclusions can be drawn.

One interesting aspect of the search for ultrahigh energy cosmic rays is that there are hints that there may be correlated events, i.e. that they may be pointing back to sources. For the very highest energy cosmic rays the ambient magnetic fields would not be expected to prevent their moving in essentially straight lines from their point of origin to the detectors, so that their detection might indicate their sources. This would be very helpful in trying to identify the mechanisms by which they are produced. Searches for these sources are expected to be an important component of future searches for ultrahigh energy cosmic rays.

3. High Energy Gamma Ray Astronomy

A question related to that of the sources of the highest energy cosmic rays is that of the sources of the highest energy gamma rays. These will also scatter from the cosmic microwave background photons, so are sharply attenuated at high energies, but they can provide important information about their sources. The Milagro detector has been operating in the Jemez mountains of New Mexico for the past several years. It is a large swimming pool instrumented with photomultipliers, and has an opaque cover. Thus it covers half the possible sky, and operates both day and night. It has identified several sources of TeV gamma rays; a Milagro sky map,[3] featuring the Cygnus region, is shown in Fig. 3.

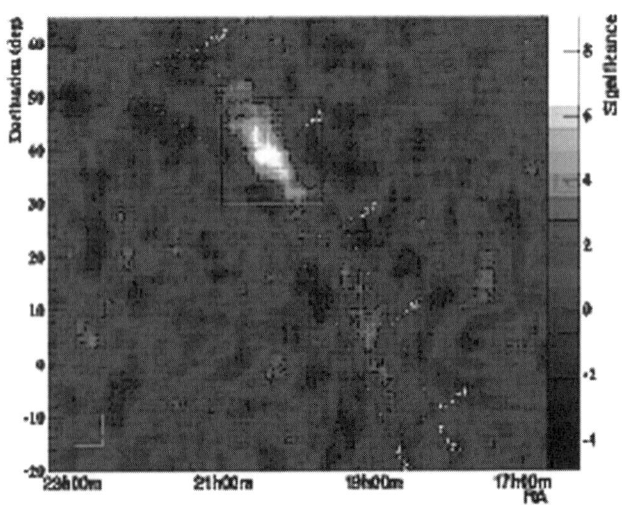

Fig. 3. Milagro sky map showing, as the brightest spot, the Cygnus region.[3]

However, its energy resolution and angular resolution are insufficient to adequately identify the sources so that they can be located for searches for optical or X-ray counterparts, or possibly even ultrahigh energy cosmic ray counterparts. Thus VERITAS, the Very Energetic Radiation Imaging Telescope Array System, is being built in southern Arizona, with funding from several countries, but from the NSF, DOE, and Smithsonian in the US. It will consist of four 12 m telescopes that will observe the air Cherenkov light produced by the ultrahigh energy gamma rays interacting with the Earth's atmosphere, and will be sensitive to energies from 10^{11} to 10^{13} eV. It is expected to be operational late in 2006. Its southern hemisphere counterpart, HESS, has already shown the worth of ultrahigh energy gamma ray astronomy, having identified and located many sources of this radiation.

4. Neutrino Astronomy

The high energy neutrinos from the Sun have been well mapped out over the past several decades by the Homestake Mine experiment, by Super-Kamiokande, and by the Sudbury Neutrino Observatory experiments, and the oscillations of these neutrinos are now well established. The existence of the low energy neutrinos has also been established by the radiochemical experiments SAGE and GALLEX/GNO. However, it is expected that the higher energy neutrino oscillations occur via the Mikaev–Schmirnov–Wolfenstein (matter-enhanced) mechanism, and the lower energy neutrino oscillations via the vacuum oscillation mechanism. The transition should take place at neutrino energies of around 1 MeV; these would be nicely amenable to detection in Borexino. This detector is currently under construction at the Gran Sasso laboratory in Italy. It is being built by an international collaboration involving seven countries, with the NSF supporting the US component. Borexino has been designed with extremely low backgrounds so that it can detect solar neutrinos with energies as low as 0.8 MeV; this would make it sensitive both to the neutrinos from the $^{7}\mathrm{Be} + \mathrm{e}^- \rightarrow {}^{7}\mathrm{Li} + \nu_e$ and $\mathrm{p} + \mathrm{e}^- + \mathrm{p} \rightarrow \mathrm{d} + \nu_e$ reactions. This is an especially interesting energy region, as not only will it extend the energy spectrum observed for solar neutrinos, but it should make it possible to see the transition from MSW to vacuum oscillations. However, there have been other possible effects suggested that might affect the neutrino spectrum in this region, such as the effect of the neutrino wave packets on the oscillations. Borexino is expected to begin taking data in 2007.

A very different entry into the field of neutrino astronomy is provided by the AMANDA/IceCube project, a detector of high energy neutrinos located at the South Pole. AMANDA has been running for several years, and is now being subsumed by IceCube, which will ultimately be an instrumented cubic kilometer of South Pole ice. The optical modules look downward for upward-traveling muons produced by interactions of neutrinos passing through the Earth and interacting either in the ice or in the Earth below the ice. The muons produce Cherenkov light, which is observed by the optical modules. The goals of IceCube are to measure the spectrum of high energy cosmic neutrinos and to search for sources. IceCube

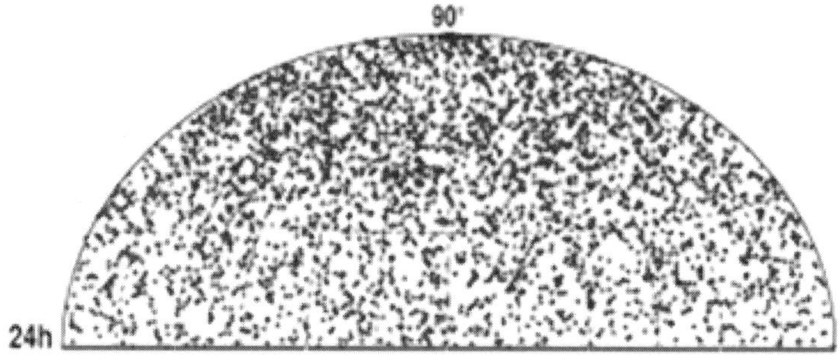

Fig. 4. AMANDA all-sky map of the neutrino sky. With these more than 3000 events, there are not yet any sources that have been detected with sufficient statistical significance.[4]

is now approximately 20% complete, but is already taking data. At present, an all-sky map of neutrinos[4] (shown in Fig. 4), actually from AMANDA, shows no statistically significant sources, but the statistics are building.

5. Synthesizing the Heaviest Elements in the Cosmos

Another of the 11 basic questions identified in "Connecting Quarks to the Cosmos" concerned the origin of the Universe's heaviest elements. One of the two processes of nucleosynthesis known to make these elements, and the process that must make all elements heavier than lead, is the rapid process, or r–process, which is characterized by an extremely high neutron density and a time scale of seconds. The pathway of this process is thought to exist roughly 20 neutrons to the neutron-rich side of stability in the chart of the nuclides, and so involves extremely unstable nuclei. The aspect of these nuclei that is critical to the ultimate r process abundances is their lifetimes.

Many of these nuclei can be made, and their lifetimes measured, at the National Superconducting Cyclotron Laboratory at Michigan State University. This facility consists of two coupled superconducting cyclotrons, which accelerate heavy ions to tens of MeV per nucleon, then smash them into target nuclei, producing a myriad of secondary nuclei. A mass analysis system then selects out the nuclei of interest, and directs them to a detector system capable of measuring their lifetimes and decay modes. These features are indicated in Fig. 5. The study of these nuclei has become a major program at that facility, with measurement of the lifetimes of some of these very neutron-rich nuclides assuming a high priority. An example of the nuclei studied is afforded by ^{78}Ni, a nucleus that is 14 neutrons to the neutron-rich side of the heaviest stable Ni isotope.

6. The Deep Underground Science and Engineering Laboratory (DUSEL)

An initiative that the NSF expects to enable some extraordinary science many years into the future is the DUSEL. Many of the experiments now being proposed

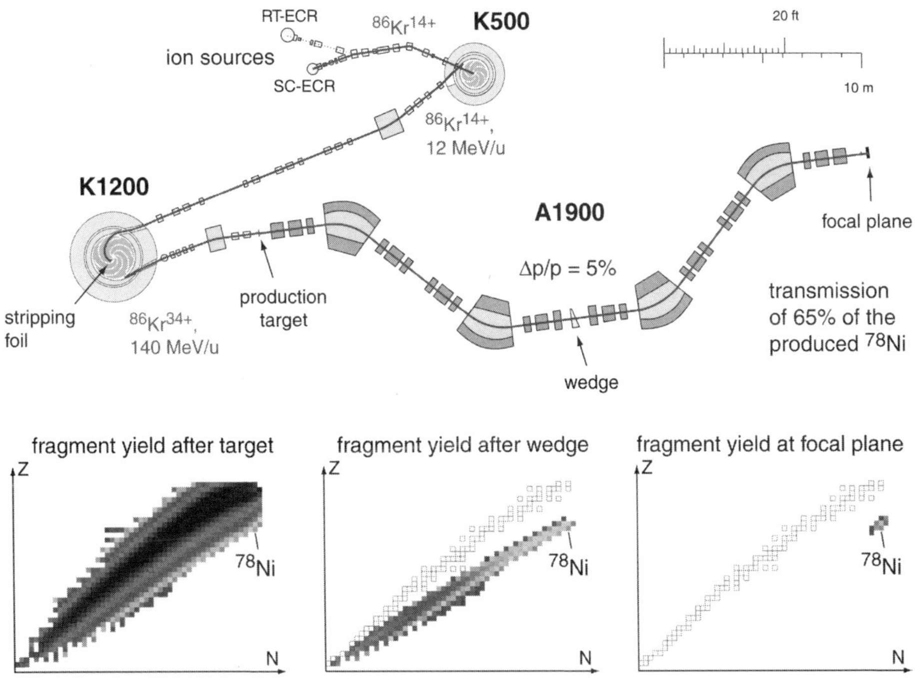

Fig. 5. Schematic of the two cyclotrons of the NSCL, along with the mass analysis system, used to study nuclei far from stability. Shown below are the yields at three stages of the analysis system. *Figure courtesy of A. Stoltz.*

require the low backgrounds that can only be achieved by siting them deep underground. The DUSEL would provide the infrastructure that would make these possible. While there are underground laboratories at several places around the world, and even in the United States, the DUSEL would have sufficient depth to match or exceed that of every other laboratory in the world, and would have the size to accommodate every experiment proposed for the foreseeable future. The DUSEL would enable next generation physics experiments such as double beta decay, dark matter, a long baseline neutrino oscillation detector (which will observe neutrinos produced at an accelerator more than 1000 km distant, and which could double as a proton decay detector), a supernova neutrino detector, and even an underground-accelerator-based effort to measure tiny, astrophysically interesting nuclear reaction cross sections. However, the DUSEL would also be aimed at addressing several major efforts in geosciences, including water transport underground and wave propagation in three dimensions. Engineering's interest in the DUSEL is centered on simply creating the laboratory, which necessitates extremes not included in any mine yet created.

The development of the DUSEL has followed a three-solicitation process. Solicitation 1 was designed to develop the science-and-engineering motivation for such a laboratory. That has taken longer than originally thought, but is essentially complete as of early 2006. While this was intended to be site-independent, it was quickly realized that this was not possible for the geosciences, as different sites afford the geoscientists different opportunities. Solicitation 2 invited proposals for support to develop conceptual designs for specific sites. A review panel met in spring 2005, and recommended that two of the sites be funded to do so. The best proposals were judged to be those from the Homestake Mine in South Dakota and the Henderson Mine in Colorado. The proponents of these two sites finished their work in June 2006. A down-select to a single site, to be funded to do a baseline analysis of the site, will be announced in early 2007. Following a year's funding for that effort, the DUSEL may then be proposed for facility construction as soon as it can be included in the NSF queue for large facilities.

7. Roadmapping the Future

There is little doubt that proposal pressure is intense in particle and nuclear astrophysics. Despite a strong growth in that budget, at least within the NSF, new ideas seem to be generated considerably more rapidly than increases in funding. Funding decisions are generally guided by studies such as the National Academy of Sciences' "Connecting Quarks to the Cosmos," and the Scientific Assessment Groups, but other studies certainly provide important guidance as well. For example, the American Physical Society initiated a recent study of neutrino physics, and the report was generated by a large fraction of that community. The long range plans of the High Energy Physics Advisory Panel and the Nuclear Sciences Advisory Committee also play an important role in guiding the decisions of the funding agencies, and aiding in the funding decisions that are made. While it is discouraging to have to reject proposals that are clearly highly meritorious, that also reflects the extraordinary strength of the field.

References

1. D. Akerib *et al.*, *Phys. Rev. Lett.* **96** (2006) 011302.
2. R. Abbasi *et al.*, *Phys. Rev. Lett.* **92** (2004) 151101.
3. A. J. Smith for the Milagro Collaboration, "Detection of Diffuse Gamma-Ray Emission from the Cygnus Region with the Milagro Gamma-Ray Observatory," in *29th Int. Cosmic Ray Conf. (ICRC 2005)*, Pune, India, 2005 (Tata Institute of Fundamental Research, 2005), p. 4.
4. G. C. Hill, for the IceCube Collaboration, "Neutrino Astronomy with IceCube and AMANDA," in *Proc. 29th Int. Cosmic Ray Conf. (ICRC 2005)*, Pune, India, 2005 (Tata Institute of Fundamental Research, 2005), p. 10.

THE DEPARTMENT OF ENERGY HIGH ENERGY PHYSICS PROGRAM

KATHLEEN TURNER
United States Department of Energy, Office of Science,
Office of High Energy Physics, SC-25,
19901 Germantown Rd, Germantown MD 20874, USA
Kathy.Turner@science.doe.gov

This paper describes the high-energy-physics program at the US Department of Energy. The mission and goals of the program are described along with the breadth of the overall program. Information on recommendations from community-based panels and committees is provided. Finally, details about the main astrophysics and cosmology projects are given.

Keywords: Agency; astrophysics; cosmology.

1. Introduction

This report describes the US Department of Energy (DOE), Office of High Energy Physics (OHEP) program, with concentration on science activities related to particle astrophysics and cosmology. The goal of the OHEP program is to explore the fundamental interactions of energy, matter, space and time. This includes understanding the unification of fundamental particles and forces, studying the mysterious dark matter that holds galaxies together and the even more mysterious dark energy that is causing the expansion of the Universe to accelerate, searching for possible new dimensions of space and investigating the nature of time itself. These goals lead to the naturally strong connections between the physics of elementary particles and the physics that determines the structure of the Universe. The OHEP program supports approximately 90% of the US high-energy-physics research. It is coordinated with the National Science Foundation (NSF), the National Aeronautics and Space Administration (NASA) and international efforts.

DOE's mission of world-class scientific research capacity is supported by the OHEP program by providing facilities and advancing our knowledge of high energy physics and related fields, including particle astrophysics and cosmology. The primary tools of the program are particle accelerators, which allow the study of fundamental interactions at the highest possible energies (see Fig. 1). The Tevatron

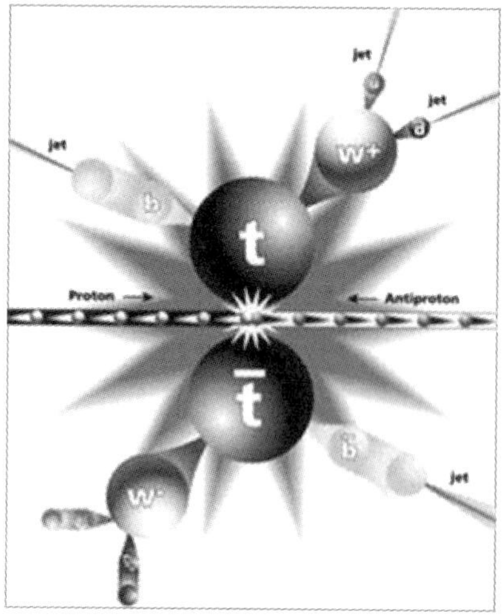

Fig. 1. Artist's conception of a proton–antiproton collision in the Fermilab Tevatron collider. A pair of top quarks is produced, which then decay into W bosons and b quarks.

collider at Fermilab in Chicago, Illinois and associated detectors study bottom and top quarks and search for the Higgs boson, extra dimensions and supersymmetry (see Fig. 2). The B-factory collider and BaBar detector at the Stanford Linear Accelerator Center (SLAC) in California studies charm and bottom quarks and searches for charge-parity violation to understand why there is more matter than antimatter in the Universe (see Fig. 3). The US is heavily involved in the construction of the next step in high energy accelerators, the Large Hadron Collider (LHC) and the

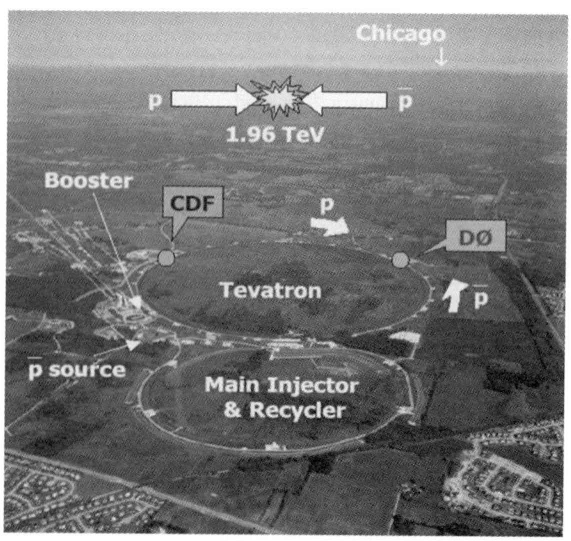

Fig. 2. Aerial photo of the Fermilab Tevatron collider and associated accelerator components.

Fig. 3. Aerial photo of the B-factory collider at SLAC.

ATLAS (see Fig. 4) and CMS detectors, which will start operating in late 2007 at the CERN laboratory in Geneva, Switzerland. Planning for an International Linear Collider (ILC) is underway. The study of neutrinos using accelerators in the US and Japan is an important and growing part of the program.

Because high energy physics is fundamentally involved in the origin and evolution of the Universe itself, the OHEP program also supports nonaccelerator studies of cosmic particles and phenomena, including experiments conducted deep underground, on the ground, on mountaintops, or in space. This includes studies of atmospheric, solar and reactor neutrinos, dark matter, dark energy and high energy cosmic and gamma rays. The major astrophysics and cosmology efforts will be described later. Currently operating experiments that OHEP is involved with are the Super-Kamiokande and KamLAND neutrino experiments in Japan, and the Cryogenic Dark Matter Search II (CDMS-II), the Sloan Digital Sky Survey (SDSS), the Supernova Cosmology Project and the Nearby Supernova Factory based in the US. Projects with OHEP involvement that are approved or under construction include the Gamma-ray Large Area Space Telescope (GLAST), the Pierre Auger array in Argentina, the Alpha Magnetic Spectrometer (AMS), the Very Energetic Radiation Imaging Telescope Array System (VERITAS) in Arizona and the Axion Dark Matter eXperiment (ADMX) at the Lawrence Livermore National Laboratory in California. Some of the proposed experiments are a reactor neutrino experiment to study neutrino mixing, the Enriched Xenon Observatory to measure neutrino mass, and mid-term and longer-term ground- and space-based dark energy experiments.

Fig. 4. View (with a man in the lower middle part) of the ATLAS detector which is under construction at CERN.

In addition to the experimental program, there is a large investment in related theoretical studies and research and development for future accelerator and detector technologies.

2. Recommendations

The DOE high-energy-physics program relies on recommendations on program direction from the High Energy Physics Advisory Panel (HEPAP), the Astronomy and Astrophysics Advisory Committee (AAAC) and the National Research Council (NRC). Recommendations related to particle astrophysics and cosmology are shown below.

The NRC's 2003 report "Connecting Quarks with the Cosmos"[1] recommended three new nonprioritized initiatives, one of which is to determine the properties of dark energy. The Committee recommended that "NASA and DOE work together to construct a wide-field telescope in space to determine the expansion history of the universe and fully probe the nature of the dark energy."

The interagency 2004 report from the National Science and Technology Council (NSTC) provided a federal cross-agency strategic plan, "The Physics of the Universe",[2] for discovery at the intersection of physics and astronomy in response to the NRC's report "Connecting Quarks with the Cosmos". The NSTC report listed dark energy measurements as its highest priority, proposing a multipronged strategy. The report recommended that NASA and DOE develop a Joint Dark Energy Mission (JDEM) and said that this "mission would best serve the scientific community if launched by the middle of the next decade." The recommendation also noted that a high-priority independent approach to studying dark energy will be made by studying the weak gravitational lensing produced by dark matter, which is a scientific goal of a ground-based Large Survey Telescope (LST), and that "NSF and DOE will begin technology development of detectors... leading to possible construction and first operations in 2012."

The 2006 NRC committee on "Elementary Particle Physics in the 21st Century"[3] laid out a "roadmap" for the field and recommended a plan in priority order. First, the US should fully exploit the opportunities for US involvement in the LHC at CERN. Secondly, the US should develop a comprehensive program to become the world-leading center for R&D for the ILC and mount a compelling bid to build it in the US. And, thirdly, the program in particle astrophysics should be expanded and the US should pursue an internationally coordinated and staged program in neutrino physics. This recommendation on particle astrophysics listed understanding dark energy and dark matter as major questions that could lead to potentially momentous implications for particle physics.

A recent (2006) report by the Dark Energy Task Force (DETF) subpanel,[4] commissioned by the AAAC and HEPAP to provide advice to NASA, DOE and NSF on the best strategies for resolving the mystery of dark energy, provides a solid analysis of how the community should move forward to understand the nature of dark energy. The report noted that dark energy could be Einstein's cosmological constant, a new exotic form of matter, or may signify a breakdown in general relativity. It said that to date there are no compelling theoretical explanations for dark energy, and therefore observational exploration must be the focus. No single technique was found to be able to answer the outstanding questions; combinations of at least two techniques, one of which is sensitive to the growth of structure, must be used. The DETF recommended a program of medium- and longer-term experiments and also said that high priority for near-term funding should be given to projects that improve our understanding of the dominant systematic effects.

The HEPAP P5 Subpanel 2006 report[5] provided recommendations on a near-term program for US high energy physics. The report recommended that DOE continue to support R&D, the SuperNova Acceleration Probe (SNAP) concept for JDEM and the Large Synoptic Survey Telescope (LSST) ground-based concept for a future LST. Furthermore, these projects should be brought to the preliminary design review stage over the next two or three years to allow sharpening of cost estimates and further development on agency and collaboration planning.

2.1. *Dark energy*

Since the discovery of dark energy in 1998, the evidence for an accelerating expansion of the Universe has become accepted in the scientific community, and the focus is now on studying its nature. The OHEP is now investigating future space- and/or ground-based telescopes for studying dark energy in cooperation with NASA and NSF and is also investigating the possibility of international partners. A plan of how to review and select which projects could go forward will be done in view of the DETF report.

The OHEP is supporting R&D for several proposed dark energy experiments. Lawrence Berkeley National Laboratory (LBNL) scientists led one of the two teams that found the first evidence of "dark energy" in 1998 and, together with 14 collaborating US and foreign institutions, they have been leading the effort to develop SNAP,[6] a next-generation space-based dark energy mission. We have been providing R&D funding for this effort since FY 2000. SNAP, which would use both supernova and weak lensing to study dark energy, is one of the concepts that may compete for the JDEM[7] mission science investigation in response to a joint Announcement of Opportunity. Several DOE laboratories are doing a small amount of R&D for proposed ground-based dark energy experiments. Fermilab is leading the effort on R&D for the proposed Dark Energy Survey (DES)[8] experiment. This project would build a new optical camera for use on the existing Blanco telescope in Chile to do a large galaxy survey in order to study dark energy. SLAC is leading the effort on R&D for a camera for the proposed LSST[9] experiment. The LSST project includes a new camera on a new telescope facility in Chile and would use weak lensing, supernovae and other methods to study dark energy.

The OHEP is providing support for continued data-taking by dark energy collaborations. This includes the Supernova Cosmology Project, which is continuing operations using ground telescope and Hubble Space Telescope measurements to collect statistics and refine their results. It also includes continued operations of the Nearby Supernova Factory, which is doing measurements of nearby supernova in order to study their properties and current dark energy effects in more detail.

In addition to the R&D efforts listed above, the President's FY 2007 budget request lists approximately $5 million in support of general R&D to be used for mid- or longer-term ground- and space-based dark energy concepts. If funding levels permit, these funds will be allocated taking into account recommendations of the DETF report.

2.2. *SDSS*

The SDSS[10] telescope (see Fig. 5) at Apache Point in New Mexico has been operating since 1998 and was recently approved for additional data-taking through summer 2008. Funding is provided by the Sloan Foundation, NSF, DOE, Japan and Germany and the experiment is led by Fermilab. The instruments include a 120-megapixel mosaic imaging camera and a 640-fiber spectrograph. Large galaxy surveys are performed and the data are used to study dark matter, dark energy and a variety of astrophysics and cosmology topics. In January 2005, the first measurement of baryon acoustic oscillations was performed with the SDSS data. The fifth public data release was done in June 2006 and the catalog now contains data on

Fig. 5. The Sloan Digital Sky Survey telescope in New Mexico.

8000 square degrees of sky, with over 1 million spectra and images of 215 million unique objects.

2.3. *CDMS-II*

The CDMS-II experiment[11] is located underground in the Soudan mine in Minnesota (see Fig. 6). Partial operations started in 2003 and full operations are expected to start in late 2006 and continue through 2007. Fermilab leads the project management. The purpose is the direct detection of weakly interacting massive particles (WIMP's), a possible form of dark matter. The experiment uses cryogenic germanium and silicon detectors to detect energy deposited by WIMP's, which are thought to constitute the dark halo of the Milky Way. The primary function of the detectors is to measure the minute phonon signals generated within a detector crystal by elastic collisions between detector nuclei and the WIMP's. In April 2005, CDMS-II set the world's lowest exclusion limits on the WIMP cross section by a factor of 10, ruling out a significant range of neutralino supersymmetric models.

2.4. *Pierre Auger Observatory*

The Pierre Auger Observatory[12] is the world's largest area cosmic ray detector, covering about 3000 sq km in Argentina (see Figs. 7–9). The scientific goal is to observe, understand and characterize the very highest energy cosmic rays. The full array is scheduled to began operations in 2007 after operating for several years with a partially completed array. The observatory consists of both surface Cherenkov detectors (about 1000 of the 1600 are currently operating) and fluorescence telescopes (18

Fig. 6. The CDMS-II detector.

Fig. 7. A water Cherenkov detector for the Pierre Auger Observatory in Argentina.

out of 24 are currently operating). This research program is being carried out by an international collaboration including scientists from the US and 19 other countries. DOE and NSF provide US funding and Fermilab leads the project management. The first scientific results were released in summer 2005 and included the cosmic ray

Fig. 8. A fluorescence telescope building, housing six telescopes, for the Pierre Auger Observatory in Argentina.

Fig. 9. Artist's conception of the Pierre Auger Observatory.

spectrum at the highest energies, results of anisotropy and point source searches, and new limits on the photon content of primary cosmic rays.

2.5. *VERITAS*

VERITAS[13] is a next-generation ground-based gamma ray observatory (see Fig. 10) which will provide the ground-based capability to study extremely energetic gamma rays, ranging in energy from 50 GeV to 50 TeV, potentially produced from a variety of astrophysical sources. The project is supported by a partnership between DOE, NSF and the Smithsonian Institution, with contributions from foreign partners. The primary scientific objectives are the detection and study of sources that could produce these gamma rays, such as black holes, neutron stars, active galactic nuclei, supernova remnants, pulsars, the galactic plane and gamma ray bursts. VERITAS will also search for dark matter candidates. The gamma rays are observed from the light they induce as they interact with the Earth's atmosphere. The imaging atmospheric Cherenkov technique, developed at the Whipple Observatory, is used to discriminate cosmic gamma rays from the cosmic ray background and to determine their energy and source direction. The study of the gamma ray sky by VERITAS and GLAST will provide observations in complementary energy ranges. VERITAS consists of four 12 m telescopes, each with 350 front aluminized glass mirror segments. The focal plane of each telescope is equipped with a detector, or camera, consisting of 499 2.5 cm photomultiplier tubes. While awaiting installation at its permanent site at the Kitt Peak National Observatory, the array has been fabricated, and will undergo engineering operations at the Whipple Observatory in 2007–2008.

Fig. 10. Artist's conception of the VERITAS observatory.

Fig. 11. Artist's conception of the GLAST telescope in orbit.

2.6. *GLAST*

The NASA GLAST[14] mission (see Fig. 11) is scheduled for launch in September 2007. The primary instrument on GLAST is the Large Area Telescope[15] (LAT) and is a partnership between DOE and NASA, with contributions from France, Italy, Japan and Sweden. The collaboration draws on the strengths of the particle-physics and high-energy-astrophysics communities. The LAT (see Fig. 12) fabrication project was managed at SLAC, which will also host the Instrument Science Operations Center (ISOC) during the data-taking phase. The LAT has finished fabrication at SLAC and is currently undergoing environmental testing at the Naval Research Laboratory.

The LAT will measure high energy gamma rays incident from space. The energy and direction will be measured in an energy range from 20 MeV to 300 GeV, over a wider field of view and with high sensitivity than any previously flown gamma ray mission. The scientific objectives of the LAT include the study of the mechanisms of particle acceleration in astrophysical environments, active galactic nuclei, pulsars and supernova remnants. They also include the resolution of unidentified galactic sources and diffuse emissions from cosmological sources, as well as determination of the high energy behavior of gamma ray bursts and transient sources. Among other topics of cosmological interest will be the information obtained on extragalactic background light generated during the epochs of star and galaxy formation in the early Universe and on dark matter.

The main components of the LAT instrument include a silicon strip track detector, a calorimeter, an anticoincidence detector and a data acquisition system. Its

Fig. 12. The LAT 4-by-4 array in the grid, before the anticoincidence detector is installed.

Fig. 13. The ISS with an artist's conception of the AMS detector in place in the middle of the picture, to the left of the main structure.

modular design consists of a 4-by-4 array of identical towers of tracker and calorimeter modules supported by a grid structure and surrounded by the anticoincidence detector. These detector modules are all integrated mechanically and thermally to the grid and electrically to the data acquisition system.

2.7. *AMS*

The AMS[16] is a 16-nation international cosmic ray particle physics and astrophysics experiment which is planned for flight on the International Space Station (ISS) (see Fig. 13). A prototype took data on flight STS-91 of the space shuttle Discovery in 1998 and verified the basic concept of the experiment. The scientific purpose is the search for dark matter, missing matter and antimatter from space.

The instrument comprises a superconducting magnet with a 0.86-tesla field and a suite of detectors, including a transition radiation detector, an eight-layer silicon tracker, a time-of-flight system, a ring-imaging Cherenkov counter and a three-dimensional electromagnetic sampling calorimeter. The fabrication of the instrument was completed in 2005 and it is currently undergoing integration and testing, which will be completed in 2007. The launch date aboard the space shuttle for deployment on the ISS is currently unknown.

3. Conclusion

In years to come, the OHEP program is looking forward to exciting new discoveries at the TeV scale, with neutrinos and in discoveries of the major constituents of the Universe: dark matter and dark energy.

References

1. "Connecting Quarks with the Cosmos: Eleven Science Questions for the New Century," a report by the National Research Council's Board on Physics and Astronomy, (2003); http://www.nap.edu/catalog/10079.html
2. "A 21st Century Frontier of Discovery: The Physics of the Universe; A Strategic Plan for Federal Research at the Intersection of Physics and Astronomy," Executive Office of the President, National Science and Technology Council, Committee on Science, (2004); http://www7.nationalacademies.org/bpa/OSTP_Q2C_Response_Draft.pdf
3. "Revealing the Hidden Nature of Space and Time: Charting the Course for Elementary Particle Physics", a report by the Committee on Elementary Particle Physics in the 21st Century (EPP2010), National Research Council (2006); prepublication available at http://www7.nationalacademies.org/bpa/EPP2010_Report_Prepub.pdf
4. "Report of the Dark Energy Task Force," subpanel (2006); http://www.nsf.gov/mps/ast/detf.jsp or http://www.science.doe.gov/hep/DETF-FinalRptJune30,2006.pdf
5. "The Particle Physics Project Prioritization Panel (P5) Subpanel Report," (2006); http://www.science.doe.gov/hep/P5InterimRptChg2June2006.pdf
6. See http://snap.lbl.gov
7. See http://www.science.doe.gov/hep/JDEM%20Reports.shtm
8. See http://www.darkenergysurvey.org
9. See http://www.lsst.org
10. See http://www.sdss.org
11. See http://cdms.berkeley.edu
12. See http://www.auger.org
13. See http://veritas.sao.arizona.edu
14. See http://glast.gsfc.nasa.gov
15. See http://www-glast.slac.stanford.edu/ or http://www-glast.stanford.edu
16. See http://ams.cern.ch/AMS

PART 2
GRAVITATIONAL THEORY

DARK ENERGY, DARK MATTER AND GRAVITY

ORFEU BERTOLAMI

Instituto Superior Técnico, Departamento de Física,
Av. Rovisco Pais 1, Lisbon, 1049-001, Portugal
orfeu@cosmos.ist.utl.pt

We discuss the motivation for high accuracy relativistic gravitational experiments in the solar system and complementary cosmological tests. We focus our attention on the issue of distinguishing a generic scalar theory of gravity as the underlying physical theory from the usual general-relativistic picture, where one expects the presence of fundamental scalar fields associated, for instance, with inflation, dark matter and dark energy.

Keywords: Dark matter; dark energy; scalar fields; gravity.

1. Introduction

Present day experimental evidence indicates that gravitational physics is in agreement with Einstein's theory of general relativity to considerable accuracy; however, there are a number of reasons, theoretical and experimental, to question the theory as the ultimate description of gravity.

On the theoretical side, difficulties arise from various corners, most stemming from the strong gravitational field regime, associated with the existence of space–time singularities and the difficulty of describing the physics of very strong gravitational fields. Quantization of gravity is a possible way to overcome these obstacles; however, despite the success of modern gauge field theories in describing the electromagnetic, weak and strong interactions, the path to describing gravity at the quantum level is still to be found. Indeed, our two foundational theories, quantum mechanics and general relativity, are not compatible with each other. Furthermore, in fundamental theories that attempt to include gravity, new long-range forces often arise in addition to the Newtonian inverse square law. Even at the classical level, and assuming the validity of the equivalence principle, Einstein's theory does not provide the most general way to establish the space–time metric. There are also important reasons to consider additional fields, especially scalar fields. Although the latter appear in unification theories, their inclusion predicts a non-Einsteinian behavior of gravitating systems. These deviations from general relativity include violations of the equivalence principle, modification of large-scale gravitational phenomena, and variation of the fundamental "constants." These predictions motivate new searches for very small deviations of relativistic gravity from general relativity and drive the need for further gravitational experiments in space. These include laser astrometric measurements,[1–4] high-resolution lunar laser ranging (LLR)[5] and long-range tracking of spacecraft using the formation flight concept, as proposed[6] to test the *Pioneer* anomaly.[7] A broader discussion on the motivations of performing fundamental physics experiments in space can be found elsewhere.[8]

On the experimental front, recent cosmological observations do lead one to conclude that our understanding of the origin and evolution of the universe based on general relativity requires that most of the energy content of the universe reside in the presently unknown dark matter and dark energy components that may permeate much, if not all of space–time. Indeed, recent cosmic microwave background radiation (CMBR) WMAP three-year data[9] indicate that our universe is well described, within the framework of general relativity, by a flat Robertson–Walker metric, meaning that the energy density of the universe is fairly close to the critical one, $\rho_c \equiv 3H_0^2/8\pi G \simeq 10^{-29}\,\mathrm{g/cm^3}$, where $H_0 \simeq 73\,\mathrm{km\ s^{-1}Mpc^{-1}}$ is the Hubble expansion parameter at present. Moreover, CMBR, supernova and large scale structure data are consistent with each other if, in the cosmic budget of energy, dark energy corresponds to about 73% of the critical density, and dark matter to about 23% and baryonic matter, the matter that we are made of, to only about 4%. Furthermore, it is generally believed that the ultimate theory that will reconcile quantum mechanics and general relativity will also allow for addressing the cosmological questions related with the origin and destiny of the universe.

It is our opinion that the crystallization of these fundamental questions is well timed, with recent progress in high-precision measurement technologies for physics experiments in space. This puts physicists in a position to realistically address crucial questions, such as the nature of dark energy and dark matter, the existence of intermediate range forces and the ultimate nature of gravity. Furthermore, given the ever-increasing practical significance of general relativity, for spacecraft navigation, time transfer, clock synchronization, weight and length standards, it is just natural to expect that the theory will be regularly tested with ever-increasing accuracy. Thus, it seems legitimate to speculate that the present state of physics represents a unique confluence of important challenges in high energy physics and cosmology together with technological advances and access to space, a conjunction that is likely to yield major discoveries.

In what follows we shall address the key issue of distinguishing a generic scalar theory of gravity, as the underlying fundamental physical theory, from the usual general-relativistic picture, where one expects the presence of fundamental scalar fields associated with inflation, dark matter and dark energy. In order to concretely discuss the matter we will consider a fairly general scalar-tensor theory of gravity as an example, and indicate how its main features can be extracted from high-resolution measurements of the parametrized post-Newtonian (PPN) parameters

β and γ. As is well known, scalar-tensor theories of gravity mimic a plethora of unification models. For instance, the graviton–dilaton system in string/M-theory can be viewed as a specific scalar–tensor theory of gravity.

Of course, one should to bear in mind that current experimental data show an impressive agreement with general relativity.[10,11] Indeed, most stringent bounds arise from the *Cassini*'s 2003 radiometric experiment[12]:

$$\gamma - 1 = (2.1 \pm 2.5) \times 10^{-5}, \tag{1}$$

$$\beta - 1 = (1.2 \pm 1.1) \times 10^{-4}, \tag{2}$$

and from limits on the strong equivalence violation parameter, $\eta \equiv 4\beta - \gamma - 3$, which are found to be $\eta = (4.4 \pm 4.5) \times 10^{-4}$, as inferred from LLR measurements.[13]

As already mentioned, in cosmology, general relativity allows for detailed predictions of the nucleosynthesis yields and of the properties of the CMBR, provided that one admits the presence of fundamental scalar fields, the inflaton, the quintessence scalar field[14] or the generalized Chaplygin gas model underlying scalar field, complex[15] or real,[16] to account for the late accelerated expansion of the universe, and in the case of some candidates for dark matter, self-interacting[17,18] or not.[19] It is worth noting that the generalized Chaplygin gas model corresponds to a unified model of dark energy and dark matter, based on the equation of state $p = -A/\rho^{\alpha}$, where p is the isotropic pressure, ρ is the energy density, and A and α are positive phenomenological constants. Its agreement with observational data has been extensively studied: CMBR,[15] supernovae,[16,23] gravitational lensing,[24] gamma ray bursts[25] and cosmic topology.[26] A fully consistent picture for structure formation in the context of the generalized Chaplygin gas model remains still an open question.[27] Another interesting cosmological issue concerns the resemblance between inflation and the late accelerated expansion of the universe, which have led to proposals where the inflaton and the quintessence scalar field are related.[28–31]

A scalar field with a suitable potential can also be the way to explain the Pioneer anomaly.[32] It is interesting to point out that scalar fields can affect stellar dynamics and hence, specific measurements of, for instance, the central temperature of stars and their luminosity can allow for setting bounds on scalar field models.[33]

2. Scalar-Tensor Theories of Gravity

In many alternative theories of gravity, the gravitational coupling strength exhibits a dependence on a field of some sort; in scalar-tensor theories, this is a function of a scalar field φ. The most general action for a scalar-tensor theory of gravity up to first order in the curvature can be written as

$$S = \frac{c^3}{4\pi G} \int d^4x \sqrt{-g} \left[\frac{1}{4} f(\varphi) R - \frac{1}{2} g(\varphi) \partial_\mu \varphi \partial^\mu \varphi + V(\varphi) + \sum_i q_i(\varphi) \mathcal{L}_i \right], \tag{3}$$

where $f(\varphi)$, $g(\varphi)$, $V(\varphi)$ are generic functions, $q_i(\varphi)$ are coupling functions and $\mathcal{L}_i$ is the Lagrangian density of the matter fields.

For simplicity, we shall consider only the theories for which $g(\varphi) = q_i(\varphi) = 1$. Hence, for a theory for which $V(\varphi)$ can be locally neglected, given that its mass is fairly small so that it acts cosmologically, the resulting effective model can be written as

$$S = \frac{c^3}{4\pi G}\int d^4x\sqrt{-\hat{g}}\left[\frac{1}{4}\hat{R} - \frac{1}{2}\partial_\mu\varphi\partial^\mu\varphi + \sum_i \mathcal{L}_i(\hat{g}_{\mu\nu} = A^2(\varphi)g_{\mu\nu})\right], \qquad (4)$$

where $A^2(\varphi)$ is the coupling function to matter and the factor that allows one to write the theory in the so-called Einstein frame.

It is shown that in the PPN limit, if one writes

$$\ln A(\varphi) \equiv \alpha_0(\varphi - \varphi_0) + \frac{1}{2}\beta_0(\varphi - \varphi_0)^2 + O(\varphi - \varphi_0)^3, \qquad (5)$$

then[34–36]

$$\gamma - 1 = -\frac{2\alpha_0^2}{1 + \alpha_0^2}, \qquad (6)$$

$$\beta - 1 = \frac{1}{2}\frac{\alpha_0^2\beta_0}{(1 + \alpha_0^2)^2}. \qquad (7)$$

Most recent bounds arising from binary pulsar PSR $B1913 + 16$ data indicate that[37]

$$\beta_0 > -4.5, \quad \alpha_0 < 0.060, \qquad (8)$$

$$\frac{\beta - 1}{\gamma - 1} < 1.1. \qquad (9)$$

These results are consistent with solar system constraints and one expects that improvement of data may allow one within a decade to achieve $|\gamma - 1| \sim 10^{-6}$, an order of magnitude better than *Cassini*'s constraint.[12] Notice that the PPN formalism for more general cases is available.[34–36] For sure, gravitational experiments in space will allow one to further constrain these models.

It is relevant to point out that scalar-tensor models have also been proposed to explain the accelerated expansion of the universe, even though not quite successfully.[38]

3. Gravitational Experiments in Space

Let us now give some examples of gravitational experiments that critically rely on space technology and that may crucially contribute to clarifying some of the discussed issues.

3.1. *Lunar laser-ranging: APOLLO facility*

The Apache Point Observatory Lunar Laser-ranging Operation (APOLLO) is a new LLR effort designed to achieve millimeter range precision and order-of-magnitude gains in the measurement of physical parameters.[5]

The major advantage of APOLLO over current LLR operations is a 3.5 m astronomical high quality telescope at a good site, the Sacramento Mountains of southern New Mexico (2780 m), with very good atmospheric quality. The APOLLO project will allow pushing LLR into the regime of millimeter range precision. For the Earth and Moon orbiting the Sun, the scale of relativistic effects is set by the ratio $(GM/rc^2) \sim v^2/c^2 \sim 10^{-8}$. Relativistic effects are small compared to Newtonian effects. The Apache Point 1-mm-range accuracy corresponds to 3×10^{-12} of the Earth–Moon distance. The impact on gravitational physics is expected to yield an improvement of an order of magnitude: the equivalence principle would result in uncertainties approaching 10^{-14}, tests of general relativity effects would be smaller than 0.1%, and estimates of the relative change in the gravitational constant would be about 0.1% of the inverse age of the universe.

Therefore, the gain in the ability to conduct even more precise tests of fundamental physics is enormous, and thus this new instrument stimulates development of better and more accurate models for the LLR data analysis at a mm level.[39]

3.2. *The LATOR mission*

The proposed Laser Astrometric Test of Relativity (LATOR)[1–4] experiment is designed to test the metric nature of gravitation, a fundamental postulate of general relativity. By using a combination of independent time series of highly accurate gravitational deflection of light in the immediate vicinity of the Sun, along with measurements of the Shapiro time delay on interplanetary scales (to a precision respectively better than 10^{-13} radians and 1 cm), LATOR will considerably improve the knowledge about relativistic gravity. Its main objectives can be summarized as follows: (i) measure the key post-Newtonian Eddington parameter γ with an accuracy of a part in 10^9, a factor 30,000 beyond the present best result, *Cassini*'s radiometric experiment[12]; (ii) perform the first measurement of gravity's nonlinear effects on light to about 0.01% accuracy, including both the traditional Eddington β parameter via gravity effect on light to about 0.01% accuracy and also the never-measured spatial metric's second order potential contribution, δ; (iii) perform a direct measurement of the solar quadrupole moment, J_2, to an accuracy of a part in 200 of its expected size; (iv) measure the "frame-dragging" effect on light due to the sun's rotational gravitomagnetic field, to 0.1% accuracy. LATOR's measurements will be able to push to unprecedented accuracy the search for relevant scalar-tensor theories of gravity by looking for a remnant scalar field. The key element of LATOR is the geometric redundancy provided by the laser-ranging and long-baseline optical interferometry.

The LATOR mission can be regarded as a 21st century version of Michelson–Morley-type experiments particularly suited to the search for effects for a scalar field in the solar system. In spite of the previous space missions exploiting radio waves for spacecraft tracking, this mission will correspond to a breakthrough in the relativistic gravity experiments, as it allows one to take full advantage of the optical techniques that have recently become available. LATOR has a number of advantages over techniques that use radio waves to measure gravitational light deflection. Indeed, optical technologies allow low bandwidth telecommunications with the LATOR spacecraft and the use of the monochromatic light enables observation of the spacecraft at the limb of the Sun. The use of narrow band filters, coronagraph optics and heterodyne detection allows suppression of background light to a level where the solar background is no longer the dominant source of noise. The short wavelength allows much more efficient links with smaller apertures, thereby eliminating the need for a deployable antenna. Finally, the use of the International Space Station enables the experiment to be above the Earth's atmosphere, the major source of astrometric noise for any ground-based interferometer. These features fully justify LATOR as a particularly fundamental mission in the search for gravitational phenomena beyond general relativity.

3.3. *A mission to test the* Pioneer *anomaly*

Pioneer 10 and *11* were launched in 1972 and 1973 to study the outer planets of the solar system. Both probes followed hyperbolic trajectories close to the ecliptic to opposite outward directions in the solar system. Due to their robust design, it was possible to determine their position with great accuracy. During the first years of the life of *Pioneer 10*, the acceleration caused by solar radiation pressure on the spacecraft was the main effect.[7] At about 20 AU (by the early 1980s) solar radiation pressure became subdominant and it was possible to identify an unaccounted for anomaly. This anomaly can be interpreted as a constant acceleration with a magnitude of $a = (8.74 \pm 1.33) \times 10^{-10}\,\mathrm{ms}^{-2}$ and is directed toward the Sun. This effect became known as the *Pioneer* anomaly. For the *Pioneer* spacecraft, it has been observed at least until 70 AU.[7] The same effect was observed in the *Pioneer 11* spacecraft.[7]

This puzzling deceleration has divided the space community in the last few years. Although, skeptics have been arguing that the most likely solution to the riddle is some unforeseen on-board generated effect such as fuel leaking from the thrusters or nonsymmetrical heat dissipation from the nuclear-powered energy sources,[40,41] the most optimistic point to the fact that this effect may signal a new force or fundamental field of nature and hence an important window for new physics.[a] The approach that has been advocated by some groups that answered the recent European Space Agency (ESA) call Cosmic Vision 2015–2025 with proposals of missions to test *Pioneer*'s anomalous acceleration is that whatever the cause of the slow down of the spacecraft, meeting the requirements of such a mission would give rise to developments that would be invaluable for building and designing noise-free spacecraft for future deep space missions. Actually, the theoretical concept of a

[a]The demonstration that the gravitational field due to the Kuiper belt is not the cause of the anomaly has recently been reanalyzed.[42] The literature is particularly rich in proposals.[6,32]

mission to verify the anomalous acceleration had been suggested earlier in a study[43] commissioned by the ESA in 2002.

A dedicated mission would rely on a simple concept, which consists in launching into deep space a geometrically symmetric[43] and spin-stabilized[44] probe whose behavior (mechanical, thermal, electromagnetic, etc.) is carefully monitored. Accurate tracking of its orbit would allow precise evaluation of the anomaly, as any deviation from the predicted trajectory would be used to examine the unmodeled anomalous acceleration.[b] The exciting possibility of using laser-ranging techniques and the flying formation concept to characterize the nature of the anomaly, and solar sailing propulsion, has been more recently discussed.[6] Particularly pleasing is the announcement that the ESA is seriously considering such an ambitious and challenging undertaking in the period 2015–2025.[46] Naturally, a mission of this nature can be particularly useful for testing the existence of any solar-system-range new interaction, as well as for exploring, for instance, the structure of the Kuiper belt.[c]

4. Discussion and Conclusions

Let us now review the main points of our discussion. It seems evident that resolving the dichotomy of dark energy–dark matter versus gravity will require a concerted effort and a whole new program of dedicated experiments in space.

It is an exciting prospect that dark matter can be directly detected in underground experiments or in the forthcoming generation of colliders. Even though it does not seem feasible to directly test the properties of dark energy, it is not impossible that indirect evidence can be found in the laboratory. The boldest proposal suggests the existence of a cutoff frequency of the noise spectrum in Josephson junctions,[49] while a more conventional approach is to investigate the effect that dark energy may have, for instance, on the variation of the electromagnetic coupling.[50–53] It follows that the characterization of dark energy and dark matter will most likely be achieved via cosmological observations, most of them to be carried out by spaceborne experiments. These encompass a large array of phenomena, such as supernovae, gamma ray bursts, gravitational lensing, cosmic shear, etc. The result of these observations will also provide increasingly detailed information on the adequacy of general relativity at cosmological scales. It is quite exciting that existing supernova data,[54] together with the latest CMBR data[9] and the recently discovered baryon acoustic oscillations,[55] are sufficiently constraining to virtually rule out,[56,57] for instance, most of the braneworld-inspired gravity models put forward to account for the accelerated expansion of the Universe. The prospect of testing some of these models through the study of the orbital motion of planets in the solar system has also been recently discussed.[58]

[b] A fairly thorough study of the main systematic effects of such a mission can be found in Ref. 45.
[c] Most recent proposals for ESA's Cosmic Vision 2015–2025 call for missions include Odissey[47] and SAGAS[48].

We have seen how the situation stands concerning scalar-tensor gravity models. Relevant results are expected within a decade from the observation of binary pulsar systems. To further test general relativity and examine the implications of its contending theories or extensions (scalar-tensor theories, braneworld models, string inspired models, etc.), a new program of gravity experiments in space is clearly needed. We have discussed how LLR can be used to improve the knowledge of relativistic gravity and pointed out how the LATOR mission and a mission to test the *Pioneer* anomaly can play a key role in the search for evidence of a remnant scalar field in the solar system, to identify new forces with ranges of a few decades of AU and, of course, to solve the *Pioneer* anomaly puzzle. It is relevant to point out that the latter type of mission, besides its technological appeal, can also be used to gather information about the vicinity of the solar system as well as to set relevant upper bounds on environmental parameters such as the density of interplanetary dust and dark matter.[42]

Acknowledgments

It is a pleasure to thank the members of the *Pioneer* Science Team for the countless discussions on the questions related to this contribution. I am particularly indebted to Jorge Páramos, Slava Turyshev, Serge Reynaud, Clovis de Matos, Pierre Toubul, Ulrich Johann and Claus Lämmerzahl for their insights and suggestions.

References

1. S. G. Turyshev, M. Shao and K. L. Nordtvedt, Jr., *Int. J. Mod. Phys. D* **13** (2004) 2035.
2. S. G. Turyshev, M. Shao and K. L. Nordtvedt, Jr., *Class. Quant. Grav.* **21** (2004) 2773.
3. S. G. Turyshev, M. Shao and K. L. Nordtvedt, Jr., in *Proc. Symposium on Relativistic Astrophysics* (Stanford Univerity, Dec. 2004), eds. P. Chen, E. Bloom, G. Madejski and V. Petrosian. SLAC-R-752, Stanford e-Conf #C041213, paper #0306. Eprint: http://www.slac.stanford.edu/econf/C041213/; gr-qc/0502113.
4. LATOR Collab. (S. G. Turyshev *et al.*), *ESA Spec. Publ.* **588** (2005) 11 [gr-qc/0506104].
5. T. W. Murphy Jr *et al.*, "The Apache Point Observatory Lunar Laser-Ranging Operation (APOLLO)," *Proc. 12th International Workshop on Laser Ranging* (Matera, Italy, Nov. 2000), in press, 2002, http://www.astro.washington.edu/tmurphy/apollo/matera.pdf
6. *Pioneer* Collab. (H. Dittus *et al.*), *ESA Spec. Publ.* **588** (2005) 3 gr-qc/0506139.
7. J. D. Anderson *et al.*, *Phys. Rev. D* **65** (2002) 082004.
8. O. Bertolami *et al.*, *Acta Astronaut.* **59** (2006) 490.
9. D. N. Spergel *et al.*, astro-ph/0603449.
10. C. M. Will, in *Proc., 100 Years of Relativity: Spacetime Structure — Einstein and Beyond*, ed. Abhay Ashtekar (World Scientific, Singapore), "Was Einstein Right? Testing Relativity at the Centenary," gr-qc/0504086.

11. O. Bertolami, J. Páramos and S. G. Turyshev, in *Proc. 359th WE-Heraeus Seminar: Lasers, Clock, and Drag-Free: Technologies for Future Exploration in Space and Gravity Tests*, (University of Bremen, ZARM, Bremen, Germany, 30 May–1 June 2005), General theory of relativity: Will it survive the next decade? gr-qc/0602016.
12. B. Bertotti, L. Iess and P. Tortora, *Nature* **425** (2003) 374.
13. J. G. Williams, S. G. Turyshev and D. H. Boggs, *Phys. Rev. Lett.* **93** (2004) 261101.
14. E. J. Copeland, M. Sami and S. Tsujikawa, hep-th/0603057.
15. M. C. Bento, O. Bertolami and A. A. Sen, *Phys. Rev. D* **66** (2002) 043507.
16. O. Bertolami *et al.*, *Mon. Not. R. Astron. Soc.* **353** (2004) 329.
17. M. C. Bento *et al.*, *Phys. Rev. D* **62** (2000) 041302.
18. M. C. Bento, O. Bertolami and R. Rosenfeld, *Phys. Lett. B* **518** (2001) 276.
19. O. Bertolami and F. Nunes, *Phys. Lett. B* **452** (1999) 108.
20. M. C. Bento, O. Bertolami and A. A. Sen, *Phys. Rev. D* **67** (2003) 063511.
21. M. C. Bento, O. Bertolami and A. A. Sen, *Phys. Lett. B* **575** (2003) 172.
22. M. C. Bento, O. Bertolami and A. A. Sen, *Gen. Relat. Gravit.* **35** (2003) 2063.
23. M. C. Bento *et al.*, *Phys. Rev. D* **71** (2005) 063501.
24. P. T. Silva and O. Bertolami, *Astrophys. J.* **599** (2003) 829.
25. O. Bertolami and P. T. Silva, *Mon. Not. R. Astron. Soc.* **365** (2006) 1149.
26. M. C. Bento *et al.*, *Phys. Rev. D* **73** (2006) 043504.
27. M. C. Bento, O. Bertolami and A. A. Sen, *Phys. Rev. D* **70** (2004) 083519.
28. P. J. E. Peebles and A. Vilenkin, *Phys. Rev. D* **59** (1999) 063505.
29. K. Dimopoulos and J. W. F. Valle, *Astropart. Phys.* **18** (2002) 287.
30. R. Rosenfeld and J. A. Friemann, *J. Cosmol. Astropart. Phys.* **0509** (2005) 3.
31. O. Bertolami and V. Duvvuri, *Phys. Lett. B* **640** (2006) 121.
32. O. Bertolami and J. Páramos, *Class. Quant. Grav.* **21** (2004) 3309.
33. O. Bertolami and J. Páramos, *Phys. Rev. D* **71** (2005) 023521.
34. T. Damour and G. Esposito-Farèse, *Phys. Rev. Lett.* **70** (1993) 2220.
35. T. Damour and G. Esposito-Farèse, *Phys. Rev. D* **54** (1996) 1474.
36. T. Damour and G. Esposito-Farèse, *Phys. Rev. D.* **58** (1998) 042001.
37. G. Esposito-Farèse, *AIP Conf. Proc.* **736** (2004) 32 [gr-qc/0409081].
38. O. Bertolami and P. J. Martins, *Phys. Rev. D* **61** (2000) 064007.
39. J. G. Williams, S. G. Turyshev and T. W. Murphy Jr., *Int. J. Mod. Phys. D* **13** (2004) 567.
40. J. I. Katz, *Phys. Rev. Lett.* **83** (1999) 1892.
41. L. K. Scheffer, *Phys. Rev. D* **67** (2003) 084201.
42. O. Bertolami and P. Vieira, *Class. Quant. Grav.* **23** (2006) 4625.
43. O. Bertolami and M. Tajmar, *Rep. ESA CR(P)* **4365** (2002)dec.
44. J. D. Anderson, M. M. Nieto and S. G. Turyshev, *Int. J. Mod. Phys. D* **11** (2002) 1545.
45. O. Bertolami and J. Páramos, *Int. J. Mod. Phys. D* **16** (2007) 1611.
46. European Space Agency Cosmic Vision: Space Science for Europe 2015–2025, BR-247, Oct. 2005.
47. B. Christo *et al.*, gr-qc/0711.2007.
48. P. Wolf *et al.*, gr-qc/0111.0304.
49. C. Beck and M. C. Mackey, *Phys. Lett. B* **605** (2005) 295.
50. K. A. Olive and M. Pospelov, *Phys. Rev. D* **65** (2002) 085044.
51. E. J. Copeland, N. J. Nunes and M. Pospelov, *Phys. Rev. D* **69** (2004) 023501.
52. M. C. Bento, O. Bertolami and N. M. C. Santos, *Phys. Rev. D* **70** (2004) 107304.
53. O. Bertolami *et al.*, *Phys. Rev. D* **69** (2004) 083513.

54. Supernova Search Team Collab. (A. G. Riess *et al.*), *Astrophys. J.* **607** (2004) 665.
55. D. Eisenstein *et al.*, *Astrophys. J.* **633** (2005) 560.
56. R. Maartens and E. Majerotto, *Phys. Rev. D* **74** (2006) 023004, astro-ph/0603353.
57. M. C. Bento *et al.*, *Phys. Rev. D* **73** (2006) 103521.
58. L. Iorio, *J. Cosmol. Astropart. Phys.* **0601** (2006) 8.

OBSERVABLE CONSEQUENCES OF STRONG COUPLING IN THEORIES WITH LARGE DISTANCE MODIFIED GRAVITY

GIA DVALI

Center for Cosmology and Particle Physics,
Department of Physics, New York University,
New York, NY 10003, USA
gd23@nyu.edu

In this talk, we review theories that modify gravity at cosmological distances, and show that any such theory must exhibit a *strong coupling phenomenon*. We show that all consistent theories that modify the dynamics of the spin-2 graviton on asymptotically flat backgrounds, automatically have this property. Due to the strong coupling effect, modification of the gravitational force is source-dependent, and for lighter sources sets in at shorter distances. This universal feature makes modified gravity theories predictive and potentially testable by precision gravitational measurements at scales much shorter than the current cosmological horizon.

Keywords: Modified gravity; strong coupling; dark energy.

1. Introduction

The observed accelerated expansion of the Universe[1,2] could result[3] from a breakdown of the standard laws of gravity at very large distances, caused in particular, by extra dimensions becoming gravitationally accessible at distances of the order of the current cosmic horizon.[4]

In this talk we shall formulate some very general properties of large distance modified gravity theories, which follow from the consistency requirements based on 4D effective field theory considerations. These arguments allow one to give a simple parametrization and display some very general *necessary* properties of such theories, even without knowing their complete explicit form.

In this talk we shall be interested in theories that, (1) are ghost-free, (2) admit the weak field linearized expansion on asymptotically flat backgrounds, and (3) modify Newtonian dynamics at a certain crossover scale, r_c. Motivated by cosmological considerations, one is tempted (and usually has to) to take r_c of the order of the current cosmological horizon $H_0^{-1} \sim 10^{28}\,\text{cm}$, but in most of our discussions we shall keep r_c as a free parameter.

There exists a unique class of ghost-free linearized theories of a spin-2 field with the above properties. These theories have a tensorial structure of the Pauli–Fierz massive graviton, and are described by the equation

$$\mathcal{E}_{\mu\nu}^{\alpha\beta} h_{\alpha\beta} - m^2(\Box)(h_{\mu\nu} - \eta_{\mu\nu} h) = T_{\mu\nu}. \tag{1}$$

Here,

$$\mathcal{E}_{\mu\nu}^{\alpha\beta} h_{\alpha\beta} = \Box h_{\mu\nu} - \Box \eta_{\mu\nu} h - \partial^\alpha \partial_\mu h_{\alpha\nu} - \partial^\alpha \partial_\nu h_{\alpha\mu} + \eta_{\mu\nu} \partial^\alpha \partial^\beta h_{\alpha\beta} + \partial_\mu \partial_\nu h \tag{2}$$

is the linearized Einstein tensor, and as usual $h \equiv \eta_{\alpha\beta} h^{\alpha\beta}$. $T_{\mu\nu}$ is a conserved source. The gravitational coupling constant is set equal to 1, for simplicity. The difference of the above theory from the Pauli–Fierz massive graviton is that, in our case, $m^2(\Box)$ is not a constant, but a more general operator which depends on $\Box$. In order to modify Newtonian dynamics at scale r_c, $m^2(\Box)$ must dominate over the first Einsteinian term at the scales $r \gg r_c$, or equivalently for the momenta $k \ll r_c^{-1}$. This implies that to the leading order

$$m^2(\Box) \simeq r_c^{2(\alpha-1)} \Box^\alpha, \tag{3}$$

with $\alpha < 1$.

The class of theories of interest can thus be parametrized by a single continuous parameter α. As we shall see in a moment, unitarity further restricts α from below by demanding $\alpha > 0$. The limiting case $\alpha = 0$ corresponds to the massive Pauli–Fierz graviton.

Currently, the only theory that has a known ghost-free generally covariant nonlinear completion is $\alpha = 1/2$, which is a five-dimensional brane world model (DGP model).[4] (See the talk by Deffayet[5] for a detailed analysis of this model.)

The limiting case of massive graviton $\alpha = 0$, unfortunately, has no known consistent nonlinear completion, and it is very likely that such completions are impossible,[7,8] at least in theories with a finite number of states.

The nonlinear completions for theories with other values of α have not been studied. It is not the purpose of the present work to discover such completions. Instead, we shall display some necessary conditions that such completions (if they exist) must satisfy. As we shall see, the effective field theory considerations combined with consistency assumptions give us a powerful tool for predicting some of the general properties of the theories, even without knowing all the details of the underlying dynamics.

The central role in our consideration is played by the *strong coupling* phenomenon, first discovered in Ref. 9, both for massive gravity ($\alpha = 0$) and for the DGP model ($\alpha = 1/2$). We shall show that this phenomenon must be exhibited by theories with arbitrary values of α from the unitarity range $0 < \alpha < 1$, and that this fact predicts the following modification of the gravitational potential for an arbitrary localized source:

$$\delta \sim \left(\frac{r}{r_c}\right)^{2(1-\alpha)} \sqrt{\frac{r}{r_g}}, \tag{4}$$

where $r_g \equiv 2G_N M$ is the usual gravitational (Schwarszchild) radius of the source, and G_N is Newton's constant. Because of this predictive power, the theories of modified gravity are extremely restrictive and potentially testable by the whole spectrum of precision gravitational measurements, presented at this meeting. These include improved accuracy laser ranging[10,11] and optical interferometry[12,13] experiments, as well as measurements based on cold atomic interferometers,[14] and other measurements of the inverse-square law.[15,16] From this point of view, precision gravitational measurements, both space- and Earth-based, are extremely important for our studies, since, contrary to naive cosmological intuition, a window to new gravitational physics could also be opened by measurements at Earth- and satellite-accessible scales.

Our analysis complements and solidifies the earlier study of Refs. 17 and 18, in which the corrections (4) where already suggested.

2. Unitarity Constraints

We shall first derive the unitarity lower bound on α. For this, consider a one-graviton exchange amplitude between the two conserved sources ($T_{\mu\nu}$ and $T'_{\mu\nu}$):

$$(\text{Amplitude})_h \propto \frac{T'_{\mu\nu}T^{\mu\nu} - \frac{1}{3}T^{\mu}_{\mu}T'^{\nu}_{\nu}}{\Box - m^2(\Box)}. \tag{5}$$

The scalar part of the (Euclidean) graviton propagator then has the form

$$\Delta(k^2) = \frac{1}{k^2 + m^2(k^2)}. \tag{6}$$

As said above, modification of gravity at large distances implies that the denominator is dominated by the second term, for $k \to 0$. To see how unitarity constrains such a behavior, let the spectral representation of the propagator $\Delta(k^2)$ be

$$\Delta(k^2) = \int_0^\infty ds \frac{\rho(s)}{k^2 + s}, \tag{7}$$

where $\rho(s)$ is a bounded spectral function. The absence of the negative norm states demands that $\rho(s)$ be a semipositive definite function. Evaluating (7) for the small k and using parametrization (3), we get

$$\frac{1}{r_c^{2(\alpha-1)}k^{2\alpha}} = \int_0^\infty ds \frac{\rho(s)}{k^2 + s}. \tag{8}$$

Nonnegativity of $\rho(s)$ implies that α cannot be negative. In the opposite case, $\Delta(k^2)$ would be zero for $k^2 = 0$, which is impossible for nonnegative $\rho(s)$. Thus, from unitarity, it follows that $\alpha = 0$ is the lowest possible bound.

3. Extra Polarizations

A graviton satisfying Eq. (1), just as in the Palui–Fierz case, propagates five degrees of freedom. These include two spin-2 helicities, two spin-1 helicities and one spin-0 helicity. The extra helicities (especially spin-0) play the central role in the strong coupling phenomenon, and it is therefore useful to separate the "new" states from the "old" helicity-2 states.

We shall now integrate the three extra helicities out and write down the effective equation for the two remaining tensorial ones. For this we shall first rewrite Eq. (1) in the manifestly gauge-invariant form, using the Stückelberg method. We can rewrite $h_{\mu\nu}$ in the form

$$h_{\mu\nu} = \hat{h}_{\mu\nu} + \partial_\mu A_\nu + \partial_\nu A_\mu, \tag{9}$$

where the Stückelberg field A_μ is the massive vector that carries two extra helicity-1 polarizations. The remaining extra helicity-0 state resides partially in A_μ and partially in $\hat{h}_{\mu\nu}$. Written in terms of $\hat{h}_{\mu\nu}$ and A_μ,

$$\mathcal{E}^{\alpha\beta}_{\mu\nu}\hat{h}_{\alpha\beta} - m^2(\Box)(\hat{h}_{\mu\nu} - \eta_{\mu\nu}\hat{h} + \partial_\mu A_\nu + \partial_\nu A_\mu - 2\eta_{\mu\nu}\partial^\alpha A_\alpha) = T_{\mu\nu}, \tag{10}$$

Eq. (1) becomes manifestly invariant under the gauge transformation

$$\hat{h}_{\mu\nu} \to \hat{h}_{\mu\nu} + \partial_\mu\xi_\nu + \partial_\nu\xi_\mu, \quad A_\mu \to A_\mu - \xi_\mu, \tag{11}$$

where ξ_μ is the gauge parameter. Note that the first Einstein's term is unchanged under the replacement (9), due to its gauge invariance. We now wish to integrate out A_μ through its equation of motion,

$$\partial^\mu F_{\mu\nu} = -\partial^\mu(\hat{h}_{\mu\nu} - \eta_{\mu\nu}\hat{h}), \tag{12}$$

where $F_{\mu\nu} \equiv \partial_\mu A_\nu - \partial_\nu A_\nu$. Before solving for A_μ, note that by taking a divergence from Eq. (12) we get the following constraint on $\hat{h}_{\mu\nu}$:

$$\partial^\mu\partial^\nu\hat{h}_{\mu\nu} - \Box\hat{h} = 0, \tag{13}$$

which means that $\hat{h}_{\mu\nu}$ is representable in the form

$$\hat{h}_{\mu\nu} = \tilde{h}_{\mu\nu} + \eta_{\mu\nu}\frac{1}{3}\Pi_{\alpha\beta}\tilde{h}^{\alpha\beta}, \tag{14}$$

where $\Pi_{\alpha\beta} = \frac{\partial_\alpha\partial_\beta}{\Box} - \eta_{\alpha\beta}$ is the transverse projector. $\tilde{h}_{\mu\nu}$ carries two degrees of freedom. Notice that, since the last term in (14) is gauge-invariant, under the gauge transformations (11), $\tilde{h}_{\mu\nu}$ shifts in the same way as $\hat{h}_{\mu\nu}$.

Integrating out A_μ, through Eq. (12), then expressing $\hat{h}_{\mu\nu}$ in terms of $\tilde{h}_{\mu\nu}$ through (14), and choosing the gauge appropriately, we can write the resulting effective equation for $\tilde{h}_{\mu\nu}$ in the form

$$\left(1 + \frac{m^2(\Box)}{\Box}\right)\mathcal{E}^{\alpha\beta}_{\mu\nu}\tilde{h}_{\alpha\beta} = T_{\mu\nu}, \tag{15}$$

which has a tensorial structure identical to the linearized Einstein equation. This latter fact indicates that $\tilde{h}_{\mu\nu}$ indeed propagates only helicity-2 polarizations, characteristic of the massless graviton.

Because $\tilde{h}_{\mu\nu}$ propagates only two degrees of freedom, the one-particle exchange amplitude between the two conserved sources $T_{\mu\nu}, T'_{\mu\nu}$ mediated by $\tilde{h}$ is

$$(\text{Amplitude})_{\tilde{h}} \propto \frac{T'_{\mu\nu}T^{\mu\nu} - \frac{1}{2}T^{\mu}_{\mu}T'^{\nu}_{\nu}}{\Box - m^2(\Box)}, \tag{16}$$

which continuously recovers the massless graviton result in the limit $m^2 \to 0$. This fact, however, *does not* avoid the well-known vDVZ discontinuity,[19] because (16) is only a part of a full one-particle exchange amplitude. This becomes immediately clear if we notice that the metric exitation that couples to the conserved $T_{\mu\nu}$ is $\hat{h}_{\mu\nu}$ (or, equivalently, $h_{\mu\nu}$), which depends on $\tilde{h}_{\mu\nu}$ through (14). The full physical amplitude is generated by the latter combination of $\tilde{h}_{\mu\nu}$ and is equal to (5), which clearly exhibits the vDVZ-type discontinuity in the $m^2(\Box) \to 0$ limit.

4. The Strong Coupling of Longitudinal Gravitons, and the Concept of r_*

We have seen that in the considered class of theories the graviton should contain three extra *longitudinal* polarizations, which lead to vDVZ discontinuity in linearized theory. Hence, any theory of modified gravity that remains weakly coupled at the solar system distances is automatically ruled out by the existing gravitational data.

Thus, the only point that could save such theories is the breakdown of the linearized approximation at the solar system distances, i.e. the *strong coupling phenomenon.*[9] In the other words, the only theories of modified gravity that can be compatible with observations are the strongly coupled ones. Any attempt at curing the strong coupling will simply rule out the theory.

Fortunately, the same extra polarization that creates a worrisome discontinuity in the gravitational amplitude also provides the strong coupling which invalidates the unwanted result at solar system scales. We shall now generalize the results of Ref. 9 to theories with arbitrary α.

Consider a theory of the graviton that in the linearized approximation satisfies Eq. (1). We shall assume that this theory has a generally covariant nonlinear completion. The nonlinear completion of the first term is the usual Einstein's tensor. The completion of the second term is unknown for general α. In fact, extrapolation from the only known case of $\alpha = 1/2$ tells us that if a completion exists, it will probably require going beyond four dimensions.[4] Discovering such a completion is beyond our program. In fact, knowing the explicit form of this completion is unnecessary for our analysis, provided that it meets some general requirements listed above. Then, we will be able to derive general properties of the strong coupling and the resulting predictions.

The propagator for the graviton satisfying Eq. (1) has the form

$$D_{\mu\nu;\alpha\beta} = \left(\frac{1}{2}\tilde{\eta}_{\mu\alpha}\tilde{\eta}_{\nu\beta} + \frac{1}{2}\tilde{\eta}_{\mu\beta}\tilde{\eta}_{\nu\alpha} - \frac{1}{3}\tilde{\eta}_{\mu\nu}\tilde{\eta}_{\alpha\beta}\right)\frac{1}{\Box + m^2(\Box)}, \tag{17}$$

where

$$\tilde{\eta}_{\mu\nu} \equiv \eta_{\mu\nu} + \frac{\partial_\mu\partial_\nu}{m^2(\Box)} = \eta_{\mu\nu} + r_c^{2(\alpha-1)}\frac{\partial_\mu\partial_\nu}{\Box^\alpha}. \tag{18}$$

The existence of the terms that are singular in r_c^{-1} is the most important fact. This singularity is precisely the source of the strong coupling observed in Ref. 9 for $\alpha = 1$ and $\alpha = 1/2$ cases. The terms that are singular in r_c^{-1} come from the additional, helicity-0 state of the resonance graviton. In the Stückelberg language given in (11), this 0-helicity polarization resides partially in A_μ and partially in the trace of $\hat{h}_{\mu\nu}$. If we denote the canonically normalized longitudinal polarization by χ, then, ignoring the spin-1 helicity vector, the full metric fluctuation can be represented as

$$h_{\mu\nu} = \tilde{h}_{\mu\nu} - \frac{1}{6}\eta_{\mu\nu}\chi + r_c^{2(\alpha-1)}\frac{\partial_\mu\partial_\nu}{3\Box^\alpha}\chi. \tag{19}$$

Notice that $\tilde{h}_{\mu\nu}$ is the same as in (14).

The same state is responsible for the extra attraction that provides a factor of 1/3 in the one-graviton exchange amplitude, as opposed to 1/2 in the standard gravity, leading to vDVZ discontinuity.

The strong coupling of the longitudinal gravitons, however, has a profound effect on discontinuity. The effect of the longitudinal gravitons becomes suppressed near the gravitating sources, where the linearized approximation breaks down. Due to the strong coupling effects, the gravitating sources of mass M, on top of the usual Schwarzchild gravitational radius $r_g \equiv 2G_N M$, acquire the second physical radius, which we shall call r_*. Breakdown of the linearized approximation near gravitating sources was first noticed in the context of $\alpha = 0$ theory in Ref. 21, and the underlying strong coupling dynamics was uncovered in Ref. 9. The key point of the present discussion is that both the strong coupling and the resulting r_* scale are properties of theories with arbitrary α.

For the longitudinal graviton χ, the latter radius plays a role somewhat similar to the one played by r_g for the transverse graviton $\tilde{h}_{\mu\nu}$. Namely, due to the strong coupling, at $r = r_*$ the nonlinear self interactions of χ become important, and the expansion in series of $G_N(r_g)$ breaks down.

The concept of r_* plays the central role in any large distance modified theory of gravity, as we shall now discuss. Consider a localized static gravitating source $T_{\mu\nu} = \delta_\mu^0\delta_\nu^0 M\delta(r)$ of gravitational radius r_g. Then, sufficiently far from the source, the linearized approximation should be valid, and the metric created by the source can be found in one graviton exchange approximation to linear order in G_N:

$$h_{\mu\nu} = \frac{\delta^0_\mu \delta^0_\nu - \frac{1}{3}\left(\eta_{\mu\nu} + r_c^{2(\alpha-1)} \frac{\partial_\mu \partial_\nu}{\Box^\alpha}\right)}{\Box - r_c^{2(1-\alpha)} \Box^\alpha} \frac{r_g}{2} \delta(r). \tag{20}$$

The term in the numerator, which is singular in $1/r_c$, vanishes when convoluted with any conserved test source $T'_{\mu\nu}$, in accordance with (5). Hence, at the distances below r_c, the metric has an $\frac{r_g}{r}$ form, but with a *wrong* (scalar-tensor type) tensorial structure, manifesting vDVZ discontinuity. However, in nonlinear interactions, the singular-in-$1/r_c$ terms no longer vanish and in fact washout the linear effects, due to the strong coupling. The scale of the strong coupling can be figured out by generalizing the analysis of Ref. 9 and 17. The straightforward power counting then shows that the leading singularity in r_c is of the order of $r_c^{4(1-\alpha)}$,[a] and comes from the trilinear interaction of the longitudinal gravitons. This vertex has a momentum dependence of the form

$$(r_c k)^{4(1-\alpha)} k^2. \tag{21}$$

Then, the scale r_* corresponds to a distance from the source for which the contribution from the above trilinear vertex becomes as important as the linear one given by (20). It is obvious that the corresponding distance is given by

$$r_* = \left(r_c^{4(1-\alpha)} r_g\right)^{\frac{1}{1+4(1-\alpha)}}. \tag{22}$$

For distances $r \ll r_*$ the correction to the Einsteinian metric coming from the longitudinal gravitons is suppressed by powers of r_*. The leading behavior can be fixed from the two requirements. First, $\chi(r)$ should become of order r_g/r_* at $r = r_*$, in order to match the linear regime (20) outside the r_* sphere. Secondly, the solution inside r_* must be possible to approximate by the analytic series in $r_c^{2(\alpha-1)}$. These two requirements fix the leading behavior as

$$\chi(r \ll r_*) \sim \frac{r_g}{r_*}\left(\frac{r}{r_*}\right)^{\frac{3}{2}-2\alpha}, \tag{23}$$

which yields the relative correction to the gravitational potential given by (4).

As pointed out in Ref. 17, for some interesting values of α these corrections are strong enough to be tested in precision gravitational measurement experiments. For $\alpha = 1/2$ theory this was also pointed out in Ref. 26, with the correction from the cosmological self-accelerated background[3,22] taken into account.

In fact, existing lunar ranging constraints already rule out theories with α significantly above 1/2, and the new generation of the improved accuracy measurements[10,11] will probably test the $\alpha \simeq 1/2$ case.

Notice that for $\alpha = 1/2$, (4) matches the known explicit solutions derived both in $1/r_c$ expansion[25] and exactly.[27]

[a]The identical result was obtained in Ref. 24 for $\alpha = 0$, agreeing with the result of Ref. 9.

To summarize briefly, for any gravitating source of Schwarzschild radius r_g, there is a new intermediate observable scale appearing. We call this scale r_*, and its essence is that at this scale perturbative expansions both in $1/r_c$ and in G_N break down. For $r \ll r_*$ and $r \gg r_*$ we can use r/r_c and r_g/r expansions respectively, but none can be used, in general, at $r \sim r_*$.[b] However, because for $r_c \sim H_0^{-1}$ all the sources within the observable Universe are well within their own r_*-s, Eq. (23) can be safely applied to them.

5. Conclusions

The possibility of large distance modification of gravity is a fundamental question, motivated by the dark energy problem.[3,23,29,30] In this talk we have tried to stress the important role played by the *strong coupling phenomenon* in this class of theories. In particular, it was already appreciated[9,18] that the strong coupling is the only cure that saves any such theory from being immediately ruled out by the solar system observations, due to the fact that any such ghost-free theory must contain an extra helicity-0 state and is subject to vDVZ discontinuity at the linearized level. This fact makes impossible the existence of consistent weakly coupled theories of modified gravity (relevant for cosmic observations).

Interestingly, in the class of theories considered, the same helicity-0 polarization that creates the problem at the linearized level, also invalidates it by exhibiting the strong coupling phenomenon.

Due to this strong coupling phenomenon, gravitating objects are "endowed" with a physical scale r_*, which for the longitudinal graviton plays the role of the second "horizon." At distances $r_* \sim r$ the nonlinear interactions of χ catchup with the linear part, and expansion in terms of the Newtonian coupling constant can no longer be trusted. Hence, vDVZ discontinuity is invalidated.

For scales $r \ll r_*$, the metric can be found in series of $\left(\frac{r}{r_c}\right)^{2(1-\alpha)}$ expansion (or possibly exactly). The assumption that the true solution can be approximated in such analytic series fixes the form of the leading correction to the Eisteinian metric in the form of (4).

This fact makes theories of modified gravity potentially testable by the precision gravitational experiments in the solar system, and in particular by lunar ranging experiments.

Acknowledgment

I wish to thank the organizers of the conference "From Quantum to Cosmos: Fundamental Physics in Space" for their invitation to this very exciting meeting.

[b]In other words, sources that are of the size of their own r_* probe the new cutoff-sensitive physics, and perturbative methods become useless. If we recall that in the self-accelerated branch of $\alpha = 1/2$ theory, the current Universe precisely has size of its r_*, it will become immediately clear that the perturbation spectrum derived from linearized analysis over this classical background cannot be trusted. For example, such a perturbative analysis that studies the stability of this brunch is, by default, beyond its validity.

References

1. Supernova Search Team Collab. (A. G. Riess *et al.*), *Astron. J.* **116** (1998) 1009 [astro-ph/9805201].
2. Supernova Cosmology Project Collab. (S. Perlmutter *et al.*), *Astrophys. J.* **517** (1999) 565 [astro-ph/9812133].
3. C. Deffayet, G. Dvali and G. Gabadadze, *Phys. Rev. D* **65** (2002) 044023 [astro-ph/0105068].
4. G. Dvali, G. Gabadadze and M. Porrati, *Phys. Lett. B* **485** (2000) 208 [hep-th/0005016].
5. G. Dvali and G. Gabadadze, *Phys. Rev. D* **63** (2001) 065007 [hep-th/0008054].
6. C. Deffayet, Theory and Phenomenology of DGP Gravity, talk given at this meeting.
7. D. G. Boulware and S. Deser, *Phys. Rev. D* **6** (1972) 3368.
8. G. Gabadadze and A. Gruzinov, hep-th/0312074.
9. C. Deffayet *et al.*, *Phys. Rev. D* **65** (2002) 044026 [hep-th/0106001].
10. E. Adelberger, private communication (2002).
11. T. W. Murphy, APOLLO Springs to Life: A New Push in Lunar Laser Ranging, talk given at this meeting.
12. K. Nordtvedt, LATOR: Its Science Objectives and Constructions of Mission Orbits Configuration, talk given at this meeting.
13. M. Shao, Optical Interferometry for Fundamental Physics, talk given at this meeting.
14. M. Kasevich, Navigation, Gravitation and Cosmology with Cold Atomic Sensors, talk given at this meeting.
15. E. Adelberger, Test of Gravitational Inverse-Square Law at the Dark Energy Length Scale, talk given at this meeting.
16. H. J. Paik, Inverse-Square Law Experiment in Space, talk given at this meeting.
17. G. Dvali, A. Gruzinov and M. Zaldarriaga, *Phys. Rev. D* **68** (2003) 024012 [hep-ph/0212069].
18. G. Dvali, *Phys. Scripta T* **117** (2005) 92 [hep-th/0501157].
19. H. van Dam and M. Veltman, *Nucl. Phys. B* **22** (1970) 397.
20. V. I. Zakharov, *J. Exp. Theor. Phys. Lett.* **12** (1970) 312.
21. A. I. Vainshtein, *Phys. Lett. B* **39** (1972) 393.
22. C. Deffayet, *Phys. Lett. B* **502** (2001) 199 [hep-th/0010186].
23. G. Dvali and M. Turner, astro-ph/0301510.
24. N. Arkani-Hamed, H. Georgi and M. D. Schwartz, *Ann. Phys.* **305** (2003) 96 [hep-th/0210184].
25. A. Gruzinov, astro-ph/0112246.
26. A. Lue and G. Starkman, *Phys. Rev. D* **67** (2003) 064002 [astro-ph/0212083].
27. G. Gabadadze and A. Iglesias, *Phys. Rev. D* **72** (2005) 084024 [hep-th/0407049].
28. G. Gabadadze and A. Iglesias, *Phys. Lett. B* **632** (2006) 622 [hep-th/0508201].
29. T. Damour, I. I. Kogan and A. Papazoglou, *Phys. Rev. D* **66** (2002) 104025 [hep-th/0206044].
30. S. M. Carrol *et al.*, astro-ph/0306438. See also, talks given at this meeting by S. M. Carrol, and by M. Trodden.

THEORY AND PHENOMENOLOGY OF DGP GRAVITY

CÉDRIC DEFFAYET
UMR7164, CNRS, Université Paris 7,
CEA, Observatoire de Paris, France
APC, 11 Place Marcelin Berthelot,
75005 Paris Cedex 05, France
and
UMR7095, CNRS, Université Paris 6, France
GReCO/IAP, 98 bis Boulevard Arago,
75014 Paris, France
deffayet@iap.fr

I review some aspects of the Dvali–Gabadadze–Porrati (also known as "brane-induced") model of gravity. This model provides a novel way to modify gravity at large distances and, as such, has potentially some interesting cosmological consequences, like the possibility of getting an accelerated expansion with a vanishing cosmological constant. In DGP gravity, the recovery of usual gravitational interaction at small (i.e. noncosmological) distances is rather nontrivial. This can lead to observable signature in observations made in the solar system. I discuss various aspects of the phenomenology of the model and briefly comment on the consistency of the whole approach.

Keywords: Modification of gravity; cosmic acceleration.

1. Preface

DGP gravity (named after Dvali, Gabadadze and Porrati, who introduced the model[1]), also known as "brane-induced gravity," provides a new and interesting way to modify gravity at large distances. In particular, this model is able to produce an accelerated expansion of the Universe, without the need for a nonvanishing cosmological constant.[2,3] One of its peculiarities is the way one recovers the usual gravitational interaction at small (i.e. noncosmological) distances. In this recovery, a key role is played by the nonlinearities of the theory, namely those of its scalar sector, leading to potentially observable signature in the solar system and precision tests of gravity on the solar system size. In the following, after giving a short introduction to the model (Sec. 2), we will review the observational signatures of DGP gravity (Sec. 3), before turning to discuss briefly some potential drawbacks of the approach (Sec. 4).

2. A short Introduction to DGP Gravity

The DGP model[1] we are considering is a five-dimensional brane-world model. As such it describes our four-dimensional Universe as a surface embedded into a five-dimensional bulk space–time. This means in particular that all the matter fields are thought of as being localized on this surface, while the gravitational fields are living in the whole bulk space–time. The characteristic feature of this particular brane world model lies in the gravitational dynamics. The latter is first governed by a bulk gravitational action, which is the usual action for 5D gravity, namely the Einstein–Hilbert action

$$S_{(5)} = -\frac{1}{2\kappa_{(5)}^2} \int d^5X \sqrt{g_{(5)}} R_{(5)}, \tag{1}$$

where $R_{(5)}$ is the 5D Ricci scalar computed from the 5D metric $g_{AB}^{(5)}$,[a] and $\kappa_{(5)}^2$ is the inverse third power of the reduced 5D Planck mass $M_{(5)}$. To account for the brane, one adds to this action a term of the form

$$S_{(4)} = \int d^4x \sqrt{g_{(4)}} \mathcal{L}. \tag{2}$$

In this expression, $\mathcal{L}$ is the Lagrangian density, given by

$$\mathcal{L} = \mathcal{L}_{(M)} - \frac{1}{2\kappa_{(4)}^2} R^{(4)}, \tag{3}$$

where $\mathcal{L}_{(M)}$ is a Lagrangian for brane localized matter (that is to say baryonic matter, dark matter, etc.) $\kappa_{(4)}^2$ is the inverse square of the reduced 4D Planck mass M_P, and $R^{(4)}$ is the Ricci scalar of the so-called induced metric $g_{\mu\nu}^{(4)}$ on the brane. It is this term, depending on $R^{(4)}$, that is responsible for all the peculiarities of the gravitational phenomenology of DGP gravity.[b] The induced metric $g_{\mu\nu}^{(4)}$ is the metric experienced by the matter we are made of; it is defined by

$$g_{\mu\nu}^{(4)} = \partial_\mu X^A \partial_\nu X^B g_{AB}^{(5)}, \tag{4}$$

where $X^A(x^\mu)$ define the brane position in the bulk. That is to say, X^A are bulk coordinates, and x^μ coordinates along the brane world volume. We also implicitly include in the action a suitable Gibbons–Hawking term[4] for the brane in order to have a well-defined variational problem from the sum of actions (1) and (2). The equation of motion for the DGP model thus reads

$$G_{AB}^{(5)} = \kappa_{(5)}^2 \delta\,(\text{brane})\, \delta_A^\mu \delta_B^\nu \left\{ T_{\mu\nu}^{(M)} - \frac{1}{\kappa_{(4)}^2} G_{\mu\nu}^{(4)} \right\}, \tag{5}$$

[a]Here and in the following, we adopt the following convention for indices: upper case Latin letters $A, B, \ldots$ denote 5D indices; Greek letters $\mu, \nu, \ldots$ denote 4D indices parallel to the brane.
[b]This term, although usually not considered, is expected to be present in generic brane world constructions. It is the hierarchy of scales between M_P and $M_{(5)}$ that is the real distinctive feature of DGP gravity.

where $G^{(5)}_{AB}$ is the 5D Einstein tensor, $G^{(4)}_{\mu\nu}$ is the 4D Einstein tensor (built out of the induced metric $g^{(4)}_{\mu\nu}$), and $T^{(M)}_{\mu\nu}$ is the matter energy–momentum tensor.

In this model, the gravitational potential between two static sources, separated by a distance r, interpolates between a 4D $1/r$ behavior at small distances and a 5D $1/r^2$ behavior at large distances, as shown in Ref. 1. The crossover distance r_c between the two regimes is given by

$$r_c \equiv \frac{\kappa^2_{(5)}}{2\kappa^2_{(4)}} = \frac{M_P^2}{2M^3_{(5)}}. \tag{6}$$

However, gravity does not reduce to a Newtonian potential, and in fact the story is much more complicated. This is because, in this model, from a 4D point of view, gravity is mediated by a continuum of massive gravitons (so-called Kaluza Klein modes), with no normalizable massless graviton entering into the spectrum. This is a consequence of the bulk being flat and infinite. As a result, the tensorial structure of the graviton propagator was shown to be that of a massive graviton and the model shares some properties with the so-called Pauli–Fierz gravity[5,6] describing a single massive graviton. It particular, it exhibits the van Dam–Veltman–Zakharov discontinuity (vDVZ in the following),[7–9] which refers to the fact that the limiting behavior of Pauli–Fierz theory, when the mass of the graviton is sent to zero, is not given by the theory of a massless graviton. This is at the root of the characteristic observable features of DGP gravity on the solar system distance scales. In fact, if one would stick to the linearized approximation to describe the metric in the solar system, for example, one would conclude that DGP gravity is ruled out by experiment, in the same way a theory of a massive gravity described by the (quadratic) Pauli–Fierz action is ruled out whatever the smallness of the mass of the graviton.[7–9] For example, the light bending predicted in the latter theory would differ by 25% from the one predicted in general relativity (GR in the following), the same being true for DGP gravity. So how can one reconcile DGP gravity with what is known of gravity at mesoscopic scales (small scales in comparison with cosmological scales)? The answer to this question lies in the nonlinearities of the theory. They open the way to a recovery of solutions sufficiently close to that of the usual GR for solar system scales,[10] in line with the old idea of A. Vainshtein for massive gravity.[11] This will be discussed with more details in the next section. We will first discuss the homogeneous cosmology of the model, an interesting subject on its own, which also provides interesting insights into the nonperturbative dynamics of the model.

3. Cosmology and Phenomenology of DGP Gravity

3.1. *Homogeneous cosmology*

Simple cosmological solutions of the DGP model are known exactly. The metric on the brane is that of a Friedmann–Lemaître–Robertson–Walker space–time with a

Hubble factor H which is a solution to the modified Friedmann equations[2]

$$\dot{\rho}_{(M)} = -3H\left(P_{(M)} + \rho_{(M)}\right), \tag{7}$$

$$H^2 + \frac{k}{a^2} = \left(\epsilon\frac{1}{2r_c} + \sqrt{\frac{\kappa_{(4)}^2\rho_{(M)}}{3} + \frac{1}{4r_c^2}}\right)^2, \tag{8}$$

where a is the scale factor on the brane, $k = 0, \pm 1$ parametrizes as usual the spatial curvature of the brane (that is to say, of our 4D space–time), $\rho_{(M)}$ and $P_{(M)}$ are respectively the brane localized matter energy density and pressure, and $\epsilon = \pm 1$ parametrizes two different phases according to the sign of the brane effective energy density $\rho_{\text{eff}} \equiv \rho_{(M)} - 3H^2/\kappa_{(4)}^2$. The two different ϵ's also correspond to the way the brane is embedded into the bulk (see Ref. 2 for more details).

The early time homogeneous cosmology is obtained from those equations taking the limit

$$\kappa_{(4)}^2\rho_{(M)} \gg r_c^{-2}. \tag{9}$$

It is easily seen that in this limit the evolution of the Universe is governed by the standard Friedmann equations

$$\dot{\rho}_{(M)} = -3H\left(P_{(M)} + \rho_{(M)}\right), \tag{10}$$

$$\begin{aligned} H^2 + \frac{k}{a^2} &= \frac{\kappa_{(4)}^2\rho_{(M)}}{3}, \\ &= \frac{8\pi G_N}{3}\rho_{(M)}, \end{aligned} \tag{11}$$

where G_N is Newton's constant. Interestingly, there is no sign of the vDVZ discontinuity in this limit.[c] Namely, cosmological solutions discussed above provide an explicit example of the Vainshtein[11] proposal: a nonperturbative recovery of solutions of standard general relativity in a theory which has the tensorial structure of massive gravity.

The late time cosmology differs from that derived in standard GR: let us assume that the energy density on the brane $\rho_{(M)}$ vanishes asymptotically when the scale factor diverges.[d] In this case, depending on the sign of ϵ, one goes either to a phase (for $\epsilon = -1$) where the right-hand side of Eq. (8) is proportional to the square of the matter density of the Universe (similar to the one discussed in Refs. 12 and 13) or to a phase (for $\epsilon = +1$) where it asymptotes the constant value r_c^{-2}. In the latter case, the Universe approaches a de Sitter space–time, and the expansion self-accelerates without the need for a nonvanishing cosmological constant. This is the basis of the proposal made in Refs. 2 and 3 to use DGP gravity as a way to

[c] In particular, one could have expected to see a renormalization of Newton's constant entering into the modified Friedmann equation with respect to the one entering into the action.
[d] Since the energy–momentum tensor of matter is conserved [see Eq. (7)], this would hold in DGP gravity for the kind of matter it is holding in standard cosmology.

produce cosmic acceleration by a large distance modification of gravity, rather than by assuming a nonzero cosmological constant.

Note that the condition (9) translates into a condition on the Hubble radius H^{-1} that is given by

$$H^{-1} \ll r_c. \tag{12}$$

As expected, the radius that governs the transition from standard cosmology to late time nonstandard cosmology is that given by Eq. (6). So, in order for the model to be in agreement with the known history of the Universe, one should demand that r_c be large enough, namely larger than (or of the order of) today's Hubble radius, H_0^{-1}. This translates in turn on a bound on $M_{(5)}$ given by

$$M_{(5)} < 10\text{–}100\,\mathrm{MeV}. \tag{13}$$

3.2. *Matching cosmological data with the self-accelerating solution*

One of the appealing features of the proposal[2,3] to use the DGP model to produce an accelerated expansion with no nonvanishing cosmological constant is that this proposal has the same number of parameters as the ΛCDM model, and yet can be distinguished from the latter model. Indeed, the generalized Friedmann equation (8) (with $\epsilon = +1$) can be rewritten as

$$H^2(z) = H_0^2\left(\Omega_k^2(1+z)^2 + \left(\sqrt{\Omega_{r_c}} + \sqrt{\Omega_{r_c} + \Omega_M(1+z)^3}\right)^2\right), \tag{14}$$

where z is the red shift, the Ω parameters have the standard definitions (above, we have only considered nonrelativistic matter and set Ω_Λ to zero,[e] but radiation or any kind of matter can be considered as in standard cosmology) and Ω_{r_c} is defined by

$$\Omega_{r_c}^{-1} = 4r_c^2H_0^2. \tag{15}$$

With the purpose of confronting this cosmology with data, one can further notice that the self-accelerating DGP cosmology is also exactly reproduced by standard cosmology with a dark energy component with a z-dependent equation-of-state parameter $w_X^{\mathrm{eff}}(z)$, which for a Universe containing only nonrelativistic matter is given by[3]

$$w_X^{\mathrm{eff}}(z) = \frac{1}{\left(\sqrt{\frac{4\Omega_{r_c}}{\Omega_M(1+z)^3}+4}\right)\left(\sqrt{\frac{\Omega_{r_c}}{\Omega_M(1+z)^3}} + \sqrt{\frac{\Omega_{r_c}}{\Omega_M(1+z)^3}+1}\right)} - 1. \tag{16}$$

At large red shift w_X^{eff} tends toward $-1/2$, reflecting the fact that the dominant term in Eq (14), after matter and curvature terms, redshifts as $(1+z)^{3/2}$ at large z. At low z, however, w_X^{eff} decreases toward an (Ω_k, Ω_M)-dependent asymptotic value.

[e]A nonvanishing cosmological constant can, however, easily be introduced in the model.

For a flat Universe, the latter is simply given by $-1/(1+\Omega_M)$. The self-accelerating DGP cosmology (14) has been confronted with observations by various authors and compared to fits to ΛCDM.[14–44] The outcome is that current observations can well be accommodated by this cosmology, even though ΛCDM fits the data more easily. This is well summarized in the recent work by Maartens and Majerotto,[44] which gives as best-fit parameters for (Ω_{r_c}, Ω_M) the values (0.130, 0.260) (using the Legacy SN data of Ref. 45). This translates into $r_c \sim 1.4\ H_0^{-1}$. However, those conclusions should be considered with some care[f]: in order to make comparison with observations in a consistent way, one should reinterpret standard observations in the full framework of DGP gravity, which has not been done yet to the necessary level. Here one should distinguish observations, such as those of the SNIa, that only depend on the background evolution [which is known to the necessary precision from Eq. (8)], from those observations which depend, in one way or another, on the dynamics of cosmological perturbations. The latter dynamics is currently not very well understood, this being related to the way perturbation theory works in DGP, as we will discuss with more details in the following two sections.

To end this subsection, let us mention that one can further complicate the model by introducing a cosmological constant on the brane, in the form of a nonvanishing brane tension. This leads to interesting properties for the expansion of the Universe, like the possibility[g] of having an equation of state-parameter w_X^{eff} lower than -1, while the Universe content is simply that of ΛCDM.[46,47]

3.3. *Cosmological perturbations*

It is fair to say that cosmological perturbations of brane world theories — defined as the linearization of the theory over a cosmological background — are far less well understood than those of the usual 4D GR. This is in particular because the problem is intrinsically 5D, and, as a result, one has to solve partial differential equations instead of ordinary differential equations. This holds for DGP gravity as well. A popular approach is to try to gain some insight into this problem by considering effective 4D equations, which take the form of the usual 4D Einstein equations where the effect of the bulk is encoded in part into a so-called Weyl fluid entering on the matter side of the equations. For example, in the DGP model, considering scalar perturbations and working in the so-called longitudinal gauge, where the 4D linearized line element (on the brane) reads

$$ds^2 = -(1+2\Phi)dt^2 + a^2(t)(1-2\Psi)\delta_{ij}dx^i dx^j, \tag{17}$$

with two gravitational potential Φ and Ψ, one can easily see[48] that, in the simplest case of a matter cosmic fluid with adiabatic perturbations and vanishing anisotropic

[f]This is also stressed in Ref. 44.
[g]Even for the $\epsilon = -1$ branch.

stress, the difference between Φ and Ψ is given by

$$\Phi - \Psi = \zeta(t)\delta\pi_{(\mathcal{E})}, \tag{18}$$

where ζ is a background-dependent coefficient that can be obtained from Refs. 48–50, and $\delta\pi_{(\mathcal{E})}$ is the anisotropic stress of the so-called Weyl fluid. This contrasts with 4D GR, where the vanishing of the matter anisotropic stress implies the equality between Φ and Ψ. This nonequality is in fact intimately related to the vDVZ discontinuity. In particular, one cannot consistently neglect the Weyl fluid contribution in the cosmological perturbation theory of DGP gravity without modifying the model itself. On the other hand, this Weyl fluid part is bothersome, because it does not allow one to obtain the evolution of the perturbations solving only equations along the brane; it does not have a local evolution equation. This has led several authors to neglect this contribution in order to obtain a closed system of equations on the brane. However, as should be clear from Eq. (18), this is not a consistent procedure. In fact, the same "approximation" would result in a disappearance of the vDVZ discontinuity on a flat background, already in the linear perturbation theory. This would mean that one has in fact modified the original model of DGP. Rather, one should keep the Weyl fluid contribution in order to compute cosmological perturbations.[48–50,43,51] Not surprisingly, the two approaches do not agree on the physical predictions. As we just mentioned, it is possible in some cases to get around this problem. For example, one can compute for large-enough, but subhorizon, perturbations the linear growth rate[43,51] (see also Ref. 52). Even so, however, further care has to be taken when fitting the data, in particular when aiming at precision cosmology parameter estimation, because the domain of validity of the linear perturbation theory is quite restricted, and not even clearly elucidated yet, as we will now illustrate in discussing the case of spherically symmetric solutions for nonrelativistic sources.

3.4. *Spherically symmetric solution for nonrelativistic sources*

As explained in Sec. 3.1, the exact cosmological solutions of the DGP model do not exhibit any sign of the vDVZ discontinuity. Based on this observation, we suggested that the vDVZ discontinuity of the DGP model was in fact an artifact of the linear perturbation theory over flat space–time[10] following A. Vainshtein's argument for extensions of Pauli–Fierz theory[h] for massive gravity.[11] This has been studied by expansions going beyond linear order.[53–59] Those analyses, mostly concentrated on the case of static spherically symmetric solutions on the brane, have confirmed so far the original suggestion made in Refs. 14–42 for a nonrelativistic source on the brane,

[h] Here we imply by the terminology that Pauli–Fierz theory is defined as a purely quadratic theory.

the solution given by the linearized theory breaks down below a nonperturbative distance, the Vainshtein radius, given by

$$r_v = \left(r_c^2 r_S\right)^{1/3}, \tag{19}$$

where r_S is the Schwarzschild radius defined as in ordinary GR. This scale can be coined from the similar one found by A. Vainshtein to appear for a Pauli–Fierz model,[11] hence its name. For distances much smaller than r_v it has been found that the spherically symmetric solution on the brane is close to the usual 4D Schwarzschild metric (with 4D, massless, "tensorial structure"), so that, for example, the light bending is in first approximation that given by standard general relativity, and there is no more discontinuity. In particular, one sees that r_v diverges as r_c goes to infinity (this is similar to sending the mass of the graviton to zero in the Pauli–Fierz action), the 4D parameters ($\kappa_{(4)}$, and the mass of the source) being fixed. So, in the large r_c limit, there is a well-defined expansion around the ordinary Schwarzschild solution. For the Sun, the Vainsthein radius is of the order 150 pc, and one has a basis to use this expansion to deal with solar system observables.[i] Foe example, the first correction to the Newtonian potential is obtained to be[53,58] $\epsilon(r_S r/2r_c^2)^{1/2}$, where ϵ is the same as the one appearing in the generalized Friedmann equation (8). This correction scales as $(r_S/r)(r/r_v)^{3/2}$. For r_c of the order of the Hubble radius, and the Schwarzschild radius of the Sun, one has $(r/r_v)^{3/2} \sim 10^{-11}$ at the radius of the Earth orbit. This would translate into similar corrections to the values for the PPN parameters. One can show that those corrections lead in particular to a deviation of the perihelion precession rate from the one predicted from GR that is given by $\epsilon\ 3/8r_c$, and assume a numerical value of 5 μas/year. It seems that the best prospect for detecting such a deviation from GR is the future generation of lunar ranging observations, or possibly the *BepiColumbo*. *MESSENGER* or *Cassini* missions.[58,60] Note an interesting feature of those deviations: they depend on the cosmological phase, through the dependence on ϵ, which can thus in principle be detected by observations in the solar system.[58,61]

The Vainshtein radius (19), and its extension to the case of a cosmological background,[58] sets a first limit on the validity of linear perturbation theory around a spherically symmetric static source, and as such should be taken into account when one is dealing with cosmological perturbations. Note in particular that all objects smaller than clusters of galaxies have a size much smaller than their own Vainsthein radius. However, recent works suggest that the situation is even more complicated, pointing to the need for a better understanding of the range of applicability of the linear perturbation theory.[59,62]

[i]Note, however, that this scale increases with the mass of the object, such that it is only for clusters of galaxies that it becomes of the same order of its size, while for smaller objects of the Universe, it stays well above that.

4. Discussion

Various concerns about the internal consistency of the DGP model, and its use to get an accelerated expansion with no cosmological constant, have been expressed in the literature. This is not the place to discuss in detail those issues, which are quite complex. We would just like to comment here briefly on some of them. It has in particular been argued that the self-accelerating phase was plagued by ghost-like instabilities. This was first discussed[63] in the framework of an effective theory advocated to describe the scalar sector of the model in the so-called *decoupling limit*: the 5D and 4D Planck scales are sent to infinity, keeping the ratio of the matter energy–momentum tensor to the 4D Planck mass finite, as well as the scale $\Lambda \equiv (M_P/r_c^2)^{1/3}$. In this effective theory, the sign in front of the kinetic term of small fluctuations of the scalar was found to flip from a Minkowski background to the self-accelerating de Sitter background. This was used to conclude that the latter background was unstable and suffered from a ghost instability. Note at this point that this is somehow an inconsistent statement, namely because the self-accelerating background is exactly at the would-be cutoff Λ of the scalar sector[j] studied in the decoupling limit (see also Gia Dvali's talk in this volume). So, on one side one argues that the theory cannot be trusted above Λ, and on the other side one decides to drop all terms expected to show up at the cutoff to claim that there is a ghost instability in the model. This put aside, however, the presence of the worrisome ghost has been confirmed in other works looking at the linear perturbations of the whole model over the self-accelerating background.[66–68] Can one conclude from this that the self-accelerating brane is unstable? This is far from being obvious at this stage. First, as we just said, the instability might just be due to an artificial truncation of the theory by discarding terms (quantum-generated) which become important at the scale Λ. Secondly, if one does not adopt this point of view (the exact meaning of the scale Λ has also to be clarified; see Refs. 69, 65 and 70), one should worry that the linear perturbation theory has a range of applicability which has yet to be sorted out, as we mentioned in the previous section. Lastly, it is not even clear that the instability is there at the quantum level (see Ref. 62).

Acknowledgment

We thank the organizers of the workshop "From Quantum to Cosmos: Fundamental Physics Research in Space" for their invitation, and for having organized such a nice and interesting meeting.

References

1. G. Dvali, G. Gabadadze and M. Porrati, *Phys. Lett. B* **485** (2000) 208 [hep-th/0005016].

[j]That the scalar effective theory considered in Ref. 63 is relevant has also been recently challenged; see Ref. 65.

2. C. Deffayet, *Phys. Lett. B* **502** (2001) 199 [hep-th/0010186].
3. C. Deffayet, G. R. Dvali and G. Gabadadze, *Phys. Rev. D* **65** (2002) 044023 [astro-ph/0105068].
4. G. W. Gibbons and S. W. Hawking, *Phys. Rev. D* **15** (1977) 2752.
5. M. Fierz, *Helv. Phys. Acta* **12** (1939) 3.
6. M. Fierz and W. Pauli, *Proc. Roy. Soc.* **173** (1939) 211.
7. H. van Dam and M. Veltman, *Nucl. Phys. B* **22** (1970) 397.
8. V. I. Zakharov, *J. Exp. Theor. Phys. Lett.* **12** (1970) 312.
9. Y. Iwasaki, *Phys. Rev. D* **2** (1970) 2255.
10. C. Deffayet *et al.*, *Phys. Rev. D* **65** (2002) 044026 [hep-th/0106001].
11. A. I. Vainshtein, *Phys. Lett. B* **39** (1972) 393.
12. P. Binetruy, C. Deffayet and D. Langlois, *Nucl. Phys. B* **565** (2000) 269 [hep-th/9905012].
13. P. Binetruy *et al.*, *Phys. Lett. B* **477** (2000) 285 [hep-th/9910219].
14. C. Deffayet *et al.*, *Phys. Rev. D* **66** (2002) 024019 [astro-ph/0201164].
15. P. P. Avelino and C. J. A. Martins, *Astrophys. J.* **565** (2002) 661 [astro-ph/0106274].
16. J. P. Uzan and F. Bernardeau, *Phys. Rev. D* **64** (2001) 083004 [hep-ph/0012011].
17. C. Deffayet, G. R. Dvali and G. Gabadadze, astro-ph/0106449.
18. J. S. Alcaniz, *Phys. Rev. D* **65** (2002) 123514 [astro-ph/0202492].
19. D. Jain, A. Dev and J. S. Alcaniz, *Phys. Rev. D* **66** (2002) 083511 [astro-ph/0206224].
20. J. S. Alcaniz, D. Jain and A. Dev, *Phys. Rev. D* **66** (2002) 067301 [astro-ph/0206448].
21. E. V. Linder, *Phys. Rev. Lett.* **90** (2003) 091301 [astro-ph/0208512].
22. O. Zahn and M. Zaldarriaga, *Phys. Rev. D* **67** (2003) 063002 [astro-ph/0212360].
23. T. Multamaki, E. Gaztanaga and M. Manera, *Mon. Not. R. Astron. Soc.* **344** (2003) 761 [astro-ph/0303526].
24. A. Lue, R. Scoccimarro and G. Starkman, *Phys. Rev. D* **69** (2004) 044005 [astro-ph/0307034].
25. H. J. Seo and D. J. Eisenstein, *Astrophys. J.* **598** (2003) 720 [astro-ph/0307460].
26. E. V. Linder, *Phys. Rev. D* **70** (2004) 023511 [astro-ph/0402503].
27. Z. H. Zhu, M. K. Fujimoto and X. T. He, *Astrophys. J.* **603** (2004) 365 [astro-ph/0403228].
28. C. Sealfon, L. Verde and R. Jimenez, *Phys. Rev. D* **71** (2005) 083004 [astro-ph/0404111].
29. J. S. Alcaniz and N. Pires, *Phys. Rev. D* **70** (2004) 047303 [astro-ph/0404146].
30. Z. H. Zhu and J. S. Alcaniz, *Astrophys. J.* **620** (2005) 7 [astro-ph/0404201].
31. F. Bernardeau, astro-ph/0409224.
32. J. S. Alcaniz and Z. H. Zhu, *Phys. Rev. D* **71** (2005) 083513 [astro-ph/0411604].
33. A. Shirata *et al.*, *Phys. Rev. D* **71** (2005) 064030 [astro-ph/0501366].
34. Y. S. Song, *Phys. Rev. D* **71** (2005) 024026 [astro-ph/0407489].
35. I. Sawicki and S. M. Carroll, astro-ph/0510364.
36. M. Fairbairn and A. Goobar, astro-ph/0511029.
37. U. Alam and V. Sahni, *Phys. Rev. D* **73** (2006) 084024 [astro-ph/0511473].
38. Z. K. Guo *et al.*, astro-ph/0603632.
39. M. C. Bento *et al.*, *Phys. Rev. D* **73** (2006) 103521 [astro-ph/0603848].
40. E. V. Linder, astro-ph/0604280.
41. Y. S. Song, I. Sawicki and W. Hu, astro-ph/0606286.
42. N. Pires, Z. H. Zhu and J. S. Alcaniz, *Phys. Rev. D* **73** (2006) 123530 [astro-ph/0606689].
43. A. Lue, R. Scoccimarro and G. D. Starkman, *Phys. Rev. D* **69** (2004) 124015 [astro-ph/0401515].

44. R. Maartens and E. Majerotto, *Phys. Rev. D* **74** (2006) 023004 [astro-ph/0603353].
45. P. Astier *et al.*, *Astron. Astrophys.* **447** (2006) 31 [astro-ph/0510447].
46. V. Sahni and Y. Shtanov, *J. Cosmol. Astropart. Phys.* **0311** (2003) 014 [astro-ph/0202346].
47. A. Lue and G. D. Starkman, *Phys. Rev. D* **70** (2004) 101501 [astro-ph/0408246].
48. C. Deffayet, *Phys. Rev. D* **66** (2002) 103504 [hep-th/0205084].
49. C. Deffayet, *Phys. Rev. D* **71** (2005) 023520 [hep-th/0409302].
50. C. Deffayet, *Phys. Rev. D* **71** (2005) 103501 [gr-qc/0412114].
51. K. Koyama and R. Maartens, *J. Cosmol. Astropart. Phys.* **601** (2006) 016 [astro-ph/0511634].
52. I. Sawicki, Y. S. Song and W. Hu, astro-ph/0606285.
53. A. Gruzinov, *New Astron.* **10** (2005) 311 [astro-ph/0112246].
54. A. Lue, *Phys. Rev. D* **66** (2002) 043509 [hep-th/0111168].
55. T. Tanaka, *Phys. Rev. D* **69** (2004) 024001 [gr-qc/0305031].
56. M. Porrati, *Phys. Lett. B* **534** (2002) 209 [hep-th/0203014].
57. C. Middleton and G. Siopsis, *Mod. Phys. Lett. A* **19** (2004) 2259 [hep-th/0311070].
58. A. Lue and G. Starkman, *Phys. Rev. D* **67** (2003) 064002 [astro-ph/0212083].
59. G. Gabadadze and A. Iglesias, hep-th/0407049.
60. A. Lue, *Phys. Rep.* **423** (2006) 1 [astro-ph/0510068].
61. G. Dvali, A. Gruzinov and M. Zaldarriaga, *Phys. Rev. D* **68** (2003) 024012 [hep-ph/0212069].
62. C. Deffayet, G. Gabadadze and A. Iglesias, Perturbations of Self-Accelerated Universe to appear in *J. Cosmol. Astropart. Phys.* [hep-th/0607099].
63. M. A. Luty, M. Porrati and R. Rattazzi, *J. High Energy Phys.* **0309** (2003) 029 [hep-th/0303116].
64. A. Nicolis and R. Rattazzi, *J. High Energy Phys.* **0406** (2004) 059 [hep-th/0404159].
65. G. Gabadadze and A. Iglesias, hep-th/0603199.
66. K. Koyama, *Phys. Rev. D* **72** (2005) 123511 [hep-th/0503191].
67. D. Gorbunov, K. Koyama and S. Sibiryakov, *Phys. Rev. D* **73** (2006) 044016 [hep-th/0512097].
68. C. Charmousis *et al.*, hep-th/0604086.
69. G. Dvali, hep-th/0402130.
70. C. Deffayet and J.-W. Rombouts, *Phys. Rev. D* **72** (2005) 044003 [gr-qc/0505134].

TESTING STRONG MOND BEHAVIOR IN THE SOLAR SYSTEM

JOÃO MAGUEIJO

Perimeter Institute for Theoretical Physics,
31 Caroline St N, Waterloo, N2L 2Y5, Canada

Canadian Institute for Theoretical Astrophysics,
60 St George St, Toronto, M5S 3H8, Canada

Theoretical Physics Group, Imperial College,
Prince Consort Road, London SW7 2BZ, UK
jmagueijo@perimeterinstitute.ca

JACOB BEKENSTEIN

Racah Institute of Physics,
Hebrew University of Jerusalem,
Jerusalem 91904, Israel

We summarize an interesting set of solar system predictions that we have recently derived for modified Newtonian dynamics (MOND). Specifically, we find that strong MOND behavior may become evident near the saddle points of the total gravitational potential. Whereas in Newtonian theory tidal stresses are finite at saddle points, they are expected to diverge in MOND, and to remain distinctly large inside a sizable oblate ellipsoid around the saddle point. While strong MOND behavior would be a spectacular "backyard" vindication of the theory, pinpointing the MOND bubbles in the setting of the realistic solar system may be difficult. Space missions such as the LISA Pathfinder, equipped with sensitive accelerometers, may be able to explore the larger perturbative region.

Keywords: Modified gravity; MOND; solar system tests.

1. Introduction

MOND[1] is a scheme for explaining extragalactic phenomenology without invoking dark matter. In the Lagrangian formulation of MOND[3] the physical gravitational potential Φ, which gives test particle acceleration by $\mathbf{a} = -\nabla\Phi$, is determined by the modified Poisson equation

$$\nabla \cdot \left[\tilde{\mu} \left(\frac{|\nabla\Phi|}{a_0} \right) \nabla\Phi \right] = 4\pi G \tilde{\rho}, \tag{1}$$

where $\tilde{\rho}$ is the baryonic mass density, $a_0 \approx 10^{-10}\,\mathrm{m\,s^{-2}}$ is Milgrom's characteristic acceleration, and the function $\tilde{\mu}(x)$ is required to approximate its argument for $x \ll 1$ and to approach unity for $x \gg 1$.

Are MOND effects of importance in the solar system (henceforth SS)? With the discovery of the "Pioneer anomaly,"[10,11] much speculation was directed toward a possible MONDian origin of the effect.[12] In Ref. 29 we searched for other sites deep inside the SS where strong MOND behavior might put the MOND phenomenon within the reach of spacecraft measurements. Strong MOND behavior is triggered by a low gradient in the *total* Newtonian potential Φ_N (the deep MOND regime is that where $|\nabla\Phi| \ll a_0$).

Two apparent candidates for strong MOND regions fail this criterion. Most obviously we have gravitational perturbations, such as those accounting for the nonrelativistic component of the perihelion of Mercury precession, or Neptune's influence upon Uranus' orbit. Most of these have a very low potential gradient, and would by themselves be in the MOND regime. However, the gradient of the total Φ_N is not small, so their effect falls in the Newtonian regime.

The Lagrange points are another apparent possibility for strong MOND regions. They are the five stationary points of the two-body dynamics; for example, L1 is the point between the Earth and the Sun where a test mass would be in inertial motion, moving neither toward the Sun nor the Earth. Each Lagrangian point orbits the Sun with the same frequency as the Earth, so the gradient of Φ_N at it must cancel the corresponding centrifugal acceleration, and is thus not especially small. This does not mean, as we shall see, that perturbative effects around these points are not present; however, strong MONDian behavior is certainly not expected.

By contrast, the saddle (or extremum) point (henceforth SP) of Φ_N between two gravitating bodies is evidently in the deep MOND regime, since $\nabla\Phi_N = 0$ there. One such point exists between any two gravitating bodies, potentially providing a testing ground for strong MONDian behavior. SP's are not inertial, but may be visited by free-falling test bodies; they are encased by small "bubbles" within which strong MOND effects are expected.

These examples show that another naively possible candidate is also a failure: the artificial planetary system in space (APSIS) of Ref. 6. This is made up of artificial test bodies operating in a drag-free environment with controlled experimental conditions. However, they are free-falling inside a satellite stationed on a Lagrange point (typically the Earth–Sun L2). Thus a saddle point in such a system would also *not* be a saddle point of the total gravitational potential.

2. The Formalism

In *TeVeS* the MOND behavior is driven by a dynamical (and dimensionless) scalar field ϕ such that the physical potential Φ in which a body falls is given by

$\Phi = \Phi_N + \phi$, where Φ_N is the usual Newtonian potential (inferred from the metric component g_{00}). In the nonrelativistic regime ϕ is governed by the equation

$$\nabla \cdot [\mu(kl^2(\nabla\phi)^2)\nabla\phi] = kG\tilde{\rho}, \tag{2}$$

where k is a coupling constant and l is a length scale which determines the Milgrom acceleration by $a_0 = \frac{\sqrt{3k}}{4\pi l} \approx 10^{-10}\,\mathrm{m\,s^{-2}}$ (we are setting Ξ, as defined in Ref. 4, to unity; thus we ignore the slight renormalization of the gravitational constant in TeVeS so that here $G_N = G$). In Eq. (2) μ is a free function not to be confused with Milgrom's $\tilde{\mu}$. Reference 4 proposed a particular form for it. The deep MOND regime is signalled by the low gradient of the scalar field ϕ; in this regime

$$\mu \approx \frac{k}{4\pi}\frac{|\nabla\phi|}{a_0}. \tag{3}$$

For strong gradients the μ proposed in Ref. 4 grow crudely as $(|\nabla\phi|/a_0)^{2/3}$. This has the effect of suppressing the contribution of $\nabla\phi$ to $\nabla\Phi$, thus bringing in the Newtonian regime. In spherically symmetric systems TeVeS with any μ satisfying Eq. (3) goes over into the Lagrangian MOND theory (1) with Milgrom's $\tilde{\mu}$ given by $\tilde{\mu} = (1+k/4\pi\mu)^{-1}$. Although this point is not well explored, it is quite possible that in less symmetric systems TeVeS does not go over to an exactly MOND behavior. For this reason we base this paper on the nonrelativistic limit of TeVeS, and not on Lagrangian MOND.

We need to solve Eq. (2) for a many-body source, but that equation is nonlinear, so the ϕ fields due to each body do not superpose. However, any nonlinear equation may be *formally* linearized by an appropriate change of variables. Here this is

$$\mathbf{u} = -\frac{4\pi\mu}{k}\nabla\phi \tag{4}$$

(see Ref. 17, where this technique was first suggested). We may then add the $\mathbf{u}$ due to each source (which is the Newtonian acceleration) and invert the *total* $\mathbf{u}$ at a given point to find $\nabla\phi$. It is essential that the sum of all sources is performed before inverting to find $\nabla\phi$.

This algorithm may be applied to any number of components. But note that even if a term in the sum is in the MOND regime, the overall system is not, unless the total $|\mathbf{u}|$ is much smaller than a_0. (It is because of this feature that the gravitational perturbations in the SS are non-MONDian.) However, it is also possible to have two components with fields not in the MOND regime such that their common field is MONDian in some region. Examples are the SP's in the gravitational potential of two bodies to be studied in this paper.

The only complication with the above technique is that $\mathbf{u}$ is generally not curl-free; indeed, it is rather the vector $\mathbf{u}/\mu$ which is curl-free. Thus the full set of equations for $\mathbf{u}$ is

$$\nabla \cdot \mathbf{u} = -4\pi G\tilde{\rho}, \tag{5}$$

$$\nabla \wedge \frac{\mathbf{u}}{\mu} = 0. \tag{6}$$

The first equation tells us that $\mathbf{u}$ equals the Newtonian acceleration $\mathbf{F}^{(N)} = -\nabla\Phi_N$ up to a curl, i.e. there must exist a vector field $\mathbf{h}$ such that

$$\mathbf{u} = \mathbf{F}^{(N)} + \nabla \wedge \mathbf{h}. \tag{7}$$

The second equation fixes the $\mathbf{h}$ (up to a gradient). This operation can only be performed upon the total $\mathbf{u}$, once again stressing the intrinsic nonlinearity of the theory. It can be shown that the curl term vanishes in a spherically symmetric situation, or in the quasi-Newtonian regime far away from the source.[3,4] Near the SP's neither of these conditions is satisfied and we have to evaluate $\nabla \wedge \mathbf{h}$. However, before plunging into the full problem, let us provide some orientation.

3. Heuristics of the MOND Bubbles

We now examine the MONDian saddle region under two simplifying assumptions. One is to replace what is essentially a many-body situation (the SS) by a two-body problem; the other is to drop the curl term in (7). We warn the reader that these approximations have very different fates. The former escapes largely unscathed from a proper treatment (see Sec. 7), and a two-body calculation can be easily adapted to the full-blown situation. The latter is extremely crude and the curl term evaluated in Secs. 4–6 introduces striking novelties. We present the simplified argument as a benchmark, justifying the rather laborious analysis in Secs. 4–6.

Consider two bodies at distance R with masses M and m, with $M \gg m$, so that the system's center of mass may be taken to coincide with the heavier body. To be definite we call them the Sun and the Earth, but we shall explore other couples later. Along the line linking them (the z axis), the Newtonian acceleration is

$$\mathbf{F}^{(N)} = \left(-\frac{GM}{\tilde{r}^2} + \frac{Gm}{(R-\tilde{r})^2}\right)\mathbf{e}_z, \tag{8}$$

where $\tilde{r}$ is the distance from the Sun and $\mathbf{e}_z$ is the unit vector in the direction Sun to Earth. The SP of the Newtonian potential Φ_N resides where $\mathbf{F}^{(N)} = 0$, i.e. at

$$\tilde{r} = r_s \approx R\left(1 - \sqrt{\frac{m}{M}}\right). \tag{9}$$

Around this point $\mathbf{F}^{(N)}$ increases linearly as it passes through zero, i.e.

$$\mathbf{F}^{(N)} \approx A(\tilde{r} - r_s)\mathbf{e}_z, \tag{10}$$

where

$$A = 2\frac{GM}{r_s^3}\left(1 + \sqrt{\frac{M}{m}}\right) \tag{11}$$

is the tidal stress at the SP along the Sun–Earth direction. The full tidal stress matrix is easy to compute. Let us use cylindrical coordinates centered at the SP, with the z axis pointing along the Sun–Earth direction, so that we have

$\partial F_z^{(N)}/\partial z = A$ and $\partial F_\varrho^{(N)}/\partial z = 0$. From the further condition that the divergence be zero (outside the Sun and the Earth), we have

$$\mathbf{F}^{(N)} = A\left(z\mathbf{e}_z - \frac{1}{2}\varrho\,\mathbf{e}_\varrho\right). \tag{12}$$

The region around the SP is obviously in the deep MOND regime since $|\mathbf{F}^{(N)}| \ll a_0$. Thus, regardless of the model adopted for μ, we have just on the basis of Eq. (3)

$$\mathbf{u} \approx -\frac{|\nabla\phi|}{a_0}\nabla\phi = \mathbf{F}^{(N)} + \nabla\wedge\mathbf{h}. \tag{13}$$

If we ignore the curl term and use $|\nabla\phi| \ll |\mathbf{F}^{(N)}|$, we have

$$\mathbf{F} = -\nabla\Phi \approx -\nabla\phi = \sqrt{Aa_0}\frac{z\mathbf{e}_z - \frac{\varrho}{2}\mathbf{e}_\varrho}{\left(z^2 + \frac{\varrho^2}{4}\right)^{1/4}} \tag{14}$$

and we see that, in contrast to the Newtonian theory, the tidal stresses here diverge at the SP. This feature survives the introduction of the less intuitive curl term and may be heuristically understood by applying the rule of thumb that in the deep MOND regime the square root of the Newtonian acceleration gives the physical acceleration. According to Eq. (10) Newtonian acceleration increases linearly along the line Sun–Earth, so the physical acceleration in the deep MOND regime is of the form $\pm\sqrt{Aa_0|\tilde{r} - r_s|}$, which has infinite derivative at r_s.

What is the size of the region where the tidal stresses remain anomalously high? Naively one might expect that the deep MOND regime is defined by the oblate ellipsoidal region defined by $|\mathbf{F}^{(N)}| = a_0$, which translates into a major semiaxis (in the ϱ direction) of size

$$\delta\tilde{r} = \frac{2a_0}{A} \approx \frac{a_0}{a_m}\sqrt{\frac{m}{M}}R, \tag{15}$$

where a_m is the acceleration of the smaller mass m. This expectation turns out wrong when, in Secs. 5 and 6, we bring the curl term into play. It is easier to estimate the size of the (larger) region where there are significant perturbative corrections to Newtonian theory, but where deep MOND behavior is not yet in evidence. For the model introduced in Ref. 4 Milgrom's $\tilde{\mu}$ can be estimated in the quasi-Newtonian region through formula (69) there (this formula, however, is rigorous only in the spherically symmetric case):

$$\tilde{\mu} = \frac{F^{(N)}}{F} \approx \left(1 - \frac{16\pi^3}{k^3}\frac{a_0^2}{F^2}\right). \tag{16}$$

Let us take $F \approx F^{(N)}$ and use Eq. (12). We see that departures at the level 10^{-4} from Newtonian gravity occur within a semimajor axis of size

$$\Delta\tilde{r} = \frac{800\pi^{3/2}}{k^{3/2}}\frac{a_0}{A}. \tag{17}$$

Using $k \approx 0.03$ (as suggested in Ref. 4) this is $\Delta\tilde{r} \approx 1900\,\text{km}$ for the Sun–Earth system, $\Delta\tilde{r} \approx 4.7 \times 10^6\,\text{km}$ for Sun–Jupiter, and $\Delta\tilde{r} \approx 700\,\text{km}$ for Moon–Earth. In contrast with (15), these estimates will withstand the closer scrutiny presented in Sec. 5.

4. Accounting for the Curl Term

By carrying out the curl in Eq. (6) we get

$$\nabla \ln \mu \wedge \mathbf{u} - \nabla \wedge \mathbf{u} = 0, \tag{18}$$

while squaring Eq. (4) gives

$$u^2 = \left(\frac{4\pi}{k}\right)^2 \mu^2 |\nabla\phi|^2. \tag{19}$$

In TeVeS $\mu = \mu(k|\nabla\phi|/a_0)$; thus $k^4 u^2/{a_0}^2$ is a function of μ only. Defining the dimensionless quantity

$$\kappa \equiv \frac{\partial \ln u^2}{\partial \ln \mu}, \tag{20}$$

we get $\nabla \ln \mu = \kappa^{-1} \nabla u^2/u^2$, so that Eq. (18) becomes

$$\kappa u^2 \nabla \wedge \mathbf{u} + \mathbf{u} \wedge \nabla u^2 = 0. \tag{21}$$

In systems with spherical, cylindrical or planar symmetry, $\mathbf{u}$ is necessarily collinear with $\nabla|\mathbf{u}|^2$. Then $\nabla \wedge \mathbf{u}$ must vanish everywhere (since κ would be expected to vanish only at isolated points). This agrees with the findings of Refs. 3 and 4 that $\mu\nabla\phi$ and $\mu\nabla\Phi$ are both curl-free in such situations. When the spatial symmetry is lower or nonexistent, the second term in Eq. (21) will not generally vanish, and will be of order $|\mathbf{u}|^3/L$, where L denotes the scale on which quantities vary. Thus if in a region $\kappa \gg 1$, we would expect $|\nabla \wedge \mathbf{u}|$ to be much smaller than its expected magnitude $|\mathbf{u}|/L$; this signals the quasi-Newtonian regime where $\mathbf{u}$ is nearly curl-free.

In TeVeS the manner of transition between the deep MOND and Newtonian regimes is dependent upon the form of μ. The form proposed in Ref. 4 are quite difficult to work with in our context. We shall thus replace it by the implicit expression

$$\frac{\mu}{\sqrt{1-\mu^4}} = \frac{k}{4\pi} \frac{|\nabla\phi|}{a_0}, \tag{22}$$

which satisfies the limit (3). The calculations in Ref. 29 then lead to

$$\frac{u^2}{{a_0}^2} = \frac{256\pi^4}{k^4} \frac{\mu^4}{1-\mu^4}. \tag{23}$$

Differentiating the logarithm of this we calculate that

$$\kappa = \frac{4}{1-\mu^4} = 4 + \frac{k^4}{64\pi^4} \frac{u^2}{{a_0}^2}. \tag{24}$$

In terms of the dimensionless vector field

$$\mathbf{U} \equiv \frac{k^2}{16\pi^2}\frac{\mathbf{u}}{a_0} \tag{25}$$

we may thus cast Eqs. (21) and (5) into the forms

$$\nabla \cdot \mathbf{U} = 0, \tag{26}$$

$$4(1+U^2)\,U^2\,\nabla \wedge \mathbf{U} + \mathbf{U} \wedge \nabla U^2 = 0, \tag{27}$$

where we have dropped the source of the first since we are interested only in the region near the SP. This pair of exact equations for one dimensionless vector is central to our study.

Once $\mathbf{U}$ is solved for we can recover $\nabla\phi$ by combining Eqs. (4), (23) and (25):

$$-\nabla\phi = \frac{4\pi a_0}{k}(1+U^2)^{1/4}\frac{\mathbf{U}}{U^{1/2}}. \tag{28}$$

As remarked earlier, the condition $\kappa \gg 1$ brings in the Newtonian limit. Now $\kappa \gg 1$ is equivalent to $U \gg 1$. Obviously in this case $-\nabla\phi \approx (4\pi a_0/k)\mathbf{U} = (k/4\pi)\mathbf{u}$, which tells us by Eq. (4) that $\mu \approx 1$, indeed the Newtonian limit [the same is obvious from Eq. (24)].

5. The Quasi-Newtonian Region

At this point we go over to spherical polar coordinates (r, ψ, ϕ) with the origin at the SP; accordingly

$$z = r\cos\psi, \quad \varrho = r\,\sin\psi. \tag{29}$$

So, for example, Eq. (12) takes the form

$$\mathbf{F}^{(N)} = Ar\mathbf{N}, \tag{30}$$

where

$$\mathbf{N}(\psi) \equiv N_r\mathbf{e}_r + N_\psi\mathbf{e}_\psi, \tag{31}$$

$$N_r = \frac{1}{4}[1+3\cos(2\psi)], \tag{32}$$

$$N_\psi = -\frac{3}{4}\sin(2\psi). \tag{33}$$

We define the quasi-Newtonian region as that where U^2 is of order 1 or larger so that the factor $1+U^2$ cannot be ignored in Eq. (27). The region's size may be estimated by dropping the curl term in (7) (an approximation to be justified *a posteriori*) and finding the solution to $U^2 = 1$ using (12) and (25) (in the Newtonian region $\mathbf{u} = \mathbf{F}^{(N)}$). This leads to the ellipsoid:

$$r^2\left(\cos^2\psi + \frac{1}{4}\sin^2\psi\right) = r_0^2 \equiv \left(\frac{16\pi^2 a_0}{k^2 A}\right)^2. \tag{34}$$

Equation (27) tells us that well outside of this ellipsoid the curl is suppressed by a factor of $1/r^2$ with respect to $\mathbf{F}^{(N)}$. As we show below, $\mathbf{U}$ is then neatly

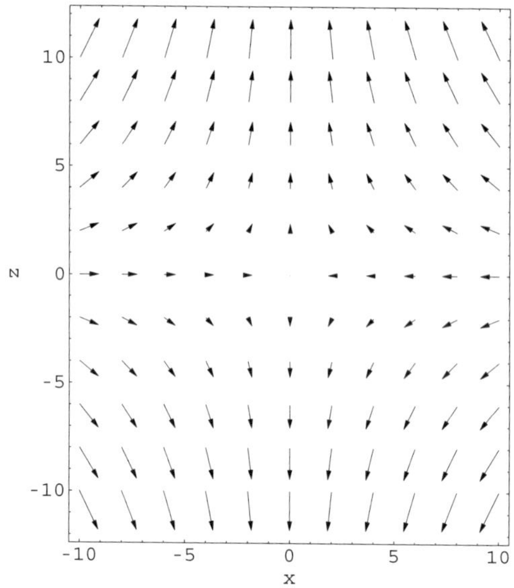

Fig. 1. The flow of $\mathbf{U}_0$ around the SP (at the origin) in a plane containing the symmetry (z) axis; for clarity all vectors have been linearly rescaled.

separated into a Newtonian component $\mathbf{U}_0$ [carrying the divergence predicted by (5) and depicted in Fig. 1] and a "magnetic" component $\mathbf{U}_2$. By definition $\mathbf{U}_2$ is solenoidal and to leading order is sourced purely by $\mathbf{U}_0$. Specifically the dynamics is approximated by

$$\mathbf{U} = \mathbf{U}_0 + \mathbf{U}_2, \tag{35}$$

$$\mathbf{U}_0 = \frac{r}{r_0}\mathbf{N}(\psi), \tag{36}$$

$$\nabla \cdot \mathbf{U}_2 = 0, \tag{37}$$

$$\nabla \wedge \mathbf{U}_2 = -\frac{\mathbf{U}_0 \wedge \nabla|\mathbf{U}_0|^2}{4|\mathbf{U}_0|^4}. \tag{38}$$

With the notation

$$\mathbf{U}_2 = U_r\mathbf{e}_r + U_\psi\mathbf{e}_\psi \tag{39}$$

Eqs. (37) and (38) become

$$\frac{1}{r^2}\frac{\partial}{\partial r}(r^2 U_r) + \frac{1}{r\sin\psi}\frac{\partial}{\partial\psi}(\sin\psi U_\psi) = 0, \tag{40}$$

$$\frac{1}{r}\left[\frac{\partial}{\partial r}(rU_\psi) - \frac{\partial U_r}{\partial\psi}\right] = \frac{s(\psi)}{r^2}, \tag{41}$$

with

$$s(\psi) \equiv -\frac{3}{8}\frac{\cos\psi\sin\psi}{\left[\cos^2\psi + \frac{\sin^2\psi}{4}\right]^2} = -\frac{12\sin 2\psi}{(5+3\cos 2\psi)^2}. \tag{42}$$

The form of Eqs. (40) and (41) suggests that both U_r and U_ψ behave as $1/r$. Accordingly we recast Eq. (39) as the ansatz

$$\mathbf{U}_2 = \frac{r_0}{r}\mathbf{B}(\psi) = \frac{r_0}{r}(F(\psi)\mathbf{e}_r + G(\psi)\mathbf{e}_\psi), \tag{43}$$

where the r dependence has been fully factored out. With this ansatz Eq. (41) collapses into

$$F' = -s = \frac{12\sin 2\psi}{(5+3\cos 2\psi)^2}, \tag{44}$$

with solution

$$F = \frac{2}{5+3\cos 2\psi} + A, \tag{45}$$

where A is a constant. Equation (40) now becomes

$$F + \frac{1}{\sin\psi}\frac{\partial}{\partial\psi}(\sin\psi\, G) = 0, \tag{46}$$

which integrates to

$$G\sin\psi = -\int F\sin\psi\, d\psi + B \tag{47}$$

where B is another constant. Performing the integral gives

$$G\sin\psi = \frac{\tan^{-1}\left(\sqrt{3} - 2\tan\frac{\psi}{2}\right) + \tan^{-1}\left(\sqrt{3} + 2\tan\frac{\psi}{2}\right)}{\sqrt{3}} + A\cos\psi + B. \tag{48}$$

To determine A and B we must discuss boundary conditions. According to Milgrom,[17] for the system (26)–(27) the normal component of $\mathbf{u}$ (or $\mathbf{U}$) must vanish on all boundaries. Parts of the symmetry axis ($\psi = 0$ as well as $\psi = \pi$) are evidently a boundary of the quasi-Newtonian region; it is obvious that N_ψ vanishes on both the North and South parts of it, where it is the normal component. Thus since $\mathbf{U}_0$ satisfies the boundary condition on the relevant pieces of the axis, so must $\mathbf{U}_2$. Accordingly we must require $G(\psi = 0) = G(\psi = \pi) = 0$, from which it follows that $A = B = -\frac{\pi}{3\sqrt{3}}$. The solutions F and G are plotted in Fig. 2. We find that $G(\pi/2) = 0$ as well. Thus on the symmetry plane ($\psi = \pi/2$) $\mathbf{U}$ is collinear with the axis.

What about the rest of the boundary? We see from Eq. (43) that $\mathbf{U}_2 \to 0$ as $r \to \infty$. Thus at large r our $\mathbf{U}$ merges with $\mathbf{U}_0$, which we know to be the limiting form of the Newtonian field as we approach the SP. It follows that our solution automatically fulfills the boundary conditions at large r. The inward part of the boundary of the quasi-Newtonian region adjoins the intermediate MOND region, where MOND effects are no longer small. Fortunately there is no need for us to set boundary conditions there; rather, the solution just described serves to set boundary conditions for the intermediate MOND region.

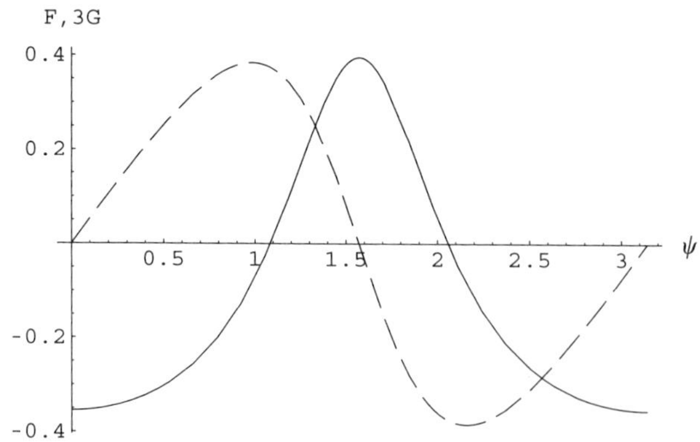

Fig. 2. The angular profile functions F (solid) and G (dashed) giving the direction of the "magnetic" field $\mathbf{B}$ in the quasi-Newtonian region; for clarity G has been multiplied by 3.

We conclude that an SP far away from the strong MOND bubble is characterized by a Newtonian component proportional to r together with a magnetic-like perturbation that falls off like $1/r$. The full physical effects in this regime may be appreciated by combining (28) with (36) and (43). We find that the extra acceleration felt by test particles is

$$\delta\mathbf{F} = -\nabla\phi \approx \frac{4\pi a_0}{k}\left(\mathbf{U}_0 + \frac{\mathbf{U}_0}{4U_0^2} + \mathbf{U}_2 + \cdots\right). \tag{49}$$

The first contribution, call it $\delta\mathbf{F}_0$, is of fully Newtonian form, and just serves to renormalize the gravitational constant, as discussed in Ref. 4. The second term was also derived in Ref. 4 [cf. Eq. (69) of Ref. 4] and is

$$\delta\mathbf{F}_1 = \frac{16\pi^3}{k^3}\frac{a_0^2}{F^{(N)2}}\mathbf{F}^{(N)} = \frac{8\pi a_0}{k}\frac{r_0}{r}\frac{\mathbf{N}(\psi)}{5+3\cos(2\psi)}. \tag{50}$$

What we have just shown is that to these two terms one should add the magnetic-like contribution

$$\delta\mathbf{F}_2 = \frac{4\pi a_0}{k}\frac{r_0}{r}\mathbf{B}(\psi), \tag{51}$$

which is of the same order of magnitude as $\delta\mathbf{F}_1$. Apart from the prefactor $4\pi a_0/k$, this term is just what was plotted in Fig. 3. In Fig. 4 we plot the angular profile $\mathbf{B}(\psi)+2[5+3\cos(2\psi)]^{-1}\mathbf{N}(\psi)$ of the total correction to the acceleration after renormalization of G. The plotted field is to be divided by r (and multiplied by $4\pi a_0 r_0/k$) to obtain the extra acceleration felt by test particles in the quasi-Newtonian region.

How do these results affect the naive expectations of Sec. 3? We have just shown that a full quantitative analysis can never neglect the "magnetic" field derived in this section. In addition the border between full and linear MONDian behavior is determined by the condition $U^2 = 1$, equivalent to the ellipsoid (34). As long as we stay well outside this ellipsoid we obtain results consistent with (16) and (17);

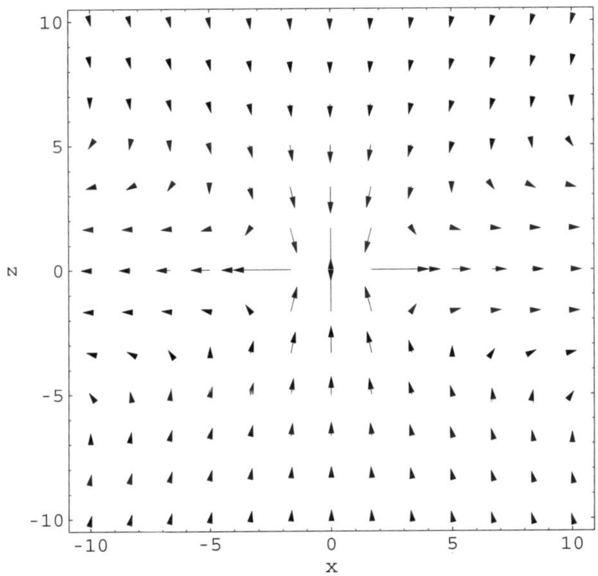

Fig. 3. The flow of $\mathbf{U}_2$ in a plane containing the z axis; coordinates are in units of r_0. For clarity the solution was cut off at $r = r_0$ (so as to avoid a divergence at the origin).

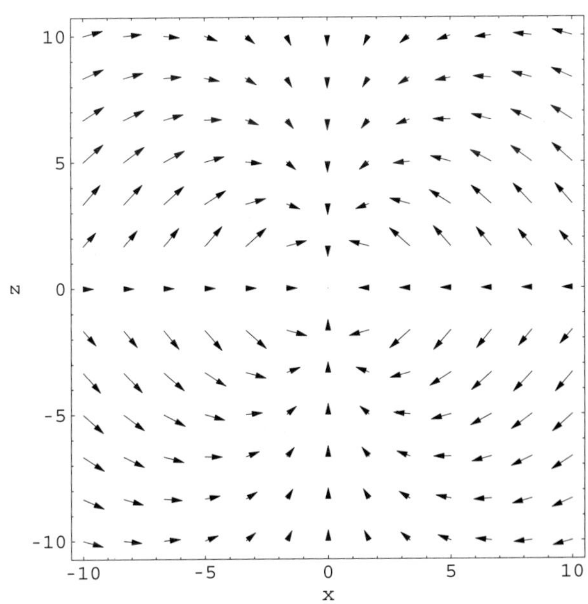

Fig. 4. The flow of $\mathbf{B}(\psi) + 2[5 + 3\cos(2\psi)]^{-1}\mathbf{N}(\psi)$. When divided by r and multiplied by $4\pi a_0 r_0/k$ this field gives the physical acceleration beyond the Newtonian one felt by test particles in the quasi-Newtonian region.

however, the order of magnitude of linear corrections outside this ellipsoid may be written as

$$\frac{\delta F}{F^{(N)}} \sim \left(\frac{4\pi}{k}\right)^3 \left(\frac{a_0}{A}\right)^2 \frac{1}{r^2} = \frac{k}{4\pi}\left(\frac{r_0}{r}\right)^2. \tag{52}$$

We learn that the highest fractional correction in this regime is achieved close to the ellipsoid (34) and is of order $k/4\pi$, around 0.0025 for $k \approx 0.03$; it then falls off as $1/r^2$ as we move away from the SP. Therefore, as long as we do not use (16) for fractional corrections larger than $k/4\pi$, we obtain qualitatively correct results (the example given in Sec. 3 satisfies this condition).

The bottom line for our predictions is that the ellipsoid (34) represents both the region where the largest linear corrections are felt and the border for the onset of full MOND behavior. For the three examples considered in Sec. 3 we have

$$r_0 \approx 383 \text{ km}, \qquad \text{Earth–Sun}, \tag{53}$$

$$r_0 \approx 9.65 \times 10^5 \text{ km}, \qquad \text{Jupiter–Sun}, \tag{54}$$

$$r_0 \approx 140 \text{ km}, \qquad \text{Earth–Moon}, \tag{55}$$

corresponding to ellipsoids with a major semiaxis of 766 km (Sun–Earth), 1.93×10^6 km (Sun–Jupiter) or 280 km (Earth–Moon). These are the relevant dimensions of the MOND bubbles.

6. The Deep MOND Region

By Eq. (24) the deep MOND regime ($\mu \ll 1$) entails $\kappa \approx 4$ or $U \ll 1$. Thus in Eq. (27) we replace $1 + U^2 \to 1$. Then, together with Eq. (26), this has a double symmetry, already noticed by Milgrom.[19] They are both invariant under $\mathbf{U} \to \text{const.} \times \mathbf{U}$ (rescaling), and under $\mathbf{x} \to \lambda\mathbf{x}$ (dilation of the coordinates). The first symmetry implies that the normalization of $\mathbf{U}$ is arbitrary [of course, the normalization is eventually fixed by taking cognizance of the sources of Eq. (26)]. The second means that a solution whose linear scale is expanded remains a solution.

In spherical polar coordinates Eqs. (26) and (27) take the forms

$$\frac{1}{r^2}\frac{\partial}{\partial r}(r^2 U_r) + \frac{1}{r\sin\psi}\frac{\partial}{\partial\psi}(\sin\psi\, U_\psi) = 0, \tag{56}$$

$$\left[\frac{4}{r}\left(\frac{\partial(rU_r)}{\partial r} - \frac{\partial U_\psi}{\partial\psi}\right) + \left(\frac{U_r}{r}\frac{\partial}{\partial\psi} - U_\psi\frac{\partial}{\partial r}\right)\right] U^2 = 0. \tag{57}$$

For a solution of these to turn into a second solution upon dilatation of the coordinates ($r \to \lambda r$), it is necessary for the r dependence of both U_r and U_ψ to be a single power. Thus we make the ansatz

$$\mathbf{U} = C\left(\frac{r}{r_0}\right)^{\alpha-2} (F(\psi)\,\mathbf{e}_r + G(\psi)\,\mathbf{e}_\psi), \tag{58}$$

with C and α dimensionless constants. The power $\alpha - 2$ is chosen for notational convenience in what follows. Substituting in Eq. (56) we obtain

$$G' + \text{ctan}(\psi)\, G + \alpha F = 0, \tag{59}$$

while substitution in Eq. (57) gives

$$F\frac{d(F^2+G^2)}{d\psi} + 2\big[\alpha G - 2F'\big](F^2+G^2) = 0. \tag{60}$$

These last constitute a coupled system of first order ordinary differential equations for $F(\psi)$ and $G(\psi)$.

These equations have several symmetries. F and G may be rescaled, i.e. multiplied by a constant (this is nothing but the scale invariance of the deep MOND regime). We also have the symmetry $\alpha \to -\alpha$, $F \to F$, $G \to -G$. Finally, the equations are parity-invariant: $\psi \to \pi - \psi$, $F \to \pm F$, $G \to \mp G$. Of course, this by itself does not compel the solutions themselves to have definite parity, i.e. $F(\psi) = \pm F(\pi - \psi)$ and $G(\psi) = \mp G(\pi - \psi)$. However, numerically we find that the only regular solutions are indeed those with definite parity, and that these only exist for a discrete sequence of αs: $\{\pm\alpha_1, \pm\alpha_2, \ldots\}$. Specifically, we find $\alpha_1 = 2$ and the approximate values $\alpha_2 \approx 3.528$, $\alpha_3 \approx 5.039$, $\alpha_4 \approx 6.545$, etc.

Seen in another way, the boundary conditions at $\psi = 0$ (see below) justify representing F as a Fourier series in $\cos(m\psi)$ and G as a Fourier series in $\sin(m\psi)$. It is only for the mentioned special αs that even and odd m modes decouple, so that we can have a solution that is a series in only odd or only even m. For other values of α the solutions mix even and odd m, but are singular at $\psi = \pi$.

As in Sec. 5, the boundary condition that the normal component of $\mathbf{U}$ vanish requires that we take $G(\psi = 0) = G(\psi = \pi) = 0$. Because C can still be adjusted, we lose no generality in requiring the corresponding boundary condition $F(\psi = 0) = F(\psi = \pi) = 1$. For were we to demand that $F(\psi = 0) \neq F(\psi = \pi)$, we would thereby introduce a jump in $\mathbf{U}$ across the plane $\psi = \pi/2$ for which there is no physical reason. Our choice of boundary conditions immediately selects a solution with definite parity, which, as mentioned earlier, are the only nonsingular ones. Regarding boundary conditions at large r, we know that there must be a match with the field in the quasi-Newtonian region. This naturally selects the particular solution with $n = 2$, since the quasi-Newtonian solution $\mathbf{U}_0$ has components with angular profiles of form $\cos 2\psi$ or $\sin 2\psi$. This logic still does not prefer positive to negative α. But to avoid a singularity at the origin [see Eq. (58)] we should select the solution with positive α, namely that for $\alpha \approx 3.528$.

The functions F and G obtained for this α are plotted in Fig. 5. These graphs are approximated at the level of 1% by the formulae

$$\begin{aligned} F(\psi) &= 0.2442 + 0.7246\cos(2\psi) + 0.0472\cos(4\psi), \\ G(\psi) &= -0.8334\sin(2\psi) - 0.0368\sin(4\psi). \end{aligned} \tag{61}$$

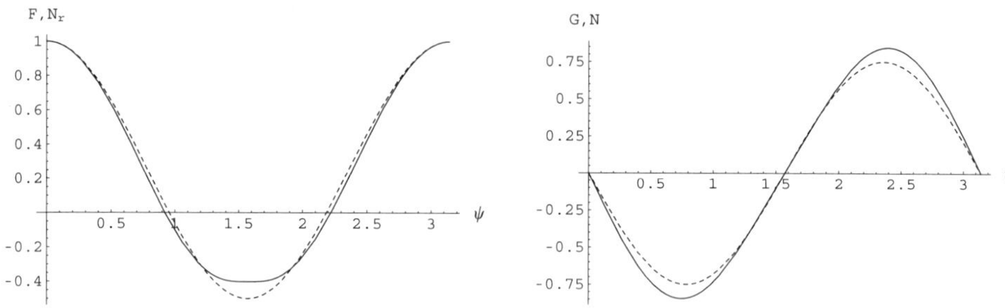

Fig. 5. The numerically determined angular profile functions F and G in the deep MOND region (solid) compared with the Newtonian profile functions N_r and N_ψ (dotted), respectively.

For comparison Fig. 5 plots also N_r and N_ψ of Eqs. (31)–(33). We see that the angular profile of the deep MOND $\mathbf{U}$ (whose flow is plotted in Fig. 6) is quite similar to that of the Newtonian $\mathbf{U}_0$ [Eq. (36) and Fig. 1]. Of course, the radial dependences of the two are quite different. Now, as mentioned earlier, in the absence of any mention of the sources in Eqs. (26) and (27), it is not possible to determine the normalization of $\mathbf{U}$. However, we may estimate C in Eq. (58) as follows. Given the similarity of the angular profiles, we may suppose that were we to extend the deep MOND $\mathbf{U}$ of Eq. (58) to the inner boundary of the Newtonian region at $r = r_0$, we should obtain $\mathbf{U}_0$. This requires that $C = 1$ and we adopt this value.

We conclude that taking the curl term into account in the deep MOND regime once again vindicates qualitatively the simplified arguments of Sec. 3, but introduces

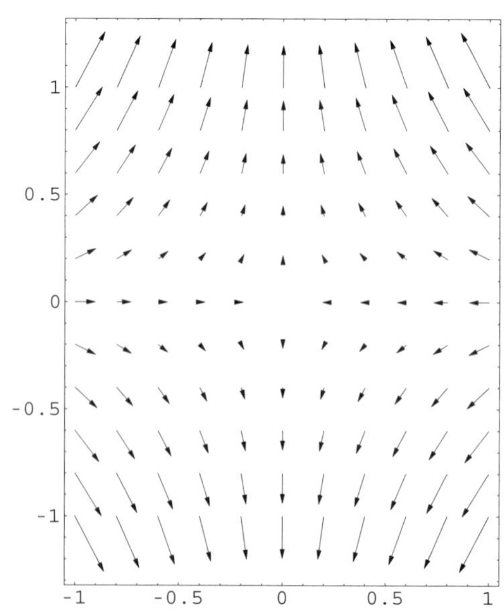

Fig. 6. The flow of the field $\mathbf{U}$ in the deep MOND regime plotted with a linear scale in units of r_0 and assuming that $C = 1$.

substantial quantitative novelties. Using (28) we find that the extra physical force is now

$$\delta \mathbf{F} \approx -\nabla\phi = \frac{4\pi a_0}{k}\frac{\mathbf{U}}{U^{1/2}}. \tag{62}$$

If we define $\mathbf{D}$ as the angular profile in the deep MOND regime (which, as we have seen, is very close to $\mathbf{N}$), then

$$\delta \mathbf{F} \approx \frac{4\pi a_0}{k}\left(\frac{r}{r_0}\right)^{\frac{\alpha-2}{2}}\frac{\mathbf{D}}{D^{1/2}}. \tag{63}$$

For $\alpha < 4$ (a condition satisfied by our solution), the tidal stresses associated with this field, i.e. its spatial derivatives, diverge at the saddle, as predicted in Sec. 3. However, the divergence is softer than in Eq. (14), where the curl term was ignored (that solution corresponds to $\alpha = 3$, which is an unphysical value, as we have seen).

Rearranging (63) with the aid of the definition of r_0 and Eq. (30), we find that the continuation of formula (52) in the deep MOND regime is

$$\frac{\delta F}{F^{(N)}} \sim \frac{k}{4\pi}\left(\frac{r_0}{r}\right)^{\frac{\alpha-4}{2}} \approx \frac{k}{4\pi}\left(\frac{r_0}{r}\right)^{-0.24}. \tag{64}$$

Hence the fractional correction to Newtonian gravity, which equals $k/4\pi \approx 0.0025$ at the ellipsoid (34), continues to grow in the strong MOND regime as we approach the saddle. (Were we to ignore the curl term, in which case $\delta F/F^{(N)} \sim 1/\sqrt{r}$, this growth would be steeper.) One implication of the growth is that the ϕ force overtakes $F^{(N)}$ in a much smaller inner region than naively expected [cf. formula (15)]. Specifically, $\delta F \approx F^{(N)}$ at

$$r \sim r_0\left(\frac{k}{4\pi}\right)^{\frac{2}{4-\alpha}} = \frac{a_0}{A}\left(\frac{k}{4\pi}\right)^{2\frac{\alpha-3}{4-\alpha}}. \tag{65}$$

This is smaller than (15) by a factor of 10^{-6}, and is essentially microscopic except for the Jupiter–Sun system. The value of $F^{(N)}$ when it becomes subdominant is not a_0 as naively expected; it is also smaller by a factor of 10^{-6}.

In summary, the full analysis reveals that there is a very large region [given by the ellipsoid (34)] inside which full MONDian effects are present. The fractional MONDian corrections to gravity in this region exceed $k/4\pi$ and are therefore significant. However, the MOND field only dominates the Newtonian field, i.e. the fractional correction becomes larger than unity, in a region far too small to be observable.

7. The Realistic Solar System

The results of Secs. 3, 5 and 6 can be used to show that the SP location for a pair of masses as determined by pure Newtonian gravity ($\mathbf{F}^{(N)} = -\nabla\Phi_N = 0$) coincides with that determined by full TeVeS ($-\nabla(\Phi_N + \phi) = 0$). In the calculations in

Sec. 6 the origin $r = 0$ is the point where $\nabla\phi = 0$, and the field configuration of $\nabla\phi$, or its surrogate $\mathbf{U}$, in the North and South hemispheres are reflections of each other (see Fig. 6). This configuration acts as a boundary condition for $\mathbf{U}$ in the quasi-Newtonian region (treated in Sec. 5). Accordingly, we expect not only the "magnetic" part $\mathbf{U}_2$, but also the $\mathbf{U}_0$, which serves as background for $\mathbf{U}_2$'s equation (38), to reflect the mentioned symmetry, and for the null points of both these fields to coincide with that of $\mathbf{U}$ of the deep MOND region. Now, as we move outward from the quasi-Newtonian region, $\mathbf{U}$ becomes dominated by $\mathbf{U}_0$, which is the pure Newtonian field. Hence the SP determined by that field (see Sec. 3) coincides with that determined by the full MOND field.

The results presented so far are "model calculations," valid under a number of simplifying assumptions not satisfied by the real SS. For example, orbits are elliptic, not circular; the barycenter of the system does not coincide with the center of M; we have a many-body problem, not a two-body problem; etc. To leading order these complications do not change the anomalous effects predicted by MOND around SP's or the size of the regions where they are felt. They do complicate the issue of locating "MOND bubbles," but since their centers coincide with the SP's of the Newtonian potential, this is in fact a Newtonian physics problem, independent of MOND dynamics.

For example, the SS barycenter is dominated by the Sun–Jupiter pair and lives just outside the solar surface, rotating with a period of approximately 11 years. But even this is a crude approximation: the relative position of the Sun and any planet depends on the configuration of the entire SS, and is chaotic. The same may be said for the location of the SP between the Sun and that planet. However, with empirical inputs and a numerical Newtonian code we can determine the location of the full set of SP's, and even predict where they will be within a few years.[20] Not only are these details in the realm of Newtonian physics, but they do not affect our conclusions on MOND effects around SP's, as long as we anchor our solutions to wherever Newtonian theory predicts the SP's to be. Indeed, we need only the result (12) as a boundary condition for our MONDian calculations.

8. Targets for the LISA Pathfinder

As stated in the introduction, the MOND effects near the Lagrange points are expected to be weak; however, this does not mean that they are beyond the reach of very sensitive equipment, such as that on board the LISA Pathfinder (LPF) mission.[14–16] Furthermore, while in transit to L1, the satellite may pass close enough to the SP to probe the quasi-Newtonian region examined in Sec. 5 (the extreme MONDian region described in Sec. 6 probably requires a dedicated mission). In the LPF mission two proof masses are suitably shielded from radiation pressure and other annoyances that prevent testing gravitational physics to a_0 accuracy in the inner parts of the SS. Naturally, the satellite itself has to bear radiation pressure, but its orbit is corrected by tracking the free-falling proof masses contained in its inside. The sensitivity to tidal stresses has been quoted as $10^{-15}\,\mathrm{s}^{-2}$ (see Refs. 15–17).

According to Eq. (11), tidal stresses at the Sun–Earth SP are of the order $A \approx 4.57 \times 10^{-11}\,\mathrm{s}^{-2}$, four orders of magnitude larger than LPF's sensitivity. The fractional corrections to Newtonian gravity contained in Eqs. (50) and (51), and plotted in Fig. 4, have a rough order of magnitude given by (52). The tidal stress corresponding to δF is thus of order $10^{-13}(r_0/r)^2\,\mathrm{s}^{-2}$ for the illustrative value $k = 0.03$ used in this paper. Therefore LPF would be sensitive to these MONDian corrections if it got to within $10r_0 \approx 3830\,\mathrm{km}$ of the saddle. This is not overly demanding; the region is the size of a planet. The MOND effects may be even apparent while LPF is in transit to L1.

In contrast, the MONDian tidal stresses felt near L1 are far too small to be within the reach of this mission. If r_L denotes L1's distance from the Sun, L1 lies at $R - r_L \approx 1.5 \times 10^6\,\mathrm{km}$ from the Earth; the saddle of the Sun–Earth potential is at $R - r_s \approx 2.6 \times 10^5\,\mathrm{km}$ from the Earth [see Eq. (9)]. Therefore L1 is $\Delta r = r_s - r_L \approx 1.24 \times 10^6\,\mathrm{km}$ away from the saddle, implying suppression of corrections to Newtonian gravity by a factor $\frac{k}{4\pi}\left(\frac{r_0}{\Delta r}\right)^2 \approx 2.4 \times 10^{-10}$. By way of contrast, the Newtonian tidal stresses at L1 are, say, for the radial component,

$$\frac{\partial F_r^{(N)}}{\partial r} \approx 8\omega_E^2 \approx 3.17 \times 10^{-13}\,\mathrm{s}^{-2}, \tag{66}$$

with ω_E being the angular frequency of the Earth's orbit. This is only two orders of magnitude above experimental sensitivity, and so the MONDian corrections to stresses in the vicinity of L1 are eight orders of magnitude too small for the quoted instrumental sensitivity.

However, "indirect" effects may possibly be detectable by LPF: effects not on its accelerometers but on its path (this comment may apply to other L1 missions). Indeed, MOND introduces a small shift to the location of L1 and its surrounding orbits. Combining Eqs. (50), (51) and (45), we obtain an extra acceleration at L1 with radial component of signed magnitude

$$\delta F = \frac{4\pi}{k}\frac{r_0}{\Delta r}\left[\frac{1}{2} - \frac{\pi}{3\sqrt{3}}\right] a_0 \approx -1.3 \times 10^{-12}\,\mathrm{m\,s^{-2}}. \tag{67}$$

Hence this extra acceleration predicted by MOND points toward the SP, i.e. away from the Sun and toward the Earth. In the usual calculation, the centrifugal acceleration at L1 is exactly balanced by the gravitational one $F^{(N)}$. This last has absolute magnitude $\omega_E^2\, r_L$ and points away from the Sun; thus $F^{(N)}$ has to point toward it. This is why L1 is closer to the Sun than the SP of the potential. With the extra force (67) to balance, L1 is further shifted toward the Sun. In view of Eq. (66), the predicted shift is approximately 4 m.

There is a similar order-of-magnitude effect on the *orbits* about L1, and while this is not the primary purpose of the LPF mission, we suggest that a careful monitoring of the spacecraft trajectory may be of interest to gravitational physics.

9. Conclusions

In this paper we have examined what might constitute "direct" detection of MOND behavior. We predicted the existence of regions displaying full MOND behavior well inside the solar system, specifically in bubbles surrounding the saddle points of the gravitational potential. If abnormally high tidal stresses are observed in these regions, this would prove MOND beyond reasonable doubt.

How general are our predictions? MOND's solid requirement is that $\tilde{\mu}(x)$ approach 1 as $x \gg 1$ and x as $x \ll 1$; the interpolating regime between these two asymptotic requirements is far less constrained. In the present work this intermediate regime translates into the quasi-Newtonian calculations presented in Sec. 5. For these we chose a reasonable form for $\mu(x)$, Eq. (22), but we should stress that the details are model-dependent. For instance, in (16) the leading correction could have been quartic in a_0/F instead of quadratic, resulting in a different power in the denominator of (38). The extra force $\delta\mathbf{F}$ would then fall off more steeply with r.

Accordingly, our calculations in the quasi-Newtonian domain are simply illustrative. We defer to a future publication a thorough study of the effect of the choice of μ (as dictated by theoretical requirements and extant observations) on planetary orbits,[28] Lagrange points, and the Pioneer anomaly.[10,11] By contrast our predictions for the interior of the ellipsoid (34), as presented in Sec. 6, are robust predictions of the MOND scenario, and of wider validity.

We thus face a dilemma. The strongest MOND effect and the theoretically more robust prediction is that made in Sec. 6 for the interior of the ellipsoid (34). However, locating it in space may be taxing, particularly since this bubble is non-inertial. In contrast, the quasi-Newtonian predictions — for example, what LISA Pathfinder might find in the vicinity of L1 — are geographically less demanding, but the predicted effects are weaker and theoretically less discriminative. Thus, observing what we predicted in Sec. 5 would support the specific model (22) there; however, failure to observe it would hardly disprove MOND in general. The interior of the ellipsoid (34) is therefore the prime experimental target for a conclusive test. But one should not despair: systems other than those examined here may naturally reveal the inner core derived in Sec. 6. For example, the movement of the saddle point through a diffuse medium — say, the rings of Saturn — could be observable.

There are other regions in the solar system where gradients of the Newtonian potential will be low, for example at the center of near-spherical objects. However, these are obviously inaccessible. By focusing on the saddle points of the gravitational potential in the solar system, we believe we have exposed the best candidates for a direct detection of strong MONDian behavior in our own backyard.

References

1. M. Milgrom, *Astrophys. J.* **270** (1983) 365, 370, 384.
2. R. H. Sanders and S. S. McGaugh, *Ann. Rev. Astron. Astrophys.* **40** (2002) 263.
3. J. D. Bekenstein and M. Milgrom, *Astrophys. J.* **286** (1984) 7.

4. J. Bekenstein, *Phys. Rev. D* **70** (2004) 083509 [Erratum, *ibid. D* **71** (2005) 069901].
5. R. H. Sanders, Astro-ph/0502222.
6. V. Sahni and Y. Shtanov, gr-qc/0606063; see also Ana Nobili's paper presented at this workshop.
7. R. H. Sanders, astro-ph/0601431.
8. See reference in Ref. [29] for a complete bibliography.
9. J. D. Bekenstein, in *Second Canadian Conference on General Relativity and Relativistic Astrophysics*, eds. A. Coley, C. Dyer and T. Tupper (World Scientific, Singapore, 1988), p. 487.
10. J. Anderson *et al.*, *Phys. Rev. Lett.* **81** (1998) 2858.
11. J. Anderson *et al.*, *Phys. Rev. D* **65** (2002) 082004.
12. A. Hellemans, A Force to Reckon with, *Scientific American*, Oct. 2005, p. 12.
13. J. R. Brownstein and J. W. Moffat, gr-qc/0511026.
14. S. Anza *et al.*, *Class. Quant. Grav.* **22** (2005) S125.
15. S. Anza *et al.*, gr-qc/0504062.
16. A. Lobo *et al.*, gr-qc/0601096.
17. M. Milgrom, *Astrophys. J.* **302** (1986) 617.
18. H. S. Zhao and B. Famaey, astro-ph/0512425.
19. M. Milgrom, *Phys. Rev. E* **56** (1997) 1148.
20. See e.g. http://www.boulder.swri.edu/hal/swift.html
21. J. Peebles, *The Large Scale Structure of the Universe* (Princeton University Press, 1980).
22. C. Skordis, astro-ph/0511591.
23. N. Zakamska and S. Tremaine, astro-ph/0506548.
24. L. Baudis, astro-ph/0511805 and astro-ph/0503549.
25. H. Zhao *et al.*, astro-ph/0509590.
26. C. Skordis *et al.*, *Phys. Rev. Lett.* **96** (2006) 011301.
27. A. Slosar, A. Melchiorri and J. Silk, *Phys. Rev. D* **72** (2005) 101301.
28. C. Talmadge *et al.*, *Phys. Rev. Lett.* **61** (1988) 1159.
29. J. Bekenstein and J. Magueijo, *Phys. Rev. D* **73** (2006) 103513.

CONSTRAINING TeVeS GRAVITY AS EFFECTIVE DARK MATTER AND DARK ENERGY

HONGSHENG ZHAO*

University of St Andrews, School of Physics and Astronomy,
KY16 9SS, Fife, UK†
hz4@st-and.ac.uk

The phenomena customarily described with the standard ΛCDM model are broadly reproduced by an extremely simple model in TeVeS, Bekenstein's[1] modification of general relativity motivated by galaxy phenomenology. Our model can account for the acceleration of the Universe seen at SNeIa distances without a cosmological constant, and the accelerations seen in rotation curves of nearby spiral galaxies and gravitational lensing of high-redshift elliptical galaxies without cold dark matter. The model is consistent with BBN and the neutrino mass between 0.05 eV to 2 eV. The TeVeS scalar field is shown to play the effective dual roles of dark matter and dark energy, with the amplitudes of the effects controlled by a μ function of the scalar field, called the μ essence here. We also discuss outliers to the theory's predictions on multiimaged galaxy lenses and outliers on the subgalaxy scale.

Keywords: Dark matter; cosmology; gravitation.

1. Introduction

As at the start of the last century, a rethinking of fundamental physics has been forced upon us by a set of experimental surprises, the only difference this time being that the whole Universe is the laboratory. Einstein's general relativity together with the ordinary matter described by the standard model of particle physics is well tested in the solar system, but fails miserably in accounting for astronomical observations from just the edge of the solar system to a cosmological distance; for example fast-rotating galaxies like ours would have been escaped the shallow gravitational potentials of their luminous constituents (stars and gas). Standard physics also cannot fully explain the cosmological observations of the cosmic acceleration seen in supernova type Ia data and the angular scales seen in the anisotropy spectrum of cosmic-microwave-background radiation (CMBR). The remedy is usually

*PPARC Advanced Fellowship.
†Member of SUPA.

to introduce two exotic components for dominating the matter-energy budget of the Universe with a split of about 25% : 74% into the Universe energy budget: dark matter (DM) as a collisionless and pressureless fluid described by perhaps SUSY physics, and dark energy (DE) as a negative pressure and nearly homogeneous field described by unknown physics.

1.1. *Challenges to dark matter and dark energy*

In spite of the success of this concordance model, the nature of DM and DE is one of the greatest mysteries of modern cosmology. For example, it has long been noted that on galaxy scales DM and baryonic matter (stars plus gas) have a remarkable correlation, and respect a mysterious acceleration scale, a_0.[2,3] The Newtonian gravity of the baryons $\mathbf{g}_b$ and the dark matter gravity $\mathbf{g}_{\rm DM}$ are correlated through an empirical relation[4,5] such that the light-to-dark ratio

$$\frac{g_b}{g_{\rm DM}} = \frac{g_{\rm DM} + \alpha g_b}{a_0}, \qquad a_0 = 1\,\text{Å}\,\sec^{-2}, \tag{1}$$

where a_0 is a dividing gravity scale, and $0 \le \alpha \le 1$ is a parameter, experimentally determined to fit rotation curves.[a] Such a tight correlation is difficult to understand in a galaxy formation theory where DM and baryon interactions enjoy huge degrees of freedom. Equally peculiar is the amplitude of DE density Λ, which is of order 10^{120} times smaller than its natural scale. It is hard to explain from fundamental physics why DE starts to dominate the Universe density only at the present epoch, hence marking the present as the turning point for the Universe from deacceleration to acceleration. This is related to the fact that $\Lambda \sim a_0^2$, where a_0 is a characteristic scale of DM. Somehow DE and DM are tuned to shift dominance when the DM energy density falls below $\frac{a_0^2}{8\pi G}$. These empirical facts should not be completely treated as random coincidences of the fundamental parameters of the Universe.

Given that the dark sector and its properties are only inferred indirectly from the gravitational acceleration of ordinary matter, one wonders if the dark sector is not just a sign of our lack of understanding of gravitational physics. Here we propose to investigate whether the roles of DM and DE could be replaced by the scalar field in a metric theory called TeVeS.

2. TeVeS Framework

TeVeS is a covariant theory proposed by Bekenstein which in the weak field limit reduces to the phenomenogically successful but noncovariant MOND theory of Ref. 6. The covariant nature of TeVeS makes it ready to be analyzed in a general setting.

[a]Note that α mimics the role of the mass to light, and hence inherits some of its uncertainty.

Just like Einstein's theory, Bekenstein's theory is a metric theory. In fact, it has two metrics. The first metric, $g_{\mu\nu}$ is minimally coupled to all the matter fields in the Universe. We shall call the frame of this metric the "matter frame" (MF). All geodesics are calculated in terms of this MF metric. For example, in a quasi-static system like a galaxy with a weak gravitational field, we can define a physical coordinate system (t, x, y, z) such that

$$d\tau^2 = g_{\mu\nu}dx^\mu dx^\nu = (1+2\Phi)dt^2 - (1-2\Phi)\left(dx^2+dy^2+dz^2\right). \tag{2}$$

Here the potential $\Phi = \Phi_b + \phi$, where the field ϕ replaces the usual role of the potential of the DM.

Another metric of TeVeS, $\tilde{g}_{\mu\nu}$, has its dynamics governed by the Einstein–Hilbert action $S_g = \frac{1}{16\pi G}\int d^4x\sqrt{-\tilde{g}}\tilde{R}$, where $\tilde{R}$ is the scalar curvature of $\tilde{g}_{\mu\nu}$. We shall call the frame of this metric the "Einstein frame" (EF). It is related to the MF metric through $g_{\mu\nu} = e^{-2\phi}(\tilde{g}_{\mu\nu} + A_\mu A_\nu) - e^{2\phi}A_\mu A_\nu$ (the notation of tildes here is opposite to that of Bekenstein), which involves the unit timelike vector field A_μ [this can often be expressed as $(\sqrt{-g_{00}}, 0, 0, 0)$ for galaxies or FRW cosmology] and a scalar field ϕ, which is governed by the action $S = \int d^4x\sqrt{-\tilde{g}}\mathcal{L}$, where, according to Refs. 1 and 7, the Lagrangian density

$$\mathcal{L} = -\Lambda + \frac{1}{16\pi G}\left[\mu_{Sk}\frac{dV}{d\mu_{Sk}} - V(\mu_{Sk})\right], \tag{3}$$

where Λ is a constant of integration equivalent to the cosmological constant, and V is a free function of μ_{Sk}, which is an implicit function of the scalar field ϕ through

$$\left(\tilde{g}^{\mu\nu} - A^\mu Ae^\nu\right)\phi_{,\mu}\phi_{,\nu} \equiv \yen = -\frac{dV}{d\mu_{Sk}}. \tag{4}$$

By selecting an expression and parameters for the scalar field Lagrangian density $\mathcal{L}$ or potential V, one picks out a given TeVeS theory.[b] Note that in all the above $G \equiv (1 - K/2)G_\oplus$ is a to-be-determined bare gravitational constant related to the usual experimentally determined value, $G_\oplus \approx 6.67\times10^{-11}$, through the coupling constant K of TeVeS (Skordis, private communication).

3. Connecting Galaxies with Cosmology

Bekenstein's original proposal was to construct the Lagrangian density with $\mathcal{L}$ as a one-to-one function of μ_{Sk}.[7] Such a one-to-one construction has the drawback that the Lagrangian necessarily has unphysical "gaps" such that a sector is reserved for spacelike systems (for example, from dwarf galaxies to the solar system in $0 < \mu_{Sk} < \mu_0$) and a disconnected sector is reserved for timelike systems (for example, an expanding Universe in $\mu_{Sk} > 2\mu_0$). While viable mathematically, such a disconnected Universe would not permit galaxies to collapse out of the Hubble

[b]This notation is directly related to Bekenstein's by $\mu_{Sk} = 8\pi\mu/k$, $\yen = \frac{y}{kl^2}$.

expansion. The particular function that Bekenstein used also result in an interpolation function, to be computed from a nontrivial implicit function of the scalar field strength, which is found to overpredict observed rotation curve amplitudes when the gravity is of order a_0.[8]

In an effort to reconnect galaxies with the expanding Universe, Ref. 4 proposed constructing the Lagrangian as a one-to-one function of the scalar field ϕ through ¥, where ¥ > 0 in galaxies and ¥ < 0 for cosmic expansion. This way allows a smooth transition from the edge of galaxies where ¥ ~ 0 to the Hubble expansion. Zhao and Famaey also suggested extrapolating the Lagrangian for galaxies to predict cosmologies so as to minimize any fine-tuning in TeVeS. The counterpart of the Zhao–Famaey model in the DM language would be Eq. (1), which they used to fit to rotation curves, and found that both $\alpha = 1$ and $\alpha = 0$ give reasonable fits, with some preference for the former.

Our aim here is to check whether the suggestions of Ref. 4 lead to reasonable galaxy rotation curves and cosmologies. To minimize fine-tuning, we consider an extremely simple Lagrangian density governing the scalar field

$$\mathcal{L}(¥) = \int_0^{¥} \frac{d¥_1}{8\pi G_\oplus} \frac{\sqrt{|¥|_1}}{a_0 \exp(-\phi_0)}, \qquad \Lambda = 0. \tag{5}$$

In quasi-static systems $\sqrt{|¥|} = |\nabla\phi| \exp(-\phi)$, where the constant ϕ_0 is the present day cosmological value of the scalar field ϕ. With this, the Poisson equation reduces to $-\nabla\left(\frac{|\nabla \exp(\phi_0 - \phi)|}{a_0}\nabla\phi\right) = 4\pi G_\oplus \rho = -\nabla \cdot \mathbf{g}_b$. So in spherical approximations we have

$$\frac{g_b}{|\nabla\phi|} = \frac{|\nabla\phi|}{a(\phi)}, \qquad a(\phi) \equiv a_0 \exp(\phi - \phi_0). \tag{6}$$

Clearly, the above Lagrangian or TeVeS Poisson equation for the scalar field essentially recovers Eq. (1) in the $\alpha = 0$ case in the DM language if we identify that the scalar field $\nabla\phi \to g_{\rm DM}$, hence playing the role of gravity of the DM $g_{\rm DM}$ at the present day when $\phi = \phi_0$. Interestingly the characteristic acceleration scale $a(\phi)$ varies with the redshift together with the scalar field $\phi(t)$.

A summary of how well TeVeS/MOND or CDM fits data on all scales is given in Table 1. To illustrate, two sample fits to rotation curves of a dwarf galaxy and a high-surface-brightness spiral galaxy are shown in Fig. 1, including the possible effects of imbedding the galaxies in a large neutrino core. We also repeat the

Table 1. Comparison of the pros (+) and cons (−) of LCDM vs TeVeS on various scales.

Data			References
Rotation curves HSB/LSB	−	++	4
Lensing by ellipticals	++	±	9
Dynamics of X-ray clusters	++	±	11, 10
Hubble expansion and CMB	++	±	7, 12

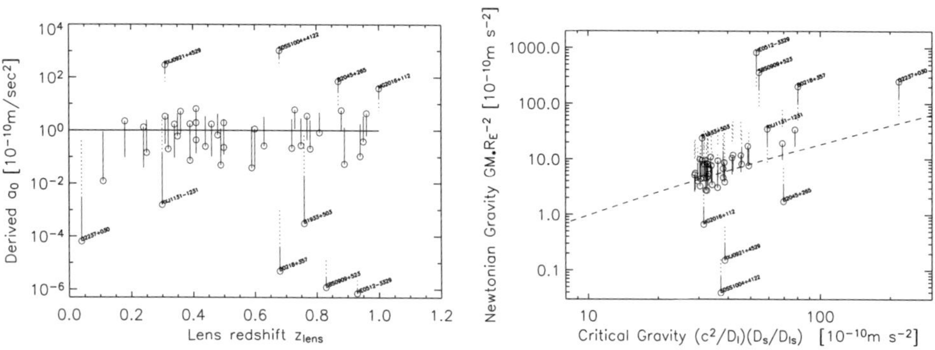

Fig. 1. Shown are the values for the TeVeS a_0 parameter derived to fit individual strong lensing Einstein radii of 50 CASTLES multiimaged systems, assuming that $M_*/L_* = 4$ (circles); also shown are the effects of raising/lowering M_*/L_* by a factor of 2 (solid vertical lines) or a factor of 4 (dotted vertical lines). A few outliers are labeled. The right panel shows that the Newtonian acceleration GM_*/R_E^2 versus the critical gravity [related to the critical surface density $(c^2/4\pi G D_l)(D_s/D_{ls})$ in GR] is the mininal local gravity for a lens to form Einstein rings. The dashed line is a prediction for point lenses in the TeVeS $\alpha = 0$ model.

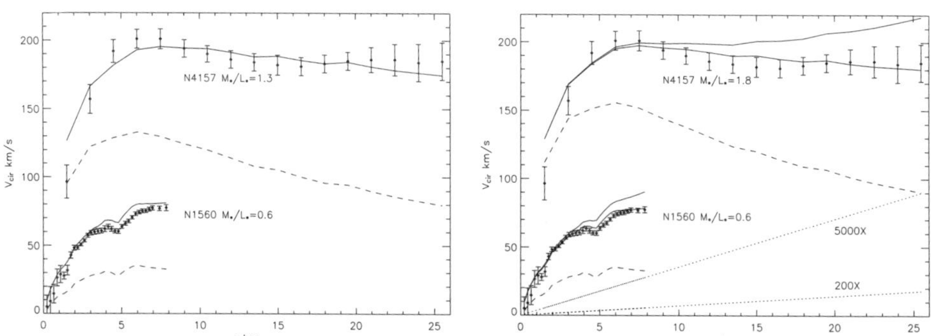

Fig. 2. The left panel shows TeVeS fits to rotation curves of a gas-rich dwarf galaxy NGC1560 and a gas-poor, larger spiral galaxy NGC4157 (solid curves), adopting $a_0 = 1.2 \times 10^{-8}$, $\alpha = 0$ μ function model without neutrinos; the Newtonian rotation curves by baryons for the assumed stellar $(M/L)_*$ are shown as well (dashed lines). The right panel is similar to the left except for assuming the $\alpha = 1$ μ function and assuming that galaxies are imbedded in a neutrino overdensity of $200 \times \frac{3H_0^2}{8\pi G} \sim 2.7 \times 10^4 M_\odot\,\mathrm{kpc}^{-3}$ or $5000 \times \frac{3H_0^2}{8\pi G} \sim 6.7 \times 10^5 M_\odot\,\mathrm{kpc}^{-3}$ (the two values bracket the typical gas density of X-ray clusters on average and in the centers). The Newtonian rotation curves of the constant neutrino cores are also shown (dotted lines).

excercise of Ref. 9, and fit the lens Einstein radii with Hernquist models in an $\alpha = 0$ modified gravity. We show in Fig. 2 that the CASTLES gravitational lenses (mostly high redshift ellipticals) are mostly consistent with TeVeS-predicted Einstein ring size within plausible uncertainties of the mass-to-light ratios. Note that the critical gravity $(c^2/D_l)(D_s/D_{ls})$ is always much stronger than 10^{-10} m/s^2 at Einstein radii of elliptical galaxy lenses, so the Einstein rings are insensitive to

MONDian effects, and hence insensitive to a_0. Some of the outliers are known to be in galaxy clusters (RXJ0921 and SDS1004), where a neutrino density core of a few times $10^{-6}M_\odot\,\mathrm{kpc}^{-3}$ might help to reduce the discrepancy. The time delay of PG1115+080 is also consistent with TeVeS prediction, with $H_0 = 70$.[13] Given that the $\alpha = 0$ model is reasonably consistent with spiral galaxy rotation curve data and Einstein rings of high redshift ellipticals, we next wish to study cosmology in this TeVeS model. The important thing to note here is that the cosmological constant Λ is set to zero, and so the zero point of the Lagarangian coincides with where the scalar field is zero.

4. Hubble Expansion and Late Time Acceleration

TeVeS is a metric theory; the uniform expanding background can be described by the FRW metric. Assuming a flat cosmology with a physical time t and a scale factor $a(t)$, we have

$$ds^2 = -dt^2 + a^2(t)\left[d\chi^2 + \chi^2\left(d\theta^2 + \sin^2\theta d\phi^2\right)\right]. \tag{7}$$

The Hubble expansion can be modeled with

$$\rho_\phi + \rho_b + \rho_r = \frac{3H^2}{8\pi G_\oplus \Gamma}, \tag{8}$$

where the first term is the scalar field effective energy density $\rho_\phi = ¥\frac{d\mathcal{L}}{d¥} - \mathcal{L} = \frac{8\sqrt{2}}{3}\exp(5\phi)(\frac{d\phi}{dt})^3(8\pi G_\oplus a_0 \exp(-\phi_0))^{-1}$ in the MF. The correction factor

$$\Gamma \equiv \frac{\exp(-4\phi)}{\left(1 + \dfrac{d\phi}{d\ln a}\right)^2} \approx \exp(4\phi_{\mathrm{BBN}} - 4\phi), \tag{9}$$

such that the expansion rate is very close to that LCDM at the epoch of BBN where the radiation density ρ_r dominates, i.e. no corrections at BBN. Note that TeVeS will mimic DM and DM if we identify

$$\rho_\phi\Gamma \to \rho_\Lambda, \qquad (\rho_b + \rho_r)(\Gamma - 1) \to \rho_{\mathrm{DM}}, \tag{10}$$

where in the LCDM framework the Hubble expansion is normally modeled with

$$\rho_\Lambda + \rho_{\mathrm{DM}} + \rho_b + \rho_r = \frac{3H^2}{8\pi G_\oplus}, \qquad H = \frac{da}{adt}. \tag{11}$$

Using the mirror-imaged $\alpha = 0$ Lagrangian of the scalar field, we derive the following second order ODE for the scalar field ϕ:

$$\frac{d}{dt}\left[\mu_s\left(\frac{d\phi}{dt}\right)\right] = (\rho_b + \rho_r)a^3, \quad dt = \frac{d\ln a}{H}, \quad \mu_s \equiv \frac{\exp(5\phi + \phi_0)a^3}{\sqrt{2}\pi G_\oplus a_0}\left(\frac{d\phi}{dt}\right). \tag{12}$$

Note the similarity of this 1D equation to the 3D Poisson equation.

We can integrate the above equation to solve for ϕ as a function of $\ln a$ or the physical synchronous time t. We note that at the current epoch $\phi \sim \phi_0$, $\mu_s \sim \frac{s}{a_0} \sim \frac{\rho_b}{\rho_\phi} \sim (\frac{s^2}{8\pi G_\oplus \rho_b})^{-1/2}$, where $s \sim \sqrt{g_{tt}\phi_{,t}\phi_{,t}}$.

We then aim to test if the cosmology specified by the above non-fine-tuned Lagrangian in TeVeS could match the behavior of an LCDM Universe. We solve the equations numerically by iteration of the bare constant G. Assuming a value for G, the initial ϕ and $\frac{d\phi}{d\ln a}$ are then set by the fact that $G_{\rm eff} \approx G_\oplus = 6.67 \times 10^{-11}$ at BBN in order to be consistent with the number of relativistic degrees of freedom at a temperature of 1 MeV. The parameters A and K are determined by the boundary condition at the present day such that we recover the normalization in the MONDian Poisson equation in galaxies and in the solar system. Typically the scalar field tracks the matter density, and $\mathcal{L}$ and ϕ are slow-varying functions of the redshift. We then iterate the parameter G such that the sound horizon angular size at LSS ($z = 1000$–1100) matches that of LCDM. The parameters typically converge in 20–30 iterations. The Hubble constant and cosmic acceleration come out without any tuning.

In Fig. 3(a) we show a model with a present day matter density $w_b = 0.024$. This is consistent with the baryon density at BBN, and hence there is no nonbaryonic matter in the present model. This model invokes *neither the cosmological constant nor DM*. The resulting model has $H_0 = 77$ km/s/Mpc. The expansion history is almost the same as for LCDM a slight difference exists in the energy density (hence the expansion rate) in the future $a > 1$.

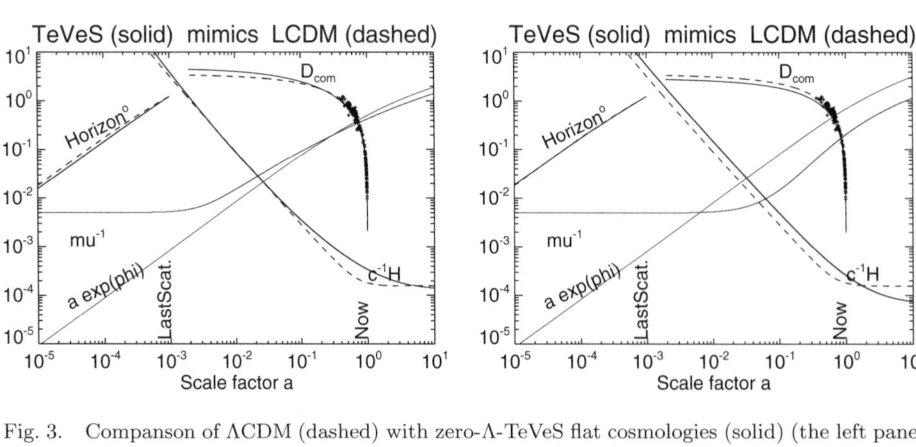

Fig. 3. Companson of ΛCDM (dashed) with zero-Λ-TeVeS flat cosmologies (solid) (the left panel assumes a zero mass for neutrinos and a μ essence with $\alpha = 0$; the right panel assumes 2 eV neutrinos and an $\alpha = 1$ model). Shown is the comoving distance $D_{\rm com}$ versus the physical scale factor a in a log–log diagram overplotted with SNIa data (small symbols) up to redshift 2. Likewise, the horizon and the Hubble parameter H in units of $\text{Mpc}^{-1}c$ in two theories are shown. The evolution of the scalar field ϕ and μ can be inferred from (thin solid lines) $a\exp(\phi)$ and μ^{-1}, with the cutoff of $\mu^{-1} = 0.005$ being adopted for numerical reasons.

To understand whether the above explanation for late acceleration and DM is unique we have also run models with a more general Lagrangian:

$$\mathcal{L}(¥) = \int_0^{¥} \frac{d¥}{8\pi G_\oplus} \frac{s}{1 - \alpha s}, \quad s \equiv \frac{\sqrt{|¥|}}{a_0 \exp(-\phi_0)}, \quad \Lambda = 0. \tag{13}$$

This is so constructed that we recover Eq. (1) for any value of α by identifying DM gravity with the scalar field $\nabla\phi$. This whole sequence of models is largely consistent with DM phenomenology on galaxy scales, with a slight preference for $\alpha = 1$ models in galaxies. Models with nonzero α also have interesting effects on the solar system. For example, a model with $\alpha = 0.2$ would predict [see Eq. (1)] a constant, non-Keperlian acceleration of $a_P = a_0\alpha^{-1} \sim 6 \times 10^{-10}\,\mathrm{ms}^{-2}$ in the solar system, consistent with the Pioneer anomaly (although a nongravitational origin is hard to exclude). Such a constant gravity would cause a gravitational redshift of 10^{-13} $(D/100\,\mathrm{AU})$ between the solar system bodies of separation D, which could be testable with experiments with accurate clocks in the future (see these proceedings). Calculating the Hubble expansion for models with increasing α, we are able to match LCDM in all cases in terms of BBN, LSS, SNeIa distances, and late acceleration. For all these models we have also varied initial conditions and found that the solutions are stable. The acceleration continues into the far future with $b = a\exp(\phi) \to cst$. The amplitude of the modification function μ also decreases with expansion. Compared to the model with $\alpha = 0$, however, larger α drives up the present day Hubble constant, unless the present day matter density w_b is also increased. This is effectively achieved by allowing for relatively massive neutrinos. For example, for $\alpha = 0.2$ this would require the matter density parameter w_b to be twice the nominal value 0.024, implying the need to include massive neutrinos of 0.8 eV. For $\alpha = 1$ this would require 2 eV neutrinos, as needed for explaining galaxy cluster data[10,11] and the CMB.[7] The latter model is shown in Fig. 3(b).

5. Conclusion

In summary, we have focused on one very specific model in the Bekenstein theory. We have shown that it may be possible to satisfy some of the most stringent cosmological observations without the need to introduce/fine-tune dark matter or dark energy. The TeVeS scalar field μ function (called the μ essence here) can be fixed by galaxy rotation curves, and it predicts the right amount of cosmic acceleration, the size of the horizon at $z = 1000$, and the present Hubble constant without fine-tuning. The ultimate test of the model should come from simulating the evolution of linear perturbations on this background and the CMB. By fitting galaxy cluster data and the third peak of the CMB we could break the degeneracy of models of different α, and constrain the neutrino mass.

Acknowledgments

I acknowledge the numerous discussions with Constantinous Skordis, David Mota and Benoit Famaey.

References

1. J. Bekenstein, *Phys. Rev. D* **70** (2004) 3509.
2. M. Milgrom, *Astrophys. J.* **270** (1983) 365.
3. S. S. McGaugh, *Phys. Rev. Lett.* **95** (2005) 171302.
4. H. S. Zhao and B. Famaey, *Astrophys. J.* **638** (2006) L9.
5. G. W. Angus, B. Famaey and H. S. Zhao, *Mon. Not. R. Astron. Soc.* **371** (2006) 138.
6. J. Bekenstein and M. Milgrom, *Astrophys. J.* **286** (1984) 7.
7. C. Skordis *et al.*, *Phys. Rev. Lett.* **96** (2006) 1301.
8. B. Famaey and J. Binney, *Mon. Not. R. Astron. Soc.* **363** (2005) 603.
9. H. S. Zhao *et al.*, *Mon. Not. R. Astron. Soc.* **368** (2006) 171.
10. E. Pointecouteau and J. Silk, *Mon. Not. R. Astron. Soc.* **364** (2005) 654.
11. R. H. Sanders, *Mon. Not. R. Astron. Soc.* **342** (2003) 901.
12. L. M. Diaz-Rivera, L. Samushia and B. Ratra, *Phys. Rev. D* **73** (2006) 3503.
13. H. S. Zhao and B. Qin, *Chin. J. Astron. Astrophys.* **6** (2006) 141.

COSMIC ACCELERATION AND MODIFIED GRAVITY

MARK TRODDEN

Department of Physics, Syracuse University,
Syracuse, NY 13244-1130, USA
trodden@physics.syr.edu

I briefly discuss some attempts to construct a consistent modification to general relativity (GR) that might explain the observed late-time acceleration of the Universe and provide an alternative to dark energy. I describe the issues facing extensions to GR, illustrate these with a specific example, and discuss the resulting observational and theoretical obstacles.

Keywords: Cosmology; gravity.

1. Introduction

Approaches to the late-time acceleration of the Universe may be divided into three broad classes. First, it is possible that there is some as-yet-undiscovered property of our existing model of gravity and matter that leads to acceleration at the current epoch. One might include in this category the existence of a tiny cosmological constant and the possibility that the backreaction of cosmological perturbations might cause self-acceleration.

Second is the idea that there exists a new dynamical component to the cosmic energy budget. This possibility, with the new source of energy density modeled by a scalar field, is usually referred to as *dark energy*.

Finally, it may be that curvatures and length scales in the observable Universe are only now reaching values at which an infrared modification of gravity can make itself apparent by driving self-acceleration.[1–19] It is this possibility that I will briefly describe in this article.

While I will mention a number of different approaches to modified gravity, I will concentrate on laying out the central challenges to constructing a successful modified gravity model, and on illustrating them with a particular simple example. Detailed descriptions of some of the other possible ways to approach this problem can be found in the excellent contributions of Sean Carroll, Cedric Deffayet, Gia Dvali, and John Moffat.

2. The Challenge

Although, within the context of general relativity (GR), one does not think about it too often, the metric tensor contains, in principle, more degrees of freedom than the usual spin-2 *graviton* (see Sean Carroll's talk in these proceedings for a detailed discussion on this).

The reason why one does not hear of these degrees of freedom in GR is that the Einstein–Hilbert action is a very special choice, resulting in second-order equations of motion, which constrain away the scalars and the vectors, so that they are nonpropagating. However, this is not the case if one departs from the Einstein–Hilbert form for the action. When using any modified action (and the usual variational principle) one inevitably frees up some of the additional degrees of freedom. In fact, this can be a good thing, in that the dynamics of these new degrees of freedom may be precisely what one needs to drive the accelerated expansion of the Universe. However, there is often a price to pay.

The problems may be of several different kinds. First, there is the possibility that along with the desired deviations from GR on cosmological scales, one may also find similar deviations on solar system scales, at which GR is rather well tested. Second is the possibility that the newly activated degrees of freedom may be badly behaved in one way or another — either having the wrong sign kinetic terms (ghosts), and hence being unstable, or leading to superluminal propagation, which may lead to other problems.

These constraints are surprisingly restrictive when one tries to create viable modified gravity models yielding cosmic acceleration. In the next few sections I will describe several ways in which one might modify the action, and in each case provide an explicit, clean, and simple example of how cosmic acceleration emerges. However, I will also point out how the constraints I have mentioned rule out these simple examples, and mention how one must complicate the models to recover viable models.

3. A Simple Model: $f(R)$ Gravity

The simplest way one could think of to modify GR is to replace the Einstein–Hilbert Lagrangian density by a general function $f(R)$ of the Ricci scalar R.[6–25]

$$S = \frac{M_{\rm P}^2}{2}\int d^4x\sqrt{-g}\,[R + f(R)] + \int d^4x\sqrt{-g}\,\mathcal{L}_m[\chi_i, g_{\mu\nu}], \tag{1}$$

where $M_{\rm P} \equiv (8\pi G)^{-1/2}$ is the (reduced) Planck mass and $\mathcal{L}_m$ is the Lagrangian density for the matter fields χ_i.

Here, I have written the matter Lagrangian as $\mathcal{L}_m[\chi_i, g_{\mu\nu}]$ to make explicit that in this frame — the *Jordan* frame — matter falls along geodesics of the metric $g_{\mu\nu}$.

The equation of motion obtained by varying the action (1) is

$$(1 + f_R)\,R_{\mu\nu} - \frac{1}{2}g_{\mu\nu}\,(R + f) + (g_{\mu\nu}\Box - \nabla_\mu\nabla_\nu)\,f_R = \frac{T_{\mu\nu}}{M_{\rm P}^2}, \tag{2}$$

where I have defined $f_R \equiv \partial f/\partial R$.

Further, if the matter content is described as a perfect fluid, with energy–momentum tensor,

$$T^m_{\mu\nu} = (\rho_m + p_m)U_\mu U_\nu + p_m g_{\mu\nu}, \tag{3}$$

where U^μ is the fluid rest-frame four-velocity, ρ_m is the energy density and p_m is the pressure, then the fluid equation of motion is the usual continuity equation.

When considering the background cosmological evolution of such models, I will take the metric to be of the flat Robertson–Walker form, $ds^2 = -dt^2 + a^2(t)d\mathbf{x}^2$. In this case, the usual Friedmann equation of GR is modified to become

$$3H^2 - 3f_R(\dot{H} + H^2) + \frac{1}{2}f + 18f_{RR}H(\ddot{H} + 4H\dot{H}) = \frac{\rho_m}{M_{\rm P}^2} \tag{4}$$

and the continuity equation is

$$\dot{\rho}_m + 3H(\rho_m + p_m) = 0. \tag{5}$$

When supplied with an equation-of-state parameter w, the above equations are sufficient to solve for the background cosmological behavior of the space–time and its matter contents. For appropriate choices of the function $f(R)$ it is possible to obtain late-time cosmic acceleration without the need for dark energy, although evading bounds from precision solar system tests of gravity turns out to be a much trickier matter, as we shall see.

While one can go ahead and analyze this theory in the Jordan frame, it is more convenient to perform a carefully chosen conformal transformation on the metric, in order to render the gravitational action in the usual Einstein–Hilbert form of GR.

Following the description in Ref. 26, consider the conformal transformation

$$\tilde{g}_{\mu\nu} = \Omega(x^\alpha)g_{\mu\nu}, \tag{6}$$

and construct the function $r(\Omega)$, which satisfies

$$1 + f_R[r(\Omega)] = \Omega. \tag{7}$$

Defining a rescaled scalar field by $\Omega \equiv e^{\beta\phi}$, with $\beta M_{\rm P} \equiv \sqrt{2/3}$, the resulting action becomes

$$\begin{aligned}\tilde{S} = \frac{M_{\rm P}}{2}\int d^4x\sqrt{-\tilde{g}}\,\tilde{R} + \int d^4x\sqrt{-\tilde{g}}\left[-\frac{1}{2}\tilde{g}^{\mu\nu}(\partial_\mu\phi)\partial_\nu\phi - V(\phi)\right] \\ + \int d^4x\sqrt{-\tilde{g}}\,e^{-2\beta\phi}\mathcal{L}_m[\chi_i, e^{-\beta\phi}\tilde{g}_{\mu\nu}],\end{aligned} \tag{8}$$

where the potential $V(\phi)$ is determined entirely by the original form (1) of the action and is given by

$$V(\phi) = \frac{e^{-2\beta\phi}}{2}\{e^{\beta\phi}r[\Omega(\phi)] - f(r[\Omega(\phi)])\}. \tag{9}$$

The equations of motion in the Einstein frame are much more familiar than those in the Jordan frame, although there are some crucial subtleties. In particular, note that in general, test particles of the matter content χ_i do not freely fall along geodesics of the metric $\tilde{g}_{\mu\nu}$.

The equations of motion in this frame are those obtained by varying the action with respect to the metric $\tilde{g}_{\mu\nu}$,

$$\tilde{G}_{\mu\nu} = \frac{1}{M_{\rm P}^2}(\tilde{T}_{\mu\nu} + T_{\mu\nu}^{(\phi)}), \tag{10}$$

with respect to the scalar field ϕ,

$$\tilde{\Box}\phi = -\frac{dV}{d\phi}(\phi), \tag{11}$$

and with respect to the matter fields χ_i, described as a perfect fluid.

Once again, I will specialize to consider background cosmological evolution in this frame. The Einstein frame line element can be written in the familiar FRW form as

$$ds^2 = -d\tilde{t}^2 + \tilde{a}^2(\tilde{t})d\mathbf{x}^2, \tag{12}$$

where $d\tilde{t} \equiv \sqrt{\Omega}\,dt$ and $\tilde{a}(t) \equiv \sqrt{\Omega}\,a(t)$. The Einstein frame matter energy–momentum tensor is then given by

$$\tilde{T}^m_{\mu\nu} = (\tilde{\rho}_m + \tilde{p}_m)\tilde{U}_\mu\tilde{U}_\nu + \tilde{p}_m\tilde{g}_{\mu\nu}, \tag{13}$$

where $\tilde{U}_\mu \equiv \sqrt{\Omega}\,U_\mu$, $\tilde{\rho}_m \equiv \rho_m/\Omega^2$ and $\tilde{p}_m \equiv p_m/\Omega^2$.

3.1. *A simple example*

For definiteness and simplicity focus on the simplest correction to the Einstein–Hilbert action; $f(R) = -\mu^4/R$, where μ is a new parameter with units of [mass].

The field equation for the metric is then

$$\left(1+\frac{\mu^4}{R^2}\right)R_{\mu\nu} - \frac{1}{2}\left(1-\frac{\mu^4}{R^2}\right)Rg_{\mu\nu} + \mu^4[g_{\mu\nu}\nabla_\alpha\nabla^\alpha - \nabla_{(\mu}\nabla_{\nu)}]R^{-2} = \frac{T^m_{\mu\nu}}{M_{\rm P}^2}. \tag{14}$$

The constant curvature vacuum solutions, for which $\nabla_\mu R = 0$, satisfy $R = \pm\sqrt{3}\mu^2$. Thus, there exists a constant curvature vacuum solution which is de Sitter space. We will see that the de Sitter solution is, in fact, unstable, albeit with a very long decay time, $\tau \sim \mu^{-1}$.

The time–time component of the field equations for this metric is

$$3H^2 - \frac{\mu^4}{12(\dot{H}+2H^2)^3}(2H\ddot{H} + 15H^2\dot{H} + 2\dot{H}^2 + 6H^4) = \frac{\rho_M}{M_{\rm P}^2}. \tag{15}$$

As I have discussed, one may now transform to the *Einstein frame*, where the gravitational Lagrangian takes the Einstein–Hilbert form and the additional degree of freedom appears as a fictitious scalar field ϕ, with potential

$$V(\phi) = \mu^2 M_{\rm P}^2 e^{-2\beta\phi}\sqrt{e^{\beta\phi}-1}, \tag{16}$$

shown in Fig. 1.

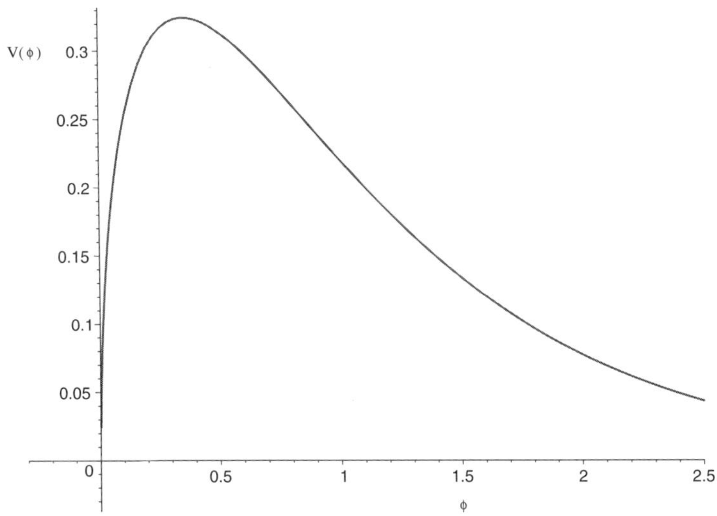

Fig. 1. The Einstein frame potential $V(\phi)$.

Denoting with a tilde all quantities except ϕ in the Einstein frame, the relevant Einstein frame cosmological equations of motion are

$$3\tilde{H}^2 = \frac{1}{M_{\rm P}^2}[\rho_\phi + \tilde{\rho}], \tag{17}$$

$$\phi'' + 3\tilde{H}\phi' + \frac{dV}{d\phi}(\phi) - \frac{1-3w}{\sqrt{6}M_{\rm P}}\tilde{\rho}_M = 0, \tag{18}$$

where a prime denotes $d/d\tilde{t}$, and where

$$\tilde{\rho}_M = \frac{C}{\tilde{a}^{3(1+w)}} \exp\left(-\frac{1-3w}{\sqrt{6}}\frac{\phi}{M_{\rm P}}\right), \tag{19}$$

with C a constant, and

$$\rho_\phi = \frac{1}{2}\phi'^2 + V(\phi). \tag{20}$$

Finally, note that the matter frame Hubble parameter H is related to that in the Einstein frame $\tilde{H} \equiv \tilde{a}'/\tilde{a}$ by

$$H = \sqrt{p}\left(\tilde{H} - \frac{\phi'}{M_{\rm P}\sqrt{6}}\right). \tag{21}$$

How about cosmological solutions in the Einstein frame? Ordinarily, Einstein gravity with a scalar field with a minimum at $V = 0$ would yield a Minkowski vacuum state. However, here this is no longer true. Even though $V \to 0$ as $\phi \to 0$, this corresponds to a curvature singularity and so is not a Minkowski vacuum. The other minimum of the potential, at $\phi \to \infty$, does not represent a solution.

Focusing on vacuum solutions, i.e. $P_M = \rho_M = 0$, the beginning of the Universe corresponds to $R \to \infty$ and $\phi \to 0$. The initial conditions we must specify are the initial values of ϕ and ϕ', denoted as ϕ_i and ϕ'_i. There are then three qualitatively distinct outcomes, depending on the value of ϕ'_i:

(1) *Eternal de Sitter.* There is a critical value of $\phi'_i \equiv \phi'_C$ for which ϕ just reaches the maximum of the potential $V(\phi)$ and comes to rest. In this case the Universe asymptotically evolves to a de Sitter solution (ignoring spatial perturbations). As we have discovered before (and as is obvious in the Einstein frame), this solution requires tuning and is unstable.
(2) *Power law acceleration.* For $\phi'_i > \phi'_C$, the field overshoots the maximum of $V(\phi)$. Soon thereafter, the potential is well approximated by $V(\phi) \simeq \mu^2 M_P^2 \exp(-\sqrt{3/2}\phi/M_P)$, and the solution corresponds to $a(t) \propto t^2$ in the matter frame. Thus, the Universe evolves to late-time power law inflation, with observational consequences similar to dark energy with equation-of-state parameter $w_{DE} = -2/3$.
(3) *Future singularity.* For $\phi'_i < \phi'_C$, ϕ does not reach the maximum of its potential and rolls back down to $\phi = 0$. This yields a future curvature singularity.

What about including matter? As can be seen from (18), the major difference here is that the equation of motion for ϕ in the Einstein frame has a new term. Furthermore, since the matter density is much greater than $V \sim \mu^2 M_P^2$ for $t \ll 14\,\text{Gyr}$, this term is very large and greatly affects the evolution of ϕ. The exception is when the matter content is radiation alone ($w = 1/3$), in which case it decouples from the ϕ equation due to conformal invariance.

Despite this complication, it is possible to show that the three possible cosmic futures identified in the vacuum case remain in the presence of matter.

Thus far, the dimensionful parameter μ is unspecified. By choosing $\mu \sim 10^{-33}\,\text{eV}$, the corrections to the standard cosmology become important only at the present epoch, explaining the observed acceleration of the Universe without recourse to dark energy.

Clearly, the choice of correction to the gravitational action can be generalized. Terms of the form $-\mu^{2(n+1)}/R^n$, with $n > 1$, lead to similar late-time self-acceleration, which can easily accommodate current observational bounds on the equation-of-state parameter.

Now, as I mentioned in the introduction, any modification of the Einstein–Hilbert action must, of course, be consistent with the classic solar system tests of gravity theory, as well as numerous other astrophysical dynamical tests. We have chosen the coupling constant μ to be very small, but we have also introduced a new light degree of freedom. As shown by Chiba,[27] the simple model above is equivalent to a Brans–Dicke theory with $\omega = 0$ in the approximation where the potential was neglected, and would therefore be inconsistent with experiment[28] (but see Refs. 29–31 for suggestions that the conformally transformed theory may not be the correct way to analyze deviations from GR).

To construct a realistic $f(R)$ model requires a more complicated function, with more than one adjustable parameter in order to fit the cosmological data and satisfy solar system bounds. Examples can be found in Refs. 13 and 32.

4. Extensions: Higher-Order Curvature Invariants

It is natural to consider generalizing the action of Ref. 6 to include other curvature invariants. There are, of course, any number of terms that one could consider, but for simplicity, focus on those invariants of lowest mass dimension that are also parity-conserving:

$$\begin{aligned} P &\equiv R_{\mu\nu}\, R^{\mu\nu}, \\ Q &\equiv R_{\alpha\beta\gamma\delta}\, R^{\alpha\beta\gamma\delta}. \end{aligned} \tag{22}$$

We consider actions of the form

$$S = \int d^4x\sqrt{-g}\,[R + f(R,P,Q)] + \int d^4x\,\sqrt{-g}\,\mathcal{L}_M, \tag{23}$$

where $f(R,P,Q)$ is a general function describing deviations from GR.

It is convenient to define

$$f_R \equiv \frac{\partial f}{\partial R}, \quad f_P \equiv \frac{\partial f}{\partial P}, \quad f_Q \equiv \frac{\partial f}{\partial Q}, \tag{24}$$

in terms of which the equations of motion are

$$\begin{aligned} R_{\mu\nu} &- \frac{1}{2}g_{\mu\nu}R - \frac{1}{2}g_{\mu\nu}f + f_R R_{\mu\nu} + 2f_P R^{\alpha}{}_{\mu}R_{\alpha\nu} + 2f_Q R_{\alpha\beta\gamma\mu}R^{\alpha\beta\gamma}{}_{\nu} \\ &+ g_{\mu\nu}\,\Box f_R - \nabla_\mu\nabla_\nu f_R - 2\nabla_\alpha\nabla_\beta[f_P R^{\alpha}{}_{(\mu}\delta^{\beta}{}_{\nu)}] + \Box(f_P R_{\mu\nu}) \\ &+ g_{\mu\nu}\,\nabla_\alpha\nabla_\beta(f_P R^{\alpha\beta}) - 4\nabla_\alpha\nabla_\beta[f_Q R^{\alpha}{}_{(\mu\nu)}{}^{\beta}] = 8\pi G\,T_{\mu\nu}. \end{aligned} \tag{25}$$

It is straightforward to show that actions of the form (23) generically admit a maximally symmetric solution: $R =$ a nonzero constant. However, an equally generic feature of such models is that this de Sitter solution is unstable. In the CDTT model the instability is to an accelerating power law attractor. This is a possibility that we will also see in many of the more general models under consideration here.

Since we are interested in adding terms to the action that explicitly forbid flat space as a solution, I will, in the same way as in Ref. 6, consider inverse powers of the above invariants and, for simplicity, specialize to a class of actions with

$$f(R,P,Q) = -\frac{\mu^{4n+2}}{(aR^2 + bP + cQ)^n}, \tag{26}$$

where n is a positive integer (taken to be unity), μ has dimensions of mass, and a, b, and c are dimensionless constants. In fact, for general n the qualitative features of the system are as for $n = 1$.[14]

4.1. *Another simple example*

For the purposes of this short aside, I will focus on a specific example — actions containing modifications involving only $P \equiv R_{\mu\nu} R^{\mu\nu}$, with the prototype being $f(P) = -m^6/P$, where m is a parameter with dimensions of mass.

It is easy to see that there is a constant curvature vacuum solution to this action given by $R^{(P)}_{\text{const}} = (16)^{1/3} m^2$. However, we would like to investigate other cosmological solutions and analyze their stability.

From (25), with the flat cosmological ansatz, the analog of the Friedmann equation becomes

$$3H^2 - \frac{m^6}{8(3H^4 + 3H^2\dot{H} + \dot{H}^2)^3}[\dot{H}^4 + 11H^2\dot{H}^3 + 2H\dot{H}^2\ddot{H} + 33H^4\dot{H}^2 + 30H^6\dot{H} + 6H^3\dot{H}\ddot{H} + 6H^8 + 4H^5\ddot{H}] = 0. \quad (27)$$

Asymptotic analysis of this equation (substituting in a power law ansatz and taking the late-time limit) yields two late-time attractors with powers $v_0 = 2 - \sqrt{6}/2 \simeq 0.77$ and $v_0 = 2 + \sqrt{6}/2 \simeq 3.22$.

However, in order to obtain a late-time accelerating solution ($p > 1$), it is necessary to give accelerating initial conditions ($\ddot{a} > 0$), otherwise the system is in the basin of attraction of the nonaccelerating attractor at $p \simeq 0.77$. (This type of behavior is generic in some other modified gravity theories.[33])

While I have given a simple example here, cosmologically viable models are described in Ref. 34.

What about the other constraints on these models? It has been shown[35] that solar system constraints, of the type I have described for $f(R)$ models, can be evaded by these more general models whenever the constant c is nonzero. Roughly speaking, this is because the Schwarzschild solution, which governs the solar system, has vanishing R and P, but nonvanishing Q.

More serious is the issue of ghosts and superluminal propagation. It has been shown[36,37] that a necessary but not sufficient condition that the action be ghost-free is that $b = -4c$, so that there are no fourth derivatives in the linearized field equations. What remained was the possibility that the second derivatives might have the wrong signs, and also might allow superluminal propagation at some time in a particular cosmological background. It has recently been shown that in an FRW background with matter, the theories are ghost-free, but contain superluminally propagating scalar or tensor modes over a wide range of parameter space.[38,39] It is certainly necessary to be ghost-free. Whether the presence of superluminally propagating modes is a fatal blow to the theories remains to be seen.

5. Conclusions

Given the immense challenge posed by the accelerating Universe, it is important to explore every option in order to explain the underlying physics. Modifying gravity may be one of the more radical proposals, but it is not one without precedent as an explanation for unusual physics. However, it is an approach that is tightly constrained both by observation and theoretical obstacles.

In the brief time and space allowed, I have tried to give a flavor of some attempts to modify GR to account for cosmic acceleration without dark energy. I have focused on two of the directions in which I have been involved and have chosen to present simple examples of the models, which clearly demonstrate not only the cosmological effects, but also how constraints from solar system tests and theoretical consistency apply.

There are a number of other proposals for modified gravity and, while I have had neither time nor space to devote to them here, others have discussed some of them in detail at this meeting.

There is much work ahead, with significant current effort, my own included, devoted to how one might distinguish between modified gravity, dark energy and a cosmological constant as competing explanations for cosmic acceleration.

Acknowledgments

I would like to thank the organizers of the Q2C conference, and in particular Slava Turyshev, for their hard work and dedication in running such a stimulating meeting. I would also like to thank my many coauthors on the work discussed here for such productive and enjoyable collaborations, and for allowing me to reproduce parts of our work in this article. This work was supported in part by the NSF under grant PHY-0354990, by the Research Corporation, and by funds provided by Syracuse University.

References

1. G. R. Dvali, G. Gabadadze and M. Porrati, *Phys. Lett. B* **485** (2000) 208 [hep-th/0005016].
2. C. Deffayet, *Phys. Lett. B* **502** (2001) 199 [hep-th/0010186].
3. C. Deffayet, G. R. Dvali and G. Gabadadze, *Phys. Rev. D* **65** (2002) 044023 [astro-ph/0105068].
4. K. Freese and M. Lewis, *Phys. Lett. B* **540** (2002) 1 [astro-ph/0201229].
5. G. Dvali and M. S. Turner, astro-ph/0301510.
6. S. M. Carroll *et al.*, *Phys. Rev. D* **70** (2004) 043528 [astro-ph/0306438].
7. S. Capozziello, S. Carloni and A. Troisi, astro-ph/0303041.
8. D. N. Vollick, *Phys. Rev. D* **68** (2003) 063510 [astro-ph/0306630].
9. E. E. Flanagan, *Phys. Rev. Lett.* **92** (2004) 071101 [astro-ph/0308111].
10. E. E. Flanagan, *Class. Quant. Grav.* **21** (2003) 417 [gr-qc/0309015].
11. D. N. Vollick, *Class. Quant. Grav.* **21** (2004) 3813 [gr-qc/0312041].
12. M. E. Soussa and R. P. Woodard, *Gen. Relativ. Gravit.* **36** (2004) 855 [astro-ph/0308114].
13. S. Nojiri and S. D. Odintsov, *Gen. Relativ. Gravit.* **36** (2004) 1765 [hep-th/0308176].
14. S. M. Carroll *et al.*, *Phys. Rev. D* **71** (2005) 063513 [astro-ph/0410031].
15. N. Arkani-Hamed *et al.*, *J. High Energy Phys.* **0405** (2004) 074 [hep-th/0312099].
16. G. Gabadadze and M. Shifman, *Phys. Rev. D* **69** (2004) 124032 [hep-th/0312289].
17. J. W. Moffat, astro-ph/0403266.

18. T. Clifton, D. F. Mota and J. D. Barrow, *Mon. Not. R. Astron. Soc.* **358** (2005) 601 [gr-qc/0406001].
19. S. M. Carroll *et al.*, astro-ph/0607458.
20. J. D. Barrow and A. C. Ottewill, *J. Phys. A* **16** (1983) 2757.
21. J. D. Barrow and S. Cotsakis, *Phys. Lett. B* **214** (1988) 515.
22. J. D. Barrow and S. Cotsakis, *Phys. Lett. B* **258** (1991) 299.
23. G. Magnano and L. M. Sokolowski, *Phys. Rev. D* **50** (1994) 5039 [gr-qc/9312008].
24. A. Dobado and A. L. Maroto, *Phys. Rev. D* **52** (1995) 1895.
25. H. J. Schmidt, *Astron. Nachr.* **311** (1990) 165 [gr-qc/0109004].
26. G. Magnano and L. M. Sokolowski, *Phys. Rev. D* **50** (1994) 5039 [gr-qc/9312008].
27. T. Chiba, *Phys. Lett. B* **575** (2003) 1 [astro-ph/0307338].
28. B. Bertotti, L. Iess and P. Tortora, *Nature* **425** (2003) 374.
29. V. Faraoni, gr-qc/0607016.
30. S. Capozziello and A. Troisi, *Phys. Rev. D* **72** (2005) 044022 [astro-ph/0507545].
31. S. Capozziello, A. Stabile and A. Troisi, gr-qc/0603071.
32. P. Zhang, *Phys. Rev. D* **73** (2006) 123504 [astro-ph/0511218].
33. D. A. Easson *et al.*, *Phys. Rev. D* **72** (2005) 043504 [astro-ph/0506392].
34. O. Mena, J. Santiago and J. Weller, *Phys. Rev. Lett.* **96** (2006) 041103 [astro-ph/0510453].
35. I. Navarro and K. Van Acoleyen, *Phys. Lett. B* **622** (2005) 1 [gr-qc/0506096].
36. T. Chiba, *J. Cosmol. Astropart. Phys.* **0503** (2005) 008 [gr-qc/0502070].
37. I. Navarro and K. Van Acoleyen, *J. Cosmol. Astropart. Phys.* **0603** (2006) 008 [gr-qc/0511045].
38. A. De Felice, M. Hindmarsh and M. Trodden, astro-ph/0604154.
39. G. Calcagni, B. de Carlos and A. De Felice, hep-th/0604201.

A MODIFIED GRAVITY AND ITS CONSEQUENCES FOR THE SOLAR SYSTEM, ASTROPHYSICS AND COSMOLOGY

J. W. MOFFAT

Perimeter Institute for Theoretical Physics,
31 Caroline St. North, Waterloo, Ontario, N2L 2Y5, Canada

Department of Physics, University of Waterloo,
Waterloo, Ontario, N2L 3G1, Canada
john.moffat@utoronto.ca

A relativistic modified gravity (MOG) theory leads to a self-consistent, stable gravity theory that can describe the solar system, galaxy and clusters-of-galaxies data, and cosmology.

Keywords: Gravitation; astrophysics; cosmology.

1. Introduction

A relativistic modified gravity (MOG) called scalar-tensor-vector gravity (STVG) describes a self-consistent, stable gravity theory that contains Einstein's general relativity in a well-defined limit.[1] The theory has an extra degree of freedom, a vector field called a "phion" field whose curl is a skew-symmetric field that couples to matter ("fifth force"). The space–time geometry is described by a symmetric Einstein metric. An alternative relativistic gravity theory called metric-skew-tensor gravity (MSTG) has also been formulated[2] in which the space–time is described by a symmetric metric, and the extra degree of freedom is a skew-symmetric second rank tensor field. These two theories yield the same weak field consequences for physical systems.

The classical STVG theory allows the gravitational coupling "constant" G and the coupling of the phion field and its effective mass to vary with space and time as scalar fields.

A MOG should explain the following physical phenomena:

(1) Galaxy rotation curve data;
(2) Mass profiles of X-ray clusters;
(3) Gravitational lensing data for galaxies and clusters of galaxies;

(4) The cosmic microwave background (CMB), including the acoustical oscillation power spectrum data;
(5) The formation of protogalaxies in the early Universe and the growth of galaxies;
(6) N-body simulations of galaxy surveys;
(7) The accelerating expansion of the Universe.

We seek a unified description of solar system, astrophysical and large-scale cosmological data without exotic nonbaryonic dark matter. Dark matter in the form of particles has until now not been discovered in spite of large-scale experimental efforts.[3] The accelerating expansion of the Universe should be explained by the MOG theory without postulating a cosmological constant.

2. Action and Field Equations

Our MOG action takes the form[1]

$$S = S_{\mathrm{Grav}} + S_\phi + S_S + S_M, \tag{1}$$

where

$$S_{\mathrm{Grav}} = \frac{1}{16\pi}\int d^4x\sqrt{-g}\left[\frac{1}{G}(R+2\Lambda)\right], \tag{2}$$

$$S_\phi = -\int d^4x\sqrt{-g}\left[\omega\left(\frac{1}{4}B^{\mu\nu}B_{\mu\nu} + V(\phi)\right)\right], \tag{3}$$

and

$$S_S = \int d^4x\sqrt{-g}(\mathcal{F}_1 + \mathcal{F}_2 + \mathcal{F}_3), \tag{4}$$

where

$$\mathcal{F}_1 = \frac{1}{G^3}\left(\frac{1}{2}g^{\mu\nu}\nabla_\mu G\nabla_\nu G - V(G)\right), \tag{5}$$

$$\mathcal{F}_2 = \frac{1}{G}\left(\frac{1}{2}g^{\mu\nu}\nabla_\mu\omega\nabla_\nu\omega - V(\omega)\right), \tag{6}$$

$$\mathcal{F}_3 = \frac{1}{\mu^2 G}\left(\frac{1}{2}g^{\mu\nu}\nabla_\mu\mu\nabla_\nu\mu - V(\mu)\right). \tag{7}$$

We have chosen units with $c = 1$, and ∇_μ denotes the covariant derivative with respect to the metric $g_{\mu\nu}$. We adopt the metric signature $\eta_{\mu\nu} = \mathrm{diag}(1, -1, -1, -1)$, where $\eta_{\mu\nu}$ is the Minkowski space–time metric, and $R = g^{\mu\nu}R_{\mu\nu}$, where $R_{\mu\nu}$ is the symmetric Ricci tensor. Moreover, $V(\phi)$ denotes a potential for the vector field ϕ^μ, while $V(G), V(\omega)$ and $V(\mu)$ denote the three potentials associated with the three scalar fields $G(x), \omega(x)$ and $\mu(x)$, respectively. The field $\omega(x)$ is dimensionless and Λ denotes the cosmological constant. Moreover,

$$B_{\mu\nu} = \partial_\mu\phi_\nu - \partial_\nu\phi_\mu. \tag{8}$$

The field equations and the test particle equations of motion are derived in Ref. 1.

The action for the field $B_{\mu\nu}$ is of the Maxwell–Proca form for a massive vector field ϕ_μ. It can be proved that this MOG possesses a stable vacuum and the Hamiltonian is bounded from below. Even though the action is not gauge-invariant, it can be shown that the longitudinal mode ϕ_0 [where $\phi_\mu = (\phi_0, \phi_i)$ $(i = 1, 2, 3)$] does not propagate and the theory is free of ghosts. Similar arguments apply to the MSTG theory.[2,a]

3. Modified Newtonian Acceleration Law and Galaxy Dynamics

The modified acceleration law can be written as[1]

$$a(r) = -\frac{G(r)M}{r^2}, \tag{9}$$

where

$$G(r) = G_N\left[1 + \sqrt{\frac{M_0}{M}}\left(1 - \exp\left(\frac{-r}{r_0}\right)\left(1 + \frac{r}{r_0}\right)\right)\right] \tag{10}$$

is an *effective* expression for the variation of G with respect to r, and G_N denotes Newton's gravitational constant. A good fit to a large number of galaxies has been achieved with the parameters[5]

$$M_0 = 9.60 \times 10^{11}\, M_\odot, \quad r_0 = 13.92 \text{ kpc} = 4.30 \times 10^{22} \text{ cm}. \tag{11}$$

In the fitting of the galaxy rotation curves for both LSB and HSB galaxies, using photometric data to determine the mass distribution $\mathcal{M}(r)$,[5] only the mass-to-light ratio $\langle M/L \rangle$ is employed, once the values of M_0 and r_0 are fixed universally for all LSB and HSB galaxies. Dwarf galaxies are also fitted with the parameters[5]

$$M_0 = 2.40 \times 10^{11}\, M_\odot, \quad r_0 = 6.96 \text{ kpc} = 2.15 \times 10^{22} \text{ cm}. \tag{12}$$

By choosing universal values for the parameters $G_\infty = G_N(1 + \sqrt{M_0/M})$, $(M_0)_{\text{clust}}$ and $(r_0)_{\text{clust}}$, we are able to obtain satisfactory fits to a large sample of X-ray cluster data.[6]

4. Solar System and Binary Pulsar

Let us assume that we are in a distance scale regime for which the fields G, ω and μ take their approximate renormalized constant values:

$$G \sim G_0(1 + Z), \quad \omega \sim \omega_0 A, \quad \mu \sim \mu_0 B, \tag{13}$$

where G_0, ω_0 and μ_0 denote the "bare" values of G, ω and μ, respectively, and Z, A and B are the associated renormalization constants. We obtain from the equations of motion of a test particle the orbital equation (we reinsert the speed of light, c)[1]

$$\frac{d^2u}{d\phi^2} + u = \frac{GM}{c^2J^2} - \frac{K}{c^2J^2}\exp\left(\frac{-r}{r_0}\right)\left[1 + \left(\frac{r}{r_0}\right)\right] + \frac{3GM}{c^2}u^2, \tag{14}$$

[a]For a detailed discussion on possible instabilities and pathological behavior of vector-gravity theories, see Ref. 4.

where $u = 1/r$, $K = G_N\sqrt{MM_0}$ and J denotes the orbital angular momentum. Using the large r weak field approximation, we obtain the orbit equation for $r \ll r_0$:

$$\frac{d^2u}{d\phi^2} + u = N + 3\frac{GM}{c^2}u^2, \tag{15}$$

where J_N denotes the Newtonian value of J and

$$N = \frac{GM}{c^2 J_N^2} - \frac{K}{c^2 J_N^2}. \tag{16}$$

We can solve Eq. (15) by perturbation theory and find for the perihelion advance of a planetary orbit

$$\Delta\omega = \frac{6\pi}{c^2 L}(GM_\odot - K_\odot), \tag{17}$$

where $J_N = (GM_\odot L/c^2)^{1/2}$, $L = a(1-e^2)$, and a and e denote the semimajor axis and the eccentricity of the planetary orbit, respectively.

For the solar system $r \ll r_0$ and from the running of the effective gravitational coupling constant, $G = G(r)$, we have $G \sim G_N$ within the experimental errors for the measurement of Newton's constant, G_N. We choose for the solar system

$$\frac{K_\odot}{c^2} \ll 1.5\,\text{km} \tag{18}$$

and use $G = G_N$ to obtain from (17) a perihelion advance of Mercury in agreement with GR. The bound (18) requires that the coupling constant ω vary with distance in such a way that it is sufficiently small in the solar system regime and determines a value for M_0 that is in accord with the bound (18).

For terrestrial experiments and orbits of satellites, we see also that $G \sim G_N$, and for $K_\oplus$ sufficiently small, we then achieve agreement with all gravitational terrestrial experiments, including Eötvös free-fall experiments and "fifth force" experiments.

For the binary pulsar PSR 1913+16, the formula (17) can be adapted to the periastron shift of a binary system. Combining this with the STVG gravitational wave radiation formula, which will approximate closely the GR formula, we can obtain agreement with the observations for the binary pulsar. The mean orbital radius for the binary pulsar is equal to the projected semimajor axis of the binary, $\langle r\rangle_N = 7\times 10^{10}\,\text{cm}$, and we choose $\langle r\rangle_N \ll r_0$. Thus, for $G = G_N$ within the experimental errors, we obtain agreement with the binary pulsar data for the periastron shift when

$$\frac{K_N}{c^2} \ll 4.2\,\text{km}. \tag{19}$$

For a massless photon we have

$$\frac{d^2u}{d\phi^2} + u = 3\frac{GM}{c^2}u^2. \tag{20}$$

For the solar system, using $G \sim G_N$ within the experimental errors gives the light deflection,

$$\Delta_\odot = \frac{4G_N M_\odot}{c^2 R_\odot}, \tag{21}$$

in agreement with GR.

5. Pioneer Anomaly

The radio tracking data from the Pioneer 10 and 11 spacecraft during their travel to the outer parts of the solar system have revealed an anomalous acceleration. The Doppler data obtained at distances r from the Sun between 20 and 70 astronomical units (AU) showed the anomaly as a deviation from Newton's and Einstein's gravitational theories. The anomaly is observed in the Doppler residual data, as the differences of the observed Doppler velocity from the modeled Doppler velocity, and can be represented as an anomalous acceleration directed toward the Sun, with an approximately constant amplitude over the range of distance, $20\,\mathrm{AU} < r < 70\,\mathrm{AU}$[7–10]:

$$a_P = (8.74 \pm 1.33) \times 10^{-8}\,\mathrm{cm\,s^{-2}}. \tag{22}$$

After a determined attempt to account for all *known* sources of systematic errors, the conclusion has been reached that the anomalous acceleration toward the Sun could be a real physical effect that requires a physical explanation.[7–10,b]

We can rewrite the acceleration in the form

$$a(r) = -\frac{G_N M}{r^2}\left\{1 + \alpha(r)\left[1 - \exp\left(\frac{-r}{\lambda(r)}\right)\left(1 + \frac{r}{\lambda(r)}\right)\right]\right\}. \tag{23}$$

We postulate a gravitational solution that the Pioneer 10/11 anomaly is caused by the difference between the running of $G(r)$ and the Newtonian value, G_N. So the Pioneer anomalous acceleration directed toward the center of the Sun is given by

$$a_P = -\frac{\delta G(r) M_\odot}{r^2}, \tag{24}$$

where

$$\delta G(r) = G_N \alpha(r)\left[1 - \exp\left(\frac{-r}{\lambda(r)}\right)\left(1 + \frac{r}{\lambda(r)}\right)\right]. \tag{25}$$

Lacking at present a solution for the variations of $\alpha(r)$ and $\lambda(r)$ in the solar system, we adopt the following parametric representations of the "running" of $\alpha(r)$ and $\lambda(r)$:

$$\alpha(r) = \alpha_\infty\left(1 - \exp\left(\frac{-r}{\bar{r}}\right)\right)^{b/2}, \tag{26}$$

$$\lambda(r) = \lambda_\infty\left(1 - \exp\left(\frac{-r}{\bar{r}}\right)\right)^{-b}. \tag{27}$$

Here, $\bar{r}$ is a nonrunning distance scale parameter and b is a constant.

[b]It is possible that a heat transfer mechanism from the spacecraft transponders could produce a nongravitational explanation for the anomaly.

In Ref. 11, a best fit to the acceleration data extracted from Fig. 4 of Ref. 10 was obtained using a nonlinear least-squares fitting routine including estimated errors from the Doppler shift observations.[8] The best fit parameters are

$$\begin{aligned} \alpha_\infty &= (1.00 \pm 0.02) \times 10^{-3}, \\ \lambda_\infty &= 747 \pm 1\,\mathrm{AU}, \\ \bar{r} &= 4.6 \pm 0.2\,\mathrm{AU}, \\ b &= 4.0. \end{aligned} \tag{28}$$

The small uncertainties in the best fit parameters are due to the remarkably low variance of residuals corresponding to a reduced χ^2 per degree of freedom of 0.42, signalling a good fit. An important result obtained from our fit to the anomalous acceleration data is that the anomalous acceleration kicks in at the orbit of Saturn.

Fifth force experimental bounds plotted for $\log_{10}\alpha$ versus $\log_{10}\lambda$ are shown in Fig. 1 of Ref. 12 for fixed values of α and λ. The updated 2003 observational data for the bounds obtained from the planetary ephemerides are extrapolated to $r = 10^{15}\,\mathrm{m} = 6,685\,\mathrm{AU}$.[13] However, this extrapolation is based on using fixed universal values for the parameters α and λ. Since known reliable data from the ephemerides of the outer planets end with the data for Pluto at a distance from the Sun, $r = 39.52\,\mathrm{AU} = 5.91 \times 10^{12}\,\mathrm{m}$, we could claim that for our range of values $47\,\mathrm{AU} < \lambda(r) < \infty$, we predict $\alpha(r)$ and $\lambda(r)$ values consistent with the *unextrapolated* fifth force bounds.

A consequence of a variation of G and $GM_\odot$ for the solar system is a modification of Kepler's third law:

$$a_{PL}^3 = G(a_{PL})M_\odot \left(\frac{T_{PL}}{2\pi}\right)^2, \tag{29}$$

where T_{PL} is the planetary sidereal orbital period and a_{PL} is the physically measured semimajor axis of the planetary orbit. For given values of a_{PL} and T_{PL}, (29) can be used to determine $G(r)M_\odot$.

For several planets, such as Mercury, Venus, Mars and Jupiter, there are planetary ranging data, spacecraft tracking data and radiotechnical flyby observations available, and it is possible to measure a_{PL} directly. For a distance-varying $GM_\odot$ we derive[14,15]

$$\left(\frac{a_{PL}}{\bar{a}_{PL}}\right) = 1 + \eta_{PL} = \left[\frac{G(a_{PL})M_\odot}{G(a_\oplus)M_\odot}\right]^{1/3}. \tag{30}$$

Here, it is assumed that $GM_\odot$ varies with distance such that η_{PL} can be treated as a constant for the orbit of a planet. We obtain

$$\eta_{PL} = \left[\frac{G(a_{PL})}{G(a_\oplus)}\right]^{1/3} - 1. \tag{31}$$

The results for $\Delta\eta_{PL}$ due to the uncertainty in the planetary ephemerides are presented in Ref. 11 for the nine planets and are consistent with the solar ephemerides.

The validity of the bounds on a possible fifth force obtained from the ephemerides of the outer planets Uranus, Neptune and Pluto is critical in the exclusion of a parameter space for our fits to the Pioneer anomaly acceleration. Beyond the outer planets, the theoretical prediction for $\eta(r)$ approaches an asymptotic value:

$$\eta_\infty \equiv \lim_{r\to\infty} \eta(r) = 3.34 \times 10^{-4}. \tag{32}$$

We see that the variations ("running") of $\alpha(r)$ and $\lambda(r)$ with distance play an important role in interpreting the data for the fifth force bounds. This is in contrast with the standard nonmodified Yukawa correction to the Newtonian force law with fixed universal values of α and λ and for the range of values $0 < \lambda < \infty$, for which the equivalence principle and lunar laser ranging and radar ranging data to planetary probes exclude the possibility of a gravitational and fifth force explanation for the Pioneer anomaly.[16–18]

A study of the Shapiro time delay prediction in our MOG is found to be consistent with time delay observations and predicts a measurable deviation from GR for the outer planets Neptune and Pluto.[19]

6. Gravitational Lensing

The bending angle of a light ray as it passes near a massive system along an approximately straight path is given to lowest order in v^2/c^2 by

$$\theta = \frac{2}{c^2}\int |a^\perp| dz, \tag{33}$$

where $\perp$ denotes the perpendicular component to the ray's direction, dz is the element of length along the ray, and a denotes the acceleration.

From (20), we obtain the light deflection

$$\Delta = \frac{4GM}{c^2 R} = \frac{4G_N\overline{M}}{c^2 R}, \tag{34}$$

where

$$\overline{M} = M\left(1 + \sqrt{\frac{M_0}{M}}\right). \tag{35}$$

The value of $\overline{M}$ follows from (10) for clusters as $r \gg r_0$ and

$$G(r) \to G_\infty = G_N\left(1 + \sqrt{\frac{M_0}{M}}\right). \tag{36}$$

We choose for a cluster $M_0 = 3.6 \times 10^{15}\, M_\odot$ and a cluster mass $M_{\rm clust} \sim 10^{14}\, M_\odot$, and obtain

$$\left(\sqrt{\frac{M_0}{M}}\right)_{\rm clust} \sim 6. \tag{37}$$

We see that $\overline{M} \sim 7M$ and we can explain the increase in the light bending without exotic dark matter.

For $r \gg r_0$ we get

$$a(r) = -\frac{G_N \overline{M}}{r^2}. \tag{38}$$

We expect to obtain from this result a satisfactory description of lensing phenomena using Eq. (33).

7. Modified Friedmann Equations in Cosmology

We shall base our results for the cosmic microwave background (CMB) power spectrum on our MOG without a second component of cold dark matter (CDM). Our description of the accelerating universe[20,21] is based on Λ_G in Eq. (62) derived from our varying gravitational constant.

We adopt a homogeneous and isotropic Friedmann–Lemaître–Robertson–Walker (FLRW) background geometry with the line element

$$ds^2 = dt^2 - a^2(t)\left(\frac{dr^2}{1-kr^2} + r^2 d\Omega^2\right), \tag{39}$$

where $d\Omega^2 = d\theta^2 + \sin^2\theta d\phi^2$ and $k = 0, -1, +1$ for a spatially flat, open and closed Universe, respectively. Due to the symmetry of the FLRW background space–time, we have $\phi_0 \equiv \phi \neq 0$, $\phi_i = 0$ and $B_{\mu\nu} = 0$.

We define the energy–momentum tensor for a perfect fluid by

$$T^{\mu\nu} = (\rho + p)u^\mu u^\nu - p g^{\mu\nu}, \tag{40}$$

where $u^\mu = dx^\mu/ds$ is the 4-velocity of a fluid element and $g_{\mu\nu}u^\mu u^\nu = 1$. Moreover, we have

$$\rho = \rho_m + \rho_\phi + \rho_S, \quad p = p_m + p_\phi + p_S, \tag{41}$$

where ρ_i and p_i denote the components of density and pressure associated with the matter, the ϕ^μ field and the scalar fields G, ω and μ, respectively.

The modified Friedmann equations take the form[1]

$$\frac{\dot{a}^2(t)}{a^2(t)} + \frac{k}{a^2(t)} = \frac{8\pi G(t)\rho(t)}{3} + f(t) + \frac{\Lambda}{3}, \tag{42}$$

$$\frac{\ddot{a}(t)}{a(t)} = -\frac{4\pi G(t)}{3}[\rho(t) + 3p(t)] + h(t) + \frac{\Lambda}{3}, \tag{43}$$

where $\dot{a} = da/dt$ and

$$f(t) = \frac{\dot{a}(t)}{a(t)}\frac{\dot{G}(t)}{G(t)}, \tag{44}$$

$$h(t) = \frac{1}{2}\left(\frac{\ddot{G}(t)}{G(t)} - \frac{\dot{G}^2(t)}{G^2(t)} + 2\frac{\dot{a}(t)}{a(t)}\frac{\dot{G}(t)}{G(t)}\right). \tag{45}$$

From (42) we obtain

$$\rho a^3 = \frac{3}{8\pi G} a\left(\dot{a}^2 + k - a^2 f - \frac{1}{3}a^2\Lambda\right). \tag{46}$$

This leads, by differentiation with respect to t, to the expression

$$\dot{\rho} + 3\frac{d\ln a}{dt}(\rho + p) + \mathcal{I} = 0, \tag{47}$$

where

$$\mathcal{I} = \frac{3a^2}{8\pi G}(2\dot{a}f + a\dot{f} - 2\dot{a}h). \tag{48}$$

An approximate solution to the field equations for the variation of G in Ref. 1 in the background FLRW space–time is given by

$$\ddot{\mathcal{G}} + 3H\dot{\mathcal{G}} + V'(\mathcal{G}) = \frac{1}{2}G_N\mathcal{G}^2\left(\rho - 3p + \frac{\Lambda}{4\pi G_N\mathcal{G}}\right), \tag{49}$$

where $\mathcal{G}(t) = G(t)/G_N$ and $H = \dot{a}/a$. A solution for $\mathcal{G}$ in terms of a given potential $V(\mathcal{G})$ and for given values of ρ, p and Λ can be obtained from (49).

The solution for $\mathcal{G}$ must satisfy a constraint at the time of big bang nucleosynthesis.[22,23] The number of relativistic degrees of freedom is very sensitive to the cosmic expansion rate at 1 MeV. This can be used to constrain the time dependence of G. Measurements of the ^{4}He mass fraction and the deuterium abundance at 1 MeV lead to the constraint $G(t) \sim G_N$. We impose the condition $\mathcal{G}(t) \to 1$ as $t \to t_{\rm BBN}$, where $t_{\rm BBN}$ denotes the time of the big bang nucleosynthesis. Moreover, *locally in the solar system* we must satisfy the observational bound from the Cassini spacecraft measurements[24]:

$$\left|\frac{\dot{G}}{G}\right| \leq 10^{-12}\ \mathrm{yr}^{-1}. \tag{50}$$

We shall now impose the approximate conditions at the epoch of recombination:

$$2\dot{a}f + a\dot{f} \sim 2\dot{a}h, \tag{51}$$

$$\frac{d}{dt}\left(\frac{\dot{G}}{G}\right) < 2\frac{\dot{a}}{a}\frac{\dot{G}}{G}. \tag{52}$$

We find from (45) and (52) that $f \sim h$, and from the condition (51) we obtain

$$\dot{f} \equiv \frac{d\Lambda_G}{dt} \sim 0, \tag{53}$$

where

$$\Lambda_G = \frac{\dot{a}}{a}\frac{\dot{G}}{G}. \tag{54}$$

By setting the cosmological constant $\Lambda = 0$, we get the generalized Friedmann equations

$$\frac{\dot{a}^2}{a^2} + \frac{k}{a^2} = \frac{8\pi G\rho}{3} + \Lambda_G, \tag{55}$$

$$\frac{\ddot{a}}{a} = -\frac{4\pi G}{3}(\rho + 3p) + \Lambda_G. \tag{56}$$

We now have from (47), (48) and (51) at the epoch of recombination $\mathcal{I} \sim 0$ and

$$\dot{\rho} + 3\frac{d\ln a}{dt}(\rho + p) \sim 0. \tag{57}$$

We adopt the equation of state, $p(t) = w\rho(t)$, and derive from (57) the approximate solution for $\rho(t)$

$$\rho(t) \sim \rho(t_0)\left(\frac{a_0}{a(t)}\right)^{3(1+w)}, \tag{58}$$

where $a/a_0 = 1/(1+z)$ and z denotes the redshift. For the matter and radiation densities ρ_m and ρ_r, we have $w = 0$ and $w = 1/3$, respectively. This gives

$$\rho_m(t) \sim \rho_m(t_0)(1+z)^3, \quad \rho_r(t) \sim \rho_r(t_0)(1+z)^4. \tag{59}$$

Let us expand $G(t)$ in a power series:

$$G(t) = G_{\rm eff}(t_r) + (t - t_r)\dot{G}(t_r) + (t - t_r)^2\ddot{G}(t_r) + \cdots, \tag{60}$$

where $t \sim t_r$ is the time of recombination and $G_{\rm eff}(t_r) = G_N(1+Z) = \text{const}$. We write the generalized Friedmann equation for flat space, $k = 0$, in the approximate form

$$H^2 = \frac{8\pi G_{\rm eff}\rho_m}{3} + \Lambda_G, \tag{61}$$

where

$$\Lambda_G = H\frac{\dot{G}}{G} > 0 \tag{62}$$

and $\dot{\Lambda}_G \sim 0$. It follows from (61) that for a spatially flat Universe

$$\Omega_m + \Omega_G = 1, \tag{63}$$

where

$$\Omega_m = \frac{8\pi G_{\rm eff}\rho_m}{3H^2}, \quad \Omega_G = \frac{\Lambda_G}{H^2}. \tag{64}$$

We shall postulate that the matter density ρ_m is dominated by the baryon density, $\rho_m \sim \rho_b$, and we have

$$\Omega_m \sim \Omega_{b\,\rm eff}, \tag{65}$$

where

$$\Omega_{b\,\rm eff} = \frac{8\pi G_{\rm eff}\rho_b}{3H^2}. \tag{66}$$

Thus, we assume that *the baryon–photon fluid dominates matter before recombination and at the surface of last scattering without a CDM fluid component.*

From the current value, $H_0 = 7.5 \times 10^{-11}\,\text{yr}^{-1}$, and (62) and (64), we obtain for $\Omega_G \sim 0.7$

$$\left|\frac{\dot{G}}{G}\right| \sim 5 \times 10^{-11}\,\text{yr}^{-1}, \tag{67}$$

valid at cosmological scales for redshifts $z > 0.1$. In the *local solar system* and for the binary pulsar PSR 1913+16 for $z \sim 0$, the experimental bound is

$$\left|\frac{\dot{G}}{G}\right| < 5 \times 10^{-12}\,\mathrm{yr}^{-1}. \tag{68}$$

We can explain the accelerated expansion of the Universe deduced from supernova measurements in the range $0.1 < z < 1.7$ using the cosmologically scaled value of $\dot{G}/G$ in (67) with Einstein's cosmological constant, $\Lambda = 0$.

8. Acoustical Peaks in the CMB Power Spectrum

Mukhanov[25,26] has obtained an analytical solution to the amplitude of fluctuations in the CMB power spectrum for $l \gg 1$:

$$l(l+1)C_l \sim \frac{B}{\pi}(O+N). \tag{69}$$

Here, O denotes the oscillating part of the spectrum, while the nonoscillating contribution can be written as the sum of three parts

$$N = N_1 + N_2 + N_3. \tag{70}$$

The oscillating contributions can be calculated from the formula

$$O \sim \sqrt{\frac{\pi}{r_h l}}\left[A_1 \cos\left(lr_p + \frac{\pi}{4}\right) + A_2 \cos\left(2lr_p + \frac{\pi}{4}\right)\right] \exp\left(-\left(\frac{l}{l_s}\right)^2\right), \tag{71}$$

where r_h and r_p are parameters that determine predominantly the heights and positions of the peaks, respectively. A_1 and A_2 are constant coefficients given in the range $100 < l < 1200$ for $\Omega_m \sim \Omega_{b\,\mathrm{eff}}$ by

$$A_1 \sim 0.1\xi \frac{(\mathcal{P} - 0.78)^2 - 4.3}{(1+\xi)^{1/4}} \exp\left(\frac{1}{2}(l_s^{-2} - l_f^{-2})l^2\right), \tag{72}$$

$$A_2 \sim 0.14 \frac{(0.5 + 0.36\mathcal{P})^2}{(1+\xi)^{1/2}}, \tag{73}$$

where

$$\mathcal{P} = \ln\left(\frac{lI}{200(\Omega_{b\,\mathrm{eff}})^{1/2}}\right), \tag{74}$$

and I is given by the ratio

$$\frac{\eta_x}{\eta_0} \sim \frac{I}{z_x^{1/2}} = 3\left(\frac{\Omega_G}{\Omega_{b\,\mathrm{eff}}}\right)^{1/6}\left(\int_0^y \frac{dx}{(\sinh x)^{2/3}}\right)^{-1} \frac{1}{z_x^{1/2}}. \tag{75}$$

Here, η_x and z_x denote a conformal time $\eta = \eta_x$ and a redshift in the range $\eta_0 > \eta_x > \eta_r$ when radiation can be neglected and $y = \sinh^{-1}(\Omega_G/\Omega_{b\,\mathrm{eff}})^{1/2}$. To determine η_x/η_0, we use the exact solution for a flat dust-dominated Universe with a constant Λ_G:

$$a(t) = a_0\left(\sinh\left(\frac{3}{2}\right)H_0 t\right)^{2/3}, \tag{76}$$

where a_0 and H_0 denote the present values of a and the Hubble parameter H.

The l_f and l_s in (72) denote the finite thickness and Silk damping scales, respectively, given by

$$l_f^2 = \frac{1}{2\sigma^2}\left(\frac{\eta_0}{\eta_r}\right)^2, \quad l_s^2 = \frac{1}{2(\sigma^2 + 1/(k_D\eta_r)^2)}\left(\frac{\eta_0}{\eta_r}\right)^2, \tag{77}$$

where

$$\sigma \sim 1.49\times 10^{-2}\left[1 + \left(1 + \frac{z_{\mathrm{eq}}}{z_r}\right)^{-1/2}\right], \quad k_D(\eta) = \left(\frac{2}{5}\int_0^\eta d\eta c_s^2\frac{\tau_\gamma}{a}\right)^{-1/2}, \tag{78}$$

and τ_γ is the photon mean-free time.

A numerical fitting formula gives[25,26]

$$\mathcal{P} \sim \ln\left(\frac{l}{200(\Omega_{b\,\mathrm{eff}}^{0.59})}\right), \quad r_p = \frac{1}{\eta_0}\int d\eta c_s(\eta). \tag{79}$$

Moreover,

$$\xi \equiv \frac{1}{3c_s^2} - 1 = \frac{3}{4}\left(\frac{\rho_b}{\rho_\gamma}\right), \tag{80}$$

where $c_s(\eta)$ is the speed of sound:

$$c_s(\eta) = \frac{1}{\sqrt{3}}\left[1 + \xi\left(\frac{a(\eta)}{a(\eta_r)}\right)\right]^{-1/2}. \tag{81}$$

We note that ξ does not depend on the value of G_{eff}. For the matter-radiation Universe,

$$a(\eta) = \bar{a}\left[\left(\frac{\eta}{\eta_*}\right)^2 + 2\left(\frac{\eta}{\eta_*}\right)\right], \tag{82}$$

where for radiation-matter equality $z = z_{\mathrm{eq}}$

$$\frac{z_{\mathrm{eq}}}{z_r} \sim \left(\frac{\eta_r}{\eta_*}\right)^2 + 2\left(\frac{\eta_r}{\eta_*}\right), \tag{83}$$

and $\eta_{\mathrm{eq}} = \eta_*(\sqrt{2} - 1)$ follows from $\bar{a} = a(\eta_{\mathrm{eq}})$.

For the nonoscillating parts, we have

$$N_1 \sim 0.063\xi^2 \frac{(\mathcal{P} - 0.22(l/l_f)^{0.3} - 2.6)^2}{1 + 0.65(l/l_f)^{1.4}} \exp\left(-\left(\frac{l}{l_f}\right)^2\right), \tag{84}$$

$$N_2 \sim \frac{0.037}{(1+\xi)^{1/2}} \frac{(\mathcal{P} - 0.22(l/l_s)^{0.3} + 1.7)^2}{1 + 0.65(l/l_f)^{1.4}} \exp\left(-\left(\frac{l}{l_s}\right)^2\right), \tag{85}$$

$$N_3 \sim \frac{0.033}{(1+\xi)^{3/2}} \frac{(\mathcal{P} - 0.5(l/l_s)^{0.55} + 2.2)^2}{1 + 2(l/l_s)^2} \exp\left(-\left(\frac{l}{l_s}\right)^2\right). \tag{86}$$

Mukhanov's formula[25,26] for the oscillating spectrum is given by

$$C(l) \equiv \frac{l(l+1)C_l}{[l(l+1)C_l]_{\mathrm{low}\, l}} = \frac{100}{9}(O + N), \tag{87}$$

where we have normalized the power spectrum by using for a flat spectrum with a constant amplitude B

$$[l(l+1)C_l]_{\mathrm{low}\, l} = \frac{9B}{100\pi}. \tag{88}$$

We adopt the parameters

$$\Omega_{bN} \sim 0.04, \quad \Omega_{b\mathrm{eff}} \sim 0.3, \quad \Omega_G \sim 0.7, \quad \xi \sim 0.6, \tag{89}$$

and

$$r_h = 0.03, \quad r_p = 0.01 \quad l_f \sim 1580, \quad l_s \sim 1100, \tag{90}$$

where $\Omega_{bN} = 8\pi G_N \rho_b / 3H^2$.

The fluctuation spectrum determined by Mukhanov's analytical formula is displayed in Fig. 1 for the choice of cosmological parameters given in (89) and (90).

The role played by CDM in the standard scenario is replaced in the modified gravity theory by the significant deepening of the gravitational potential well by the effective gravitational constant, $G_{\mathrm{eff}} \sim 7G_N$, which traps the nonrelativistic baryons before recombination. The deepening of the gravitational well reduces the baryon dissipation due to the photon coupling pressure, and the third and higher peaks in the acoustic oscillation spectrum are not erased by finite thickness and baryon drag effects. The effective baryon density, $\Omega_{b\,\mathrm{eff}} = (1+Z)\Omega_{bN} \sim 7\Omega_{bN} \sim 0.3$, dominates the fluid before recombination, *and we fit the acoustical power spectrum data without a CDM fluid component.* For $t < t_{\mathrm{dec}}$, where t_{dec} denotes the time of matter-radiation decoupling, luminous baryons and photons are tightly coupled, and for photons the dominant collision mechanism is scattering by nonrelativistic electrons due to Thompson scattering. It follows that luminous baryons are dragged along with photons and perturbations at wavelength $\lambda_w < \ell_s$ will be partly erased where ℓ_s is the proper Silk length, given by $\ell_s \sim 3.5\,\mathrm{Mpc}\,\Omega_{b\,\mathrm{eff}}^{-1/2}$.[31,32] We have $\ell_s \sim 6\,\mathrm{Mpc}$ for $\Omega_{b\,\mathrm{eff}} \sim 0.3$ compared to $\ell_s \sim 18\,\mathrm{Mpc}$ for $\Omega_{bN} \sim 0.04$. The Silk

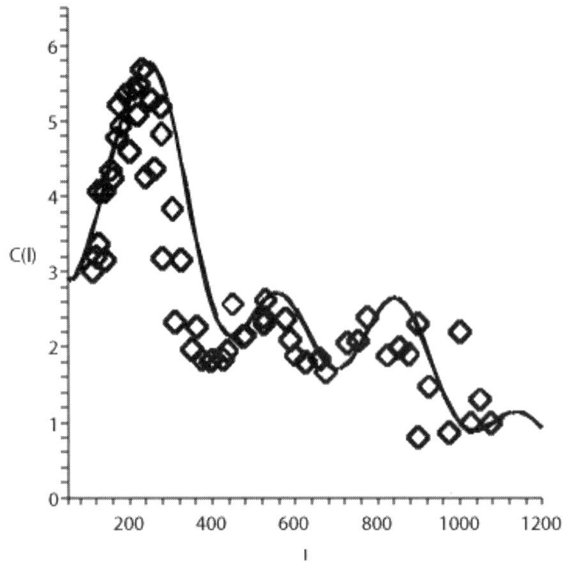

Fig. 1. The solid line shows the result of the calculation of the power spectrum acoustical oscillations, $C(l)$, and the $\diamond$'s correspond to the WMAP, Archeops and Boomerang data in units $\mu K^2 \times 10^{-3}$, as presented in Refs. 26–29.

mass is reduced by more than an order of magnitude.[c] Thus, sufficient baryonic perturbations should survive before $t \sim t_{\rm dec}$ to explain the power spectrum without collisionless dark matter.

Our predictions for the CMB power spectrum for large angular scales corresponding to $l < 100$ will involve the integrated Sachs–Wolfe contributions obtained from the modified gravitational potential.

9. Conclusions

We have demonstrated that a modified gravity theory[1] can lead to a satisfactory fit to the galaxy rotation curve data, mass profiles of X-ray cluster data, the solar system and the binary pulsar PSR 1913+16 data. Moreover, we can provide an explanation for the Pioneer 10/11 anomalous acceleration data, given that the anomaly is caused by gravity. We can fit satisfactorily the acoustical oscillation spectrum obtained in the cosmic-microwave-background data by employing the analytical formula for the fluctuation spectrum derived by Mukhanov.[25,26]

Λ_G obtained from the varying gravitational constant in our MOG replaces the standard cosmological constant Λ in the concordance model. Thus, the accelerating expansion of the Universe is obtained from the MOG scenario.

An important problem to investigate is whether an N-body simulation calculation based on our MOG scenario can predict the observed large scale galaxy surveys.

[c]Note that there will be a fraction of dark baryonic matter before decoupling.

The formation of protogalaxy structure before and after the epoch of recombination and the growth of galaxies and clusters of galaxies at later times in the expansion of the Universe have to be explained.

We have succeeded in fitting in a unified picture a large amount of data over 16 orders of magnitude in distance scale from Earth to the surface of last scattering some 13.7 billion years ago, using our modified gravitational theory without exotic dark matter. The data fitting ranges over four distance scales: the solar system, galaxies, clusters of galaxies and the CMB power spectrum data at the surface of last scattering.

Acknowledgments

This work was supported by the Natural Sciences and Engineering Research Council of Canada. I thank Joel Brownstein, Martin Green and Justin Khoury for helpful discussions.

References

1. J. W. Moffat, *J. Cosmolog. Astropart. Phys.* **3** (2006) 004 [gr-qc/0506021].
2. J. W. Moffat, *J. Cosmolog. Astropart. Phys.* **5** (2005) 003 [astro-ph/0412195].
3. L. Baudis, *Int. J. Mod. Phys. A* **21** (2006) 1925 [astro-ph/0511805].
4. M. A. Clayton, gr-qc/0104103.
5. J. R. Brownstein and J. W. Moffat, *Astrophys. J.* **636** (2006) 721 [astro-ph/0506370].
6. J. R. Brownstein and J. W. Moffat, *Mon. Not. Roy. Astron. Soc.* **367** (2006) 527 [astro-ph/0507222].
7. J. D. Anderson *et al.*, *Phys. Rev. Lett.* **81** (1998) 2858 [gr-qc/9808081].
8. J. D. Anderson *et al.*, *Phys. Rev. D* **65** (2002) 082004 [gr-qc/0104064].
9. S. G. Turyshev, M. M. Nieto and J. D. Anderson, *EAS Publication Series* **20** (2006) 243 [gr-qc/0510081].
10. M. M. Nieto and J. D. Anderson, *Class. Quant. Grav.* **22** (2005) 5343 [gr-qc/0507052].
11. J. R. Brownstein and J. W. Moffat, *Class. Quant. Grav.* **23** (2006) 3427 [gr-qc/0511026].
12. S. Reynaud and M. M. Jaekel, *Int. J. Mod. Phys. A* **20** (2005) 2294 [gr-qc/0501038].
13. E. Fischbach (2005), private communication.
14. E. Fischbach and C. L. Talmadge, *The Search for Non-Newtonian Gravity* (Springer, Heidelberg, New York, 1999).
15. C. Talmadge *et al.*, *Phys. Rev. Lett.* **61** (1988) 1159.
16. C. Stubbs *et al.*, *Phys. Rev. Lett.* **58** (1987) 1070.
17. E. G. Adelberger, B. R. Heckel and A. E. Nelson, *Ann. Rev. Nucl. Part. Sci.* **53** (2003) 77 [hep-ph/0307284].
18. C. W. Will, gr-qc/0510072.
19. J. W. Moffat, gr-qc/0605141.
20. S. Perlmutter *et al.*, *Astrophys. J.* **517** (1999) 565 [astro-ph/9812133].
21. A. G. Riess *et al.*, *Astron. J.* **116** (1998) 1009 [astro-ph/9805201].
22. R. Bean, S. Hansen and A. Melchiorri, *Phys. Rev. D* **64** (2001) 103508 [astro-ph/0104162].
23. C. J. Copi, A. N. Davis and L. M. Krauss, *Phys. Rev. Lett.* **92** (2004) 171301 [astro-ph/0311334].

24. B. Bertotti, L. Iess and P. Tortora, *Nature* **425** (2003) 374.
25. V. Mukhanov, *Int. J. Theor. Phys.* **43** (2004) 623 [astro-ph/0303072].
26. V. Mukhanov, *Physical Foundations of Cosmology* (Cambridge University Press, 2005).
27. G. Hinshaw *et al.*, astro-ph/0603451.
28. D. N. Spergel *et al.*, astro-ph/0603449.
29. M. Tristram *et al. Astron. Astrophys.* **436** (2005) 785 [astro-ph/0411633].
30. W. C. Jones *et al.*, astro-ph/0507494.
31. J. Silk, *Astrophys. J.* **151** (1968) 459.
32. T. Padmanabhan, *Structure Formation in the Universe* (Cambridge University Press, 1993), p. 172.

LONG RANGE GRAVITY TESTS AND THE PIONEER ANOMALY

SERGE REYNAUD

Laboratoire Kastler Brossel, Université Pierre et Marie Curie, case 74, Campus Jussieu, F75252 Paris cedex 05, France*
reynaud@spectro.jussieu.fr

MARC-THIERRY JAEKEL

Laboratoire de Physique Théorique,† Ecole Normale Supérieure, 24 rue Lhomond, F75231 Paris cedex 05, France
jaekel@lpt.ens.fr

Experimental tests of gravity performed in the solar system show a good agreement with general relativity. The latter is, however, challenged by the Pioneer anomaly, which might be pointing at some modification of gravity law at ranges of the order of the size of the solar system. As this question could be related to the puzzles of "dark matter" or "dark energy," it is important to test it with care. There exist metric extensions of general relativity which preserve the well-verified equivalence principle while possibly changing the metric solution in the solar system. Such extensions have the capability to preserve compatibility with existing gravity tests while opening free space for the Pioneer anomaly. They constitute arguments for new mission designs and new space technologies as well as for having a new look at data of already-performed experiments.

Keywords: General relativity; gravity tests; Pioneer anomaly.

1. Introduction

The commonly heard assertion that gravity tests show a good agreement with general relativity (GR) has to be understood as a set of more detailed statements.[1,2] It first implies that the gravitational field may be identified with the metric tensor $g_{\mu\nu}$ in a Riemannian space–time, as a consequence of the fact that the equivalence principle is one of the most accurately verified properties of nature. It then means that this metric tensor appears to have a form close to that predicted by GR, as shown by the confrontations of observations with the family of more general

*CNRS, ENS, UPMC.
†CNRS, ENS, UPMC.

PPN solutions. This second statement can be put in the alternative form of a good agreement of the gravity force law with the prediction of GR, deviations being predicted by unification models but not observed to date.[3,4]

Besides these successes, GR is challenged by observations performed at various scales. First, anomalies are known to affect the rotation curves of galaxies. They are commonly accounted for by introducing "dark matter" to reproduce these curves.[5,6] Further anomalies have been detected more recently in the relation between redshifts and luminosities, showing an acceleration of cosmic expansion. They are usually interpreted as being due to the presence of some "dark energy."[7,8] Both components of the "dark side" of the Universe have no known origin and are not observed through other means than the gravitational anomalies they have been designed to cure. As long as this situation is lasting, the related anomalies may as well be interpreted as long range deviations from GR.[9–11]

The Pioneer anomaly constitutes a new piece of information in this puzzling context, which might already reveal an anomalous behavior of gravity at scales of the order of the size of the solar system.[12,13] Though a number of mechanisms have been considered for this aim,[14–19] the anomaly has up to now escaped all attempts at explanation as a systematic effect generated by the spacecraft itself or its environment. The importance of the Pioneer anomaly for space navigation already justifies it to be submitted to further scrutiny. Meanwhile its potential impact on fundamental physics cannot be underestimated, since the possibility exists that the Pioneer anomaly be the first hint of a long range modification of gravity law.[20–26]

These questions are reviewed in the present paper, with emphasis put on the key issue of the compatibility of the Pioneer anomaly with other gravity tests.

2. Gravity Tests in the Solar System

GR provides us with an excellent description of gravitational phenomena in the solar system. In order to discuss the meaning of this common statement, we first recall the basic features of this theoretical description and then briefly review the experimental evidences supporting it.

In order to apply the principle of relativity to accelerated motions, Einstein introduced what is now called the equivalence principle.[27,28] A weak form of this principle is expressed by the universality of free fall, a central property of the theory of gravitation since Galileo and Newton which acquired with Einstein a geometrical significance, gravitation fields being identified with the metric tensor $g_{\mu\nu}$ in a Riemannian space–time. Ideal atomic clocks measure the proper time $\int ds$ along their trajectory in space–time with $ds^2 \equiv g_{\mu\nu} dx^\mu dx^\nu$. Meanwhile freely falling motions are the geodesics of this Riemannian space–time, which are also the curves which extremize the integral $\int ds$.

The equivalence principle is one of the best-tested properties of nature. Potential violations are usually parametrized by a relative difference η in the accelerations a_1

and a_2 undergone by two test bodies of different compositions in free fall at the same location and with the same velocity. Modern experiments constrain the parameter η to stay below the 10^{-12} level. They test the principle at distances ranging from the millimeter in laboratory experiments (Ref. 4 and references therein) to the sizes of the Earth–Moon[29] or Sun–Mars orbit.[30]

The geometrical interpretation is the very core of GR, but it is not sufficient to fix the latter theory. In order to do that, it is necessary to write also the equations determining the metric tensor from the distribution of energy and momentum in space–time or, in other words, to fix the form of the coupling between curvature and stress tensors. Among the curvature tensors available in Riemannian geometry, the Einstein tensor $E_{\mu\nu} \equiv R_{\mu\nu} - \frac{1}{2} g_{\mu\nu} R$ is defined from the Ricci tensor $R_{\mu\nu}$ and the scalar curvature R so that it has a null covariant divergence, $D^\mu E_{\mu\nu} \equiv 0$. This geometrical property has to be compared with the physical property $D^\mu T_{\mu\nu} \equiv 0$, which expresses conservation of energy and momentum as the condition of null divergence of the stress tensor $T_{\mu\nu}$. Note that the latter relation is a necessary and sufficient condition for motions of test masses to follow geodesics.

GR corresponds to a simple proportionality relation between the two tensors $E_{\mu\nu}$ and $T_{\mu\nu}$, the constant being determined from the Newton gravitation constant G_N and the velocity of light c:

$$E_{\mu\nu} = \frac{8\pi G_N}{c^4} T_{\mu\nu}. \tag{1}$$

This Einstein–Hilbert equation[31–33] is tested through comparisons of its predictions with observations or experiments. To this end, the metric tensor in the solar system is first deduced by solving (1). In the simple case where the gravity source, i.e. the Sun, is described as a pointlike motionless mass M, the metric can be written as an expansion in terms of the Newton potential ϕ:

$$\begin{aligned} ds^2 &= g_{00} c^2 dt^2 + g_{rr} \left(dr^2 + r^2 (d\theta^2 + \sin^2\theta d\varphi^2) \right), \\ g_{00} &= 1 + 2\phi + 2\phi^2 + \cdots, \quad g_{rr} = -1 + 2\phi + \cdots, \\ \phi &\equiv -\frac{\kappa}{r}, \quad \kappa \equiv \frac{G_N M}{c^2}, \quad |\phi| \ll 1. \end{aligned} \tag{2}$$

Spherical coordinates have been used (t and r are time and radius; θ and φ are colatitude and azimuth angles) with the Eddington gauge convention of isotropic spatial coordinates. κ is the gravitational radius of the Sun, ~1.5 km.

GR is usually tested through its confrontation with the enlarged family of parametrized post-Newtonian (PPN) metric tensors introduced by Eddington[34] and then developed by several physicists[35–39]:

$$g_{00} = 1 + 2\alpha\phi + 2\beta\phi^2 + \cdots, \quad g_{rr} = -1 + 2\gamma\phi + \cdots. \tag{3}$$

The three parameters α, β and γ are constants, the first of which can be set to unity by fixing the Newton constant G_N. Within the PPN family, GR thus corresponds to $\gamma = \beta = 1$. The anomalies $\gamma - 1$ or $\beta - 1$ of these Eddington parameters affect motions, i.e. the geodesics associated with the metric (3), and they can therefore be measured by comparing observations with predictions.

Experiments that have been performed for more than four decades have led to more and more constraining bounds on these anomalies. For example, Doppler ranging on Viking probes in the vicinity of Mars[30] and deflection measurements using VLBI astrometry[40] or radar ranging on the Cassini probe[41] have given smaller and smaller values of $|\gamma - 1|$, with presently a bound of a few 10^{-5}. Analysis of the precession of planet perihelions[42] and of the polarization by the Sun of the Moon orbit around the Earth[43] have led to determinations of linear superpositions of β and γ, resulting now in $|\beta - 1|$ smaller than a few 10^{-4}.

An alternative way to test GR has been to check the r dependence of the Newton potential, i.e. also of the component g_{00} in (2). Hypothetical modifications of its standard expression, predicted by unification models, are usually parametrized in terms of an additional Yukawa potential depending on two parameters, the range λ and the amplitude α measured with respect to Newton potential.[3] The presence of such a Yukawa correction has been looked for on a large range of distances. The accuracy of short range tests has recently been improved, as gravity experiments were pushed to smaller distances[44–47] and as Casimir forces, which become dominant at the submillimeter range, were more satisfactorily taken into account.[48–51] On the other side of the distance range, long range tests of the Newton law are performed by following the motions of planets or probes in the solar system. They also show an agreement with GR with a good accuracy for ranges of the order of the Earth–Moon[29] or Sun–Mars distances.[52–54] When the whole set of results is reported on a global figure (see Fig. 1 in Ref. 55, reproduced by courtesy of Coy *et al.*[56]), it appears that windows remain open for violations of the standard form of Newton force law at short ranges, below the millimeter, as well as long ones, of the order of or larger than the size of the solar system.

To sum up this discussion, tests of gravity confirm its metric interpretation and provide strong evidence for gravitation theory being very close to GR. A few exceptions exist, among which are notably the anomalous observations recorded on Pioneer probes. We will see below that this contradiction between Pioneer observations and other gravity tests may be resolved in an extended framework, where deviations from GR may show a scale dependence. It is precisely the merit of Newton force law tests to shed light on this possibility of a scale dependence, with any specific experiment being only sensitive to a given range of distances. The issue of scale dependence has to be considered with great attention, especially in the context recalled in the Introduction where questions arise about the validity of GR at galactic or cosmic scales. As will be recalled in forthcoming sections, scale dependence is also a natural consequence of radiative corrections to GR to be taken into account.

3. The Pioneer Anomaly

After the discussions of the previous section, it is clear that the gravity laws have to be tested at all possible scales. It is of particular interest to study the largest scales attainable by man-made instruments, in an attempt to bridge the gap between experiments made on Earth or in its vicinity and the much larger galactic and cosmic scales. The best example of such a strategy to date is the NASA decision to extend the Pioneer 10 and 11 missions after their primary periods with the aim, among others, of testing the laws of gravity at large heliocentric distances.[57,58] When considered as a gravity test, the extended Pioneer missions were the largest-scaled test ever carried out, and they failed to confirm the known laws of gravity.

The anomaly was recorded on deep space navigation (DSN) tracking data from the Pioneer 10 and 11 probes.[59] An up-link radio signal is emitted from Earth at a DSN station, is then received and sent back by the probe, and the down-link radio signal is finally received on Earth at the same or another DSN station. For probes equipped with range measurement capabilities (which was not the case for Pioneer 10 and 11), the ranging observable is defined as half the time elapsed on Earth from the emission time to the reception time. For the Pioneer 10 and 11 probes, the tracking technique was based on the measurement of the Doppler shift, a proper observable defined as the ratio of cycle counting rates of reference clocks located at emission and reception stations.[60] The same information can be encoded in a Doppler velocity v, with the ratio of received to emitted frequencies written as

$$\frac{f}{f_0} \equiv \frac{1-\frac{v}{c}}{1+\frac{v}{c}}. \tag{4}$$

The observable v represents a relative velocity of the probe with respect to the station, with relativistic and gravitational effects taken into account in the definition (4) and perturbations due to transmission media effects properly accounted for.[13]

These Doppler tracking data were analyzed during the travel of the Pioneer 10 and 11 probes to the outer parts of the solar system. When the probes had reached a quieter environment, after flying by Jupiter and Saturn, a precise comparison of tracking data with predictions of GR showed that the observed Doppler velocity departed from the calculated Doppler velocity. The velocity was thus showing an anomaly δv varying linearly with elapsed time (see Fig. 8 of Ref. 13),

$$\delta v \equiv v_{\text{observed}} - v_{\text{modeled}} \simeq -a_P(t - t_{\text{in}}), \tag{5}$$

with a_P an anomalous acceleration directed toward the Sun and having an approximately constant amplitude over a large range of heliocentric distances (AU = astronomical unit)

$$a_P = (0.87 \pm 0.13)\,\text{nm}\,\text{s}^{-2}, \quad 20\,\text{AU} \lesssim r_P \lesssim 70\,\text{AU}. \tag{6}$$

It is worth emphasizing that the Pioneer anomaly has been registered on the two deep space probes showing the best navigation accuracy. Other anomalous observations have been reported for the Ulysses and Galileo probes, but they were not as reliable as for Pioneer probes.[13] For other probes like Voyager 1 and 2 and Cassini, the navigation accuracy was not sufficient. In other words, the Pioneer gravity test has been performed twice with identical probes on similar trajectories — but escape directions opposite in the solar system — and the same result. This is not an impressive statistics when we compare it to the large number of tests confirming GR. In particular, when the possibility of an artefact on board the probe or in its environment is considered, this artefact could be the same on the two probes. However, no satisfactory explanation of this kind has been found to date, though intensive efforts have been devoted to this aim. The extensive analysis of Anderson *et al.*,[13] published after years of cross-checks, has been confirmed by an independent analysis.[61] Such independent reanalyses of the data remain an important tool for confirming or invalidating the existence of the anomaly and they now experience a revival thanks to recently recovered data covering the whole period of the Pioneer 10 and 11 missions, from their launch to the last data point.[62–65]

4. A Key Question: Is the Pioneer Anomaly Compatible with Other Gravity Tests?

In this context, the question of the compatibility of the observed Pioneer anomaly with other gravity tests acquires the status of a key issue. If there exist gravity theories where a Pioneer-like anomaly can take a natural place, it is indeed of the first importance to consider these theories with great care because, as stated in the Introduction, the anomaly could be the first hint of a modification of gravity at large scales, with potentially a tremendous impact on galactic and cosmic physics. But if there exist no such theories, the Pioneer anomaly may remain an interesting curiosity with a potentially large impact on navigation in the solar system, but probably lesser importance for fundamental physics.

At this point, it is worth repeating that tests of the equivalence principle (EP) have shown it to be preserved at a very high accuracy level, better than 10^{-12} in laboratory experiments as well as in tracking of the motion of Moon on its orbit around the Earth. This is in any case a much higher accuracy than the EP violation which would be needed to account for the Pioneer anomaly: the standard Newton acceleration at 70 UA is of the order of $1\,\mu\mathrm{m\,s}^{-2}$ while the Pioneer anomaly is of the order of $1\,\mathrm{nm\,s}^{-2}$. Should the anomaly be interpreted in terms of an EP violation, the latter would be of the order of 10^{-3}. This does not contradict the possibility of EP violations which are predicted by unification models[66–68] and looked for in space experiments with an excellent precision, such as MICROSCOPE[69] and STEP.[70] But such violations are expected to occur at a lower level than needed to affect the Pioneer anomaly and we will therefore restrict our attention to a confrontation of GR with alternative metric theories of gravity.

In this well-established metric interpretation, the precise form of the coupling between space–time curvature and gravity sources can still be discussed.[71] Like the other fundamental interactions, gravitation may be treated within the framework of field theory.[72–74] Radiative corrections due to its coupling to other fields then naturally lead to embedding GR within a larger class of theories.[75–77] Modifications are thus expected to appear,[78–81] in particular, though not only, at large length scales.[82–85] This suggests to consider GR as an effective theory of gravity valid at the length scales for which it has been accurately tested but not necessarily at smaller or larger scales. Note that, in contrast to GR,[86] the fourth order theories which are a natural extension of GR show renormalizability as well as asymptotic freedom at high energies.[87] This is a strong argument for extending the gravitation theory at scales not already constrained by experiments, for instance using renormalization group trajectories.[88] Renormalizability of these theories, however, comes with a counterpart, i.e. the problem of ghosts, but it has been argued that this problem does not constitute a definitive dead end for an effective field theory valid in a limited scale domain.[89] In particular, the departure from unitarity is expected to be negligible at ordinary scales tested in the present day Universe.[90]

In the following, we will briefly review the main features of a phenomenological framework which has recently been developed for the purpose of answering the question of the compatibility of the Pioneer anomaly with other gravity tests.[21–23] It will be presented below as covering the whole spectrum of metric extensions of GR which remain in the vicinity of GR. In particular, it will be shown to include as particular cases the PPN extensions as well as the already-invoked modifications of Newton force law. Let us stress that this larger family of theories is not just an *ad hoc* extension showing the nice property of allowing a place for the Pioneer anomaly. It emerges in a natural manner as the extension of GR induced by radiative corrections due to the coupling of gravity with other fields,[91] this idea having been explored before it was noticed that it led to Pioneer-like anomalies.[21–23]

5. Post-Einsteinian Metric Theories of Gravity

In order to present the extensions of GR in a simple manner, we start with the linearized version of gravitation theory.[21,22] We will then present some salient features of the nonlinear theory.[23,26]

In the linearized treatment, the metric field is represented as a small perturbation $h_{\mu\nu}$ of Minkowski metric $\eta_{\mu\nu}$:

$$g_{\mu\nu} = \eta_{\mu\nu} + h_{\mu\nu}, \quad \eta_{\mu\nu} = \mathrm{diag}(1, -1, -1, -1), \quad |h_{\mu\nu}| \ll 1. \tag{7}$$

The field $h_{\mu\nu}$ is a function of position x in space–time or, equivalently in Fourier space, of wave vector k. Gauge-invariant observables of the metric theory are given by curvature tensors. In the linearized theory, i.e. at first order in $h_{\mu\nu}$, Riemann, Ricci, scalar and Einstein curvatures have simple expressions in the momentum representation (they are given in Ref. 22).

These curvature fields are similar to the gauge-invariant electromagnetic fields of electrodynamics so that, while being supported by its geometrical interpretation, GR shows essential similarities to other field theories.[73,75] This suggests that GR may be considered as the low energy effective limit of a more complete unified theory[80,81] which should describe the coupling of gravity with other fields. In any case, this theory should contain radiative corrections to the graviton propagator, leading to a modification of the Einstein–Hilbert equation (1) and to a momentum dependence of the coupling between curvature and stress tensors. In the weak field approximation, the Einstein tensor, which is divergenceless, has a natural decomposition on the two sectors corresponding to different conformal weights,[91] i.e. also on traceless (conformal weight 0) and traced components (conformal weight 1).

When considering the isotropic and stationary situation with a pointlike and motionless Sun of mass M, the general coupling between curvature and stress tensors is thus described by two running constants, $\tilde{G}^0$ and $\tilde{G}^1$, which depend on the spatial wave vector $\mathbf{k}$ and live in the two sectors (0) and (1). Solutions to the extended gravitation equations (given in Refs. 21 and 22), written anew with spatial isotropic coordinates, depend on two potentials:

$$g_{00} = 1 + 2\Phi_N, \quad g_{rr} = -(1 - 2\Phi_N + 2\Phi_P). \tag{8}$$

These two potentials obey Poisson equations with running constants $\tilde{G}_N$ and $\tilde{G}_P$ given by linear combinations of $\tilde{G}^0$ and $\tilde{G}^1$:

$$\begin{aligned} -\mathbf{k}^2\Phi_a[\mathbf{k}] &= \tilde{G}_a[\mathbf{k}]\frac{4\pi M}{c^2}, \quad a = N, P, \\ \tilde{G}_N \equiv \frac{4\tilde{G}^0 - \tilde{G}^1}{3}&, \quad \tilde{G}_P \equiv \frac{2(\tilde{G}^0 - \tilde{G}^1)}{3}. \end{aligned} \tag{9}$$

The standard Einstein equation is recovered when the running constants $\tilde{G}^0$ and $\tilde{G}^1$ are momentum-independent and equal to each other, i.e. also when

$$\begin{aligned} \left[\tilde{G}_N\right]_{\text{st}} \equiv G_N&, \quad \left[\tilde{G}_P\right]_{\text{st}} = 0, \\ \left[\Phi_N(r)\right]_{\text{st}} \equiv \phi(r)&, \quad \left[\Phi_P(r)\right]_{\text{st}} = 0. \end{aligned} \tag{10}$$

The two potentials Φ_a will be written as sums of these standard expressions and anomalies, which have to remain small:

$$\Phi_a(r) \equiv \left[\Phi_a(r)\right]_{\text{st}} + \delta\Phi_a(r), \quad |\delta\Phi_a(r)| \ll 1. \tag{11}$$

This linearized form of the extended theory is quite useful for introducing the ideas in terms of an effective field theory of gravitation. It is, however, not sufficient to deal with the general relation between metric and curvature tensors which involves nonlinear expressions. It is no more satisfactory for the general discussion of gravity tests as some of them also involve nonlinearity of the gravitation theory.

It turns out that the extended theory may in fact be given a full nonlinear formulation, discussed in great detail in Ref. 23. Most formulas are thus written more conveniently in terms of Schwarzschild coordinates[92]:

$$\begin{aligned} ds^2 &= \bar{g}_{00}(\bar{r})c^2dt^2 + \bar{g}_{rr}(\bar{r})d\bar{r}^2 - \bar{r}^2(d\theta^2 + \sin^2\theta d\varphi^2), \\ \bar{g}_{\mu\nu}(r) &\equiv \left[\bar{g}_{\mu\nu}(r)\right]_{\text{st}} + \delta\bar{g}_{\mu\nu}(r), \quad |\delta\bar{g}_{\mu\nu}(r)| \ll 1, \end{aligned} \tag{12}$$

with the standard GR solution treated exactly,

$$[\bar{g}_{00}]_{\text{st}} = 1 - 2\frac{\kappa}{\bar{r}} = -\frac{1}{[\bar{g}_{rr}]_{\text{st}}}, \tag{13}$$

and the anomalous metric dealt with at first order. It is possible to define in the nonlinear theory two potentials, $\delta\bar{\Phi}_N$ and $\delta\bar{\Phi}_P$, which generalize (11) while taking into account the nonlinear corrections involving powers of $\kappa/\bar{r}$.

We do not reproduce here the corresponding calculations but emphasize a few salient features of the results. First, the phenomenological freedom of the extended framework is represented by the two functions $\delta\bar{g}_{00}(r)$ and $\delta\bar{g}_{rr}(r)$, which contain the same information, through the appropriate transformations,[23] as $\delta g_{00}(r)$ and $\delta g_{rr}(r)$, or $\delta\Phi_N(r)$ and $\delta\Phi_P(r)$, or $\delta G_N[\mathbf{k}]$ and $\delta G_P[\mathbf{k}]$. They can also be described by the Einstein curvatures E_0^0 and E_r^r, which no longer vanish outside the source.[23] The PPN family is recovered as a particular case which already shows an anomalous behavior of Einstein curvatures nonnull apart from the gravity source. Once more, the post-Einsteinian metric theory is nothing but an extension of this anomalous behavior with more general dependences of the curvatures versus the distance r to the Sun. In loose words, the post-Einsteinian metric theories can be thought of as an extension of the PPN metric with PPN parameters no longer constants but now functions of r.

6. Phenomenological Consequences

The new phenomenological framework is characterized by the two functions $\delta g_{00}(r)$ and $\delta(g_{00}g_{rr})(r)$. The first function represents an anomaly of the Newton potential which has to remain small to preserve the good agreement between GR and gravity tests performed on planetary orbits.[55,56] Meanwhile, the second sector represents an extension of PPN phenomenology with a scale-dependent Eddington parameter γ. It opens an additional phenomenological freedom with respect to the mere modification of the Newton potential, and this freedom opens the possibility of accommodating a Pioneer-like anomaly besides other gravity tests.[21–23]

Recent publications have forced us to be more specific about the relation between the Pioneer anomaly and modifications of the Newton potential, i.e. anomalies in the first sector according to the terminology of the preceding paragraph. Interpreting the Pioneer anomaly in such a manner requires that δg_{00} vary roughly as r at the large radii explored by Pioneer probes. If this dependence also holds at smaller radii,[13] or if the anomaly follows a simple Yukawa law,[55] one deduces that it cannot have escaped detection in the more constraining tests performed with Martian probes.[52,53] Brownstein and Moffat have explored the possibility that the linear dependence holds at distances explored by Pioneer probes while being cut at the orbital radii of Mars.[24] Other authors[93,94] have in contrast argued that the ephemeris of outer planets were accurate enough to discard the presence of the required linear dependence in the range of distances explored by the Pioneer probes. This argument has been contested by the authors of Ref. 24 and the conflict remains to be settled.

The authors of Refs. 93 and 94 have pushed their claim one step farther by restating their argument as an objection to the very possibility of accounting for the Pioneer anomaly in any viable metric theory of gravity. This claim is clearly untenable, because it only considers metric anomalies in the first sector while disregarding those in the second sector. At this point, we want to repeat that the discussion of the compatibility of metric anomalies with observations performed in the solar system has to be done carefully, accounting for the presence of the two sectors as well as for possible scale dependences. This question has already been discussed in Refs. 22 and 23 for the cases of deflection experiments on electromagnetic sources passing behind the Sun.[40,41] It has a particularly critical character for the ranging experiments which involve directly the Shapiro time delay.[96]

The second potential, $\delta(g_{00}g_{rr})(r)$, naturally produces an anomaly on Doppler tracking of probes with escape trajectories in the outer solar system. This Pioneer-like anomaly can be calculated by taking into account the perturbations on probe motions as well as on light propagation between stations on Earth and probes. The time derivative of the Doppler velocity thus computed can be written as a Doppler acceleration a and the anomaly evaluated as the difference of the values obtained in the extended and standard theories:

$$\delta a \equiv a_{\text{extended}} - a_{\text{standard}}, \quad a \equiv \frac{dv}{dt}. \tag{14}$$

The result of the calculation given in Refs. 21 and 22 was unfortunately corrupted by a mistake. The mistake has been corrected in a recent publication,[26] which also contains the evaluation of an annually modulated anomaly coming out, as the secular anomaly, as a natural consequence of the presence of an anomalous metric in the second sector. As observations of such annual anomalies are reported in Ref. 13, this situation certainly pleads for pushing this study farther and comparing the theoretical expectations with the newly recovered Pioneer data.[62,64,65]

More generally, these data will make available a lot of information on the status of the probes as well as on the Doppler tracking details, for the whole duration of the Pioneer 10 and 11 missions, from their launch to the last data points. The recovery is now completed at JPL[97] and the upcoming data analysis planned as an international effort.[98] Numerous open questions can potentially be solved by this new analysis. The systematics can certainly be much better controlled while several important properties of the force — direction, time variation of the secular anomaly, annual or diurnal modulations, spin contribution, etc. — can be more precisely characterized. Then, the availability of early data may make it possible to confirm whether or not the anomaly arises at Pioneer 11 at the Saturn encounter, as is suggested by Fig. 7 of Ref. 13. Finally, the data will be confronted with the detailed predictions now available for a variety of theoretical proposals.

If we follow the line of thought presented in this paper, the confrontation of data with extended metric theories of gravitation is of particular interest, as the anomaly observed on the trajectories of the Pioneer 10 and 11 probes may well be a first hint of a modification of gravity law in the outer part of the solar system. This possibility would have such a large impact on fundamental physics, astrophysics and cosmology that it certainly deserves further investigations. The evaluations presented in Ref. 26 will allow one to address these questions in a well-defined theoretical framework. It is only after a quantitative comparison, taking into account the details known to be important for data analysis,[13] that it will be possible to know whether the post-Einsteinian phenomenological framework shows the capability of fitting the Pioneer observations.

When using the corrected expression for the secular anomaly, identification with the observed Pioneer anomaly now points to a quadratic dependence of the second potential with radius. This corresponds to a constant curvature (see the evaluations in Ref. 23) with an unexpectedly large value in the outer solar system (see Ref. 26). This quadratic dependence may have to be cut off at distances exceeding the size of the solar system as well as in the inner solar system in order to pass Shapiro tests on Martian probes. As already stressed, it is of crucial importance to check that the modification of GR needed to produce the Pioneer anomaly does not spoil its agreement with other gravity tests. At the same time, this study can lead to Pioneer-related anomalies, produced by the same metric anomalies, but to be looked for in other kinds of experiments or, in some cases, by taking a new look at data of already-performed experiments.

The second potential, $\delta\Phi_P$, has a direct effect on the propagation of light rays. It affects the Eddington deflection experiments as well as the ranging experiments which are sensitive to the Shapiro time delay. These experiments can in fact be described as determining the Eddington parameter γ, with the new feature that the latter can now depend on the heliocentric distance (more discussions in Ref. 22 for deflections amplified near occultation). The results are reduced to PPN ones when γ is a constant. Otherwise, they show that deflection or ranging tests can reveal the presence of $\delta\Phi_P$ in the vicinity of the Sun through a space dependence of the parameter γ. Such a potential dependence might already be looked for through a reanalysis of existing data, such as VLBI measurements,[40] the Cassini experiment,[41] or HIPPARCOS data.[99] It may also be studied in the future through higher accuracy Eddington tests, made possible by the global mapping of deflection over the sky in the GAIA project,[100] or by the high accuracy LATOR mission.[101] For this kind of tests, the goal can be described as a construction of the dependence of the deflection versus the elongation of the ray with respect to the Sun. This function directly probes the space dependence of the second potential,[22] and its unambiguous experimental determination will either produce a clear signature of a deviation of GR or put improved constraints on the existence of the second potential at heliocentric distances smaller than 1 AU.

The presence of $\delta\Phi_P$ can also be sensed in planetary tests. In particular, the perihelion precession of planets has been evaluated in the nonlinear theory.[23] The expression given there, written as an anomaly with respect to GR and truncated after leading ($\propto e^0$) and subleading ($\propto e^2$) orders in the eccentricity e of the planetary orbit, shows that the perihelion precession can be used as a sensitive probe of the value and variation of the second potential. Note that the second potential could in principle be present at the long distances explored by the Pioneer probes, but not at the smaller distance corresponding to the radius of the Mars orbit. This means that it would be extremely interesting to track with accuracy the motions of small bodies which may have significant radial velocities while being at large heliocentric distances. This possibility of testing GR by following small bodies can be considered as a further fundamental challenge for GAIA.[100]

Generally speaking, the eccentricity of the orbits plays a key role in Pioneer-related anomalies. It takes large values for Pioneer-like probes which sense $\delta\Phi_P$, whereas it is zero for circular orbits which do not. This suggests making a dedicated analysis of the intermediate situation, not only for the two categories of bound and unbound orbits, but also for the flybies used to bring Pioneer-like probes from the former category to the latter. It would be worth studying planetary probes on elliptical orbits, for example on transfer orbits from Earth to Mars or Jupiter. Another natural target for such a study could be LISA with its three craft on slightly elliptical orbits.[102]

Finally, there are strong motivations for new missions designed to study the anomaly and try to understand its origin.[103] A cheaper and quicker alternative could be to fly dedicated passenger instruments on planetary missions with different primary purposes. In the meantime, a wise strategy is to develop and validate enabling technologies, such as laser and radio techniques for ranging, accelerometers for controlling the deviation from geodesic motion, and accurate clocks on board for measuring separately the two components of the metric.

Acknowledgments

Thanks are due, for discussions to the members of the Deep Space Gravity Probe team (H. Dittus *et al.*),[103] of the Pioneer Anomaly Investigation Team (S. G. Turyshev *et al.*),[98] and of the Groupe Anomalie Pioneer, in particular F. Bondu, P. Bouyer, B. Christophe, J.-M. Courty, B. Foulon, S. Léon, A. Lévy, G. Metris and P. Touboul.

References

1. C. M. Will, *Theory and Experiment in Gravitational Physics* (Cambridge University Press, 1993).
2. C. M. Will, *Living Rev. Rel.* **4** (2001) 4; http://relativity.livingreviews.org
3. E. Fischbach and C. Talmadge, *The Search for Non-Newtonian Gravity* (Springer-Verlag, Berlin, 1998).
4. E. G. Adelberger, B. R. Heckel and A. E. Nelson, *Ann. Rev. Nucl. Part. Sci.* **53** (2003) 77.
5. A. Aguirre *et al.*, *Class. Quant. Grav.* **18** (2001) R223.
6. A.G. Riess *et al.*, *Astron. J.* **116** (1998) 1009.

7. S. Perlmutter *et al.*, *Astrophys. J.* **517** (1999) 565.
8. S. Perlmutter, M. S. Turner and M. White, *Phys. Rev. Lett.* **83** (1999) 670.
9. R. H. Sanders and S. S. McGaugh, *Annu. Rev. Astron. Astrophys.* **40** (2002) 263.
10. A. Lue, R. Scoccimaro and G. Starkman, *Phys. Rev. D* **69** (2004) 044005.
11. S. M. Carroll *et al.*, *Phys. Rev. D* **70** (2004) 043528.
12. J. D. Anderson *et al.*, *Phys. Rev. Lett.* **81** (1998) 2858.
13. J. D. Anderson *et al.*, *Phys. Rev. D* **65** (2002) 082004.
14. J. D. Anderson, M. M. Nieto and S. G. Turyshev, *Int. J. Mod. Phys. D* **11** (2002) 1545.
15. J. D. Anderson *et al.*, *Mod. Phys. Lett. A* **17** (2003) 875.
16. M. M. Nieto and S. G. Turyshev, *Class. Quant. Grav.* **21** (2004) 4005.
17. S. G. Turyshev, M. M. Nieto and J. D. Anderson, *Adv. Space Res.* **39** (2007) 291.
18. M. M. Nieto, *Phys. Rev. D* **72** (2005) 083004.
19. M. M. Nieto, S. G. Turyshev and J. D. Anderson, *Phys. Lett. B* **613** (2005) 11.
20. O. Bertolami and J. Paramos, *Class. Quant. Grav.* **21** (2004) 3309.
21. M.-T. Jaekel and S. Reynaud, *Mod. Phys. Lett. A* **20** (2005) 1047.
22. M.-T. Jaekel and S. Reynaud, *Class. Quant. Grav.* **22** (2005) 2135.
23. M.-T. Jaekel and S. Reynaud, *Class. Quant. Grav.* **23** (2006) 777.
24. J. R. Brownstein and J. W. Moffat, *Class. Quant. Grav.* **23** (2006) 3427.
25. C. Lämmerzahl, O. Preuss and H. Dittus, in *Lasers, Clocks, and Drag-Free: Technologies for Future Exploration in Space and Tests of Gravity* (Springer, Berlin, 2006), p. 75.
26. M.-T. Jaekel and S. Reynaud, *Class. Quant. Grav.* **23** (2006) 7561.
27. A. Einstein, *Jahrbuch der Radioaktivität und Elektronik* **4** (1907) 411.
28. A. Einstein, *Ann. Phys.* **35** (1911) 898.
29. J. G. Williams, X. X. Newhall and J. O. Dickey, *Phys. Rev. D* **53** (1996) 6730.
30. R. W. Hellings *et al.*, *Phys. Rev. Lett.* **51** (1983) 1609.
31. A. Einstein, *Sitz. der Preuss. Akad. der Wissenschaften zu Berlin* (1915) 844.
32. D. Hilbert, *Nachr. von der Gesellshaft der Wissenshaften zu Göttingen* (1915) 395.
33. A. Einstein, *Ann. Phys.* **49** (1916) 769.
34. A. S. Eddington, *The Mathematical Theory of Relativity* (Cambridge University Press, 1957).
35. H. P. Robertson in *Space Age Astronomy* (Academic, 1962).
36. D. H. Ross and L. I. Schiff, *Phys. Rev.* **141** (1966) 1215.
37. K. Nordtvedt, *Phys. Rev.* **169** (1968) 1014, 1017.
38. C. M. Will and K. Nordtvedt, *Astrophys. J.* **177** (1972) 757.
39. K. Nordtvedt and C. M. Will, *Astrophys. J.* **177** (1972) 775.
40. S. S. Shapiro *et al.*, *Phys. Rev. Lett.* **92** (2004) 121101.
41. B. Bertotti, L. Iess and P. Tortora, *Nature* **425** (2003) 374.
42. C. Talmadge *et al.*, *Phys. Rev. Lett.* **61** (1988) 1159.
43. K. Nordtvedt, gr-qc/0301024.
44. C. D. Hoyle *et al.*, *Phys. Rev. Lett.* **86** (2001) 1418.
45. C. D. Hoyle *et al.*, *Phys. Rev. D* **70** (2004) 042004.
46. J. Long *et al.*, *Nature* **421** (2003) 922.
47. J. Chiaverini *et al.*, *Phys. Rev. Lett.* **90** (2003) 151101.
48. M. Bordag, U. Mohideen and V. M. Mostepanenko, *Phys. Rep.* **353** (2001) 1.
49. A. Lambrecht and S. Reynaud, in *Poincaré Seminar 2002*, eds. B. Duplantier and V. Rivasseau (Birkhaüser-Verlag, Basel, 2003), p. 109.
50. R. S. Decca *et al.*, *Phys. Rev. D* **68** (2003) 116003.
51. F. Chen *et al.*, *Phys. Rev. A* **69** (2004) 022117.

52. R. D. Reasenberg *et al.*, *Astrophys. J.* **234** (1979) L219.
53. J. D. Anderson *et al.*, *Astrophys. J.* **459** (1996) 365.
54. N. I. Kolosnitsyn and V. N. Melnikov, *Gen. Relativ. Gravit.* **36** (2004) 1619.
55. M.-T. Jaekel and S. Reynaud, *Int. J. Mod. Phys. A* **20** (2005) 2294.
56. J. Coy *et al.*, private communication (2003).
57. R. O. Fimmel, W. Swindell and E. Burgess, NASA Publication SP-349/396 (NASA, Washington, 1977); electronic version at http://history.nasa.gov/SP-349/sp349.htm
58. R. O. Fimmel, J. A. Van Allen and E. Burgess, NASA Publication SP-446 (NASA, Washington, 1980).
59. S. W. Asmar *et al.*, *Radio Sci.* **40** (2005) RS2001.
60. T. D. Moyer, *Formulation for Observed and Computed Values of Deep Space Network Data Types for Navigation* (Wiley, New York, 2003); http://descanso.jpl.nasa.gov
61. C. Markwardt, gr-qc/0208046; http://lheawww.gsfc.nasa.gov/users/craigm/atdf
62. S. G. Turyshev, M. M. Nieto and J. D. Anderson, *EAS Publ. Ser.* **20** (2006) 243.
63. M. M. Nieto and J. D. Anderson, *Class. Quant. Grav.* **22** (2005) 5343.
64. S. G. Turyshev *et al.*, *Int. J. Mod. Phys. D* **15** (2006) 1.
65. V. T. Toth and S. G. Turyshev, *Can. J. Phys.* **84** (2006) 1063.
66. T. Damour, *Class. Quant. Grav.* **13** (1996) A33.
67. T. Damour, F. Piazza and G. Veneziano, *Phys. Rev. D* **66** (2002) 046007.
68. J. M. Overduin, *Phys. Rev. D* **62** (2000) 102001.
69. MICROSCOPE Site @ CNES, http://smsc.cnes.fr/MICROSCOPE
70. STEP Site @ Stanford Univ., http://einstein.stanford.edu/STEP
71. S. Weinberg, *Gravitation and Cosmology* (Wiley, New York, 1972).
72. W. E. Thirring, *Ann. Phys.* **16** (1961) 96.
73. R. P. Feynman, *Acta Phys. Pol.* **24** (1963) 711.
74. S. Weinberg, *Phys. Rev. B* **138** (1965) 988.
75. R. Utiyama and B. De Witt, *J. Math. Phys.* **3** (1962) 608.
76. S. Deser and P. van Nieuwenhuizen, *Phys. Rev. D* **10** (1974) 401.
77. D. M. Capper, M. J. Duff and L. Halpern, *Phys. Rev. D* **10** (1974) 461.
78. K. S. Stelle, *Phys. Rev. D* **16** (1977) 953.
79. K. S. Stelle, *Gen. Relativ. Gravit.* **9** (1978) 353.
80. A. D. Sakharov, *Dokl. Akad. Nauk SSSR* **177** (1967) 70.
81. R. J. Adler, *Rev. Mod. Phys.* **54** (1982) 729.
82. T. Goldman *et al.*, *Phys. Lett. B* **281** (1992) 219.
83. C. Deffayet *et al.*, *Phys. Rev. D* **65** (2002) 044026.
84. G. Dvali, A. Gruzinov and M. Zaldarriaga, *Phys. Rev. D* **68** (2003) 024012.
85. G. Gabadadze and M. Shifman, *Phys. Rev. D* **69** (2004) 124032.
86. G. 't Hooft and M. Veltman, *Ann. Inst. H. Poincaré A* **20** (1974) 69.
87. E. S. Fradkin and A. A. Tseytlin, *Nucl. Phys. B* **201** (1982) 469.
88. O. Lauscher and M. Reuter, *Class. Quant. Grav.* **19** (2002) 483.
89. J. Z. Simon, *Phys. Rev. D* **41** (1990) 3720.
90. S. W. Hawking and T. Hertog, *Phys. Rev. D* **65** (2002) 103515.
91. M.-T. Jaekel and S. Reynaud, *Ann. Phys.* **4** (1995) 68.
92. C. W. Misner, K. S. Thorne and J. A. Wheeler, *Gravitation* (Freeman, New York, 1972).
93. L. Iorio and G. Giudice, *New Astron.* **11** (2006) 600.
94. K. Tangen, *Phys. Rev. D* **76** (2007) 042005.
95. J. W. Moffat, *Class. Quant. Grav.* **23** (2006) 6767.
96. I. I. Shapiro, *Rev. Mod. Phys.* **71** (1999) S41.
97. Contact Slava G. Turyshev <turyshev@jpl.nasa.gov>.

98. See the pages of the Pioneer Anomaly Investigation Team at the International Space Sciences Institute, http://www.issi.unibe.ch/teams/Pioneer
99. HIPPARCOS Site @ ESA, http://www.rssd.esa.int/Hipparcos
100. GAIA Site @ ESA, http://www.rssd.esa.int/Gaia
101. LATOR Collab. (S. G. Turyshev *et al.*), *ESA Spec. Publ.* **588** (2005) 11.
102. LISA Site @ ESA, http://www.rssd.esa.int/Lisa
103. Pioneer Explorer Collab. (H. Dittus *et al.*), *ESA Spec. Publ.* **588** (2005) 3.

PART 3

GRAVITATIONAL EXPERIMENT

EXPERIMENTAL GRAVITY IN SPACE — HISTORY, TECHNIQUES AND PROSPECTS

RONALD W. HELLINGS
Physics Department, Montana State University,
Bozeman MT 59717, USA
hellings@physics.montana.edu

This is more an after-dinner talk than it is a presentation of research — nothing new, no references, incomplete. It is, rather, a somewhat personal view of: (1) the history of experimental gravity in space, (2) a discussion on the theoretical basis for the two major thrusts in the field — tests of post-Newtonian gravity and gravitational astronomy, and (3) a couple of brief comments on two of the future space opportunities that are discussed at this meeting, LATOR and LISA. My judgment is that we have done a lot to provide an experimental basis for the theory of gravity, and that we can do even more.

Keywords: History; PPN; gravitational astronomy.

1. History

In 1970, after three years as a graduate student at UCLA, I decided that I would like to do my PhD research in general relativity. I went to talk with the UCLA professor who had given an excellent graduate relativity course to see if there was a possibility of doing some work with him. His answer was that relativity was something a physicist could do for a hobby, but no one could make a living at it. Disappointed, I sought a second opinion from a professor at Caltech, who described an exciting emerging field of experimental GR, where the demand would be high and a well-educated graduate could write his own ticket for his future career. The truth, of course, lay somewhere in between.

Still, I chose to go to Montana State University to work with Ken Nordtvedt, who was just then collaborating with Cliff Will on developing the parametrized post-Newtonian formalism for testing relativistic gravity. I worked on an alternative vector-metric theory of gravity for my dissertation and have managed to make a career in the field, so I guess the Caltech advice was good. But my UCLA professor's advice was also very reasonable for the time. Let me explain why.

We all know of the early corroboration of Einstein's 1916 GR observational predictions — the gravitational redshift of spectral emission lines in the Sun, the explanation of the well-known anomalous perihelion shift of Mercury, and the striking results of Eddington's 1919 eclipse expedition that confirmed the gravitational bending of light to about 20%. These results made Einstein a household name. However, what is less well known is the fact that by the 1950's, the situation was far less clear. The perihelion precession persisted, but the gravitational red shift of solar emission lines repeatedly failed to agree with GR, due essentially to the complicated environment in the photosphere of the Sun where Doppler and pressure broadening made the lines hard to center and superposition of lines from varying depths made it hard to model. And then, most significantly, subsequent eclipse observations measured deflections from $\frac{3}{4}$ to $1\frac{1}{2}$ times the GR value. The experimental support for GR had been badly eroded.[a] It seemed like the theory would never have any observable effects in practical applications, and belief in the theory seemed to be more a matter of religion than of science. During my UCLA years, while I was working part-time as a trajectory engineer at the Rockwell Space Division, another trajectory engineer who knew me very well took me aside confidentially one day to ask if I really believed in relativity. As late as 1978, when the first GPS satellites were launched, the needed relativistic corrections to the modeling software were included as an option which could be turned on or off, so as not to place the US military at the mercy of ivory-tower theorists.

It was the space age that changed everything — access to space away from drag and other nongravitational forces, precise clocks and frequency standards, precise laser and radar ranging techniques, and digital computing to analyze trajectories. The two decades of the 1960's and 1970's changed GR from an impractical theory with no observable consequences to an essential engineering technique, verified to parts in a thousand. Let me outline what seem to me to have been the important steps during this time frame:

1960: Pound and Rebka test the gravitational red shift using the Mossbauer effect on Fe^{57} γ-ray emission and absorption in the 74-foot elevator shaft in the Jefferson Tower at Harvard, verifying the Einstein prediction to $\pm 1\%$.

1961: Brans and Dicke publish their scalar-tensor theory of gravity, a complete and viable Lagrangian-based alternative to GR, predicting results for light-bending and perihelion precession that differ from those of GR.

1961: The Fermi School on Evidence for Gravitational Theories is held in Varenna, Italy. This is the first international conference dedicated to experimental gravitation. Although most of the lectures discuss theoretical alternatives to GR, there are talks by Weber on methods for detecting gravitational waves and by Bertotti giving a theoretical framework for observational tests of relativity.

1963: Shapiro publishes a paper deriving the gravitational time delay for a signal passing near the sun, a fourth "classical" test of relativity.

[a]These elements of history are documented in: Clifford M. Will, *Was Einstein Right?* (Harper-Collins, 1986).

1967: The first radar ranges to Mercury verify the gravitational time delay to $\pm 20\%$.

1968: Weber publishes results from his aluminum bar detector experiment. Though there seem to be gravitational wave pulses detected, analysis of his data by others puts the detections in question. (Subsequent, more sensitive detectors also fail to see pulses at the amplitude claimed by Weber.)

1968: Isaacson publishes a paper showing how to define energy in gravitational waves. There had been a long-standing suggestion that wave-like solutions to Einstein's equations were just coordinate anomalies. This paper shows that they cannot be transformed away.

1968: Nordtvedt publishes a paper explaining how many alternative gravity theories predict a violation of the strong equivalence principle, in which self-gravity does not couple to an external gravitational field in the same way that matter-energy does. This effect should be testable with precise ranging to the Moon.

1969: A plate of corner-cube reflectors is placed on the Moon by the astronauts of Apollo 11. Subsequent laser ranging to the Moon has produced a nearly-40-year record of observations, now at an accuracy of ± 1.5 cm. The strong equivalence principle is now verified to an accuracy of 5×10^{-4}.

1970: A spacecraft ranging to the Mariner 6 and 7 spacecraft during an interplanetary cruise at solar conjunction measures the gravitational time delay to $\pm 3\%$, limited by unmodeled nongravitational forces on the spacecraft.

1971: At a summer "Workshop on Experimental Gravity" at Montana State University, Will and Nordtvedt work out the parametrized post-Newtonian (PPN) formalism. This is a form of the space–time metric with PPN parameters as coefficients of the potentials, allowing observations to be related to the parameters. The form is valid for a broad class of theories of gravitation, different theories corresponding to different values of the parameters.

1972: A spacecraft ranging to the Mariner 9 spacecraft in orbit around Mars determines PPN parameter γ to $\pm 2\%$. Since Mariner 9 is orbiting Mars, it is anchored to a body with no significant nongravitational forces acting on it. The remaining error is a result of scintillation of the radio tracking signal due to the plasma in the solar corona.

1972: The Fermi School on Experimental Gravitation is held in Varenna, Italy. The PPN formalism is presented by Will; Weber and Braginsky debate the sensitivity of bar gravitational wave antennas; Dicke suggests that a large measured quadrupole moment for the Sun means that GR does not correctly predict the perihelion shift of Mercury; Anderson explains solar system data analysis techniques and results.

1974: Hulse and Taylor discover PSR1913+16, the binary pulsar. Four years later, the change in frequency due to loss of orbital energy to gravitational radiation is observed. This indirect proof of the existence of gravitational waves earns Hulse and Taylor the Nobel Prize in 1994.

1976: The Viking Landers land on Mars. Anchored directly to the planet, with 5 m-accuracy ranging systems and concurrent dual-band ranging from the Viking orbiters to calibrate the effects of solar plasma, the six years of Viking range data (1976–1982) provide a measure of PPN parameter γ that agrees with the prediction of GR to $\pm 10^{-3}$.

1976: Vessot launches a hydrogen maser on the suborbital Gravity Probe-A rocket flight. By comparing the spaceborne clock with clocks on the Earth, the gravitational red shift is verified to $\pm 10^{-4}$.

1979: Using the Voyager spacecraft and the Earth as two free masses, a search is made for LF gravitational waves (with periods from 10 s to 1000 s). The amplitude sensitivity is 3×10^{-14}, corresponding to a signal-to-noise ratio for plausible sources of 10^{-7} (meaning that signals should be too weak by seven orders of magnitude to be observed).

There was little progress in experimental gravitation in space for the following two decades. There were several disappointments with new missions to Mars (the Soviet Phobos lander in 1988, JPL's Mars Observer in 1992 and Mars Climate Orbiter in 1998), so the main activity consisted in mining the existing planetary data for new effects — testing Moffat's theory, looking for a variable gravitational constant, looking for a deviation from the inverse-r-squared law of gravity, etc. On the other hand, ground-based experiments began to flourish. The laboratory gravitational-wave projects LIGO and VIRGO were proposed, funded, and built. A flurry of activity in testing the equivalence principle and the inverse-r-squared law at laboratory distances was motivated by Fischbach's detection of a signal for an equivalence-principle-violating fifth force in Eötvös's original data.

Then came the new century, and, with it, another significant improvement in the knowledge of the PPN parameters. In contrast to the ~1300 Viking and Mariner 9 ranging points, we now have ~150,000 Pathfinder, Odyssey, and Mars Global Surveyer ranges, with a higher-frequency X-band ranging system that is an order of magnitude less sensitive to plasma delays. The result of this reinstrumenting of the solar system laboratory is a 30-year stretch of planetary data, with recent accuracies reaching down to about ± 1 m. Additional lunar laser data have accumulated as well, but there has also been one surprising breakthrough in PPN parameter γ due to an experiment using the Cassini spacecraft.

Beginning in the late 1990's, Bertotti, Iess, and Tortura began to point out that the gravitational time delay did not require knowing the absolute delay, but could be measured by noting the change in the delay that would be seen as a spacecraft moved quickly along its path on the far side of the Sun, as seen from the Earth. This was important, because the Cassini spacecraft had no ranging system on board, but had a very precise dual-frequency Doppler tracking system. Finally, in 2002, when the Cassini spacecraft passed behind the Sun, a careful data-taking session was used to produce the apparent Doppler shifts shown in Fig. 1.

The Cassini experiment gave the result $\gamma - 1 = (2.1 \pm 2.3) \times 10^{-5}$. Preliminary results from the newly reinstrumented solar system give $\beta - 1 = (1.0 \pm 1.1) \times 10^{-4}$. The 21st century data have already produced an order-of-magnitude improvement in our knowledge of the PPN parameters.

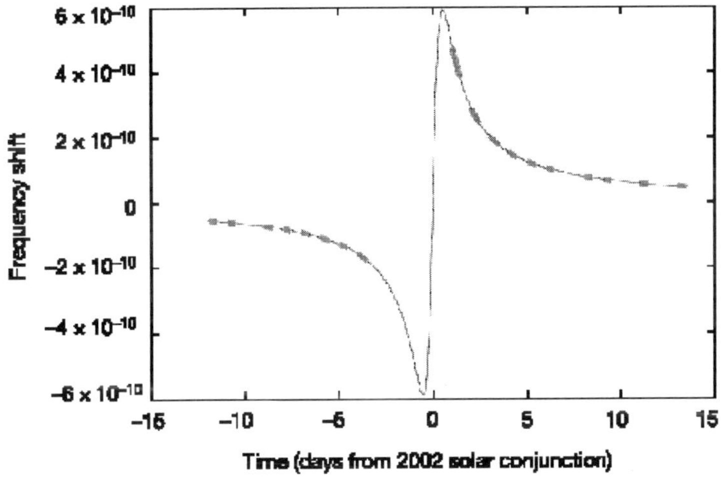

Fig. 1. Relative Doppler shifts $\Delta f/f$ as a function of time for the 2002 Cassini solar conjunction experiment. The small dots are the data; the line is the theoretical prediction of GR.

2. PPN Tests

In this section, we develop the theory of the PPN tests. The PPN metric is a metric made up of terms that arise in GR and in other theories of relativistic gravitation. At the present accuracy of solar system measurements, the form of the PPN metric applicable to solar system dynamics is

$$g_{00} = 1 - 2\frac{GM}{x}\left[1 - \frac{J_2}{x^2}P_2(\hat{\mathbf{k}}\cdot\hat{\mathbf{x}})\right] - 2\sum_a \frac{Gm_a}{x_a} + 2\beta\frac{G^2M^2}{x^2} + \alpha_1\frac{GM}{x}\frac{w}{c}, \quad (1)$$

$$g_{0k} = \alpha_1\frac{GM}{x}\frac{w^k}{c}, \quad (2)$$

$$g_{ij} = -\left(1 - 2\gamma\frac{GM}{x} - 2\gamma\sum_a\frac{Gm_a}{x_a} + 2\delta\frac{G^2M^2}{x^2}\right)\delta_{ij}. \quad (3)$$

Here, M is the mass of the Sun, m_a is the mass of planet a, x is the distance between the field point and the Sun, x_a is the distance between the field point and planet a, J_2 is the quadrupole moment of the Sun, $\hat{\mathbf{k}}$ is a unit vector normal to the Sun's equator, $\hat{\mathbf{x}}$ is a heliocentric unit vector toward the field point, $\mathbf{w} = (-354.44, 28.34, 34.08)$ km/s is the velocity of the solar system relative to the mean rest frame of the universe as determined by the COBE dipole signature, and $\beta, \gamma, \alpha_1, \delta$ are PPN parameters.

Using this form of the metric, gravitational geodesic equations are derived. (These are too complicated and not instructive enough to reproduce here.) To find the positions of the solar system bodies, the geodesic equations are integrated, using estimated values for the constants in Eqs. (1)–(3) and for the initial positions and velocities for each body. Using these positions, the times of flight of tracking signals can be modeled by integrating the null geodesic condition,

$$ds^2 = g_{00}dt^2 + g_{0k}dtdx^k + g_{ij}dx^i dx^j, \tag{4}$$

from the time and position of emission to the time and position of reception. The result of this integration is the light propagation equation

$$\tau = |\mathbf{x}_r - \mathbf{x}_e| + (1+\gamma)GM \ln\left[\frac{x_r + \mathbf{x}_r \cdot \hat{\mathbf{n}}}{x_e + \mathbf{x}_e \cdot \hat{\mathbf{n}}}\right], \tag{5}$$

where we assume that the gravitational field of bodies besides the Sun may be neglected, where $\mathbf{x}_r$ and $\mathbf{x}_e$ are positions of reception and emission, respectively, and $\hat{\mathbf{n}}$ is a unit vector oriented along the path of the signal propagation. Given the known reception time and position of a signal, the time of flight of the signal will determine the time (and position) of emission, and the time (and position) of emission will determine the time of flight. The propagation equation must therefore be iterated to arrive at a self-consistent predicted value for τ.

Currently, the solar system data include the following:

- 45,893 optical transit circle observations (α, δ pairs, $\pm 1''$) of the Sun, Moon, and planets;
- 636 radar ranges (± 1 km) to Mercury;
- 544 radar ranges (± 1 km) to Venus;
- 23 closure points (differenced radar ranges to the same surface point) to Mercury;
- 1272 Viking range points (± 10 m) to Mars (1976–1982);
- 89 Pathfinder Lander range points (± 2 m) to Mars (6/1997–9/1997);
- 89,949 Mars Global Surveyer (MGS) range points (± 2 m) to Mars (1999–2003);
- 38,524 Odyssey range points (± 2 m) to Mars (2002–2003);
- 16,250 laser ranging normal points (± 1.5 cm) to Moon (1969–01/2006).

The model of an observation, such as a range point to a Viking Lander, includes a host of parameters whose values must be determined from the observations. These include the initial positions and velocities of the planets, the masses of the planets and several individual asteroids, the mean densities of about 300 additional asteroids, divided into three taxonomic families, the PPN parameters, and parameters that are particular to each data set, such as the position of the lander on Mars, the Mars physical ephemeris (rotation), the position of antennas on Earth, etc. When *a priori* values are assumed for these parameters, the model of an observation and the actual observation still do not exactly coincide, for two reasons — random measurement error and incorrect *a priori* values for the parameters. The goal of solar system data analysis is to adjust the parameters so that the differences between prediction and observation are reduced to the random measurement errors only. The differences, the residuals, are represented via

$$y_i = \tau_{i,\text{observed}} - \tau_{i,\text{computed}} = \frac{\partial \tau_i}{\partial q_1}\delta q_1 + \frac{\partial \tau_i}{\partial q_2}\delta q_2 + \frac{\partial \tau_i}{\partial q_3}\delta q_3 + \cdots + \nu_i, \tag{6}$$

where q_α are the parameters and ν_i are the measurement noises. Thus each parameter has its own signature $(\partial\tau_i/\partial q_\alpha)$ which may be used in conjunction with the other signatures to soak up trends in the data. The amount of each signature needed gives the correction to the parameter required to best fit the data. The fit is accomplished using linear least-squares formalism. The uncertainty in the value for a parameter is determined partly using the formal uncertainty produced by the least-squares covariance matrix and partly from numerical experiments with the solutions designed to root out effects of possible systematic errors.

3. Gravitational Astronomy

In this section, we develop the ideas of gravitational wave detection. Let us begin by writing the metric for the space–time in the wave zone of a binary star. For a binary of angular frequency ω, with initial position in its orbit ϕ_0 and with an orbit plane whose normal makes an angle i with the propagation vector from the source to the detector, the binary produces two polarizations of the gravitational wave:

$$\bar{h}_+ = (1 + \cos^2 i)\mathcal{A}\cos(2\omega t + 2\phi_0), \tag{7}$$

$$\bar{h}_\times = (-2\cos i)\mathcal{A}\sin(2\omega t + 2\phi_0). \tag{8}$$

In these equations, $\mathcal{A}$ is the amplitude of the wave, falling off as $1/r$. The line element in the presence of the weak gravitational wave is

$$ds^2 = dt^2 - (\eta_{ij} + h_{ij})dx^i dx^j. \tag{9}$$

In a coordinate system oriented with the z axis along propagation vector $\hat{\mathbf{k}}$, the spatial part of the metric takes the familiar form

$$(h_{ij}) = \begin{pmatrix} \bar{h}_+ & \bar{h}_\times & 0 \\ \bar{h}_\times & -\bar{h}_+ & 0 \\ 0 & 0 & 1 \end{pmatrix}. \tag{10}$$

The detector reference frame, however, is not likely to be the same as the source reference frame, so in the detector coordinates the gravitational wave takes on the more general form

$$h_{ij}(t) = h_+(t, \mathbf{x})\epsilon^+_{ij}(\hat{\mathbf{n}}) + h_\times(t, \mathbf{x})\epsilon^\times_{ij}(\hat{\mathbf{n}}), \tag{11}$$

where $\hat{\mathbf{n}} = -\hat{\mathbf{k}}$ is a unit vector in the direction of the source. The $\epsilon^{+/\times}_{ij}$ are related to the h_{ij} by coordinate rotations, and the amplitudes depend on the position of the detector as

$$h_{ij}(t, \mathbf{x}) = h_{ij}(t - \hat{\mathbf{n}} \cdot \mathbf{x}), \tag{12}$$

the relationship between the phase of $\bar{h}_{+/\times}$ and $h_{+/\times}$ depending on the position of the detector. In order to detect a wave, the distance between two free masses must be monitored. The tracking signal will travel a null path in the local geometry, satisfying

$$dt^2 = (\eta_{ij} + h_{ij})dx^i dx^j = ds^2 \left(1 + h_{ij}\frac{dx^i}{ds}\frac{dx^j}{ds}\right). \tag{13}$$

For small amplitudes, the square root may be approximated, giving an integrated time of flight from emission of the tracking signal to its reception of

$$\ell \equiv \int_e^r dt \approx \int_e^r ds + \frac{1}{2}\int_e^r h_{ij}\frac{dx^i}{ds}\frac{dx^j}{ds}ds, \tag{14}$$

or

$$\ell = s + \frac{1}{2}\hat{s}^i\hat{s}^j \int_e^r h_{ij}ds = s + \frac{1}{2}\hat{s}^i\hat{s}^j\left[H_{ij}(r) - H_{ij}(e)\right], \tag{15}$$

where $\hat{s}^i$ are the components of a unit vector along the tracking signal and $H_{ij}(t) \equiv \int h_{ij}(t)dt$ is the antiderivative of h_{ij}. The change in the time of flight produced by the gravitational wave is

$$\Delta\ell = \ell - s = \frac{1}{2}\hat{s}^i\hat{s}^j[H_{ij}(t_r - \hat{\mathbf{n}}\cdot\mathbf{x}_r) - H_{ij}(t_e - \hat{\mathbf{n}}\cdot\mathbf{x}_e)]. \tag{16}$$

For a signal received at $(t_r = t, \mathbf{x}_r = 0)$ the emission would have occurred at $(t_r = t - s, \mathbf{x}_r = s\hat{\mathbf{s}})$, so Eq. (16) becomes

$$\Delta\ell = \frac{1}{2}\hat{s}^i\hat{s}^j\left\{H_{ij}[t - (1 + \hat{\mathbf{n}}\cdot\hat{\mathbf{s}})s] - H_{ij}[t]\right\}. \tag{17}$$

This shows that the change in apparent distance between the two free masses is proportional to the *integral* of the gravitational wave. To find a quantity that is proportional to the gravitational wave itself, we must take the derivative of Eq. (17) to get the apparent change in radial velocity:

$$\frac{d(\Delta\ell)}{dt} = \Delta\frac{d\ell}{dt} = \frac{1}{2}\hat{s}^i\hat{s}^j\{h_{ij}[t - (1 + \hat{\mathbf{n}}\cdot\hat{\mathbf{s}})s] - h_{ij}[t]\}. \tag{18}$$

Thus we see that the effect of a gravitational wave on the signal linking two free masses is to change the apparent velocity between the bodies, with the wave amplitude appearing twice in the time series — once at the moment when the wave strikes the receiver ($h_{ij}[t]$) and once when the signal emitted as the wave strikes the emitter gets down to the receiver ($h_{ij}[t - (1 + \hat{\mathbf{n}}\cdot\hat{\mathbf{s}})s]$). In the limit where the wavelength is large compared to the light time between the masses, the two terms in Eq. (18) overlap and cancel, leaving

$$\frac{d(\Delta\ell)}{dt} \approx -\frac{1}{2}\hat{s}^i\hat{s}^j(1 + \hat{\mathbf{n}}\cdot\hat{\mathbf{s}})s\dot{h}_{ij}[t], \tag{19}$$

where the dot denotes time differentiation. If we integrate Eq. (19) we get back to an expression that is proportional to the wave amplitude itself:

$$\frac{\Delta\ell}{\ell} \approx -\frac{1}{2}\hat{s}^i\hat{s}^j(1 + \hat{\mathbf{n}}\cdot\hat{\mathbf{s}})h_{ij}[t], \tag{20}$$

Equation (18) gives the general response of a tracking signal to the passage of a gravitational wave, with Eq. (20) applying in the limit $\lambda \ll s$.

Pulsar timing detectors use a pulsar and the Earth as free masses and use the pulsar pulses as a tracking signal. Because of the light time between emission and reception of the pulse (~1000 yrs), the two-pulse structure in Eq. (14) will not become apparent over the lifetime of the observer. However, researchers have used the fact that the second term in Eq. (18) will occur at the same time regardless of the direction to the pulsar to do a cross-correlation of timing signals from pairs of pulsars to dig into the timing noise to search for a gravitational background.

Doppler detectors use a spacecraft and the Earth as free masses and use the round-trip Doppler as the tracking signal. Two versions of Eq. (18) are combined, producing a three-pulse response to the gravitational wave. [Since the immediate second term in Eq. (18) for the uplink and the delayed first term for the downlink occur at the same time, there are three pulses and not four.] With very high radio frequencies and dual frequencies to calibrate charged particles in the beam, the plasma scintillation noise has been reduced, but the experiments are still limited by troposphere noise and by jitter in the frequency standard.

Space interferometers like LISA use three free-flying spacecraft as end masses and use lasers to track the relative positions of the spacecraft. Since lasers are typically less stable than the frequency standards used for Doppler detectors, the signals in two different arms are combined and subtracted to eliminate laser phase jitter. If the two arms could be maintained at equal lengths, a simple subtraction of two round-trip Doppler signals could be used. However, when the arms are unequal, a set of one-arm signals may be combined with appropriate time lags to form time series in which laser phase jitter is canceled exactly. In this time-delay interferometry (TDI), each one-way signal is modeled using Eq. (18).

4. Prospects

Available techniques for experimental gravity in space, especially techniques using spaceborne lasers, continue to expand and improve. Of the projects that have been proposed, two seem particularly feasible and scientifically interesting. Both of them are discussed in this symposium.

LATOR is a proposed mission that should improve the determination of γ by nearly four orders of magnitude. Such breakthroughs are rare in science. At the projected accuracy of a few parts in 10^9, LATOR will get down to a sensitivity below the point where many models of string-motivated quantum gravity predict that effects of the new theories will appear.

The mission consists of two pieces — a set of two laser transmitter spacecraft in heliocentric orbit and an optical interferometer. The interferometer is currently envisioned to fly on board the Space Station but could also possibly be flown in a dedicated Earth-orbiting satellite. The two satellites are separated by about 1°, as seen from the Earth. As they pass behind the Sun, their laser signals are gravitationally bent by different amounts, each being measured to an accuracy of 10^{-12}

radians. A recent covariance study of LATOR sensitivity predicts the following accuracies:

- γ to $\pm 7.3 \times 10^{-10}$;
- δ to $\pm 1.8 \times 10^{-4}$;
- J_2 to $\pm 1.3 \times 10^{-9}$.

We may thus expect a breakthrough in tests of relativistic gravity, determining γ to an interesting level of accuracy and producing the first measurement of a post-Newtonian parameter, δ.

LISA is a planned mission that will detect gravitational waves in the LF band. It consists of three free-flying drag-free spacecraft that track each other with lasers. Using TDI, a set of Michelson-type interferometers can be formed that cancel laser phase jitter and produce a sensitivity sufficient to see a number of known and unknown sources. LISA will be able to detect a handful of known sources, interacting white dwarf binaries and others, providing a clean physics experiment in which detection verifies the theory and nondetection falsifies the theory. The SNR for these sources should reach 100. In addition, there will probably be a confusion background of close, compact white dwarf binaries and thousands of individual separable sources, providing a catalog for use in stellar evolution studies, especially common envelope binary studies. LISA's most exciting observation will be the inspiral and coalescence of binary massive black holes, formed by the merger of colliding galaxies containing massive black holes in their nuclei. Estimates have been made suggesting that several to several hundred such observations will occur over the LISA lifetime, and SNR's should reach the thousands. Finally, there is the chance of seeing the inspiral of compact solar-mass objects into massive black holes, where the wave forms will probe and trace out the space–time around the hole — the closest thing one can imagine to "seeing" a black hole.

5. Conclusion

Space experiments and observations have been the critical element in our present verification of general relativity, and we seem to be poised on the threshold of new spaceborne observations that will deepen our understanding of the fundamental laws of the Universe — tests of weak gravity capable of seeing nonlinear effects, tests of strong gravity via the gravitational waves produced by massive black binary inspiral and coalescence, dedicated space cosmology missions that will refine our theories of dark energy and dark matter. It is an exciting time to be working in such an exciting field.

PROBING SPACE–TIME IN THE SOLAR SYSTEM: FROM CASSINI TO BEPICOLOMBO

LUCIANO IESS
Università La Sapienza, Rome, Italy
Luciano.Iess@uniroma1.it

SAMI ASMAR
Jet Propulsion Lab, California Institute of Technology,
4800 Oak Grove Dr., Pasadena, CA 91109-8099, USA
sami.asmar@jpl.nasa.gov

Spacecraft radio science techniques can be used for precision solar system tests of relativistic gravity, as was demonstrated by the measurement of the Doppler shift of radio signals with the Cassini mission. Similar experiments are planned for the BepiColombo mission to Mercury. Recent theoretical developments based on string theory and inflationary cosmologies link the validity of general relativity to the expansion of the Universe and indicate that violations may be within the reach of future, precise experiments. In spite of the uncertainty of the theoretical scenarios, the motivations for further tests of gravitational theories are stronger then ever: string theory, new cosmological observations, the hypotheses of dark matter and dark energy, all point to the need for a new and more profound understanding of the Universe and its laws, including the laws of gravity. This paper describes experiments for probing space–time in the solar system with the Cassini and BepiColombo missions, and discusses the experimental limitations of microwave systems used for these tests, including attitude motion and nongravitational accelerations of the spacecraft, propagation noise, and mechanical noise of ground antenna.

Keywords: Relativistic gravity; radio science; spacecraft tracking.

1. Introduction

Is general relativity (GR) the ultimate theory of gravity? In spite of the striking experimental success, questions about the theory's range of validity arise from the derivation of the field equations, since there is an attractive but arbitrary criterion of mathematical beauty and simplicity, as well as from the failure to reconcile it with quantum mechanics. The simplest among the proposed modifications entails the inclusion of a scalar term in the field equations. This additional term, considered by Einstein himself, would affect the long range behavior of gravity, with consequences for the motion of bodies from planetary to cosmological scales. Modifications of Newton–Einstein laws of gravity at short spatial scales have also been investigated, for example by adding a Yukawa-like potential to the Newtonian potential of a point mass.

At cosmological scales, GR has generally been used in models of the Universe's evolution, but the available astronomical data are not accurate enough to support GR over alternative theories. Rather, the suggestion of dark energy to explain the recent observations of the accelerated expansion of the Universe is often associated with a cosmological, scalar term in the field equations. It is not unlikely that future, more accurate observations and models of the Universe at large distances will require modifications of GR. Theoretical attempts to link inflationary cosmological models to string theory have also been made, as briefly indicated in the next section. In these models, current violations of GR are traced to the small and decreasing scalar field remnant of the inflation.

After the recent theoretical and observational developments, the motivations for further, more accurate tests of GR are strong. Unfortunately, no reliable prediction exists on the level where a violation would occur in any feasible experiment. Given this uncertainty, any test that improves the current results is significant. The most accurate test of GR so far has been carried out with the Cassini spacecraft on its way to Saturn.[4] The experiment was proposed after the original selection of the investigations and made use of existing radio science instrumentation without requiring any additional ground or spacecraft resources other than antenna tracking time. Future planetary and astrometry missions, such as BepiColombo and GAIA, will offer additional opportunities to carry out improved gravity experiments with little additional cost, although only dedicated and therefore much more expensive missions will be able to provide improvements of several orders of magnitude.

Radio science investigations of the relativistic gravity rely on precision range and range-rate tracking of interplanetary spacecraft. The Ka band (32–34 GHz) uplink and downlink technology introduced into deep space tracking systems with the Cassini mission, and uniquely available at one of NASA's Deep Space Network (DSN) stations, has proven to be a powerful tool for scientific research, especially when combining the Ka band with the X band ($\sim$8 GHz) to form a multifrequency link system that reduces the Doppler noise to a remarkably small 1 μm/s, at 1000 s integration time, after tropospheric calibration, overcoming susceptibility to interplanetary plasma.[1] Largely relying on existing ground and spacecraft instrumentation, radio science tests of relativistic gravity can be hosted on any dynamically stable spacecraft. Brief descriptions follow of the Cassini solar conjunction experiment and the gravitational measurements planned for BepiColombo, the mission to Mercury of the European Space Agency (ESA).

2. The Cassini Experiment

In the summer of 2002 the Cassini spacecraft was tracked by the antennas of NASA's Deep Space Network across a superior solar conjunction period in an experiment aiming to measure the effect of solar gravity on the propagation of photons.[4,6] This experiment confirmed the predictions of GR with an experimental error of 23 parts per million. Like other classical tests of GR, such as the deflection and time delay of photons, the Cassini test is affected only by the space components of the solar metric.[9,16] In the framework of the PPN formalism, appropriate to solar system experiments, all metric theories of gravity are classified according to the value assumed by ten parameters appearing as coefficients of the expansion of the metric in terms of small potentials of the order of the dimensionless Newtonian gravitational potential $\epsilon \approx GM/c^2$. In the simplest formulation (the Eddington–Robertson–Schiff form), only two parameters are needed, β and γ, controlling the time and space components, respectively, of the metric tensor and affecting all solar system tests carried out so far. In GR, β and γ have a unit value.

Although all experiments, including the Cassini solar conjunction experiment, agreed with the predictions of GR, theoretical arguments support the existence of violations. In a seminal work, Damour and Nordtvedt[5] suggested the existence of a small scalar field, a remnant of the inflation, which would modify the metric tensor and generate a nonzero value for $\beta - 1$ and $\gamma - 1$. Remarkably, the violations of GR would be time-dependent, with the largest deviations occurring near the big bang, while $\beta - 1$ and $\gamma - 1$ approach zero as time approaches infinity. GR would then become an asymptotic theory, an attractor to which any theory of gravity would tend as the expansion of the Universe proceeds. Although this theoretical scenario is very attractive for its attempt to link GR, cosmology and string theory, no precise prediction is made regarding the level at which a violation of GR may occur. The suggestion that experimental results may deviate from GR values at levels of 10^{-7}–10^{-5} can be regarded as purely speculative. Nonetheless, the motivation for further, more precise tests of gravitational theories is stronger than ever: an experimental violation of Einstein's theory would be a landmark discovery in physics, whose consequences in fundamental physics and cosmology would be as unpredictable as profound.

Until Cassini, the most accurate tests of GR have been based on two classes of experiments: the gravitational deflection of radio waves using VLBI and the time delay of radio waves as they propagate between the Earth and an interplanetary spacecraft. A photon passing at a small distance b from the Sun (whose gravitational radius is R_g) is deflected by an amount

$$\theta = 2(1+\gamma)\frac{GM}{bc^2} = 4 \times 10^{-6}(1+\gamma)\frac{R_g}{b} \text{ rad.} \tag{1}$$

By comparing the angular separation of 541 radio sources near and far from solar conjunction, VLBI measurements have been able to constrain $\gamma - 1$ to the value $(-1.7 \pm 4.5) \times 10^{-4}$.[12]

The accuracy of this determination, based upon a large set of observations, improved by a factor of 2 the results of the time delay experiment carried out in 1978 using ranging signals to the Viking landers on the surface of Mars.[10] The Viking experiment exploited the accurate measurement of the two-way propagation time of a modulated signal between a ground antenna and the Mars landers. Ranging measurements were carried out by modulating an S band carrier (2.1 GHz) with a set of square waves (the ranging tones) which are received by the landers and coherently retransmitted back to the Earth. A correlation of the received signal with a suitably phase-advanced replica of the transmitted signal provides the round-trip light-time measurement with an error of a few meters. As integration times of several minutes are required, the ranging system uses the Doppler measurement of the range rate to advance the phase of the transmitted signal prior to the correlation. By exploiting a range rate measurement, consider the Doppler-aided phase advance as if the ground station is in motion at a speed equal to the actual range rate, *de facto* electronically freezing the distance between the ground station and the lander to the value at the start of the integration. For a signal propagating close to the Sun, the two-way Newtonian propagation time is increased by an amount approximately equal to[16]

$$\Delta t \simeq 2(1+\gamma)\frac{GM}{bc^3}\ln\frac{4r_e r}{b^2} \simeq (1+\gamma)\left(240 + 20\ln\frac{(b/R)^2}{r/r_e}\right)\mu\text{s}. \tag{2}$$

In this equation, b is the distance of closest approach of the beam to the Sun, r and r_e the distance of the spacecraft and the Earth from the Sun, and is the solar radius. For a beam grazing the Sun, the delay amounts to about 460 μs (138 km), a quantity which must be compared to the accuracy of the ranging system of tens of nanoseconds (a few meters). Unfortunately the Newtonian light time is not accessible and, in this experiment, only the variation of the delay across the conjunction is effectively measured. The variations are affected by the second term in the above equation, which is not only smaller but has a much faster decay with the distance b. However, the limiting factor in the Viking experiment was the effect of the solar corona, a dispersive and turbulent medium which induces a significant noise in range measurements. Nonetheless the time delay test resulted in a determination of $\gamma - 1$ compatible with GR within an experimental error of 1×10^{-3}, an outstanding result unmatched for almost 25 years.

Note that Cassini did not measure the travel time of photons to and from the spacecraft but exploited precision measurements of the spacecraft's radial velocity. The radial velocity (a more precise description is "the two-way range rate") is obtained by measuring the phase of the received carrier of the radio signal. Phase measurements are extremely accurate due to the use of highly stable frequency standards in the generation of the uplink signal and in the detection of the received phase. In addition, both the ground and spacecraft electronics are especially designed to preserve the phase coherence of the radio link. Although the Cassini spacecraft supported range measurements accurate to 1–2 m in the X band channel (7.2–8.4 GHz), they are far less useful than direct range rate measurements for the detection of the relativistic effect.

Due to the orbital motion of the spacecraft and the Earth during the conjunction period, the closest approach distance changes with time, inducing a relative frequency shift of the carrier approximately equal to

$$\frac{\Delta\nu}{\nu} = \frac{d\Delta t}{dt} \simeq -4(1+\gamma)\frac{GM}{bc^3}\frac{db}{dt} = -(20\ \mu\text{s})(1+\gamma)\frac{1}{b}\frac{db}{dt}. \tag{3}$$

During the 2002 experiment, the Cassini spacecraft was en route to Saturn (its final destination) at a heliocentric distance of 7.4 AU. The quantity db/dt is therefore very close to the Earth's orbital velocity and the relative frequency shift is of the order of 10^{-10}–10^{-9}, depending on the distance of the radio beam from the Sun. This quantity is almost five orders of magnitude larger than the measurement noise of the Cassini radio system.

All experiments based upon the propagation radio signals are strongly affected by the solar corona, turbulent plasma causing random variations of the optical path. As the refractive index of the coronal plasma in the microwave region of the spectrum is inversely proportional to the square of the carrier frequency (magnetic corrections are negligible beyond 2–3 solar radii), the Cassini experiment exploited a multifrequency radio link in X and Ka bands for a nearly complete cancellation of the plasma effects.[3,13,14] Due to its immunity to plasma noise, the Cassini experiment resulted in an improved confirmation of GR, with $\gamma - 1 = (2.1 \pm 2.3) \times 10^{-5}$.

3. Measurement Techniques

The crucial component of the plasma noise cancellation system is a Ka band radio link (34 GHz uplink, 32.5 GHz downlink) especially developed for Cassini radio science experiments. The 34 m DSN station in California (designated DSS-25) underwent significant upgrades to meet stringent radio science requirements for precision Doppler investigations, including the separation of the transmitting and receiving feeds to account for aberration effects.

On board the Cassini spacecraft, the key instrument is a Ka band Translator (KaT) that coherently retransmits the uplink signal back to the ground station. Provided by the Italian Space Agency, the KaT was especially designed for high phase stability with an Allan deviation of 4×10^{-16} over time scales of 1000 s. This ensured that instrumental noise contribution was lower than the overall end-to-end system performance specification of $\sim 3 \times 10^{-15}$, over the same time scale, after full calibration of the interplanetary plasma and the Earths troposphere.

The error budget of the Cassini experiment is dominated by two sources — propagation effects and mechanical noise of the ground antenna. While no special attempt was made to improve the antenna thermomechanical stability, dedicated instrumentation was developed for the calibration of the tropospheric propagation effect. An Advanced Media Calibration (AMC) system included a precise water vapor radiometer, along with a suite of instruments for meteorological measurements that enabled the team to reduce the residual noise after plasma compensation by about a factor of 3. The final Allan deviation of the frequency residuals at 1000 s integration time was 1.5×10^{-14}, a nearly constant value independent of the solar elongation angle. A detailed analysis of the error budget is found in Ref. 1.

Doppler data acquired in the period of 6 June to 7 July 2002 were used in the data analysis. Range data were not used since they were available only at the X band link and were strongly affected by solar plasma. The conjunction occurred on 21 June 2002, with a minimum impact parameter of 1.6 solar radii; the spacecraft was ~7.4 AU away from the Sun. In this period, the spacecraft was kept in a quiet dynamical and thermal state. The only forces with potential impact on the experiment were the solar radiation pressure and the anisotropic thermal emission from the three radioisotope thermal generators (RTG's), which was modeled as thermal thrust, constant in the spacecraft frame. The corresponding nongravitational accelerations were small due to the spacecraft's large mass, with its largest component oriented along the axis of the high gain antenna and estimated at 3×10^{-9} m/s^2, almost two orders of magnitude bigger than the solar radiation pressure.

A total of 12 quantities were used in the orbital fit, namely the six components of the spacecraft state vector at a reference epoch, the three components of the RTG-induced acceleration, the specular and diffuse reflectivity of the spacecraft high gain antenna, and the PPN parameter γ. The determination of the spacecraft position was remarkably accurate, with formal errors of 3.5 km, 46 km and 35 km along the three axes of the J2000 frame,[13] an unprecedented result in planetary navigation from Doppler data only and across a solar conjunction.

These Cassini radio science experiments made the first-ever use of a two-way Ka band and multifrequency radio link over interplanetary distances. Their successful conclusion has indicated the benefits of the novel instrumentation in spacecraft tracking, both for scientific uses and for interplanetary navigation. After Cassini, several future missions will be endowed with Ka band radio systems, notably BepiColombo, the ESA mission to Mercury, and the Jovian orbiter Juno, NASA's second mission of the New Frontier program.

4. The BepiColombo Experiment

The motion of Mercury about the Sun has been the subject of a long and fascinating scientific investigation. The anomalous precession of its perihelion was difficult to explain in the framework of classical physics and the orbit of Mercury gave GR its first experimental success. Solving the equation of motion in a relativistic formalism in 1932, Einstein and de Sitter computed a perihelion drift incredibly close to the observed one. GR has since passed many experimental tests, but Mercury, being the solar system object most affected by relativistic accelerations, has always been considered for improved tests of GR. In particular, a mission to Mercury, both after the arrival at the planet and in its cruise phase that includes superior solar conjunctions, makes for a unique laboratory to further investigate relativistic

Table 1. Current accuracies of selected PPN parameters and the values expected in the future from the BepiColombo MORE experiment. Metric theories of gravity with no preferred frame effects are assumed. Current accuracies are based on Refs. 4, 17 and 18.

Parameter	Present accuracy	MORE
γ	2×10^{-5}	2×10^{-6}
β	1×10^{-4}	2×10^{-6}
η	5×10^{-4}	8×10^{-6}
$J_2^{\odot}$	4×10^{-8}	2×10^{-9}
$\dot{G}/G$	9×10^{-13} yr^{-1}	3×10^{-13} yr^{-1}

gravity via radio science techniques. The scientific relevance of gravitational tests at Mercury led also to the proposal of dedicated space missions.[2]

The European Space Agency has selected a science team to conduct the Mercury Orbiter Radioscience Experiment (MORE), which includes relativistic gravity. MORE can repeat previous tests with much-improved accuracy and explore new aspects of gravitational theories. MORE's scientific objectives are to test general relativity and alternative theories of gravity to the levels shown in Table 1, and to determine the gravity field and internal structure of Mercury. The main observable quantities are the time of flight (range) by measuring the time delay, and the Doppler shift (range rate) of radio waves propagating to and from the spacecraft in a coherent, two-way radio link. Overly simplifying the complexities of the problem, Doppler data will provide the determination of the Mercury's gravity field and the Mercury-centric position of the spacecraft, while the ephemerides of Mercury in the solar system baricentric frame will be mainly derived from range measurements. In the data analysis software under development, both data sets will be used in a consistent, global orbital solution, which includes the gravity field of the planet to at least degree and order 25, the Love number k_2, the solar J_2, the time variation of the Newton's gravitational constant G, a number of PPN parameters and the Nordtvetdt parameter η.

As a major instrumental development over Cassini, MORE will use a novel and highly accurate ranging system, capable of measuring the two-way time of flight of radio waves with an accuracy of 0.7 μs (20 cm). This goal will be attained by the use of higher frequency modulating tone (20 MHz) and an accurate calibration of all time delays introduced by the on-board and ground electronics. Crucial for attaining the goals of the investigation is also an on-board accelerometer, accurate to 1×10^{-6} cm/s$^2/\sqrt{\text{Hz}}$ over a 10^{-4}–10^{-1} Hz bandwidth. Indeed, most BepiColombo radio science measurements will be carried out while the spacecraft is in orbit around Mercury, where the large nongravitational accelerations would impair the orbit determination if not appropriately accounted for in the data analysis. The illustration can be seen in Fig. 1.

An important scientific goal of the MORE experiment is the generation of Mercury's ephemerides with an accuracy of 10 m or less, depending on the tracking

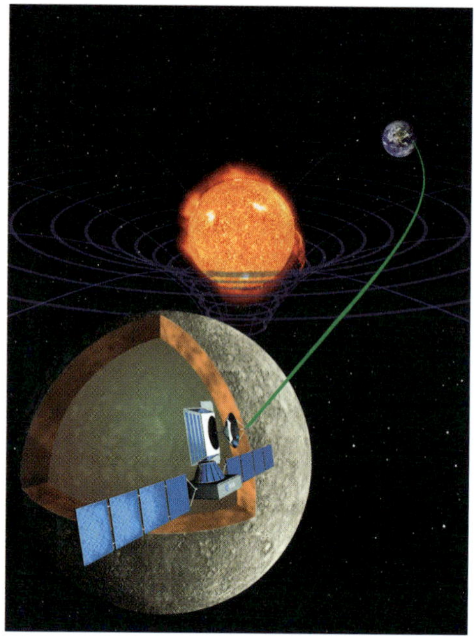

Fig. 1. Illustration of the BepiColombo relativistic gravity experiment during its mission of exploring Mercury (NASA JPL).

geometry. This data set will result in an excellent estimation of the precession of both the perihelion and the node of the planet's heliocentric orbit. The latter will provide a direct, model-independent determination of the solar gravitational quadrupole moment J_2. Known J_2, the relativistic contribution to precession of the perihelion, controlled by the PPN parameter β, will be unambiguously measured. Although the focus of past classical tests of GR has been on the parameters β and γ, MORE offers the opportunity to determine other parameters of the PPN expansion, most notably α_1 and α_2, related to the existence of preferred frames.

The accurate tracking of BepiColombo will determine whether Sun–Mercury and the Sun–Earth pair fall with the same acceleration in the combined gravitational field of Jupiter and the other planet. Mercury and the Earth have indeed different gravitational binding energies, which may contribute to their gravitational masses in violation to the strong equivalence principle. The fraction of the gravitational compactness factor (the ratio of the gravitational binding energy to the rest energy Mc^2) contributing to the body's gravitational mass is Nordtvedt parameter η. The MORE experiment is expected to improve significantly the determination of η, therefore carrying out a new test of the strong equivalence principle (see Table 1).

Some theories of gravity predict that the locally measured Newtonian gravitational "constant" will vary over (cosmological) time scales. Observational constraints on any time variation of G are discussed by Will.[15,16] Ranging to the Viking landers provided the then-best constraint on $\dot{G}/G$; lunar laser ranging gives the best current bound, $|\dot{G}/G| < 9 \times 10^{-13}$ per year. The MORE experiment is expected to improve current bounds by a factor of 3 or more, to $< 3 \times 10^{-13}$ per year.

The expected accuracies of the MORE experiment have been largely based upon the conceptual design of the required on-board and ground instrumentation and detailed orbital simulations using realistic error models.[7,8] This preparatory work, stimulated by the European Space Agency, indicates that the nominal, one-year, BepiColombo mission can attain the accuracies shown in Table. 1. These values were obtained assuming that gravity is described by a metric theory, with no preferred frame effects ($\alpha_1 = \alpha_2 = 0$). The improved value of $d(\ln G)/dt$ over values given in Ref. 8 reflects recent changes in the design and performances of the accelerometer (with a reduction of thermal effects) and the ranging system (with decreased long-term transponder delays).

5. Summary

With the steadily increasing accuracy in space experiments (e.g. Gravity Probe B, Gaia mission, etc.) we are approaching the regime at which violations of Einstein's theory may be assessed and understood. Gravity Probe B is measuring the Lense–Thirring precession predicted by general relativity, but its measurements may also lead to a better determination of γ. BepiColombo MORE will provide a low-cost opportunity to carry out a full set of tests of post-Newtonian gravity with unprecedented accuracy. As a result of the accurate determination of Mercury's ephemerides, the solar quadrupole coefficient J_2 and the time variation of the gravitational constant G and the Nordtvedt parameter η will also be estimated with improved accuracies. Theoretically, a new paradigm rooted in the role of string theory in early cosmology has recently emerged. The equivalence principle and general relativity are generically and jointly violated due to a small scalar field with its own dependence on time and space, which is expected to affect all the different physical interactions. While a real, computable theory seems at this time difficult to achieve, we at least have a firm prediction that $\gamma < 1$ based on general considerations; disproving this constraint would constitute a major advance.

From Viking to Cassini, BepiColombo and beyond, radio science techniques continue to make significant contributions to relativistic gravity.

Acknowledgments

Part of this work was performed at the Jet Propulsion Laboratory, California Institute of Technology, under contract with the National Aeronautics and Space Administration.

References

1. S. W. Asmar *et al.*, *Radio Sci.* **40** (2005) RS2001.
2. P. L. Bender *et al.*, *Adv. Space Res.* **9** (1989) 113.

3. B. Bertotti, G. Comoretto and L. Iess, *Astron. Astrophys.* **269** (1993) 608.
4. B. Bertotti, L. Iess and P. Tortora, *Nature* **425** (2003) 374.
5. T. Damour and K. Nordtvedt, *Phys. Rev. Lett.* **70** (1993) 2217.
6. L. Iess *et al.*, *Class. Quant. Grav.* **16** (1999) 1487.
7. L. Iess and G. Boscagli, *Planet. Space Sci.* **49** (2001) 1597.
8. A. Milani *et al.*, *Phys. Rev. D* **66** (2002) 082001.
9. C. W. Misner, K. S. Thorne and J. A. Wheeler, *Gravitation* (W. H. Freeman, San Francisco, 1973), chaps. 39 and 40.
10. R. D. Reasenberg *et al.*, *Astrophys. J. Lett.* **234** (1979) L219.
11. I. I. Shapiro, *Phys. Rev. Lett.* **13** (1964) 789.
12. S. S. Shapiro *et al.*, *Phys. Rev. Lett.* **92** (2004) 121101.
13. P. Tortora *et al.*, *J. Guidance Contr. Dynam.* **27** (2004) 251.
14. P. Tortora, L. Iess and R. G. Herrera, in *Proc. 2003 IEEE Aerospace Conference* (Big Sky, Montana, USA, 8–15 Mar. 2003).
15. C. M. Will, *Theory and Experiment in Gravitational Physics*, revised edn. (Cambridge University Press, 1993).
16. C. M. Will, *Living Reviews in Relativity* **9** (2006) 3; e-print available at http://www.Livingreviews.org/lrr-2006-3.
17. T. L. Duvall, Jr. *et al.*, *Nature* **310** (1984) 22.
18. J. G. Williams, S. G. Turyshev and D. H. Boggs, *Phys. Rev. Lett.* **93** (2004) 261101.

APOLLO: A NEW PUSH IN LUNAR LASER RANGING

T. W. MURPHY, JR, E. L. MICHELSON and A. E. ORIN

Center for Astrophysics and Space Sciences,
University of California, San Diego,
9500 Gilman Drive, La Jolla, CA 92093-0424, USA
tmurphy@physics.ucsd.edu

E. G. ADELBERGER, C. D. HOYLE and H. E. SWANSON

Center for Experimental Nuclear Physics and Astrophysics,
Box 354290, University of Washington,
Seattle, WA 98195-4290, USA

C. W. STUBBS and J. B. BATTAT

Department of Physics, Harvard University,
Cambridge, MA 02138, USA

APOLLO (the Apache Point Observatory Lunar Laser-ranging Operation) is a new effort in lunar laser ranging that uses the Apollo-landed retroreflector arrays to perform tests of gravitational physics. It achieved its first range return in October 2005, and began its science campaign the following spring. The strong signal (> 2500 photons in a ten-minute period) translates to one-millimeter random range uncertainty, constituting at least an order-of-magnitude gain over previous stations. One-millimeter range precision will translate into order-of-magnitude gains in our ability to test the weak and strong equivalence principles, the time rate of change of Newton's gravitational constant, the phenomenon of gravitomagnetism, the inverse-square law, and the possible presence of extra dimensions. An outline of the APOLLO apparatus and its initial performance is presented, as well as a brief discussion on future space technologies that can extend our knowledge of gravity by orders of magnitude.

Keywords: Gravity; laser ranging; APOLLO.

1. Introduction

Lunar laser ranging (LLR) has a long history of providing many of our strongest tests of gravity.[1–3] It currently provides the best tests of the following gravitational parameters, at the indicated levels of precision:

- Strong equivalence principle (SEP) to 4×10^{-4},
- Weak equivalence principle (WEP) to 10^{-13},

- Time rate of change of the gravitational constant ($\dot{G}/G$) to 10^{-12} per year,
- Gravitomagnetism (basis of frame dragging) to 0.1%,
- Geodetic precession to 0.35%,
- Test of $1/r^2$ to 10^{-10} times the strength of gravity, at $\sim 10^9$ length scales.

LLR thus far has not seen deviations from the expectations of general relativity. Recent astrophysical observations, together with theoretical developments in gravity, motivate an invigorated experimental campaign to seek departures from general relativity. For example, the apparent acceleration of the expansion of the Universe may indicate a lack of understanding of gravity on cosmological scales. This includes ideas relating to the departure of gravitons from our four-dimensional space–time into extra dimensions.[4,5] As another example, the existence of dark matter for explaining rotation curves of galaxies can be countered with modifications to gravitation[6] that solar system tests of gravitation should be sensitive to at some level. In general, any scalar-field modifications to general relativity — such as those that generically accompany quantum gravity models or covariant forms of dark-matter-hiding modified gravity[7] — violate the equivalence principle and introduce secular changes to the fundamental constants of nature. Many of the best tests of these effects are accomplished via LLR.

The state of the art in 2005 was 2-cm-range precision, usually accomplished in an observing period lasting a few tens of minutes, and collecting 5–50 photons of returned laser energy. The two LLR stations that routinely acquire data — in Texas (MLRS[8]) and France (OCA[9]) — typically see a return rate to the larger Apollo 15 array of 0.002 and 0.01 photons per pulse, respectively. At a 10 Hz pulse repetition rate, this corresponds to one photon every 50 and 10 seconds, respectively.

The LLR error budget is typically dominated by uncertainty associated with the tilt of the retroreflector array normal relative to the line of sight. These tilts — up to about 7° in each axis — are caused by "optical" librations of the Moon, due to the elliptical and inclined nature of the lunar orbit. The result is a range of distances to the array: there is in general a nearest corner and a farthest corner. Even if the orientation is known precisely, it is impossible to determine from which part of the array a particular photon was reflected. Thus an uncertainty in the range measurement is introduced, with a peak-to-peak uncertainty in the ballpark of $a \tan 6° \approx 0.1a$, where a is the array dimension of roughly 1 m. In a root-mean-square sense, the resulting 30–50-mm-range uncertainty can be averaged to 1 mm uncertainty by gathering 900–2500 photons. This number is well outside the grasp of the MLRS or OCA stations.

A new lunar ranging apparatus, APOLLO (the Apache Point Observatory Lunar Laser-ranging Operation), has begun operation in southern New Mexico on a mountaintop at an elevation of 2780 m. Using a 3.5 m telescope aperture and taking advantage of good atmospheric image quality ("seeing"), APOLLO is capable of receiving multiple photons per pulse. A basic outline of the new system and its features is presented here, along with a summary of initial performance. This is followed by a brief discussion on the prospects for improving tests of gravity in the solar system beyond the capabilities of APOLLO.

2. Link Equation

The return photon rate is dominated by the signal loss from divergence of the outgoing and return beams. The outgoing beam, even if perfectly collimated, is limited by the atmosphere to a divergence of one-to-a-few arcseconds, with one arcsecond corresponding to 1.8 km on the lunar surface. The small corner cubes constituting the Apollo arrays return a beam with an effective divergence of 7–10 arcsec,[10] or about 15 km on the Earth's surface. If Φ is the atmospheric divergence, ϕ the corner-cube divergence, D the diameter of the collecting telescope, and $d = 38$ mm the diameter of the n corner cubes in the array, then the link efficiency is approximately

$$\epsilon \approx \eta^2 f Q \left(\frac{nd^2}{r^2\Phi^2}\right)\left(\frac{D^2}{r^2\phi^2}\right). \tag{1}$$

Here, r is the distance to the Moon, η is the telescope/atmospheric transmission efficiency (experienced both ways), f is the receiver throughput — dominated by a narrow-band filter, and Q is the detector efficiency. With 1 arcsec seeing, 40% telescope efficiency, 25% receiver efficiency, 30% photon detection efficiency, and 115 mJ pulses, the APOLLO photon rate would be $\sim$5 photons per pulse on the smaller Apollo arrays ($n = 100$), or a photon rate of approximately 100 Hz. Speckle interference of the beam pattern on the Moon makes the photon return rate highly variable, with 50% of the pulses returning 1–10 detected photons (based on an average of 6).

So far APOLLO has achieved return rates as high as 0.6 photons per pulse (12 Hz) sustained over short intervals, with as many as 8 photons per pulse recorded, and a sustained average rate of 0.25 photons per pulse, or 5 Hz over 10 minutes. At the return rates seen thus far, we obtain the necessary photon number for millimeter precision on a time scale of 10 minutes. More discussion of achieved performance appears in Sec. 4.

A significant advantage of a high return rate is that even at 10% of the observed peak rate, we detect about one photon per second. This rate is high enough to apply feedback for optimization of the optical system — a practical impossibility for previous-generation LLR stations.

3. APOLLO System Overview

3.1. *Telescope*

The 3.5 m telescope at the Apache Point Observatory is a world-class astronomical telescope run of the Astrophysical Research Consortium, composed member universities. The University of Washington (through which APOLLO has access) has a 28% share of the total observing time. The telescope is typically scheduled into half-night allotments, though some measurement campaigns in the past have utilized frequent one-hour blocks of time. APOLLO will largely operate on the latter arrangement, using a distributed arrangement of time blocks of one hour or less. The median image quality observed at the telescope is 1.1 arcsec.

3.2. *APOLLO laser*

APOLLO utilizes a commercial mode-locked, cavity-dumped Nd:YAG laser.[11] The laser has a pulse energy of 115 mJ, a repetition rate of 20 Hz, and a pulse width of < 100 ps FWHM (full width at half-maximum) at 532 nm. The 7 mm Gaussian profile beam is expanded to 17 mm via a pair of lenses whose separation can be adjusted to set the beam divergence. A negative lens diverges the beam into a cone matching the telescope focal path, such that the telescope recollimates the beam into the sky, filling most of the primary mirror. Various tests of beam divergence indicate that the output beam is collimated to well under 1 arcsec, meaning that atmospheric seeing dominates the outgoing beam divergence.

3.3. *APD arrays*

APOLLO uses integrated avalanche photodiode (APD) arrays fabricated by MIT Lincoln Lab, having 30-μm-diameter elements arranged in a square pattern with 100 μm spacing (Fig. 1). We place the array at the focal plane of our optical receiver, in such a way as to oversample the point spread function across a handful of elements (e.g. via 0.35 arcsec pixels). A microlens array in front of the detector recovers a

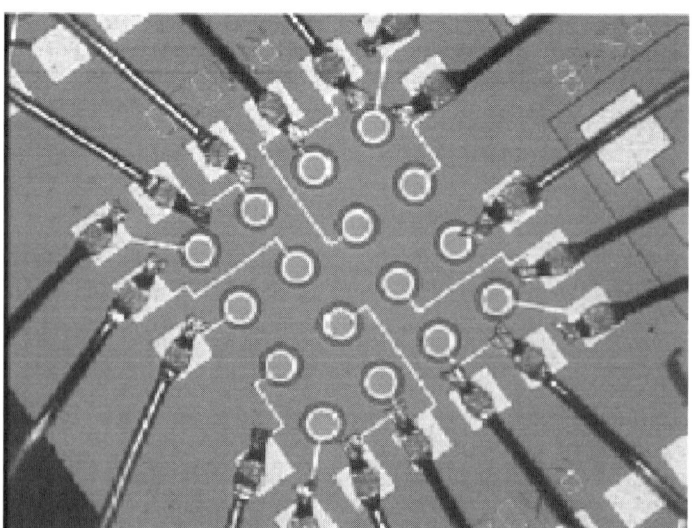

Fig. 1. 4×4 APD array employing 30-μm-diameter active areas on 100 μm centers — each capable of measuring the arrival time of a single photon to ~ 50 ps precision.

nearly 100% fill factor, compared to the 7% fill factor of the bare array. An array format provides a variety of advantages: not only does the oversampling parcel out photons to individual detector elements — thereby allowing each photon to be individually time-tagged — but two-dimensional spatial information is preserved, so that real-time guiding and beam alignment/focus diagnostics can be employed. We have measured the timing performance of the detectors, finding that the intrinsic response to a photon has a jitter — or variability — of less than 50 ps. But the radial location of the incident photon within the detector element introduces a separate spread that is 60 ps in magnitude.

3.4. *Time-zero fiducial*

The pulse departure time is measured via photons reflected from a corner cube located at the exit aperture of the telescope and mechanically referenced to the spatially "fixed" intersection of telescope drive axes. Thus APOLLO performs a *differential* measurement between the local corner cube and the lunar reflector arrays. The corner cube photons share the *same optical path* as the lunar photons, except for the highly reflective coatings on the front and rear surfaces of the rotating transmit/receive switch that attenuate the reference signal to a few photons. Because we are using an array detector, we can accept a reference signal strength of a few photons per pulse without biasing our result. Thus *virtually every* shot is "calibrated," in contrast to today's single-photon ranging systems, in which 1 out of every 5–10 pulses has a fiducial measurement. A fast photodiode detecting the laser fire provides a strong-signal timing "anchor" with tested 15 ps performance. Over short time scales (minutes), the photodiode provides the start reference (with known offset relative to the corner cube fiducial), while the corner cube return is relied upon for tracking slow variations in path length or electronic drifts associated with environmental variations. A side benefit of this scheme is that we continuously monitor the overall time resolution of the system, so that we may separate the effect of the finite size of the lunar retroreflector array.

3.5. *APOLLO timing system*

The timing scheme for APOLLO is based on a Global Positioning System (GPS)-disciplined clock[12] and frequency multiplier providing a highly stable 50 MHz frequency reference. The RMS jitter of the clock over the ~2.5 s round-trip travel time of the pulse is about 7 ps. Long-term stability of the ~ 2.5 s interval (over month-long time scales) is provided by reference to GPS — itself referenced to the ground-based atomic clock ensemble. Time-to-digital converters (TDC's) having 25 ps resolution and 15 ps RMS jitter[13] measure the precise time interval between APD detection events and the next clock pulse. A digital counter system continuously keeps track of the number of clock pulses between events in a redundant way, accommodated on a single Programmable Logic Device (PLD) within a custom unit called the APOLLO Command Module (ACM). The ACM also carries out a

Table 1. Single-photon random error budget.

Expected statistical error	RMS error (ps)	One-way error (mm)
Laser pulse (95 ps FWHM)	40	6
APD avalanche initiation location	60	9
APD intrinsic jitter	50	7
TDC jitter	15	2.2
50 MHz frequency reference	7	1
APOLLO system total	89	13
Lunar retroreflector array	80–230	12–35
Total error per photon	120–245	18–37

fast (1 kHz) calibration of the TDC units with precise start/stop pulses spaced at integer multiples of 20.00 ns apart (±10 ps).

3.6. *Net timing performance*

The expected statistical RMS timing errors are shown in Table 1. The range given for the retroreflector spread represents typical values, assuming a lunar libration angle of 5°–8°, and a reflector spanning 0.5–0.8 m. The retroreflector array generally dominates the error budget, so that $\sim$ 1000 photons are typically required to achieve a 1 mm statistical error. At a photon rate of 0.1 photons per shot, this photon number may be achieved in less than 10 minutes. Initial engineering operations of the APOLLO apparatus at the telescope support the total random uncertainty estimate presented in Table 1.

4. APOLLO Project Status

The summer of 2005 saw most of the hardware and software come together in an integrated system at the observatory. Though the system was not complete, enough was in place to provide opportunities to practice ranging, allowing significant system evolution — mostly on the software front. With the placement of the microlens array in front of the detector in October 2005, we achieved our first unambiguous range results, reaping about 2400 photons in a period of less than 30 minutes. In reference to Table 1, this is more than enough photons to provide statistical averaging at the 1 mm level. The McDonald Laser Ranging System collected a similar number of lunar return photons over a three-year period, from 2000 to 2002. Also of note is that this initial success was achieved *near full moon*, when other stations are unable to acquire the range signal against the lunar background.

At present, APOLLO has accomplished much in its first half-year of operation:

- $>$ 2000 lunar return photons within 10 minutes (on two occasions);
- Peak rates of $>$ 0.6 photons per pulse over short intervals (on two occasions);
- As many as 8 return photons have been seen in a single pulse (plus many 7's, 6's, etc.);

- About half of the return photons in strong runs arrive in multiphoton packets;
- Full-moon ranging does not represent a noticeable challenge;
- Typical acquisition time for each reflector is less than one minute.

An example run is shown in Fig. 2. While the performance has not yet hit the anticipated level of 1–5 photons per pulse, we have demonstrated the capability of collecting sufficient numbers of photons to achieve 1 mm precision on time scales of less than 10 minutes. The eventual goal for APOLLO is to routinely achieve millimeter precision on each of the four available reflectors in a period of less than an hour, and eventually within 30 minutes.

APOLLO range measurements in October, November, December, and January were processed by the analysis group at JPL. A solution for the APOLLO station position was found that resulted in range deviations at the 0.1 ns level, corresponding to 1–2 cm. This level of imprecision was not inconsistent with the knowledge of our system performance at that time. First, the GPS-disciplined clock providing the timing backbone for APOLLO was not suited to its position within the main laser/timing enclosure, as this enclosure rotates with the telescope. Such motion disrupted the oscillator lock (likely via a sudden reorientation of the thermal gradient within the temperature-controlled oven), effectively producing centimeter-level inaccuracies in the range measurement. The second problem was that the fiducial returns from the corner cube were not yet interleaved between the lunar gates, so

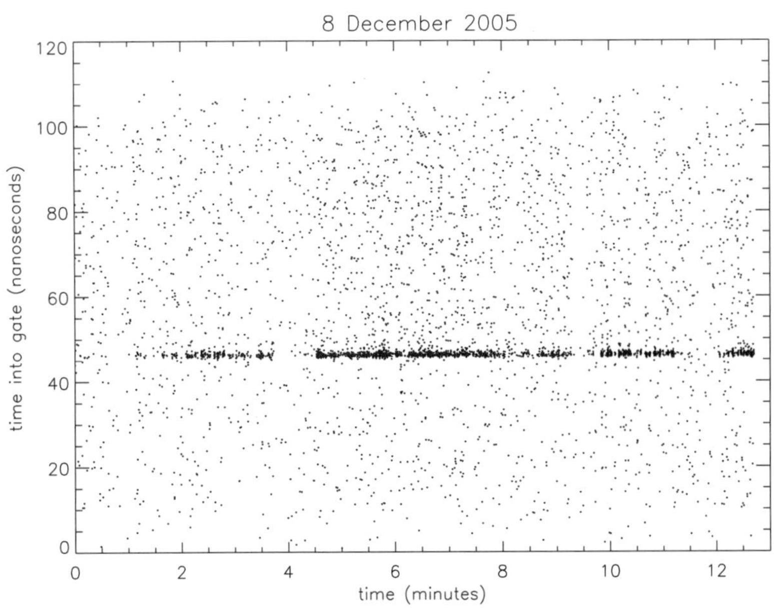

Fig. 2. Example Apollo 15 time series showing photon return time (vertically) within the range gate. The lunar return is evident against the background photons. The width is consistent with the temporal spread of the reflector array.

that we did not have a real-time differential range measurement. Both issues were addressed in March 2006, so that the April 2006 acquisitions represent what we believe to be the first *unbiased, differential* measurements from APOLLO.

APOLLO is poised at the edge of a data campaign unlike anything else in the history of lunar ranging. Early work on the 2.7 m telescope at McDonald Observatory approached single-photon-per-pulse performance, but at 0.3 Hz repetition rate and 4 ns pulse width. The APOLLO return rate is at least two orders of magnitude higher than currently operating LLR stations (helped some by a higher repetition rate), so that order-of-magnitude gains in physics seem feasible. Project status updates are available on the APOLLO website.[14]

5. Future Directions

The 1 mm range precision of the APOLLO campaign will translate into order-of-magnitude gains in understanding fundamental gravity. But it is not clear that the LLR technique can be pushed much further, because effects of site displacement and meteorological influences will likely impose limits at this level. Dramatic improvements in precision are still possible within the solar system environment, involving the implementation of new space hardware.

In the near term, a return to the Moon may be accompanied by new ranging hardware. Current lunar ranging stations are limited primarily by the uncertainty imposed by the tilt of the tightly populated retroreflector array. A new sparse array would allow the resolution of individual corner cubes, so that relatively inexpensive improvements in the ground station (shorter-pulse lasers, better detectors) could dramatically improve range performance. At present, modest improvements in the ground apparatus would be wasted in the face of the reflector uncertainty (see Table 1). A sparse array (> 20 cm corner cube separation) does not need to be rigid or even uniform, as the individual corner cube positions can be determined via postprocessing of the return data. Strong-signal bias would also cease to be a problem, as the timing of each received photon unambiguously reveals the identity of the element involved, so that one no longer must rely on stability of the centroid.

Asynchronous transponder technology is likely to play a major role in future tests of gravity within the solar system. The $1/r^4$ scaling relationship of passive reflectors [as in Eq. (1)] becomes prohibitive for ranges beyond 10^9 m. Laser altimeters are already in effect transponders, and the altimeter on the MESSENGER spacecraft has recently been successfully used in this capacity both to demonstrate ranging capability on solar system scales, and to establish the spacecraft distance to ± 20 cm.[15] Clever implementation of optical communications devices can permit a "piggybacked" transponder capability.[16] A test of communications/transponder technology on the Moon would have the following two major benefits:

- The resulting $1/r^2$ link efficiency would open up LLR to a wide host of satellite laser ranging (SLR) stations, permitting one to average over site displacements and meteorological conditions, thus overcoming the 1 mm barrier a single station is likely to experience.

- The technology test would be of direct relevance and advantage to more ambitious optical communications and/or transponder missions in the future.

Examples of future missions that would rely on interplanetary transponder capabilities are groupings of satellites that span the solar system to measure the effects of space–time curvature (thus constraining the Eddington parameter, γ, to 10^{-8} or better), and devices placed on solar system bodies to constrain the strong equivalence principle the time rate of change of the gravitational constant, etc. to better precision than APOLLO expects to reach via traditional LLR techniques.

References

1. J. G. Williams, X. X. Newhall and J. O. Dickey, *Phys. Rev. D* **53** (1996) 6730.
2. J. G. Williams, S. G. Turyshev and D. H. Boggs, *Phys. Rev. Lett.* **93** (2004) 261101.
3. J. G. Williams, S. G. Turyshev and D. H. Boggs (2005) [gr-qc/0507083].
4. G. Dvali, A. Gruzinov and M. Zaldarriaga, *Phys. Rev. D* **68** (2003) 024012.
5. A. Lue and G. Starkman, *Phys. Rev. D* **67** (2003) 064002.
6. M. Milgrom, *Astrophys. J.* **270** (1983) 365.
7. J. D. Bekenstein, *Phys. Rev. D* **70** (2004) 083509.
8. http://www.csr.utexas.edu/mlrs/
9. E. Samain *et al.*, *Astron. Astrophys. Suppl. Ser.* **130** (1998) 235.
10. J. O. Dickey *et al.*, *Science* **265** (1994) 482.
11. http://www.continuumlasers.com/products/pulsed_Leopard_series.asp
12. http://www.symmttm.com/products_gps_xl-dc.asp
13. http://www.phillipsscientific.com/phisci1.htm
14. http://physics.ucsd.edu/~tmurphy/apollo
15. D. E. Smith *et al.*, *Science* **311** (2006) 53.
16. S. Merkowitz, *Int. J. Mod. Phys. D* **16** (2007) 2151.

ASYNCHRONOUS LASER TRANSPONDERS: A NEW TOOL FOR IMPROVED FUNDAMENTAL PHYSICS EXPERIMENTS

JOHN J. DEGNAN

Sigma Space Corporation, 4801 Forbes Blvd, Lanham MD 20706, USA
John.Degnan@sigmaspace.com

Since 1964, the NASA Goddard Space Flight Center (GSFC) has been using short pulse lasers to range to artificial satellites equipped with passive retroreflectors. Today, a global network of 40 satellite laser ranging (SLR) stations, under the auspices of the International Laser Ranging Service (ILRS), routinely tracks two dozen international space missions with few-millimeter precision using picosecond pulse lasers in support of Earth science. Lunar laser ranging (LLR) began in 1969, shortly after NASA's Apollo 11 mission placed the first of five retroreflector packages on the Moon. An important LLR data product has been the verification of Einstein's equivalence principle and other tests of general relativity. In 1975, the University of Maryland used a laser ranging system to continuously transfer time between two sets of atomic clocks — one set on the ground and the other in an aircraft — to observe the predicted relativistic effects of gravity and velocity on the clock rates. Two-way asynchronous laser transponders promise to extend these precise ranging and time transfer capabilities beyond the Moon to the planets, as evidenced by two successful experiments carried out in 2005 at distances of 24 and 80 million km respectively.

Keywords: Laser; interplanetary ranging; time transfer; transponder; relativity.

1. Heritage: Satellite and Lunar Laser Ranging

Laser ranging to passive retroreflectors on Earth-orbiting satellites was first demonstrated at the NASA Goddard Space Flight Center on 31 October 1964.[1] The basic measurement of this single-ended instrument is both simple and unambiguous. The outgoing laser pulse starts a highly precise timer, is reflected by the satellite, and the return signal stops the timer. One then multiplies the time interval by the speed of light, correcting for satellite signature (impulse response) and atmospheric propagation delay effects, to compute a range to the satellite center of mass. Today, an international network of approximately 40 satellite laser ranging (SLR) stations routinely tracks two dozen space missions in Earth orbit. Over the past four decades, the ranging precision has improved from a few meters to one or two millimeters, and the subcentimeter absolute accuracy is presently limited, not by the instrumentation, but by uncertainties in the atmospheric propagation model and pulse spreading by the satellite target arrays. For more information on SLR, the reader is referred to a series of review articles devoted to SLR history and science applications,[2] SLR hardware,[3] and mathematical models.[4]

Since its inception in 1998, the International Laser Ranging Service (ILRS), an official service of the International Association for Geodesy (IAG), has set mission tracking policy and managed the daily operations of the international SLR network. The global distribution of ILRS stations is shown in Fig. 1, and, as will be demonstrated later, most of these stations are potentially capable of supporting future centimeter ranging and subnanosecond time transfer to the other planets within the solar system.

A select few of the ILRS stations have successfully tracked one or more of the five retroreflectors placed on the Moon by the manned US Apollo 11, 14, and 15 and two unmanned Soviet Lunakhod missions to the Moon. Most of the operational lunar laser ranging (LLR) data over the past four decades has come from three sites — the NASA/University of Texas station at McDonald Observatory, the French CERGA station in the coastal Mediterranean town of Grasse, and the NASA/University of Hawaii station at the top of Mt Haleakala in Maui. Unfortunately, the last-mentioned site was decommissioned in 1992 due to NASA funding cuts. It is important to note that, even with meter class telescopes located at

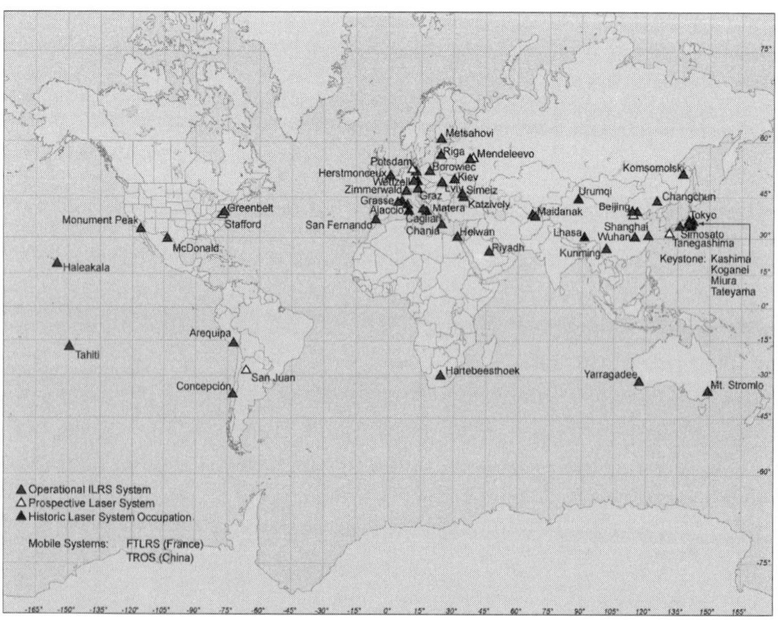

Fig. 1. Global distribution of the ILRS satellite-laser-ranging network.

mountaintop sites with excellent atmospheric "seeing" and with moderately high subnanosecond pulse energies on the order of 100–200 mJ, LLR systems typically detect one single photon return from the lunar arrays out of every 10–20 laser fires, or roughly one photon per second at typical 10–20 Hz fire rates. This low signal photon return rate makes the extraction of the signal from background noise difficult, except when the sunlit lunar surface is outside the receiver field of view (FOV). On the other hand, LLR observers have also found it necessary to offset their pointing from prominent lunar features in order to guide their narrow laser beam successfully to the target. The net consequence of these two constraints is to limit lunar tracking to temporal periods which are far from both "Full Moon" and "New Moon." In spite of these limitations, LLR has proved invaluable to a number of important scientific endeavors in the fields of lunar physics and general relativity.[5]

Under the APOLLO (Apache Point Observatory Lunar Laser-ranging Operation) program in New Mexico, activities have been underway to produce multiphoton lunar ranging returns through the use of larger, 3.5-m-diameter telescopes and more powerful lasers,[6] and the first lunar returns were reported in October 2005. Returns from both the strongest (Apollo 15 with 300 retroreflectors) and weakest (Apollo 11 with 100 retroreflectors) lunar targets were obtained, including some successful experimental sessions near Full Moon. During the best run reported to date, 420 returns were detected out of 5000 attempts for an 8.4% return rate. Nevertheless, the conventional SLR technique of ranging to passive retroreflectors is unlikely to be useful for targets much beyond the Earth–lunar distance (384,000 km, or 0.0026 AU). This is due to the R^{-4} dependence of the received signal strength from a passive target, where R is the target range.

Laser transponders can overcome the distance limitations of conventional SLR and LLR systems ranging to passive targets. Transponders consist of two active terminals — each with its own laser, telescope, and timing receiver.[7] As a result, the signal strength falls off only as R^{-2}, and this greatly extends the range over which precise measurements can be made.[8] Transponders on the Moon would greatly ease the laser power and pointing requirements and make LLR accessible to the least capable SLR stations throughout the lunar cycle, thereby eliminating the current temporal outages near New and Full Moon. More importantly, transponders could permit subcentimeter interplanetary ranging and subnanosecond time transfer in the very near future, thereby providing a powerful new tool for improved fundamental physics experiments.

2. The Natural Synergism Between Laser Ranging and General Relativity

Einstein often explained his relativistic theories in terms of light propagation in an inertial frame. Thus, in the 1960's, there was a natural synergy between the burgeoning field of SLR and the relativity community. It is not surprising then that a group of early SLR practitioners at NASA/GSFC and elsewhere formally teamed up with theoretical and experimental relativists at various universities to consider and plan definitive tests of Einstein's predictions. The Lunar Retroreflector Experiment (LURE) team successfully lobbied NASA Headquarters for the placement of a retroreflector array on the lunar surface during the first Apollo Moon landing. In addition to vastly improving our physical knowledge of the Moon and its interactions with Earth, LLR has provided some of the most quantitative and convincing tests to date of Einstein's predictions regarding general relativity.[5] For example, LLR has not only verified the equivalence principle but also tightly constrained both the β parameter in the Robertson–Walker metric and the time rate of change in the gravitational "constant," G.

On 22 November 1975, Professor Carroll Alley of the University of Maryland and several of his graduate students (including the author) performed a unique laser ranging experiment to directly demonstrate the effects of the gravitational red shift and time dilation on atomic clocks.[9,10] In this experiment, ultrashort laser pulses were generated by a modelocked Nd:YAG laser, located in a trailer on the grounds of the Patuxent Naval Air Station near Washington, D.C., and were used to transfer time at a 30 pps rate between an ensemble of clocks on the ground and an identical ensemble of clocks in the aircraft. The aircraft, which was equipped with both a passive retroreflector and a high speed detector, flew in an elliptical racetrack pattern in the vicinity of the ground station for approximately 15 hours. The laser transmitter and receive telescope were pointed at the aircraft via a manually operated joystick controller. The ground clocks recorded the laser pulse time of departure and the time of arrival of the reflected pulse. The midpoint of these two time marks is coincident with the pulse arrival time recorded by the aircraft clock ensemble. The time transfer method is summarized by the Minkowski space–time diagram in Fig. 2(a). The time lines of the ground and airborne clock ensembles are represented by the lower and upper horizontal lines respectively. Light

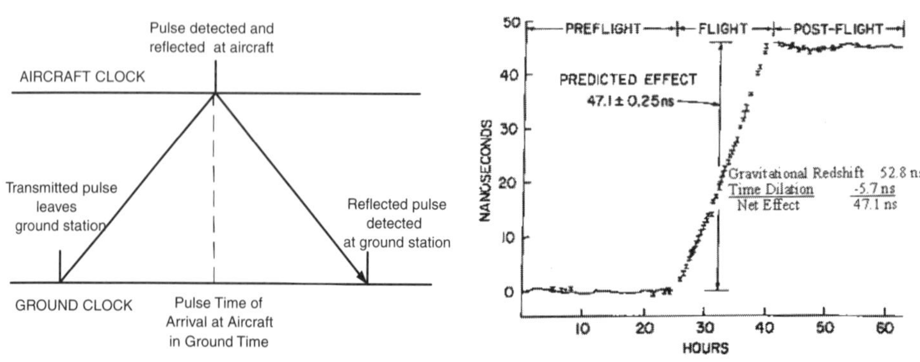

(a) (b)

Fig. 2. University of Maryland Atomic Clock Experiment: (a) Minkowski space–time diagram; (b) cumulative time offset between the airborne and ground ensembles of atomic clocks.

pulse propagation to and from the aircraft is indicated by "photon world lines" with slopes of ±45° relative to the horizontal. The experimental results are summarized in Fig. 2(b). During the preflight period of 0–25 hours, the synchronized clocks in both ensembles ran at the same rate. During the flight period between 25 and 40 hours, the clocks ran at different rates, resulting in a total time offset of 47.1 nsec. The 52.8 nsec accumulated offset due to aircraft clocks running faster in the lower gravitational field ("gravitational red shift") was partially offset by aircraft velocity effects ("time dilation") on the order of 5.7 nsec, both in good agreement with the predictions of relativity. Following the flight, the rates of the two clock ensembles were again equal, and there was no further accumulation of time offset. Unlike prior aircraft relativity experiments where clock ensembles could only be compared before and after the flight to obtain a total offset, the laser time transfer permitted the instantaneous rates to be monitored and compared throughout the flight. In fact, the aircraft altitude increased three times during the 15-hour experiment as fuel was expended with corresponding increases in the aircraft clock rate due to the weaker gravitational field at the higher altitudes.[9]

3. Interplanetary Ranging and Time Transfer via Laser Transponders

There are two types of transponders: echo and asynchronous. The Minkowski space–time diagrams for echo and asynchronous transponders are shown in Figs. 3(a) and 3(b) respectively. In an Earth–Moon echo transponder, for example, a pulse emitted from the Earth terminal at time t_{E1} is detected by the lunar terminal at time t_{M1}, which then generates a response pulse at time t_{M2} subsequently detected on Earth at time t_{E2}. The delay between the received and the transmitted pulse at the lunar terminal, t_d, would be either known *a priori* through careful calibration or controlled via active electronics and would be subtracted from the observed roundtrip time before computing the target range. Alternatively, the delay can be measured locally by a timer at the lunar terminal and transmitted to the Earth terminal via a communications link. The signal return rate at the primary station is then equal to the fire rate of the laser multiplied by the joint probability that

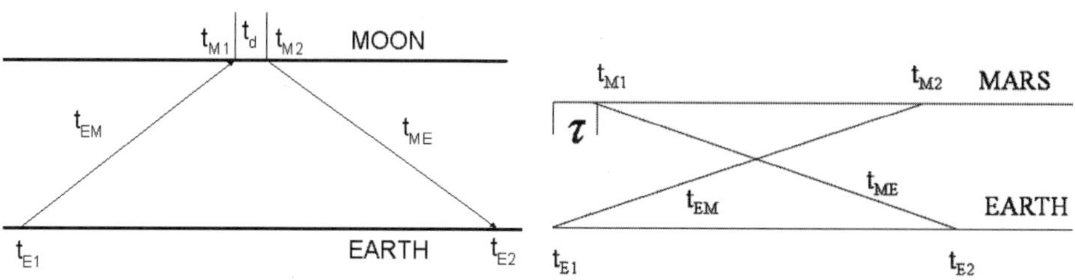

(a) (b)

Fig. 3. Timing diagrams for (a) echo and (b) asynchronous transponders.

pulses are detected at both ends of the link. Thus, the simple echo approach works very well when the roundtrip time-of-flight is relatively short and there is a high probability of detection at both ends of the link, i.e. when both the uplink and the downlink signal are reasonably strong and pointing uncertainties are small relative to the transmitter divergence. This approach should work very well over Earth–Moon or shorter links. However, in interplanetary links where the light transit time is relatively long (several minutes to hours) and the probability of detection is potentially small at one or both ends of the link, it is worth considering the asynchronous laser transponder.

In an asynchronous transponder, the two terminals independently fire pulses at each other at a known laser fire rate, as illustrated by the Minkowski diagram in Fig. 3(b). For an Earth–Mars link, for example, the Earth terminal records the times of departure of its own transmitted pulses (t_{E1}) as well as the times of arrival of pulses from Mars (t_{E2}), and vice versa. In a high SNR system with good pointing, the pulses arrive at roughly the laser fire rate, whereas in low SNR or photon-counting systems, the pulses may arrive intermittently.[7] The departure and arrival times measured at each terminal are then communicated to, and properly paired at, an Earth-based processor, which then calculates a range and clock offset between the two terminals for each set of two-way measurements occurring within a reasonably short time interval. The relevant equations are

$$R = \frac{2}{c}\left[(t_{E2} - t_{E1}) + (t_{M2} + t_{M1})\right] \tag{1}$$

for the interterminal range at the time when the two photon world lines cross in Fig. 2(b) and

$$\tau = \frac{(t_{E2} - t_{E1}) - (t_{M2} - t_{M1})}{2(1 + \dot{R}/c)} \tag{2}$$

for the corresponding time offset between the pulses departing from each terminal, where $\dot{R}/c$ is a correction for the range rate between the two terminals.

For a more extensive discussion on the theory of laser transponders, background noise and error sources, proposed methods for terminal and signal acquisition, and detailed analyses of an Earth–Mars link, the reader is referred to a comprehensive article previously published by the author.[7]

In late May 2005, NASA/GSFC conducted the first successful two-way transponder experiments at a wavelength of 1064 nm with a laser altimeter on board the Messenger spacecraft, which is currently en route to Mercury. From a distance of about 24 million km (0.17 AU), the Messenger spacecraft performed a raster scan of Earth while firing its Q-switched Nd:YAG laser at an 8 Hz rate. Simultaneously, a ground-based Q-switched Nd:YAG laser at GSFC's 1.2 m telescope was aimed at the Messenger spacecraft. The repetition rate of the ground laser was much higher (240 Hz) in order to ensure that at least one pulse would arrive at the spacecraft when the altimeter range gate was open. During the few-second periods when the

Messenger raster scan happened to pass over the Earth station, pulses were successfully exchanged between the two terminals.[11] The pulse times of departure and arrival at the two terminals were then used to estimate the Earth–spacecraft range with decimeter precision,[12] orders of magnitude better than could be achieved with the spacecraft microwave Doppler data.

In late September 2005, the same GSFC team successfully transmitted hundreds of Q-switched laser pulses to the Mars Orbiter Laser Altimeter (MOLA), an instrument on the Mars Global Surveyor (MGS) spacecraft in orbit about Mars. Because the MOLA laser was no longer operable, this was limited to a one-way uplink experiment. The instrument parameters for the two GSFC experiments are summarized in Table 1.

The MLA and MOLA experiments demonstrate that decimeter interplanetary ranging is within the state of the art and can be achieved with modest laser powers and telescope apertures. It must be stressed, however, that these were experiments of opportunity rather than design. First of all, since the spacecraft had no ability to lock onto the opposite terminal or even the Earth image, the spaceborne lasers and receiver FOV's were scanned across the Earth terminal, providing only a few seconds of data. Secondly, near-infrared (NIR) detectors at 1064 nm are far less sensitive than the photon-counting visible detectors at 532 nm typically used in SLR and LLR. As a result, the laser energies and telescope apertures in the MLA and MOLA experiments, although modest, were significantly higher than would be necessary for a dedicated deep space transponder mission. Furthermore, range precision was limited to roughly a decimeter or two by the long laser pulse widths (6 nsec) and comparable receiver bandwidths which were designed for altimetry rather than precise ranging. This is to be compared with the 2–3 mm RMS single shot precision achieved by many modern SLR systems, which operate at the 532 nm green wavelength and use picosecond lasers, detectors, and timers.

Table 1. Summary of key instrument parameters for recent deep space transponder experiments at 1064 nm. The EA (PA) product is the laser energy (average laser power) at one terminal multiplied by the telescope receive area at the opposite terminal.

Experiment	MLA (cruise)		MOLA (Mars)
Range (10^6 km)	24.3		~80.0
Wavelength, nm	1064		1064
	Uplink	Downlink	Uplink
Pulsewidth, nsec	10	6	5
Pulse energy, mJ	16	20	150
Repetition rate, Hz	240	8	56
Laser power, W	3.84	0.16	8.4
Full divergence, μrad	60	100	50
Receive area, m^2	0.042	1.003	0.196
EA product, J-m^2	0.00067	0.020	0.0294
PA product, W-m^2	0.161	0.160	1.64

Achieving subcentimeter precision over interplanetary distances would allow significantly more accurate (~three orders of magnitude) tests of the strong equivalence principle, much tighter constraints on important relativistic metrics (e.g. β and γ) than is currently provided by LLR or microwave radar ranging to the planets, and a convincing measurement of the Shapiro time delay.

4. Technology Roadmap: Where Do We Go from Here?

Space applications of lasers are currently of great interest to NASA. Laser altimeters have already mapped the surfaces of Earth, Mars, and the Moon from orbit. Another altimeter (MLA) is en route to Mercury and will provide a second opportunity for a transponder experiment as it flies by Venus in June 2007. State-of-the-art sensors, such as high resolution cameras, 3D imaging lidars, synthetic aperture radars (SARs) and hyperspectral imagers, generate unprecedented amounts of data. Their potential use elsewhere in the solar system would exceed the data-handling capabilities of conventional microwave communications systems and require the use of higher bandwidth laser communications systems to download the data to Earth. Interferometric approaches to laser ranging, such as the Laser Interferometer Space Antenna (LISA), have been proposed for the detection of gravitational waves and have the ambitious goal of detecting range variations between satellites with 10 pm resolution over distances of 5 million km (although knowledge of the absolute range would be no better than 10 m).

Fortunately, there is substantial overlap among the various ancillary space-based subsystems needed to support the aforementioned applications such that technology development can be leveraged to benefit multiple users. For example, with the exception of altimetry, there is a common need to point to, acquire, and lock onto the opposite terminal, maintain accurate time on board, and store and download the science and housekeeping data. The laser transmitter and receiver, on the other hand, are tailored to the specific need or application. For example, LISA and lasercom use low peak power CW and quasi-CW (i.e. modulated) beams respectively. However, accurately measuring the relative phases of incoming beams in LISA to detect small relative motions induced by gravitational waves or maintaining a low bit error rate (BER) in a lasercom link requires a high signal-to-noise ratio (SNR), and hence the beams must be highly collimated in order to make optimum use of the available laser power. Since narrow beams complicate and extend the duration of the terminal acquisition process, either the divergence of the primary laser beam is temporarily increased or a separate laser beacon is used for acquisition. If terminal lock is lost in the absence of a separate beacon, the acquisition sequence must be restarted with a possible interruption or loss of the primary data product. Transponders make ideal beacons since they utilize ultrashort pulses, whose high peak powers are readily seen against the optical background — even with relatively low pulse energies and high beam divergences.[7] Furthermore, they provide valuable navigation and timing information to the mission. For example, a transponder can potentially measure the Earth-to-spacecraft (or spacecraft-to-spacecraft) distance with an absolute accuracy of a few mm, synchronize the spacecraft clock to GPS time via its time transfer capabilities, and even measure the local gravity field via its effect on spacecraft motion[2,4] or on the spaceborne clock rate as discussed in Sec. 2.

Most of the hardware necessary for mm accuracy ranging between the planets has already been demonstrated in the laboratory or in field SLR systems, and many key components have been space-qualified. Space qualification of the laser transmitter and picosecond timer needs to be demonstrated, but the technologies do not appear to present any difficult hurdles. The current technology trend in SLR is away from the traditional low repetition rate (5–20 Hz), high energy (100 mJ) systems toward the kHz rate, low energy (< 1 mJ) photon-counting systems as pioneered by NASA's SLR2000 system.[13,14] The photon-counting approach results in a much smaller and lighter spaceborne package, and it has been shown[7] analytically that low power ($\sim$100 mW), kHz rate transmitters, pumped by CW diode lasers, can support an Earth–Mars link throughout its synodic cycle, corresponding to interplanetary ranges between 0.52 and 2.52 AU.

The best satellite ranging results (< 3 mm single shot RMS) to date have been achieved with very short laser pulsewidths on the order of 10–20 picoseconds generated by relatively large modelocked Nd:YAG lasers. Passively Q-switched Nd:YAG microchip lasers are capable of producing multi-kHz trains of sub-nanosecond (typically 250–800 psec) pulses with up to several hundred milliwatts of average power from a very compact package pumped by CW diode lasers and should readily produce cm accuracy results with available space-qualifiable timers. Space-qualified CW pump diodes and passive Q switches are already available and have been used successfully in several NASA and ESA space experiments. Furthermore, the small size and monolithic construction of microchip lasers makes them relatively insensitive to misalignment, vacuum or thermal gradient effects as compared to larger, multicomponent lasers which have already performed successfully in spaceborne altimeters such as MOLA, GLAS, and MLA. If the need for still shorter pulses is warranted by future improvements in compact space-qualified timers, kHz trains of ultrashort pulses (< 50 ps) with sufficient average power can be generated efficiently in a very compact and rugged package via a CW diode-pumped fiber laser amplifier seeded by either an ultrashort pulse diode laser or a microchip laser passively Q-switched by a semiconductor saturated absorber mirror (SESAM).[15]

Compact, space-qualified multistop picosecond timers can be constructed using either CMOS time-to-digital converter (TDC) chips or field programmable gate arrays (FPGA's). The best TDC chips are manufactured by ACAM in Germany and have a timing resolution of 25 psec (< 4 mm range). Radiation-hardened versions of a similar but older TDC chip (100 psec resolution) have been successfully developed for use in the lidar docking system on ESA's Automated Transport Vehicle (ATV) for the International Space Station (ISS). Sigma Space Corporation has recently developed an alternative multichannel multistop timing system for a UAV-based lidar that utilizes a commercial FPGA and has a timing resolution of 92 psec (14-mm-range resolution) and a dead time of less than 2 nsec.[16] Space-qualified FPGA counterparts are readily available.

State-of-the-art SLR systems in, Europe presently use compensated single photon avalanche diodes (C-SPADs), and the Czech Technical University in Prague has been building space-qualified versions in anticipation of interplanetary transponders.[17] NASA SLR systems worldwide use microchannel plate photomultiplier tubes (MCP/PMT's). The fast rise time (140 psec), high quantum efficiency (40%), zero dead time, inherent vacuum compatibility, and low dark count (~30 kHz) of recently developed microchannel plate photomultipliers equipped with GaAsP photocathodes make them especially attractive for the most advanced photon-counting SLR systems and for transponders as well. Furthermore, they can be configured with multiple anodes, such as a quadrant, thereby allowing the tube to provide fine pointing corrections while simultaneously providing high resolution timing pulses as in NASA's developmental SLR2000 system.[13,14] Because they are vacuum-compatible by design, past PMT models have readily made the transition to space qualification on numerous NASA and ESA missions.

Space-qualified versions of other components required to build a complete transponder already exist. These include:

- CCD cameras for capturing the planetary image during initial acquisition.
- Accurate atomic clocks (e.g. Rb, Cs) routinely fly on GPS satellites.
- Dual wedge Risley Prism for introducing transmitter point ahead[7] — although it did not fly, the beam alignment mechanism (BAM) in the GLAS instrument on ICESat was fully space-qualified.
- Gimballed Two-Axis Telescope Pointing System — developed by Ball Aerospace for US military lasercom experiments and by ESA for intersatellite lasercom. A similar capability will be required for LISA.

The transponder can share these components with a collocated lasercom or interferometric system such as LISA. The manner in which these various components are combined to form a complete transponder system is discussed elsewhere.[7]

When one of the transponder or lasercom terminals is Earth-based, we must consider the effects of atmospheric turbulence on the uplink and downlink beam propagation. These effects include beam spreading, short term beam wander, and scintillation (signal fading).[4] It is important that the impact of the atmosphere on the link can be accurately assessed and tested, but end-to-end laboratory or field experiments which can convincingly simulate all aspects of these complex systems are both difficult to envision and expensive to implement. Fortunately, atmospheric transmission and turbulence effects on the uplink and downlink beams are the same whether the uplink beam is being reflected from a passive high altitude satellite in Earth orbit, as in SLR or LLR, or transmitted from a distant transponder or lasercom terminal in deep space. It has therefore been suggested[18] that, in preparing for an interplanetary mission, one might use existing SLR satellites to simulate the interplanetary link, complete with atmosphere. Each station must be located within the reflected return spot of the other station, however, and, for the existing satellite constellation, this implies an interstation distance of a few hundred meters or less.

The GSFC is currently implementing a joint transponder/lasercom experiment in which their 1.2 m telescope facility (Station A) ranges to satellites in the infrared (1064 nm) while their 40 cm aperture photon-counting SLR2000 system (Station B), in close proximity, ranges to the same satellite in the green (532 nm). Each station is equipped with receiver channels at 1064 and 532 nm, so they can detect their own reflected pulses as well as those from the other station. The experiment is self-calibrating since the transponder measures the length of the dogleg defined by the path Station A – satellite – Station B while the original single-ended ranging systems measure the Station A – satellite and Station B – satellite distances. Ground surveys typically define the interstation vector to better than 2 mm. Two-station SLR provides an accurate way to test the ranging and time transfer algorithms. Automated acquisition of the "Earth station" (A) by the "remote terminal" (B) can be demonstrated by either turning off or ignoring the closed ranging loop at 532 nm while it searches for the reflected light at 1064 nm. The ability to lock Station A onto the satellite via a closed single-ended ranging loop at 1064 nm ensures a steady source of photons from the Earth station for the remote terminal to find and lock onto, thereby simulating a distant laser source locked onto a solar-illuminated Earth image. It has been shown[18] that the equivalent lasercom or transponder link distance, R_T, is given by the equation

$$R_T(h, \theta_A, \sigma_s) = R_R^2(h, \theta_A)\sqrt{\frac{4\pi}{\sigma_s}\left(\frac{T_B^{\sec\theta_B}}{T_A^{\sec\theta_A}}\right)} \simeq R_R^2(h, \theta_A)\sqrt{\frac{4\pi}{\sigma_s}\frac{1}{T_A^{\sec\theta_A}}}, \tag{3}$$

where R_R is the slant range to the target satellite, h is the satellite altitude above sea level, σ_s is the optical cross-section of satellite retroreflector array, T_A and T_B are the zenith transmissions of Earth and planetary atmospheres respectively, and θ_A and θ_B are the zenith angles of the opposite terminal when viewed from Earth and planetary terminals respectively. The final approximation holds if the remote transponder or lasercom terminal is in interplanetary cruise phase, in orbit, or sitting on the surface of a planet or moon with little or no atmosphere ($T_B \sim 1$). Since the SLR satellites are normally tracked over the range $0° < \theta_A < 70°$, Eq. (3) defines a maximum and minimum simulated transponder range for each satellite. These are indicated by the lower curves, connected by squares, in Fig. 4 for our selected satellites. In the plots, we have assumed a value $T_A = 0.7$ corresponding to the one-way zenith transmission for a standard clear atmosphere at 532 nm. The upper curves, connected by circles, are plots of the minimum and maximum interplanetary distances of the Moon and other planets from Earth.

Figure 4 shows that a dual station ranging experiment to the lowest of the SLR satellites, Champ, provides a weaker return than a two-way lunar transponder. Low

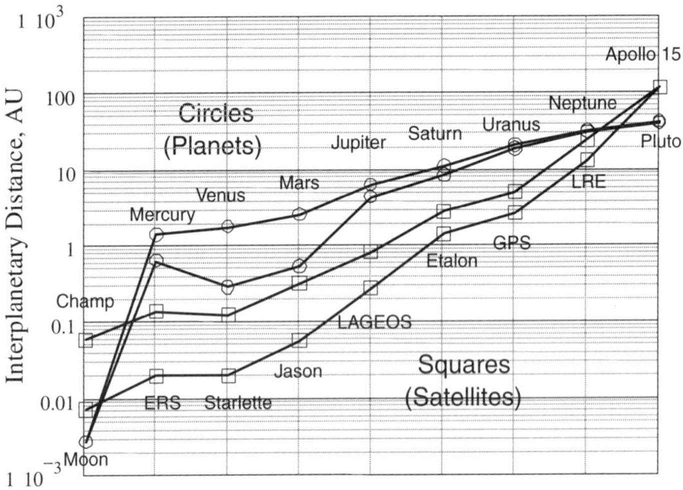

Fig. 4. The minimum and maximum distances from the Earth to the Moon and the eight planets are illustrated by the two top curves (circles) in the figure. The minimum and maximum transponder ranges simulated by the various SLR satellites are indicated by the two bottom curves (squares).

elevation angle experiments to Jason are comparable to a Venus or Mars link when they are closest to Earth. Experiments to the LAGEOS and Etalon satellites would simulate ranging to Mercury, Venus, and Mars throughout their synodic cycles while experiments to GPS and LRE (at 25,000 km) would simulate links up to and beyond Jupiter and Saturn. Dual station experiments to the Apollo 15 reflector on the lunar surface would simulate transponder links to over 100 AU, well beyond Pluto's orbit at 40 AU.

The nine SLR satellites in Fig. 4 were chosen based on their ability to simulate different transponder ranges and because pulse spreading by the target array is minimal. The latter property improves the precision of the measured transponder range and also provides a reasonably high fidelity facsimile of the outgoing optical pulse train from a ground-based lasercom transmitter.

5. Summary

Based on the recent successful experiments to the Messenger and MGS spacecraft, the space-qualified technology for decimeter accuracy interplanetary laser transponders is clearly already available, and more compact, space-qualified, subcentimeter accuracy systems can be made available within the next few years with very modest technology investments that also would benefit interplanetary laser communications and LISA.

The ability of laser transponders to simultaneously measure range, transfer time between distant clocks, and indirectly monitor the local gravity field at the spacecraft makes it a very useful tool for fundamental physics studies within the solar system. Other transponder applications include:

Solar system, lunar, and planetary science:

- Solar physics: gravity field, internal mass distribution and rotation.
- Few-mm accuracy lunar ephemerides and librations.
- Improved ranging accuracy and temporal sampling over current LLR operations to Apollo retroreflectors on the Moon with small, low energy ground stations.
- Decimeter-to-millimeter accuracy planetary ephemeredes.
- Mass distribution within the asteroid belt.

Lunar and planetary mission operations:

- Decimeter or better accuracy spacecraft ranging.
- Calibration/validation/backup for DSN microwave tracking.
- Subnanosecond transfer of GPS time to interplanetary spacecraft for improved synchronization of Earth/spacecraft mission operations.
- The transponder can serve as an independent self-locking beacon for collocated laser communications systems.

The ILRS has established a Transponder Working Group, which is presently developing hardware and software guidelines for member stations interested in participating in future transponder experiments. The ILRS is expected to support upcoming experiments such as the MLA Venus flyby in June 2007 and one-way differential range measurements to a visible detector on board NASA's Lunar Reconnaissance Orbiter, scheduled for launch in 2008.

Plans are well underway at the GSFC to utilize the retroreflector arrays on international SLR spacecraft to simulate interplanetary transponder and lasercom links via two-station SLR. Sigma has designed an improved array simulator for a possible "piggyback" on a future GPS or GEO satellite, which will permit the participating stations to be separated by up to a few kilometers.

References

1. H. H. Plotkin *et al.*, *Proc. IEEE* **53** (1965) 301.
2. J. J. Degnan, Thirty years of satellite laser ranging, in *Proc. Ninth International Workshop on Laser Ranging Instrumentation* (Canberra, Australia, 1994), p. 1.
3. J. J. Degnan, *IEEE Trans. Geosci. Remote Sens.* **GE-23** (1985) 398.
4. J. J. Degnan, Millimeter accuracy satellite laser ranging: A review, in *Contributions of Space Geodesy to Geodynamics: Technology*, eds. D. E. Smith and D. L. Turcotte (AGU Geodynamics Series, 25, 1993), p. 133.
5. J. O. Dickey *et al.*, *Science* **265** (1994) 482.
6. T. W. Murphy Jr. *et al.*, APOLLO: Meeting the millimeter goal, in *Proc. 14th International Workshop on Laser Ranging* (San Fernando, Spain, 2004).
7. J. J. Degnan, *J. Geodynamics* **34** (2002) 551.
8. J. J. Degnan, *Surv. Geophys.* **22** (2001) 431.
9. C. O. Alley, Proper time experiments in gravitational fields with atomic clocks, aircraft, and laser light pulses, in *Quantum Optics, Experimental Gravity, and Measurement Theory*, eds. P. Meystre and M. O. Scully (Plenum, 1983), p. 363.

10. N. Calder, *Einstein's Universe* (Greenwich House, New York, 1982).
11. X. Sun *et al.*, Laser ranging between the Mercury laser altimeter and an earth-based laser satellite tracking station over a 24 million kilometer distance, *OSA Annual Meeting Abstracts* (Tucson, AZ, Oct. 16–20, 2005).
12. D. E. Smith *et al.*, *Science* **311** (2006) 53.
13. J. J. Degnan and J. McGarry, SLR2000: Eyesafe and autonomous satellite laser ranging at kilohertz rates, in *Proc. Laser Radar Ranging and Atmospheric Lidar Techniques* (London, UK, Sep. 24–26, 1997), *SPIE Proc.* **3218** (1998) 63.
14. J. McGarry *et al.*, Early satellite ranging results from SLR2000, in *Proc. 14th International Workshop on Laser Ranging* (San Fernando, Spain, June 6–10, 2004).
15. G. J. Spuhler *et al.*, *J. Opt. Soc. Am. B* **16** (1999) 376.
16. J. J. Degnan *et al.*, "Second-Generation, Scanning, 3D Imaging Lidars Based on Photon-Counting," presented at IGARSS 2006 Conference (Denver, CO, Jul. 31–Aug. 4, 2006).
17. I. Prochaska *et al.*, SPAD detector package for space born applications, in *Proc. 13th Int. Workshop on Laser Ranging*, eds. R. Noomen *et al.* (NASA/CP-2003-212248, 2003).
18. J. J. Degnan, Laser transponders for high accuracy interplanetary laser ranging and time transfer, in *Lasers, Clocks, and Drag-Free: Exploration of Relativistic Gravity in Space*, eds. H. Dittus, C. Lammerzahl and S. Turyshev (Springer, New York, 2006), p. 231.

LASER RANGING FOR GRAVITATIONAL, LUNAR AND PLANETARY SCIENCE

STEPHEN M. MERKOWITZ*, PHILIP W. DABNEY,
JEFFREY C. LIVAS, JAN F. MCGARRY, GREGORY A. NEUMANN
and THOMAS W. ZAGWODZKI

NASA Goddard Space Flight Center, Greenbelt MD 20771, USA
**Stephen.M.Merkowitz@nasa.gov*

More precise lunar and Martian ranging will enable unprecedented tests of Einstein's theory of general relativity as well as lunar and planetary science. NASA is currently planning several missions to return to the Moon, and it is natural to consider if precision laser ranging instruments should be included. New advanced retroreflector arrays at carefully chosen landing sites would have an immediate positive impact on lunar and gravitational studies. Laser transponders are currently being developed that may offer an advantage over passive ranging, and could be adapted for use on Mars and other distant objects. Precision ranging capability can also be combined with optical communications for an extremely versatile instrument. In this paper we discuss the science that can be gained by improved lunar and Martian ranging along with several technologies that can be used for this purpose.

Keywords: Lunar ranging; general relativity; Moon; Mars.

1. Introduction

Over the past 35 years, lunar laser ranging (LLR) from a variety of observatories to retroreflector arrays placed on the lunar surface by the Apollo astronauts and the Soviet Luna missions has dramatically increased our understanding of gravitational physics along with Earth and Moon geophysics, geodesy, and dynamics. During the past few years, only the McDonald Observatory (MLRS) in Texas and the Observatoire de la Côte d'Azur (OCA) in France have routinely made lunar range measurements. A new instrument, APOLLO, at the Apache Point facility in New Mexico is expected to become operational within the next year, with somewhat increased precision over previous measurements.[1]

Setting up retroreflectors was a key part of the Apollo missions, so it is natural to ask if future lunar missions should include them as well. The Apollo retroreflectors are still being used today, and the 35 years of ranging data has been invaluable for scientific as well as other studies, such as orbital dynamics. However, the available

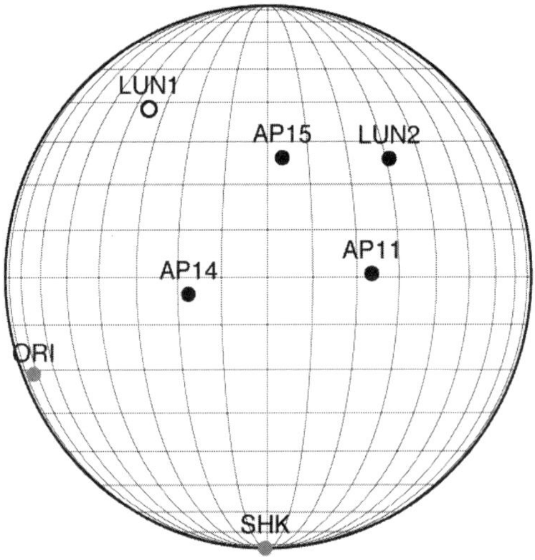

Fig. 1. Location of the lunar retroreflector arrays. The three Apollo arrays are labeled AP and the two Luna arrays are labeled LUN. ORI and SHK show the potential locations of two additional sites that would aid in strengthening the geometric coverage.

retroreflectors all lie within 26° latitude of the equator, and the most useful ones within 24° longitude of the sub-Earth meridian, as shown in Fig. 1. This clustering weakens their geometrical strength. New retroreflectors placed at locations other than the Apollo sites would enable more detailed studies, particularly those that rely on the measurement of lunar librations. In addition, more advanced retroreflectors are now available that will reduce some of the systematic errors associated with using the Apollo arrays.

In this paper we discuss the possibility of putting advanced retroreflectors at new locations on the lunar surface. In addition, we discuss several active LLR instruments that have the potential for greater precision and can be adapted for use on Mars. These additional options include laser transponders and laser communication terminals.

2. Gravitational Science from Lunar Ranging

Gravity is the force that holds the Universe together, yet a theory that unifies it with other areas of physics still eludes us. Testing the very foundation of gravitational theories, like Einstein's theory of general relativity, is critical for understanding the nature of gravity and how it relates to the rest of the physical world.

The equivalence principle, which states the equality of gravitational and inertial mass, is central to the theory of general relativity. However, nearly all alternative theories of gravity predict a violation of the equivalence principle. Probing the validity of the equivalence principle is often considered the most powerful way to search for new physics beyond the Standard Model.[2] A violation of the equivalence principle would cause the Earth and Moon to fall at different rates toward the Sun, resulting in a polarization of the lunar orbit. This polarization shows up in LLR as a displacement along the Earth–Sun line with a 29.53 d synodic period. The current limit on the equivalence principle is given by LLR: $\Delta(M_G/M_I)_{\mathrm{EP}} = (-1.0 \pm 1.4) \times 10^{-13}$.[3]

General relativity predicts that a gyroscope moving through curved space–time will precess with respect to the rest frame. This is referred to as geodetic or de Sitter precession. The Earth–Moon system behaves as a gyroscope with a predicted geodetic precession of 19.2 msec/year. This is observed using LLR by measuring the lunar perigee precession. The current limit on the deviation of the geodetic procession is $K_{\mathrm{gp}} = (-1.9 \pm 6.4) \times 10^{-3}$.[3] This measurement can also be used to set a limit on a possible cosmological constant: $\Gamma < 10^{-26}\,\mathrm{km}^{-2}$.[4]

It is also useful to look at violations of general relativity in the context of metric theories of gravity. Post-Newtonian parametrization (PPN) provides a convenient way to describe simple deviations from general relativity. The PPN parameters are usually denoted as γ and β; γ indicates how much space–time curvature is produced per unit mass, while β indicates how nonlinear gravity is (self-interaction). γ and β are identically one in general relativity. Limits on γ can be set from geodetic precession measurements, but the best limits come from measurements of the gravitational time delay of light, often referred to as the Shapiro effect. Ranging measurements to the Cassini spacecraft set the current limit on γ, $(\gamma - 1) = (2.1 \pm 2.3) \times 10^{-5}$,[5] which combined with LLR data provides the best limit on β: $(\beta - 1) = (1.2 \pm 1.1) \times 10^{-4}$.[3]

The strength of gravity is given by Newton's gravitational constant G. Some scalar-tensor theories of gravity predict some level of time variation in G. This will lead to an evolving scale of the solar system and a change in the mass of compact bodies due to a variable gravitational binding energy. This variation will also show up on larger scales, such as changes in the angular power spectrum of the cosmic microwave background.[6] The current limit on the time variation of G is given by LLR: $\dot{G}/G = (4 \pm 9) \times 10^{-13}$/year.[3]

The above effects are the leading gravitational limits that have been set by LLR, but many more effects can be studied using LLR data at various levels. These include gravitomagnetism (frame-dragging), $1/r^2$ force law, and even tests of Newton's third law.[7]

3. Lunar Science from Lunar Ranging

Several areas of lunar science are aided by LLR. First, the orientation of the Moon can be used for geodetic mapping. The current IAU rotation model, with respect to which images and altimetry are registered, has errors at the level of several hundred meters. A more precise model, DE403,[8] is being considered that is based on LLR and dynamical integration, but will require updating since it uses data only through 1994. Errors in this model are believed to be several meters. Further tracking will quantify the reliability of this and future models for lunar exploration.

Second, LLR helps provide the ephemeris of the Moon and solar system bodies. The position of the lunar center of mass is perturbed by planetary bodies, particularly Venus and Jupiter, at the level of hundreds of meters to more than 1 km. LLR is an essential constraint on the development of planetary ephemerides and navigation of spacecraft.

LLR can also be used to study the internal structure of the Moon, such as the possible detection of a solid inner core. The second-degree tidal lunar Love numbers are detected by LLR, as well as their phase shifts. From these measurements, a fluid core of 20% the Moon's radius is suggested. A lunar tidal dissipation of $Q = 30 \pm 4$ has been reported to have a weak dependence on tidal frequency. Evidence for the oblateness of the lunar fluid-core/solid-mantle boundary may be reflected in a century-scale polar wobble frequency. The lunar vertical and horizontal elastic tidal displacement Love numbers h_2 and l_2 are known to no better than 25% of their values, and the lunar dissipation factor Q and the gravitational potential tidal Love number k_2 no better than 11%. These values have been inverted jointly for structure and density of the core,[9,10] implying a semiliquid core and regions of partial melt in the lunar mantle, but such inversions require stochastic sampling and yield probabilistic outcomes.

4. Rationale for Additional Lunar Ranging Sites

While a single Earth ranging station may in principle range to any of the four usable retroreflectors on the near side from any longitude during the course of an observing day, these observations are nearly the same in latitude with respect to the Earth–Moon line, weakening the geometric strength of the observations. Additional observatories improve the situation somewhat, but of stations capable of ranging to the Moon, only Mt Stromlo in Australia is not situated at similar northern latitudes. The frequency and quality of observations vary greatly with the facility and power of the laser employed. Moreover, the reflector cross-sections differ substantially. The largest reflector, Apollo 15, has 300 cubes and returns only a few photons per minute to MLRS. The other reflectors have 100 cubes or less, and proportionately smaller rates. Stations and reflectors are unevenly represented, so that in recent years most ranging has occurred between one ground station and one reflector. Over the past six years, 85% of LLR data have been taken from MLRS and 15% from OCA. 81% of these were from the Apollo 15 reflector, 10% from Apollo 11, 8% from Apollo 14, and about 1% from Luna 2.[11] The solar noise background and thermal distortion makes ranging to some reflectors possible only around the quarter-Moon phase. The APOLLO instrument should be capable of ranging during all lunar phases.

The first LLR measurements had a precision of about 20 cm. Over the past 35 years, the precision has increased only by a factor of 10. The new APOLLO instrument has the potential to gain another factor of 10, achieving mm level precision, but this capability has not yet been demonstrated.[12] Poor detection rates were a major limiting factor in past LLR. Not every laser pulse sent to the Moon results in a detected return photon, leading to poor measurement statistics. MLRS typically collects less than 100 photons per range measurement, with a scatter of about 2 cm. The large collecting area of the Apache Point telescope and the efficient avalanche photodiode arrays used in the APOLLO instrument should result in thousands of detections (even multiple detections per pulse) leading to a potential statistical uncertainty of about 1 mm. Going beyond this level of precision will likely require new lunar retroreflectors or laser transponders that are more thermally stable and are designed to reduce the error associated with the changing orientation of the array with respect to the Earth due to lunar librations.

Several tests of general relativity and aspects of our understanding of the lunar interior are currently limited by present LLR capabilities. Simply increasing the precision of the LLR measurement, either through ground station improvement or through the use of laser transponders, will translate into improvements in these areas.

Additional ranging sites will also help improve the science gained through LLR. The structure and composition of the interior require dynamic measurements of the lunar librations, while tests of general relativity require the position of the lunar center of mass. In all, six degrees of freedom are required to constrain the geometry of the Earth–Moon system (in addition to Earth orientation). A single ranging station and reflector is insufficient to accurately determine all six, even given the rotation of the Earth with respect to the Moon.

To illustrate the importance of adding high-cross-section reflectors (or transponders) near the lunar limb, we performed an error analysis based on the locations of two observing stations that are currently operating and the frequency with which normal points have been generated over the last 10 years. While data quality has improved over the years, it has reached a plateau for the last 15 years or so. The presently operating stations are comparable in quality, and we assume an average of 2.5 cm for all observations. The normal point accumulation is heavily weighted toward Apollo 15, and a negligible number of returns are obtained from Lunakhod 2. We anticipate that ranging to a reflector with a 4$\times$ higher cross-section than Apollo 15 would approach 1 cm quality, simply by the increased return rate.

The model assumes a fixed Earth–Moon geometry to calculate the sensitivity of the position determination jointly with lunar rotation along three axes parallel to Earth's X–Y–Z coordinate frame at a moment in time when the Moon lies directly along the positive X axis. The Z axis points North and Y completes the right-hand system. Partial derivatives of range are calculated with respect to perturbations in position and orientation, where orientation is scaled from radians to meters by an equatorial radius of 1738 km. The analysis makes no prior assumptions regarding the dynamical state of the Moon.

The normal equations are weighted by the frequency of observations at each pair of ground stations and reflectors over the last 10 years. We then replace some of the observations with ranges to one or more new reflectors. The results are given in Table. 1.

Table 1. Additional ranging sites and stations increase the precision of the normal points for the same measurement precision due to better geometrical coverage. The precision in meters on the six degrees of freedom of a typical normal point is shown for several possible ranging scenarios.

X	Y	Z	Rot$X$	Rot$Y$	Rot$Z$	
0.265	6.271	23.294	15.958	0.179	0.225	MLRS and OCA
0.263	3.077	23.305	7.611	0.174	0.140	25% of observations to ORI
0.259	2.840	23.271	4.692	0.114	0.198	25% of observations to SHK
0.259	2.969	23.291	4.850	0.116	0.086	Both ORI and SHK
0.030	2.501	2.902	4.244	0.050	0.078	Both ORI and SHK, with 25% additional observations from Mt Stromlo

The addition of one or more reflectors would improve the geometrical precision of a normal point by a factor of 1.5 to nearly 4 at the same level of ranging precision. Such improvements directly scale to improvements in measurement of ephemeris and physical librations. The uncertainty in the Moon–Earth distance (X) is highly correlated with the uncertainty in position relative to the ecliptic (Z) when all stations lie at similar latitudes. An advanced reflector with a high cross-section would enable southern hemisphere ground stations such as Mt Stromlo to make more frequent and precise observations. The geometric sensitivity to position is dramatically improved by incorporating such a ground station, as shown in the last row of Table 1.

5. Retroreflectors

Five retroreflector arrays were placed on the Moon in the period 1969–1973. Three were placed by US astronauts during the Apollo missions (11, 14, and 15), and two were sent on Russian Lunokhod landers. The Apollo 11 and 14 arrays consist of 100 fused silica "circular opening" cubes (diameter 3.8 cm each) with a total estimated lidar cross-section of 0.5 billion square meters. Apollo 15 has 300 of these cubes and therefore about three times the lidar cross-section and is the lunar array with the highest response. Because the velocity aberration at the Moon is small, the cube's reflective face angles were not intentionally spoiled (deviated from 90°).

The two Lunokhod arrays consist of 14 triangular cubes, each side 11 cm. Shortly after landing, the Lunokhod 1 array ceased to be a viable target — no ground stations have since been able to get returns from it. It is also very difficult to get returns from Lunokhod 2 during the day. The larger size of the Lunokhod cubes makes them less thermally stable, which dramatically reduces the optical performance when sunlit.

Since 1969, multiple stations successfully ranged to the lunar retroreflectors. Some of these stations are listed in Table 2 along with their system characteristics. However, there have only been two stations continuously ranging to the Moon since the early 1970's: OCA in Grasse, France, and MLRS in Texas. The vast majority of their lunar data come from the array with the highest lidar cross-section — Apollo 15.

Table 2. Lunar-retroreflector-capable laser ranging stations and their expected return rate from the Apollo 15 lunar array. Link calculations use 1 billion meters squared for Apollo 15's cross-section, a mount elevation of 30°, and a standard clear atmosphere (transmission = 0.7 at the Zenith) for all but Apache Point, where transmission = 0.85 at the zenith. The laser divergence was taken to be 40 rad for MLRS and 20 arcsec for the other systems. The detector quantum efficiency was assumed to be 30% for all systems.

System	Telescope aperture (m)	Pulse energy exiting system (J)	Laser fire rate (Hz)	System transmission	Apollo 15 photoelectrons/min link calculation
MLRS	0.76	60	10	0.5	4
OCA (France)	1.54	60	10	0.22	20
Matera (Italy)	1.5	22	10	0.87	60
Apache Point	3.5	115	20	0.25	1728

The difficulty in getting LLR data is due to the distance to the Moon coupled with the $1/r^4$ losses in the signal, and the technology available at the ground stations. MLRS achieves an expected return rate from Apollo 15 of about one return per minute. Increasing the lidar cross-section of the lunar arrays by a factor of 10-would correspond to a factor-of-10 increase in the return data rate. This can be achieved by making arrays with 10 times more cubes than Apollo 15 or by changing the design of the cubes. One possibility is to increase the cube size. The lidar cross-section of a cube with a diameter twice that of an Apollo cube would be 16 times larger. However, simply making solid cubes larger increases their weight by the ratio of the diameter cubed. Additional size also adds to thermal distortions and decreases the cube's divergence: a very narrow divergence will cause the return spot to completely miss the station due to velocity aberration. Spoiling can compensate

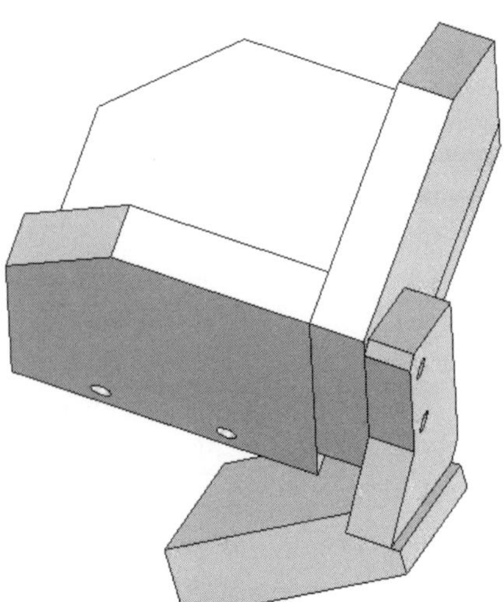

Fig. 2. Hollow retroreflectors can potentially be used to build large cross-section lightweight arrays.

for the velocity aberration but reduces the effective lidar cross-section. Changing the design of the cubes, such as making them hollow, may be a better alternative. For example, 300 unspoiled 5 cm beryllium hollow cubes would have a total mass less than that of Apollo 15 but would have a $3\times$ higher lidar cross-section.

An option being investigated at Goddard is to replace solid glass cubes with hollow cubes which weigh much less than their solid counterparts. Thermal distortions are less, especially in hollow cubes made of beryllium, so the cubes can be made larger without sacrificing optical performance. Hollow cubes (built by PLX) flew on the Japanese ADEOS satellite and on the Air Force Relay Mirror Experiment, but are generally not used on satellites for laser ranging. This is due in part to the lack of optical performance test data on these cubes under expected thermal conditions, but also because of early investigations which showed that hollow cubes were unstable at high temperatures. Advances in adhesives and other techniques for bonding hollow cubes make it worthwhile to reinvestigate them. Testing that was done for Goddard by ProSystems showed that hollow cubes (with faces attached via a method that is being patented by ProSystems) can survive thermal cycles from room temperature to 150 degrees Celsius. Testing has not yet been done at cold temperatures. Preliminary mechanical analysis indicates that the optical performance of hollow beryllium cubes would be more than sufficient for laser ranging.

6. Satellite Laser Ranging Stations

Satellite laser ranging began in 1964 at NASA's Goddard Space Flight Center. Since then it has grown into a global effort, represented by the International Laser Ranging Service (ILRS),[13] of which NASA is a participant. The ILRS includes ranging to Earth-orbiting artificial satellites and ranging to the lunar reflectors, and is actively working toward supporting asynchronous planetary transponder ranging.

The ILRS lunar-retroreflector-capable stations have event timers with precisions of better than 50 picoseconds, and can tie their clocks to UTC to better than 100 nanoseconds. Most have arc-second tracking capabilities and large aperture telescopes (> 1 m). Their lasers have very narrow pulse widths (< 200 psec) and most have high energy per pulse (> 50 mJ). All have the ability to narrow their transmit beam divergence to less than 50 rad. The detectors have a relatively high quantum efficiency ($> 15\%$). All current LLR systems range at 532 nm.

Clearly, there is more than one way to increase the laser return rate from the Moon. One is to deploy higher response retroreflector arrays or transponders on the Moon. Another is to increase the capability of the ground stations. A third is to add more lunar-capable ground stations. A combination of all these options would have the biggest impact.

The recent development of the Apache Point system, APOLLO,[1] shows what a significant effect improving the ground station can make. Apache Point can theoretically achieve a thousand returns per minute from Apollo 15 versus the few-per-minute return rate from MLRS (see Table 2). Apache Point does this by using a very large aperture telescope, a somewhat higher laser output energy and fire rate, and a judicious geographical location (where the astronomical seeing is very good). Other areas that could also improve ground station performance are higher-quantum-efficiency single photon detectors ($> 30\%$ QE at 532 nm), higher-repetition-rate lasers (kilohertz versus tens of hertz), and the use of adaptive optics to maintain tight beam control.

Higher-cross-section lunar retroreflectors may make it possible to use NASA's next generation of satellite laser ranging stations (SLR2000) for LLR. The prototype SLR2000 system is currently capable of single photon asynchronous laser transponder ranging, and will participate in both a two-way asynchronous transponder experiment in 2007 and the one-way laser ranging to the Lunar Reconnaissance Orbiter (LRO) in 2008–2009. Approximately ten SLR2000 stations are expected to be built and deployed around the world in the coming decade. Adding ten LLR stations to the existing few would dramatically increase the volume of data as well as giving the data a wide geographical distribution. The global distribution of the new SLR2000 stations would be very beneficial to data collection from an asynchronous transponder on the Moon.

7. Laser Transponders

Laser transponders are currently being developed for satellite ranging, but they can also be deployed on the lunar surface. Transponders are active devices that detect an incoming signal, respond with a known or predictable response signal, and are used to either determine the existence of the device or positioning parameters, such as range and/or time. For extraterrestrial applications, a wide range of electromagnetic radiation, such as radio frequency (RF), is used for this signal. To date, most spacecraft have been tracked using RF signals, particularly in the S and X bands of the spectrum. NASA and several other organizations routinely track Earth-orbiting satellites using optical satellite laser ranging (SLR). Laser transponders have approximately a $1/r^2$ link advantage over direct ranging loss of $1/r^4$, essentially because the signal is propagating in only one direction before being regenerated. In fact, it is generally considered that ranging beyond lunar distances is not practical using direct optical ranging to cube-corner reflectors. Laser transponders are in general more energy- and mass-efficient than RF transponders since they can work at single photon detection levels with much smaller apertures and beam divergences. A smaller beam divergence has the added benefit that there is less chance of interference with other missions, as well as making the link more secure should that be necessary. With the development and inclusion of laser communications for spaceflight missions, it is logical to include an optical transponder that uses the same optomechanical infrastructure such that it has minimal impact on the mission resources.

The simplest conceptual transponder is the synchronous or echo transponder. An echo transponder works by sending back a timing signal with a fixed delay from the receipt of the base station signal. This device has the potential for the lowest complexity and autonomous operations with no RF or laser-based communications channel. To enable this approach, an echo pulse must be created with a fixed offset delay that has less than 500 ps jitter from the arrival of the Earth station signal. This is very challenging given the current state of the art in space-qualifiable lasers. Furthermore, several rugged and simple laser types would be excluded as candidates due to the lack of precision control of the pulse generation. The synchronous/echo transponder has a total link probability that is the joint probability of each direction's link probability (approximately the product of each).

Asynchronous laser transponders (ALTs) have been shown analytically[14] and experimentally[15] to provide the highest link probability since the total link is the root sum square of each one-way link probability. Furthermore, they allow the use of free-running lasers on the spacecraft that operate at their most efficient repetition rates, are simpler, and potentially more reliable. Figure 3 shows a conceptual asynchronous laser transponder using an existing NASA SLR ground station that

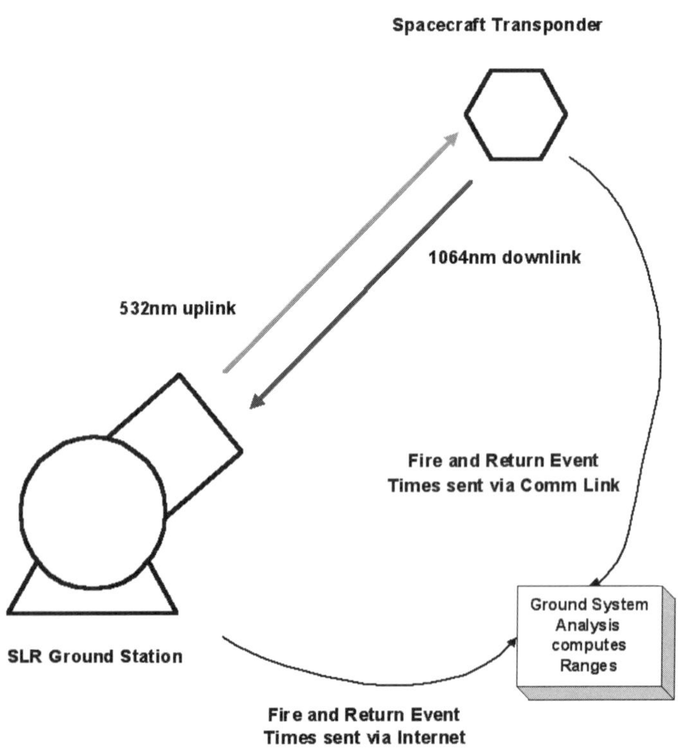

Fig. 3. In an asynchronous laser transponder system the ground and remote stations fire independently of each other, recording the pulse transmission and detection times. The data from the two sites are then combined to calculate the range.

is already precisely located and calibrated in the solar reference frame, and a spacecraft transponder that receives green photons (532 nm) and transmits near-infrared (NIR) photons (1064 nm). This diagram shows the spacecraft event times being downlinked on the RF (S band) channel, but this could be done on the laser communication channel if one exists. This dual wavelength approach is being explored for reasons of technical advantage at the ground station, but may also be used to help remove atmospheric effects from the range data (due to its wavelength-dependent index of refraction). Using the same wavelength for each direction is also possible. Expressions for recovering the range parameters from an asynchronous measurement can be found in Ref. 14 and in the parameter retrieval programs developed by Gregory Neumann for the Earth-MLA asynchronous transponder experiments.[15,16]

An ALT will likely have systematic errors that will limit its long-term accuracy. A retroreflector array located near the ALT's lunar site should allow the study and calibration of the ALT's systematic errors. Performing this experiment on the Moon would be particularly important should this technology be adapted for Mars or other bodies where retroreflectors cannot be used.

Recently, two interplanetary laser transponder experiments were successfully performed from the NASA Goddard Geophysical and Astronomical Observatory (GGAO) SLR facility. The first utilized the nonoptimized Mercury Laser Altimeter (MLA) on the Messenger spacecraft and the second utilized the Mars Orbiting Laser Altimeter (MOLA) on the Mars Orbiter spacecraft. The Earth-MOLA experiment was a one-way link that set a new distance record of 80 M-km for detected signal photons. The Earth-MLA experiment was a two-way experiment that most closely resembled the proposed asynchronous laser transponder concept. This experiment demonstrated the retrieval of the clock offset, frequency drift, and range of 24 M-km using a small number of detected two-way events. Such experiments have proven the concept of being able to point both transceivers, detect the photons, and retrieve useful parameters at low-link margins.

The Lunar Reconnaissance Orbiter (LRO) mission includes a GSFC-developed laser ranger that will provide a one-way ranging capability. In this case the clock is assumed to be stable enough over one lunar orbit. The result is a range profile that is extremely precise but far less accurate than what a two-way asynchronous transponder would provide with its full clock solution.

An ALT conceptual design was developed as part of the LRO laser ranger trade study. Link analyses performed on this design showed that it is possible to make more than 500 two-way range measurements per second using a 20 mm aperture and a 10 microjoule/pulse, 10 kHz laser at the Moon and the existing eye-safe SLR2000 telescope located at the GGAO. The ALT was not selected due to the need for a very-high-readiness design for the LRO, but the analysis did show its feasibility. It was also shown that many of the international SLR systems could participate with nominal receiver and software upgrades, thereby increasing the ranging coverage. The ALT increases the tracking/ranging availability of spacecraft since the link margins are higher than for direct ranging to reflector arrays.

8. Communication Terminal

A communications terminal conceptually represents the most capable kind of transponder ranging system and is at the other end of the spectrum in terms of complexity from the echo transponder. In general, a communications link of some kind is necessary for operating and recovering data from a spacecraft or remote site. There are several potential benefits if the communications link can be made part of the ranging system, including savings in weight, cost, and complexity over implementations that use separate systems for each requirement.

As with other types of transponder systems, the active terminals for a full-duplex communications system mean that the loss budget for the ranging/communications link scales as $1/r^2$ instead of $1/r^4$, which is a substantial advantage. The communications link need not be symmetric in terms of the data rate to achieve this benefit. Very often the uplink is relatively low-bandwidth for command and control, and the downlink is at a higher data rate for dumping data. The ground antenna can be made substantially larger than the remote antenna, both to make the receiver more sensitive and as a way to reduce the mass and the pointing and tracking requirements for the spacecraft, since a smaller antenna has a larger beam.

Forward error correction (FEC) is a technique that can improve the link budget and hence the range of the system. It is a signal processing technique that adds more bits to the communications data stream through an algorithm that generates enough redundancy to allow these bits to be used to detect errors. There are a wide variety of possible FEC algorithms that can be used, but it is possible to get link budget gains of the order of 8 dB at the cost of 7% overhead on the data rate even at data rates of several Gbits/s. Gains of the order of 10 dB and higher are available at lower data rates, but at the cost of higher overhead. Generally the FEC algorithm may be optimized for the noise properties of the link if they are known.

A synchronous communications terminal must maintain a precise clock to be able to successfully recover the data. A remote terminal will recover the clock from the incoming data stream and phase-lock a local oscillator. All modern wide area terrestrial communications networks use synchronous techniques, so the techniques and electronics are well known and generally available. The advantage of having a stable reference clock that is synchronized to the ground terminal is that long times (and therefore long distances) may be measured with a precision comparable to that of the clock simply by counting bits in the data stream and carefully measuring the residual timing offset at the ground station. A maximal-length pseudorandom code can be used to generate a pattern with very simple cross-correlation properties that may be used to unambiguously determine the range, and the synchronous nature of the signal plus any framing or FEC structure imposed on the data stream means that even long times may be measured with the same precision as the clock.

For optical communications terminals, it is almost as cost-effective to run at a high data rate as it is at a low data rate. The data rate might then reasonably be chosen for the timing precision instead of for the data downlink requirements.

For example, a 10 Gbps data rate has a clock period of 100 picoseconds, which translates to a submillimeter distance precision with some modest averaging — just based on the clock. Specialized modulation formats such as phase-shift keying offer the possibility of optical phase-locking in addition to electrical phase-locking, which may allow further increases in precision.

Spacecraft or satellites may be used as repeaters or amplifiers much as they are in terrestrial telecom applications, further extending the reach. Multiple communications terminals distributed around an object, such as a planet, offer the ability to measure more complicated motion than just a range and a change in range. A high-data-rate terminal might also be used as part of a communications network in space. In addition to serving as a fixed point for high precision ranging, it could provide various communications functions, such as switching and routing.

9. Conclusions

LLR has made great advances in the past 35 years. However, the amount of light returned by the current retroreflectors is so little that only the largest ranging stations can be used for this purpose; poor detection statistics remain the leading source of error. Thermal and orientation effects will ultimately limit range measurement to the Apollo retroreflectors. Measurements of the lunar librations are also limited by the poor geometric arrangement of the visible retroreflectors.

More precise range measurements to retroreflectors placed at sites far from the existing arrays will greatly improve the gravitational and lunar science discussed above. A number of improvements (such as a higher cross-section) can be made to the retroreflector designs to realize these gains. This natural extension to the Apollo instruments is likely to produce a solid incremental improvement to these scientific studies for many years to come.

To make a much larger leap in ranging accuracy, a laser transponder or communications terminal will most likely be required. The robust link margins will enable the use of much smaller ground stations, which would provide for more complete time and geometric coverage as more ranging stations could be used. An active system will also not be susceptible to the libration-induced orientation errors.

An active laser ranging system can be considered a pathfinder for a Mars instrument, as it is likely to be the only way to exceed the meter level accuracy of current ranging data to Mars. Laser ranging to Mars can be used to measure the gravitational time delay as Mars passes behind the Sun relative to the Earth. With 1 cm precision ranging, the PPN parameter γ can be measured to about 10^{-6}, ten times better than the Cassini result.[17] The strong-equivalence-principle polarization effect is about 100 times larger for Earth–Mars orbits than for the lunar orbit. With 1 cm precision ranging, the Nordtvedt parameter, $\eta = 4\beta - \gamma - 3$, can be measured to between 6×10^{-6} and 2×10^{-6} for observations ranging between one and ten years.[18] Combined with the time delay measurements this leads to a measurement of the PPN parameter β to the 10^{-6} level. Mars ranging can also be used in combination with lunar ranging to get more accurate limits on the time variation of the gravitational constant.

The ephemeris of Mars itself is known to meters in-plane, but hundreds of meters out-of-plane.[19] Laser ranging would get an order-of-magnitude better estimate, significant for interplanetary navigation. Better measurements of Mars' rotational dynamics could provide estimates of the core size.[20] The elastic tidal Love number is predicted to be less than 10 cm, within reach of laser ranging. There is also an unexplained low value of Q, inferred from the secular decay of Phobos' orbit, that is a constraint on the present thermal state of the Mars interior.[21] Laser ranging to Phobos would help solve this mystery.

References

1. T. W. Murphy Jr. *et al.*, in *12th Int. Workshop on Laser Ranging* (Matera, Italy, 13–17 Nov. 2000).
2. T. Damour, *Class. Quant. Grav.* **13** (1996) A33.
3. J. G. Williams, S. G. Turyshev and D. H. Boggs, *Phys. Rev. Lett.* **93** (2004) 261101.
4. M. Sereno and P. Jetzer, *Phys. Rev. D* **73** (2006) 063004.
5. B. Bertotti, L. Iess and P. Tortora, *Nature* **425** (2003) 374.
6. J.-P. Uzan, *Rev. Mod. Phys.* **75** (2003) 403.
7. K. Nordtvedt, *Class. Quant. Grav.* **18** (2001) L133.
8. E. M. Standish *et al.*, JPL Planetary and Lunar Ephemerides, DE403/LE403, Jet Propulsion Laboratory internal report JPL IOM 314.10-127 (1995).
9. A. Khan *et al.*, *J. Geophys. Res.* **109** (2004) E09007.
10. A. Khan and K. Mosegaard, *Geophys. Res. Lett.* **32** (2005) L22203.
11. P. Shelus, private communication.
12. J. G. Williams, S. G. Turyshev and T. W. Murphy Jr, *Int. J. Mod. Phys. D* **13** (2004) 567 [gr-qc/0311021].
13. M. R. Pearlman, J. J. Degnan and J. M. Bosworth, *Adv. Space Res.* **30** (2002) 135.
14. J. J. Degnan, *J. Geodyn.* **34** (2002) 551.
15. D. E. Smith *et al.*, *Science* **311** (2006) 53.
16. G. A. Neumann *et al.*, in *15th Int. Laser Ranging Workshop* (Canberra, Australia 16–20 Oct. 2006).
17. S. G. Turyshev *et al.*, in *The 2004 NASA/JPL Workshop on Physics for Planetary Exploration* (2004) [gr-qc/0411082].
18. J. D. Anderson *et al.*, *Astrophys. J.* **459** (1996) 365.
19. A. S. Konopliv *et al.*, *Icarus* **182** (2006) 23.
20. W. M. Folkner *et al.*, *Science* **278** (1997) 1749.
21. B. G. Bills *et al.*, *J. Geophys. Res.* **110** (2005) E07004.

SPACE-BASED TESTS OF GRAVITY WITH LASER RANGING

SLAVA G. TURYSHEV* and JAMES G. WILLIAMS

Jet Propulsion Laboratory,
California Institute of Technology,
4800 Oak Grove Drive,
Pasadena, CA 91109, USA
**turyshev@jpl.nasa.gov*

Existing capabilities of laser ranging, optical interferometry, and metrology, in combination with precision frequency standards, atom-based quantum sensors, and drag-free technologies, are critical for space-based tests of fundamental physics; as a result of the recent progress in these disciplines, the entire area is poised for major advances. Thus, accurate ranging to the Moon and Mars will provide significant improvements in several gravity tests, namely the equivalence principle, geodetic precession, PPN parameters β and γ, and possible variation of the gravitational constant G. Other tests will become possible with the development of an optical architecture that allows one to proceed from meter to centimeter to millimeter range accuracies on interplanetary distances. Motivated by anticipated accuracy gains, we discuss the recent renaissance in lunar laser ranging and consider future relativistic gravity experiments with precision laser ranging over interplanetary distances.

Keywords: Tests of gravity; lunar laser ranging; fundamental physics.

1. Introduction

Because of a much higher transmission data rate and, thus, larger data volume delivered from large distances, higher communication frequency is very important for space exploration. Higher frequencies are less affected by the dispersion of delay in the solar plasma, thus allowing a more extensive coverage, when deep space navigation is concerned. Presently, the highest frequency implemented at the NASA Deep Space Network is the 33 GHz frequency of the Ka band. There is still a possibility of moving to even higher radiofrequencies, say to $\sim$60 GHz; however, besides being very challenging technologically, this would put us closer to the limit that the Earth's atmosphere imposes on signal transmission. Beyond these frequencies radio communication with distant spacecraft will be inefficient. The next step is switching to optical communication.

Lasers — with their spatial coherence, narrow spectral emission, high power, and well-defined spatial modes — are highly useful for many space applications. While in free space, optical laser communication (lasercomm) would have an advantage as opposed to the conventional radiocommunication methods. Lasercomm would provide not only significantly higher data rates (on the order of a few Gbps), it would also allow a more precise navigation and attitude control. In fact, precision navigation, attitude control, landing, resource location, three-dimensional imaging, surface scanning, formation flying and many other areas are thought of in terms of laser-enabled technologies. Here we investigate how near-future free-space optical technologies might spur progress in gravitational and fundamental physics experiments performed in the solar system.

This paper focuses on current and future optical technologies and methods that will advance fundamental physics research in the context of solar system exploration. Section 2 discusses the current state and the future performance expected with the new lunar laser ranging (LLR) technology. Section 3 addresses the possibility of improving tests of gravitational theories with the Apache Point Observatory Lunar Laser-ranging Operation (APOLLO) — the new LLR station in New Mexico. We investigate possible improvements in the accuracy of the tests of relativistic gravity with the anticipated reflector/transponder installations on the Moon. We also discuss the next logical step — interplanetary laser ranging — and address the possibility of improving tests of fundamental physics with laser ranging to Mars. We close with a summary and recommendations.

2. LLR Contribution to Fundamental Physics

2.1. *History and scientific background*

LLR is the living legacy of the Apollo program; in fact, it is the only continuing investigation from the Apollo era and the longest-running experiment in space. Since its initiation by the Apollo 11 astronauts in 1969, LLR has strongly contributed to our understanding of the Moon's internal structure and the dynamics of the Earth–Moon system. The data provide for unique, multidisciplinary results in the areas of lunar science, gravitational physics, Earth sciences, geodesy and geodynamics, solar system ephemerides, and terrestrial and celestial reference frames.

The placement of the retroreflector arrays on the Moon was motivated by the strong scientific potential of LLR. The first deployment of an LLR package on the lunar surface took place during the Apollo 11 mission in the summer of 1969, making LLR a reality.[1] Additional packages were set up by the Apollo 14 and 15 astronauts. The goal was to place arrays at three lunar locations to study the Moon's motion. Two French-built retroreflector arrays were on the Lunokhod 1 and 2 rovers placed on the Moon by the Soviet Luna 17 and Luna 21 missions, respectively.[a]

[a]The Lunokhod 1 location is poorly known and the retroreflector cannot be used currently.

While some early efforts were brief and demonstrated capability, most of the scientific results came from long observing campaigns at several observatories. Today, with several tens of satellite laser ranging (SLR) stations around the world, only two of them routinely range to the Moon. One of the presently operating stations is the McDonald Laser Ranging System (MLRS) (http://www.csr.utexas.edu/mlrs/) in Texas, USA.[2] The other is at the Observatoire de la Côte d'Azur (OCA) (http://www.obs-nice.fr/) in France.[3,4] Both stations operate in a multiple-target mode, observing targets other than the lunar retroreflectors.

The LLR effort at the McDonald Observatory in Texas has been carried out from 1969 to the present. The first sequence of observations was made from the 2.7 m telescope. In 1985 ranging operations were moved to the MLRS and in 1988 the MLRS was moved to its present site. The current 0.76 m MLRS has the advantage of a shorter laser pulse and improved range accuracy over the earlier, 2.7 m system, but the pulse energy and aperture are smaller. OCA began its accurate observations in 1984, which have continued to the present, though first detections were demonstrated earlier. Although originally built to operate as a lunar-only station, the operation is now divided among the four lunar retroreflectors, the two LAGEOS targets, and the several high altitude spacecraft (Glonass, Etalon, and GPS).

2.2. *Design of the experiment*

LLR measures the range from an observatory on the Earth to a retroreflector on the Moon. The geometry of the Earth, Moon, and orbit is shown in Fig. 1. For the Earth and Moon orbiting the Sun, the scale of relativistic effects is set by the ratio $(GM/rc^2) \simeq v^2/c^2 \sim 10^{-8}$. The mean distance of the Moon is 385,000 km, but there is considerable variation owing to the orbital eccentricity and perturbations due to Sun, planets, and the Earth's J_2 zonal harmonic. The solar perturbations are thousands of kilometers in size and the lunar orbit departs significantly from an ellipse. The equatorial radii of the Earth and Moon are 6378 km and 1738 km,

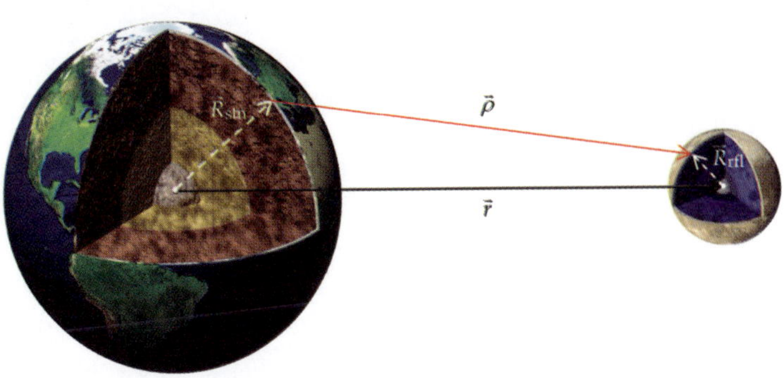

Fig. 1. Lunar laser ranging accurately measures the distance between an observatory on the Earth and a retroreflector on the Moon. Illustration not to scale.

respectively, so that the lengths and relative orientations of the Earth–Moon vector, the station vector, and the retroreflector vector influence the range. Thus, not only is there sensitivity of the range to anything which affects the orbit, there is also sensitivity to effects at the Earth and Moon.

In addition to the lunar orbit, the range from an observatory on the Earth to a retroreflector on the Moon depends on the position in space of the ranging observatory and the targeted lunar retroreflector. Thus, orientation of the rotation axes and the rotation angles of both bodies are important, with tidal distortions, plate motion, and relativistic transformations also coming into play. These various sensitivities allow the ranges to be analyzed to determine many scientific parameters. To extract the gravitational physics information of interest, it is necessary to accurately model a variety of effects.[5,6]

Each LLR measurement is the round-trip travel time of a laser pulse between an observatory on the Earth and one of the four corner cube retroreflector arrays on the Moon. To range the Moon, the observatories fire a short laser pulse toward the target array. The lasers currently used for the ranging operate at 10 Hz, with a pulse width of about 200 psec; each pulse contains $\sim 10^{18}$ photons. Under favorable observing conditions a single reflected photon is detected every few seconds for most LLR stations and in less than 1 sec for the Apache Point.[7] Such a low return rate is due to huge attenuation during the round trip of the pulse. The outgoing narrow laser beam must be accurately pointed at the target. The beam's angular spread, typically a few arcsec, depends on atmospheric seeing, so the spot size on the Moon is a few kilometers across. The amount of energy falling on the array depends inversely on that spot area.

The beam returning from the Moon cannot be narrower than the diffraction pattern for a corner cube; this pattern has a sixfold shape that depends on the six combinations of ways that light can bounce off the three orthogonal reflecting faces. Green laser light ($0.53\,\mu$m) with Apollo corner cubes gives 5 arcsec for the angular diameter of the central diffraction disk. The larger Lunokhod corner cubes would give half that diffraction pattern size. Thermal distortions, imperfections, and contaminating dust can make the size of the beam larger than the diffraction pattern.

The returning pulse illuminates an area around the observatory which is a few tens of kilometers in diameter ($\sim$10 km for green light). The observatory has a very sensitive detector which records single photon arrivals. The power received by the telescope depends directly on the telescope's collecting area and inversely on the returning spot area. Velocity-caused aberration of the returning beam is roughly 1". At the telescope's detector both a diaphragm restricting the field of view and a (few angstroms) narrow-bandpass filter reduce background light. When the background light is high the small diaphragm reduces the interference to increase the signal-to-noise ratio. When the seeing is poor the image size increases and this requires a larger diaphragm. Color and spatial filters are used to eliminate much of the background light.

A normal point is the standard form of an LLR datum used in the analysis. It is the result of a statistical combination of the observed transit times of several individual photons arriving at the observing detector within a relatively short time, typically less then a few tens of minutes.[4,8–10] Photons from different laser pulses have similar residuals with respect to the expected round-trip flight time and are separated from the widely scattered randomly arriving background photons. The resulting "range" normal point is the round-trip light time for a particular firing time. By October 2006, there were more than 16,250 normal points collected.

2.3. *LLR and tests of fundamental physics*

LLR data have been acquired from 1969 to the present. LLR has remained a viable experiment with fresh results over 37 years because the data accuracies have improved by an order of magnitude. The International Laser Ranging Service (ILRS) (http://ilrs.gsfc.nasa.gov/index.html) archives and distributes LLR data and related products, and supports laser ranging activities.

The measured round-trip travel times Δt are two-way (i.e. round-trip), but in this paper equivalent ranges in one-way length units are $c\Delta t/2$. The conversion between time and length (for distance, residuals, and data accuracy) uses 1 nsec = 15 cm. The ranges of the early 1970s had accuracies of approximately 25 cm (see Fig. 2). By 1976 the accuracies of the ranges had improved to about 15 cm. Accuracies improved further in the mid-1980s; by 1987 they were 4 cm, and the present accuracies are ~2 cm. One immediate result of lunar ranging was the great improvement in the accuracy of the lunar ephemeris,[11] lunar science,[12] and results of various tests of gravitational phenomena.[6,11]

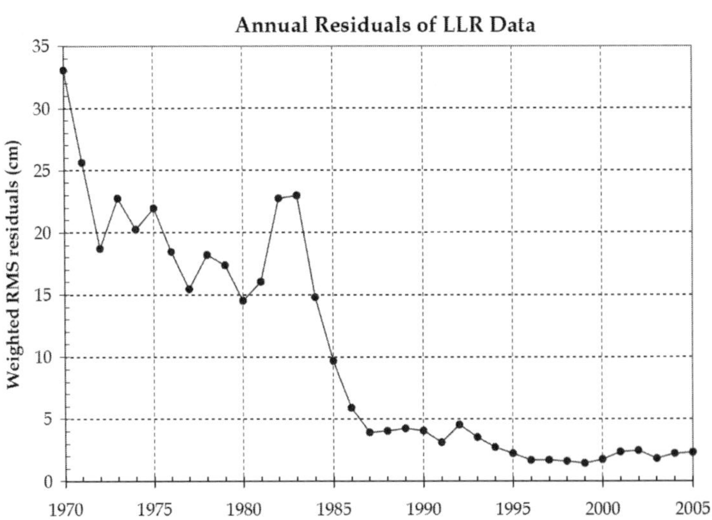

Fig. 2. Weighted rms residuals (observed–computed Earth–Moon distance) annually averaged.

LLR offers very accurate laser ranging (the weighted rms residual is currently ~ 2 cm or $\sim 5\times10^{-11}$ in fractional accuracy) to retroreflectors on the Moon. Analysis of these very precise data contributes to many areas of fundamental and gravitational physics. Thus, these high-precision studies of the Earth–Moon system moving in the gravitational field of the Sun provide the most sensitive tests of several key properties of weak-field gravity, including Einstein's strong equivalence principle (SEP), on which general relativity rests (in fact, LLR is the only current test of the SEP). LLR data yield the strongest limits to date on variability of the gravitational constant (the way gravity is affected by the expansion of the Universe), and the best measurement of the de Sitter precession rate.

Expressed in terms of the combination of the PPN parameters $\eta = 4\beta - 3 - \gamma$ (γ and β are the two Eddington parameters that are both equal to 1 in general relativity, and thus $\eta = 0$ in that theory), a violation of the equivalence principle leads to a radial perturbation of lunar orbit $\delta r \sim 13\eta \cos D\,\mathrm{m}$.[13–15] LLR investigates the SEP by looking for a displacement of the lunar orbit along the direction to the Sun. The Equivalence Principle can be split into two parts: the weak equivalence principle tests the sensitivity to composition and the SEP checks the dependence on mass. There are laboratory investigations of the weak equivalence principle (at the University of Washington) which are about as accurate as LLR.[16,17] LLR is the dominant test of the SEP. In our recent analysis of LLR data we used 16,250 normal points through 11 January, 2006 to test the EP of $\Delta(M_G/M_I)_{\mathrm{EP}} = (-0.8 \pm 1.3) \times 10^{-13}$, where $\Delta(M_G/M_I)$ signifies the difference between gravitational-to-inertial mass ratios for the Earth and the Moon. This result corresponds to a test of the SEP of $\Delta(M_G/M_I)_{\mathrm{SEP}} = (-1.8 \pm 1.9) \times 10^{-13}$, with the SEP violation parameter $\eta = 4\beta - \gamma - 3$ found to be $\eta = (4.0 \pm 4.3) \times 10^{-4}$.

The determined value for the combination of PPN parameters η can be used to derive the value for the PPN parameter, β. Thus, in combination with the recent value of the space curvature parameter $\gamma_{\mathrm{Cassini}}$ $[\gamma - 1 = (2.1 \pm 2.3) \cdot 10^{-5}]$, derived from Doppler measurements to the Cassini spacecraft,[18] the nonlinearity parameter β can be determined by applying the relationship $\eta = 4\beta - 3 - \gamma_{\mathrm{Cassini}}$. Then the PPN parameter β is determined at the level of $\beta - 1 = (1.0 \pm 1.1) \times 10^{-4}$.[b]

A recent limit on a change of G comes from LLR. This is the second-most-important gravitational physics result that LLR provides. General relativity does not predict a changing G, but some other theories do, thus testing for this effect is important. The current LLR $\dot{G}/G = (6 \pm 7) \times 10^{-13}\,\mathrm{yr}^{-1}$ is the most accurate limit obtained using LLR data. The $\dot{G}/G$ uncertainty is 107 times smaller than the inverse age of the Universe, $t_0 = 13.73$ Gyr, with the value for the Hubble constant $H_0 = 73.4\,\mathrm{km/sec/Mpc}$ from the WMAP3 data.[19] The uncertainty for $\dot{G}/G$ is improving rapidly because its sensitivity depends on the square of the data

[b]The LLR result for the space curvature parameter γ as determined from the EIH (Einstein–Infeld–Hoffmann) equations is less accurate than the results derived from other measurements.

span. The parameter $\dot{G}/G$ benefits from the long time span of LLR data and has experienced the biggest improvement over the past few years.

LLR has also provided the only accurate determination of the geodetic precession. Reference 6 reports a test of geodetic precession which, expressed as a relative deviation from general relativity, is $K_{\rm gp} = -0.0005 \pm 0.0047$. The GP-B satellite should provide improved accuracy over this value. LLR also has the capability of determining PPN β and γ directly from the point mass orbit perturbations. A future possibility is detection of the solar J_2 from LLR data combined with the planetary ranging data. Also possible are dark matter tests, looking for any departure from the inverse square law of gravity, and checking for a variation of the speed of light. The accurate LLR data have been able to quickly eliminate several suggested alterations of physical laws. The precisely measured lunar motion is a reality that any proposed laws of attraction and motion must satisfy.

Results for all relativistic parameters obtained from the JPL analysis are shown in Table 1. The realistic errors are comparable with those obtained in other recent investigations.[6,11]

LLR expands our understanding of the precession of the Earth's axis in space, the induced nutation, Earth orientation, the Earth's obliquity to the ecliptic, the intersection of the celestial equator and the ecliptic, lunar and solar solid body tides, lunar tidal deceleration, lunar physical and free librations, the structure of the Moon and energy dissipation in the lunar interior, and the study of core effects. LLR provides accurate retroreflector locations useful for lunar surface cartography and geodesy. It helps determine Earth station locations and motions, the mass of the Earth–Moon system, lunar and terrestrial gravity harmonics, and tidal Love numbers.

For a general review of LLR see Ref. 20. A comprehensive paper on tests of gravitational physics is Ref. 5. Our recent paper describes the model improvements needed to achieve millimeter-level accuracy for LLR.[21] Also, Refs. 6 and 11 have the most recent JPL LLR results for gravitational physics.

3. Future Laser-Ranging Tests of Gravity

It is essential that the acquisition of new LLR data continues in the future. Centimeter-level accuracies are now achieved, and a further improvement is

Table 1. Determined values for the relativistic quantities and their realistic errors.

Parameter	Results
SEP parameter η	$(4.0 \pm 4.3) \times 10^{-4}$
PPN parameter $\gamma - 1$	$(4 \pm 5) \times 10^{-3}$
PPN parameter: $\beta - 1$ from point mass	$(-2 \pm 4) \times 10^{-3}$
$\beta - 1$ from $\eta = 4\beta - 3 - \gamma_{\rm Cassini}$	$(1.0 \pm 1.1) \times 10^{-4}$
Time-varying gravitational constant, $\dot{G}/G$ (yr^{-1})	$(6 \pm 7) \times 10^{-13}$
Geodetic precession, $K_{\rm gp}$	-0.0005 ± 0.0047

expected. Analyzing improved data would allow a correspondingly more precise determination of gravitational physics parameters and other parameters of interest. In addition to the existing LLR capabilities, there are two near-term possibilities: the construction of new LLR stations and the emerging field of interplanetary laser ranging that has recently demonstrated its readiness for deployment in space.

3.1. *APOLLO facility*

LLR has remained a viable experiment, with fresh results over 37 years, because the data accuracies have improved by an order of magnitude. A future LLR station should provide another-order-of-magnitude improvement. APOLLO is a new LLR effort designed to achieve millimeter-range precision and corresponding order-of-magnitude gains in measurements of fundamental physics parameters. Using a 3.5 m telescope the APOLLO facility is pushing LLR into the regime of stronger photon returns with each pulse, enabling millimeter-range precision to be achieved.[7,21,22]

The high accuracy LLR capability that was recently installed at the Apache Point[21,22] should provide major opportunities. As expected, the Apache Point 1-mm-range accuracy corresponds to 3×10^{-12} of the Earth–Moon distance. The resulting LLR tests of gravitational physics would improve by an order of magnitude: the equivalence principle would give an uncertainty approaching 10^{-14}, tests of general relativity effects would be <0.1%, and estimates of the relative change in the gravitational constant would be 0.1% of the inverse age of the Universe. This last number is impressive considering that the expansion rate of the Universe is approximately one part in 10^{10} per year. Therefore, the gain in our ability to conduct even more precise tests of fundamental physics is enormous, and thus this new instrument stimulates development of better and more accurate models for the LLR data analysis at millimeter level. (The current status of APOLLO is discussed in Ref. 7.)

3.2. *New retroreflectors and laser transponders on the Moon*

There are two critical factors that control the progress of the LLR-enabled science — the distribution of retroreflectors on the lunar surface and their passive nature. Thus, the four existing arrays[20] are distributed from the equator to midnorthern latitude of the Moon and are placed with modest mutual separations relative to the lunar diameter. Such a distribution is not optimal; it limits the sensitivity of the ongoing LLR science investigations. The passive nature of reflectors causes signal attenuation proportional to the inverse fourth power of the distance traveled by a laser pulse. The weak return signals drive the difficulty of the observational task; thus, only a handful of terrestrial SLR stations are capable of also carrying out the lunar measurements, currently possible at centimeter level.

Return to the Moon provides an excellent opportunity for LLR, particularly if additional retroreflector arrays will be placed on the lunar surface at more widely separated locations. Due to their potential for new science investigations, these instruments are well justified.

3.2.1. *New retroreflector arrays*

Range accuracy, data span, and distributions of Earth stations and retroreflectors are important considerations for future LLR data analysis. Improved range accuracy helps all solution parameters. Data span is more important for some parameters, e.g. change in G, precession, and station motion, than others. New retroreflectors optimized for pulse spread, signal strength, and thermal effects will be valuable at any location on the Moon.

Overall, the separation of lunar three-dimensional rotation, the rotation angle and orientation of the rotation axis (also called physical librations), and tidal displacements depends on a good geographical spread of retroreflector array positions. The current three Apollo sites plus the infrequently observed Lunokhod 2 are close to the minimum configuration for separation of rotation and tides, so that unexpected effects might go unrecognized. A wider spread of retroreflectors could improve the sensitivity to rotation/orientation angles and the dependent lunar science parameters by factors of up to 2.6 for longitude and up to 4 for pole orientation. The present configuration of retroreflector array locations is quite poor for measuring lunar tidal displacements. Tidal measurements would be very much improved by a retroreflector array near the center of the disk, longitude 0 and latitude 0, plus arrays further from the center than the Apollo sites.

Lunar retroreflectors are the most basic instruments, for which no power is needed. Deployment of new retroreflector arrays is very simple: deliver, unfold, point toward the Earth, and walk away. Retroreflectors should be placed far enough away from astronaut/moonbase activity that they will not get contaminated by dust. One can think about the contribution of smaller retroreflector arrays for use on automated spacecraft and larger ones for manned missions. One could also benefit from colocating passive arrays and active transponders and use a few LLR-capable station ranging retroreflectors to calibrate the delay-vs-temperature response of the transponders (with their more widely observable strong signal).

3.2.2. *Opportunity for laser transponders*

LLR is one of the most modern and exotic observational disciplines within astrometry, being used routinely for a host of fundamental astronomical and astrophysical studies. However, even after more than 30 years of routine observational operation, LLR remains a nontrivial, sophisticated, highly technical, and remarkably challenging task. Signal loss, proportional to the inverse fourth power of the Earth–Moon distance, but also the result of optical and electronic inefficiencies in equipment, array orientation, and heating, still requires that one observe mostly single photoelectron events. Raw timing precision is some tens of picoseconds, with the out-and-back range accuracy being approximately an order of magnitude larger. Presently, we are down to sub-centimeter lunar ranging accuracies. In this day of routine SLR operations, it is sobering to realize that ranging to the Moon is many orders of magnitude harder than to an Earth-orbiting spacecraft. Laser transponders may help to solve this problem. Simple time-of-flight laser transponders offer a unique opportunity to overcome the problems above. Although there are great opportunities for scientific advances provided by these instruments, there are also design challenges as transponders require power, precise pointing, and thermal stability.

Transponders require development: optical transponders detect a laser pulse and fire a return pulse back toward the Earth.[8] They give a much brighter return signal accessible to more stations on the Earth. Active transponders would require power and would have more limited lifetimes. Transponders might have internal electronic delays that would need to be calibrated or estimated, since if these delays were temperature-sensitive that would correlate with the SEP test. Transponders can also be used to good effect in asynchronous mode,[9,10] wherein the received pulse train is not related to the transmitted pulse train, but the transponder unit records the temporal offsets between the two signals. The LLR experience can help determine the optimal location on the Moon for these devices.

3.2.3. *Improved LLR science*

LLR provides valuable lunar science, provides results on lunar orbit (ephemeris) and rotation/orientation, tests relativity, and is sensitive to information on Earth geophysics and geodesy. Many LLR science investigations benefited from an order-of-magnitude gain in the rms-weighted residuals that progressed during last 37 years from ∼25 cm in the 1970s to 2 cm in the last decade. The new lunar opportunities may be able to widen the array distribution on the lunar surface and to enable many SLR stations to achieve millimeter-level LLR ranging — a factor-of-20 gain compared to the present state.

Increased sensitivity would allow a search for new effects due to the lunar fluid core free precession, inner core influences, and stimulation of the free rotation modes. Future possibilities include detection of an inner solid core interior to the fluid core. Advances in gravitational physics are also expected. Several tides on the Earth have been measured through their profound influence on the lunar orbit. Geocentric positions of tracking stations are determined and station motions are measured. Precession and nutation of the Earth's equator/pole are measured and Earth rotation variations are strong in the data. The small number of current LLR-capable stations could be expanded if the return signal was stronger.

A wider geographic distribution of retroreflectors or transponders than the current retroreflector distribution would be a benefit; the accuracy of the lunar science parameters would increase several times. The lunar science includes the interior information: measuring tidal response, tidal dissipation, and core effects.

Gravitational physics includes the equivalence principle and limits on variation of the gravitational constant G.[6,11,21] (See discussion in Sec. 2.3.)

The small number of operating Earth stations is a major limitation on current LLR results for geophysics and geodesy. A bright transponder source on the Moon would open LLR to several dozen terrestrial SLR stations which cannot detect the current weak signals from the Moon. Resulting Earth geophysics and geodesy results would include the positions and rates for the Earth stations, Earth rotation, precession rate, nutation, and tidal influences on the orbit. From the lunar rotation and orientation at the expected millimeter accuracy, inferences can be made on the liquid lunar core and its size and oblateness. Parameters from gravitational physics could be modeled with vastly improved accuracy. A larger number of stations will add geographical diversity to the present narrow sample. The science, technology, and exploration gains from the new lunar deployment will be significant. For instance (if equipped with clocks stable at the subnanosecond scale), laser transponders may enable accurate time transfer for multiple users on the Earth. Thus, besides the classic LLR science, new investigations will be possible.

In addition to their strong return signals and insensitivity to lunar orientation effects, laser transponders are attractive due to their potential to become an increasingly important part of space exploration efforts. Laser transponders on the Moon can be a prototype demonstration for later laser ranging to Mars and other celestial bodies to give strong science returns in the areas similar to those investigated with LLR. A lunar installation would provide a valuable operational experience.

3.3. *Science with ranging to Mars*

The prospects of increased space traffic on the Earth–Mars route will provide excellent science opportunities, especially for gravitational physics. In particular, the Earth–Mars–Sun–Jupiter system allows for a sensitive test of the SEP which is qualitatively different from that provided by LLR.[23] The outcome of these ranging experiments has the potential to improve the values of the two relativistic parameters — a combination of PPN parameters η (via a test of the SEP) and a direct observation of the PPN parameter γ (via Shapiro time delay or solar conjunction experiments). The Earth–Mars range would also provide for a very accurate test of $\dot{G}/G$. Below, we shall briefly address these possibilities.

3.3.1. *Planetary test of the SEP*

Accurate ranging to Mars may be used to search for a violation of the SEP. One can determine the PPN parameter η from the improved solution for the Martian orbit. By precisely monitoring the range between the two planets, the Earth and Mars, one studies their free-fall accelerations toward the Sun and Jupiter. The PPN model of this range includes terms due to violation of the SEP introduced by the possible inequality between gravitational and inertial masses.[23] Should η have a small but finite value, the Martian orbit would be perturbed by the force responsible for the violation of the SEP. If one accounts for the orbital eccentricity and inclination effects, together with the tidal interaction, the size of this range perturbation is $\delta r = 1569\ \eta$m.

If the accuracy of the Earth–Mars range reaches 1 cm, one will be able to determine the parameter η with an accuracy of $\Delta\eta \sim 1\,\text{cm}/(1569\,\text{m}) = 6.4 \times 10^{-6}$. The accuracy in determining η increases if one is able to continue ranging to Mars with this accuracy for a number of years. For instance, after 10 years (or slightly more than 5 complete Martian years), the experiment may yield η with an accuracy of $\sim 2 \times 10^{-6}$ (limited only by the noise introduced by asteroids). Because of a larger gravitational self-energy of the Sun,[23] this accuracy would provide a SEP violation test of $[\Delta(M_G/M_I)]_{\text{SEP}} \sim 10^{-13}$, which is comparable to that of the present LLR.

3.3.2. *Solar conjunction experiments*

The Eddington PPN parameter γ is another relativistic parameter that may be precisely measured with accurate Earth–Mars ranges. The measurement may be done during solar conjunctions in an experiment similar to that of the Cassini mission in 2002.[18] In the conjunction experiments one measures either Shapiro time delay of the signal that is going through the solar gravitational field or the deflection angle due to the solar gravity.[24,25] A model that describes the effects of both the deflection of light and the light-time delay explicitly depends on the parameter γ; thus the data analysis efforts and the solution are reasonably well understood.

On the limb of the Sun, the gravitational delay of a photon (Shapiro time delay) from a source on the Martian surface is about 250 microsec. This effect is inversely proportional to the solar impact parameter. With a Cassini-type Ka- and X-band communication system one can come as close to the Sun as about 4–6 solar radii, which will result in a delay of 192 microsec. If one measures this delay to 1 cm accuracy, one may determine γ accurate to 1×10^{-6}. Even a greater accuracy is possible with optical ranging.[26]

3.3.3. *Search for time variation in the gravity constant*

Similar to the LLR experiment, analysis of light travel times between the Earth and Mars would yield a stringent limit on the fractional variation of the gravitational constant $\dot{G}/G$. The uncertainty for $\dot{G}/G$ is improving rapidly because its sensitivity depends on the square of the data span. Continuing these Earth–Mars laser measurements for five years even at an accuracy of $\sim$1 cm would allow for significant reduction of the uncertainty in the $\dot{G}/G$ parameter to better than 1 part in 10^{13} per year, a limit close to the effect predicted by some theories.

Other potential locations for interplanetary laser transponders may be either on celestial bodies in the solar system (such as a asteroids with highly eccentric orbits) or space probes (such as a mission of opportunity on the BepiColombo mission to Mercury[27] or a dedicated gravity experiment proposed for the LATOR mission[28–30]).

4. Conclusions

LLR provides the most precise way to test the SEP as well as for time variation of Newton's constant. With technology improvements and substantial access to a large-aperture, high-quality telescope, the APOLLO project will take full advantage of the lunar retroreflectors and will exploit the opportunity provided by the unique Earth–Moon "laboratory" for gravitational physics. An order-of-magnitude improvement in the accuracy of the LLR tests of relativistic gravity is expected with the new APOLLO instrument. Opportunities for new reflector/transponder deployment on the Moon may provide for a significant improvement in LLR-enabled science, including lunar science, geophysics, and gravitational physics.

Laser ranging may offer very significant improvements in many areas of deep-space navigation and communication. What is critical for the purposes of fundamental physics is that, while in free space, the lasercomm allows for a very precise trajectory estimation and control to an accuracy of less than 1 cm at distances of ∼2 AU. The recent successful two-way laser transponder experiments conducted with a laser altimeter on board the Messenger spacecraft en route to Mercury, and also successful transmission of hundreds of Q-switched laser pulses to the Mars Orbiter Laser Altimeter (MOLA), an instrument on the Mars Global Surveyor (MGS) spacecraft in orbit about Mars, have demonstrated the maturity of laser ranging technologies for interplanetary applications.[31,32] In fact, the MLA and MOLA experiments demonstrated that a decimeter interplanetary ranging is within the state of the art and can be achieved with modest laser power and telescope apertures. Achieving millimeter-class precisions over interplanetary distances is within reach, thus opening a way to significantly more accurate (several orders of magnitude) tests of gravity on the solar system scales.[33]

The future deployment of laser transponders on interplanetary missions will provide new opportunities for highly improved tests of the SEP and measurements of the PPN parameters γ and β. With their anticipated capabilities, interplanetary transponders will lead to very robust advances in the tests of fundamental physics and could discover a violation or extension of general relativity, or reveal the presence of an additional long range interaction in physical laws. As such, these devices should be used in planning both the next steps in lunar exploration and the future interplanetary missions to explore the solar system.

Acknowledgments

The work described here was carried out at the Jet Propulsion Laboratory, California Institute of Technology, under a contract with the National Aeronautics and Space Administration.

References

1. P. L. Bender *et al.*, *Science* **182** (1973) 229.
2. P. Shelus *et al.*, McDonald ranging: 30 years and still going, in *Proc. 13th Int. Workshop on Laser Ranging* (Oct. 2002 7–11, Washington, DC, USA, 2003); http://cddisa.gsfc.nasa.gov/lw13/lw_proceedings.html
3. C. Veillet *et al.*, Lunar laser ranging at CERGA for the ruby period (1981–1986), in *Proc. Space Geodesy to Geodynamics: Technology*, eds. D. E. Smith and D. L. Turcotte *AGU Geodynam. Ser.* **25** (1993) 133.
4. E. Samain *et al.*, *Astron. Astrophys. Suppl. Ser.* **130** (1998) 235.
5. J. G. Williams, X. X. Newhall and J. O. Dickey, *Phys. Rev. D* **53** (1996) 6730.
6. J. G. Williams, S. G. Turyshev and D. H. Boggs, *Phys. Rev. Lett.* **93** (2004) 261101 [gr-qc/0411113].
7. T. W. Murphy Jr. *et al.*, *Int. J. Mod. Phys. D* **16** (2007) 2127.
8. J. J. Degnan, *AGU Geodynam. Ser.* **25** (1993) 133.
9. J. J. Degnan, *J. Geodynam.* **34** (2002) 551.
10. J. J. Degnan, Laser transponders for high accuracy interplanetary laser ranging and time transfer, in *Lasers, Clocks, and Drag-Free: Exploration of Relativistic Gravity in Space*, eds. H. Dittus, C. Lammerzahl and S. G. Turyshev (Springer, New York, 2006).
11. J. G. Williams, S. G. Turyshev and D. H. Boggs, Lunar laser ranging tests of the equivalence principle with the Earth and Moon, in *Testing the Equivalence Principle on Ground and in Space* (Pescara, Italy, 20–23 Sep. 2004) eds. C. Laemmerzahl, C. W. F. Everitt and R. Ruffini, to appear in *Lect. Notes Phys. Ser.* (Springer, 2006) [gr-qc/0507083].
12. J. G. Williams *et al.*, *Adv. Space Res.* **37** (2006) 67 [gr-qc/0412049].
13. K. L. Nordtvedt, J. Müller and M. Soffel, *Astron. Astrophys.* **293** (1995) L73.
14. T. Damour and D. Vokrouhlicky, *Phys. Rev. D* **53** (1996) 4177.
15. T. Damour and D. Vokrouhlicky, *Phys. Rev. D* **53** (1996) 6740.
16. S. Baeßler *et al.*, *Phys. Rev. Lett.* **83** (1999) 3585.
17. E. G. Adelberger, *Class. Quant. Grav.* **18** (2001) 2397.
18. B. Bertotti, L. Iess and P. Tortora, *Nature* **425** (2003) 374.
19. D. N. Spergel *et al.*, submitted to *Astrophys. J.* (2006) [astro-ph/0603449].
20. J. O. Dickey *et al.*, *Science* **265** (1994) 482.
21. J. G. Williams, S. G. Turyshev and T. W. Murphy Jr., *Int. J. Mod. Phys. D* **13** (2004) 567 [gr-qc/0311021].
22. T. M. Murphy *et al.*, The Apache Point Observatory Lunar Laser-Ranging Operation (APOLLO), in *Proc. 12th Int. Workshop on Laser Ranging* (Matera, Italy; Nov. 2000), http://www.astro.washington.edu/tmurphy/apollo/matera.pdf
23. J. D. Anderson *et al.*, *Astrophys. J.* **459** (1996) 365 [gr-qc/9510029].
24. I. I. Shapiro, *et al.*, *J. Geophys. Res.* **82** (1977) 4329.
25. R. D. Reasenberg *et al.*, *Astrophys. J. Lett.* **234** (1979) L219.
26. J. F. Chandler, M. R. Pearlman, R. D. Reasenberg and J. J. Degnan, Solar-system dynamics and tests of general relativity with planetary laser ranging, in *Proc. 14th Int. Laser Ranging Workshop* (San Fernando, Spain, 7–11 June, 2004); http://cddis.nasa.gov/lw14
27. L. Iess and S. Asmar, *Int. J. Mod. Phys. D* **16** (2007) 2117.
28. S. G. Turyshev, M. Shao and K. L. Nordtvedt, *Class. Quant. Grav.* **21** (2004) 2773 [gr-qc/0311020].

29. S. G. Turyshev, M. Shao and K. L. Nordtvedt, Science, technology and mission design for the laser astrometric test of relativity mission, in *Lasers, Clocks, and Drag-Free: Exploration of Relativistic Gravity in Space*, eds. H. Dittus, C. Lammerzahl and S. G. Turyshev, to be published (Springer, 2007) [gr-qc/0601035].
30. K. L. Nordtvedt, *Int. J. Mod. Phys. D* **16** (2007) 2205.
31. X. Sun *et al.*, Laser ranging between the mercury laser altimeter and an Earth-based laser satellite tracking station over a 24 million kilometer distance, *OSA Annual Meeting Abstracts* (Tucson, AZ, USA; 16–20 Oct., 2005).
32. D. E. Smith *et al.*, *Science* **311** (2006) 53.
33. J. J. Degnan, *Int. J. Mod. Phys. D* **16** (2007) 2137.

INVERSE-SQUARE LAW EXPERIMENT IN SPACE

HO JUNG PAIK*, VIOLETA A. PRIETO† and M. VOL MOODY‡

Department of Physics, University of Maryland,
College Park, MD 20742-4111, USA
**hpaik@physics.umd.edu*
†violeta@glue.umd.edu
‡mvmoody@mail.umd.edu

The objective of ISLES (Inverse-Square Law Experiment in Space) is to perform a null test of Newton's law in space with a resolution of 10 ppm or better at a 100 μm distance. ISLES will be sensitive enough to detect the axion, a dark matter candidate, with the strongest allowed coupling and probe large extra dimensions of string theory down to a few micrometers. The experiment will be cooled to <2 K, which permits superconducting magnetic levitation of the test masses. This soft, low-loss suspension, combined with a low-noise SQUID, leads to extremely low intrinsic noise in the detector. To minimize Newtonian errors, ISLES employs a near-null source, a circular disk of large diameter-to-thickness ratio. Two test masses, also disk-shaped, are suspended on the two sides of the source mass at a nominal distance of 100 μm. The signal is detected by a superconducting differential accelerometer.

Keywords: Newton's law; extra dimensions; axion.

1. Objective of Research

The objective of ISLES (Inverse-Square Law Experiment in Space) is to test Newton's inverse-square ($1/r^2$) law to better than one part in 10^5 at a 100 μm range. Figure 1 shows the sensitivities (2σ) of ISLES and its ground experiment versus the existing limits for the $1/r^2$ law at $\lambda \leq 1$ mm, where the total potential is written as

$$V(r) = -G\frac{m_1 m_2}{r}\left[1 + \alpha \exp\left(-\frac{r}{\lambda}\right)\right]. \qquad (1)$$

The line and the shaded region represent violations predicted by higher-dimensional string theory and the axion theory, respectively. The expected resolution of ISLES represents an improvement by over six orders of magnitude beyond the limits obtained by Chiaverini *et al.*,[1] Long *et al.*,[2] and Hoyle *et al.*[3]

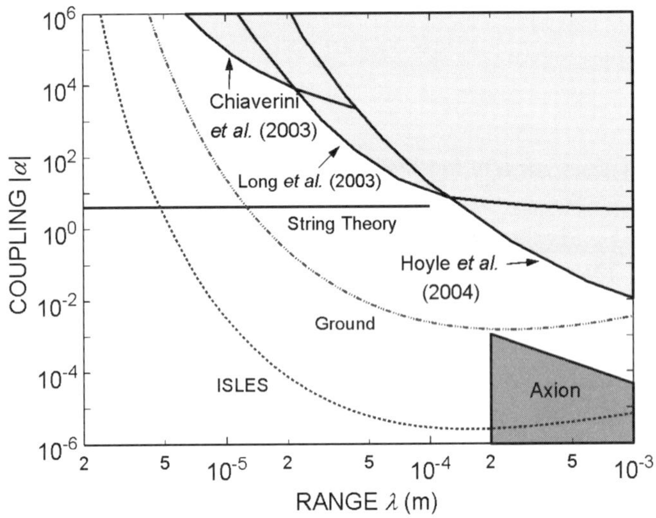

Fig. 1. Sensitivity of ISLES versus the existing limits.

1.1. *Test of string theories*

String theory is defined in terms of a fundamental scale M_*. If there are n compact dimensions with radii $R_1, R_2, \ldots, R_n$, Gauss's law implies that the Planck mass $M_{\rm Pl}$ is related to M_* by

$$M_{\rm Pl}^2 \approx M_*^{2+n} R_1 R_2 \ldots R_n. \tag{2}$$

As we probe distances shorter than one of the radii R_i, a new dimension opens up and changes the r dependence of the gravitational force law. For $r > R_i$, the deviation is found to be Yukawa-type. In particular, when extra dimensions are compactified on an n torus, the strength of the potential is $\alpha = 2n$.[4,5] The string-theory-derived deviation shown in Fig. 1 corresponds to $n = 2$.

One theoretically well-motivated value for M_* is 1 TeV, which solves the gauge hierarchy problem. For two large dimensions of similar size, one obtains $R_1 \approx R_2 \approx$ 1 mm.[6] Cosmological and astrophysical constraints give a bound $M_* > 100$ TeV,[7,8] which corresponds to $R_1 \approx R_2 < 1\,\mu$m. While this is beyond the reach of our experiment, there are cosmological assumptions going into these bounds.

1.2. *Search for the axion*

The Standard Model of particle physics successfully accounts for all existing particle data; however, it has one serious blemish — the strong CP problem. Strong interactions are such that parity (P), time reversal (T), and charge conjugation (C) symmetries are automatically conserved in perturbation theory. However, nonperturbative effects induce violations of P and CP, parametrized by a dimensionless angle θ. The *a priori* expectation of this parameter is of the order of unity, but no such violations of P or CP have been observed in strong interactions.

Peccei and Quinn[9] developed an attractive resolution of this problem. One ramification of their theory is the existence of a new light-mass boson, the *axion*.[10,11] The axion mediates a short-range mass–mass interaction. The upper bound $\theta \leq 3\times10^{-10}$ corresponds to a violation of the $1/r^2$ law at the level of $|\alpha| \approx 10^{-3}$ at $\lambda = 200\,\mu$m, which is within the reach of ISLES. The axion could also solve a major open question in astrophysics: the composition of dark matter. It is one of the strongest candidates for cold dark matter.[12]

2. Principle and Design of the Experiment

To maximize the masses that can be brought to $100\,\mu$m from each other, flat disk geometry is used for both the source and test masses. An infinite plane slab is a Newtonian null source. We approximate this by using a circular disk of large diameter-to-thickness ratio. Figure 2 shows the configuration of the source and test masses with associated coils. Two disk-shaped test masses are suspended on two sides of the source and are coupled magnetically to form a differential accelerometer.

As the source mass is driven at frequency $f_S/2$ along the symmetry axis, the first-order Newtonian fields arising from the finite diameter of the source mass are

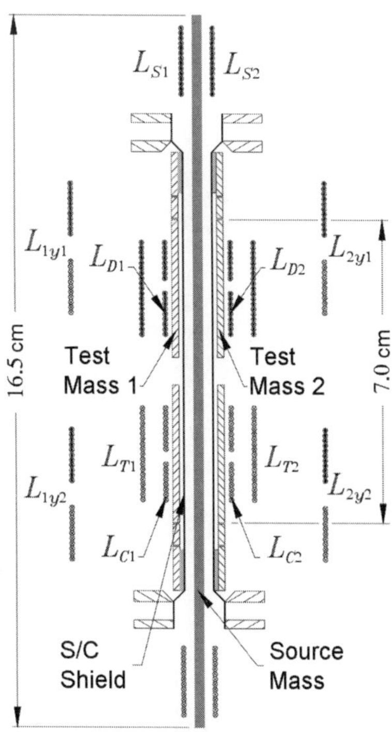

Fig. 2. Configuration of the source and test masses with associated coils.

canceled upon differential measurement, leaving only a second-order error at f_S. By symmetry, the Yukawa signal also appears at f_S. The second harmonic detection, combined with the common-mode rejection of the detector, reduces vibration coupling by over 200 dB.

2.1. *Overview of the apparatus*

Figure 3 shows an expanded cross section of the ground experiment. To eliminate differential contraction and provide good electromagnetic shielding, the entire housing is fabricated from niobium (Nb). The source mass is made out of tantalum (Ta), which closely matches Nb in thermal contraction. It is suspended by cantilever springs and driven magnetically. The test masses are also made out of Ta and suspended by cantilever springs. A thin Nb sheet provides electrostatic and magnetic shielding between the source and each test mass.

The experiment is suspended from the top of the cryostat via three rubber tubes. Voice-coil transducers, incorporated into each vertical leg of the suspension, are used to shake the instrument for balance and calibration. By varying the magnitude and phase of the current through the coils, vertical acceleration or tilt in any direction is applied to the instrument. The tilt is sensed with a two-dimensional optical lever

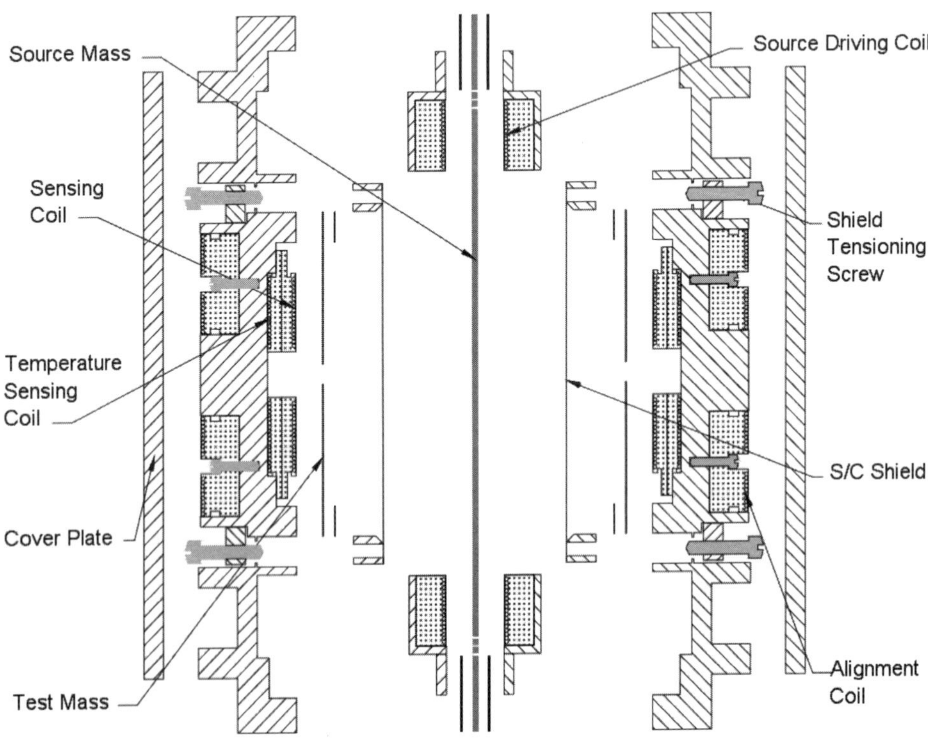

Fig. 3. Expanded cross section of the ground experiment.

consisting of a laser, a beam splitter, an x–y photodiode, and a planar mirror mounted at the top of the instrument.

The cryostat has a cold plate and copper can, which will be cooled to below 2 K by pumping on liquid helium through a capillary.

2.2. *Source and test masses*

The source is a disk 1.65 mm thick by 165 mm in diameter, with mass 590 g. The source mass, cantilever springs, and rim are machined out of a single plate of Ta. Ta is chosen for its high density (16.6 g cm^{-3} compared to 8.57 g cm^{-3} for Nb), which increases the signal, and its relatively high critical field ($H_c = 0.070$ T at 2 K). The test masses are identical Ta disks 250 μm thick by 70 mm in diameter. Their dynamic mass is $m = 16.7$ g. The mechanical resonance frequency of the test mass is 11 Hz.

The equilibrium spacing between the source and each test mass is 150 μm. These are shielded from each other by means of a 12.5-μm-thick Nb shield. The source mass is driven magnetically by sending AC currents to the two source coils. The design allows a source displacement of $\pm 87\,\mu$m.

The differential acceleration signals expected from the Yukawa force with $|\alpha| = 10^{-3}$ and $\lambda = 200\,\mu$m are plotted in Fig. 4 as a function of the source mass position. The small Newtonian term arising from the finite source mass diameter is also shown. The source mass looks like an "infinite plane slab" to the test mass due to its proximity. The Yukawa signal is almost purely second harmonic to the source motion. Its rms amplitude, corresponding to a $\pm 87\,\mu$m displacement, is 2.6×10^{-14} m s^{-2}.

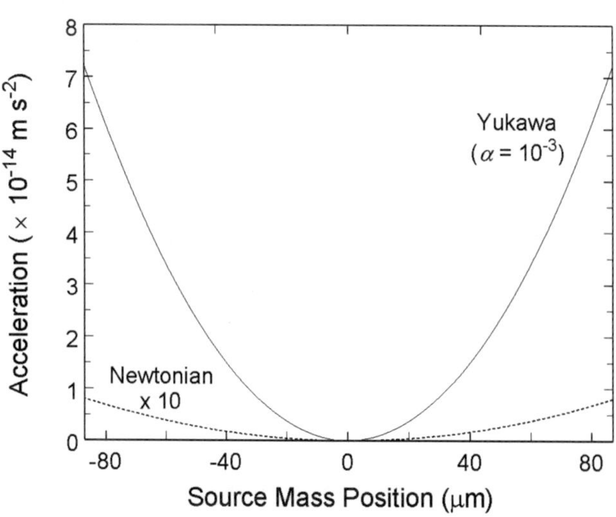

Fig. 4. Newtonian and Yukawa signals versus source position.

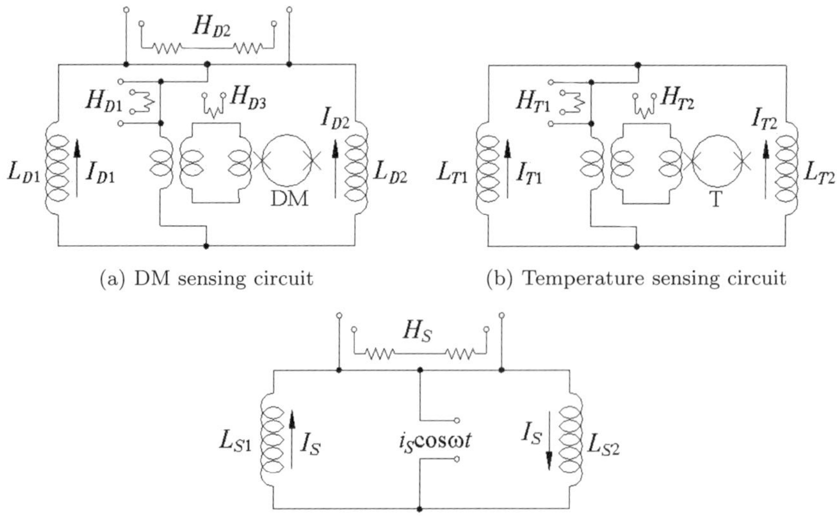

(a) DM sensing circuit

(b) Temperature sensing circuit

(c) Source driving circuit

Fig. 5. Superconducting circuits for the detector and source.

2.3. *Superconducting circuits*

Figure 5(a) is the differential mode (DM) sensing circuit. A persistent current is stored through the loop that is the *parallel* combination of L_{D1} and L_{D2} and the transformer primary. Another current is stored in the loop that is the *series* combination of L_{D1} and L_{D2}, which in turn tunes the ratio I_{D2}/I_{D1}. For redundancy, the CM circuit (not shown) is designed to be physically identical but with a different current configuration.

Figure 5(b) is the temperature sensing circuit. The coils L_{T1} and L_{T2} are mounted directly on the Nb housing (see Fig. 2), making them sensitive only to temperature variation. The output of this circuit is used to compensate for the temperature sensitivity of the differential accelerometer.

Figure 5(c) shows the superconducting circuit for the source. A large persistent current, $I_S \approx 5$ A, is stored in the main loop. The source is then driven by sending a small current, $i_S \cos \omega t$, across the loop. This reduces the magnetic crosstalk between the source and the detector.

3. Experimental Errors

3.1. *Metrology errors*

For the thin disk geometry of the source, errors in the thickness are dominant over density inhomogeneity. Due to the cylindrical symmetry of the test masses, the linear taper of the source produces a second-order error and azimuthal asymmetry is averaged out. Radial thickness variation is the dominant error source. Due to the null nature of the source, test mass metrology is unimportant, except for the suspension spring. The source mass is lapped to meet the required dimensional tolerances and is stress-relieved.

3.2. *Intrinsic noise of the detector*

The intrinsic power spectral density of a superconducting differential accelerometer can be written[13,14] as

$$S_a(f) = \frac{8}{m}\left[\frac{k_B T \omega_D}{Q_D} + \frac{\omega_D^2}{2\eta\beta} E_A(f)\right], \tag{3}$$

where m is the mass of each test mass, $\omega_D = 2\pi f_D$ and Q_D are the DM (angular) resonance frequency and quality factor, β is the electromechanical energy coupling coefficient, η is the electrical energy coupling coefficient of the SQUID, and $E_A(f)$ is the input energy resolution of the SQUID. For our ground experiment, we find that $S_a^{1/2}(f) = 8.4 \times 10^{-12}$ m s^{-2} Hz$^{-1/2}$ at $f = 0.1$ Hz. We have chosen $f_S = 0.1$ Hz since the seismic noise dips at that frequency and the SQUID noise is close to the white noise level.

3.3. *Other noise*

Modulation of the penetration depth of a superconductor with temperature gives rise to temperature sensitivity in a superconducting accelerometer.[13] Extrapolating from the low-frequency noise spectrum measured during the initial cooldown of the experiment, we find a temperature-induced noise of 2.4×10^{-12} m s^{-2} Hz$^{-1/2}$ at $f = 0.1$ Hz. This will be reduced by a factor of 10 by the temperature compensation.

The displacement of the source induces a platform tilt of 1.2×10^{-6} rad at $f_S/2$. The tilt modulates the Earth's gravity and produces an unacceptably high level of linear acceleration as well as angular acceleration. We will cancel the source-driven tilt by a factor of 10^2 with a feedback loop. The tilt is measured with the optical tilt sensor and the signal is fed back to the voice-coil actuators to null the tilt signal.

Magnetic crosstalk between the source and the detector is an important error source. The entire housing is machined out of Nb and a Nb shield is provided between the source and each test mass. Further rejection by the frequency discrimination provides a comfortable margin. The Casimir force[15] is not important in our experiment since gaps between conducting planes are $> 10\,\mu$m.[16]

3.4. *Expected resolution of the ground experiment*

Table 1 combines all the errors. To reduce the random noise to the levels listed, a 10^6 s integration was assumed. By equating the total error with the expected Yukawa signal, we compute the minimum detectable $|\alpha|$. Figure 1 shows the 2σ error plotted as a function of λ. The best resolution of the ground experiment is $|\alpha| \approx 1 \times 10^{-3}$ at $\lambda = 150\,\mu$m. The experiment will probe extra dimensions down to $10\,\mu$m when two extra dimensions are compactified on a square torus.

Table 1. Error budget (1σ) of the ground experiment.

Error source	Error ($\times 10^{-15}$ m s^{-2})
Metrology	1.4
Random (10^6 s integration)	
Intrinsic	8.4
Temperature	2.4
Seismic	0.5
Source dynamic	17.4
Gravity noise	< 0.1
Magnetic coupling	< 0.1
Electrostatic coupling	< 0.1
Total	19.5

4. Progress of the Ground Experiment

4.1. *Construction of the apparatus*

Two 1.75-mm-thick Ta sheets were "two-face-grounded" commercially to improve the surface flatness. A measurement with dial indicators showed the surface to be flat to 10 μm, which is an acceptable tolerance for the initial experiment. A source mass was produced from one of these sheets by wire EDM and was then heat-treated to relieve stress. Four test masses were constructed, also with wire EDM, from a 0.25-mm-thick Ta sheet.

Two Nb housings that held the source mass, test masses, and various coils together were machined. Superconducting shields for the test masses were fabricated by diffusion-bonding 12.5μm Nb sheets to Nb rings. All the superconducting coils were wound and tested.

The entire apparatus was assembled and integrated with the cryostat. Figure 6 shows the instrument suspended from the cryostat insert. The exposed coil and a similar one on the opposite side (not shown in Fig. 3) are used to cancel the recoil of the platform in response to source mass motion.

4.2. *Test of the apparatus*

The apparatus has been cooled down several times and several mechanical problems have been discovered and corrected. One of the source coils partially came off from the Macor coil form upon cooling, touching the source mass. To reduce the driving current and thus the magnetic crosstalk, the source coils had been wound with five layers of Nb–Ti coils. The resulting thick layer of epoxy cracked due to differential contraction. The test masses also touched the Nb shields due to stresses put on them by the stretching of the shields.

We have now rewound all the failed superconducting coils. The source coils have been rewound using two layers of wire. We have also inserted capacitor plates to monitor the source position *in situ* relative to the superconducting shields. We are now reassembling the apparatus, with increased spacing between the test masses and the shields, to ensure that the test masses remain free at low temperature.

Fig. 6. The apparatus integrated with the cryostat.

The resonance frequencies of the Nb shields were measured to be ~ 1 kHz, with Q's in excess of 10^5. This result is very encouraging, since the modulated Casimir force from the source mass will be sufficiently attenuated by the stiff shields.

The He^4 cold plate was tested while monitoring the temperature sensing circuit. It reached a steady operating temperature of 1.6 K. The temperature remained stable for approximately 10 h.

5. The Space Experiment

The response of a test mass m to the force $F(\omega)$ goes as

$$x(\omega) = \frac{F(\omega)/m}{\omega_0^2 - \omega^2 + \jmath\omega/\tau}, \tag{4}$$

where ω_0 and τ are the resonance frequency and relaxation time of the mode, and ω is the signal frequency. In order to maximize $x(\omega)$, both ω_0 and ω must be reduced and/or a resonance experiment ($\omega = \omega_0$) must be conducted. At the same time, τ must be maximized to reduce the Brownian motion noise of the test mass.

A promising soft suspension with low dissipation is superconducting magnetic levitation. Levitation in $1g$, however, requires a large magnetic field, which tends to couple to the measurement degree of freedom through metrology errors and stiffen the mode. The situation improves dramatically in space. The g level is reduced by more than six orders of magnitude, so the test masses can be supported with much weaker magnetic springs, permitting the realization of both the lowest resonance frequency and the lowest dissipation. With such a weak magnetic coupling to the test masses, the gas-damping limit of Q may be achievable.

ISLES combines the advantages of the low-g environment of space with the superconducting accelerometer technology to achieve the vastly improved sensitivity. It employs a circular disk of large diameter-to-thickness ratio as the near-null source, as in the ground experiment. Two test masses, also disk-shaped, are *magnetically* suspended near opposite faces of the source at a nominal distance of 100 μm. The space experiment is expected to improve the resolution of the $1/r^2$ law by two to three orders of magnitude over the ground experiment. It will probe the extra dimensions down to 5 μm and search for the axion with sensitivity sufficient to detect the particle with strength within two orders of magnitude from the upper limit.

Acknowledgments

This work is partially supported by NASA under grant NAG32874 and by NSF under grant PHY0244966. We are grateful for useful discussions with Don Strayer at the Jet Propulsion Laboratory.

References

1. J. Chiaverini *et al.*, *Phys. Rev. Lett.* **90** (2003) 151101.
2. J. C. Long *et al.*, *Nature* **421** (2003) 922.
3. C. D. Hoyle *et al.*, *Phys. Rev. D* **70** (2004) 042004.
4. E. G. Floratos and G. K. Leontaris, *Phys. Lett. B* **465** (1999) 95.
5. A. Kehagias and K. Sfetsos, *Phys. Lett. B* **472** (2000) 39.
6. N. Arkani-Hamed, S. Dimopoulos and G. Dvali, *Phys. Rev. D* **59** (1999) 086004.
7. S. Cullen and M. Perelstein, *Phys. Rev. Lett.* **83** (1999) 268.
8. L. J. Hall and D. Smith, *Phys. Rev. D* **60** (1999) 085008.
9. R. D. Peccei and H. Quinn, *Phys. Rev. Lett.* **38** (1977) 1440.
10. S. Weinberg, *Phys. Rev. Lett.* **40** (1978) 223.
11. F. Wilczek, *Phys. Rev. Lett.* **40** (1978) 279.
12. M. S. Turner, *Phys. Rep.* **197** (1990) 67.
13. H. A. Chan and H. J. Paik, *Phys. Rev. D* **35** (1987) 3551.
14. M. V. Moody, E. R. Canavan and H. J. Paik, *Rev. Sci. Instrum.* **73** (2002) 3957.
15. H. B. G. Casimir, *Proc. K. Ned. Akad. Wet.* **51** (1948) 793.
16. S. K. Lamoreaux, *Phys. Rev. Lett.* **78** (1997) 5.

LASER ASTROMETRIC TEST OF RELATIVITY: SCIENCE, TECHNOLOGY AND MISSION DESIGN

SLAVA G. TURYSHEV* and MICHAEL SHAO†

Jet Propulsion Laboratory, California Institute of Technology,
4800 Oak Grove Drive, Pasadena, CA 91109, USA
**turyshev@jpl.nasa.gov*
†mshao@huey.jpl.nasa.gov

The Laser Astrometric Test of Relativity (LATOR) experiment is designed to explore the general theory of relativity in close proximity to the Sun — the most intense gravitational environment in the solar system. Using independent time series of highly accurate measurements of the Shapiro time delay (interplanetary laser ranging accurate to 3 mm at ~2 AU) and interferometric astrometry (accurate to 0.01 picoradian), LATOR will measure gravitational deflection of light by the solar gravity with an accuracy of one part in a billion — a factor of ~30,000 better than what is currently available. LATOR will perform a series of highly accurate tests in its search for cosmological remnants of the scalar field in the solar system. We present the science, technology and mission design for the LATOR mission.

Keywords: Tests of general relativity; interferometric astrometry; laser ranging.

1. Motivation

Recent remarkable progress in observational cosmology has again subjected the general theory of relativity to a test by suggesting a non-Einsteinian model of the Universe's evolution. From the theoretical standpoint, the challenge is even stronger — if the gravitational field is to be quantized, general relativity will have to be modified. Additionally, recent advances in the scalar-tensor extensions of gravity have intensified searches for very small deviations from Einstein's theory, at the level of three-to-five orders of magnitude below the level currently tested by experiment.

The Laser Astrometric Test of Relativity (LATOR) is a proposed space-based experiment to significantly improve the tests of relativistic gravity.[1,2] LATOR is designed to address the questions of fundamental importance by searching for a cosmologically evolved scalar field that is predicted by modern theories, notably the string theory. It will also test theories that attempt to explain the small acceleration rate of the Universe (so-called dark energy) via modification of gravity at very large distances, notably brane-world theories.

Section 1 of this paper discusses the theoretical framework, science motivation and objectives for LATOR. Section 2 provides an overview of the mission and optical designs for LATOR. Section 3 discusses the next steps in mission development.

1.1. *The PPN formalism*

Generalizing on Eddington's phenomenological parametrization of the gravitational metric tensor field, a method called the parametrized post-Newtonian (PPN) formalism has been developed.[3–7] This method represents the gravity tensor's potentials for slowly moving bodies and weak interbody gravity, and it is valid for a broad class of metric theories, including general relativity as a unique case. The several parameters in the PPN metric expansion vary from theory to theory, and they are individually associated with various symmetries and invariance properties of the underlying theory. An ensemble of experiments can be analyzed using the PPN metric, to determine the unique value for these parameters, and hence the metric field itself. In locally Lorentz-invariant theories the expansion of the metric field for a single, slowly rotating gravitational source in PPN coordinates[3–7] is given by

$$\begin{aligned} g_{00} &= 1 - 2\frac{M}{r}Q(r,\theta) + 2\beta\frac{M^2}{r^2} + \mathcal{O}(c^{-6}), \\ g_{0i} &= 2(\gamma+1)\frac{[\mathbf{J}\times\mathbf{r}]_i}{r^3} + \mathcal{O}(c^{-5}), \\ g_{ij} &= -\delta_{ij}\left[1 + 2\gamma\frac{M}{r}Q(r,\theta) + \frac{3}{2}\delta\frac{M^2}{r^2}\right] + \mathcal{O}(c^{-6}), \end{aligned} \tag{1}$$

where M and $\mathbf{J}$ are the mass and angular momentum of the Sun, $Q(r,\theta) = 1 - J_2\frac{R^2}{r^2}\frac{3\cos^2\theta - 1}{2}$, with J_2 being the quadrupole moment of the Sun and R being its radius; r is the distance between the observer and the center of the Sun; and β, γ, δ are the PPN parameters (in general relativity $\beta = \gamma = \delta = 1$). The M/r term in the g_{00} component is the Newtonian limit; the terms multiplied by the post-Newtonian parameters β, γ are post-Newtonian terms. The term multiplied by the post-post-Newtonian parameter δ also enters into the calculation of the relativistic light deflection.[8]

The most precise value for the PPN parameter γ is at present given by the Cassini mission[9] as $\gamma - 1 = (2.1 \pm 2.3) \times 10^{-5}$. Using the recent Cassini result[9] on the PPN γ, the parameter β was measured as $\beta - 1 = (0.9 \pm 1.1) \times 10^{-4}$ from lunar laser ranging (LLR).[10] The PPN parameter δ has not yet been measured, though its value can be inferred from other measurements.

The Eddington parameter γ, whose value in general relativity is unity, is perhaps the most fundamental PPN parameter, in that $\frac{1}{2}(1-\gamma)$ is a measure, for example, of the fractional strength of the scalar gravity interaction in scalar-tensor theories of gravity.[11] Within perturbation theory for such theories, all other PPN parameters to all relativistic orders collapse to their general relativistic values in proportion to $\frac{1}{2}(1-\gamma)$. Thus, a measurement of the first order light deflection effect at the level of accuracy comparable with the second order contribution would provide the crucial information for separating alternative scalar-tensor theories of gravity from general relativity,[3] and also for probing possible ways of gravity quantization and testing modern theories of cosmological evolution[14,16,18] (see Sec. 1.2). The LATOR mission is designed to directly address this issue with an unprecedented accuracy.

1.2. *Motivations for precision gravity experiments*

Recent theoretical findings suggest that the present agreement between general relativity and experiment might be naturally compatible with the existence of a scalar contribution to gravity. Damour and Nordtvedt, in Ref. 14,[a] have found that a scalar-tensor theory of gravity may contain a "built-in" cosmological attractor mechanism toward general relativity. These scenarios assume that the scalar coupling parameter $\frac{1}{2}(1-\gamma)$ was of order 1 in the early Universe, and show that it then evolves to be close to, but not exactly equal to, zero at the present time.[1] Reference 14 estimates the likely order of magnitude of the left-over coupling strength at the present time which, depending on the total mass density of the Universe, can be given as $1-\gamma \sim 7.3 \times 10^{-7}(H_0/\Omega_0^3)^{1/2}$, where Ω_0 is the ratio of the current density to the closure density and H_0 is the Hubble constant in units of 100 km/s/Mpc. Compared to the cosmological observations a lower bound of $(1-\gamma) \sim 10^{-6}$–10^{-7} can be derived for the present value of the PPN parameter γ.

Reference 18 estimated $\frac{1}{2}(1-\gamma)$ within the framework compatible with string theory and modern cosmology, confirming the previous result.[14] This analysis discusses a scenario where a composition-independent coupling of dilaton to hadronic matter produces detectable deviations from general relativity in high-accuracy light deflection experiments in the solar system. This work assumes only some general properties of the coupling functions and then only assumes that $1-\gamma$ is of order 1 at the beginning of the controllably classical part of inflation. It was shown[18] that one can relate the present value of $\frac{1}{2}(1-\gamma)$ to the cosmological density fluctuations. For the simplest inflationary potentials favored by the WMAP mission[20] (i.e. $m^2\chi^2$) it was found that the present value of $1-\gamma$ could be just below 10^{-7}. In particular, within this framework the value of $\frac{1}{2}(1-\gamma)$ depends on the model taken for the inflation potential; thus for $V(\chi) \propto \chi^2$, with χ being the inflation field, the level of the expected deviations from general relativity is $\sim 0.5 \times 10^{-7}$.[18] Note that these predictions are based on the work on scalar-tensor extensions of gravity which are consistent with, and indeed often part of, present cosmological models.

The analyses above motivate new searches for small deviations of relativistic gravity in the solar system by predicting that such deviations are currently present in the range from 10^{-5} to 5×10^{-8} for $\frac{1}{2}(1-\gamma)$, i.e. for observable post-Newtonian deviations from general relativity predictions and, thus, should be easily detectable

[a] See also Ref. 16 for nonmetric versions of this mechanism, together with Ref. 18 for the recent summary of a dilaton-runaway scenario.

with LATOR. This would require measurement of the effects of the next post-Newtonian order ($\propto G^2$) of light deflection resulting from gravity's intrinsic non-linearity. An ability to measure the first order light deflection term at the accuracy comparable with the effects of the second order is of the utmost importance and a major challenge for 21st century fundamental physics.

There are now multiple evidences indicating that 70% of the critical density of the Universe is in the form of a "negative pressure" dark energy component; there is no understanding as to its origin and nature. The fact that the expansion of the Universe is currently undergoing a period of acceleration now seems rather well tested: it is directly measured from the light curves of several hundred type Ia supernovae,[21–23] and independently inferred from observations of CMB by the WMAP mission[20] and other CMB experiments.[24,25] Cosmic speed-up can be accommodated within general relativity by invoking a mysterious cosmic fluid with large negative pressure, dubbed dark energy. The simplest possibility for dark energy is a cosmological constant; unfortunately, the smallest estimates for its value are 55 orders of magnitude too large. Most of the theoretical studies operate in the shadow of the cosmological constant problem, the most embarrassing hierarchy problem in physics. This fact has motivated other possibilities, most of which assume that $\Lambda = 0$, with the dynamical dark energy being associated with a new scalar field. However, none of these suggestions is compelling and most have serious drawbacks.

Given the challenge of this problem, a number of authors considered the possibility that cosmic acceleration is not due to some kind of stuff, but rather arises from new gravitational physics (see the discussion in Refs. 26–28). In particular, some extensions to general relativity in a low energy regime[27] were shown to predict an experimentally consistent evolution of the Universe without the need for dark energy. These models are expected to produce measurable contribution to the parameter γ in experiments conducted in the solar system also at the level of $1 - \gamma \sim 10^{-7} - 5 \times 10^{-9}$, thus further motivating the relativistic gravity research. Therefore, the PPN parameter γ may be the only key parameter that holds the answer to most of the questions discussed above.[b]

In summary, there are a number of theoretical and experimental reasons to question the validity of general relativity; LATOR will address these challenges.

2. Overview of LATOR

The LATOR experiment uses the standard technique of time-of-fight laser ranging between two microspacecraft whose lines of sight pass close by the Sun, and also a long-baseline stellar optical interferometer (placed above the Earth's atmosphere) to accurately measure deflection of light by the solar gravitational field in

[b]Also, an anomalous parameter δ will most likely be accompanied by a "γ mass" of the Sun which differs from the gravitational mass of the Sun and therefore will show up as anomalous γ.

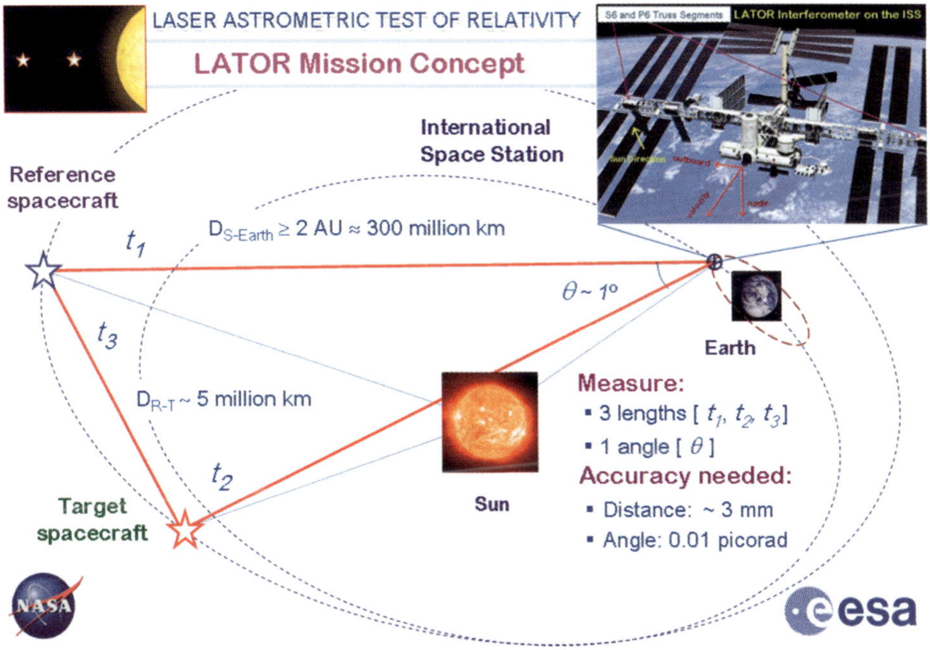

Fig. 1. The overall geometry of the LATOR experiment.

extreme proximity to the Sun.[1] Figure 1 shows the general concept for the LATOR missions including the mission-related geometry, experiment details and required accuracies.

2.1. *Mission design and anticipated performance*

LATOR is a Michelson–Morley-type experiment designed to test the pure tensor metric nature of gravitation — a fundamental postulate of Einstein's theory of general relativity.[1,2] With its focus on gravity's action on light propagation it complements other tests which rely on the gravitational dynamics of bodies. LATOR relies on a combination of independent time series of highly accurate measurements of the gravitational deflection of light in the immediate proximity to the Sun along with measurements of the Shapiro time delay on the interplanetary scales (to a precision respectively better than 0.01 prad and 3 mm). Such a combination of observables is unique and enables LATOR to significantly improve tests of relativistic gravity.

The schematic of the LATOR experiment is given in Fig. 1. Two spacecraft are injected into a heliocentric solar orbit on the opposite side of the Sun from the Earth. The triangle in the figure has three independent quantities but three arms are monitored with laser metrology. Each spacecraft is equipped with a laser ranging system that enables a measurement of the arms of the triangle formed by the two spacecraft and the ISS. According to Euclidean rules this determines a specific angle at the interferometer; LATOR can measure this angle directly. In particular, the laser beams transmitted by each spacecraft are detected by a long-baseline ($\sim$ 100 m) optical interferometer on the ISS. The actual angle measured at the interferometer is compared to the angle calculated using Euclidean rules and three side measurements; the difference is the non-Euclidean deflection signal (which varies in time during spacecraft passages), which contains the scientific information. This built-in redundant-geometry optical truss eliminates the need for drag-free spacecraft for high-accuracy navigation.[1]

The uniqueness of LATOR comes with its geometrically redundant architecture that enables it to measure the departure from Euclidean geometry ($\sim 8 \times 10^{-6}$ rad) caused by the solar gravity field, to a very high accuracy. This departure is shown as a difference between the calculated Euclidean value for an angle in the triangle and its value directly measured by the interferometer. This discrepancy, which results from the curvature of the space–time around the Sun and can be computed for every alternative theory of gravity, constitutes LATOR's signal of interest. The precise measurement of this departure constitutes the primary mission objective.

LATOR's primary mission objective is to measure the key PPN parameter γ with an accuracy of a part in 10^9. When the light deflection in solar gravity is concerned, the magnitude of the first order effect as predicted by general relativity for the light ray just grazing the limb of the Sun is $\sim$ 1.75 arcsec. The effect varies inversely with the impact parameter. The second order term is almost six orders of magnitude smaller, resulting in $\sim$ 3.5 microarcsec (μas) light deflection effect, and falls off inversely as the square of the light ray's impact parameter.[1,29–31] The relativistic frame-dragging term is ± 0.7 μas, and the contribution of the solar quadrupole moment, J_2, is sized as 0.2 μas (using the theoretical value of the solar quadrupole moment $J_2 \simeq 10^{-7}$). The small magnitudes of the effects emphasize the fact that, among the four forces of nature, gravitation is the weakest interaction; it acts at very long distances and controls the large-scale structure of the Universe, thus making the precision tests of gravity a very challenging task.

The first order effect of light deflection in the solar gravity caused by the solar mass monopole is $\alpha_1 = 1.75$ arcsec, which corresponds to an interferometric delay of $d \simeq b\alpha_1 \approx 0.85$ mm on a $b = 100$ m baseline. Using laser interferometry, we are currently able to measure distances with an accuracy (not just precision but accuracy) of ≤ 1 pm. In principle, the 0.85 mm gravitational delay can be measured with 10^{-10} accuracy versus 10^{-4} available with current techniques. However, we use a conservative estimate for the delay of 5 pm which would produce the measurement of γ to an accuracy of 1 part in 10^9 (rather than 1 part in 10^{-10}), which would already be a factor-of-30,000 accuracy improvement when compared to the recent Cassini result.[9] Furthermore, we have targeted an overall measurement accuracy of 5 pm per measurement, which for $b = 100$ m translates to the accuracy of 0.05 prad $\simeq 0.01$ μas. With four measurements per observation, this yields an accuracy of $\sim 5.8 \times 10^{-9}$ for the first order term. The second order light deflection is approximately 1700 pm and, with a 5 pm accuracy and $\sim$ 400 independent data points, it could be measured with an accuracy of ~ 1 part in 10^4, including the first-ever measurement of the PPN parameter δ. The frame-dragging effect would be measured with $\sim$ 1-part-in 10^3 accuracy and the solar quadrupole moment can be modestly measured to 1 part in 200, all with respectable signal-to-noise ratios. Recent covariance studies performed for the LATOR mission[34,35] confirm the design performance parameters and also provide valuable recommendations for further mission developments.

We shall now consider the LATOR mission architecture.

2.2. *Mission architecture: evolving light triangle*

The LATOR mission architecture uses an evolving light triangle formed by laser ranging between two spacecraft (placed in $\sim$1 AU heliocentric orbits) and a laser transceiver terminal on the ISS, via European collaboration. The objective is to measure the gravitational deflection of laser light as it passes in extreme proximity to the Sun (see Fig. 1). To that extent, the long-baseline ($\sim$100 m) fiber-coupled optical interferometer on the ISS will perform differential astrometric measurements of the laser light sources on the two spacecraft as their lines of sight pass behind the Sun. As seen from the Earth, the two spacecraft will be separated by about 1°, which will be accomplished by a small maneuver immediately after their launch.[1,36,37] This separation would permit differential astrometric observations to accuracy of $\sim 10^{-13}$ radians needed to significantly improve measurements of gravitational deflection of light in the solar gravity.

To achieve the primary objective, LATOR will place two spacecraft into a heliocentric orbit, to provide conditions for observing the spacecraft when they are behind the Sun as viewed from the ISS (see Figs. 2 and 3). Figure 2 shows the trajectory and the occultations in more detail. The figure on the right is the spacecraft position in the solar system, showing the Earth's and LATOR's orbits relative to the Sun. The epoch of this figure shows the spacecraft passing behind the Sun as viewed from the Earth. The figure on the left shows the trajectory when the spacecraft would be within 10° of the Sun as viewed from the Earth. This period of 280 days will occur once every 3 years, provided that proper maneuvers are performed. Two similar periodic curves give the Sun–Earth–probe angles for the two spacecraft, while the lower smooth curve gives their angular separation as seen from the Earth.

An orbit with a 3:2 resonance with the Earth uniquely satisfies the LATOR orbital requirements.[1,2] For this orbit, 13 months after the launch, the spacecraft are within $\sim 10^\circ$ of the Sun, with first occultation occurring 15 months after launch.[1] At this point, LATOR is orbiting at a slower speed than the Earth, but as LATOR approaches its perihelion, its motion in the sky begins to reverse and the spacecraft is again occulted by the Sun 18 months after launch. As the craft slow down and move out toward the aphelion, their motion in the sky reverses again, and it is occulted by the Sun for the third and final time 21 months after launch.

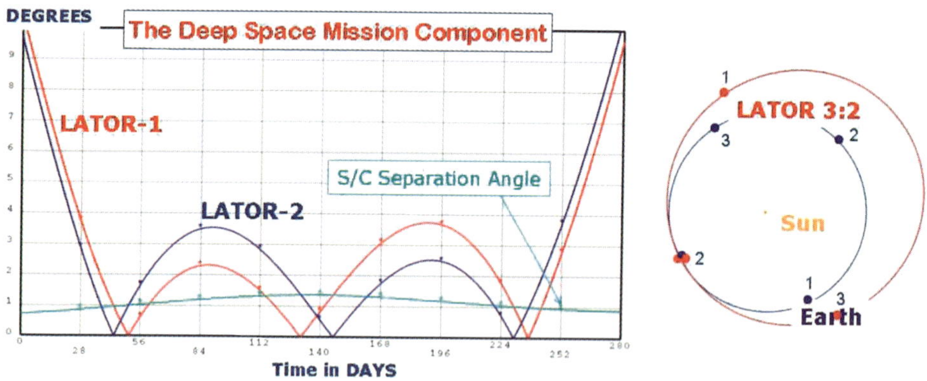

Fig. 2. *Left*: The Sun–Earth–probe angle during the period of three occultations (two periodic curves) and the angular separation of the spacecraft as seen from the Earth (lower smooth line). The time shown is days from the moment when one of the spacecraft is at 10 distance from the Sun. *Right*: View from the North Ecliptic of the LATOR spacecraft in a 3:2 resonance. The epoch is taken near the first occultation.

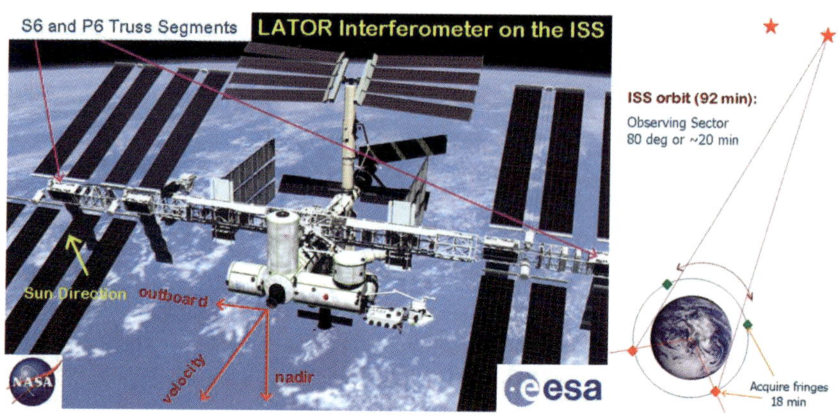

Fig. 3. *Left*: Location of the LATOR interferometer on the ISS. To utilize the inherent ISS Sun-tracking capability, the LATOR optical packages will be located on the outboard truss segments P6 and S6 outward. *Right*: Signal acquisition for each orbit of the ISS; a variable baseline allows for resolving fringe ambiguity.

2.3. *Observing sequence*

As a baseline design for the LATOR orbit, the two spacecraft will be launched on the same launch vehicle. Almost immediately after the launch there will be a 30 m/s maneuver that separates the two spacecraft on their 3:2 Earth resonant orbits (see Fig. 2). This sequence will be initiated at the beginning of the experiment period, after the ISS' emergence from the Earth's shadow (see Fig. 3). It assumes that boresighting of the spacecraft attitude with the spacecraft transmitters and receivers has already been accomplished. This sequence of operations is focused on establishing the ISS-to-spacecraft link. The interspacecraft link is assumed to be continuously established after final deployment (at $\sim 15^\circ$ off the Sun), since the spacecraft never lose the line of sight with each another.

The laser beacon transmitter at the ISS is expanded to have a beam divergence of 30 arcsec in order to guarantee illumination of the LATOR spacecraft. After re-emerging from the Earth's shadow this beam is transmitted to the craft and reaches them in about 18 minutes. At this point, the LATOR spacecraft acquire the expanded laser beacon signal. In this mode, a signal-to-noise ratio (SNR) of 4 can be achieved with 30 s of integration. With attitude knowledge of 10 arcsec and an array field of view of 30 arcsec, no spiral search is necessary. Upon signal acquisition, the receiver mirror on the spacecraft will center the signal and use only the center quad array for pointing control. Transition from acquisition to tracking should take about 1 minute. Because of the weak uplink intensity, at this point, tracking of the ISS station is done at a very low bandwidth. The pointing information is fed forward to the spacecraft transmitter pointing system and the transmitter is turned on. The signal is then retransmitted down to the ISS.

Each interferometer station or laser beacon station searches for the spacecraft laser signal. The return is heterodyned by using an expanded bandwidth of 300 MHz. In this case, the solar background is the dominant source of noise, and an SNR of 5 is achieved with 1 s integration.[2] Because of the small field of view of the array, a spiral search will take 30 s to cover a 30 arcsec field. Upon acquisition, the signal will be centered on the quad cell portion of the array and the local oscillator frequency locked to the spacecraft signal. The frequency band will then be narrowed to 5 kHz. In this regime, the solar background is no longer the dominant noise source and an SNR of 17.6 can be achieved in only 10 msec of integration. This will allow one to have a closed loop pointing bandwidth of greater than 100 Hz and be able to compensate for the tilt errors introduced by the atmosphere. The laser beacon transmitter will then narrow its beam to be diffraction limited ($\sim$1 arcsec) and to point toward the LATOR spacecraft. This completes the signal acquisition phase, and the entire architecture is in-lock and transmits scientific signal. This procedure is re-established during each 92-minute orbit of the ISS (see Ref. 2 for details).

2.4. *Principles of optical design*

A single aperture of the interferometer on the ISS consists of three 20-cm-diameter telescopes (see Fig. 4 for a conceptual design). One of the telescopes, with a very-narrow-bandwidth laser line filter in front and with an InGaAs camera at its focal plane, sensitive to the 1064 nm laser light, serves as the acquisition telescope to locate the spacecraft near the Sun. The second telescope emits the directing beacon to the spacecraft. Both spacecraft are served out of one telescope by a pair of piezocontrolled mirrors placed on the focal plane. The properly collimated laser light ($\sim$10 W) is injected into the telescope focal plane and deflected in the right direction by the piezoactuated mirrors.

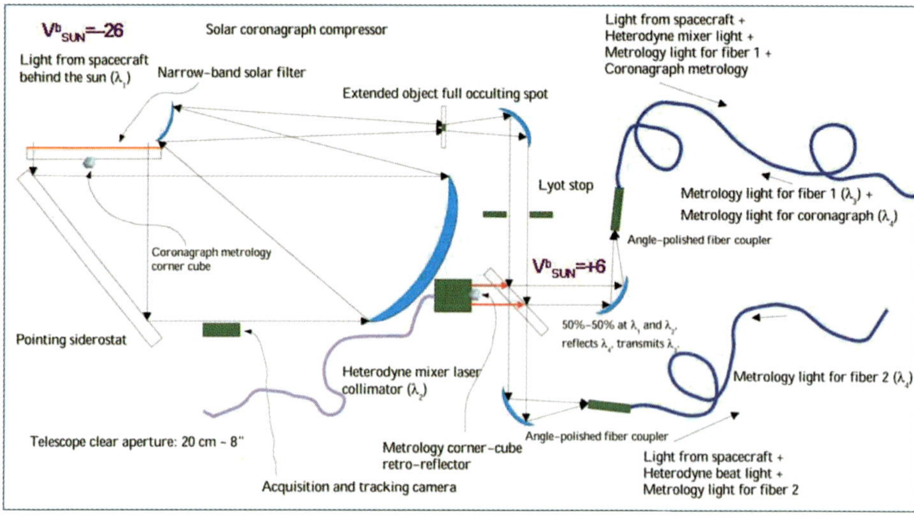

Fig. 4. Basic elements of optical design for the LATOR interferometer. The laser light (together with the solar background) is going through a full aperture ($\sim 20\,\text{cm}$) narrow band-pass filter with $\sim 10^{-4}$ suppression properties. The remaining light illuminates the baseline metrology corner cube and falls onto a steering flat mirror, where it is reflected to an off-axis telescope with no central obscuration (needed for metrology). It then enters the solar coronograph compressor by first going through a 1/2 plane focal plane occulter and then coming to a Lyot stop. At the Lyot stop, the background solar light is reduced by a factor of 10^6. The combination of a narrow band-pass filter and a coronograph enables solar luminosity reduction from $V = -26$ to $V = 4$ (as measured at the ISS), thus permitting the LATOR precision observations.

The third telescope is the laser-light-tracking interferometer input aperture, which can track the two spacecraft at the same time. To eliminate beam walk on the critical elements of this telescope, two piezoelectric X–Y–Z stages are used to move two single-mode fiber tips on a spherical surface while maintaining the focus and beam position on the fibers and other optics. Dithering at a few Hz is used to make the alignment to the fibers and the subsequent tracking of the two spacecraft completely automatic. The interferometric tracking telescopes are coupled together by a network of single-mode fibers whose relative length changes are measured internally by a heterodyne metrology system to an accuracy of less than 5 pm.

The spacecraft are identical in construction and contain a 1 W, stable (2 MHz per hour; $\sim$500 Hz per second), small cavity fiber-amplified laser at 1064 nm. Nearly 75% of the power of this laser is pointed to the Earth through a 15-cm-aperture telescope and its phase is tracked by the interferometer. With the available power and the beam divergence, there are enough photons to track the slowly drifting phase of the laser light. The remaining part of the laser power is diverted to another telescope, which points toward the other spacecraft. In addition to the two transmitting telescopes, each spacecraft has two receiving telescopes. The receiving telescope, which points toward the area near the Sun, has laser line filters and a simple knife-edge coronagraph to suppress the Sun's light to 1 part in 10^4 of the light level of the light received from the space station. The receiving telescope that points to the other spacecraft is free of the Sun light filter and the coronagraph.

The spacecraft also carry a small (2.5 cm) telescope with a CCD camera. This telescope is used to initially point the spacecraft directly toward the Sun so that their signal may be seen at the space station. One more of these small telescopes may also be installed at right angles to the first one, to determine the spacecraft attitude, using known, bright stars. The receiving telescope looking toward the other spacecraft may be used for this purpose part of the time, reducing hardware complexity. Star trackers with this construction were demonstrated many years ago and they are readily available. A small RF transponder with an omnidirectional antenna is also included in the instrument package to track the spacecraft while they are on their way to assuming the orbital position needed for the experiment.

The LATOR experiment has a number of advantages over techniques that use radio waves to measure gravitational light deflection. Advances in optical communications technology allow low bandwidth telecommunications with the LATOR spacecraft without having to deploy high gain radio antennae needed to communicate through the solar corona. The use of the monochromatic light enables the observation of the spacecraft almost at the limb of the Sun, as seen from the ISS. The use of narrowband filters, coronagraph optics and heterodyne detection will suppress background light to a level where the solar background is no longer the dominant noise source. In addition, the short wavelength allows much more efficient links with smaller apertures, thereby eliminating the need for a deployable antenna. Finally, the use of the ISS will allow conducting of the test above the Earth's atmosphere — the major source of astrometric noise for any ground-based interferometer. This fact justifies LATOR as a space mission.

3. Conclusions

The LATOR mission aims to carry out a test of the curvature of the solar system's gravity field with an accuracy better than 1 part in 10^9. The LATOR experiment benefits from a number of advantages over techniques that use radio waves to study the light propagation in the solar vicinity. The use of monochromatic light enables the observation of the spacecraft almost at the limb of the Sun. The use of narrowband filters, coronagraph optics, and heterodyne detection will suppress background light to a level where the solar background is no longer the dominant noise source. The short wavelength allows much more efficient links with smaller apertures, thereby eliminating the need for a deployable antenna. Advances in optical communications technology allow low bandwidth telecommunications with the LATOR spacecraft without having to deploy high gain radio antennae needed to communicate through the solar corona. Finally, the use of the ISS not only makes the test affordable, but also allows conducting of the experiment above the Earth's atmosphere — the major source of astrometric noise for any ground-based interferometer. This fact justifies the placement of LATOR's interferometer node in space.

LATOR is envisaged as a partnership between NASA and ESA wherein the two partners are essentially equal contributors, while focusing on different mission elements: NASA provides the deep space mission components and interferometer design, while building and servicing infrastructure on the ISS is an ESA contribution.[2,38] The NASA focus is on mission management, system engineering, software management, integration (of both the payload and the mission), the launch vehicle for the deep space component, and operations. The European focus is on interferometer components, the initial payload integration, optical assemblies and testing of the optics in a realistic ISS environment. The proposed arrangement would provide clean interfaces between familiar mission elements.

Acknowledgments

The work described here was carried out at the Jet Propulsion Laboratory, California Institute of Technology, under a contract with the National Aeronautics and Space Administration.

References

1. S. G. Turyshev, M. Shao and K. L. Nordtvedt, *Class. Quant. Grav.* **21** (2004) 2773 [gr-qc/0311020].
2. S. G. Turyshev, M. Shao and K. L. Nordtvedt, Science, Technology and Mission Design for the Laser Astrometric Test of Relativity, in *Proc. Lasers, Clocks and Drag-Free Key Technologies: Future High Precision Tests of General Relativity,* eds. H. Dittus, C. Lämmerzahl and S. G. Turyshev (Springer-Verlag, 2006), p. 429, gr-qc/0601035.
3. K. L. Nordtvedt, *Phys. Rev.* **169** (1968) 1017.
4. K. L. Nordtvedt, *Phys. Rev.* **170** (1968) 1186.
5. K. L. Nordtvedt, *Astrophys. J.* **320** (1987) 871.
6. C. M. Will and K. L. Nordtvedt, *Astrophys. J.* **177** (1972) 757.
7. C. M. Will, *Theory and Experiment in Gravitational Physics* (Cambridge University Press, 1993).
8. K. L. Nordtvedt, *Class. Quant. Grav.* **13** (1996) A11.
9. B. Bertotti, L. Iess and P. Tortora, *Nature* **425** (2003) 374.
10. J. G. Williams, S. G. Turyshev and D. H. Boggs, *Phys. Rev. Lett.* **93** (2004) 261101 [gr-qc/0411113].
11. T. Damour and G. Esposito-Farese, *Phys. Rev. D* **53** (1996) 5541 [gr-qc/9506063].
12. T. Damour and G. Esposito-Farese, *Phys. Rev. D* **54** (1996) 1474 [gr-qc/9602056].
13. T. Damour and G. Esposito-Farese, *Phys. Rev. Lett.* **70** (1993) 2220.
14. T. Damour and K. L. Nordtvedt, *Phys. Rev. Lett.* **70** (1993) 2217.
15. T. Damour and K. L. Nordtvedt, *Phys. Rev. D* **48** (1993) 3436.
16. T. Damour and A. M. Polyakov, *Gen. Relativ. Gravit.* **26** (1994) 1171.
17. T. Damour and A. M. Polyakov, *Nucl. Phys. B* **423** (1994) 532.
18. T. Damour, F. Piazza and G. Veneziano, *Phys. Rev. Lett.* **89** (2002) 081601 [gr-qc/0204094].
19. T. Damour, F. Piazza and G. Veneziano, *Phys. Rev. D* **66** (2002) 046007 [hep-th/0205111].
20. C. L. Bennett *et al.*, *Astrophys. J. Suppl.* **148** (2003) 1 [astro-ph/0302207].
21. S. Perlmutter *et al.*, *Astrophys. J.* **517** (1999) 565 [astro-ph/9812133].

22. A. G. Riess *et al.*, *Astron. J.* **116** (1998) 1009.
23. J. L. Tonry *et al.*, *Astrophys. J.* **594** (2003) 1 [astro-ph/0305008].
24. N. W. Halverson *et al.*, *Astrophys. J.* **568** (2002) 38 [astro-ph/0104489].
25. C. B. Netterfield *et al.*, *Astrophys. J.* **571** (2002) 604 [astro-ph/0104460].
26. P. J. E. Peebles and B. Ratra, *Rev. Mod. Phys.* **75** (2003) 559 [astro-ph/0207347].
27. S. M. Carroll *et al.*, *Phys. Rev. D* **70** (2004) 043528 [astro-ph/0306438].
28. S. M. Carroll *et al.*, *Phys. Rev. D* **68** (2003) 023509 [astro-ph/0301273].
29. R. Epstein and I. I. Shapiro, *Phys. Rev. D* **22** (1980) 2947.
30. E. Fischbach and B. S. Freeman, *Phys. Rev. D* **22** (1980) 2950.
31. G. W. Richter and R. A. Matzner, *Phys. Rev. D* **26** (1982) 1219.
32. G. W. Richter and R. A. Matzner, *Phys. Rev. D* **26** (1982) 2549.
33. G. W. Richter and R. A. Matzner, *Phys. Rev. D* **28** (1983) 3007.
34. J. E. Plowman and R. W. Hellings, *Class. Quant. Grav.* **23** (2006) 309 [gr-qc/0505064].
35. K. L. Nordtvedt, LATOR's Measured Science Parameters and Mission Configuration, in *Proc. Lasers, Clocks and Drag-Free Key Technologies: Future High Precision Tests of General Relativity,* eds. H. Dittus, C. Lämmerzahl and S. G. Turyshev (Springer-Verlag, 2006), p. 501.
36. S. G. Turyshev, M. Shao and K. L. Nordtvedt, eConf C041213 #0306 (2004), gr-qc/0502113.
37. S. G. Turyshev, M. Shao and K. L. Nordtvedt, *Int. J. Mod. Phys. D* **13** (2004) 2035 [gr-qc/0410044].
38. S. G. Turyshev *et al.*, *ESA Spec. Publ.* **588** (2005) 11 [gr-qc/0506104].

LATOR: ITS SCIENCE PRODUCT AND ORBITAL CONSIDERATIONS

KENNETH NORDTVEDT

Northwest Analysis: 118 Sourdough Ridge Road,
Bozeman MT 59715, USA
knordtvedt@bresnan.net

In a LATOR mission to measure the non-Euclidean relationship between three sides and one angle of a light triangle near the Sun, the primary science parameter, to be measured to part-in-10^9 precision, is shown to include not only the key parametrized post-Newtonian (PPN) γ, but also the Sun's additional mass parameter, M_Γ, which appears in the spatial metric field potential. M_Γ may deviate from the Sun's well-measured gravitational mass due to post-Newtonian features of gravitational theory not previously measured in relativistic gravity observations. Under plausible assumptions, M_Γ is a linear combination of the Sun's gravitational and inertial masses.

If LATOR's two spacecraft lines of sight are kept close to equal and opposite relative to the Sun during the mission's key measurements of the light triangle, it is found that the navigational requirements for the spacecraft positions are greatly relaxed, eliminating the need for on-board drag-free systems. Spacecraft orbits from the Earth to achieve the equal and opposite passages by the Sun's, line of sight are illustrated.

Keywords: High-precision gravity tests; laser ranging and interferometery; LATOR.

1. Overview

LATOR is a mission concept to measure the three sides and one angle of a light triangle in the solar system. With at least one side of the triangle passing close by the Sun, the triangle is distorted by gravity from Euclidean structure, and through this distortion provides a high-precision test of general relativity.[1,2] The most important science parameter measured by LATOR is the strength of the first-order light deflection; the mission's goal is a part-in-10^9 measurement of the parametrized post-Newtonian (PPN) "Eddington" coefficient γ, which scales this deflection and whose deviation from one is a signal (and measure) of modification of theory from the pure tensor gravity of general relativity.[3–5] PPN γ is not estimated in isolation by this measurement; it is found that a new mass parameter for the Sun, sensitive to the second post-Newtonian level of theory, contributes to the measured science parameter together with γ.[6]

The three sides of the light triangle are measured by round-trip, transponded laser ranging: one of the angles of the triangle is measured by a laser interferometer which is located near the Earth and which looks at the angle between the lines of sight of two spacecraft which pass behind the Sun as seen from the interferometer (see the paper by Shao in this issue). Figure 1 illustrates a preferred mission configuration with the Sun enclosed by the light triangle, resulting in the doubling of the science signal due to addition of the light deflections occurring on two sides of the light triangle.

An additional advantage from keeping the two spacecraft lines of sight equal and opposite with respect to the Sun is that this desensitizes the science signal from uncertainties in the location of the light triangle with respect to the Sun. No drag-free systems on the spacecraft are then needed to preserve high-precision knowledge of the spacecraft position during the period of data measurements in the mission. A mission scenario is illustrated in Fig. 2, which achieves the equal and opposite motions of spacecraft lines of sight past the Sun; there is a 1.5-year version of the mission and a 2.5-year version.

A sequence of views of the passing spacecrafts as seen from the interferometer near the Earth is given in Fig. 3. Preferred polar passages of the spacecraft are shown in which the light deflection signal due to the Sun's quadrupole moment gravity field better "separates" from that due to the Sun's monopolar gravity field. The mission could go forth using equatorial passages of the spacecraft with some sacrifice in measurement precision for PPN γ, if technical and cost considerations ruled against slight out-of-ecliptic orbital changes.

If light traveled according to the geometrical laws of Euclid and at constant speed, the laser ranging times from the three sides of the light triangle would fix the angle the interferometer sees between the two spacecraft lines of sight. This

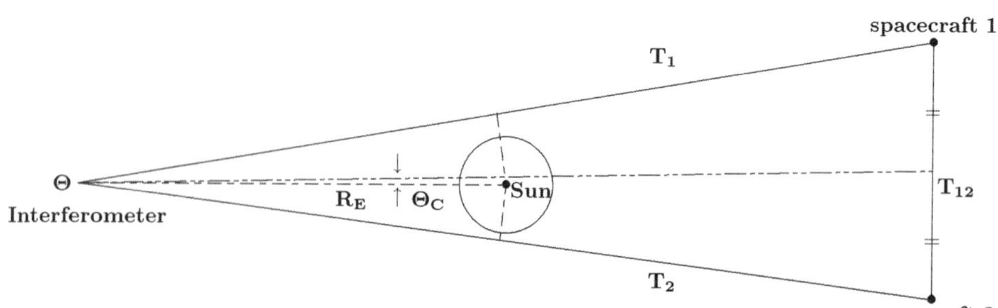

Fig. 1. Light triangle in the LATOR mission: A laser interferometer at distance R_E from the Sun measures angle Θ between lines of sight to two spacecraft passing close by the Sun's line of sight, and the round-trip laser ranging times T_1, T_2, T_{12} are also measured. Deviations from the Euclidean relationship between these four observables probe general relativity's predicted effects of the Sun's gravity on the light trajectories. If the average angular location Θ_C of the spacecraft is kept close to the Sun's center, then only modest navigational requirements for spacecraft locations are needed, eliminating the need for drag-free systems.

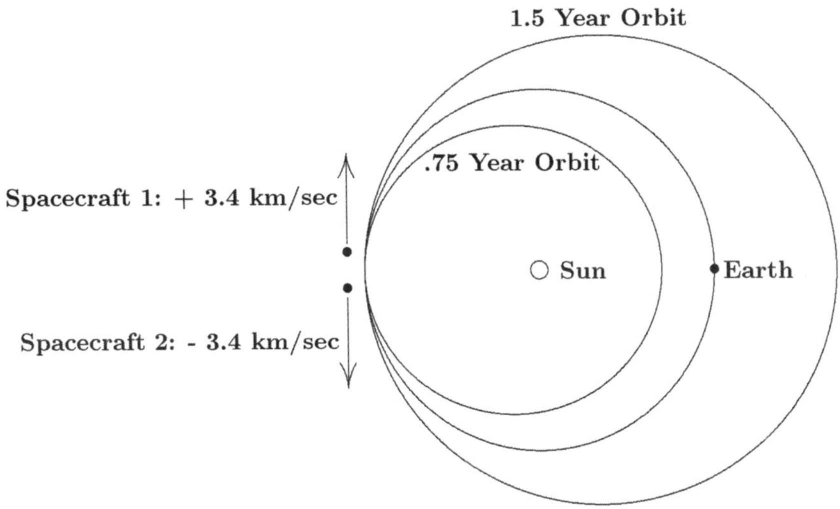

Fig. 2. To achieve equal and opposite passages of two spacecraft lines of sight as viewed from an interferometer near the Earth, spacecraft are put into 1.5-year and 0.75-year orbits by being given escape velocities of about 3.4 km/s parallel and antiparallel to the Earth's motion around the Sun. A 2.5-year mission could be achieved with smaller spacecraft escape velocities, ±2.0 km/s, which produce 5/4-year and 5/6-year orbits.

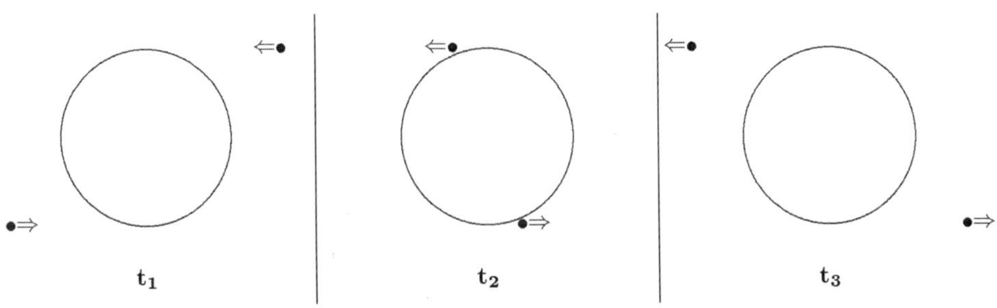

Fig. 3. Polar passages of two spacecraft by the Sun: At successive times t_1, t_2, t_3 the angular positions of two spacecraft passing at equal and opposite distances from the Sun are shown for polar passages. Polar passage has the advantage of better "separating" gravity's monopolar deflection angle signal from that due to the Sun's gravitational quadrupole moment, the latter changing sign during the data-collecting period if passages are sufficiently off the solar equator.

defined angle from three observables, Θ_{Euc}, then establishes the benchmark from which the actual measured angle can be compared.

$$\sin\Theta_{\text{Euc}} = \frac{1}{2T_1T_2}\sqrt{2T_{12}^2\,(T_1^2+T_2^2) - T_{12}^4 - (T_1^2-T_2^2)^2}. \quad (1)$$

The leading order deflection signal which contributes to the interferometer's measured angle, the fourth observable, can be expressed as

$$\Theta - \Theta_{\text{Euc}} = (1+\gamma^*)\left[\frac{GM_G(\text{Sun})}{c^2R_E}\right]\left(\frac{1}{\sin(\Theta/2+\Theta_C)} + \frac{1}{\sin(\Theta/2-\Theta_C)}\right) + \cdots, \quad (2)$$

with the variables Θ, Θ_C, and R_E shown in Fig. 1, and $GM_G(\text{Sun})/c^2$ being a solar system parameter measured from other missions of the past along with sufficient knowledge of the time-dependent R_E to render the dimensionless package $GM_G(\text{Sun})/c^2R_E$ sufficiently known for LATOR's needs. The gravity theory parameter γ^* is to be estimated by fitting the time-dependent data from the mission's four observables during the passages of the two spacecraft's lines of sight past the Sun. The detailed content of γ^* is discussed in the next section.

2. Light Coordinate Speed Function and the Science Signal

In metric tensor gravity the coordinate speed-of-light function is obtained from the null geodesic principle

$$\sqrt{g_{\mu\nu}dx^\mu dx^\nu} = 0, \tag{3}$$

so the spatial metric potentials contribute to light's speed on an equal footing with the temporal metric potentials. If these two main components of the metric tensor field are expressed to leading order, there results

$$g_{oo} = 1 - 2\frac{GM_G}{c^2R} + \cdots, \tag{4}$$

$$g_{ab} = -\delta_{ab}\left(1 + 2\gamma\frac{GM_\Gamma}{c^2R}\right) + \cdots. \tag{5}$$

M_G is the Sun's gravitational mass, while M_Γ is a new mass parameter of the Sun which may differ from the gravitational mass by the order of the gravitational binding energy of the Sun (fractionally $4 \cdot 10^{-6}$ of the Sun's mass-energy), just as the Sun's gravitational mass may differ from its inertial mass-energy by this order.[6] The possible differences of these mass parameters are probes of the next relativistic order of the metric tensor fields, as is outlined in App. A. The light coordinate speed function is then

$$c(R,\hat{c}) = 1 - \left(1 + \frac{\gamma M_\Gamma}{M_G}\right)\frac{GM_G}{c^2R} + \cdots. \tag{6}$$

The additional, unstated terms of the speed-of-light function arise from higher order metric potentials due to gravity's intrinsic nonlinearity and to the motion of the Sun in the solar system barycentric frame. The leading order contributions to the deflection angles of the two sides of the light triangle illustrated in Fig. 1 can then be put into the form of Eq. (1). using the parameters defined in that figure,

$$\begin{aligned}\Theta - \Theta_{\text{Euc}} = &\left(1 + \frac{\gamma M_\Gamma}{M_G}\right)\left[\frac{GM(G)_{\text{Sun}}}{c^2R_E}\right]\\ &\times\left(\frac{1}{\sin(\Theta/2 + \Theta_C)} + \frac{1}{\sin(\Theta/2 - \Theta_C)}\right) + \cdots,\end{aligned} \tag{7}$$

with R_E being the coordinate distance from the interferometer near the Earth to the Sun's gravity field center, and $R_E\ \sin(\Theta/2 \pm \Theta_c)$ are, in lowest order, the distances of closest approach to the Sun of the two sides of the light triangle.

The primary science signal measures the package $\gamma M_\Gamma/M_G$ rather than γ alone! This is true, in fact, for any future experiment which seeks to measure PPN γ at levels of precision exceeding the fractional gravitational binding energy of the pertinent gravity field sources. γ will always appear together with the indicated mass ratios of key bodies.[a]

It has been shown that a PPN metric tensor field expansion which fulfills "extended Lorentz invariance" must have mass parameters M_Γ which can be expressed as linear combinations of the bodies' gravitational and inertial masses.[6] "Extended Lorentz invariance" means that the first post-Newtonian order, local gravitational fields remain Lorentz-invariant under arbitrary rescaling due to more distant mass distributions.

$$M_G + \gamma M_\Gamma = (1+\gamma)M_I. \tag{8}$$

This renders the coefficient of first order light deflection in Eq. (6) as $1+\gamma^* = (1+\gamma)M_I(\text{Sun})/M_G(\text{Sun})$. The ratio of inertial to gravitational mass of the Sun can be inferred to sufficient precision from lunar-laser-ranging measurements of the same ratio for the Earth, or the Sun's ratio can directly be measured through improved solar system ranging experiments between the inner planets. In this case PPN γ becomes directly measured by itself.

PPN γ may have an anomalous value due to rescaling from the solar system's unique place in the cosmos. See App. B for a more detailed discussion. Just as Newton's G is rescaled due to the gravitational potential of distant matter, γ is also rescaled due to second post-Newtonian order structure of the spatial components of the metric tensor gravity field. We know from observations of the microwave background radiation of the Universe that the cosmological gravity potential has part-in-10^5 variations about the average; the observed "great attractor" of 10^{16} solar masses at about 65 mega-parsec distance produces here a Newtonian potential of about this size. A part-in-10^9 measurement of γ should therefore probe the unique combination of second post-Newtonian order parameters derived in the appendix at the part-in-10^4 level.

3. Locating the Light Triangle with Respect to the Sun

In Eqs. (2) and (7), which express the leading order light deflection signal, the "common mode" location angle of the two spacecraft, Θ_C, can be considered the variable which locates the light triangle transversely with respect to the Sun. Knowledge of the other five degrees of freedom for location and orientation of the light triangle with respect to the Sun is not so critical. Acquiring knowledge of Θ_C, however, would seem necessary in order to gain sufficiently accurate knowledge of the impact parameters of the light triangle sides which pass close by the Sun. This could be

[a]The same thing occurs for high precision measurements of PPN β; this parameter will always manifest itself in post-Newtonian effects together with body mass ratios such as M_β/M_G, etc.

a mission stopper. But if the common mode angle Θ_C could be kept very small by equal and opposite spacecraft passages of the Sun, the location issue for the light triangle can be managed more easily. For a small common mode angle, the trigonometric expression in Eqs. (2) and (7) can be expanded as

$$\frac{1}{\sin(\Theta/2+\Theta_C)}+\frac{1}{\sin(\Theta/2-\Theta_C)}=\frac{2}{\sin(\Theta/2)}\left(1+4\left(\frac{\Theta_c}{\Theta}\right)^2+\cdots\right). \tag{9}$$

The contribution from the leading quadratic Θ_C expansion term can be kept below five parts in 10^{10} if $X_C = 2R_E\Theta_c$ is less than 21 *kilometers*. This requires only modest along-track navigational accuracies of 30 *kilometers* for each of the spacecraft.

On the other hand, if only one side of the light triangle passed close by the Sun to primarily produce the science signal, the distance of closest approach to the Sun for that side needs to be known to five parts in 10^{10}, or, equivalently, about 70 *centimeters* for the spacecraft position in its along-track direction. This would require fitting for this location (and velocity) at one point in the mission, costing precision of estimation for the science parameters in the least squares fit of the data, and then having a drag-free system on board capable of measuring $3\cdot 10^{-14}g$ nongravitational steady-state accelerations on the spacecraft — a daunting and expensive challenge for the mission.

Achievement of equal and opposite passages of the two spacecraft lines of sight is obtained by putting one spacecraft into a 1.5-year orbit and the other spacecraft into a 0.75-year orbit. 1.5 years later the spacecraft will be opposite the Earth from the Sun with lines of sight passing in the desired fashion. This is illustrated in Fig. 2 with the view of these passages from the interferometer shown in Fig. 3.

Appendix A. M_G and M_Γ as Probes of the Next Post-Newtonian Order Metric Gravity

For a collection of mass elements with negligible gravitational self-energies, and locally Lorentz-invariant gravity, the second post-Newtonian order expansion of the temporal component of the metric tensor gravity field is

$$g_{00} = 1 - 2U(\mathbf{r},t) + 2\beta U(\mathbf{r},t)^2 + (4\beta-2)\sum_{i\neq j}\frac{G^2 m_i m_j}{c^4 r_i r_{ij}} + (2\gamma+1)\sum_i \frac{G m_i v_i^2}{c^4 r_i}, \tag{A.1}$$

with $r_i = |\mathbf{r}-\mathbf{r}_i|$ and $r_{ij} = |\mathbf{r}_i - \mathbf{r}_j|$; γ and β are the two Eddington PPN coefficients ($\gamma=\beta=1$ in general relativity); and the dimensionless Newtonian potential function is

$$U(\mathbf{r},t) = \sum_i \frac{G m_i}{c^2 r_i}. \tag{A.2}$$

If many mass elements are collected into an equilibrium massive body, and the total contributions to the $1/R$ potentials collected with R being the distance from the assembled massive body, then contributions from the $4\beta - 2$ and $2\gamma + 1$ terms in Eq. (A.1) add to the linear Newtonian potential, giving in total

$$g_{00} = 1 - 2\frac{GM_G}{c^2R} + \cdots, \tag{A.3}$$

with

$$M_G = \sum_i m_i \left(1 + \frac{v_i^2}{2c^2} - \sum_{j\neq i} \frac{Gm_j}{2c^2 r_{ij}}\right) - (4\beta - 3 - \gamma) \sum_{i,j\neq i} \frac{Gm_i m_j}{2c^2 r_{ij}} \tag{A.4}$$

$$= M_I - (4\beta - 3 - \gamma) \sum_{i,j\neq i} \frac{Gm_i m_j}{2c^2 r_{ij}}. \tag{A.5}$$

So, unless $4\beta - 3 - \gamma = 0$ the gravitational mass of a body will differ from its inertial mass-energy.[7]

A similar investigation can be made of the spatial components of the metric tensor gravity field.[8] The second post-Newtonian order expansion in this case is

$$g_{ab} = -\delta_{ab}\left(1 + 2\gamma\, U(\mathbf{r},t) + \frac{3}{2}\delta\, U(\mathbf{r},t)^2 - \kappa \sum_{i,j\neq i} \frac{G^2 m_i m_j}{c^4 r_i r_{ij}} \right.$$
$$\left. - \gamma \sum_i \frac{Gm_i}{c^4 r_i^3} (\mathbf{v}_i \cdot \mathbf{r}_i)^2 \right) - 2(\gamma + 1) \sum_i \frac{Gm_i}{c^4 r_i} (\mathbf{v}_i)_a (\mathbf{v}_i)_b\,, \tag{A.6}$$

with the two additional PPN parameters, δ and κ, being introduced ($\delta = \kappa = 1$ in general relativity). The potential terms with coefficients involving γ are fixed from requirements of the local Lorentz invariance of the metric tensor fields. As was done for the temporal component of the field, collecting many mass elements into an equilibrium massive body, and identifying the total contributions to the $1/R$ potential of the spatial metric, the massive body's mass parameter in the spatial metric becomes

$$\gamma M_\Gamma = \gamma M_I - \left(\frac{3\kappa - 2 - \gamma}{3}\right) \sum_{i,j\neq i} \frac{Gm_i m_j}{2c^2 r_{ij}}. \tag{A.7}$$

Appendix B. Renormalization of G and PPN Parameters by Distant Bodies

The expansion of the temporal component of the metric tensor field given by Eq. (A.1) also results in renormalization of Newton's gravitational coupling parameter G due to distant matter.[6] This is derived by dividing the general collection of bodies in Eq. (A.1) into contributions from local bodies plus contributions from all the distant (spectator) bodies:

$$U(\mathbf{r},t) = \sum_{\text{local } i} \frac{Gm_i}{c^2 r_i} + U_S, \quad \text{with } U_S = \sum_{\text{spectators } s} \frac{Gm_s}{c^2 R_s}. \tag{B.1}$$

The gravitational potential of the spectator bodies also rescales the local proper time and space coordinates

$$t^* = t\,(1 - U_S)\,, \tag{B.2}$$

$$\mathbf{r}^* = \mathbf{r}\,(1 + \gamma U_S)\,. \tag{B.3}$$

Inserting these factors into Eq. (A.1), and recalling that the tensor transformation rule is

$$g^*_{00} = \left(\frac{\partial t}{\partial t^*}\right)^2 g_{00}\,, \tag{B.4}$$

the temporal metric tensor component becomes in the local region and to linear order

$$g^*_{00} = 1 - 2\sum_{\text{local } i} \frac{G^* m_i}{c^2 r^*_i}\,, \tag{B.5}$$

with the renormalized G^*:

$$\frac{G^*}{G} = 1 - (4\beta - 3 - \gamma)U_S. \tag{B.6}$$

Renormalization of PPN γ due to distant matter is obtained in a similar fashion using the spatial components of the metric tensor field given in Eq. (A.6).[8] These metric field components become locally after the renormalizations and to linear order

$$g^*_{ab} = -\delta_{ab}\left(1 + 2\gamma^* \sum_{\text{local } i} \frac{G^* m_i}{c^2 r^*_i}\right), \tag{B.7}$$

with renormalized γ^* given by

$$\left(\frac{G^*}{G}\right)\gamma^* = \gamma + \left(\frac{3\delta - \kappa - 2\gamma^2}{2}\right) U_S. \tag{B.8}$$

From the metric gravity expression for the light speed function, Eq. (1), the expression for the coordinate speed of light, sufficiently complete for use to model the LATOR mission, becomes

$$c(\mathbf{R}, \hat{c}, \mathbf{v}_S) = 1 - (1 + \gamma^*)U(\mathbf{R},t) + \beta^* U(\mathbf{R},t)^2 + \mathbf{h}\cdot\hat{c}. \tag{B.9}$$

γ^* has previously been defined in Eq. (6), and $U(\mathbf{R},t)$ is the Sun's Newtonian gravitational potential including the multipole contributions due to nonsphericity,

$$U(\mathbf{R},t) = \frac{GM_G}{c^2 R}\left(1 + J_2\left(\frac{a}{R}\right)^2 P_2(\theta) + \cdots\right), \tag{B.10}$$

with θ being the latitude of the location with respect to the solar equator. β^* is a collection of nonlinear and linear order PPN coefficients,

$$\beta^* = \frac{4\beta - 3\delta + 4\gamma + 6\gamma^2 - 2}{4}, \tag{B.11}$$

and $\mathbf{h}_a = g_{oa}$ is the "gravitomagnetic"-vector-potential part of the metric tensor field due to the moving matter of the Sun. This motion comes from two sources — the motion of the Sun's center of mass in the barycentric inertial frame and the motion of matter due to rotation of the Sun about its axis. The Sun's gravitational potential is time-dependent by virtue of the velocity of that body in the solar system,

$$R - R(t) = |\mathbf{R} - \mathbf{R}_S(t)|, \tag{B.12}$$

$$\mathbf{h} = \frac{4\gamma + 3}{2}\frac{GM_G}{c^3 R}\mathbf{v}_S + \frac{1}{2}\frac{GM_G}{c^3 R^3}\mathbf{R}\mathbf{R}\cdot\mathbf{v}_S + (1+\gamma)\frac{G\mathbf{J}_S x\mathbf{R}}{c^3 R^3}, \tag{B.13}$$

with $\mathbf{v}_S$ and $\mathbf{J}_S$ being the Sun's velocity and angular momentum, respectively. The effect on the speed of light due to the solar-angular-momentum contribution to the gravitomagnetic potential [the last term in Eq. (30)] is a maximum of $4 \cdot 10^{-7}$, the dominant monopolar effect. So a part-in-10^9 mission design would be able to modestly measure the contribution to the deflection of a single side of the light triangle. But our preferred configuration of two sides equally and oppositely located with respect to the Sun produces about equal deflections for the two sides and in senses which lead to nullification of the effect on the triangle's angle at the interferometer. The contribution from the motion of the Sun in the solar system is a maximum of $4.4 \cdot 10^{-8}$, the leading effect. So this will have to be included in what can be called the "housekeeping" corrections to the predicted signal along with similarly sized corrections due to the changing location of the Sun's gravitational potential during the transit time of the light past the Sun. Other housekeeping tasks include evaluation of the small, but observable, contributions to the light triangle distortions due to the other masses in the solar system.

Measurement of the β^* coefficient of second post-Newtonian order alteration in the light speed, and hence in the ranging and deflection of lines of sight, can be considered a secondary and modest explicit measurement of second order features of the metric tensor gravitational field. This contribution to the interferometer's angular anomaly measurements will depend on the inverse square of the light's distance of closest approach to the Sun, so it will be separable from the chief science signal, which depends on only the inverse first power of closest approach distance.

Unless independent determination of the Sun's gravitational quadrupole moment strength J_2 reaches sufficient precision from other observations, this parameter will have to be simultaneously estimated along with the chief science signal γ^* in the least squares fit of the data. The latitudinal dependence of the quadrupole field's effect on the light speed can be used to advantage; high latitude trajectories of the spacecraft as they pass the Sun's line of sight will produce a quadrupole field deflection signal which, by changing sign, easily orthogonalizes from the chief science signal from the monopolar field, so fitting for J_2 in such scenarios will produce only minimal degradation in the estimation precision for the chief science signal.

References

1. S. Turyshev, M. Shao and K. Nordtvedt, in *Proc. 359th W. E. Hereaus Seminar on Lasers, Clocks, and Drag-Free: Technologies for Future Exploration in Space and Tests of Gravity* (ZARM, Bremen, Germany; May 30–June 1, 2005), eds. H. Dittus, C. Laemmerzahl and S. Turyshev (Springer-Verlag), to appear.
2. S. Turyshev, M. Shao and K. Nordtvedt, *Class. Quant. Grav.* **21** (2004) 2773.
3. C. Will and K. Nordtvedt, *Astrophys. J.* **177** (1972) 757.
4. T. Damour and K. Nordtvedt, *Phys. Rev. Lett.* **70** (1993) 2217.
5. T. Damour and K. Nordtvedt, *Phys. Rev. D* **48** (1993) 3436.
6. K. Nordtvedt, *Astrophys. J.* **297** (1984) 390.
7. K. Nordtvedt, *Phys. Rev.* **170** (1968) 1186.
8. K. Nordtvedt, *Astrophys. J.* **407** (1993) 758.

SATELLITE TEST OF THE EQUIVALENCE PRINCIPLE: OVERVIEW AND PROGRESS

JEFFERY J. KOLODZIEJCZAK
NASA / Marshall Space Flight Center,
National Space Science and Technology Center,
320 Sparkman Drive, XD 12, Huntsville, AL 35805, USA
Jeffery.Kolodziejczak-1@nasa.gov

JOHN MESTER
Stanford University,
W. W. Hansen Experimental Physics Laboratory,
MC: 4085, STEP, Stanford, CA 94305-4085, USA
mester@step.stanford.edu

STEP, the Satellite Test of the Equivalence Principle, is reviewed and the current status of the project is discussed. This space-based experiment will test the universality of free fall and is designed to advance the present state of knowledge by over five orders of magnitude. The international STEP collaboration is pursuing a development plan to improve and verify the technology readiness of key systems. We discuss recent advances with an emphasis on accelerometer fabrication and tests. Critical technologies successfully demonstrated in flight by the Gravity Probe B mission also contribute to progress.

Keywords: Equivalence principle; instrumentation; gravity; fundamental physics.

1. Introduction

For centuries, the apparent equivalence between inertial mass and gravitational mass has been a fruitful source of both experimental ingenuity and theoretical inspiration in physics. This paper briefly describes the motivation for equivalence principle tests and some background on previous experiments. This leads to a rationale for the Satellite Test of the Equivalence Principle (STEP), a cryogenic SQUID-based drop tower test on an orbiting drag-free platform. We describe STEP and discuss its current status, as well as plans for demonstrating technical readiness to begin flight system development.

Two cornerstone theories of physics, general relativity and the Standard Model of strong weak and electromagnetic interactions, have proved difficult to unify.

This is due in part to a lack of experimental data for distinguishing a multitude of theories. Unifying theories often contain additional couplings or fields which lead to equivalence principle violations resulting from currently unknown forces or from a more complex gravitational coupling. A sensitive equivalence principle test would provide a rare data point in the chasm between general relativity and the Standard Model which might be a guidepost for theoretical progress.

Advancement beyond the equivalence principle state-of-the-art by five orders of magnitude may be compared with a similar increase in particle accelerator energies. Breakthroughs in understanding in going from 1 GeV to 100 TeV encompass practically everything in the Standard Model beyond quantum electrodynamics, including electroweak processes, W and Z bosons, quantum chromodynamics, quarks and gluons. A factor of 10^5 in astronomical sensitivity is comparable to the difference between the human eye and the Hubble Space Telescope. The leap in sensitivity offered by STEP could provide similar illumination.

For this reason, the National Academy of Sciences stated[1]: "Improvement by a factor of around 10^5 could come from an equivalence principle test in space... at these levels, null experimental results provide important constraints on existing theories, and a positive signal would make for a scientific revolution."

Since Galileo's free fall experiment, as described by his student Vincenzo Viviani, the measurement of the equivalence principle has taken many forms. Ground experiment has reached a 3×10^{-13} violation sensitivity level.[2] Other non-orbiting concepts include a stratospheric balloon experiment[3] with a sensitivity goal of a few $\times 10^{-15}$ and an atom interferometer[4] aiming toward $< 10^{-15}$ sensitivity. The limitations of seismic motion and gravity gradient torques constrain the Earth-bound experimental limits. To achieve higher precision requires a space-based platform with drag-free control to reduce effects of aerodynamic disturbances. The ONERA team developing the Microscope experiment predicts 10^{-15} sensitivity in an experiment performed with a warm capacitive readout.[5] The Galileo Galilei (GG) space experiment[6] has the goal of reaching 10^{-17}. Much higher sensitivity is possible in a cryogenic environment using a SQUID readout.

STEP probes for an equivalence principle violation at the level of a part in 10^{18} by measuring the position difference between pairs of test masses at a level where the acceleration of one relative to the other is determined to 4×10^{-19} g over a $\sim$ 100 ksec averaging time interval. During this time the common acceleration of the two test masses relative to the spacecraft is determined to 2×10^{-15} g.

2. The STEP Instrument

STEP simultaneously performs four independent orbiting drop tower tests. Each test compares the relative motion of two test masses. The test mass pairs form individual differential accelerometers (DA's). Four DA's form the instrument payload which attaches to a spacecraft.

Table 1. Candidate test mass materials.

Material	Z	N	Baryon number $\left(\frac{N+Z}{\mu}-1\right)\times 10^3$	Lepton number $\frac{N-Z}{\mu}$	Coulomb parameter $\frac{Z(Z-1)}{\mu(N+Z)^{\frac{1}{3}}}$
Be	4	5	−1.3518	0.11096	0.64013
Si	14	14.1	0.8257	0.00387	2.1313
Nb	41	52	1.0075	0.11840	3.8462
Pt	78	117.116	0.18295	0.20051	5.3081

2.1. *Test masses*

The core of any EP experiment is test masses of differing materials. Some distinguishing characteristics among candidate materials are shown in Table 1. These characteristics are derived using a general approach defined by Damour.[7] Under this general picture, lepton number, baryon number and electrostatic energy provide the means of gauging the influence of currently unknown coupling. STEP will use three materials in different combinations in the four test mass pairs.

Each STEP test mass is cylindrically symmetric. This allows the placement of one test mass inside the other, which reduces gravity gradient effects, provides the dimensional constraint to produce a periodic signal, and a workable layout for the acceleration and displacement detection components. The belted cylinder design enables the masses to behave gravitationally, like spheres. The shape also reduces the gravity gradient effect by minimizing multipole moments of inertia up to sixth order.[8]

Processes for test mass manufacturing are subject to tight tolerances. This combined with the stringent density homogeneity requirements limits the material options. Manufacturing feasibility has been demonstrated using all the materials listed in Table 1. A niobium thin film coating provides the needed superconductivity for bearing and readout purposes, and a gold overcoat layer reduces the patch effect caused by surface oxides.

2.2. *Differential accelerometer*

The STEP instrument employs test masses in pairs in order to directly measure the acceleration of one test mass relative to the other. The four pairs thereby define four independent differential accelerometers. A difference in acceleration between the two test masses indicates that they are falling toward the Earth at different rates. The design of the instrument therefore has a primary requirement that it provide a signal in which a specific difference in observed Earth-ward acceleration between the two test masses has no plausible interpretation in terms of known forces. A null signal then implies that EP is preserved at the specified level of sensitivity, and a differential detection would imply an equivalence violation.

This requirement is met by keeping all disturbances which could imitate a difference in acceleration toward the Earth below readout sensitivity and providing a readout signal of adequate fidelity to meet the part-in-10^{-18} sensitivity requirement. There are several key DA technologies associated with meeting this requirement, including the test masses discussed above. Superconducting bearings constrain motion to one dimension while minimizing friction. SQUID sensors provide a stable signal referenced to fundamental physical constants. An electrostatic positioning system measures and controls the test mass position to enable experiment initialization and charge measurement. A charge management system controls electrostatic forces to the necessary levels. Further description of these subsystems follows.

Each accelerometer contains an inner and an outer superconducting bearing to constrain motion of the test masses to one direction. A laser photolithography process patterns Nb thin film meander circuits onto cylindrical quartz substrates to within 5 arcsec of the cylinder's axis. Persistent currents injected into these circuits during DA operation generate a radially repulsive force on the facing test-mass-bearing surfaces.

SQUID's provide low intrinsic noise and the highest possible sensitivity for the STEP readout. The circuit shown in Fig. 1 employs two SQUID's for each DA. Each test mass is placed between two superconducting coils subject to a persistent current. Inductance between the coils and the superconducting surfaces of the test mass changes as a test mass moves toward one coil and away from the other. A current approximately proportional to the displacement is thus generated in a third parallel coil as a result of the inductance change. This third coil is shared by the two test mass subsystems and is coupled to a DC SQUID. When properly balanced, the current sensed by this SQUID is proportional to the difference in displacement of the two masses. The second SQUID provides a common mode displacement signal for the test mass pair by coupling to a separate inductor in which the two current components sum rather than difference. This common mode signal provides a reference for spacecraft translation control. Balancing the accelerometer consists of adjusting the relative supercurrents in the respective test mass circuit components to null the differential signal in the presence of an external excitation. The DA design accommodates one part-in-10^4 common mode rejection in the differential signal chain.

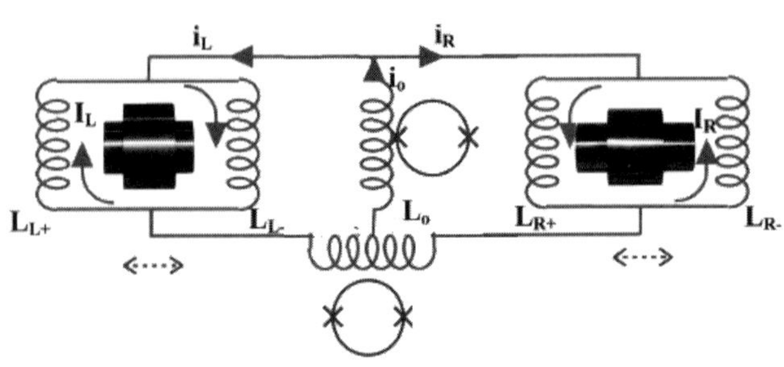

Fig. 1. SQUID circuit.

The test mass position can be electrostatically controlled and measured by controlling and sensing the voltages on various nonsuperconducting thin film electrodes deposited onto nonbearing components surrounding the test masses. The measured capacitance provides a position readout. Charge is also measured using these electrodes by applying a periodic voltage to excite an oscillation in the test mass and then capacitively measuring the amplitude of the oscillation at the excitation frequency.

The test mass will acquire charge primarily through radiation in space. Control of this charge requires a UV light source coupled with a bias electrode. The UV stimulates photoelectric emission with a net current either from test mass to electrode or from electrode to test mass, depending on the bias, thus enabling control of the test mass potential in either the positive or the negative direction.

2.3. *Payload and spacecraft*

Environmental disturbance rejection is achieved primarily at the payload and spacecraft levels through several technologies. The four DA's are assembled into a quartz block which provides a stable measurement platform which becomes part of a probe assembly. A cryoperm blanket and superconducting shielding provides magnetic field stability. A liquid helium Dewar provides thermal stability and the necessary cyrogenic environment for the instrument. The thermal environment of the instrument and electronics is further controlled through active systems. Tide control is maintained using a helium permeable aerogel filler in the Dewar. The spacecraft provides a drag-free control system for aerodynamic disturbance control, as well as stable power, communications and data handling.

The quartz block provides dimensional stability between separate DA's. This ensures that the combined common mode signal used for drag-free control has adequate fidelity and enables precise gravity gradient measurement. It also enables combined use of differential data from different DA's for EP violation analysis. The orientations of the four DA's in the quartz block are transverse to the Sun line and are clocked sequentially by 90°.

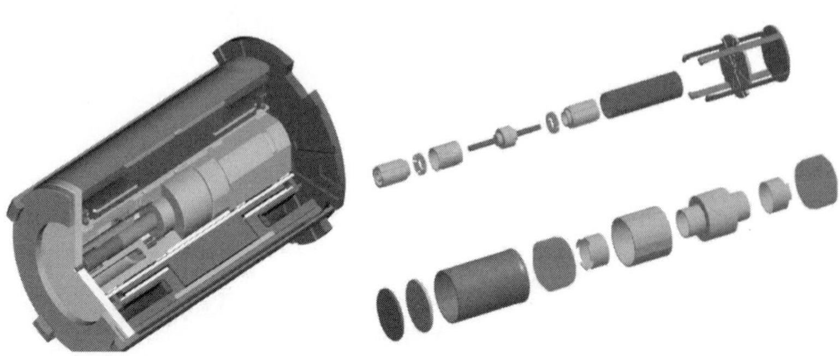

Fig. 2. Differential accelerometer (*left*) and exploded view of differential accelerometer (*right*).

The Dewar supports operation of the superconducting instrumentation and magnetic shielding, reduces the thermal noise of the test masses, and provides the source of boil-off gas for the thrusters. An aerogel material which fills the entire Dewar reservoir mitigates the possibility of disturbances resulting from liquid helium motion. The Dewar also provides a stable, high-capacity thermal sink for the active thermal control system needed to maintain the required SQUID performance stability.

The force of aerodynamic drag on the spacecraft introduces a common mode signal on the test mass pairs. The magnitude of this effect must be reduced to the point where its contribution to the differential signal is negligible. A drag-free control system provides this capability. In its fine mode, proportional thrusters, fed by the Dewar boil-off, will null the common mode signal to below 2×10^{-15} g over the measurement time. This combined with the DA's 10^{-4} common mode rejection reduces the signal below the differential mode noise level.

3. The STEP Mission

A mission to test the equivalence principle in space with cryogenic instrumentation was proposed by Worden and Everitt[9] following interaction with Chapman and Hanson.[10] The technologies necessary for STEP have matured over time with funding from NASA, partially as a result of GP-B development of shared technologies, and the National Science Foundation in the US. STEP has grown into an international collaboration including the institutions listed in Table 2 and has further benefited from the support of ESA, CNES (France), DLR (Germany), ASI (Italy) and PPARC (UK).

The essential data set for STEP is the comparison of accelerations between the test masses. A time sequence of these differential acceleration measurements performed under a specific set of conditions represents an equivalence principle measurement. The satellite will orbit in a near-polar Sun-synchronous orbit while rolling around an axis roughly pointed toward the Sun to minimize thermal variations with roll frequency zero to $\sim 10^{-3}$ Hz. This orientation will minimize the gravity gradient force experienced by the test masses and will result in EP violation signal modulated at the orbit-minus-roll frequency. This strategy permits optimization of sensitivity by operating the experiment at a roll frequency where noise and systematics are at a minimum.

Table 2. STEP international collaboration.

Stanford University, PI Francis Everitt	
Marshall Space Flight Center	University of Birmingham, UK
ESTEC	FCS Universität, Jena, Germany
Imperial College, London, UK	Institute des Hautes Études Scientifiques, UK
ONERA, Paris, France	PTB, Braunschweig, Germany
Rutherford Appleton Laboratory, UK	University of Strathclyde, UK
Universitá di Trento, Italy	ZARM Universität, Bremen, Germany

Satellite operations will consist of three main phases: setup, initial calibration and experiment operation. The setup includes electronic and drag-free control (DFC) system initialization. DFC is initialized in a coarse mode using position sensing based on capacitance measurements. The SQUID's will also be calibrated against the capacitance readout to recheck ground calibration.

The calibration period begins when DFC fine mode is achieved using the SQUID common mode position sensing. During this phase the supercurrents will be adjusted to produce no response in the differential channel to applied longitudinal accelerations. Transverse acceleration is then applied to calibrate any misalignment coupling to the differential signal. Finally, the centers of mass are aligned using the gravity gradient signal as a reference.

The experiment operation consists of a series of tests lasting 100–500 ksec. Tests will vary potential systematic error sources. Temperature, roll rate, electrode bias, and test mass charge represent some of the possible variable parameters for these studies.

4. Development Status

STEP was selected in the 2000 NASA Small Explorer (SMEX) competition to undertake a concept study. Subsequent to completion of the study, reviewers determined that the STEP risk of exceeding SMEX cost and schedule constraints was too high to begin the next phase, which is preliminary flight system design. For this reason, the STEP team has begun a development program to bring STEP to the necessary technical readiness level by demonstrating performance of critical STEP-unique subsystems in a relevant environment.

The completion of the science data collection phase of the GP-B mission has advanced the state of readiness of a number of technologies shared with STEP. The GP-B Dewar, SQUID's, electrostatic gyroscope suspension system, drag free control system, and UV charge control system provide confidence that similar designs for

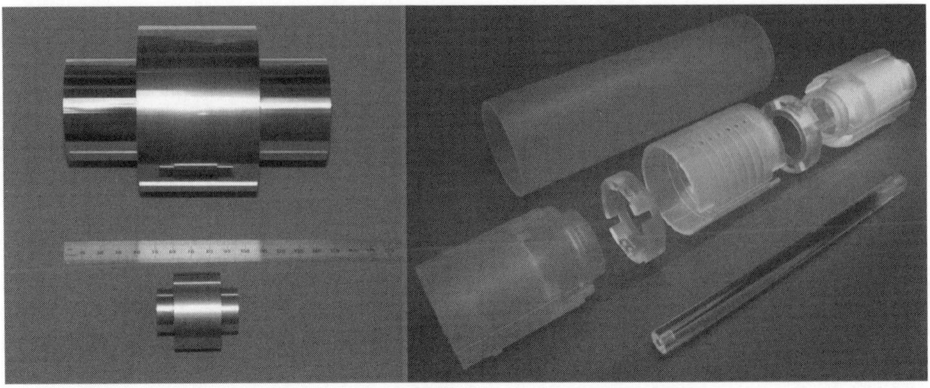

Fig. 3. Outer and inner test masses (*left*) and quartz parts for the inner accelerometer (*right*).

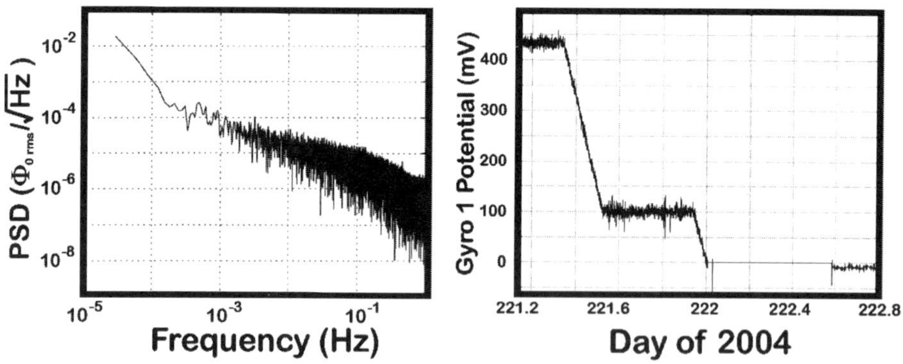

Fig. 4. Two examples of GP-B flight–proven technology. On-orbit GP-B SQUID noise (*left*), and a UV charge control profile which shows two discharges indicated by the ramps (*right*).

STEP will not uncover unexpected surprises. Figure 4 illustrates this point, using SQUID noise and UV charge control as examples. These systems performed at or near required levels for GP-B and the remaining design issues for STEP are well understood.

The focus of current STEP team efforts is to reach the necessary point in development of items requiring long procurement times, to address the significant lessons learned from GP-B and demonstrate performance of STEP-unique subsystems. The most significant long-lead item for STEP is the Dewar, which also requires a suitable application of aerogel technology, as discussed earlier. Aerogel component validation testing was successful,[11] but establishing a source and prototype demonstration tasks remain.

Team members at Stanford and MSFC are defining verifiable requirements early in the technology development to ensure that it is possible to clearly trace the results of a specific test or demonstration to its impact on the flight mission. This is especially important for STEP because the 10^{-18} performance will not be directly measured on the ground. Demonstration of error budget fidelity and specification of relationships to design requirements are necessary before preliminary flight system design can begin.

One of the more significant lessons from GP-B, which could impact STEP, is to fully understand and carefully schedule the time required to initialize and tune the experiment on-orbit. GP-B took roughly twice the allocated time to complete all the tasks needed to begin its science data collection phase. The GP-B initialization process was considerably more complex than STEP because of gyroscope spinup and the risk involved in the loss of suspension while the gyroscopes were spinning. STEP is not subject to these critical operations; however, the GP-B knowledge gained in establishing gains, biases, filter parameters and alignments to the necessary precision will be captured in a ground integrated test facility (ITF). This facility, based on the GP-B ITF, integrates vehicle dynamics, instrument characteristics, and hardware-in-the-loop capabilities. The test bed supports the development of flight algorithms and operations procedures which minimize the risk of delay in the start of equivalence principle experiments.

The most important component of the STEP technology development plan has the goal of demonstrating performance of a prototype differential accelerometer in a 1 g cryogenic vacuum. The plan to achieve this goal includes a number of intermediate stages designed to reduce as much risk as possible at each stage. These steps build on the significant progress that was made prior to and during the SMEX study, which included proof-of-concept tests of a SQUID-based differential accelerometer,[12] progress in caging mechanism development by ZARM, charge control testing by Stanford and Imperial College, and, most significantly, building and testing a prototype accelerometer magnetic bearing.

The steps identified to reach this goal include several stages of brass-board development for the inner accelerometer using Macor parts, manufacture of an engineering model, and a technology demonstration test. A similar outer accelerometer development process with brass-board and EM stages will be followed by a technology demonstration test of the differential accelerometer in a cryogenic vacuum. An accelerometer test facility (ATF), developed to suspend the test mass and manage disturbances, provides the test environment. PTB-manufactured niobium and beryllium test masses will be used in the tests, along with flightlike quartz accelerometer components with thin films applied using coating processes developed at Stanford for flight application. The brass-board stages include incremental testing which improves the likelihood of success by enabling design modifications between stages. So far, the Stanford accelerometer development team has completed a plastic prototype to define optimum cable routing, and an initial brass-board assembly for room-temperature testing of the EPS electrode configuration. Currently, a second brass-board development cycle will culminate in a cyrogenic SQUID test. Quartz parts and PTB test mass for the engineering model of the inner accelerometer are complete and have successfully undergone acceptance metrology. Coating and assembly processes are underway which will lead to a test in the first half of 2007. Similar steps for the outer accelerometer will result in a test one year later. Successful execution of this plan will provide the required confidence to enter the preliminary flight design phase.

5. Achieving Technology Readiness for Transition to a Flight Program

One key to gaining support to begin the development of a mission is to demonstrate not only that the technology can achieve the required performance, but that the technology has reached a maturity where the project management team can have confidence in planning a realistic design, manufacturing and test schedule. The result of this confidence is the ability to obtain a consensus within the funding agency that the risk of exceeding performance, cost and schedule constraints is acceptably small. The STEP technology development plan is designed to demonstrate readiness through the process of test, and internal and external peer review. One MSFC contribution to the development is to help define a process by which the technology readiness level of STEP may be independently assessed and certified.

For this purpose, a robust plan is to assemble an independent certification board prior to each milestone test for a readiness review. This review enumerates the test requirements and demonstrates that these requirements are satisfied. It also defines the success criteria for the test, which must be agreed upon by the board members. Following the successful completion of the test, the board reconvenes for a certification review, where the test results are measured against the success criteria. At this point the board decides whether or not to certify the technical readiness level of the test article.

The certification process is a means of reducing much of the cost and schedule risk, but only partially mitigates the inherent risk in an instrument that requires the actual space environment to fully test its performance. Another approach is to fly a pathfinder mission; however, the cost of this for STEP would be prohibitive given the relatively small size of the actual mission. However the ONERA mission, MicroScope, represents a pathfinder for the STEP EPS system. A successful certification process builds a convincing case that STEP is ready to proceed as a flight program.

6. Conclusions

GP-B has flight-proven many technologies needed for STEP. A technology development effort is surmounting the remaining challenges in the areas of differential accelerometer manufacture and test, and helium tide control. The STEP team is progressing in charge control, error analysis, drag-free control and operational optimization to support the mission needs. The STEP collaboration now includes the Marshall Space Flight Center to further exploit the benefits of the GP-B experience. Assuming adequate funding, the STEP team will be ready to transition to a flight program in 2008 with a projected launch date in 2012.

References

1. The National Academy of Sciences, *Connecting Quarks with the Cosmos: Eleven Science Questions for the New Century* (National Academies Press, 2003), p. 162.
2. E. G. Adelberger, *Class. Quant. Grav.* **18** (2001) 2397.
3. I. I. Shapiro *et al.*, Flight Definition of an Experiment to Test the Equivalence Principle in an Einstein Elevator, in *Proc. 2nd Pan-Pacific Basic Workshop on Microgravity Sciences*, FP-1080 (AIAA, 2001).
4. S. Dimopoulos *et al.*, gr-qc/0610047.
5. P. Touboul *et al.*, *C. R. Acad. Sci. Paris* **2** (série V) (2001) 1271.
6. A. M. Nobili *et al.*, *Phys. Lett. A* **318** (2003) 172.
7. T. Damour, *Class. Quant. Grav.* **13** (1996) 33.
8. N. A. Lockerbie, *Class. Quant. Grav.* **19** (2002) 2063.

9. P. Worden and C. W. F. Everitt, Tests of the Equivalence of Gravitational and Inertial Mass based on Cryogenic Techniques, in *Proc. Int. Sch. Physics* (Enrico Fermi, 1972), p. 382.
10. P. Chapman and A. Hanson, in *Proc. Conf. Experimental Tests of Gravitation*, (Caltech, JPL TM #33-449, 1970), p. 228.
11. S. Wang *et al.*, *Class. Quant. Grav.* **18** (2001) 2551.
12. P. Worden, "Measurement and Control of Disturbing Forces on an Equivalence Principle Experiment," in *Proc. Int. Symp. Experimental Gravitational Physics*, (Guangzhou, China), ed. P. Michelson (World Scientific, 1987).

TESTING THE PRINCIPLE OF EQUIVALENCE IN AN EINSTEIN ELEVATOR

I. I. SHAPIRO*, E. C. LORENZINI†, J. ASHENBERG,
C. BOMBARDELLI‡ and P. N. CHEIMETS

Harvard–Smithsonian Center for Astrophysics,
60 Garden Street, Cambridge, MA 02138, USA
**ishapiro@cfa.harvard.edu*

V. IAFOLLA, D. M. LUCCHESI, S. NOZZOLI
and F. SANTOLI

Istituto di Fisica dello Spazio Interplanetario (IFSI–INAF),
Via Fosso del Cavaliere 100, 00133 Rome, Italy

S. GLASHOW

Boston University, Department of Physics,
590 Commonwealth Avenue, Boston, MA 02215, USA

Improving the level of accuracy in testing the principle of equivalence (PE) requires reliably extracting a very small signal from an instrument's intrinsic noise and the noise associated with the instrument's motion. In fact, the spin velocity required to modulate a PE-violating signal produces a relatively high level of motion-related noise and modulation of gravity gradients at various frequencies. In the test of the PE in an Einstein elevator under development by our team, the differential acceleration detector free-falls while spinning around a horizontal axis inside an evacuated, comoving capsule released from a stratospheric balloon. The accuracy goal of the experiment is to test the PE at an accuracy of a few parts in 10^{15}, a limit set by the expected white-noise sources in our detector. The extraction of a very small signal from the prevailing noise sources is necessary for the experiment to succeed. In this paper, we discuss different detector configurations and describe a particular design that is able to provide a remarkable attenuation and frequency separation of the effects of motion and gravity gradients with respect to a PE-violating signal. Numerical simulations of the detector's dynamics in

†Presently Professor at the University of Padova, Department of Mechanical Engineering, Padova, Italy.
‡Presently Researcher at the European Space Agency, ESTEC, Noordwijk, The Netherlands.

the presence of relevant perturbations, realistic errors, and construction imperfections show the merits of this configuration for the differential acceleration detector.

Keywords: Mechanical instruments; experimental tests of general relativity; principle of equivalence.

1. Introduction

Although its name is modern, the principle of equivalence (PE) and related aspects of inertia have engaged scientists for well over two millennia. Aristotle is the first famous natural philosopher known to have grappled with these issues. His qualitative views can be summarized in mathematical language[1]: a body in motion obeys the law $v = P/M$, where v is the body's velocity, P the "motive power," and M the effect of the medium through which the body moves. Aristotle's view apparently held sway for the better part of a millennium, despite the obvious problem it creates: the velocity becomes infinite as the medium, and its effects, disappear. A solution to this problem,[2] again in more modern language, was argued in the fifth century by Ionnes Grammaticus ("John the Grammarian," a.k.a. John Philoponus), reputedly an Alexandrian of Greek origin[2]: $v = P - M$. There is no evidence of which we are aware that Grammaticus did experiments, such as simultaneously dropping from the same height objects of different materials (as opposed to different masses, which he did consider) and observing which, if either, reached the ground first. Throughout the Dark and Middle Ages various thinkers, such as the scholars Avempace and Al-Bitrogi, tackled these problems with no notable advances flowing from their thoughts.

Galileo, the father of modern science, is widely credited with putting the PE on a strong experimental basis through the simultaneous dropping of disparate stones from the top of the leaning tower of Pisa. This experiment has been discredited,[1] apparently by his own writing while he was in Pisa. Only later in Padova did Galileo adopt a more modern version of the PE. Newton, about a century later, first placed the test of the PE on a quantitative basis, confirming it at a level of about one part in 10^3. The last century saw dramatic gains in accuracy,[3,4] up to the most recent laboratory torsion experiments[5] and the lunar laser ranging (LLR) measurements,[6] which verified the PE to about a few parts in 10^{13}. LLR provided the first measurements sensitive to the gravitational binding energy contribution to a body's mass.

Now, funding allowed, we are poised to make further significant gains with balloon and possible spacecraft experiments. Why might one expect such experiments to be important? With the discovery of the present acceleration in the Universe's expansion,[7] a revolution in the most fundamental laws of physics is in the wind. No one knows how deep or far-reaching it will be. But it is manifest that this acceleration poses profound problems. The most straightforward way of "explaining" this acceleration, based on presently accepted particle-physics concepts, leads to an overestimate of this acceleration by a factor of about 10^{120}, a big number even by astronomical standards. Testing the PE to substantially higher accuracy will, at least, pose more useful constraints on a new theory that can accommodate this facet of the expanding Universe. Detection of a violation would be earthshaking.

In the remainder of this paper, we discuss briefly a balloon test of the PE whose accuracy is expected to be at the few-parts-in-10^{15} level — a hundredfold higher accuracy than currently achieved in ground and space experiments. Our main emphasis is on a discussion of the choice of detector configuration that minimizes the susceptibility of the experiment to systematic errors. We do not here discuss the details of the detector design.

2. Brief Description of the Experiment

Our strategy is to test the PE in an Einstein elevator in which a differential accelerometer (i.e., the detector) free-falls inside a co-falling, evacuated capsule released from a high-altitude balloon[8] (see Fig. 1). The accuracy goal of the experiment is a few parts in 10^{15}, a limit set by the expected white-noise sources of our detector.

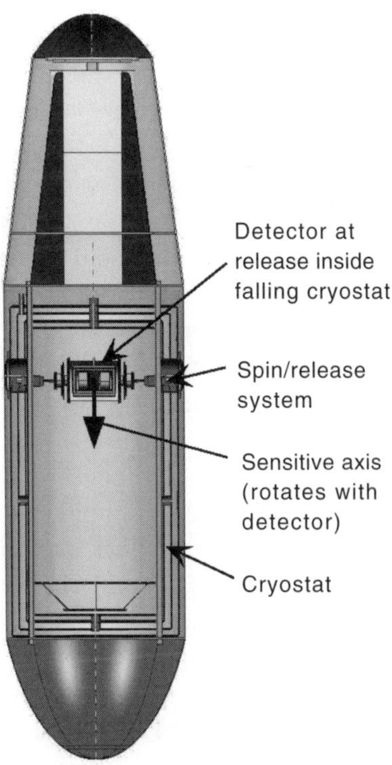

Fig. 1. Schematic of a vertical free-fall experiment with a detector shown at release inside the capsule.

In this vertical free-fall scenario, the detector is spun about a horizontal axis to modulate the PE signal and released inside a co-moving capsule shortly after the capsule is released from a high-altitude balloon. The detector's release mechanism is patterned after a roasting spit. The detector is attached along its spin axis to rotating release plates at each end. The contact is provided by three hemispherical protrusions at each end of the axis which are placed on the vertices of an equilateral triangle and fit into matching hemispherical holes on the ends of the detector's spin axis. At the desired spin velocity and release conditions, the plates holding the protrusions are quickly retracted and the spinning detector freely falls within the co-moving capsule.

The free-fall time is dependent upon the capsule's release altitude, its ballistic coefficient, and its length. We can assume conservatively a free-fall time of 25 s within a capsule of modest length (i.e., a 2-m-long internal experimental space)[9] released from an altitude of 40 km. The release of the capsule from the balloon implies a change of acceleration from 1 g to 0 g that will force relatively large oscillations of the detector's proof masses that need to be damped out before the measurement phase takes place. The damping is accomplished by drastically reducing (by a factor of about 10^5), through resistive feedback, the quality factor of the detector.[10] A damping phase, with a conservative maximum duration of 5 s, is therefore included in our simulation model.

At release, there are errors in the orientation of the spin axis with respect to the local gravity vector and there are angular velocity errors. Orientation errors with respect to the horizontal plane (defined as perpendicular to the local gravity vector) are important in both ground and orbital experiments because they ultimately produce gravity-gradient components at the spin frequency that are not distinguishable from a PE-violating signal. Angular-velocity errors perpendicular to the symmetry axis (ideally coincident with the spin axis) determine the amplitude of the precession cone that the spin axis describes about the angular-momentum vector. Obviously, in an orbital experiment, the amplitude of the precession cone is not determined by release conditions but rather by the accuracy with which the spacecraft is set to spin and later controlled.

3. Detector Configurations

Testing the PE, at an accuracy substantially higher than the present state of the art, requires resolving a very small signal out of not only the (white) intrinsic noise of the detector and its preamplifier but also the colored noise associated with the detector motion and gravity gradients. The white-noise sources are considered in most papers dealing with gravitational tests[11–13] but only a few papers have touched upon the effects of motion-related noise.[14,15] In PE tests, in either orbital or vertical free fall, the need for signal modulation requires spinning the platform hosting the instrument at a frequency that is as distinct as possible from the frequencies of disturbances. However, the rotational motion itself produces a comparatively high level of motion-related noise that needs to be considered, and further, gravity gradients produce noise components at once and twice the spin frequency.

We have followed the approach of first developing a good-fidelity dynamics model of the motion of the detector proof masses and the platform hosting them. Then, we have simulated the response of different detector configurations under realistic conditions of errors and imperfections with the goal of extracting a known PE-violating signal from the noisy output. We found that extracting the PE-violating signal from the motion-related noise is very strongly dependent on the configuration of the acceleration detector, its location inside the hosting platform, and (obviously) the magnitude of construction and centering errors. Another key issue is the orientation of the spin axis with respect to the gravity vector. If the spin axis is not orthogonal to the gravity vector, then the off-diagonal terms of the gravity-gradient tensor produce harmonics in the detector's output at the spin frequency. As shown below, detectors with different designs have very different sensitivities to gravity-gradient force components and the requirements on these can be strongly relaxed if an appropriate configuration is selected. The goal of our study was to develop a detector design with low sensitivity to the effects of the damaging gravity-gradient forces and torques, and of the motion of the platform hosting the detector.

An accelerometer measures the displacement between two bodies (i.e., the case and the proof mass) to obtain the acceleration of the case. In a differential accelerometer, there are three bodies: two proof masses and the case. Since the goal is to measure the differential acceleration of the proof masses, this information can be obtained by measuring the displacement of each proof mass with respect to the case and differencing the two measurements.

We can classify differential accelerometers by the type of motion of the proof masses. Accelerometers can be designed to exploit a purely linear motion, a combination of linear and rotational motion or, in the design that we devised, a purely rotational motion of the proof masses. Differential accelerometers under development[16] and prototype instruments built[17] thus far have used only the first two options. Also, differential accelerometers have been designed thus far with one mass purely of one material and the other purely of a different material.

Figure 2 shows a schematic of each of the three types. The purely rotational configuration, in turn, can be designed with pivot axes orthogonal to the spin axis [see Fig. 2(c)] or, as shown in the following, with pivot axes parallel to the spin axis.

In this paper we show in some detail the responses of different detector configurations by focusing in particular on the differences between the translational and purely rotational configurations. We first introduce briefly the mathematical models of those configurations and then analyze their responses in free fall. These classes of detectors are not exhaustive of all the possible geometrical options, but they do capture the characteristics of large families of differential-accelerometer designs.

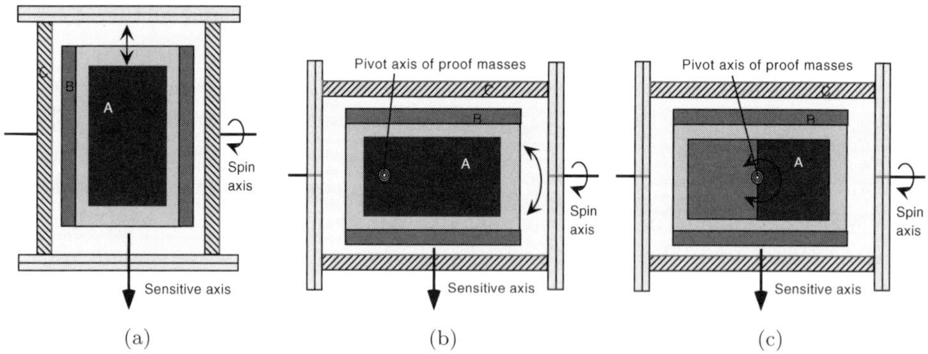

Fig. 2. Detector configurations grouped by the type of relative motion of the proof masses: (a) purely translational, (b) rotational–translational, and (c) purely rotational with the pivot axis perpendicular to the spin-axis. The bodies are: A = first proof mass; B = second proof mass; and C = case. The sensitive axis indicates the direction of a PE-violating acceleration that will cause a relative motion (indicated by the double arrow) of the proof masses. The pivot axis is the axis about which the proof mass rotates; the pivot axis is perpendicular to the page in panels (b) and (c).

4. Translational Detectors

4.1. *Mathematical model*

In this detector configuration the two proof masses can slide with respect to each other while the instrument package (housing the detector) rotates about an axis orthogonal to the sensitive axis [see Fig. 2(a)].

As we shall show, the choice of the ratio between the precession frequency and the spin rate is the key to being able to extract a PE-violating signal from the dynamics-related noise. A precession frequency that is too low (such as for an instrument package that is too close to spherical) produces an incomplete sinusoid in the precessional motion, over the 20 s integration time for the experiment, leading to a large "leakage" in the Fourier content of the data. At the opposite end, a precession frequency that is too high produces strong accelerations of the proof masses (if they are not centered at the center of mass of the package or spacecraft) that can easily saturate a supersensitive accelerometer.

We call $\boldsymbol{\omega_A}$, $\boldsymbol{\omega_B}$, $\boldsymbol{\omega_C}$ the angular velocities of the three bodies A, B (the proof masses), and C (the case), respectively, and $\boldsymbol{\delta r_A}$, $\boldsymbol{\delta r_B}$, $\boldsymbol{\delta r_C}$ the error-position vectors defining the center of mass of each body with respect to the common center of mass (CM) of the three bodies, where $\boldsymbol{\delta r_i} = (\delta_{ix}, \delta_{iy}, \delta_{iz})$, m_i is the mass of body i, $[I_i]$ is the moment-of-inertia tensor of body i, and $i = A, B, C$. For purely translational motion of the proof masses, the three angular-velocity vectors are coincident and we call the common vector $\boldsymbol{\omega}$ for simplicity. The angular-momentum vector of this system is[19]

$$\begin{aligned} H = {} & [I_A]\omega + [I_B]\omega + [I_C]\omega + m_A\delta r_A \times (\delta\dot{r}_A + \omega \times \delta r_A) \\ & + m_B\delta r_B \times (\delta\dot{r}_B + \omega \times \delta r_B) + m_C\delta r_C \times (\delta\dot{r}_C + \omega \times \delta r_C). \end{aligned} \tag{1}$$

The magnitudes of the error-position vectors δ are very small *for proof masses located very close to the instrument package's CM* because they consist solely of construction and centering errors. The error contributions to the angular momentum are of order δ^2 and, consequently, the terms involving the cross products in Eq. (1) are negligible with respect to the other terms. The time derivative of the angular momentum taken in the detector's body principal-axes frame yields the Euler equations of rigid-body motion:

$$\begin{aligned} I_x^{\rm tot}\dot{\omega}_x + (I_z^{\rm tot} - I_y^{\rm tot})\omega_y\omega_z &= Q_x, \\ I_y^{\rm tot}\dot{\omega}_y + (I_x^{\rm tot} - I_z^{\rm tot})\omega_x\omega_z &= Q_y, \\ I_z^{\rm tot}\dot{\omega}_z + (I_y^{\rm tot} - I_x^{\rm tot})\omega_x\omega_y &= Q_z. \end{aligned} \tag{2}$$

The x, y and z axes are the body C principal axes, with x the sensitive axis (i.e., along the symmetry axis of the detector), and y and z the transverse axes, with z the spin axis. $I_x^{\rm tot} = I_{Ax}+I_{Bx}+I_{Cx}$, $I_y^{\rm tot} = I_{Ay}+I_{By}+I_{Cy}$ and $I_z^{\rm tot} = I_{Az}+I_{Bz}+I_{Cz}$ are the principal (axial) moments of inertia of the instrument package and $\mathbf{Q}$ represents the external, perturbing torque.

The equations of the relative motion (along the sensitive body axis, x) of the proof masses with respect to the accelerometer case (body C) are as follows[19]:

$$\begin{aligned} &\ddot{x}_A - (\omega_y^2 + \omega_z^2 - g_{xx})x_A + \frac{k_A}{m_{AC}}x_A + \frac{k_B}{m_C}x_B + \frac{\xi_A}{m_{AC}}\dot{x}_A + \frac{\xi_B}{m_C}\dot{x}_B \\ &\quad + [(\dot{\omega}_y - g_{xz})\delta_{Az} - (\dot{\omega}_z + g_{xy})\delta_{Ay} + (\omega_z\delta_{Az} + \omega_y\delta_{Ay})\omega_x - (\omega_z^2 + \omega_y^2 - g_{xx})\delta_{Ax}] \\ &\quad = a_x^{\rm PEV}, \end{aligned} \tag{3}$$

$$\begin{aligned} &\ddot{x}_B - (\omega_y^2 + \omega_z^2 - g_{xx})x_B + \frac{k_B}{m_{BC}}x_B + \frac{k_A}{m_C}x_A + \frac{\xi_B}{m_{BC}}\dot{x}_B + \frac{\xi_A}{m_C}\dot{x}_A \\ &\quad + [(\dot{\omega}_y - g_{xz})\delta_{Bz} - (\dot{\omega}_z + g_{xy})\delta_{By} + (\omega_z\delta_{Bz} + \omega_y\delta_{By})\omega_x - (\omega_z^2 + \omega_y^2 - g_{xx})\delta_{Bx}] \\ &\quad = 0, \end{aligned} \tag{4}$$

where $m_{AC} = \frac{m_A m_C}{m_A + m_C}$, $m_{BC} = \frac{m_B m_C}{m_B + m_C}$ are reduced masses; k_A, k_B are the respective elastic stiffness constants of each proof mass suspension; ξ_A, ξ_B are the damping coefficients; δ_{Aj}, δ_{Bj} $(j = x, y, z)$ are the components of the distance errors from the CM of each proof mass to the common CM; and $a_x^{\rm PEV}$ is the PE-violating acceleration along the sensitive body axis x. Equations (3) and (4) can be further simplified if $m_A, m_B \ll m_C$, as is often the case.

The g's are the components of the gravity-gradient tensor expressed in body axes. Their expressions[19] can be derived by transforming the gravity-gradient tensor from the Earth-fixed reference frame $[X, Y, Z]$ into the body axes $[x, y, z]$ by one of the Euler rotation sequences. We adopted the 2–1–3 Euler's rotation sequence to avoid any discontinuity about a horizontal spin axis.

Equations (3) and (4) show clearly the similar roles played by the gravity-gradient-tensor components and the rotational dynamics of the detector. They both influence the accelerometer's output if the CM distance errors of the proof masses are different from zero. Centering errors between the proof mass CM's (e.g., $\delta_{Ax} - \delta_{Bx}$) will produce differential accelerations while identical displacement errors (e.g. $\delta_{Ax} = \delta_{Bx} \neq 0$) will produce common-mode accelerations. The latter accelerations are less worrisome than the former, but they must be kept sufficiently small to avoid saturating the detector's response. If we consider now a simplified model of an axisymmetric detector, modeled as a rigid body, the solution of the simplified Euler equations describing its (disturbance-free) dynamics is

$$\begin{aligned} \omega_x &= \omega_t \sin(\Omega_p t), \\ \omega_y &= \omega_t \cos(\Omega_p t), \\ \omega_z &= \omega_s, \end{aligned} \tag{5}$$

where ω_s is the spin angular rate (rad/s), Ω_p the body precession rate, and ω_t the transverse angular velocity perpendicular to the spin-axis of the detector. The body precession rate is determined by the inertia distribution of the instrument package. For a body with axisymmetric mass distribution, $\Omega_p = (1 - I_a/I_t)\omega_s$, where I_a and I_t are the axial and transverse moments of inertia of the instrument package.

For practical purposes, ω_t may be considered the transverse-angular-velocity error of the detector's release mechanism in vertical free-fall experiments. For experiments using a spinning spacecraft, ω_t is determined by both the accuracy with which the orientation of the spin axis is controlled and the angular misalignment between the principal body axis of the spacecraft (close to the spin axis) and the physical body axis along which the accelerometers are placed.

Substitution of Eq. (5) into the relative motion of the proof masses [Eqs. (3) and (4)], after neglecting gravity gradients, elasticity, and damping terms, yields the acceleration a_x of a proof mass along the sensitive axis (x) associated with the rigid-body dynamics of the instrument package or spacecraft:

$$\begin{aligned} a_x = &\left[\omega_s^2 + \frac{1}{2}\omega_t^2(1 + \cos(2\Omega_p t))\right]\delta_x \\ &+ \frac{1}{2}\omega_t^2 \sin(2\Omega_p t)\delta_y - \omega_t(\omega_s - \Omega_p)\cos(\Omega_p t)\delta_z, \end{aligned} \tag{6}$$

where $\delta_x, \delta_y, \delta_z$ are the components of the distance vector from this proof mass CM to the CM of the instrument package or spacecraft. Equation (6) is quite useful because it shows that the rigid-body motion of the package (or spacecraft) contributes two harmonics to the accelerometer output: once and twice the body precession rate, Ω_p. Typically $\omega_s \gg \omega_t$ and, consequently, the first harmonic (at the precession frequency) is much larger than the second harmonic. Moreover, the amplitude of the first harmonic depends on the angular rate difference $(\omega_s - \Omega_p)$, which becomes relatively large for retrograde precession, in which the precession frequency is negative (we recall that retrograde precession applies to bodies that spin about a maximum principal axis of inertia and prograde precession to minimum-inertia-axis spinners).

The amplitude of the rigid-body-motion harmonics may be large with respect to the expected limit on (or detection of) a PE-violating signal (especially if the accelerometer is located far from the instrument package's CM) and, consequently, the frequency placement of the two harmonics associated with the rigid-body dynamics is the key to the ability to extract a PE-violating signal from the prevailing dynamics-related noise.

4.2. *Numerical results for translational detectors*

We now focus our attention on the dynamics response of a translational detector for a vertical free-fall experiment. Based on a large number of simulations, we adopted a spin frequency, $f_s = 0.5\,\text{Hz}$, that strikes a good balance between a sufficient number of signal cycles (i.e., 10 for an integration time of 20 s) in the output data and relatively low spin-related accelerations. Figure 3 shows results for a spin frequency, $f_s = 0.5\,\text{Hz}$, and a value of the precession frequency of 0.1 Hz. The resonant (elastic) frequency of the detector proof masses of 3 Hz clearly appears in the numerical results. The resonant frequency in a vertical free-fall experiment (which has a very limited duration) must be of the order of a few Hz to allow quick damping of the oscillations after release. This damping phase is only a few seconds long in Fig. 3.

We assumed a spin-axis orientation error at release of 0.1 deg (which would require a rather complex stabilization of the capsule before release) in this simulation to limit the gravity-gradient error contribution at the signal frequency, a transverse angular-velocity error at release of 0.1 deg/s (which is reasonably easy to achieve), centering errors of 1 micron between the CM's of the proof masses, and 10 micron errors between each proof mass CM and the CM of the instrument package.[20] Each of these errors was assumed to be present in each Cartesian component.

Figure 3 shows clearly the first precession-related harmonic at 0.1 Hz (the second harmonic at 0.2 Hz would appear more prominently for larger velocity errors at release) and the harmonic at 1 Hz (i.e., at $2f_s$) that is associated with a diagonal component of the gravity-gradient tensor.

Selection of the precession frequency is very important. Ideally, the precessional motion should produce a few complete cycles over the duration of the experiment and be at a frequency lower than the signal frequency, which implies a package with a minimum axis of inertia along the spin axis. Such minor-axis spinners (with prograde precessions) are preferable, but attention must be paid to the placement of the two harmonic peaks associated with the precession dynamics. Specifically, the precession frequency should be smaller than the signal frequency in such a way that the frequency of the second harmonic produced by precession does not "overlap" with f_s, as in Fig. 3.

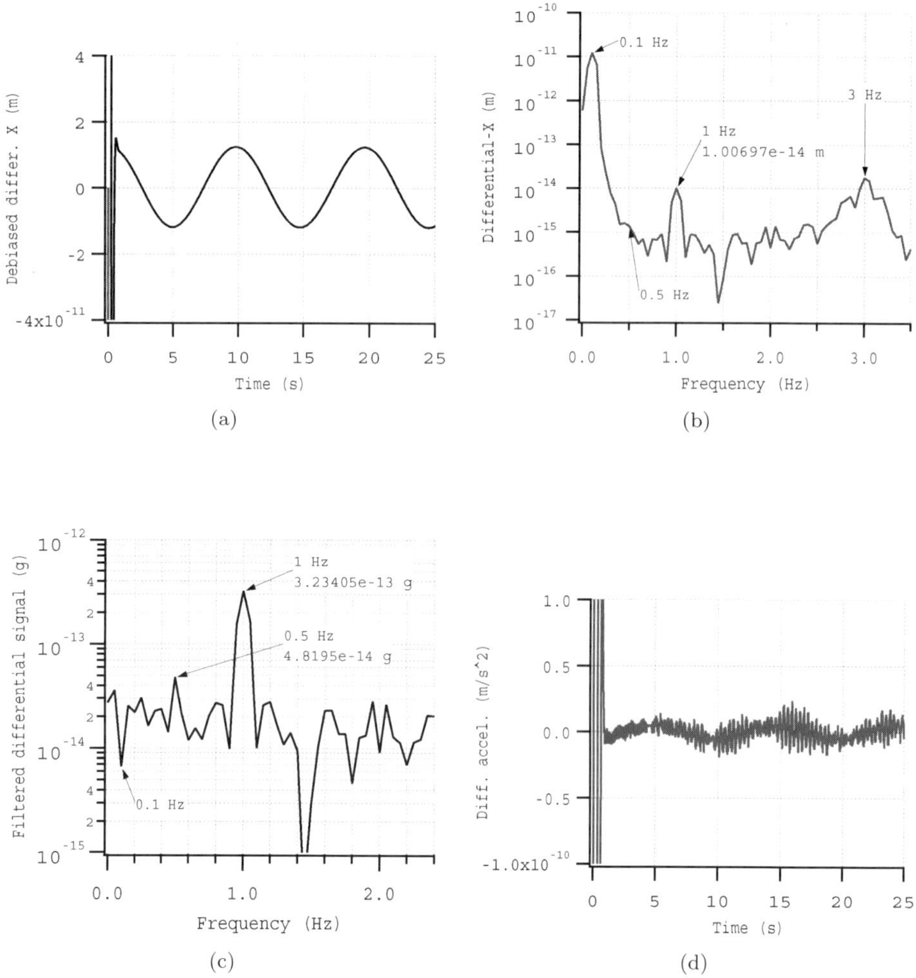

Fig. 3. Simulated dynamic response of a translational detector for a free-fall time = 25 s, spin frequency = 0.5 Hz, resonant (elastic) frequency = 3 Hz, and falling package precession frequency = 0.1 Hz (i.e., prograde precession). The detector's construction and release errors (see text) are: $\delta_{Ax} = \delta_{Ay} = \delta_{Az} = 10\ \mu$m; $\delta_{Bx} = \delta_{By} = \delta_{Bz} = 11\ \mu$m; $\omega_{x0} = \omega_{y0} = 0.1$ deg/s; spin axis angle with respect to horizontal plane = 0.1 deg; moment-of-inertia asymmetry of package $(I_{Cx} - I_{Cy})/I_{Cy} = 0.01$; amplitude of initial oscillations of proof masses = 50 μm. The panels are: (a) differential displacement along the sensitive axis x between the proof masses with constant bias removed; (b) spectrum of the differential displacement taken over the time window 5–25 s; (c) spectrum of differential acceleration along x, with the two precession-related harmonics filtered out, showing the recovered signal harmonic at 0.5 Hz (see text); (d) differential acceleration between the proof masses; (e) plot of ω_y vs ω_x that shows the amplitude of the precession of the angular velocity vector; and (f) fluctuations of the spin velocity about the average value of 3.14 rad/s (i.e., 0.5 Hz). Note that multiplication factors on the axes of panels (a), (d), (e), and (f) apply to all the values on the relevant axis.

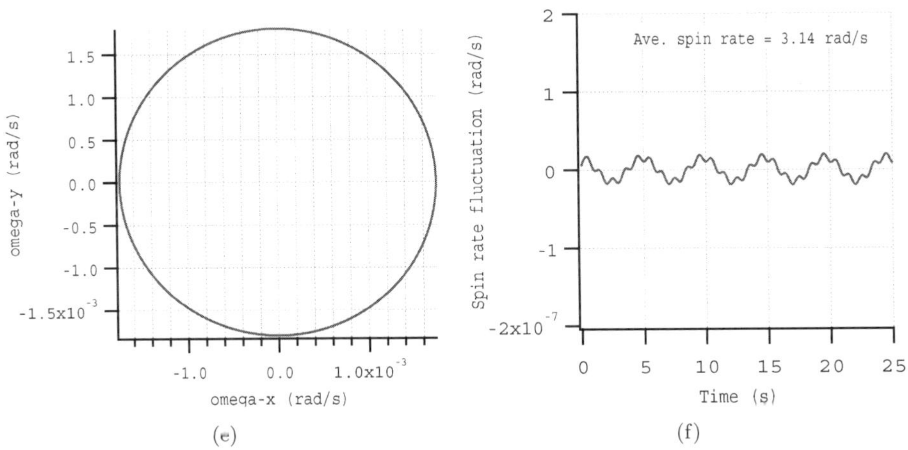

Fig. 3. (*Continued*)

Although the instrument package is a minor-axis spinner, spin-axis conversion is not an issue here because the time scale of the damping phase is only a few seconds long and the energy dissipation is very small.

Retrograde precessions in which the precession frequency is higher than the spin frequency can lead to saturation of the accelerometer (for sensitivity values of interest) because of the amplification of the first precession-related harmonic of Eq. (6). Moreover, numerical results indicate that the dynamics noise level near the signal frequency is too high for fast (and retrograde) precessions to allow the extraction of a small PE-violating signal.

In conclusion, the results of our analysis for a translational configuration in vertical free fall with 20 s integration time, reasonably realistic errors at release, and detector imperfections, are that a PE-violating signal no smaller than several parts in 10^{14} could be extracted (see Fig. 3) from the motion-related noise even under the best conditions with a ratio of precession to spin frequency, $f_p/f_s \sim 0.2$. A signal of 5×10^{-14} g could be extracted with an amplitude error of about 5%, after filtering the two harmonic peaks associated with the precession dynamics of the instrument package. These peaks, although separated from the signal frequency, are sufficiently strong to affect it.

Results for rotational–translational detectors are similar to those of translational detectors. Since our goal is to detect violation signals of a few parts in 10^{15}, we examine below a different configuration: the purely rotational differential accelerometer.

5. Purely Rotational Detectors

An accelerometer with a purely rotational configuration will respond only to torques and, consequently, a sensing mass made of one homogeneous material would not be able to sense linear accelerations (except for errors between the pivot axis and the CM) but only rotational accelerations. To sense a PE violation, we make a proof mass of two different materials (with the two materials on opposite sides of the pivot axis) so that a PE violation will generate a torque about the pivot axis. Such a proof mass will be highly insensitive to any linear acceleration, inclusive of those associated with gravity-gradient forces, as explained below, but will react to torques.

The sensitivity to common-mode rotational motion is removed by making the second proof mass (see Fig. 4) of one material and having the resonant frequency (of its rotational motion) ideally equal to that of the first proof mass, which is made from a combination of the two materials being tested. The cross sections of the proof masses are annuli. Differences in the density of the two materials can be accommodated by hollowing a portion of the denser material (a detail not shown in Fig. 4 for simplicity). Thus, the first proof mass will be sensitive to PE violations while the second mass acts as the dynamics reference that removes the effects of common-mode rotational motion.

This design is insensitive to gravity-gradient forces because these forces produce a torque, about the pivot axis, proportional to the product $\delta_i \delta_j$, where δ_i and δ_j are the small distance errors between each proof mass CM and the pivot axis. The torques resulting from the gravity-gradient forces will therefore be negligible when compared to our target sensitivity for detecting any PE violation. For similar reasons, the angular accelerations associated with the precession frequency are also substantially smaller than for the previously described configurations.

The other dominant perturbation to be considered is the gravity-gradient torque about the pivot axis due to construction inequalities of the moments of inertia of

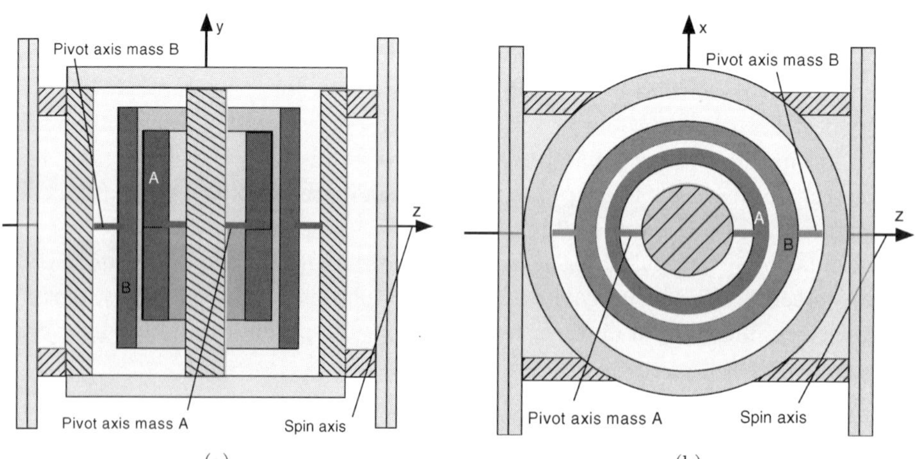

Fig. 4. Schematic of a purely rotational detector configuration with pivot axes of proof masses parallel to the detector's spin axis: (a) longitudinal cross section; and (b) transverse cross section [through a plane perpendicular to the corresponding plane for (a)].

each proof mass. This torque about the pivot axis, which is associated with the (error) difference in moments of inertia of a proof mass, is modulated by the spin at a frequency that depends on the orientation of the pivot axis with respect to the spin axis: for a pivot axis perpendicular to the spin axis, as in Fig. 2(c), the modulation frequency is equal to the spin frequency; for a pivot axis parallel to the spin axis, as shown in Fig. 4, the modulation frequency is twice the signal frequency. The latter option is far preferable to the former, because a dominant perturbation will then be well separated in frequency from the PE-violating signal.

5.1. *Mathematical model*

In the following, we show the equations for a geometrical arrangement in which the pivot axes of the proof masses are parallel to the spin axis, z (see Fig. 4). In the principal body frame, z is the spin axis, y is along the longitudinal axes of the proof masses at rest, and x is the direction along which the relative motion of the proof masses is sensed. The Euler equations that describe the attitude of the whole detector are as follows[19]:

$$\begin{aligned}
I_x^{\text{tot}}\dot{\omega}_x + (I_z^{\text{tot}} - I_y^{\text{tot}})\omega_y\omega_z + (I_{Bz}\dot{\theta}_B + I_{Az}\dot{\theta}_A)\omega_y &= Q_x,\\
I_y^{\text{tot}}\dot{\omega}_y + (I_x^{\text{tot}} - I_z^{\text{tot}})\omega_x\omega_z - (I_{Bz}\dot{\theta}_B + I_{Az}\dot{\theta}_A)\omega_x &= Q_y,\\
I_z^{\text{tot}}\dot{\omega}_z + (I_y^{\text{tot}} - I_x^{\text{tot}})\omega_x\omega_y + (I_{Bz}\ddot{\theta}_B + I_{Az}\ddot{\theta}_A) &= Q_z,
\end{aligned} \tag{7}$$

where $\dot{\theta}_A$ and $\dot{\theta}_B$ are the angular velocities of proof masses A and B, respectively, with respect to the accelerometer case (body C), $\ddot{\theta}_A$ and $\ddot{\theta}_B$ are the corresponding angular accelerations, and the other quantities are defined as for Eq. (2). The equation for the rotation phase, angle θ_A, of proof mass A with respect to the differential accelerometer case is[19]

$$\begin{aligned}
&\ddot{\theta}_A + E_1\dot{\theta}_A + E_2\theta_A + \dot{\omega}_z + E_3(\dot{\omega}_x - \omega_y\omega_z + g_{yz}) + E_4(\dot{\omega}_y + \omega_x\omega_z + g_{xz})\\
&\quad + E_5(\omega_x^2 - \omega_y^2 - g_{xx} + g_{yy}) + E_6(\omega_x\omega_y - g_{xy}) = E_7(\tau_{Az}^{\text{PEV}} + \tau_{Az}^{gg}),
\end{aligned} \tag{8}$$

where

$$E_1 = \frac{\xi_A}{\tilde{I}_{Az}}, \quad E_3 = -\frac{m_A}{\tilde{I}_{Az}}\delta_{Ax}\delta_{Az}, \quad E_4 = -\frac{m_A}{\tilde{I}_{Az}}\delta_{Ay}\delta_{Az},$$

$$E_2 = \frac{m_A}{\tilde{I}_{Az}}\left(\frac{K_{\theta_{Az}}}{m_A} - \delta_{Axy}^2\omega_z^2 - \delta_{Ax}^2\omega_x^2 - 2\delta_{Ax}\delta_{Ay}\omega_x\omega_y - \delta_{Ay}^2\omega_y^2\right),$$

$$E_5 = -\frac{m_A}{\tilde{I}_{Az}}\delta_{Ax}\delta_{Ay}, \quad E_6 = -\frac{m_A}{\tilde{I}_{Az}}(\delta_{Ay}^2 - \delta_{Ax}^2), \quad E_7 = \frac{1}{\tilde{I}_{Az}},$$

$$\tilde{I}_{Az} = I_{Az} + m_A\delta_{Axy}^2, \quad \delta_{Axy} = \sqrt{\delta_{Ax}^2 + \delta_{Ay}^2},$$

and where $K_{\theta Az}$ is the elastic stiffness constant of the torsional suspension of proof mass A (about the pivot axis, z), τ_{Az}^{PEV} is the PE-violating torque, and τ_{Az}^{gg} is the Earth's gravity-gradient torque about z.

The equation for the rotational motion of proof mass B with respect to the accelerometer case is obtained by replacing the subscript A with B in Eq. (8) and its coefficients, and setting $\tau_{Bz}^{\mathrm{PEV}} = 0$. In deriving Eq. (8) we have assumed that $I_A, I_B \ll I_C$, which is indeed valid for our detector design. For clarity in plotting the results (see below), rotations and rotational accelerations of the proof masses about z have been converted to linear displacements and accelerations along x by multiplying the relevant angular variables by $l/4$, where l is the length (typically about 10 cm) along the symmetry axis of the inner proof mass. The location of the displacement pickup plates of the detector determines this factor.

Perturbing torques are important in a purely rotational configuration that responds to torques. Major perturbing torques are those associated with the Earth's gravity gradients acting on the accelerometer case and, more importantly, on the proof masses themselves. Higher-order torques can be made negligibly small by shaping the proof masses[21] and sizing the drop capsule appropriately.

The components of the Earth's gravity-gradient torque (associated with the difference between principal moments of inertia of a proof mass) are

$$\begin{aligned}
\tau_x^{gg} &= \frac{3}{2}\frac{GM}{R^3}(I_y - I_z)\sin(2\phi)\cos\psi, \\
\tau_y^{gg} &= \frac{3}{2}\frac{GM}{R^3}(I_z - I_x)\sin(2\phi)\sin\psi, \\
\tau_z^{gg} &= \frac{3}{2}\frac{GM}{R^3}(I_y - I_x)\cos^2(\phi)\cos(2\psi),
\end{aligned} \tag{9}$$

where GM is the Earth's gravitational constant, R the distance from the Earth's center, ϕ the (elevation) angle of the spin axis with respect to the horizontal plane, and ψ the spin angle (where $\psi = \omega_s t$, with t the time).

The attitude dynamics of the whole instrument package is influenced by the three components of the torque, whereas proof masses A and B react only to the component about the pivot axis. The strength of this gravity-gradient-torque component (acting on each proof mass) depends on construction imperfections because each proof mass is designed to have an ellipsoid of inertia as spherical as possible. More importantly, the torque component about the spin axis (z) is modulated at $2f_s$ [see Eq. (9)] and therefore does not overlap with any PE-violating signal. For this reason the pivot axis of the proof masses in the configuration that we selected is parallel to the spin axis. In this configuration, the frequency separation insures that relatively high construction errors of the proof masses can be tolerated (see below).

5.2. *Numerical results for purely rotational detectors*

Figure 5 shows numerical results for the preferred configuration with the pivot axis parallel to the spin axis. A signal of $1 \times 10^{-15}\,g$ was recovered, with its amplitude almost unaltered, by simple frequency analysis without filtering the output data of the harmonics associated with the precession dynamics.

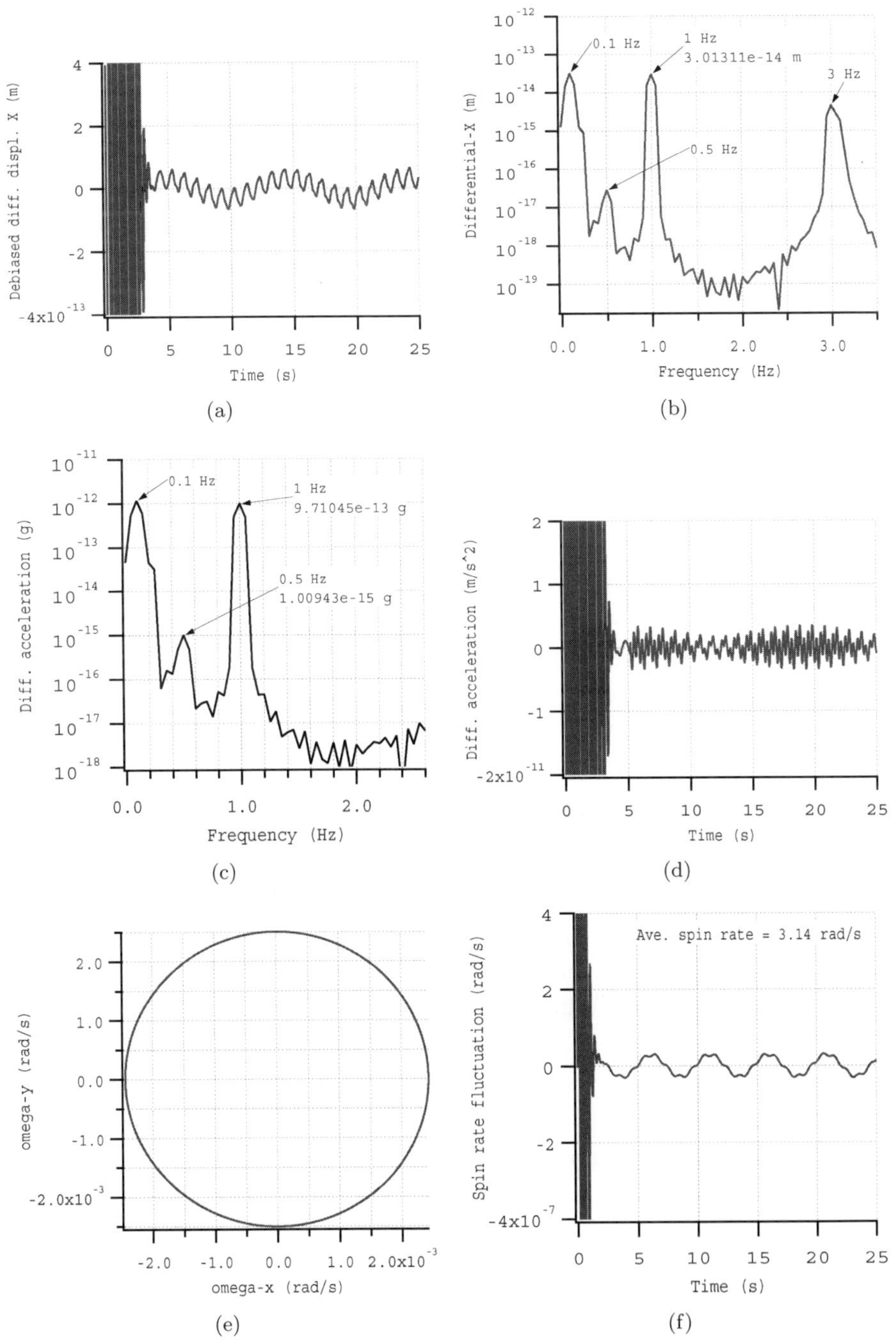

Fig. 5. Simulated dynamic response of a purely rotational detector. Same conditions as in Fig. 3 except for different error values, as follows: $\delta_{Ax} = \delta_{Ay} = \delta_{Az} = 10\,\mu$m; $\delta_{Bx} = \delta_{By} = 8\,\mu$m, $\delta_{Bz} = 12\,\mu$m; spin axis elevation angle = 1 deg; inertial asymmetry of proof mass $(I_x - I_y)/I_y = 10^{-4}$ (this error is not important for translational detectors). See Fig. 3 for an explanation of the panels. Note that in panel (c) a simulated PE-violating acceleration of $1 \times 10^{-15}\,g$, included in the simulation, was recovered almost unaltered.

This PE-violating signal strength is smaller than the accuracy goal for our experiment of a few parts in 10^{15}. The numerical results clearly show that the purely rotational configuration of the detector has made the experiment much less sensitive to detector dynamics, gravity gradients, and the elevation angle of the spin axis (which in this simulation was 1 deg). Thus, a detector can be designed with moderate requirements of centering errors and axial symmetries. Likewise, the release mechanism has to satisfy a moderate requirement (of order 1 deg) on the orientation of the spin axis with respect to the plane normal to the gravity vector. Thus, for a rotational detector configuration, the accuracy of the experiment is no longer limited by the detector's motion and gravity-gradient effects, but rather by white-noise sources like the Brownian noise of the detector and its preamplifier. These latter noise sources were not included in the simulations so as not to confuse the motion-related issues, but they do determine the target accuracy of our experiment to be a few parts in 10^{15}.

6. Conclusions

We have analyzed the free-fall dynamics of different configurations of differential acceleration detectors to test the principle of equivalence (PE) and have investigated the ability to extract a small PE-violating signal from the motion-related noise. The analysis shows that differential detectors using translational and rotational–translational motions of proof masses are too sensitive to motion-related noise, gravity gradients associated with the location, and centering errors of the proof masses, and orientation of the spin axis with respect to the gravity vector to attain an accuracy of about a few parts in 10^{15}. However, a configuration that makes use of purely rotational motion of the proof masses dramatically reduces the errors due to gravity gradients and attitude motion of the hosting platform. We conclude that our proposed detector design leads to a substantial simplification over other possible designs of the requirements for a PE experiment in vertical free fall, thus significantly increasing its likelihood of success.

Acknowledgments

We would like to acknowledge the financial support from NASA's Glenn Research Center through grant NAG3-2881 to the Smithsonian Astrophysical Observatory, and from the Italian Space Agency through contract I/R/098/02 to the Institute of Interplanetary Space Physics, IFSI–INAF.

References

1. E. A. Moody, *J. Hist. Ideas* **12** (1951) 163, 375.
2. M. R. Cohen and I. E. Drabkin, *A Source Book in Greek Science* (Harvard University Press, 1958).
3. C. M. Will, *Theory and Experiment in Gravitational Physics* (Cambridge University Press, 1981).

4. T. M. Niebauer, M. P. McHugh and J. E. Faller, *Phys. Rev. Lett.* **59** (1986) 609.
5. S. Baessler *et al.*, *Phys. Rev. Lett.* **83** (1999) 3585.
6. J. G. Williams, X. X. Newhall and J. O. Dickey, *Phys. Rev. D* **53** (1996) 6730.
7. A. Clocchiati *et al.*, *Astrophys. J.* **642** (2006) 1 and references therein.
8. E. C. Lorenzini *et al.*, *Il Nuovo Cimento B* **109** (1994) 1195.
9. I. I. Shapiro *et al.*, Flight definition of an experiment to test the equivalence principle in an Einstein elevator, *Proc. 2nd Pan Pacific Basin Workshop on Microgravity Sciences*, paper FP-1080 (1–4 May 2001, Pasadena, California; Association of Pacific Rim Universities, 2001).
10. V. Iafolla *et al.*, General relativity accuracy test (GReAT): New configuration for the differential accelerometer, *Proc. 35th COSPAR Scientific Assembly* (18–25 July 2004, Paris, France, 2004).
11. R. P. Giffard, *Phys. Rev. D* **14** (1976) 2478.
12. H. J. Paik, *J. Astronaut. Sci.* **29** (1981) 1.
13. V. B. Braginsky and A. B. Manukin, *Measurements of Weak Forces in Physics Experiments* (University of Chicago Press, 1974).
14. P. Worden, J. Mester and R. Torii, *Class. Quant. Grav.* **18** (2001) 2543.
15. P. Touboul *et al.*, *Acta Astronaut.* **50** (2002) 433.
16. J. Mester *et al.*, *Class. Quant. Grav.* **18** (2001) 2475.
17. V. Iafolla *et al.*, Development of a high-sensitivity differential accelerometer to be used in the experiment to test the equivalence principle in an Einstein elevator, *Proc. XXVIII Recontres de Moriond: Gravitational Waves and Experimental Gravity* (22–29 Mar. 2003, Les Arcs, France, 2003).
18. E. C. Lorenzini *et al.*, Detector Configurations for Equivalence Principle Tests with Strong Separation of Signal from Noise, *XXVIII Spanish Relativity Meeting: A Century of Relativity Physics* (ERE 2005, Oviedo, Spain), *AIP Conf. Proc.* **841**, 502, eds. L. Mornas and J. Diaz Alonso (Melville, New York, 2006).
19. I. I. Shapiro *et al.*, Test of the Equivalence Principle in an Einstein Elevator, Annual Report #2, NASA Grant NAG3-2881 (Mar. 2005).
20. I. I. Shapiro *et al.*, Test of the Equivalence Principle in an Einstein Elevator, Final Report, NASA Grant NAG8-1780 (Apr. 2004).
21. I. I. Shapiro *et al.*, Test of the Equivalence Principle in an Einstein Elevator, Final report, NASA Grant NAG3-2881 (July 2007).

A LABORATORY TEST OF THE EQUIVALENCE PRINCIPLE AS PROLOG TO A SPACEBORNE EXPERIMENT

ROBERT D. REASENBERG* and JAMES D. PHILLIPS†

Smithsonian Astrophysical Observatory,
Harvard–Smithsonian Center for Astrophysics,
Cambridge, MA, USA
**reasenberg@cfa.harvard.edu*
†*jphillips@cfa.harvard.edu*

To test the equivalence principle (EP) to an accuracy of at least $\sigma(\Delta g)/g = 5 \times 10^{-14}$, we are developing a modern Galilean experiment. In our principle-of-equivalence measurement (POEM), we directly examine the relative motion of two test mass assemblies (TMA) that are freely falling. Such an experiment tests both for a possible violation of the weak equivalence principle (WEP) and for new forces that might mimic a WEP violation. For the terrestrial version of the experiment, there are three key technologies. A laser gauge measures the separation of the TMA to picometer accuracy in a second as they fall freely in a comoving vacuum chamber. The motion system launches the TMA from their kinematic mounts inside the chamber and keeps the chamber on a trajectory that mimics free fall until the chamber nears the bottom of its motion. It then "bounces" the chamber back to upward motion in preparation for a new launch of the TMA. A capacitance gauge system measures an additional four degrees of freedom of the motion of each TMA. The resulting estimate of the rotation around and translation along the horizontal axes is used to correct systematic errors. We describe the status of POEM and discuss recent progress.

Keywords: WEP; general relativity.

1. Introduction

One needs look no further than these proceedings to find reasons for more stringent testing of the weak equivalence principle (WEP). The WEP is central to the present accepted theory of gravity, and some theorists argue it is the aspect of general relativity that would most easily manifest a breakdown. The evidence that leads to postulating dark energy may be telling us that we need a new gravity theory. Attempts to create a quantum theory of gravity show a failure of the equivalence principle. Finally, the least well tested of the fundamental forces is gravity, which governs the large-scale structure of the Universe. An experiment that tests for a possible violation of the WEP is also sensitive to new forces.

To test the WEP to an accuracy of at least $\sigma(\Delta g)/g = 5 \cdot 10^{-14}$, we are developing a Galilean principle-of-equivalence measurement (POEM) in which we directly examine the relative motion of a pair of test masses that are freely falling. The test mass assemblies (TMA) will be in free fall in a comoving vacuum chamber for about 0.8 s per "toss," i.e. motion both up and down along a vertical path of about 90 cm. These brief periods of free fall repeat at intervals of a little over a second.

Figure 1 shows the principal components of the first generation measurement system. Inside the vacuum chamber is a single pair of TMA resting on shelves separated by 0.5 m. Each TMA contains a sample of a test substance (A or B) and a corner-cube retroreflector. Conditioned laser light entering at the lower right reaches the beamsplitter, illuminating the optical cavity formed by the two retroreflectors, and is then passed to the detector (upper left). The compensator plate makes possible the alignment of the cavity in the presence of imperfect retroreflectors and a wedged beamsplitter. The vacuum chamber is attached to a cart that rides on a vertical track. The position of the cart is driven by a linear motor inside the track and position feedback is provided by a linear incremental encoder with 20 micron markings. The position loop is closed by a DSP-based controller and PWM amplifier, operating together at 20 kHz.

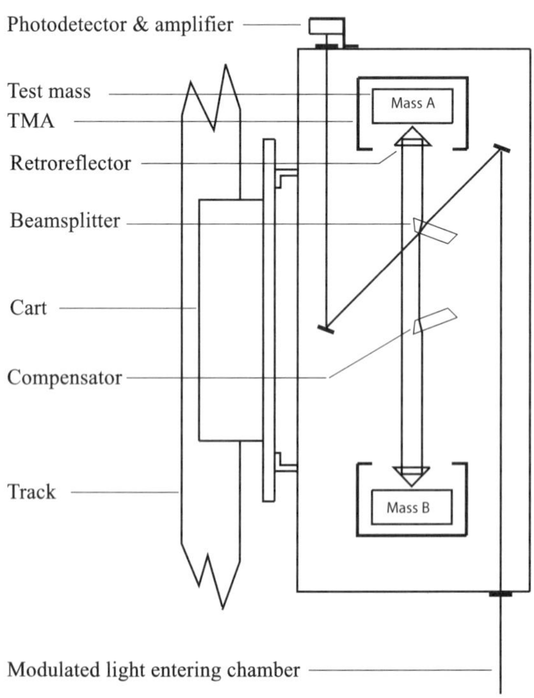

Fig. 1. Principal components of the measurement system.

We are developing POEM in three stages (Gen-I, Gen-II, Gen-III), with possible further development to be defined after the measurement system is working. The measurement system is being designed both for the control of systematic error and, where applicable, to be easily transformed to be space-based. There are three key technologies for POEM: the laser gauge, the capacitance gauge, and the motion system. These will be discussed after a brief description of POEM.

2. POEM

Figure 2 shows the "probe," which comprises all of the components inside the chamber, the chamber's top flange, and the detector for the laser gauge (mounted on top of the flange). One can see the TMA sitting on the shelves as well as the other components of the interferometer. Each TMA appears as a cylinder of 44.5 mm diameter and 36.5 mm height, exclusive of the ball-end tungsten rods that stick out each end and engage the TMA kinematic support. Figure 3 shows an exploded view of a TMA. The retroreflector sits on raised pads in the base plate and is held in place by a flexible ring that has three grooved feet to sit on the retroreflector directly above the raised pads. The ring is loaded by screws from above. The small

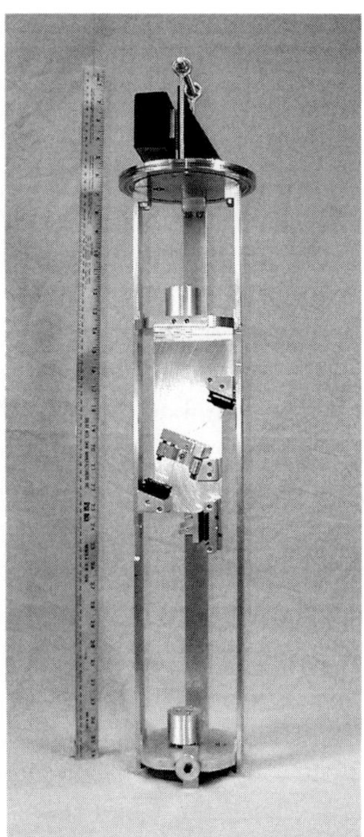

Fig. 2. POEM probe. Some support elements have been removed to make the optics and optical mounts visible. The scale is 36″ long and has both centimeter and inch markings.

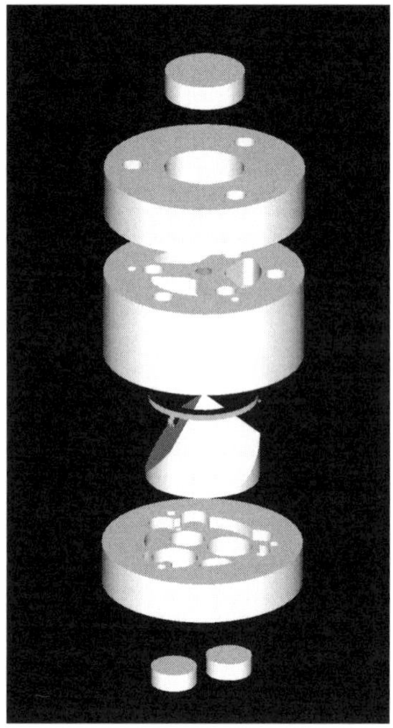

Fig. 3. Exploded view of a TMA.

cylinder at the top is a screw for adjusting the location of the center of mass along the cylindrical axis.

The moving vacuum chamber offers three advantages over the more usual fixed chamber. First, there is no need for mechanisms inside the chamber to drive the motion of the TMA and the TMA-observing devices during each toss. Thus, the vacuum chamber contains no motors, precision bearings, or wall-penetrating shafts that drive high speed motion to sub-mm accuracy. Second, there is no need for a slide (cart plus rails) inside the chamber to guide the motion of the TMA and the TMA-observing devices. The slide would need to provide smooth motion with transverse noise of under about 10 nm. A slide based on ball bearings will not suffice. An air slide, a precision teflon lubricated steel on steel slide, or other low-vibration slide would need to operate in the chamber. The parts that would need to move include most of the probe below the vacuum flange and the cart — about a third of the weight of the present moving assembly. In order to control vibration, the rail inside the chamber would need to be about a third as stiff as the one outside the chamber, and so would be of nearly the same size. Third, the moving chamber can be relatively small and easily opened. A nonmoving chamber would be large and would have to be lifted off with an overhead hoist and therefore be located in a high-bay area. Notwithstanding these advantages, the moving-vacuum-chamber approach has drawbacks. It entails moving tens of kilograms at speeds approaching 5 m/s, which implies large forces and the potential for significant amounts of vibrational energy from which the measurement must be protected. Further, the pump for sustaining (but not for creating) the vacuum must ride with the chamber, although this pump can be of relatively low mass.

The chamber motion repeats in about 1.3 s. The chamber is sent upward at 5 m/s. During the upper portion of the motion, the linear motor and its control system serve to enforce a free-fall trajectory, overcoming friction. At the bottom of the free-fall portion of the motion, the chamber encounters a "bouncer" that passively applies an upward force, absorbs the energy of the falling chamber, and returns the chamber to upward motion in about 0.4 s with a minimum of force required from the motor. The bouncer will be discussed in Sec. 3.1.

In the Gen-II version of POEM, there will be a larger probe (and vacuum chamber) to hold two pairs of TMA (Fig. 4) with a lateral separation of 7 cm. This will permit a double difference observable that cancels many systematic errors. Prominent among these is the gravity gradient, including the vertical component, $dg/dz = 3 \cdot 10^{-7}$ g/m. Additionally, there are small components that are time-dependent, including those due to ground-water variation and parked cars on the nearby street.

The Gen-II dual-measurement probe will be modified for Gen-III. With a Gen-III science goal of $\sigma(\Delta g)/g = 5 \cdot 10^{-14}$ and a TMA mass that is 30% test substance, we require a measurement accuracy of $\sigma(\Delta g)/g = 1.5 \cdot 10^{-14}$ for the TMA. This requirement, when combined with the vertical gravity gradient, implies a requirement for absolute distance measurement with an uncertainty under 0.05 μm. The required laser gauge will be discussed in the next section.

The science goal of $\sigma(\Delta g)/g = 5 \cdot 10^{-14}$ requires careful attention to systematic error. In the Gen-III version of POEM, systematic error is mitigated by a series of interchanges: (1) left–right robotic interchange of TMA on a time scale of perhaps 10 minutes, (2) top–bottom interchange of TMA between runs, and (3) manual interchange of test substance halfway through the experiment if we find that it is needed. These interchanges substantially reduce the biasing due to the gravity gradient and several small effects.

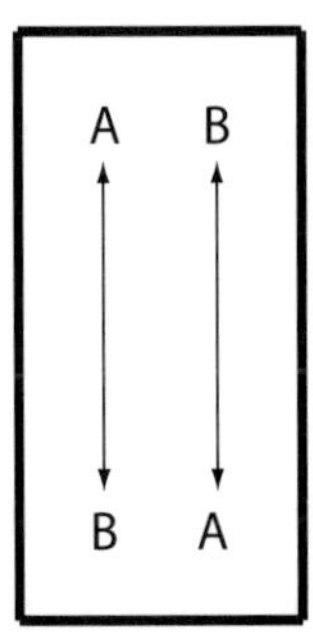

Fig. 4. Locations of the test substances (A and B) inside the TMA in Gen-II and Gen-III. The acceleration difference (left–right) adds any WEP-violating signal and cancels biasing effects.

Some of the error sources depend on parameters that we estimate principally from calibration maneuvers, which we introduce to reduce the impact on measurement time of obtaining these estimates. These error sources include beam alignment and rotation of the TMA in the presence of an offset between the center of mass and the optical reference point (ORP).[a]

We enrich the data set by exaggerating the normally small displacements during calibration maneuvers distributed throughout a run:

(i) At its launch point below the chamber, the laser beam is tipped and tilted forward, back, left, and right of its nominal (vertical) direction.
(ii) Each TMA is pushed (again in four directions) by applying high voltage to the capacitance-gauge electrodes. These pushes may be given midway through the free-fall period, yielding corresponding changes in lateral velocity.
(iii) Each TMA is rotated around two horizontal axes by applying high voltage to the capacitance plates. These rotation maneuvers form the basis for determining Δz, the vertical CM-ORP separation of each TMA, with $\sigma(\Delta z) < 0.03\ \mu$m. Δz is important because it acts with the (measured) TMA rotation rate to produce an apparent TMA acceleration. While the CM-ORP separation also contributes to the uncorrected gravity-gradient acceleration, its effect cancels in the difference after a top–bottom swap. Similarly, the calibration maneuvers provide the basis for correcting the bias due to beam misalignment (cosine error) and to beam walk on imperfect optics.

In a perfect retroreflector cavity, translation of the input beam yields no change in S, the round trip optical path. Beam tilt would cause a cosine error. In our (real) system, the relative lateral motions and rotations of the TMA, vacuum chamber, and incoming beam, combined with angular errors in the retroreflectors and the wedged beamsplitter and compensator, yield errors that mimic an acceleration (Δg) of a few $10^{-11}\,\mathrm{m\,s^{-2}}$, which is comparable to the single-toss error from the laser gauge. To substantially remove these effects, we will produce and analyze the enriched data set resulting from the calibration maneuvers that exaggerate the 12 perturbations of the measurement system (per pair of TMA) described above. The perturbations will be observed to better-than-required accuracy using the capacitance gauges and a pair of "quad cells" that look at the incoming beam at the bottom and top of the chamber. The extra parameters associated with calibration maneuvers (e.g. Δz) will be estimated along with other experiment parameters.

Coriolis force can produce TMA acceleration far above the intended accuracy level. To address this systematic error, we will use a four-channel capacitance gauge

[a]The ORP is defined as the point around which a small rotation of the TMA does not cause a change in optical path. For a hollow retroreflector, the ORP is the geometric apex, and even a large rotation around the ORP causes no change in the optical path. For a solid retroreflector rotated around the ORP by θ, the lowest order correction to the optical path is proportional to θ^4.

for each TMA.[b] The POEM error budget requires that the transverse velocity be measured to 33 nm/s in each toss and that the bias in the average of these measurements be under 0.25 nm/s. In order to keep the stability requirement for the capacitance gauge at a reasonable level, we require that the TMA transverse velocity be under 10 μm/s. Since the TMA will pick up the chamber's transverse velocity at the time of launch, when the chamber will be moving upward at nearly 5 m/s, we require that the rail guiding the motion have angular deviations from vertical under $2 \cdot 10^{-6}$ rad. This, in turn, requires both a straight rail and careful leveling, which can be adjusted based on the TMA trajectories, as measured by the capacitance gauge.

2.1. *Tracking frequency laser gauge*

For POEM, we use the tracking frequency laser gauge (TFG) that we developed more than 15 years ago for POINTS,[1] an affordable space-based astrometric optical interferometer with a nominal single-measurement accuracy of 2 μas for a pair of bright stars separated by 90 ± 3 deg. Because of the close connection among size, weight, complexity, and cost, we kept the baseline length at 2 m and thus required high precision in the metrology in order to achieve the nominal single-measurement astrometric accuracy. The mission requirement was for a single metrology leg (laser gauge) to have an accuracy of 2 pm on time scales from 1 to 100 minutes. We could find no existing laser gauge that would meet that requirement. In particular, we looked at and rejected the standard precision laser gauge, the heterodyne interferometer.

As will become apparent, the TFG offers four advantages over the heterodyne gauge.[2] First, it is intrinsically free of the nm-scale cyclic bias that plagues the heterodyne gauges. It has only one beam and thus it cannot be subject to problems associated with separating beams of different frequencies. Second, it naturally operates not only in a nonresonant interferometer (Michelson, Mach–Zehnder, etc.), but also in a resonant cavity. Thus, additional accuracy is accessible when needed. However, operation in a nonresonant interferometer requires unequal paths. Third, the TFG can be used to measure absolute distance with little additional effort. Fourth, the TFG can suppress some additional errors associated with polarization sensitivity and, when used in a cavity, with beam alignment.

The TFG is an application of Pound–Drever–Hall locking, and Fig. 5 shows one realization. An optical signal from the variable frequency source (VFS) of an adjustable wavelength λ_{VFS} is phase modulated at a frequency f_m and introduced into the measurement interferometer, whose length, L, is to be determined. When λ_{VFS} is equal to λ_0, the wavelength at the intensity extremum with $N\lambda_0 = 2L$, the optical signal emerging from the interferometer is phase modulated but not

[b]The capacitance gauge is being developed in a collaboration with Winfield Hill of the Rowland Institute at Harvard.

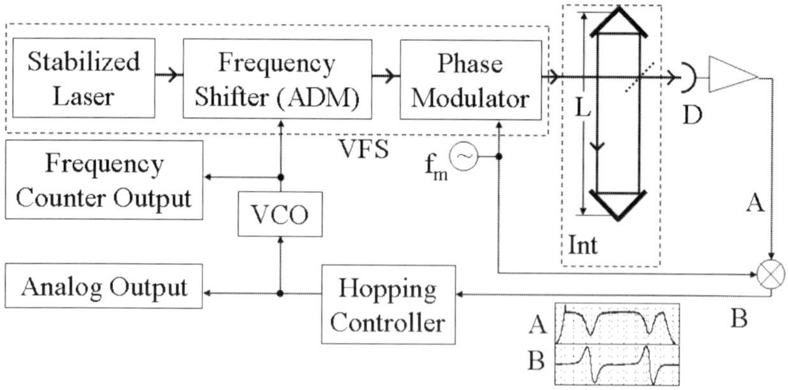

Fig. 5. TFG block diagram, classic realization.

amplitude modulated. When λ_{VFS} is away from λ_0, the optical signal emerging from the interferometer is both phase and amplitude modulated, resulting in an electrical signal at the detector output at f_m with a magnitude and sign that indicate the offset. Synchronous detection at f_m and filtering yield a signal that is used to control λ_{VFS}, driving it back to λ_0. The corresponding optical frequency shift is measured by a frequency counter.

In any realization of the TFG, the usable range of the VFS will be limited either by the VFS or by the frequency counter. For most applications, the corresponding distance range would be too small. Therefore, we have introduced a nonlinear aspect into the TFG loop controller. It detects that it is running out of the frequency shifter's range and hops to a mode at the far end of the range, shifting the optical frequency by an integer multiple of the free spectral range, $\Phi = c/2L$. The hop is fast enough (about $1\,\mu$s) to be "unobserved" by the linear portion of the loop controller. We have demonstrated a rate of $5 \cdot 10^4$ hops/s, which corresponds to a linear velocity of 16 mm/s, using our HeNe version operating in a low finesse resonant interferometer for which $\Phi = 300$ MHz.

If the precision of the TFG were limited by photon-counting statistics, then for $1\,\mu$W of HeNe (633 nm) power detected from a Michelson interferometer, the incremental distance uncertainty would be 0.06 pm after 1 s. The current TFG is limited by technical noise: $\sigma = 10$ pm at either 1 or 10 samples per second.

The hopping provides an easy means of measuring absolute distance. By measuring the frequency shift before and after a hop, the TFG measures the free spectral range, Φ, of the measurement interferometer corresponding to the current length L. The estimate of L is then $Kc/2\delta F$, where δF is the frequency difference after hopping K fringes. The precision of the absolute distance measurement is

$$\sigma_T(L) = \frac{2}{\eta}\sqrt{\frac{\tau}{T}}\,\sigma_\tau(\delta L)\,, \tag{1}$$

where T and t are the integration times for absolute (L) and incremental (δL) distance, respectively. η is the fractional bandwidth, $\Delta F/F$, where F is the initial laser frequency.

One complication with this scheme comes from fluctuations in L due, for example, to vibration. There are two ways around this problem. Either use two lasers to read simultaneously or hop at a rate that reaches a portion of the disturbance spectrum where the noise is acceptably low. For sufficiently rapid hopping, vibration has this characteristic. For applications in air, path disturbances due to turbulence also diminish with frequency. The TFG does hop fast (50 kHz demonstrated) and the next generation TFG, which will be based on distributed feedback (DFB) lasers, will hop faster. Unlike most narrow-linewidth laser systems, the DFB laser provides rapid frequency shift for hopping without the use of an acousto-optic (AO) modulator. AO devices have a limited frequency-shift range, are bulky, prone to failure, require a free-space beam with attendant alignment issues, and have a time delay due to the acoustic propagation, which limits the achievable servo bandwidth and therefore the hop rate.

How large can η be? In the initial realization of the TFG, ΔF is limited to 500 MHz by the ADM. Since $\Phi = 300$ MHz in the test rig and the HeNe operates at 633 nm, $\eta = 300\,\text{MHz}/470\,\text{THz} = 0.6 \cdot 10^{-6}$. We are developing the SL-TFG, a realization that uses DFB semiconductor lasers. In the SL-TFG, we could shift frequency both upward and downward from nominal to achieve $\Delta F = 2$ GHz, limited by our frequency counter and its internal dividers. (See Sec. 4.) For the DFB laser operating at 1550 nm, $\eta = 2\,\text{GHz}/200\,\text{THz} = 10^{-5}$.[c] A frequency counter with a greater range would permit an increase in η. For POEM, this must be a dual-channel counter with very small differential timing jitter.

2.2. *Capacitance gauge*

Coriolis force can produce TMA acceleration a_c far above the intended accuracy level: $a_c = 2\,v_{\text{e-w}}|\omega|\cos(\text{latitude})$ where $v_{\text{e-w}}$ is the east–west component of the TMA velocity in the lab frame and the Earth rotation rate is $|\omega| = 7.292 \times 10^{-5}$/s. To address this and other sources of systematic error, we will use a four-channel (later five-channel) capacitance gauge for each TMA. The POEM error budget requires that the transverse velocity be measured to 33 nm/s in each toss and that the bias in the average of these measurements be under 0.25 nm/s.

Figure 6 is a block diagram of the POEM capacitance gauge. It is of an unusual design because the TMA can neither be grounded nor connected directly to the sense amplifier. Instead, it is capacitively connected via an encircling cylindrical plate. Drive signals, applied through a tightly coupled transformer to plates that are segments of a cylinder, are at multiples of 2 kHz and in the range 10–18 kHz. These signals are kept small to limit their contribution to the vertical acceleration

[c]The SL-TFG is discussed briefly in the last section.

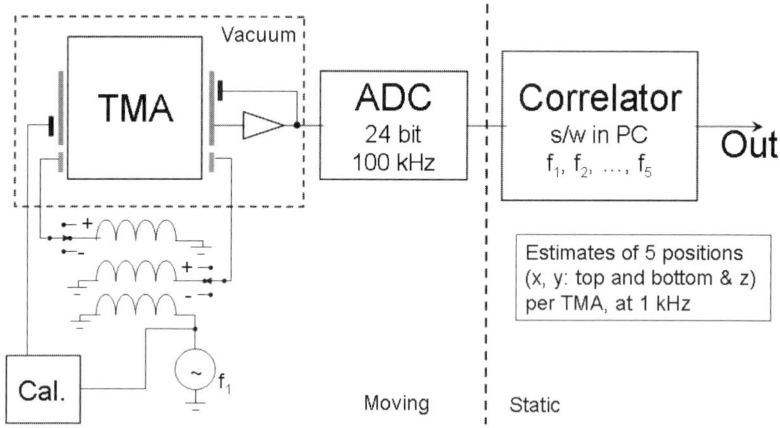

Fig. 6. Block diagram of the POEM capacitance gauge. The diagram shows the hardware associated with only one of the four drive signals.

of the TMA through fringing fields. With the nominal drive level of 0.1 V rms, we expect a position sensitivity better than 8 nm in 1 s in each channel. When the TMA is centered, the capacitive gaps are nominally 1 mm for both the drive and pickup plates. The optimized system has a single large pickup electrode and four pairs of small drive electrodes.

In the initial configuration, there are four pairs of drive plates covering two axes near both the top and the bottom of the TMA. The combination of displacement signals picked up by the encircling cylindrical plate is amplified and digitized before being sent to a PC for separate extraction in a software correlator. With the ADC trigger and the drive signals derived from the same stable oscillator, all 0.5 ms batches of data will have the same (integer) number of cycles of each of the drive signals, making possible a clean separation. The measurement intervals are a multiple of 0.5 ms.[d]

We use capacitive feedback around the sense amplifier to avoid phase shifts and a frequency-dependent response. Both the feedback capacitor and the calibration capacitor are machined into the assembly and must be mechanically stable. In order to keep the stability requirement for the capacitance gauge at a reasonable level,[e] we set the full scale at 10 μm and thus require that the TMA transverse velocity be under 10 μm/s. Since the TMA will pick up the chamber's transverse velocity at the time of launch, when the chamber will be moving upward at nearly 5 m/s,

[d]The motion system, discussed below, includes a high power (a few kW) PWM driver operating at 20 kHz. There is bound to be pickup in the CG from the lines to the motor. However, if the CG clock and the clock in the motor controller run accurately at their nominal speeds, then the 20 kHz signals from the motor controller will be well removed in the correlation process.

[e]The stability requirement is defined for this purpose as the ratio of the largest signal that can be processed to long-term change in that signal due to drift, e.g. from changing gain or bias. We take the stability requirement to be $4 \cdot 10^4$, based on engineering judgment.

we require that the rail guiding the motion have angular deviations from vertical under $2 \cdot 10^{-6}$. This, in turn, requires both a straight rail and careful leveling, which can be adjusted based on the capacitance gauge measurements.

3. Motion System

A key feature of POEM is that the experiment is conducted in a vacuum chamber that is placed in free fall with the TMA. In order to have a precision experiment based on a free fall time of 3/4 s, it is necessary to have a large number of repetitions, which suggests a need for rapid recycling. To meet this objective, we built a "bouncer" that catches the falling vacuum chamber and returns it to upward motion with little loss of energy or shock to the instrument inside the chamber. Once the chamber is again moving upward at the required speed (nearly 5 m/s), the TMA must be launched.

3.1. *Torsion bar bouncer*

It is our intention to run the laser gauge continuously, including during the bounce period when vertical acceleration reaches about 5 g. For this reason, to ease the job of the motor control servo, and to limit probe vibration so that the TMA are launched with minimal transverse velocity, it is essential that the bouncer produces minimal shock and vibration. This requirement rules out many obvious candidate designs. Having the chamber contact a cable (or flexible band) under tension results in a force on the chamber that initially grows linearly with cable deflection. We use a 1/4 inch steel cable, 62 cm long. It has a 0.1 kg mass but only 0.05 kg "effective mass" as seen by the falling chamber, which now has a mass of about 40 kg. (This will increase with Gen-II and Gen-III.) In addition, the cable probably flexes on initial contact, further reducing the shock to the moving system. Thus, as the chamber makes contact with the cable, there is a small bump as it loses one part in 800 of its velocity. The change of velocity is spread over a short interval of time by the flexing of the cable.

The next question is: How to store the energy of the falling chamber so that it can be returned to upward motion? Coil springs have internal modes that are problematic. Our initial design,[3] which used a ton of lead, a 5:1 lever, and a long cable running over pulleys, had two problems. First, on initial contact with the falling chamber, the cable experienced a longitudinal acceleration that excited an oscillation with the lead. Second, friction internal to the cable as it ran over the pulleys caused the system to fail to meet the efficiency requirement. The new bouncer uses torsion bars connected to the cable through stiff levers. Internal torsional modes of the bars are above 1 kHz and, to the extent excited, cannot contain much of the system's total energy since the bar's moment of inertia is very small compared to $M_{\text{chamber}} R_{\text{lever}}^2$. Tests show that bouncer energy loss is small and masked by uncertainty in the losses in the present slide. There is no sign of the torsion-bar bouncer introducing vibration, but the critical test awaits a quieter system based on an air-bearing slide.

3.2. *Air-bearing slide*

The present commercial slide uses instrument grade track rollers running on small (1/4 inch) but well-supported rails. At full speed, we find micron scale vibration, strongest in the 100–200 Hz band, yielding transverse velocity of 1 mm/s scale. We are in the process of implementing an air-bearing slide, as described in our original plans for this experiment.

We plan to use porous graphite bearings running on a granite beam. The undulations of the slope of the granite surface will need to under $2 \cdot 10^{-6}$ in the region where the cart is traveling when the TMA are launched. On linear scales shorter than the diameter of the air bearings (nominally 2 inch), the requirement can be relaxed because of the averaging that will take place. On large scales (say, over 1 m), the requirement can again be relaxed. Micron scale transverse displacement of the chamber, either after or well before TMA launch, has negligible effect on the intended results.

Such surface quality requirements are well within the capabilities of the large-optics industry. However, costs there are high, in part because the facilities are intended for making complex (aspherical) surfaces, not just the required flat. Fortunately, the requirements are just within the capability of the precision granite industry, which operates at significantly lower cost. As of this writing, we are preparing to place an order for a granite beam. Should we eventually wish to further reduce the transverse motion of the chamber due to the shape of the granite beam, we would consider connecting each air bearing to the moving assembly through an actuator (e.g. a PZT pusher) and using an inertial sensor to measure the required correction.

4. Key Technology Status and Conclusion

We have had a working TFG since the early 1990's. In the last few years, we have increased the frequency shift range, added the hopping capability, and demonstrated the measurement of absolute distance with the HeNe-based version. The HeNe TFG has operated successfully in the moving chamber, measuring the distance between the TMA. In this mode, it has tracked motion of several-μm amplitude and several-ms duration, consistent with brief episodes of TMA liftoff caused by the vibration, which was substantial at the time.

The standard frequency counter has dead time of up to 100 ms following each gate interval. We wish to be able to operate the TFG with samples as short as 0.1 ms although, for POEM, we anticipate 10 ms as a normal frequency measurement interval. Having dead time degrades the data set in three ways. First, only a portion of the available time is spent taking data; in some cases a small portion. Second, if there were no dead time, then certain errors made by the counter in sample n

would be made with the opposite sign in sample $n + 1$. Such "perfectly correlated noise" has a substantially smaller effect on estimates of the amplitudes of slowly varying quantities than does ordinary noise.[4] Third, a bias results from a mismatch of gate times for the two channels of the POEM counter for noncontiguous measurements.

A dual-channel frequency counter was built for POEM by Jim MacArthur, head of the Electronics Instrument Design Lab in Harvard University's Physics Department. This counter provides for contiguous measurements (no dead time) and precise synchronization between channels of the boundaries between measurement intervals. It operates to 200 MHz and contains $2\times, 4\times$, and $8\times$ dividers that permit a maximum count rate (each channel) of 400, 800, and 1200 MHz, respectively.

For the past few years, we have been working with DFB semiconductor lasers operating the 1550 nm communications band. Because of the large market for devices operating in this band, there are a wide variety of rugged, moderately priced components available. We have purchased a cavity with a finesse of 300, and locked a DFB laser to it. This cavity comprises a (45 mm OD $\times$ 150 mm) cylinder of Zerodur with an (13 mm) axial bore. Mirrors (one flat and one concave on its inner face) are optically contacted to the ends of the cylinder. They have identical v-coatings on their inner faces, providing 99.2% reflectivity. Further, they are wedged and have identical AR coatings on the outer faces to prevent these pieces of fused quartz from acting as spurious etalons.

We have locked two DFB lasers to adjacent orders of the cavity and measured the characteristics of the beat note. So far, we have only used a servo of rather low gain. We have since increased the servo gain substantially, and new results are pending.

The capacitance gauge is nearly completed. The architecture has long been established. We have received a preliminary version of the electrode assemblies and modified the probe to mount these around the TMA. All electronic components have been designed and are in various stages of fabrication, except for the preamp, for which only a basic design exists. We still need to decide how to package the electronics and connect through the probe vacuum flange. Most of the custom electronics will ride on or in the vacuum chamber.

The motion system is in the midst of an upgrade to reduce vibration. The new bouncer, based on a pair of torsion bars, has shown the high mechanical efficiency that we expected. Unlike the previous bouncer, it is left–right symmetric. Because of the higher efficiency of the new bouncer, the motor servo can be (and has been) tuned less aggressively. This results in a lower level of motor-driven vibration, yet the moving assembly still follows the intended trajectory within tens of μm.

We have measured the vibration in the present slide and found it too high. For this reason, we have begun to work on the long-intended upgrade to an air-bearing slide to replace the present commercial system that has wheels running along steel rails. According to the present plan, the air-bearing slide will use a granite beam and porous graphite bearings. Preliminary designs have been completed and no serious problem has been identified. Vendors have been found for all of the major components, and they are able to meet the requirements at a reasonable price. The hardware to make clean, dry compressed air has been delivered.

In summary, the SAO principle-of-equivalence measurement (POEM) is a Galilean test of the WEP divided into three developmental generations. The goal for the Gen-III version of the experiment is $\sigma(\Delta g)/g = 5 \cdot 10^{-14}$ for several pairs of substances. All Gen-I components are working and being tuned or modified for better performance. Work on some components originally described as part of Gen-II has started. These include the capacitance gauge, which is nearly finished, and the air slide, for which a preliminary design has been completed. The measurement system is being designed both for the control of systematic error and, where applicable, to be easily translated into a space-based version, for which we anticipate $\Delta g/g = 10^{-16}$.

Acknowledgments

We thank Kelzie Beebe (Harvard), Alexandru Ene (Harvard), Elizabeth Gould (Worcester Polytechnic), and Glen Nixon (Purdue), and high school student Alex McCaouley for skilful laboratory work. We thank colleagues Jim Faller (JILA), Robert Kimberk (SAO, Central Engineering), Tim Niebauer (Micro-g Corp.), and Doug Robertson (NGS/NOAA) for helpful discussions. We gratefully acknowledge support from the National Aeronautics and Space Administration through grant NNC04GB30G, and from the Smithsonian Institution directly and through the SAO IR&D program.

References

1. R. D. Reasenberg *et al.*, *Astron. J.* **32** (1998) 1731.
2. J. D. Phillips and R. D. Reasenberg, *Rev. Sci. Instr.* **76** (2005) 064501.
3. R. D. Reasenberg and J. D. Phillips, *Class. Quant. Grav.* **18** (2001) 2435.
4. R. D. Reasenberg, *Am. Inst. Aeronaut. Astronaut.* **10** (1972) 942.

EXPERIMENTAL VALIDATION OF A HIGH ACCURACY TEST OF THE EQUIVALENCE PRINCIPLE WITH THE SMALL SATELLITE "GALILEO GALILEI"

ANNA M. NOBILI

Department of Physics, University of Pisa and INFN,
Largo Bruno Pontecorvo 3,
56127 Pisa, Italy
nobili@dm.unipi.it

GIAN LUCA COMANDI

Department of Physics, University of Bologna and INFN,
viale C. Berti Pichat, 6/2,
40127 Bologna, Italy

SURESH DORAVARI and FRANCESCO MACCARRONE

Department of Physics, University of Pisa and INFN,
Largo Bruno Pontecorvo 3,
56127 Pisa, Italy

DONATO BRAMANTI and ERSEO POLACCO

Istituto Nazionale dif Fisica Nucleare,
Largo Bruno Pontecorvo 3,
56127 Pisa, Italy

The small satellite "Galileo Galilei" (GG) has been designed to test the equivalence principle (EP) to 10^{-17} with a total mass at launch of 250 kg. The key instrument is a differential accelerometer made up of weakly coupled coaxial, concentric test cylinders rapidly spinning around the symmetry axis and sensitive in the plane perpendicular to it, lying at a small inclination from the orbit plane. The whole spacecraft spins around the same symmetry axis so as to be passively stabilized. The test masses are large (10 kg each, to reduce thermal noise), their coupling is very weak (for high sensitivity to differential effects), and rotation is fast (for high frequency modulation of the signal). A 1 g version of the accelerometer ("Galileo Galilei on the Ground" — GGG) has been built to the full scale — except for coupling, which cannot be as weak as in the absence of weight, and a motor to maintain rotation (not needed in space due to angular momentum conservation). GGG has proved: (i) high Q; (ii) auto-centering and long term stability; (iii) a sensitivity to EP testing which is close to the target sensitivity of the GG experiment provided that the physical properties of the experiment in space are going to be fully exploited.

Keywords: General relativity; space physics; experimental gravity.

1. Introduction

Laboratory tests of the equivalence principle (EP) allow the experimental results to be checked beyond question. The best such results have been obtained in a remarkable series of experiments using slowly rotating torsion balances[1] that have found no violation to 10^{-12} and slightly better.

It is known that an experiment performed inside a spacecraft orbiting at low altitude around the Earth can aim at improving these results, in the gravitational field of the Earth, by several orders of magnitude. The first proposed satellite experiment is STEP,[2] aiming at an EP test to 1 part in 10^{18}. STEP requires a cryogenic accelerometer, an actively controlled three-axis stabilized or slowly rotating spacecraft, with a total mass at launch of about 1 ton. A scaled-down, noncryogenic version of STEP has been designed, named μSCOPE,[3] to fly in 2009–2010. Abandoning cryogenics has allowed the total μSCOPE mass to be reduced to one-fourth of the STEP's mass, for an expected EP test to 1 part in 10^{15}. The concept of the STEP and μSCOPE experiments is outlined in Fig. 1.

"Galileo Galilei" (GG)[4] is a proposed space experiment to test the EP at room temperature with a total mass at launch close to that of μSCOPE ($\simeq$250 kg) but aiming at an EP test competitive with STEP, namely to 1 part in 10^{17}. We can convincingly argue that ultimately this is made possible by one single change in the experiment design from that of STEP and μSCOPE (see Fig. 2). Such an apparently simple change and its far-reaching consequences for EP testing in space are outlined in Sec. 2. Section 3 describes the GGG prototype[6–8] that we have built with the support of INFN (Istituto Nazionale di Fisica Nucleare), and reports the experimental results and current sensitivity. Finally, Sec. 4 shows the implications of the sensitivity obtained in the lab for the target of the space mission by analyzing the physical properties of the space environment and how the experiment, because of the way it has been designed, will benefit from it.

In February 2006 the GG mission was included in the National Aerospace Plan (PASN) of ASI (Agenzia Spaziale Italiana) for the next three years (see *ASI Plan*,[9] p. 47).

2. Peculiarities of the GG Experiment in Space

The GG experiment is designed to measure the relative acceleration of two test masses in free fall in the gravity field of the Earth. Thus GG tests the universality of free fall (UFF), whereby all bodies fall with the same acceleration regardless of their mass and composition, which is a direct consequence of the EP. A space mission can reach a sensitivity much higher than a ground experiment because,

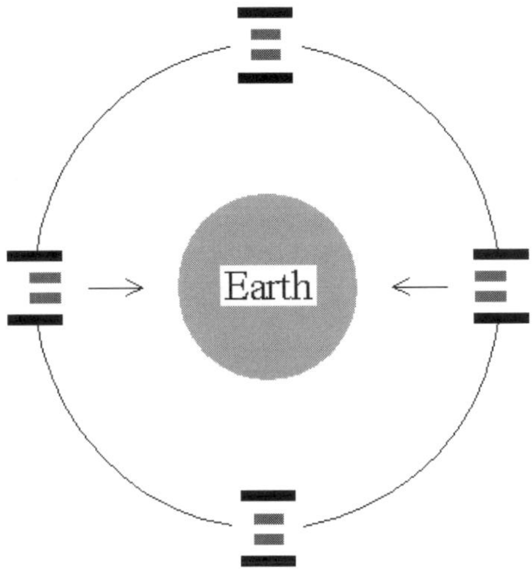

Fig. 1. Concept of the STEP and μSCOPE space experiments to test the equivalence principle. The figure (not to scale) shows a section in the plane along the symmetry axis of two coaxial, concentric test cylinders made of different materials in orbit around the Earth, the symmetry axis lying in the orbit plane. Along this axis the cylinders are weakly coupled, while coupling is stiff in the other directions. Together they form a 1D accelerometer sensitive along the symmetry axis. If the EP were violated — hence one cylinder would be attracted by the Earth more than the other — a differential acceleration would appear as indicated by the arrows, namely at the orbital frequency of the spacecraft. In order to separate the frequency of the signal from the orbital one, at which many disturbances occur, and to up-convert the signal to a higher frequency for reduction of $1/f$ noise, the whole satellite enclosing the accelerometer is actively rotated around an axis perpendicular to the orbit plane (hence to the sensitive/symmetry axis). In so doing, the sensitive axis, and the test cylinders around it, are physically rotated with respect to the Earth as if the accelerometer were still and the Earth were rotating around it. The physical system is a forced oscillator, the forcing effect being at the rotation frequency of the spacecraft with respect to the center of the Earth.

in the case of EP violation, test bodies in low Earth orbit are subjected to an acceleration from the Earth which is more than a thousand times larger than that due to the Sun acting on torsion balances on the ground. Another main advantage of space is weightlessness, which makes it possible to use extremely weak suspensions resulting in a large response to EP violation.

In GG two test masses of different composition are arranged to form a differential accelerometer (Fig. 3, right). The test masses (10 kg each) are concentric, coaxial, hollow cylinders. These two masses are mechanically coupled by attaching them at their top and bottom to two ends of a coupling arm by using flexible lamellae. The coupling arm is made up of two concentric tubes similarly attached at their midpoints to a single shaft. This assembly preserves the overall symmetry of the apparatus, when the two parts of the arm are taken together. The masses are mechanically coupled through the balance arm such that they are free to move in the transverse XY plane, and all of them taken together form the physical system. The masses oscillate in a two-dimensional harmonic potential defined by

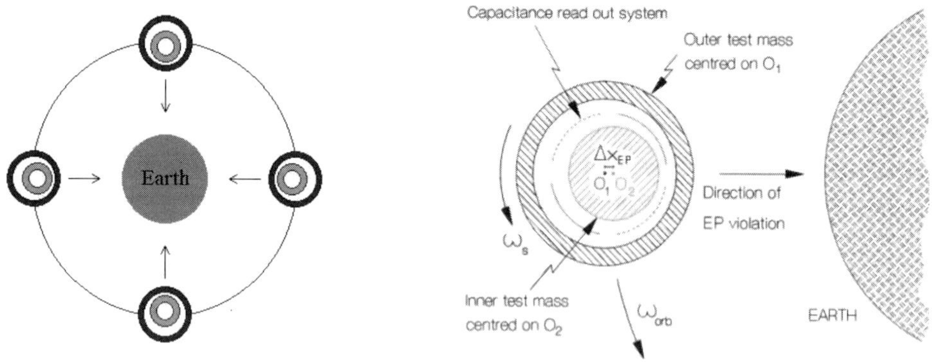

Fig. 2. (*Left*): Concept of the GG space experiment to test the equivalence principle. As compared to the STEP and μSCOPE scheme shown in Fig. 1, the symmetry axis of the test cylinders is simply turned by $90°$ to become perpendicular to the orbit plane. In order to sense an EP violation effect from the Earth the cylinders are now weakly coupled in the orbit plane and stiff along the symmetry axis, thus forming a 2D accelerometer. The figure (not to scale) shows a section across the symmetry axis, which is also the symmetry axis of a capacitance bridge readout placed between the test cylinders and of the whole satellite (neither of them shown). By making it the axis of the maximum moment of inertia and the spin axis of the whole satellite, the satellite is passively stabilized without requiring active attitude control. If EP were violated, and hence one cylinder would be attracted by the Earth more than the other, all along the orbit their centers of mass would be displaced toward the center of the Earth, as indicated by the arrows. (*Right*): Enlargement from the previous figure showing (not to scale) the test cylinders and the capacitance readout at one given position along the orbit. They all spin at angular velocity ω_s while orbiting the Earth at ω_{orb}. In the case of EP violation the centers of mass of the test bodies are displaced from one another by a vector $\Delta\mathbf{x}_{\mathrm{EP}}$. Under the (differential) effect of this new force the test masses, which in this plane are weakly coupled by mechanical suspensions, reach equilibrium at a displaced position where the new force is balanced by the weak restoring force of the suspension, while the bodies rotate independently around O_1 and O_2 respectively. The vector of this relative displacement has a constant amplitude (for zero orbital eccentricity) and points to the center of the Earth (the source mass of the gravitational field). The signal is therefore modulated by the capacitors at their spinning frequency with respect to the center of the Earth $\omega_{s\oplus} \equiv \omega_s - \omega_{\mathrm{orb}}$.

the suspension springs at the ends of the balance arm while free-falling around the Earth. A differential acceleration of the masses would thus give rise to a displacement of the equilibrium position in the XY plane. The displacement of the test masses is sensed by two sets of capacitance plates located between the test cylinders, one set for each orthogonal direction (X and Y). Each set of capacitance plates is placed in an AC bridge configuration such that a displacement of the masses causes an imbalance of the bridge and is thus converted into a voltage signal. When the physical system is mechanically well balanced it is insensitive to "common mode" accelerations. In addition, the capacitance bridges are predominantly sensitive to differential displacements. Thus, the differential nature of the accelerometer is ensured both by the dynamics of the physical system and by the displacement transducer.

The goal of testing the EP to 1 part in 10^{17} in the gravitational field of the Earth requires the detection of a differential acceleration $a_{\mathrm{EP}} \simeq 8.4 \cdot 10^{-17}\,\mathrm{m/s^2}$.

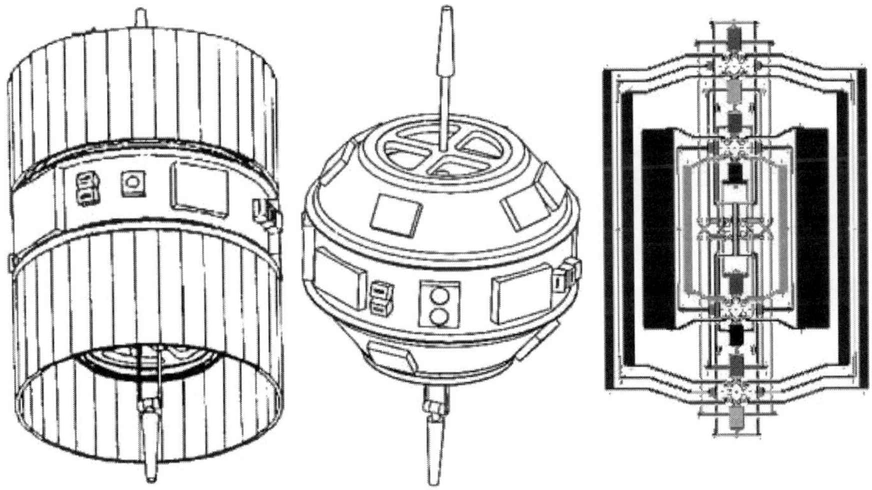

Fig. 3. The GG satellite with solar panels (*left*) and without (*center*). GG is a compact 1-m-diameter structure in the shape of a spinning top stabilized passively by one-axis rotation. Its total mass is 250 kg and its orbit is low (520 km altitude), almost circular and almost equatorial. Inside the "spinning top" (center) — through an intermediate corotating attenuation stage known as PGB — is located the key instrument (*right*) for testing the equivalence principle in the gravitational field of the Earth. It consists of four test cylinders (10 kg each), one inside the other, forming two differential accelerometers: the inner one for EP testing (cylinders of different composition; shown in green and blue respectively) and the outer one for zero check (cylinders made of the same material; both shown in brown). In each accelerometer the two test cylinders are coupled to form a beam balance, as described in Sec. 2. Note that: (i) the whole system is symmetric around the spin axis as well as top/down; (ii) the two accelerometers are both centered at the center of mass of the spacecraft (unlike other proposed space experiments) in order to reduce tidal effects and improve reliability of the zero check; (iii) mechanical suspensions provide electric grounding and passive discharging; (iv) cryogenics is not required.

To achieve this sensitivity it is necessary: (i) that the test masses are very weakly coupled to each other (otherwise the displacement signal resulting from such a tiny acceleration is too small to detect); (ii) that the signal (at the orbital frequency) is up-converted to a higher frequency, the higher the better, to reduce $1/f$ noise. In the GG accelerometer, once unlocked in orbit, the target acceleration signal, $a_{\rm EP}$, would generate a displacement $\Delta x_{\rm EP} \simeq 0.6\,{\rm pm}$ pointing to the center of the Earth. As shown in Fig. 2 (right), by spinning the satellite and the enclosed accelerometer, with its displacement transducer, around their common symmetry axis, the EP violation displacement signal is *modulated* at the spin frequency of the system relative to the center of the Earth: $\omega_{s\oplus} \equiv \omega_s - \omega_{\rm orb}$.

It is to be noted that this signal modulation could in principle be achieved by spinning the displacement transducer *only*, and not the test cylinders themselves (though it would not be wise), which means that there is no forcing of the coupled cylinders due to rotation. Instead, in Fig. 1, where the rotation axis is perpendicular to the sensitive axis of the accelerometer, the whole accelerometer must rotate — faster than it orbits around the Earth — in order to up-convert the signal to a higher frequency. Therefore, it will necessarily respond as a forced oscillator, with natural angular frequency ω_n, forced at the spin frequency $\omega_{s\oplus}$, just as if the accelerometer is sitting still and the Earth is rotating around it at $\omega_{s\oplus}$. This fact limits the spin frequency to be smaller than the natural coupling one ($\omega_{s\oplus} < \omega_n$) because a forcing signal at a frequency higher than the natural one would be attenuated by the factor $(\omega_{s\oplus}/\omega_n)^2$. Instead, very sensitive EP tests require *both* weak coupling (i.e. small ω_n) *and* fast rotation (i.e. high $\omega_{s\oplus}$). With its novel design (see Fig. 2), GG can satisfy both these needs, a property which is unique to this experiment since the limitation reported above holds also on the ground for EP tests with rotating torsion balances.

Once the spacecraft has been given the required rate of rotation at the beginning of the mission (2 Hz with respect to the center of the Earth), no motor or ball bearings are needed inside the satellite. In fact, all parts of the apparatus and the satellite corotate around a common symmetry axis. Since the satellite is not constrained to spin slowly, a spin speed which optimizes the stability of the satellite can be chosen. In this way the spacecraft is also passively stabilized by rotation around its symmetry axis and no active attitude control is required for the entire duration of the space mission. This passive stabilization results in a reduction in the total mass, complexity, cost and (last but not least) acceleration noise on the sensitive accelerometer (Fig. 3, right), which is the heart of the experiment.

Due to the very weak coupling between the masses and rapid spin, the GG system is a rotor in the supercritical regime and supercritical rotors are known to auto-center even if fabrication and mounting errors give rise to departures from ideal cylindrical symmetry. The only disadvantage of spinning at frequencies above the natural oscillation frequencies of the rotor is the onset of whirl motions. These occur at the natural frequencies of the system as orbital motion of the masses around the equilibrium position. Whirl arises due to losses in the suspensions (the smaller the losses, the slower the growth rate of whirl) and needs to be damped to prevent instability. With a Q of at least 20,000, which laboratory tests have shown to be achievable, whirl growth is so slow that experimental runs can be performed between successive damping cycles, thus avoiding any disturbance from damping forces.

The largest disturbing acceleration experienced by the accelerometer is due to the effect of residual air drag acting on the spacecraft and not on test masses suspended inside it. This inertial acceleration, resulting from air drag and in general from nongravitational forces acting on the spacecraft, is in principle the same on both the test bodies. Ideally, common mode effects should not produce any differential signal to compete with the target differential signal of an EP violation. In reality, they can only be partially rejected. In the GG space experiment the strategy chosen is for air drag to be partially compensated by the drag-free control system, and partially rejected by the accelerometer itself. Drag compensation requires the spacecraft to be equipped with appropriate thrusters and a control system to force the spacecraft to follow the motion of an undisturbed test mass inside it at (and close to) the frequency of the signal.

Realistic error budget and numerical simulations of the GG experiment carried out within Phase A mission studies funded by ASI[4] are consistent with an EP test to 1 part in 10^{17}.

The novel design of GG has allowed us to build the full scale "Galileo Galilei on the Ground" (GGG) prototype of the satellite experiment, in which the basic physical principles as well as all the associated technology are tested.

3. GGG Prototype: Design, Experimental Results and Sensitivity

GGG[5–8] mimics the design of GG in every possible way within the constraints set by local gravity. At 1 g, unlike in space, (i) the test masses (10 kg each) need to be supported against local gravity, which breaks the symmetry of the accelerometer along the Z axis; (ii) thicker suspensions are needed, which reduce the time period of the natural oscillation of the balance; (iii) bearings (ball bearings in our case) and a motor are needed to maintain a constant rotation speed, which inevitably conveys some noise to the test bodies; (iv) only the accelerometer rotates, everything else around being a potential source of disturbance (primarily axis tilts).

As shown in Fig. 4, GGG consists of two concentric, coaxial hollow cylinders suspended from a balancing arm as in a beam balance, the beam being vertical and coinciding with the common symmetry axis. This assembly is supported by a hollow shaft, at the midpoint of the balancing arm, through a third cardanic joint. The suspension is arranged such that the masses are free to move in the horizontal plane confined by a weak harmonic potential given by the spring constants of the three suspensions. If a force in the horizontal plane acts on one of the masses, and not on the other, it results in a deviation of the masses from their equilibrium position which is sensed by two sets of capacitance plates, as in GG. If a gravitational interaction with an external body results in such a displacement (and other "classical" differential couplings are ruled out), it would constitute a violation of the EP.

In the GGG experiment, the Sun is the source mass of an expected EP violation and therefore the deviation (if any) would occur at a period of 24 h. If $\eta \equiv \Delta a/a$ is the difference in acceleration between the two suspended masses, normalized to the average acceleration that they experience toward the Sun, their displacement from the equilibrium position is given by $\Delta x = \eta a/\omega_n^2$, where ω_n is the natural frequency of oscillation of the masses relative to each other in the harmonic potential of the balance. The centers of mass of the suspended bodies coincide (unlike ordinary balances), as in GG, to minimize classical tidal effects. The shaft is supported on ball bearings such that the accelerometer can be rotated around the symmetry axis to modulate the displacement signal, as in GG. The microstepping motor, driven by very stable clock, is weakly coupled to the rotor in order to minimize the noise conveyed to the rotor from the motor.

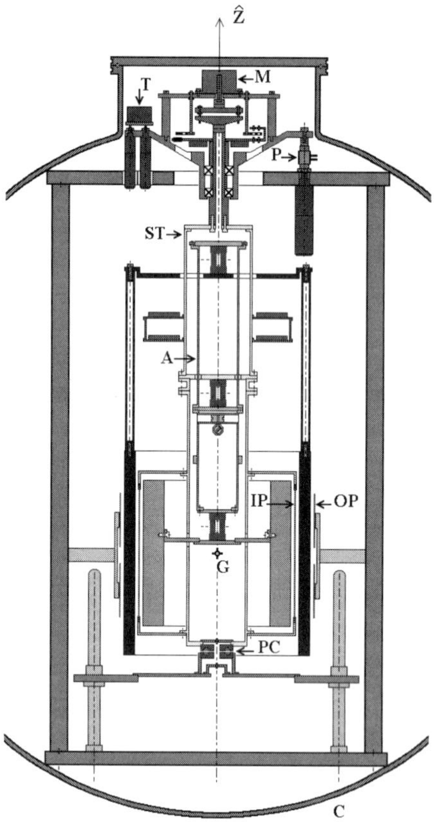

Fig. 4. Section through the spin axis $\hat{Z}$ of the GGG differential accelerometer inside the vacuum chamber. The drawing is to scale and the inner diameter of the vacuum chamber is 1 m. C — vacuum chamber; M — motor; x — ball bearings; ST — suspension tube; A — coupling (balance) arm, located inside the suspension tube, with its three laminar cardanic suspensions (in red); G — center of mass of the two-cylinder system (in blue the outer cylinder, in green the inner one; 10 kg each). IP are the internal capacitance plates of the differential motion detector, OP are the outer ones for whirl control and PC is the contactless inductive power coupler providing power to the electronics inside the rotor. T and P, at the top of the rotor, are the tilt meter and three PZT's (at 120° from one another; only one is shown) for automated control of low frequency terrain tilts.

Figure 5 shows, with a systematic series of measurement runs, the property of the GGG test cylinders to auto-center to reach a well-defined position of physical equilibrium which — for the given apparatus — is independent from initial conditions, as expected theoretically. Figure 6 reports some of the Q values measured with GGG, indicating that the values required for GG — with thinner suspensions — are realistic.

Figure 7 reports, in $m/\sqrt{\text{Hz}}$, residual relative displacements of the test cylinders in the horizontal, nonrotating plane, as measured between June 2005 and October 2006. We have acquired the ability to perform long runs during which

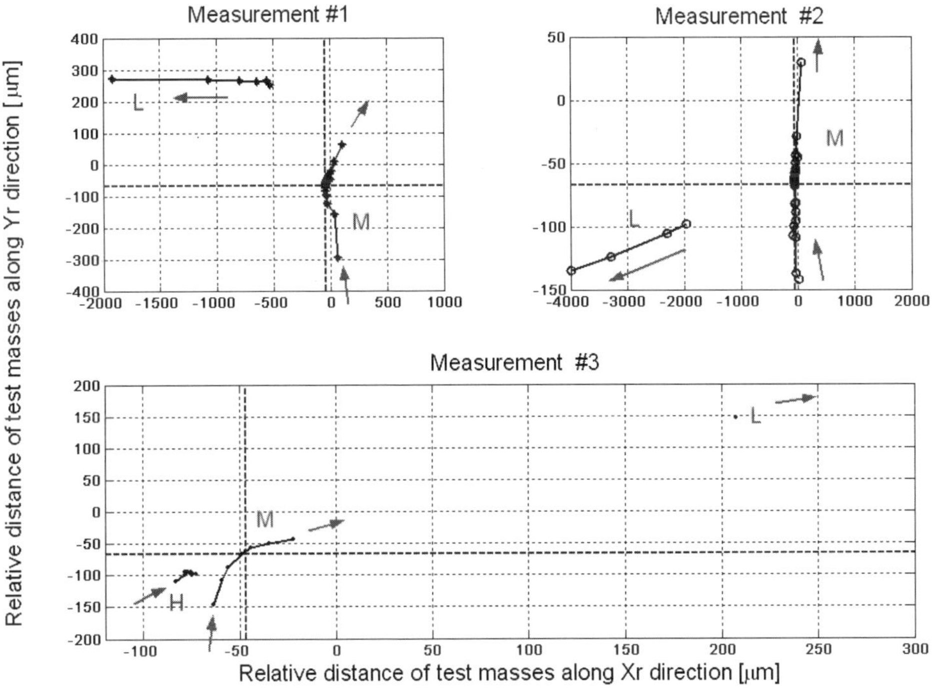

Fig. 5. Experimental evidence for auto-centering of the test cylinders in supercritical rotation. In the horizontal plane of the rotor, $X_r Y_r$, the centers of mass of the test cylinders approach each other as the spin frequency increases (along red arrow) from below the first resonance (L), to between the two resonances (M), to above both resonances (H). The equilibrium position reached is always the same (determined by the intersection of the two dashed lines), independently of initial conditions, as predicted theoretically. Each data point refers to a run of several hours.

whirl (at $\simeq 0.076\,\mathrm{Hz}$, 13.2 s natural period) is accurately controlled and is not a limitation. For an EP experiment in the field of the Sun, i.e. at $1.16 \times 10^{-5}\,\mathrm{Hz}$, the best result is $\simeq 2 \times 10^{-6}\,\mathrm{m}/\sqrt{\mathrm{Hz}}$, amounting — in 3.8 days of integration time — to $\simeq 3.5 \times 10^{-9}\,\mathrm{m}$. With a natural frequency of 0.076 Hz, and an acceleration from the Sun of $6 \times 10^{-3}\,\mathrm{m/s^2}$, this means a sensitivity $\eta_{\odot 3.8\,\mathrm{days}} \simeq 1.3 \times 10^{-7}$, currently limited by the 16-bit ADC converter and by the residual noise due to terrain tilts and temperature variations. Increasing the natural period would improve the sensitivity as the period squared. We are working on that, though it is not easy at 1 g. Longer integration times are feasible (e.g. 10 months, with an improvement by a factor 10); a spin frequency closer to the nominal 2 Hz value of the GG satellite will be tested.

4. Relevance of GGG Current Sensitivity for the Experiment in Space

Let an accelerometer with the capabilities of GGG as of today be flown in a GG satellite (520 km altitude, $1.75 \times 10^{-4}\,\mathrm{Hz}$ orbital frequency). From Fig. 7,

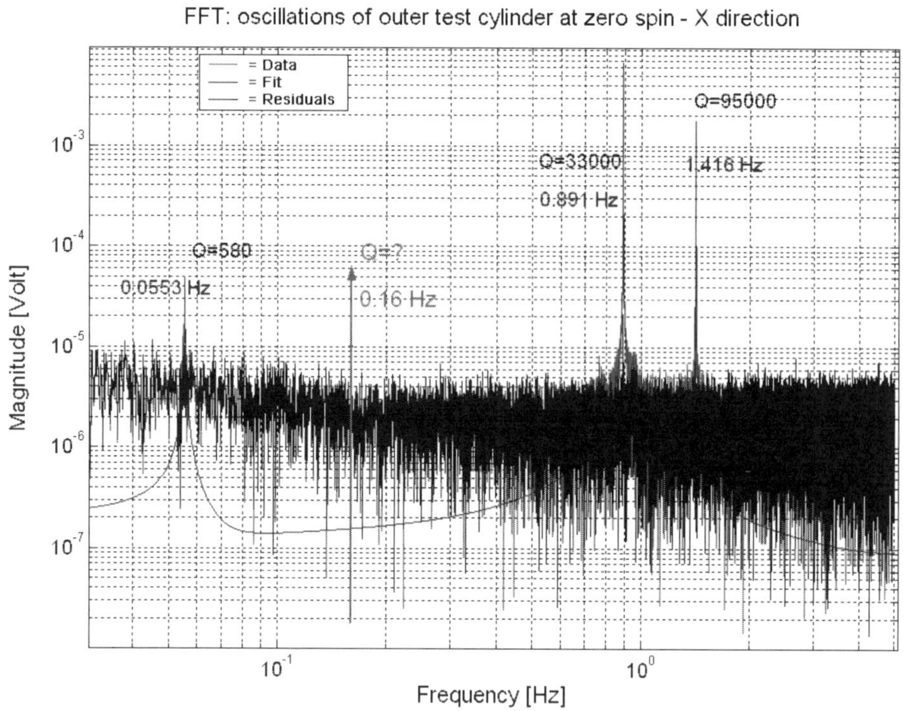

Fig. 6. Quality factors of the GGG accelerometer at its three natural frequencies, obtained by exciting them (at zero spin) and measuring the oscillation decay. In a later assembly, at the natural frequency of 0.08 Hz we have measured $Q = 3970$. In supercritical rotation the relevant Q is measured from the growth of whirl at the natural differential frequency of the test cylinders; at a spin frequency of 0.16 Hz (green line) we have measured 3020.

at 1.75×10^{-4} Hz it is sensitive to $10^{-6}\,\mathrm{m}/\sqrt{\mathrm{Hz}}$, and hence — in 10 months of integration time, very compatible with the GG mission's duration — to relative displacements of 280 pm. With a natural frequency of 0.076 Hz (13.2 s period), and an acceleration from the Earth of $8.4\,\mathrm{m/s^2}$, this means an EP test in the field of the Earth to 7.5×10^{-12}, obtained by performing in space just as GGG in the lab today.

It is to be noted that in space the accelerometer will have *both* a lower platform noise *and* a better sensitivity. A *lower platform noise* than in GGG is due to the absence of terrain tilts, motor and bearings, resulting in an improvement by about a factor of 50. A *better sensitivity* requires us to weaken the coupling of the test cylinders; this can be done because the largest acceleration on GG (due to residual air drag) is 10^8 times smaller than local gravity on GGG. In fact, GG is designed with a natural period of 545 s[4] versus 13.2 s now in GGG, which means an improvement by a factor of 1700. For these improvements to take effect, the ADC converter will also need to be improved; as for thermal effects, they will be less

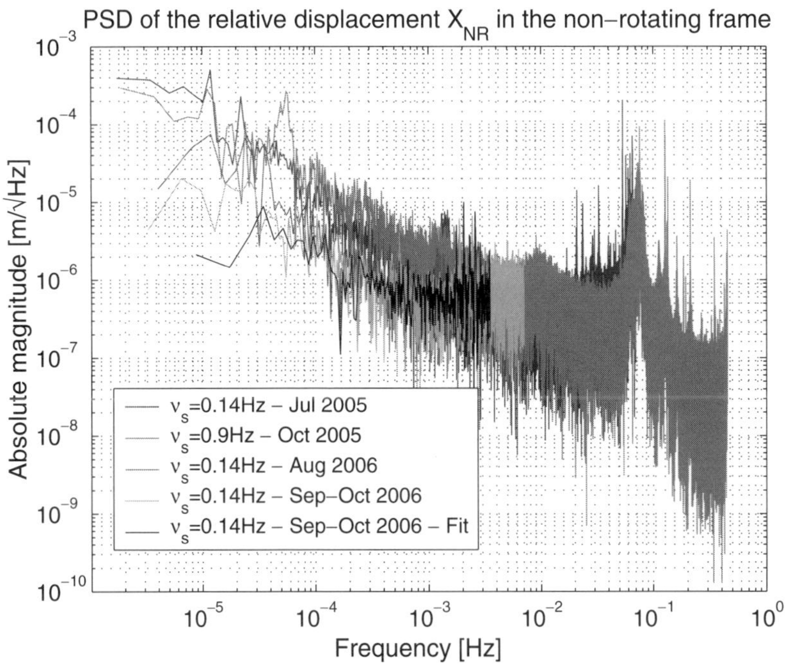

Fig. 7. Power spectral density (in $\mathrm{m}/\sqrt{\mathrm{Hz}}$) of the relative displacements of GGG test cylinders in the horizontal, nonrotating plane (X_{nr} direction). Whirl at the frequency $\simeq 0.076$ Hz is clearly well controlled. The best result (shown in black) was obtained after subtracting the displacements due to thermal expansions. The thermal expansion was modeled as a linear function of temperature. An EP violation signal from the Sun would occur at $1.16 \cdot 10^{-5}$ Hz (1 day). The orbital frequency of GG, at which an EP violation from the Earth would occur in space (see Fig. 2), is 1.75×10^{-4} Hz (5700 s).

severe because of the common rotation of the whole satellite, but require appropriate choices (e.g. carbon fiber structure, thermal insulation) already taken into account in GG mission studies.[4]

With such a reduced platform noise and an improved sensitivity made possible in space, the sensitivity reported above for GGG (7.5×10^{-10}) gets close to the GG mission target of testing the EP to 10^{-17}, which requires one to detect relative displacements of 0.6 pm at the satellite orbital frequency of 1.75×10^{-4} Hz.

Acknowledgment

GG studies have been funded by ASI and the GGG experiment by INFN.

References

1. S. Baeßler *et al.*, *Phys. Rev. Lett.* **83** (1999) 3585.
2. STEP website, http://einstein.stanford.edu/STEP/step2.html
3. μCROSCOPE website, http://www.onera.fr/dmph/accelerometre/index.html

4. "*Galileo Galilei*" (*GG*) *Phase A Study Report* (A. M. Nobili *et al.*) ASI Nov. 1998, 2nd edn., Jan. 2000 (see also GG website, http://eotvos.dm.unipi.it/nobili).
5. G. L. Comandi *et al.*, *Phys. Lett. A* **318** (2003) 213.
6. http://eotvos.dm.unipi.it/nobili/comandi_thesis (G. L. Comandi, PhD thesis, University of Pisa, 2004).
7. G. L. Comandi *et al.*, *Rev. Sc. Ins.* **77** (2006) 034501.
8. G. L. Comandi *et al.*, *Rev. Sc. Ins.* **77** (2006) 034502.
9. ASI National Aerospace Plan, 2006, http://www.asi.it/html/ita/news/20060124_pasn.pdf

PROBING GRAVITY IN NEO'S WITH HIGH-ACCURACY LASER-RANGED TEST MASSES

A. BOSCO, C. CANTONE, S. DELL'AGNELLO*, G. O. DELLE MONACHE, M. A. FRANCESCHI, M. GARATTINI and T. NAPOLITANO

Laboratori Nazionali di Frascati (LNF)
dell'Istituto Nazionale di Fisica Nucleare (INFN),
Frascati (Rome) 00044, Italy
**Simone.DellAgnello@lnf.infn.it*

I. CIUFOLINI

University and INFN, Lecce, I-73100, Italy

A. AGNENI, F. GRAZIANI, P. IALONGO, A. LUCANTONI, A. PAOLOZZI, I. PERONI and G. SINDONI

School of Aerospace Engineering,
University "La Sapienza," Rome, 00184, Italy

G. BELLETTINI and R. TAURASO

Department of Mathematics,
University "Tor Vergata," Rome, 00133, Italy

E. C. PAVLIS

University of Maryland,
Baltimore & NASA Goddard, MD 21250, USA

D. G. CURRIE

University of Maryland,
College Park, MD 20742, USA

D. P. RUBINCAM

NASA Goddard Space Flight Center,
Greenbelt, MD 20771, USA

D. A. ARNOLD

94 Pierce Road, Watertown,
MA 02472-3035, USA

R. MATZNER
University of Texas at Austin, Austin, TX 78712, USA

V. J. SLABINSKI
US Naval Observatory,
Washington DC, 20392-5420, USA

Gravity can be studied in detail in near Earth orbits NEO's using laser-ranged test masses tracked with few-mm accuracy by ILRS. The two LAGEOS satellites have been used to measure frame dragging (a truly rotational effect predicted by GR) with a 10% error. A new mission and an optimized, second generation satellite, LARES (I. Ciufolini PI), is in preparation to reach an accuracy of 1% or less on frame dragging, to measure some PPN parameters, to test the $1/r^2$ law in a very weak field and, possibly, to test select models of unified theories (using the perigee). This requires a full thermal analysis of the test mass and an accurate knowledge of the asymmetric thermal thursts due to the radiation emitted by the Sun and Earth. A Space Climatic Facility (SCF) has been built at INFN-LNF (Frascati, Italy) to perform this experimental program on LAGEOS and LARES prototypes. It consists of a 2 m × 1 m cryostat, simulators of the Sun and Earth radiations and a versatile thermometry system made of discrete probes and an infrared digital camera.

The SCF commissioning is well underway. A test of all its subsystems has been successfully completed on August 4, 2006, using a LAGEOS 3 × 3 retroreflector array built at LNF. This prototype has been thermally modeled in detail with a commercial simulation software. We expect to demonstrate the full functionality of the SCF with the thermal characterization of this LAGEOS array by the beginning of September 2006.

Keywords: Gravitomagnetism; climatic test; thermal analysis.

1. Probing Gravity in NEO's with LAGEOS

The LAGEOS[a] I and II satellites have been launched, respectively, in 1976 (by NASA) and 1992 (NASA-ASI) into orbits with high inclinations ($i = 109.9°$ and $52.65°$), low eccentricities ($e = 0.004$ and 0.014) and large semimajor axes ($a = 12{,}270$ and $12{,}163$ km). They are high-accuracy, passive, spherical test masses, whose orbit is tracked with < 1 cm precision by the 40+ stations of ILRS (International Laser Ranging Service) scattered all over the Earth. They have a weight of about 400 kg, a 60 cm diameter and 426 fused silica cube corner retroreflectors (CCR's) for the satellite laser-ranging measurement (SLR). The primary purpose of LAGEOS I was space geodesy. Later it was shown that a pair of these satellites with supplementary inclinations was a good tool for experimental tests of general relativity.[1]

The LAGEOS data were used in 1998 for the first-ever measurement[2] of the phenomenon of dragging of inertial frames by a central rotating mass (the Earth

[a]LAser GEOdynamics Satellite.

in this case) acting on its orbiting satellite. This effect was predicted by Einstein (who named it "frame dragging"), Lense and Thirring in 1916–1918. For its formal similarity to electromagnetism, it is also referred to as the Earth "gravitomagnetism." Frame dragging can be observed also with pointlike spinning bodies (i.e. gyroscopes): this is the goal of the Gravity Probe B mission, launched in 2004, which ended its data-taking in 2006. GP-B is an active, high-technology satellite (probably one of the most sophisticated ever), aimed at a one-time-only measurement of frame dragging with an accuracy $\leq 1\%$.

Recently, a larger set of LAGEOS data (about 11 years), in conjunction with a much-improved determination of the Earth geopotential field (mainly due to the two GRACE satellites), was used to remeasure the frame dragging effect in NEO's with 10% accuracy.[3] The measured value of the frame dragging precession of the two combined orbital nodes, $\dot{\Omega}_{\rm FD} = 47.9\,{\rm mas/yr}$ (milliarc-sec/yr), is in good agreement with the GR prediction, $\dot{\Omega}_{\rm FD} = 48.2\,{\rm mas/yr}$. For the LAGEOS altitude $h \sim 6000\,{\rm km}$, this amounts to a nodal precession of about $2\,{\rm m/yr}$.

In the next few years, the knowledge of the geopotential is expected to improve thanks to the GRACE, CHAMP and GOCE missions. Since LAGEOS has a virtually limitless orbit lifetime ($\sim$ 1 million years), the nongravitational perturbations (NGP's) will become an important experimental error on $\dot{\Omega}_{\rm FD}$. Among NGP's, the main sources of error are nonconservative thermal forces due to the varying and asymmetric space climatic conditions. These contribute to $\sigma(\dot{\Omega}_{\rm FD})$ with a few %. A very detailed error analysis and error budget can be found in Ref. 4.

2. The New LARES Mission

This paper describes the focussed R&D effort which is being carried out by the LARES Collaboration, within the infrastructure of the Frascati National Laboratory of INFN (LNF) near Rome, Italy, to address the significant issue of the thermal NGP's. This work builds upon an extensive analysis on this subject performed by several authors in the past. At LNF, experimental measurements in a NEO space climatic facility will be done for the first time. This program has two main goals:

- Climatically characterize LAGEOS prototypes to reduce NGP's errors on $\dot{\Omega}_{\rm FD}$. A significant improvement can be reached on this, but we have to accept the unavoidable limitation that we are testing prototypes and validating models, not the original satellites (but we are trying to get hold of the original engineering models).
- Design a new mission and build a fully characterized satellite, which avoids as much as possible the weaknesses of LAGEOS and is capable of reaching an accuracy $\sigma(\dot{\Omega}_{\rm FD}) \leq 1\%$. Such a follow-up mission, LARES,[b] has been considered since the late 1990's. Because SLR is a consolidated technique, the LAGEOS

[b] LAser RElativity Satellite.

data analysis is mature, and thanks to the SCF, the time is right to launch a modern, second generation test mass.

These measurements will make the LAGEOS nodes more robust observables under the effect of thermal NPG's, but LARES will be needed to get the ultimate accuracy, both for physics and for space geodesy.[c] Unlike LAGEOS, the LARES perigee will be usable in the analysis, in addition to the node (which is much less perturbed by NGP's). LARES can be the beginning of the implementation of a high-accuracy SLR *constellation*.

The construction and testing of LARES was formally proposed to INFN in mid-2004. In 2005, the mission was approved by INFN for R&D work in 2006 and 2007. The SCF was then built and completed by July 2006. In 2006 the Collaboration proposed the launch to space agencies. A proposal was submitted to the joint ASI-INFN committee for the qualification flights of the new ESA launcher, VEGA.

3. Thermal NGP's on LAGEOS

Laboratory measurements of the thermo-optical properties of the LAGEOS retro-reflectors have been advocated for many years by the leading experts in this field (Rubincan, Farinella, Slabinski, etc.). Tests like these in a NEO SCF were not conducted on either LAGEOS I or LAGEOS II. Due to their larger temperature asymmetry and emissivity, the CCR's give rise to a thermal drag perturbation far much larger than that of the aluminum structure of the satellite. The orbital perturbations depend significantly on the Yarkovsky effects — specifically the diurnal Yarkovsky effect, the seasonal Yarkovsky effect (also known as *thermal drag* and first understood, in the case of LAGEOS, by Rubincam) and the Yarkovsky–Schach effect. The seasonal Yarkovsky effect is due to the Earth infrared radiation, while the Yarkovsky–Schach effect is due to the modulation of the solar radiation by the Earth shadow. The magnitude of these effects depends upon the spin axis orientation, the spin rate and the thermal properties of the satellites. Among the thermal properties, of particular importance and concern for the orbital dynamics of LAGEOS and LARES, is the *thermal relaxation time* of the CCR's ($\tau_{\rm CCR}$). There are semianalytical approaches to the calculation of $\tau_{\rm CCR}$ and empirical ones based on the analysis of the orbit residuals. Both have limitations. In the literature, estimates vary by over 300% from 1625 s to about 7070 s.

The frame dragging effect on the node of each LAGEOS satellite (the *single* node, not the special linear combination of the two nodes used in the most recent analysis[3]) is about 31 mas/yr.[1] Taking the central value of the above wide range of $\tau_{\rm CCR}$, the thermal drag effects on the nodes turn out to have a very-long-period, secular-like amplitude $\dot{\Omega}_{\rm TD} = 1$–$2\,{\rm mas/yr}$ (the subscript TD indicates the *thermal drag* contribution to $\dot{\Omega}_{\rm FD}$). The size of this effect is, on the node, mainly due to the

[c] At the 2005 ILRS conference in Eastbourne, UK, it was pointed out that one of the next frontiers is to aim for mm-level SLR accuracy.

thermal drag during the satellite eclipses by the Earth. Indeed, when the satellite orbits are not entering the Earth shadow there is a reduction of the *nodal* thermal drag by a factor of the order e.

Let us now consider the uncertainty on the prediction for τ_{CCR} and let us take, for example, $\sigma(\tau_{\mathrm{CCR}})/\tau_{\mathrm{CCR}} \sim 250\%$. First of all, the orbital effects of the thermal drag are periodical and are, thus, averaged out or fitted for[2] over very long periods at the level of 90% and only a residual factor, $R_{\mathrm{TD}} = 10\%$, remains. Second, the long period nodal perturbations of the thermal drag during the eclipse season are linearly proportional to τ_{CCR}. Therefore

$$\frac{\sigma(\dot{\Omega}_{\mathrm{TD}})}{\dot{\Omega}_{\mathrm{TD}}} = \frac{\sigma(\tau_{\mathrm{CCR}})}{\tau_{\mathrm{CCR}}} \times R_{\mathrm{TD}} = 250\% \times 10\% = 25\%. \tag{1}$$

For $\dot{\Omega}_{\mathrm{TD}} = 1$–$2\,\mathrm{mas/yr}$ (see above), one then gets

$$\sigma(\dot{\Omega}_{\mathrm{TD}}) = 1\text{–}2\,\mathrm{mas/yr} \times 25\% = 0.25\text{–}0.5\,\mathrm{mas/yr}. \tag{2}$$

Finally, the relative uncertainty in the measurement of the frame dragging effect from the thermal relaxation time *only* is

$$\frac{\sigma(\dot{\Omega}_{\mathrm{TD}})}{\dot{\Omega}_{\mathrm{FD}}} = \frac{0.25\text{–}0.5\,\mathrm{mas/yr}}{31\,\mathrm{mas/yr}} = 0.8\text{–}1.6\%. \tag{3}$$

Clearly, for LARES, it will be critical to have an accurate measurement of τ_{CCR} to use in the NASA GEODYNE orbit determination program used for data analysis. The SCF has been designed to achieve $\sigma(\tau_{\mathrm{CCR}})/\tau_{\mathrm{CCR}} \leq 10\%$ for the LARES retroreflectors, which, in the baseline design, are the same as the LAGEOS ones. Since τ_{CCR} is the basic observable which governs all thermal forces, this will put the climatic NGP's well under control for the LARES goals.

4. The LNF Space Climatic Facility

A schematic view of the SCF is shown in Fig. 1. The size of the steel cryostat is approximately 2 m length by 1 m diameter. The inner copper shield is painted with Aeroglaze Z306 black paint (0.95 emissivity and low outgassing properties) and is kept at $T = 77\,\mathrm{K}$ with liquid nitrogen. When the SCF is cold, the vacuum is typically in the 10^{-6} mbar range.

A support fixture on the ceiling holds the prototype spacecraft in front of the Earth infrared simulator (inside the SCF). The solar simulator is outside, behind a quartz window (40 cm diameter, 4 cm thickness), which is transparent to the solar radiation up to 3000 nm. A side flange with a germanium window allows one to take thermograms of the prototypes with a FLIR infrared digital camera.

The Earth simulator is a 30-cm-diameter disk painted with Aeroglaze Z306, kept at the appropriate temperature (250 K) and distance from the satellite prototype in order to provide the CCRs with the same viewing angle in orbit ($\sim 60°$ for LAGEOS). The sun simulator (from www.ts-space.co.uk) provides a 40-cm-diameter beam with a close spectral match to the AM0 standard of 1 Sun in space

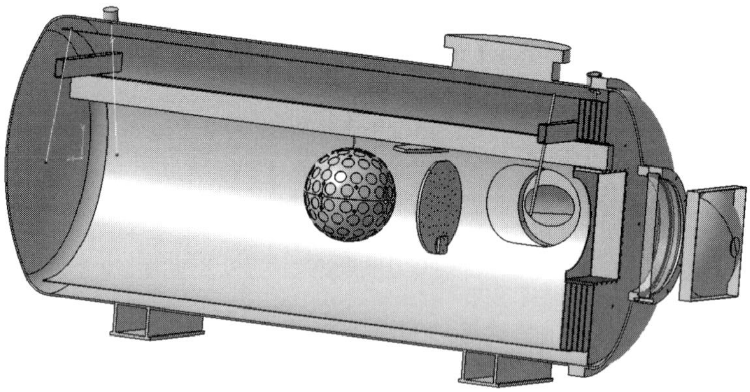

Fig. 1. Sketch of the LNF Space Climatic Facility.

(1366.1 W/m^2), with a uniformity of ±5% over an area of 35 cm diameter. The spectrum is formed from the output of two sources, namely an HMI arc lamp (UV-V), together with a tungsten filament lamp (Red-IR). The quartz halogen lamp (with the tungsten filament) has a power of 12 KW, while the metal halide lamp has 6 KW power. These two sources are filtered such that when the two beams are combined with a beam splitter/filter mirror, the resulting spectrum is a good match to AM0 in the range of 400–1800 nm (see Fig. 2). The spectrum has also been measured up to $\lambda = 3000$ nm (important for $\dot{\Omega}_{\mathrm{TD}}$) and found to be in reasonable agreement with AM0 (see Fig. 3).

The absolute scale of the solar simulator intensity is established by exposing the beam to a reference device, the *solarimeter*, which is a standard www.epply.com

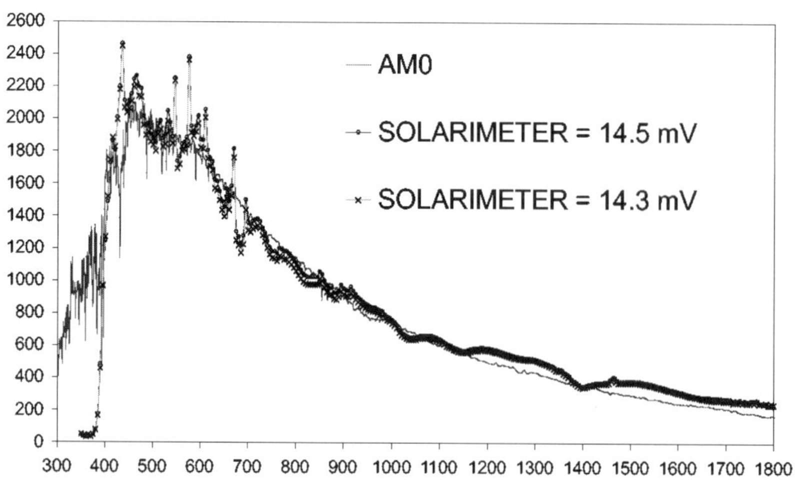

Fig. 2. AM0 spectrum (W/m^2/nm × 10^3) as a function of wavelength (nm) and two (almost undistinguishable) spectra measured with the SCF simulator, for two different values of the lamp currents and solarimeter readings. Typical currents are around 36 A (tungsten) and 29 A (HMI).

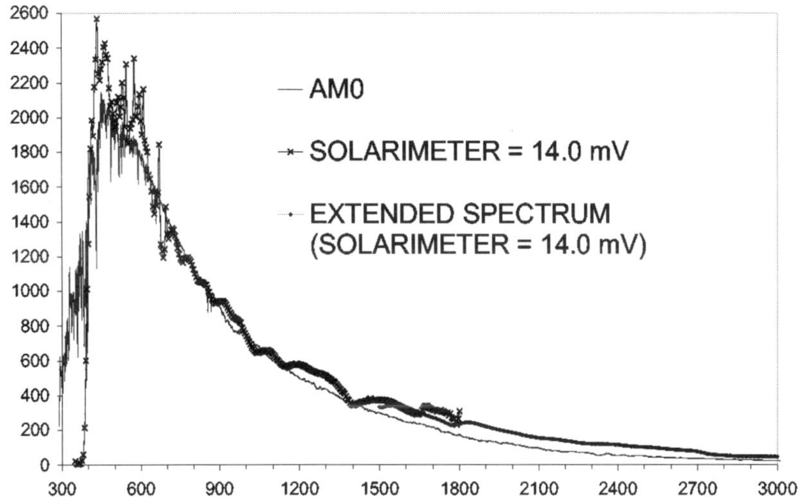

Fig. 3. Measured extended solar simulator spectrum ($W/m^2/nm \times 10^3$) for $\lambda > 1500$ nm.

thermopile. The solarimeter is basically a calibrated blackbody, accurate and stable over 5+ years to ±2%. It is used over long times to adjust the power of the lamps and compensate for their ageing. During continuous operation, the beam intensity is monitored and controlled by means of a feedback PID photodiode which reads a portion of the beam with a small optical prism.

4.1. *The LAGEOS "matrix" prototype*

An array of 3 × 3 CCR's has been built at LNF, following the traditional LAGEOS CCR mounting configuration (see Fig. 4). This matrix contains nine LAGEOS-type CCRs, KEL-F plastic mounting rings, Al retainer rings and screws (three per CCR).

5. Thermal Simulations and Experimental Measurements

The simulations have been performed with a commercial specialized satellite software by C&R–Tech (www.crtech.com), Thermal Desktop (geometric thermal modeler) + RadCad (radiation analysis module) + Sinda-Fluint (solver) + orbital simulator, indicated with TRS in the rest of the paper. We expect to iterate several times between SCF measurements and TRS simulations.

The overall strategy of the program is described in the following:

(i) Hold the average temperature of the Al body of the prototype, $T(\text{AL})$, to the expected value of 300 K[5] and measure in the SCF:

  (a) emissivity (ϵ) and reflectivity (ρ) of CCR's and Al retainer rings;
  (b) τ_{CCR} and, similarly, for the Al retainers, τ_{AL};
  (c) surface temperature distribution (i.e. thermal forces).

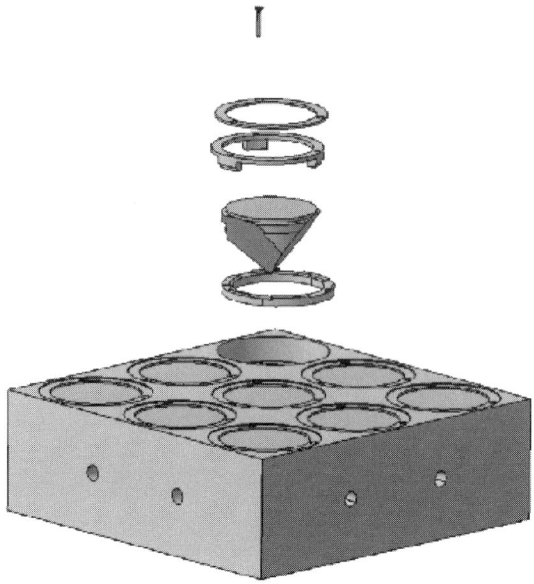

Fig. 4. LAGEOS CCR assembly. The assembly elements facing outside (and, therefore, causing thermal thrusts) are the CCR's and the Al retainer rings (screws can be neglected).

(ii) Repeat all of the above for $T(\mathrm{AL})$ different from 300 K; we are also considering changing ϵ and ρ of the Al body by modifying the surface finish, in both a uniform and a nonuniform way.

(iii) Tune the TRS models to the SCF data for "static" climatic conditions, in which the Sun and Earth radiations are turned on and off alternatively.

(iv) Use the validated TRS models to predict the LAGEOS and LARES behavior along their full orbits using the TRS orbital simulator.

(v) The sequence of the prototype simulations and measurements will be:

(a) Test the LNF LAGEOS matrix in detail; a LAGEOS I sector from NASA-GSFC may become available for testing at the SCF in September 2006.

(b) Use the matrix results to simulate and optimize the design of LARES in order to reduce the thermal forces wrt LAGEOS.

(c) Build a LARES prototype and test it in the SCF.

(vi) Test the effect of satellite spin, first in the simulation and then in the SCF.

5.1. *Simulation results for the matrix*

τ_{CCR} has been estimated from TRS for various climatic conditions and values of $T(\mathrm{AL})$. For example, Fig. 5 shows the temperature variation of the front face of the CCR, $T(\mathrm{CCR})$, in the case of illumination by the Sun for $T(\mathrm{AL}) = 300\,\mathrm{K}$. Figure 6 shows τ_{CCR} vs $T_{\mathrm{avg}} = \frac{1}{2}(T_{t=\infty} + T_{t=0})$. $T(\mathrm{CCR})$ is a strong function of the CCR

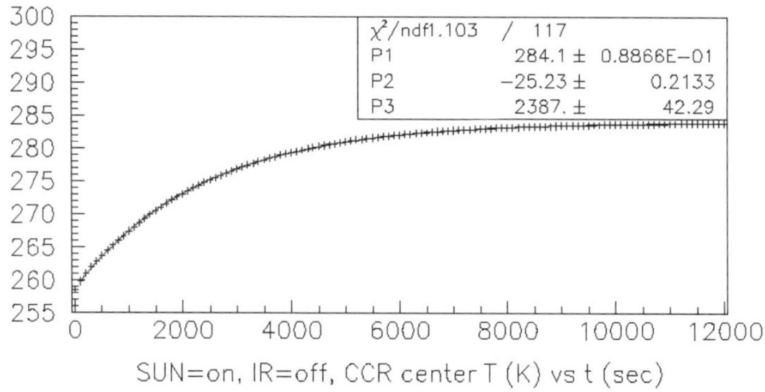

Fig. 5. LAGEOS matrix. Exponential fit to T(CCR) vs time in the simulation when the Sun is turned on at t > 0, $\alpha_{\rm Sun} = 15\%$ and T(AL) $= 300\,$K. P3 $= \tau_{\rm CCR}$ in seconds. This data assume a temperature accuracy of 0.5 K.

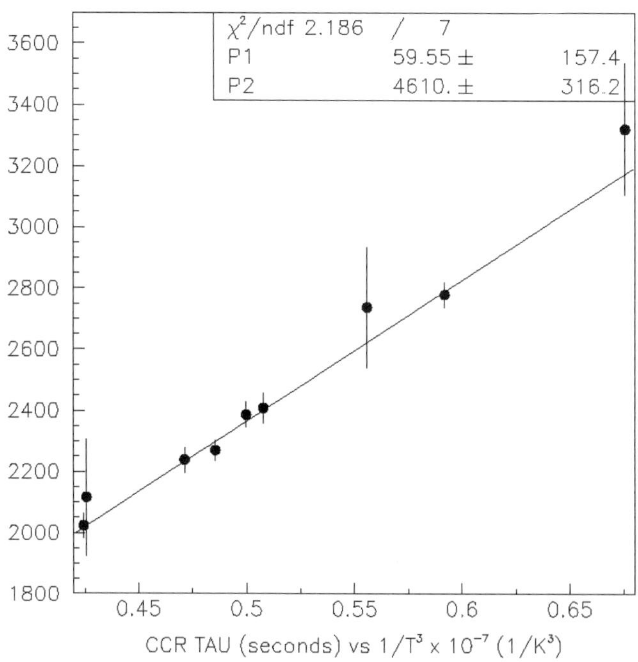

Fig. 6. LAGEOS matrix. $\tau_{\rm CCR}$ vs $1/T^3_{\rm avg} \times 10^{-7}$ in the simulation for $\alpha_{\rm Sun} = 15\%$. Each point is a different set of conditions in terms of the Sun and Earth radiations, angle of exposure and value of T(AL). For example, the first point is for T(AL) $= 320\,$K, Sun = on, IR = off; the last point is for T(AL) $= 280\,$K, Sun = on, IR = on. All other points are for T(AL) $= 300\,$K.

solar absorptivity, $\alpha_{\rm Sun}$. Figure 7 shows the effect of a change of $\alpha_{\rm Sun}$ from 15% (adopted in Ref. 5) to 1.5%.

We also studied the effect of ageing of the satellite aluminum surface, by varying its emissivity from ϵ(AL) $= 0.05$ (the value for LAGEOS II before launch[5]), 0.2

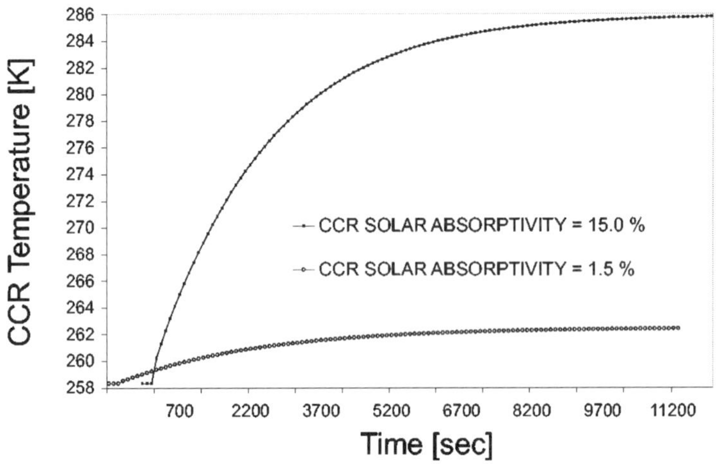

Fig. 7. LAGEOS matrix. Warm-up time of the CCR for $\alpha_{\mathrm{Sun}} = 15\%$ (top) and 1.5% (bottom).

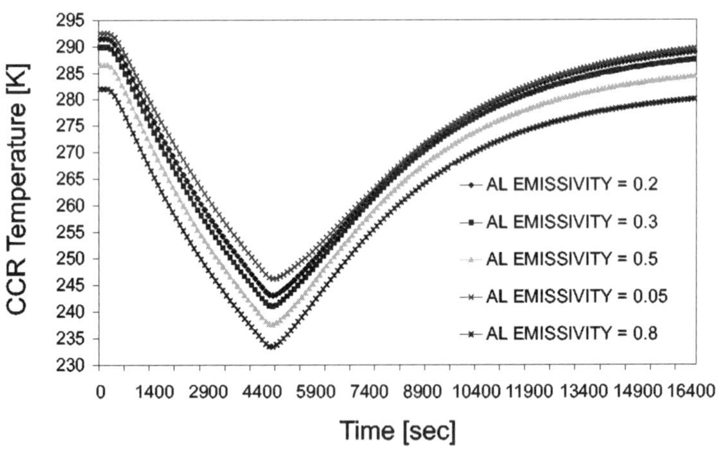

Fig. 8. LAGEOS matrix. Effect of Al ageing on $T(\mathrm{CCR})$ for $\alpha_{\mathrm{Sun}} = 15\%$. Sun off between 200 s and 4700 s (Earth shadow). The higher the value of $\epsilon(\mathrm{AL})$, the lower the $T(\mathrm{CCR})$ curve.

(the value for LAGEOS I before launch[5]) to 0.3, 0.5 and 0.8. This causes $T(\mathrm{AL})$ to change wrt 300 K (the value for no ageing). $T(\mathrm{CCR})$ is also changed by the ageing of the aluminum (see Fig. 8), but the variation of τ_{CCR} (i.e. the shape of $T(\mathrm{CCR})$ vs time) is not significant within our target accuracy, $\sigma(\tau_{\mathrm{CCR}})/\tau_{\mathrm{CCR}} \leq 10\%$.

5.2. *Parametric model of the LAGEOS thermal forces*

A full thermal analysis of LAGEOS is in progress, but it will be completed after the detailed analysis of the matrix. However, in the meantime, a simplified and parametric model of the thermal forces experienced by LAGEOS has been done, which shows the capability of the TRS software and some of the basic features of the thermal NGP's.

The simulated SCF configuration is: (1) satellite pole facing the Sun and Earth simulators, (2) steady state with both simulators turned on at $t = 0$, (3) Sun turned off between $t = 0$ and 4500 s, (4) zero thermal conductance between the Al retainer screws and the Al satellite body. This configuration can be easiily implemented in the SCF and it mimics, approximately, the satellite passage through the Earth shadow and a satellite spin directed along the ecliptic plane. The thermal thrusts are estimated in a parametrized way, using a single CCR in a cavity of an aluminum block held fixed at 300 K. For each row, the single CCR is illuminated by the solar lamp at the appropriate angle and the thermal thrust is computed from the software. The contribution of all CCR's in a row, of all rows and of the two hemispheres, is then summed. The results are shown in Fig. 9 (same climatic conditions as in Fig. 8).

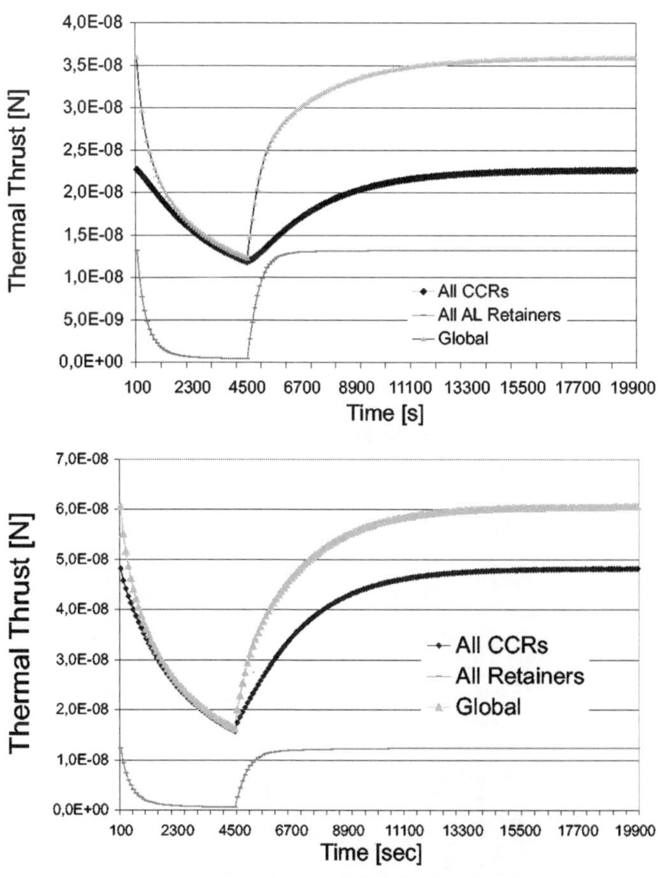

Fig. 9. LAGEOS parametric model. Estimate of the thermal thrusts on the satellite due to the SCF Sun and Earth simulators for $\alpha_{\rm Sun} = 1.5\%$ (top plot) and $\alpha_{\rm Sun} = 15\%$ (bottom plot).

6. Proposal of a New LARES Design

An original design has been developed at LNF[6] to strongly suppress thermal forces. This "shell-over-the-core" design consists of two outer aluminum half-shells, which host the CCR's, and an inner massive spheroidal core, which provides an area–mass ratio less than or equal to the LAGEOS value. The basic idea is to mount CCR's from the inside on the shells, which in turn are screwed over the core, leaving a vacuum gap in between. Since the Al retainer rings will be replaced by retainer seats machined directly from the Al shells, this CCR "back-mounting" option will entirely remove the significant thermal forces due to the Al rings. In addition, some significant thermal radiation released in the gap by warm, illuminated CCR's can thus propagate to colder CCR's in a dark region. This must be aided by a proper choice of the thermo-optical parameters of the inner core, which has yet to be optimized. The goal is to make $T(\mathrm{CCR})$ more uniform than for LAGEOS. A full thermal simulation is in an advanced stage and a 1:2 scale prototype has been built at LNF to test the effectiveness of this new design at the SCF (see. Fig. 10).

7. Completion of the SCF "System Test"

The last major component of the SCF, the solar simulator, has been delivered to LNF at the beginning of July 2006. The following month has been devoted to a system test of the whole apparatus and of the main procedures using the LAGEOS matrix. The test included the combined operation of: cryostat (cooled down to 85 K in a few hours), vacuum vessel (down to 3×10^{-6} mbar), temperature measurement with PT100 probes, temperature control of the Al matrix block with thermocoolers, use of vacuum feedthroughs and irradiation with the AM0 beam (with reduction of the beam to the matrix size with a shroud). The test has been successful and

Fig. 10. 1:2 scale prototype of the new shell-over-the-core LARES design built at LNF.

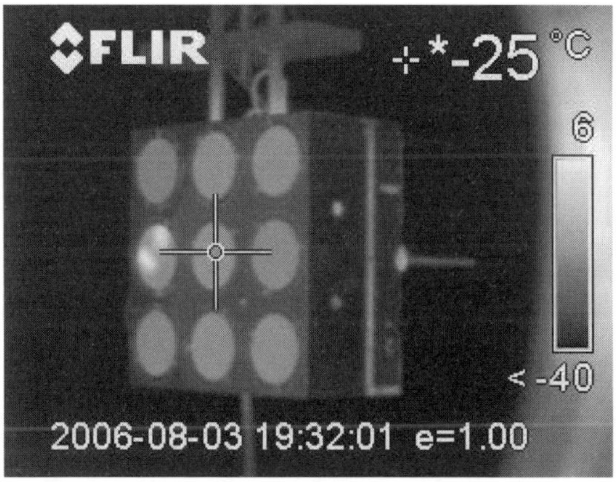

Fig. 11. Thermogram of the LAGEOS matrix taken with the IR camera through the Ge window.

has shown two problems (already solved): one of the feedthroughs did not keep the vacuum and prevented temperature control; the tungsten lamp could barely be operated at maximum power (this was because the voltage in Italy is typically 220 V instead of the 240 V in the UK, where the simulator has been built). Two brand-new feedthroughs have been purchased and delivered; the spare 10 KW tungsten lamp has been swapped with a 12 KW lamp at no exta cost. Figure 11 shows a thermogram of the matrix taken during the cooldown of the SCF with the IR camera.

The digital IR camera has been extensively tested separately and its performance found to be within specs. In an inside test, in air and at room temperature, it was used to estimate the infrared emissivity and reflectivity of the CCR's and the aluminum of the LAGEOS matrix [ϵ(CCR) ~ 0.82, ϵ(AL) ~ 0.15] with a-few-% accuracy.

8. Other Applications of the SCF

An optical test is being set up at LNF for the measurement of far field diffraction patterns of retroreflector arrays in absolute units. Ultimately, this test will be done with the prototypes inside the SCF, thus merging, to a large extent, the thermal and the optical facilities.

Preliminary optical calculations of the expected laser performance of LARES have been carried out. These are based on the baseline assumption that the outer surface of the new satellite will look like LAGEOS, with the diameter scaled from 60 to 30 cm. LARES will have 102 CCR's, versus the 426 of LAGEOS. The calculations indicate that the laser return will be 1–1.5 CCR's and that the expected ranging fluctuations of LAGEOS and LARES will be similar (the smaller radius compensates for the larger number of CCR's).

The LNF group proposes to use these two facilities for the thermal and laser characterization of CCR arrays foreseen on future GNSS[d] constellations (GPS-3 and GALILEO), in close collaboration with ILRS, NASA-GSFC and UMCP.

Answering a call for proposals for the 2006–2008 study by ASI, LNF has also proposed to participate in the design and test of the laser-ranged test masses for the Deep Space Gravity Probe (DSGP) mission, which is being conceived to accurately study the Pioneer effect, as well as to perform important (inter)planetary science investigations. With minor upgrades, the SCF is capable of performing test measurements for SLR in the outer solar system for DSGP, which is a formation of an active spacecraft and a few SLR masses. These upgrades consist in the attenuation of the AM0 solar beam with a set of appropriate wire meshes (which do not distort the spectrum) and in the adoption of cryocoolers, in order to cool down prototypes to temperatures below that of liquid nitrogen. The first of these upgrades has been suggested by the vendor of the solar simulator, which comes with built-in provisions for installing the wire meshes. The IR radiation of planets of the outer solar system can be simulated with black disks of varying size and distance from the prototypes. Note that the observed and unexplained Pioneer 10 and 11 decelerations are about a-factor-of-10 *larger* than typical LAGEOS thermal accelerations. Finally, the typical distances which are foreseen between the active DSGP spacecraft (equipped with the laser) and the SLR test masses are in the kilometer range: therefore, the expensive complication of CCR dihedral angle offsets can be avoided.

9. Conclusions

This paper describes an 18-month preparation of a Space Climatic Facility at INFN-LNF dedicated to the complete thermal and (though less advanced) optical characterization of high-accuracy laser-ranged test masses to probe gravity in NEO's. The main goal is to improve the 10% measurement of the frame-dragging effect currently achieved with LAGEOS down to an accuracy $\leq 1\%$ with the new LARES mission. This second generation satellite, in addition to LAGEOS I and II, would be very valuable also in space geodesy, to strengthen and improve the definition of the International Terrestrial Reference Frame (ITRF). This latter application of the SCF will be further expanded in the near future with the proposed test of retro-reflector arrays for GPS-3 and GALILEO in collaboration with ILRS, NASA-GSFC and UMCP.

Acknowledgments

The authors wish to thank the technicians of the LNF Cryogenics Service (Accelerator Division) and of the SSCR Mechanics Service (Research Division) who have been involved in the design, constructions and operation of the SCF: G. Bisogni,

[d]Global Navigation Satellite System.

A. Ceccarelli, G. Ceccarelli, R. Ceccarelli, A. De Paolis, E. Iacuessa, N. Intaglietta, V. Lollo, U. Martini and A. Olivieri. We wish to deeply and warmly thank the LNF Director, Prof. Mario Calvetti, Ing. Claudio Sanelli and Dr Maria Curatolo for their encouragment and for supporting the SCF with lab infrastructure funds. Special thanks to Dr Marco Ricci, of the INFN Astroparticle Physics Committe, for his patient and practical help, especially during the critical initial phase of the effort. Monumental thanks to Dr Gianfranco Giordano of LNF for allowing the LARES group to use his optical laboratory and numerous accessories.

References

1. I. Ciufolini, *Phys. Rev. Lett.* **56** (1986) 278.
2. I. Ciufolini *et al.*, *Science* **279** (1998) 2100.
3. I. Ciufolini and E. C. Pavlis, *Nature* **431** (2004) 958.
4. I. Ciufolini, E. C. Pavlis and R. Peron, *New Astron.* **11** (2006) 527.
5. V. J. Slabinski, *Cel. Mech. Dyn. Astr.* **66** (1997) 131.
6. G. Bellettini *et al.*, LNF 2005 Annual Report (2005), p. 74.

MEASUREMENT OF THE GRAVITATIONAL CONSTANT USING THE ATTRACTION BETWEEN TWO FREELY FALLING DISCS: A PROPOSAL

LEONID VITUSHKIN* and PETER WOLF†

Bureau International des Poids et Mesures, Time, Frequency and Gravimetry Section, Pavillon de Breteuil, 92312 Sèvres Cedex, France

**LVitushkin@bipm.org*

†PWolf@bipm.org

ARTYOM VITUSHKIN

Minex-Engineering Corp., 1000 Apollo Ct., Unit G, Antioch, CA 94509, USA

artyom@minex-engineering.com

The constant of gravitation, G, is the least well-known of the physical constants. A new, independent method of measurement, estimated as having a potential uncertainty at least as small as that achieved by existing methods, would be useful for an improvement in G determination. This experiment is based on the measurement of the relative motion of two freely falling test bodies (discs), caused by their gravitational attraction. The uncertainties are analyzed for two parallel tungsten discs with masses of about 30 kg. The use of test bodies with an incorporated optical system of multipass two-beam interferometers, as well as of multibeam interferometers, is proposed to measure their relative displacement. The estimations were made for laboratory experiment with free fall duration of 0.714 s. In this case, the relative displacement to be measured is about 0.1 μm. These estimates show that relative uncertainties lower than 5×10^{-5} can be obtained in G measurement in a single drop of the test bodies. The proposed experiment can be made in outer space. In space a lower uncertainty can be achieved because the time interval of the measurement of relative motion of the test bodies can be increased.

Keywords: Gravitational constant; laser interferometry; space experiments; metrology.

1. Introduction

Recent advances in laser interferometry allow for displacement measurements with subnanometer uncertainty. For example, laser displacement interferometers combined with X-ray interferometers are now being developed for the calibration of linear transducers with subnanometer uncertainty. Another scientific effort concerns the design of interferometers for the detection of gravitational waves that are extremely sensitive to the relative displacement of test bodies. In both cases the resolution of laser displacement interferometers, limited only by shot noise of photo detection, was reached.

Current experiments aimed at detecting gravitational waves (see for example Ref. 1) have reached the shot-noise-limited resolution of about $1 \times 10^{-19}\,\mathrm{m/Hz^{\frac{1}{2}}}$. Also, a shot-noise level of $1 \times 10^{-10}\,\mathrm{m/Hz^{\frac{1}{2}}}$ has been reached using the combined optical/X-ray interferometer.[2]

An essential experimental result for understanding the limitations of the accuracy of two-beam laser displacement interferometry was reported in Ref. 3. The residual nonlinearity of the interference signal of the heterodyne interferometer outlined in this paper was less than 0.02 nm.

Such developments in laser displacement interferometry will permit determination, with appropriate accuracy, of the gravitational constant G from direct measurements of the motion of two nearby free-moving test bodies caused by their proper gravitational attraction.

This new, independent method to measure the gravitational constant G will have an uncertainty, at least, not greater than that achieved in previous experiments.

In this paper we analyze the possibility of measuring G from the relative motion of two free-falling discs with parallel basal planes. We show that the gravitational attraction between two discs is greater than that between two spheres of the same mass and separation between the surfaces. Furthermore, the discs are simpler to incorporate within an optical system for measurement of their relative motion. It is worth noting that regularity of a disc can be one or two orders better than that of a sphere.[4]

The sources of uncertainties in the proposed ground-based experiment are analyzed for the reasonable characteristics of the experimental setup. This analysis allows the conclusion that if such an experiment is carried out on a spacecraft, where the time interval of the measurement of relative motion of the test bodies can be significantly increased, the uncertainty in the measurement of the gravitational constant can be diminished.

2. Basic Estimations

2.1. *Relative acceleration of test bodies due to mutual gravity attraction*

A relative acceleration of two identical spheres with the mass $M_1 = M_2 = M$ due to their gravitational attraction is described by

$$a_{GS} = 2G\frac{M}{L^2} = \frac{8\pi G R_s^3 \rho}{3(2R_s + d)^2}, \tag{1}$$

where L is the distance between the centers of the spheres, $G = 6.672 \times 10^{-11}\,\mathrm{Nm^2kg^{-2}}$, ρ is the density of the spheres and d is the distance between their surfaces. For tungsten spheres ($\rho = 19.3 \times 10^3\,\mathrm{kg/m^3}$) with $R = 0.072\,\mathrm{m}$, $d = 0.001\,\mathrm{m}$ and $M = 30.3\,\mathrm{kg}$, we obtain

$$a_{GS}(R = 0.072\,\mathrm{m},\ d = 0.001\,\mathrm{m}) = 1.9 \times 10^{-7}\,\mathrm{m\,s^{-2}}. \tag{2}$$

A numerical calculation of the relative acceleration a_{GS} of two parallel tungsten discs with the mass of 30.3 kg, the radius $R = 0.1\,\mathrm{m}$, the thickness $h = R/2 = 0.05\,\mathrm{m}$ and the distance $d = 0.001\,\mathrm{m}$ between the adjacent surfaces gives

$$a_{GS}(R = 0.1\,\mathrm{m},\ h = 0.05\,\mathrm{m},\ d = 0.001\,\mathrm{m}) = 3.97 \times 10^{-7}\,\mathrm{m\,s^{-2}}. \tag{3}$$

It was found that the optimal shape of the discs to give the maximum mutual attraction for a fixed mass is that with $h = R/2$.

2.2. *Equation of motion*

The gravitational constant may be evaluated from the relative motion equation of two free-falling discs,

$$\frac{\partial^2(\Delta z)}{\partial t^2} = \frac{\partial^2 z_2}{\partial t^2} - \frac{\partial^2 z_1}{\partial t^2} = \gamma_{zz}\Delta z + 2G\frac{M}{L(R,h,\Delta z)}, \tag{4}$$

where z_i is the z axis coordinate of the gravity center of the ith test body, γ_{zz} is the vertical gravity gradient of the Earth gravity field and $L(R,h,\Delta z)$ is a numerically calculated function of the dimensions of the discs and of the distance between the gravity centers.

For such an evaluation the distance between the discs and time intervals should be measured during a free fall. The distance–time intervals should be used for a least-squares evaluation of G. This task is somewhat similar to that of the measurement of a vertical gravity gradient using an absolute ballistic gravity gradiometer with two free-falling bodies.[5,6]

3. Required Uncertainty in Displacement Measurement

The uncertainty in the displacement measurement is crucial for the G measurement. For a fall height of 2.5 m, the free fall duration will be 0.714 s. The relative displacement of the discs with relative acceleration of $3.9 \times 10^{-7}\,\mathrm{m\,s^{-2}}$ is

$$\Delta z_G = \frac{1}{2}a_{GD}t^2 = 0.10\,\mu\mathrm{m}. \tag{5}$$

The optical path change Δs_G in a two-beam interferometer is twice Δz_G. In order to achieve relative uncertainty for a single measurement below 5 parts in 10^5, the change of the optical path should be measured with a relative uncertainty of 0.01 nm. If the multipass optical system with, for example, 48 double passes of the beam will be used in the interferometer, the total optical path change is

$$\Delta s_G = 9.5\,\mu\mathrm{m} = 18.8\lambda, \tag{6}$$

where the laser wavelength $\lambda = 0.515\,\mu\mathrm{m}$ (see Refs. 7 and 8). Use of the multipass optical system allows increased uncertainty in the measurement of the optical path change to 0.5 nm.

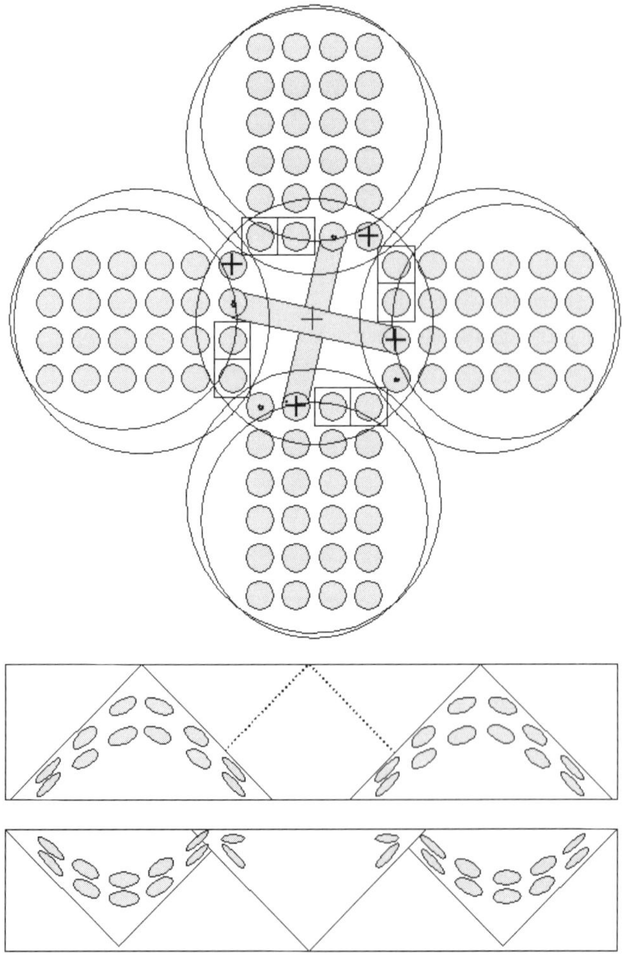

Fig. 1. Diagram of the propagating light beams in two independent multipass optical cells with 48 double passes based on the use of the shifted conical reflectors.

4. Laser Displacement Interferometer

Two independent multipass optical cells[9] based on the use of conical reflectors[10] are proposed for the interferometer (Fig. 1) used in the G measurement. This arrangement allows one to monitor relative tilts of the discs during the fall. The basic idea of the optical arrangement is described in Ref. 9. Instead of using traditional corner cubes, we propose using conical reflectors in order to avoid the deformation of the wave front on the edges of the corner cubes.

The main parts of the interferometer, including the optical elements of the reference arm, a beamsplitter and a photodetector board, can be fixed on the moving cart. This cart will be used to fix the test bodies at their initial positions and also to drop, catch and lift them back. An output beam from a laser installed on the pillar is separated by a beamsplitter in the reference and measuring arms of the interferometer.

5. Sources of the Uncertainty in the Measurement

A long list of possible measurement uncertainties, including that in the measurement of the relative displacements, time intervals, mass and dimensions of the test bodies, was analyzed.

Some of the disturbing factors acting on the test bodies are the same for the two test bodies and will therefore have no effect on their relative motion. They include tidal variations of the gravitational field, electromagnetic interactions between the test bodies and the chamber, interaction of the test bodies with the magnetic fields of the Earth, etc.

The ultimate sensitivity to absolute displacements of the mirror in a two-beam interferometer, limited solely by the shot noise of the photodetector, is given by the formula

$$\delta l_{\min} = 1.5 \times 10^{-6}\sqrt{\Delta f}\,\text{nm}, \tag{7}$$

for the laser power $P = 1\,\text{mW}$, the quantum efficiency of the photodetector $q = 0.3$ and the wavelength $\lambda = 515\,\text{nm}$. In this formula Δf is the signal frequency band. It is seen that $\delta l_{\min}$ is much lower than the uncertainty required in the G measurement.

The uncertainty in the measurement of displacement by a two-beam interferometer with the difference ΔL of the length of the interferometer arms is limited by the laser frequency instability. For the laser with $\Delta f/f = 1 \times 10^{-12}$ and $\Delta L = 1\,\text{m}$, we find the estimation of this uncertainty to be $\delta l_\lambda = 1 \times 10^{-4}\,\text{nm}$.

One of the principal sources of uncertainties in the G measurements is the inhomogeneous gravity field in the laboratory. The relative displacement of two discs caused by the vertical gravity gradient may be obtained from the formula[7]

$$\Delta z_\gamma = \frac{1}{2}\gamma_{zz}\Delta z_0(\Delta t)^2, \tag{8}$$

where Δz_0 is the initial distance between the gravity centers of the discs and M is the free fall time. For the mentioned dimensions of the discs, their initial separation of 1 mm and the vertical gravity gradient of the normal gravity field of the Earth $\gamma_{zz} = 3.086 \times 10^{-6}\,\text{s}^{-2}$, we obtain $\Delta z_\gamma = 0.79\,\mu\text{m}$.

This displacement is practically the same as that caused by mutual gravitational attraction of the test bodies. The relative displacement due to mutual gravitational attraction and that caused by the gravity gradient are opposite in sign. The gravity gradient should be measured for use in the evaluation of motion equations. This measurement can be done with the same system, but with an increased initial separation between the discs to diminish the mutual gravitational attraction. It will also increase the sensitivity of such a vertical gravity gradiometer.

If the measuring system is located on a spacecraft, the influence of the higher derivatives of gravity potential is dramatically diminished, particularly with a proper design of the mechanical and optical assembly (supporting bench), which provides the minimal inhomogeneity of the background gravity field on the axis of the measuring system. The relative tilts of the discs can cause changes in the gravitational attraction and changes of the optical path of the beam. Preliminary estimations show that relative tilts below 0.1 arcsec cause a relative uncertainty in the G measurement below 3×10^{-6}.

In order to have a relative uncertainty in the G measurement below 1×10^{-6}, the pressure of residual gas in the chamber should be below 3×10^{-9} Torr. Precautions should be taken against the electrostatic charging of the discs to avoid disturbing the relative acceleration of the test bodies. The potential difference between the discs should be not more than a few microvolts, to allow a relative uncertainty of 1×10^{-5} in the G measurement. The analysis of other known sources of uncertainty leads to the conclusion that the relative uncertainty of the G measurement below 5×10^{-5}, in a single drop of the test bodies, is possible in the setup with the discs and the above-mentioned parameters.

Those other sources of uncertainty would be the following:

- angular instability of laser radiation,
- diffraction effects,
- inhomogeneity of the material of the discs,
- uncertainties in the evaluation of gravity attraction between the test bodies in the form of the discs with incorporated optical elements,
- uncertainties in the measurement of the dimensions and mass (currently, that can be done for the dimensions and masses used in our estimations at the level below 1×10^{-6}),
- thermal instabilities,
- negligible influence of the Casimir effect,
- uncertainty in the time interval measurements.

The following improvements would diminish the uncertainty in the measurement:

- the use of larger test bodies,
- a shorter wavelength of laser radiation,
- longer time intervals of the free fall of test bodies,
- simultaneous (or subsequent) measurements using a laser radiation at different wavelengths (e.g. at 515 nm and 532 nm),
- special design of test bodies that makes it possible to reinstall the discs after rotation at 180° on a vertical axis to diminish the influence of inhomogeneity of the material of the discs,
- the use of different materials for the discs.

A longer free motion of test bodies could be obtained in an experiment on a spacecraft. If this time interval were to be increased to 10 s, the measured relative displacement of the test bodies would be increased by a factor of 200, making it

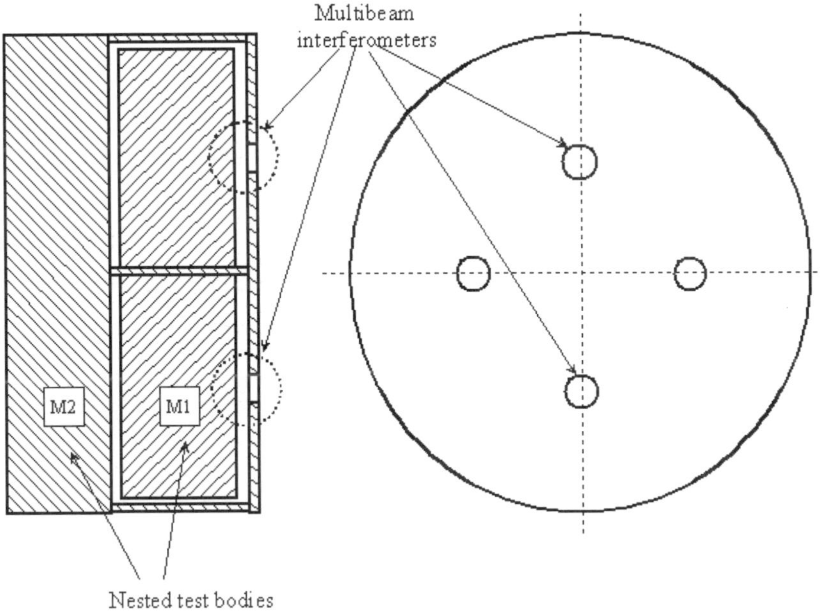

Fig. 2. Schematic configuration of the nested test bodies with four independent multibeam interferometers.

possible to diminish the masses of test bodies and to simplify the optical system. In addition, the influence of the inhomogeneity of the Earth's gravity field would be decreased.

A longer free motion time interval would also make it possible to use multibeam interferometers, of the Fabry–Perot type, for the measurement of the relative displacement of the test bodies. The relative displacement of $0.1\,\mu$m, estimated above for the experiment on the ground, is within one interference fringe corresponding to a displacement of $\lambda/2$ and could not be measured with the required uncertainty. In contrast, on a spacecraft with a possible relative displacement of $20\,\mu$m corresponding to about 78 interference fringes, multibeam interferometers can be used. A special design of the nested test bodies somewhat similar to that proposed for a vertical ballistic gravity gradiometer[5] can be used in the G measurement. In such a configuration (Fig. 2), the multibeam interferometers in reflection are formed by the reflecting surface of one test body and the optical mirrors incorporated in the thin plate fixed to the second test body. Use of four independent interferometers allows the control of relative tilts. The various tests, for example with the use of various wavelengths of laser radiation, are also possible with the independent interferometers.

6. Conclusions

We have proposed a new experiment on the measurement of the gravitational constant. The experiment is based on the measurement of the relative motion due to mutual gravity attraction of two freely falling discs with incorporated components of the optical laser interferometer for the displacement measurement. Precise measurements of the dimensions and mass of the test bodies, as well as precise computation of the function which describes their gravity attraction, are required. Currently this can be done with the relative uncertainty below 1×10^{-6} for the estimated parameters of the test bodies. The gravitational constant is then evaluated from the measured intervals of time and distance using the equation of the motion of freely falling test bodies. The vertical gravity gradient should be measured using the same experimental setup with the increased initial separation between the discs (for example, from 1 mm to 10 cm). Then the measured gravity gradient value should be included in the motion equation.

The theoretical estimations made using realistic parameters of the experimental setup (for example, for the height of free fall of 2.5 m) and based on the practical results showed that the uncertainty in the G measurement may be below 5×10^{-5}. Only actual experimentation can show how much lower the obtained uncertainty would be. Recommended in 2002 by the CODATA value $6.6742(10) \times 10^{-11}\,\mathrm{m^3 kg^{-1} s^{-2}}$ of the gravitational constant is given with the relative standard uncertainty of 1.5×10^{-4} (see Ref. 11). The smallest assigned uncertainty of 1.5×10^{-5} was reported in Ref. 12. The experiments, used in the CODATA adjustment usually based on the torsion balance or on some measurements with the suspended masses. It would be interesting to perform a new, independent determination of the gravitational constant with a potential uncertainty at the same level of uncertainty that was obtained by existing methods.

The advantage of a new method is that it is one of a few which can be performed on a spacecraft where the uncertainty in the G measurement can be diminished at least by a factor 10. In such an experiment the observation time in a single measurement, determined by the length of the relative displacement of the test bodies, can be significantly increased.

The uncertainty in G determination on a spacecraft will mainly be limited by the following factors:

- the uncertainty in mass and dimension measurement,
- the inhomogeneity of the disc material,
- misalignments in the optical system,
- misalignments in the mechanical system.

It is worth noting that the technologies developed for the realization of the proposed experiment on the ground and in space will also stimulate progress in the absolute ballistic gravity gradiometry and in the study of gravitation at short distances.

Acknowledgments

The authors are grateful to T. J. Quinn and R. Davis for the useful discussions based on their own experiences in the measurement of the gravitational constant.

References

1. Laser Interferometer Space Antenna (LISA) (P. Bender *et al.*), Pre-Phase A Report, 2nd edn., July 1998, MPQ 233 (Max-Planck-Institute für Quantenoptick, D-85748 Garching, Germany, 1998).
2. A. Yacoot, Combined optical–X-ray interferometry (COXI), in *Proc. 159 PTB Seminar: Requirements and Recent Developments in High Precision Length Metrology*, eds. H. Bosse and J. Fluegge (Fertigungsmesstechnik PTB-F-45, Braunschweig; Nov. 2001), p. 56.
3. C. Wu, J. Lawall and R. D. Deslattes, *Appl. Opt.* **38** (1999) 4089–4094.
4. Y. T. Chen and A. Cook, *Gravitational Experiments in the Laboratory* (Cambridge University Press, 1993).
5. L. F. Vitushkin, T. M. Niebauer and A. L. Vitushkin, Ballistic gradiometer for the measurement of the vertical gravity gradient: A proposal, in *Proc. IAG Symp. on Airborne Gravity Field Determination* (Calgary, Aug. 1995) p. 47.
6. T. M. Niebauer, D. van Westrum, J. M. Brown and F. J. Klopping, New absolute gradiometer, in *Proc. Workshop "IMG-2002: Instrumentation and Metrology in Gravimetry", Cahiers du Centre Européen de Géodinamique et de Séismologie*, Vol. 22 (Luxembourg, 2003), p. 11.
7. L. Vitushkin and O. Orlov, *Directors Report on the Activity and Management of the International Bureau of Weights and Measures* (*BIPM*) **4** (2003) 164.
8. L. Vitushkin and O. Orlov, *Proc. SPIE* **5856** (2005) 281.
9. A. L. Vitushkin and L. F. Vitushkin, *Appl. Opt.* **37** (1998) 162.
10. A. L. Vitushkin and L. F. Vitushkin, On the use of conical reflectors in laser displacement interferometry, in *Conference Digest of Conference on Precision Electromagnetic Measurements* (14–19 May 2000; Sydney, Australia), p. 477.
11. P. J. Mohr and B. N. Taylor, *Rev. Mod. Phys.* **77** (2005) 1.
12. J. H. Gundlach and S. M. Merkowitz, *Phys. Rev. Lett.* **85** (2000) 2869.

CONCEPT CONSIDERATIONS FOR A DEEP SPACE GRAVITY PROBE BASED ON LASER-CONTROLLED FREE-FLYING REFERENCE MASSES

ULRICH A. JOHANN

EADS Astrium GmbH, 88039 Friedrichshafen, Germany
ulrich.johann@astrium.eads.net

Concept considerations for a space mission with the objective of precisely testing the gravitational motion of a small test mass in the solar system environment are presented. In particular, the mission goal is an unambiguous experimental verification or falsification of the Pioneer anomaly effect. A promising concept is featuring a passive reference mass, shielded or well modeled with respect to nongravitational accelerations and formation flying with a rather standard deep space probe. The probe provides laser ranging and angular tracking to the reference mass, ranging to Earth via the radio-communication link and shielding from light pressure in the early parts of the mission. State-of-the-art ranging equipment can be used throughout, but requires in part optimization to meet the stringent physical budget constraints of a deep space mission. Mission operation aspects are briefly addressed.

Keywords: Pioneer anomaly; formation flying; laser ranging; deep space gravity probe.

1. Introduction

Recent developments in fundamental physics, cosmology and the analysis of the so-called Pioneer anomaly[1] have sparkled new interest in precision tests of the gravitational law on all distance scales. In particular, on the scale and in the environment of the solar system, the Pioneer anomaly presumably hints at a potential flaw in the understanding of the free fall motion of a reference mass, moving outbound from the Sun on a solar system escape trajectory. For an unambiguous verification of the anomalous motion and its precise characterization, the influence of nongravitational, nearly constant (quasi-dc) microacceleration noise on the mass has to be determined and carefully discriminated from the well-modeled gravitational motion in free fall to a bias accuracy of around 10^{-12} m/s^2 in all three coordinates.

Several novel mission and payload concepts for such a deep space gravity probe have been proposed and elaborated by the author recently.[2–5] One attractive option is based on a two-step measurement process (Fig. 1).

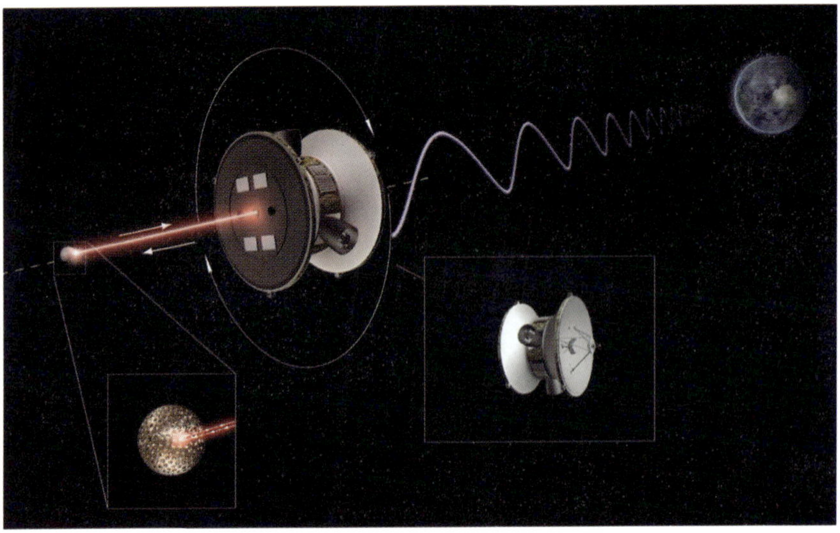

Fig. 1. Proposed scenario for a deep space gravity probe[2,4]: a passive spherical reference mass equipped with optical corner cubes (*left*) is formation-flying in the vicinity of a deep space probe which could be a primary mission probe carrying the gravity experiment package as a passenger, a dedicated spacecraft or, alternatively, the retired propulsion module with added equipment. A two-step tracking scheme based upon a radio communication link to Earth and a short range laser ranging between probe and reference mass is employed to track the latter with respect to a ground station with high precision by eliminating microacceleration noise of the probe caused by noninertial forces.

They involve a standard radio science link of a nearly classically operated noisy probe with respect to an Earth reference, but combined with radio or laser ranging between the probe and one or a few free-falling reference masses. The (nearly) free-falling reference masses are shielded or well modeled with respect to probe and space environment interaction. Consequently, they are either kept in free space in formation flying in the vicinity of the probe at a sufficient distance to minimize probe to reference mass interaction or, alternatively, they are placed inside the probe and shielded from the space environment, but then subjected to disturbing interaction with the probe itself. For the latter, the reference mass can be operated also as a test mass of an accelerometer which is coupled to the probe while measuring any nongravitational acceleration of the test mass itself. In any case, nearly constant (dc-bias) nongravitational acceleration must be monitored, removed or modeled to the quoted level of accuracy.

The requirement of operation in the dc regime or at extremely low frequency (days or weeks in terms of the period) is a distinct technical challenge as compared to similar devices for measuring or shielding very small accelerations in other space missions (GOCE, Microscope, LISA Pathfinder, LISA) operating at a low but non-dc frequency band (hours in terms of the period). This requirement is the main reason to consider formation flying reference masses external to the probe, despite their added operational complexity. The same requirement, however, imposes also operational and design constraints of similar complexity in concepts, where the reference mass is placed inside the probe. A similar concept has been proposed independently.[6]

Laser ranging lends itself as a suitable tool for monitoring the distance between the reference mass and fiducial points located in the probe, because it can be made to be essentially without bias and drift and can provide the necessary resolution with a minimum of equipment based upon proven technology. The reference mass would then be a passive sphere with well-defined surface and electromagnetic properties, equipped with equally distributed corner cube retroreflectors, similar to the Lageos satellites in Earth orbit, but much smaller (typically of 20 cm diameter). Laser metrology to read out the relative position of a reference mass inside an inertial sensor/accelerometer is presently employed for the LISA pathfinder payload. A nonpolarizing heterodyne interferometer with differential wavefront sensing for attitude measurement is complementing here the capacitive readout system at much-improved accuracy along the sensitive measurement axis and for lateral attitude angles. It provides a resolution of $10\,\mathrm{pm}/\sqrt{\mathrm{Hz}}$ and $10\,\mathrm{nrad}/\sqrt{\mathrm{Hz}}$, respectively, within the band 10^{-3}–10^{-1} Hz. Augmented by a molecular frequency reference and a thermally stable reference arm, sufficient stability may be reached for much lower or quasi-dc frequencies. A different and much simpler laser ranging technique can be employed in the case of external, formation flying reference masses. The following discussion will focus on the latter scenario.

2. Requirements on the Measurement Process

The present hypothesis is that the Pioneer anomaly is a constant or very smooth effect, featuring a constant anomalous acceleration either in the direction of the Sun, the Earth line of sight or opposite to the probe's velocity vector in the solar system coordinate frame. Which case applies cannot be discriminated on the basis of available data. The effect is constant in the sense that it is independent of time and orbital position or velocity at least during cruising phases. As such, any deviation in Doppler–interpreted as a corresponding velocity change–should accumulate linear in time. In terms of distance ranging, the effect should hence be quadratic in time. It is obvious that the required range rate or ranging precision for the measurement scales inversely with the measurement intervals (or integration time). In fact, for cruising intervals of 10 years or more, periodic short measurements about every month should be perfectly sufficient. Even outliers in the form of sudden accelerations caused for example by micrometeorites can be discriminated and tolerated. Nevertheless, a certain higher frequency measurement capability (hours) may be desirable for reliable chasing of the reference mass by the probe. The ranging budget is required to achieve a 1% accuracy goal ($\delta a_{\mathrm{RM}} = 8\cdot 10^{-12}\ \mathrm{m/s^2}$) of the reference mass acceleration a_{RM} measurement. Hence, the measurements on both legs, the Earth–probe radio link and the probe–RM laser link, respectively, must be accurate enough to unambiguously discriminate the Pioneer anomaly effective contribution from all other known effects influencing the trajectory. In addition, the modeling accuracy for the acceleration of the RM by all known gravitational and nongravitational effects must be within the δa_{RM} budget. As a consequence, the reference mass should be purely passive, with well-defined surface and electrical properties. The Pioneer anomaly effect accumulates to a Doppler shift, velocity deviation or distance deviation relative to the modeled parameter values over one day or one month, respectively, as shown in Table 1, together with the accuracy to be provided by the ranging.

These figures of course apply to the complete two-step process, comprising the radio link to Earth and the local laser ranger, and are relative values. It is evident that — even if factorized — only moderate requirements are imposed for the laser ranger in terms of measurement precision. This fact has the important consequence that the tracking can be accomplished with rather simple sensor equipment on the probe in addition to the anyway-existing radio link to Earth.

Because the probe is on a presumably close-to-Sun-radial outbound trajectory, the ranging accuracy along the Earth line of sight by far dominates the measurement accuracy and the lateral coordinates can be determined with less precision. Nevertheless, VLBI techniques for the radio link to Earth allow μarcsec precision if desired. The orientation of the laser ranging vector relative to the Earth line of sight can be easily determined to within a sufficient, few-arcsec accuracy, for example by a simple star tracker. The capabilities of current radio ranging in Deep-Space Network communication links are summarized in Table 2.

3. Ranging and Measurement Concepts

In principle, the following ranging technologies can be combined in the proposed two-step process (Table 3):

Table 1. Single bin measurement accuracy required to resolve a (smooth) Pioneer acceleration with sufficient precision of 1% as a function of the measurement interval. "Relative" here refers to the trajectory without the PA effect. The values are for the total measurement (Earth–probe–sphere). A monthly period is considered to be sufficient, greatly relaxing sensor requirements. A daily period may, however, be desirable for active tracking of the free-fall reference mass. Please note that no absolute ranging accuracy to that precision is required for the Earth–probe link, but a constant bias can be tolerated.

Measurement period	Relative velocity (m/s)	Relative distance (m)	Relative Doppler Ka (mHz)	Required ranging accuracy
1 day	70 μm/s	6 m (25 cm/h)	7 mHz	6 cm 0.7 μm/s 0.07 mHz
1 month	2100 μm/s	5300 m	200 mHz	50 m

Table 2. Present DSN radio link capabilities for the spacecraft tracking Earth–probe (*DSN Handbook*[7]).

Parameter	Measures	Accuracy (1σ)
Doppler	Range rate	$0.03\,\mathrm{mm/s}$
Range	Range	~ 1–$2\,\mathrm{m}$
Angle	Lateral angular position (right assention, declination)	$0.01°$ ($170\,\mu\mathrm{rad}$)
DDOR (VLBI)	Lateral angular position (right assention, declination)	$0.14\,\mu°$ ($2.4\,\mathrm{nrad}$)

Table 3. Ranging technologies suitable for the proposed two-step process and rendering the measurement insensitive to noisy probe effects. This paper focuses on the first option.

Earth–probe link	Probe–reference mass (RM) link
Classical bidirectional bi-wavelength radio science	Laser ranging + star tracker of reference mass with corner cubes (in shadow)
Classical bidirectional bi-wavelength radio science	Radio tracking of passive radar reflector (in shadow)
Classical bidirectional bi-wavelength radio science	Active transponder on reference mass (in shadow)
Classical bidirectional bi-wavelength radio science	Radar reflector tracking in main communication beam (in light)
Laser ranging to Earth station (ground or orbit)	Everything above

The ranging requirements between probe and formation flying reference mass (or several masses) are further intensified by the necessity to chase the latter while actively maneuvering the probe. The impact on probe operations will, however, be kept at a minimum, particularly in scenarios where the experiment is a passenger on a mission with different objectives. Obviously, the frequency of (very low Δv) correction maneuvers and the allowance of letting the reference mass drift far away, while stressing the local ranging requirements, are conjugate. A further requirement indirectly related is the necessity to shield or model any reference disturbance at sufficient precision. Among the various disturbance sources, here the dominant effects of solar light pressure and gravity or thermal interaction with the probe itself define constraints. A sphere of 25 cm diameter, 5 kg, equipped with retroreflecting corner cubes experiences at 1 AU (Earth distance to the Sun) a light pressure acceleration of about $5 \cdot 10^{-8}$ m/s^2, a figure about 100 times the Pioneer effect. To achieve the 1% measurement accuracy, a modeling of the induced acceleration to within a 10^{-4} accuracy is required, which is a challenging task, considering ageing surface properties, etc. At 10 AU, an accuracy of 1% is still necessary. Hence, flying the reference mass in the shadow of the probe would be an advantage at least in early parts of the trajectory. Assuming a probe carrying an opaque 2.5 m antenna, it will cast a core shadow (umbra) of about 260 m and 2600 m for 1 AU and 10 AU, respectively. Equipped with a dedicated Sun shield of, say, 10 m diameter, the shielded range could be extended to 1 km and 10 km, respectively. At close range, a 500 kg probe would pull the reference mass equivalent to the desired acceleration accuracy at about a 65 m distance. In case the gravity interaction can be modeled to a 1% accuracy, a lower limit of only 6.5 m could be allowed, but then thermal and electrostatic interaction with the probe become important. Figure 2 illustrates the geometry of the formation flying mission.[8,9]

The frequency of maneuvers to keep the reference mass in the range interval allowed by the above constraint, is then set primarily by the differential light pressure acceleration. For a typical 2.5-m-diameter probe, a range walk of 180 m/day (1AU) and 1.8 m/day (10 AU) would occur by this effect. Hence, at 1 AU a daily maneuver imposing a Δv of about 10 mm/s in the Sun's direction would be required or, alternatively, a continuous thrust of about 25 μN. At 10 AU that reduces to a monthly maneuver imposing a Δv of about 3 mm/s. Beyond 10 AU, the probe can be allowed to drift away in sunlight until the tracking capabilities are exhausted.

The reference mass can be centered in the probe shadow by very low thrust lateral maneuvers of the probe. It is important to note that these positioning maneuvers are not comparable with the complex and risky orbit correction maneuvers in classical missions, as only very low thrust authority is employed. A possible alternative operation mode would be to let the reference mass and the probe drift apart independently for long intervals (month), to reacquire the target with the star tracker and to actively steer the probe close, within a few meters, followed by

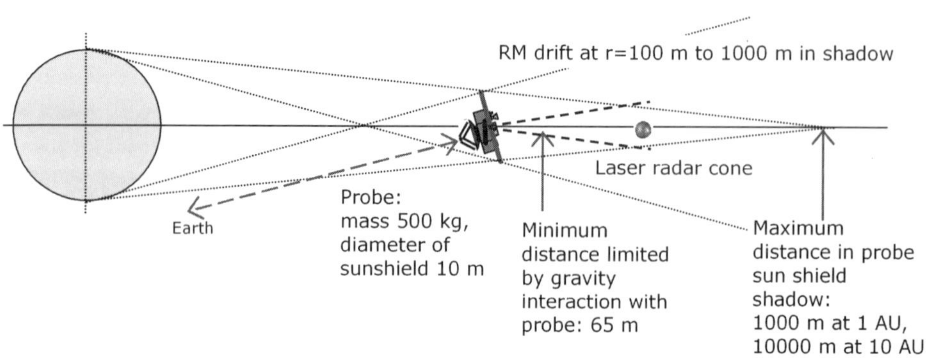

Fig. 2. Geometry of formation flying reference mass shielded from solar radiation pressure by shadowing. The sphere is tracked by a laser radar and a star sensor payload on the probe and allowed to drift in a range between about 100 m and 1000 m at 1 AU (10000 m at > 10 AU. The probe is actively chasing the sphere with very low thrust maneuvers with a period of several days, depending on Sun distance and design parameters. The laser radar pressure and the thermal radiation pressure from the probe are sufficiently small.

a very simple short range calibration. Obviously, the shadowing would be lost in that mode.

4. Suitable Technologies for Laser Ranging to the Reference Mass

Numerous laser ranging techniques exist for many applications on the ground and in space and for a large range of requirements. For the envisaged application, the selection depends on the mission scenario, in particular whether it is based upon close range continuous tracking or large range drifts. It is further driven by the least impact on probe operations (mass, power, operation modes, required actuation or AOCS maneuvers) and by simplicity, robustness and space heritage.

A (nonexhaustive) list of candidate measurement principles is:

- Pulsed time-of-flight laser radar for ranging and star tracker for relative attitude,
- Frequency chirp coherent laser radar (FMCW),
- Continuous tracking with a coherent laser heterodyne interferometer,
- Calibrated measurement of the sphere diameter (interferometric fringe contrast),
- Triangulation using three star trackers,
- Intensity ranging.

Another important requirement is the capability to periodically acquire the probe following inactive intervals. Obviously, the system robustness demands that the reference mass should not be lost. The periodic operation modes are then:

- Warm up,
- Calibration (option),
- Target acquisition,
- Target tracking (option),
- Range, range rate and angular measurement,
- Stand by.

In the following, the pulsed laser radar based upon existing technology is shown to be perfectly suitable even for the large range scenarios. A detailed trade, however, and final selection has to be incorporated into a complete system level study for the experiment. It is assumed that the ranging done by a laser radar is supported by a directional measurement using a state-of-the-art simple star tracker. The star tracker also adds significant robustness to the system, because the target can be "seen" in its field of view and in front of the star field, either illuminated by the laser or by the Sun.

One interesting option for a very simple system is a uniform (white) sphere, illuminated by a defocused laser beam, which can be broadband, but must be calibrated in power. The backscattered laser light is detected by the star tracker, which also receives a power calibration from the transmission (e.g. fiber link) and locates the sphere relative to the beam axis. For laser intensity calibrated to 10^{-3} relative accuracy and an intensity flat beam lobe within the angular accuracy of the star tracker ($\sim$ 50 μrad), a range resolution of $\delta R = 2.5 \cdot 10^{-4} \cdot R$ can be achieved. That corresponds to 25 mm at $R = 100$ m. A mode for initial calibration and recalibration, compensating for ageing target surface properties, would have to be incorporated, however.

A time-of-flight radar would require a target (sphere or disk) packed with corner cubes in order to support the link budget. Acquisition strategies typically employ scanning laser beams (spiral or rectangular patterns), defocused beams followed by reorientation and refocusing or combinations thereof. Also, the star tracker signal can be used to actively point the focused laser beam at an acquired target, if the two boresights are aligned. In that case a scanning or defocusing of the beam can be avoided, provided that the target is illuminated.

The received power for a defocused (non-diffraction-limited) transmitted beam reflected off a corner-cube-carrying sphere can be expressed as:

$$P_r = \frac{D_t^2}{\left(D_t + \dfrac{R\lambda}{d_T}\right)^2} n\rho \left(\frac{d_T}{d_b}\right)^2 P_t, \tag{1}$$

where, in the example considered here, $D_t = 0.1$ m is the transmit/receive telescope diameter, R is the range, $\lambda = 1\,\mu$m is the wavelength, $d_T = 0.02$ m is the diameter of the reflector corner cubes, $n = 19$ is the number of illuminated cubes with an average efficiency of $\rho = 0.5$, $d_b = 10$ m is the diameter of the laser-illuminated area at the location of the sphere and P_t is the transmitted power.

For a defocused beam to 10 m diameter at 10 km distance, the received power is then

$$P_r = 1 \times 10^{-6} P_t. \tag{2}$$

A suitable transmitted power of 100 mW provides therefore a reception signature of 100 nW, which is by far sufficient for localization. A focused beam (about diffraction-limited) provides

$$P_r = 2.6 \times 10^{-3} P_t \tag{3}$$

and hence 260 μW. This value is again by far sufficient for a ranging accuracy of about 10 mm in a time-of-flight laser radar (see below).

The example is in fact representing an overdesigned system, illustrating the capabilities.

5. Available Suitable Sensors

A space-qualified time-of-flight laser radar which could be employed with few (descoping) modifications has been developed by Jena Optronik together with EADS Astrium GmbH ad Rigel GmbH. It has flown successfully on several space shuttle missions, where it served as a rendezvous and docking sensor. Figure 3 illustrates the device and Table 4 summarizes the main parameters.[10] The scanning mechanism would be omitted and replaced by a refocusing mechanism in the foreseen application, further reducing power and mass budgets.

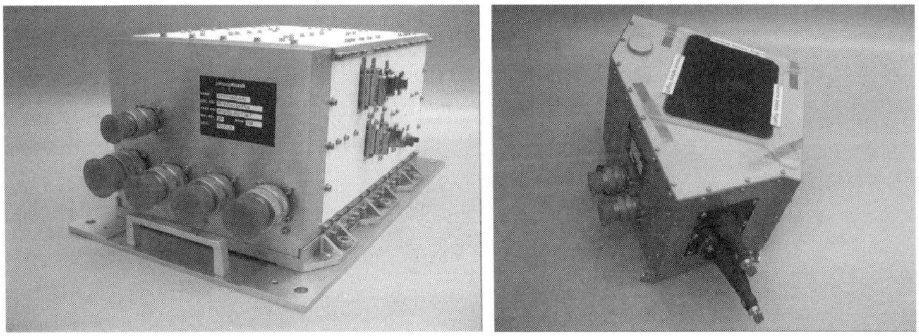

Fig. 3. Space-demonstrated laser range finder developed by Jena Optronik, Astrium GmbH and Riegel GmbH for space shuttle rendezvous and docking.

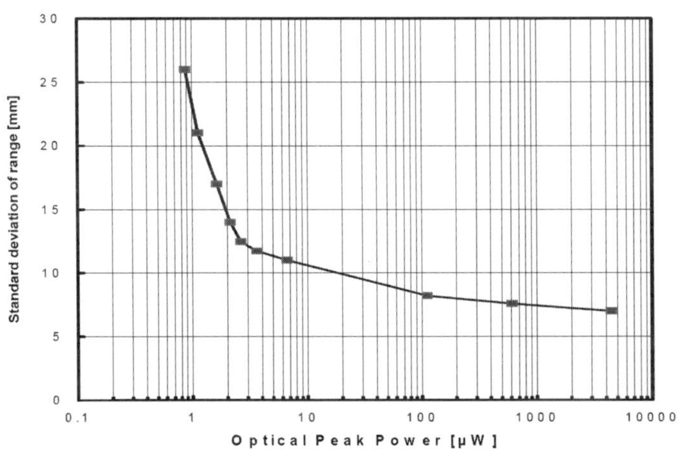

Fig. 4. Ranging performance measured as a function of received optical power. A resolution of 10 mm is obtained for 10 μW (Jena Optronik).

Table 4. Performance and budgets of the Jena Optronik laser range finder. Custom-specific modification should lead to significantly reduced mass and power budgets.

Field of view		Up to $40^\circ \times 40^\circ$
Measurement parameter		Azimuth α, elevation β, roll R, pitch P, yaw Y, time, range r
Accuracy for range (700 m – 3 m)	LOS (noise + bias) LOS bias	$< 0.1^\circ$ $< 0.1^\circ$
Power		Min: 35 W Max: 70 W
Temperature	Operational Nonoperational	35°C $\cdots$ +65°C 55°C $\cdots$ +70°C
Mechanical size	Opt. head (w/o fiber) Electronic box	270 mm × 287 mm × 196 mm 315 mm × 224 mm × 176 mm
Mass	Optical head Electronic box	6.1 kg 8.2 kg

Fig. 5. Autonomous star tracker Astro 10 (Jena Optronik).

Table 5. Performance and budgets of the Jena Optronik Astro 10 autonomous star tracker.

Dimensions	Head: diameter 185 mm, height 242 mm (including 30° baffle) E-box: 150 mm × 145 mm × 75 mm separated box design
Mass	< 960 g for optical head (without baffle) < 1180 g for electronic box < 510 g for 30° baffle (380 g for 40° baffle) < 350 g for cabling E-box/opt. head (1 m length)
Power	Star sensor: < 10.0 W at 20° interface temperature optical head
Sensor performance	LOS accuracy (BOL): $\leq$ 5 arcsec (3σ), pitch/yaw = 35 arcsec (3σ), roll, slew rate at 0.6° s^{-1}
Operating modes	Boot, stand-by, initial acquisition Attitude lock-in, high accuracy attitude, simulation
Data interface	RS 422; alternatively MIL 1553 B
Input voltage range	22–35 V

Figure 3 shows measured range resolution versus received power.

A suitable star tracker is also available from Jena Optronik, although other devices exist, which may be even more compact, lighter and less power-consuming (such as the ATC of the University of Denmark). Figure 5 and Table 5 present the main design and performance figures for the Astro 10 autonomous star tracker.[10]

6. Conclusions

Probing the gravity field on the solar system scale with high accuracy in an active space mission emerges as a scientifically and technically challenging goal in itself. We have conceptually assessed mission scenarios which the aim of testing the Pioneer anomaly with sufficient precision to unambiguously verify and eventually characterize its nature. One promising scenario has been outlined, which can be flown either in a dedicated mission or as a passenger in a deep space mission. Required ranging technology readily exists, potentially meeting the stringent constraints of a deep space mission in terms of mass, power and operational impact.

Acknowledgment

The author acknowledges the fruitful collaboration and stimulating discussions with Slava G. Turyshev of the Jet Propulsion Laboratory.

References

1. J. D. Anderson *et al.*, *Phys. Rev. D* **65** (2002) 0820041 [gr-qc/0104064].
2. U. Johann and R. Förstner, "ENIGMA," Unsolicited proposal to ESA/ESTEC, Fundamental Physics and Advanced Concepts (2003).
3. U. Johann and R. Förstner, "ENIGMA," Presentation at the First International Pioneer Anomaly Workshop (ZARM, Bremen, May 2004). See the meeting's website, http://www.Zarm.uni-bremen.de/Pioneer
4. H. J. Dittus, S. G. Turyshev and the Pioneer Anomaly Team, "A Mission to Explore the Pioneer Anomaly," in *2005 ESLAB Symposium: Trends in Space Science and Cosmic Vision 2020* (ESA/ESTEC, Noordwijk, The Netherlands, 19 Apr. 2005), http://www.congrex.nl/05a14. *ESA Publication* **SP-588** (2005) 3 [gr-qc/0506139].
5. H. J. Dittus, S. G. Turyshev and the Pioneer Anomaly Team, "A Consolidated Cosmic Vision Theme Proposal to Explore the Pioneer Anomaly," submitted to ESA FPAG (Oct. 2004).
6. K. Penanen and T. Chui, *Nucl. Phys. Proc. Supp.* **134** (2004) 211 [gr-qc/0406013].
7. DSN 810-005, Rev. E, 203; see electronic version at NASA DSMS website, http://deepspace.jpl.nasa.gov/dsndocs/810-005/203/203A.pdf
8. U. Johann and R. Förstner, "On Technologies for Pioneer Anomaly Tests," presentation at the First Pioneer Explorer Collaboration, International Space Sciences Institute (IISI) (Bern, Switzerland, 7–11 Nov. 2005). See the meeting's website, http://www.issibern.ch/teams/Pioneer/.
9. U. Johann and H. J. Dittus, "Novel Mission and Payload Concepts for a Deep Space Gravity Probe," presentation at 36th COSPAR Scientific Assembly (Beijing, China, 17–24 July 2006, http://meetings.copernicus.org/cospar2006.
10. http://www.jena-optronik.de (Apr. 2004).

PROPOSED OBSERVATIONS OF GRAVITATIONAL WAVES FROM THE EARLY UNIVERSE VIA "MILLIKAN OIL DROPS"

RAYMOND Y. CHIAO

University of California at Merced,
PO Box 2030, Merced, CA 95344, USA
rchiao@ucmerced.edu

Pairs of Planck-mass drops of superfluid helium coated by electrons (i.e. "Millikan oil drops"), when levitated in a superconducting magnetic trap, can be efficient quantum transducers between electromagnetic (EM) and gravitational (GR) radiation. This leads to the possibility of a Hertz-like experiment, in which EM waves are converted at the source into GR waves, and then back-converted at the receiver from GR waves into EM waves. Detection of the gravitational-wave analog of the cosmic microwave background using these drops can discriminate between various theories of the early Universe.

Keywords: Gravitational radiation; quantum mechanics; cosmic microwave background.

1. Forces of Gravity and Electricity Between Two Electrons

Consider the forces exerted by an electron upon another electron at a distance r away in the vacuum. Both the gravitational and the electrical force obey long-range, inverse-square laws. Newton's law of gravitation states that

$$|F_G| = \frac{Gm_e^2}{r^2}, \tag{1}$$

where G is Newton's constant and m_e is the mass of the electron. Coulomb's law of electricity states that

$$|F_e| = \frac{e^2}{r^2}, \tag{2}$$

where e is the charge of the electron. The electrical force is repulsive, and the gravitational one attactive.

Taking the ratio of these two forces, one obtains the dimensionless constant

$$\frac{|F_G|}{|F_e|} = \frac{Gm_e^2}{e^2} \approx 2.4 \times 10^{-43}. \tag{3}$$

The gravitational force is extremely small compared to the electrical force, and is therefore usually omitted in all treatments of quantum physics. Note, however, that this ratio is not strictly zero, and therefore can in principle be amplified.

2. Gravitational and Electromagnetic Radiation Powers Emitted by Two Electrons

The above ratio of the coupling constants Gm_e^2/e^2 is also the ratio of the powers of gravitational (GR) to electromagnetic (EM) radiation emitted by two electrons separated by a distance r in the vacuum, when they undergo an acceleration a relative to each other. Larmor's formula for the power emitted by a single electron undergoing acceleration a is

$$P_{\mathrm{EM}} = \frac{2}{3}\frac{e^2}{c^3}a^2. \tag{4}$$

For the case of two electrons undergoing an acceleration a relative to each other, the radiation is quadrupolar in nature, and the modified Larmor formula is

$$P'_{\mathrm{EM}} = \kappa\frac{2}{3}\frac{e^2}{c^3}a^2, \tag{5}$$

where the prefactor κ accounts for the quadrupolar nature of the emitted radiation.[a] Since the electron carries mass, as well as charge, and its charge and mass comove rigidly, two electrons undergoing an acceleration a relative to each other will also emit homologous quadrupolar gravitational radiation according to the formula

$$P'_{\mathrm{GR}} = \kappa\frac{2}{3}\frac{Gm_e^2}{c^3}a^2, \tag{6}$$

with the same prefactor of κ.

The equivalence principle demands that the lowest order of gravitational radiation be quadrupolar, and not dipolar, in nature. Hence the ratio of gravitational to electromagnetic radiation powers emitted by the two-electron system is given by the same ratio of coupling constants, viz.

$$\frac{P'_{\mathrm{GR}}}{P'_{\mathrm{EM}}} = \frac{Gm_e^2}{e^2} \approx 2.4 \times 10^{-43}. \tag{7}$$

Thus it would seem at first sight to be hopeless to try and use any two-electron system as the means for coupling between electromagnetic and gravitational radiation.

3. The Planck Mass Scale

However, the ratio of the forces of gravity and electricity of two "Millikan oil drops" (to be described in more detail below; however, see Fig. 1) need not be so hopelessly small.[1]

[a]Here $\kappa = \frac{2}{15}\frac{v^2}{c^2}$, where v is their relative speed and c is the speed of light, if $v \ll c$.

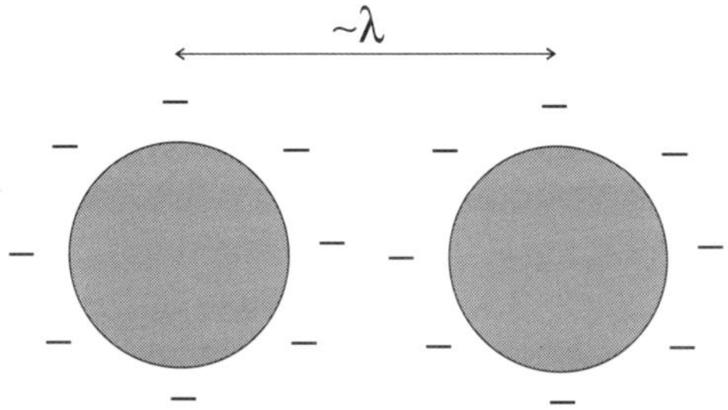

Fig. 1. A pair of levitated "Millikan oil drops" (i.e. electron-coated superfluid helium drops with Planck-scale masses) in a superconducting magnetic trap, separated by around a microwave wavelength λ.

Suppose that each Millikan oil drop contains a Planck-mass amount of superfluid helium, viz.

$$m_{\text{Planck}} = \sqrt{\frac{\hbar c}{G}} \approx 22 \text{ micrograms}, \tag{8}$$

where $\hbar$ is Planck's constant/2π, c is the speed of light, and G is Newton's constant. Planck's mass sets the characteristic scale at which quantum mechanics ($\hbar$) impacts relativistic gravity (c, G). Note that this mass scale is mesoscopic,[2] and not astronomical, in size. This suggests that it may be possible to perform some novel nonastronomical, tabletop-scale experiments at the interface of quantum mechanics and general relativity, which are accessible to the laboratory.

The ratio of the forces of gravity and electricity between the two Millikan oil drops now becomes

$$\frac{|F_G|}{|F_e|} = \frac{Gm_{\text{Planck}}^2}{e^2} = \frac{G\,(\hbar c/G)}{e^2} = \frac{\hbar c}{e^2} \approx 137. \tag{9}$$

Now the force of gravity is 137 times stronger than the force of electricity, so that instead of a mutual repulsion between these two charged objects, there is now a mutual attraction between them. The sign change from mutual repulsion to mutual attraction between these two Millikan oil drops occurs at a critical mass m_{crit}, given by

$$m_{\text{crit}} = \sqrt{\frac{e^2}{\hbar c}}\, m_{\text{Planck}} \approx 1.9 \text{ micrograms}, \tag{10}$$

whereupon $|F_G| = |F_e|$, and the forces of gravity and electricity balance each other. This is a strong hint that mesoscopic-scale quantum effects can lead to nonnegligible couplings between gravity and electromagnetism.

The critical mass m_{crit} is also the mass at which there occurs an equal amount of electromagnetic and gravitational radiation power generated upon scattering of radiation from the pair of Millikan oil drops, each member of the pair with a mass $m_{\rm crit}$ and with a single electron attached to it. Now the ratio of quadrupolar gravitational to quadrupolar electromagnetic radiation power is given by

$$\frac{P'_{\rm GR}}{P'_{\rm EM}} = \frac{Gm_{\rm crit}^2}{e^2} = 1, \tag{11}$$

where the prefactors of κ in Eqs. (5) and (6) cancel out, if the charge of the drop comoves rigidly with its mass. This implies that the scattered power from these two charged objects in the gravitational wave channel becomes equal to that in the electromagnetic wave channel. However, it should be emphasized that it has been assumed here that a given drop's charge and mass move together as a single unit, in accordance with a Mössbauer-like mode (i.e. a zero-phonon mode) of response to radiation fields, which will be discussed below. This is a purely quantum effect based on the quantum adiabatic theorem's prediction that the system will remain adiabatically, and hence rigidly, in its nondegenerate ground state during perturbations arising from externally applied radiation fields.

4. Millikan Oil Drops Described in More Detail

Let the oil of the classic Millikan oil drops be replaced with superfluid helium (^{4}He) with a gravitational mass of around the Planck-mass scale, and let these drops be levitated in a superconducting magnetic trap with Tesla-scale magnetic fields.

The helium atom is diamagnetic, and liquid helium drops have successfully been magnetically levitated in an anti-Helmholtz magnetic trapping configuration.[3] Due to its surface tension, the surface of a freely suspended, ultracold superfluid drop is atomically perfect. When an electron approaches a drop, the formation of an image charge inside the dielectric sphere of the drop causes the electron to be attracted by the Coulomb force to its own image. However, the Pauli exclusion principle prevents the electron from entering the drop. As a result, the electron is bound to the surface of the drop in a hydrogenic ground state. Experimentally, the binding energy of the electron to the surface of liquid helium has been measured using millimeter-wave spectroscopy to be 8 K,[4] which is quite large compared to the millikelvin temperature scales for the proposed experiment. Hence the electron is tightly bound to the surface of the drop.

Such a Millikan oil drop is a macroscopically phase-coherent quantum object. In its ground state, which possesses a single, coherent quantum-mechanical phase throughout the interior of the superfluid, the drop possesses a zero circulation quantum number (i.e. contains no quantum vortices), with one unit (or an integer multiple) of the charge quantum number. As a result of the drop being at ultralow temperatures, all degrees of freedom other than the center-of-mass degrees of freedom are frozen out, so that there results a zero-phonon Mössbauer-like effect, in which the entire mass of the drop moves rigidly as a single unit in response to radiation fields. Also, since it remains adiabatically in the ground state during perturbations due to these radiation fields, the Millikan oil drop possesses a quantum rigidity and a quantum dissipationlessness that are the two most important quantum properties for achieving a high conversion efficiency for gravitational-wave antennas.[5]

Note that a pair of spatially separated Millikan oil drops have the correct quadrupolar symmetry in order to couple to gravitational radiation, as well as to quadrupolar electromagnetic radiation. When they are separated by a distance on the order of a wavelength, they should become an efficient quadrupolar antenna capable of generating, as well as detecting, gravitational radiation.

5. A Pair of Millikan Oil Drops as a Transducer

Now imagine placing a pair of levitated Millikan oil drops separated by approximately a microwave wavelength inside a black box, which represents a quantum transducer that can convert GR waves into EM waves.[5] This kind of transducer action is similar to that of the tidal force of a gravity wave passing over a pair of charged, freely falling objects orbiting the Earth, which can convert a GR wave into an EM wave.[1] Such transducers are linear, reciprocal devices.

By time-reversal symmetry, the reciprocal process, in which another identical pair of Millikan oil drops converts an EM wave back into a GR wave, must occur with the same efficiency as the forward process, in which a GR wave is converted into an EM wave by the first pair of Millikan oil drops. The time-reversed process is important because it allows the generation of gravitational radiation, and can therefore become a practical source of such radiation.

This raises the possibility of performing a Hertz-like experiment, in which the time-reversed quantum transducer process becomes the source, and its reciprocal quantum transducer process becomes the receiver of GR waves in the far field of the source. Room-temperature Faraday cages can prevent the transmission of EM waves, so that only GR waves, which can easily pass through all classical matter such as the normal (i.e. dissipative) metals of which standard, room-temperature Faraday cages are composed, are transmitted between the two halves of the apparatus that serve as the source and the receiver, respectively. Such an experiment would be practical to perform using standard microwave sources and receivers, since the scattering cross-sections and the transducer conversion efficiencies of the two Millikan oil drops turn out not to be too small, as will be shown below. The Hertz-like experiment would allow the calibration of the Millikan-oil-drops receiver for detecting the gravitational-wave analog of cosmic microwave background radiation from the extremely early Big Bang.

6. Mössbauer-Like Response of Millikan Oil Drops in a Magnetic Trap to Radiation Fields

Let a pair of Millikan oil drops be levitated in a superconducting magnetic trap, where the drops are separated by a distance on the order of a microwave wavelength, which is chosen so as to satisfy the impedance-matching condition for a good quadrupolar microwave antenna. See Fig. 1.

Now let a beam of EM waves in the Hermite–Gaussian TEM_{11} mode,[6] which has a quadrupolar transverse field pattern that has a substantial overlap with that of a GR plane wave, impinge at a 45° angle with respect to the line joining these two charged objects. As a result of being thus irradiated, the pair of Millikan oil drops will appear to move in an antiphased manner, so that the distance between them will oscillate sinusoidally with time, according to an observer at infinity. Thus the apparent simple harmonic motion of the two drops relative to one another produces a time-varying mass quadrupole moment at the same frequency as that of the driving EM wave. This oscillatory motion will in turn scatter (in a linear scattering process) the incident EM wave into gravitational and electromagnetic scattering channels with comparable powers, provided that the ratio of quadrupolar Larmor radiation powers given by Eq. (11) is of the order of unity, which will be the case when the mass of both drops is on the order of the critical mass m_{crit} for the case of single electrons attached to each drop. The reciprocal process should also have a power ratio of the order of unity.

The Mössbauer-like response of Millikan oil drops will now be discussed in more detail. Imagine what would happen if one were to replace an electron in the vacuum with a single electron which is firmly attached to the surface of a drop of superfluid helium in the presence of a strong magnetic field and at ultralow temperatures, so that the system of the electron and the superfluid, considered as a single quantum entity, would form a single, macroscopic quantum ground state. Such a quantum system can possess a sizeable gravitational mass. For the case of many electrons attached to a massive drop, where a quantum Hall fluid forms on the surface of the drop in the presence of a strong magnetic field, there results a nondegenerate, Laughlin-like ground state.

In the presence of Tesla-scale magnetic fields, an electron is prevented from moving at right angles to the local magnetic field line around which it is executing tight cyclotron orbits. The result is that the surface of the drop, to which the electron is tightly bound, cannot undergo liquid-drop deformations, such as the oscillations between the prolate and oblate spheroidal configurations of the drop which would occur at low frequencies in the absence of the magnetic field. After the drop has been placed into Tesla-scale magnetic fields at millikelvin operating temperatures, both the single- and many-electron drop systems will be effectively frozen into the ground state, since the characteristic energy scale for electron cyclotron motion in Tesla-scale fields is on the order of kelvins. Due to the tight binding of the electron(s) to the surface of the drop, this would freeze out all shape deformations of the superfluid drop.

Since all internal degrees of freedom of the drop, such as its microwave phonon excitations, will also be frozen out at sufficiently low temperatures, the charge and the entire mass of the Millikan oil drop should comove rigidly as a single unit, in a Mössbauer-like response to applied radiation fields. This is a result of the elimination of all internal degrees of freedom by the Boltzmann factor at sufficiently low temperatures, so that the system stays in its ground state, and only the external degrees of freedom of the drop, consisting only of its center-of-mass motions, remain.

The criterion for this Mössbauer-like mode of response of the electron-drop system is that the temperature of the system is sufficiently low, so that the probability for the entire system to remain in its nondegenerate ground state without even a single quantum of excitation of any of its internal degrees of freedom being excited, is very high, i.e.

$$\text{Prob. of zero internal excitation} \approx 1 - \exp\left(-\frac{E_{\text{gap}}}{k_B T}\right) \to 1 \text{ as } \frac{k_B T}{E_{\text{gap}}} \to 0, \quad (12)$$

where E_{gap} is the energy gap separating the nondegenerate ground state from the lowest permissible excited states, k_B is Boltzmann's constant, and T is the temperature of the system. Then the quantum adiabatic theorem ensures that the system will stay adiabatically in the nondegenerate ground state of this quantum many-body system during perturbations, such as those due to weak, externally applied radiation fields, whose frequencies are below the gap frequency $E_{\text{gap}}/\hbar$. By the principle of momentum conservation, since there are no internal excitations to take up the radiative momentum, the center of mass of the entire system must undergo recoil in the emission and absorption of radiation. Thus the mass involved in the response to radiation fields is the mass of the whole system.

For the case of a single electron (or many electrons in the case of the quantum Hall fluid) in a strong magnetic field, the typical energy gap is given by

$$E_{\text{gap}} = \hbar\omega_{\text{cyclotron}} = \frac{\hbar e B}{mc} \gg k_B T, \quad (13)$$

an inequality which is valid for the Tesla-scale fields and millikelvin temperatures being considered here.

7. Estimate of the Scattering Cross-Section

Let $d\sigma_{a\to\beta}$ be the differential cross-section for the scattering of a mode a of radiation of an incident GR wave to a mode β of a scattered EM wave by a pair of Millikan oil drops (Latin subscripts denote GR waves, and Greek subscripts EM waves). Then, by time-reversal symmetry

$$d\sigma_{a\to\beta} = d\sigma_{\beta\to a}. \quad (14)$$

Since electromagnetic and weak gravitational fields both formally obey Maxwell's equations[7] (apart from a difference in the signs of the source density and the source current density), and since these fields obey the same boundary conditions, the solutions for the modes for the two kinds of scattered radiation fields must also have the same mathematical form. Let a and α be a pair of corresponding solutions, and b and β be a different pair of corresponding solutions to Maxwell's equations for gravitational and electromagnetic modes, respectively. For example, a and α could represent incoming plane waves which copropagate in the same direction, and b and

β scattered, outgoing plane waves which copropagate in a different direction. Then, for the case of a pair of critical-mass drops with single-electron attachment, there is an equal conversion into the two types of scattered radiation fields in accordance with Eq. (11), and therefore

$$d\sigma_{a\to b} = d\sigma_{a\to\beta}, \tag{15}$$

where b and β are corresponding modes of the two kinds of scattered radiation.

By the same line of reasoning, for this pair of critical-mass drops

$$d\sigma_{b\to a} = d\sigma_{\beta\to a} = d\sigma_{\beta\to\alpha}. \tag{16}$$

It therefore follows from the principle of reciprocity (i.e. time-reversal symmetry) that

$$d\sigma_{a\to b} = d\sigma_{\alpha\to\beta}. \tag{17}$$

To estimate the size of the total cross-section, it is easier to consider first the case of electromagnetic scattering, such as the scattering of microwaves from two Planck-mass-scale drops, with radii R and a separation r on the order of a microwave wavelength (but with $r > 2R$). See Fig. 1. Let the electrons on the Millikan oil drops be in a quantum Hall plateau state, which is known to be that of a perfectly dissipationless quantum fluid, like that of a superconductor. Furthermore, it is known that the nondegenerate Laughlin ground state is that of a perfectly rigid, incompressible quantum fluid.[8] The two drops thus behave like perfectly conducting, shiny, mirrorlike spheres, which scatter light in a manner similar to that of perfectly elastic hard-sphere scattering in idealized billiards. The total cross-section for the scattering of electromagnetic radiation from a pair of drops is therefore given approximately by the geometric cross-sectional areas of two hard spheres,

$$\sigma_{\alpha\to\mathrm{all}\,\beta} = \int d\sigma_{\alpha\to\beta} \simeq \text{Order of } 2\pi R^2, \tag{18}$$

where R is the hard-sphere radius of a drop.

However, if, as one might expect on the basis of classical intuitions, the total cross-section for the scattering of GR waves from the two-drop system is extremely small, like that of all classical matter such as the Weber bar, then by reciprocity, the total cross-section for the scattering of EM waves from the two-drop system must also be extremely small. In other words, if Millikan oil drops were to be essentially invisible to gravitational radiation, then they must also be essentially invisible to electromagnetic radiation. This would lead to a contradiction with the hard-sphere cross-section given by Eq. (18), or with any other reasonable estimate for the electromagnetic scattering cross-section of these drops, so these classical intuitions must be incorrect.

From the reciprocity principle and from the important properties of quantum rigidity and quantum dissipationlessness of these drops, one therefore concludes that for two critical-mass Millikan oil drops, it must be the case that

$$\sigma_{a\to\mathrm{all}\,b} = \sigma_{\alpha\to\mathrm{all}\,\beta} \simeq \text{Order of } 2\pi R^2. \tag{19}$$

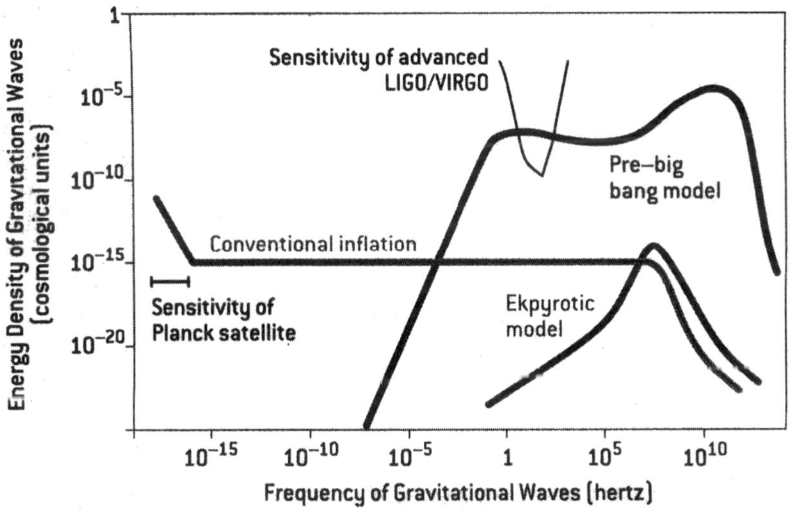

Fig. 2. Spectrum of gravitational waves from the Planck era of the Big Bang according to three different models. *Adapted from Ref. 9.*

8. Cosmic Microwave Background in Gravitational Waves

An important problem in cosmology is the detection and the measurement of the spectrum of gravitational radiation from the extremely early Universe, especially around microwave frequencies. Since gravitational radiation decouples from matter at a much earlier era of the Big Bang (i.e. the Planck era) than electromagnetic radiation, observations of these primordial gravity waves would constitute a much deeper probe of the structure of the early Universe than is the case for the usual CMB.

In particular, the string-inspired pre-Big-Bang model, the ekpyrotic model based on brane theory, and the conventional inflation model give totally different predictions as to the gravitational-wave spectrum.[9] See Fig. 2. Observations in the radio- and microwave-frequency parts of the spectrum would be decisive in determining which model (if any) is the correct one, since the positions of the maxima in the spectra predicted by the pre-Big-Bang and ekpyrotic models and their strengths are strikingly different from each other. Both models in turn yield spectra which differ greatly from the spectrum predicted by the conventional inflation model, which is extremely flat up to the microwave frequency range, where there is a cutoff, but where there are no maxima at all.

Acknowledgment

I would like to thank the organizers for inviting me to participate in NASA's recent "Quantum to Cosmos" conference.

References

1. R. Y. Chiao, Lamb Medal Lecture on Jan. 5, 2006, quant-ph/0601193.
2. C. Kiefer and C. Weber, *Ann. Phys.* (Leipzig) **14** (2005) 253.
3. M. A. Weilert *et al.*, *Phys. Rev. Lett.* **77** (1996) 4840.
4. C. C. Grimes and G. Adams, *Phys. Rev. Lett.* **36** (1976) 145.
5. R. Y. Chiao, in *Science and Ultimate Reality*, eds. J. D. Barrow, P. C. W. Davies and C. L. Harper, Jr. (Cambridge University Press, 2004), p. 254 [quant-ph/0303100].
6. A. Yariv, *Quantum Electronics*, 1st edn. (John Wiley & Sons, New York, 1967), p. 223.
7. R. M. Wald, *General Relativity* (University of Chicago Press, 1984).
8. R. B. Laughlin, *Phys. Rev. Lett.* **50** (1983) 1395.
9. G. Veneziano, *Sci. Am.* (May 2004) 64.

A ROBUST TEST OF GENERAL RELATIVITY IN SPACE

JAMES GRABER

Technology Assessment, Library of Congress,
101 Independence Ave. SE,
Washington, DC 20540, USA
jgra@loc.gov
jgraber@mailaps.org

LISA may make it possible to test the black-hole uniqueness theorems of general relativity, also called the no-hair theorems, by Ryan's method of detecting the quadrupole moment of a black hole using high-mass-ratio inspirals. This test can be performed more robustly by observing inspirals in earlier stages, where the simplifications used in making inspiral predictions by the perturbative and post-Newtonian methods are more nearly correct. Current concepts for future missions such as DECIGO and BBO would allow even more stringent tests by this same method. Recently discovered evidence supports the existence of intermediate-mass black holes (IMBHs). Inspirals of binary systems with one IMBH and one stellar-mass black hole would fall into the frequency band of proposed maximum sensitivity for DECIGO and BBO. This would enable us to perform the Ryan test more precisely and more robustly. We explain why tests based on observations earlier in the inspiral are more robust and provide preliminary estimates of possible optimal future observations.

Keywords: LISA; BBO; DECIGO; no-hair.

1. Introduction

The theme of this NASA workshop, "From Quantum to Cosmos: Fundamental Physics Research in Space," is testing fundamental physics in space, celebrating past such accomplishments and anticipating possible future achievements. One fundamental test of general relativity that apparently depends on space-based gravitational wave detectors for practical implementation is the test of the black-hole uniqueness theorems (no-hair theorems), first proposed by Ryan.[1] Testing general relativity is one of the official goals of the LISA project[2,3] and includes specifically measuring the extreme-mass-ratio inspirals (EMRIs)[4] that are necessary for performing Ryan's test. Ryan[5] concluded that LISA could perform his test to an accuracy of order 1% with data from a favorable EMRI. In Ref. 6, we reached a similar conclusion. The prospects that we will be able to perform robust and accurate versions of Ryan's test have brightened considerably due to the recent discovery of probable intermediate-mass black holes, which increases not only the number of gravitational-wave-dominated binary inspirals that are likely to be seen, but also the likelihood that we can observe them in the stages of the inspiral where the predictions are most robust and where the data are most likely to support precise and reliable tests.

If substantial numbers of IMBHs exist, as recently proposed,[7–9] it will be possible to perform a greatly enhanced Ryan test with future possible space missions such as Big Bang Observer (BBO)[10,11] or DECIGO.[12] This is because the inspiral of a stellar-mass black hole into an IMBH falls into the most sensitive band of BBO or DECIGO, where there are no interfering white-dwarf binaries, and where it will spiral through millions of cycles in less than ten years. BBO, which is optimized to find faint gravitational waves from the Big Bang itself, will be more than a thousand times as sensitive as LISA and will be able to see light IMRIs throughout the entire universe.

In this paper we briefly review the recent developments affecting our expectations of observing extreme- and intermediate-mass-ratio inspirals (EMRIs and IMRIs), and consider the eventual possibilities for performing more robust and more accurate tests of general relativity. We point out that data from early in the inspiral have some advantages over data from later stages for performing robust and accurate tests. We give order-of-magnitude estimates for the possible improvements in accuracy and for possible increases in the number of systems observed to indicate the potential rich harvest that awaits these future, more sensitive missions to test fundamental physics in space by observing black holes with gravitational waves.

2. Definition of EMRIs, Light and Heavy IMRIs

For simplicity, supermassive black holes are defined as those greater than 10^6 solar masses, stellar-mass black holes as those less than 100 solar masses, and intermediate-mass black holes (IMBHs) as those from 10^2 to 10^6 solar masses.

A classic EMRI is the inspiral of a stellar-mass black hole into a supermassive black hole. A heavy IMRI is the inspiral of an IMBH into a supermassive black hole. A light IMRI is the inspiral of a stellar-mass black hole into an IMBH.

3. Short Summary of DECIGO and BBO Proposals

BBO and DECIGO are concepts for far-more-sensitive, space-based gravitational-wave observatories to follow LISA. One of the key ideas of the DECIGO and BBO proposals is to put LIGO and VIRGO technology in space. Another key factor in these proposals is arm lengths ten times shorter than LISA, resulting in peak sensitivities at higher frequencies. BBO, in particular, is optimized to detect very weak gravitational waves from the Big Bang itself. The fact that this also makes it so useful for performing Ryan's test with light EMRIs is a bonus.

The inclusion of shorter arm lengths will make BBO/DECIGO-type systems not only more sensitive than LISA, but also sensitive to different sources. It turns out that the inspirals of light IMRIs fall right into this sensitivity band. LISA's peak sensitivity is approximately 10^{-20} strain per root Hertz from 0.003 Hz to 0.01 Hz. Proposed DECIGO and BBO systems are planned to have peak sensitivity of 10^{-23} strain per root Hz from 0.1 to 1.0 Hz, i.e. about 1000 times more sensitive in a frequency band 10–100 times higher. This band is ideally suited for observing the inspirals of light IMRIs.

4. Short Summary of Testing General Relativity by Ryan's Method

The basic observable gravitational-wave form is quasi-sinusoidal with a slowly rising frequency, called a chirp. The phase of this sinusoid (ϕ) corresponds to twice the phase of the orbiting binary. It can be recovered exactly by removing Doppler shifts for the appropriate direction and referring the LISA signal to the solar system barycenter. By matched filtering, we can determine the frequency of the chirp as a function of time with an error of less than a single cycle in the length of the filter, which can potentially be many thousands of cycles long. This frequency evolution function (FEF) (technically $d\phi/2\pi dt$ as a function of time) will be observed with this accuracy over tens or hundreds of thousands of — or even a million or more — cycles in a typical chirp observed by LISA. Hence the FEF will be known with an accuracy better than one part in 10^5 or 10^6. This is what enables us to perform precision tests of general relativity, by comparing the observed FEF to a predicted FEF.

According to the black-hole uniqueness theorems,[13–17] in general relativity the only astrophysically possible neutral black hole is a Kerr black hole, which is uniquely determined by its mass M and spin S. General relativity predicts that the magnitude of the suitably defined quadrupole moment Q of a Kerr black hole is $Q = S^2/M$. If Q is not equal to $Q = S^2/M$, general relativity is falsified.

Ryan[1] showed that one can determine the mass M, the spin S and the quadrupole moment Q from just the first four terms in the Taylor expansion of the FEF in the extreme-mass-ratio circular-orbit case. Put another way, Ryan showed that if you can measure the first three terms of this series, you can predict the fourth. Use of this decomposition of the FEF to check whether or not $Q = S^2/M$ is the test of the black-hole uniqueness theorems by Ryan's method, or the Ryan test. This is one of the easiest and cleanest tests for the correctness of general relativity, and one of the most restrictive on possible alternate theories of gravity. In principle, one needs only three numbers (M, S, Q) for this test.

Since the FEF is a convergent series [particularly far away from the innermost circular orbit (ISCO)], the first four terms are generally decreasing, and the accuracy of the test is determined by the size of the fourth (smallest) term.

Since the number of cycles is the most directly measurable feature, and the error is of order one cycle, the dominant error is of order one over the number of cycles contributed by the fourth term.

The *accuracy* of the test is determined by how precisely we can *measure* the number of cycles contributed by the first four terms of this series.

The *robustness* of the test is determined by how precisely we can *predict* the number of cycles contributed by the first four terms of this series.

5. Why Earlier Is Better?

The lack of theoretical robustness in general-relativity inspiral predictions primarily comes from uncomputed higher-order terms[18] and from the progressive failure of the adiabatic hypothesis and other simplifications made to compute these inspiral predictions.[19] It is well known that these errors and deviations get larger near the ISCO.[20–30] On the contrary, the unknown terms become less important and the approximations become more acurate as one moves earlier in the inspiral and farther from the ISCO.

Also important is how well we can isolate, observationally and theoretically, the number of cycles contributed by higher-order terms, as well as the terms of the first four orders.

Due to a prefactor of order -5, as you move earlier in the inspiral (and away from the ISCO), the number of cycles contributed by terms of order 4 and less *increases*, whereas the number of cycles contributed by terms of order 6 or higher *decreases*. Since the contributions of these higher-order terms decrease as we move away from the ISCO, it is easier to get an accurate measurement of the lower-order terms farther from the ISCO, as long as there is enough frequency sweep to cleanly separate the terms of different orders.

Thus, it is more robust to measure the inspiral at an earlier stage, somewhat removed from the ISCO, for two reasons:

First, the general relativity predictions are cleaner and more robust theoretically.

Second, the measurement of the contributions of the lower-order terms needed for Ryan's test is more precise.

We will see that light IMRIs and proposed second- and third-generation missions, i.e. DECIGO and BBO, help achieve these objectives of getting more inspiral cycles farther from the ISCO.

6. Summary of Evidence for IMBHs

A small number of nearby globular clusters and dwarf galaxies have shown dynamical evidence consistent with IMBHs.[31] A very large number of ultraluminous X-ray sources (ULXs) have been observed, on the order of several per L* galaxy.[32] If a significant fraction of ULXs are IMBHs, as now seems likely, IMBHs are approximately as numerous as L* galaxies. For supermassive black holes, it is commonly accepted that there is one in almost every L* galaxy, but they are only actively emitting X-rays about 1% of the time. If a similar ratio of IMBHs are active as ULXs at any time, that would imply that IMBHs are of order 100 times as numerous as L* galaxies.

Another, somewhat more speculative line of reasoning merely assumes that IMBHs are approximately as numerous as globular clusters,[33] since some dynamical evidence supports IMBHs in globular clusters (e.g. M15[34], and G1[35]), and some ULXs are associated with globular clusters. This also results in a ratio of IMBHs to L* galaxies of order 100 to 1.

The same type of argument can be given for dwarf galaxies in place of globular clusters. The evidence is less firm, but the expected relative numbers are again the same within an order of magnitude.

Hereafter we assume for our optimistic estimate that IMBHs are 100 times as numerous as L* galaxies, with of course large uncertainties.

Another argument for the existence of IMBHs is that almost all supermassive-black-hole formation scenarios pass through an IMBH stage.[36]

The simulations of IMBH formation in globular clusters suggest that it is a natural result of runaway core collapse and stellar collisions in the central cusp of the globular cluster. Many stellar-mass black holes are expected to be present and to be absorbed by the growing IMBH.[37] The formation of light IMRIs in globular clusters is highly likely in this scenario.

7. Why Light IMRIs Give a More Precise and Robust Ryan Test?

The overall sensitivity of the Ryan test is proportional to the number of cycles of the inspiral that are observed. The higher frequency of the BBO band and the light IMRIs, as compared to the LISA band and the heavy IMRIs and classical EMRIs, results in 100 times more cycles in the same amount of time.

As discussed in Sec. 3, the robustness and the accuracy of the Ryan test are greater earlier in the inspiral.

The classical EMRIs and the heavy IMRIs begin to get lost in the white-dwarf binary confusion noise as one moves away from the ISCO. The light IMRIs have at least two extra decades of frequency sweep before they hit that limit. Also, as they accumulate cycles more than 100 times faster, their measurement is also less likely to be impacted by the mission duration limit. Hence the light IMRIs with BBO or DECIGO are likely to permit a very substantially more robust and precise measurement than the EMRIs (or heavy IMRIs) and LISA.

8. Conclusion

We have briefly explained why earlier inspiral data are more theoretically robust. They contain a greater number of total cycles and a higher number of cycles per octave of frequency sweep. They also contain a higher ratio of predicted cycles to unpredicted cycles. Light IMRIs in the 0.1 Hz band are likely to give a very robust and precise Ryan test when BBO or DECIGO flies.

References

1. F. D. Ryan, *Phys. Rev. D* **52** (1995) 5707.
2. http://lisa.nasa.gov
3. http://sci.esa.int/home/lisa
4. T. Prince and K. Danzmann, *LISA Science Requirements Document, version 3.0, 12 May 2005*, http://www.srl.caltech.edu/lisa/documents.html
5. F. D. Ryan, *Phys. Rev. D* **56** (1997) 1845.
6. J. Graber, gr-qc/060408.
7. J. M. Miller, Present Evidence for Intermediate Mass Black Holes in ULXs and Future Prospects, to appear in *From X-Ray Binaries to Quasars: Black Hole Accretion on All Mass Scales*, eds. T. J. Maccarone, R. P. Fender and L. C. Ho (Kluwer, Dordrecht, 2006).
8. L. M. Winter, R. F. Mushotzky and C. S. Reynolds, to appear in *Astrophys. J.* [astro-ph/0512480].
9. E. J. M. Colbert and M. C. Miller, astro-ph/0402677.
10. E. S. Phinney *et al.*, *The Big Bang Observer: Direct Detection of Gravitational Sources from the Birth of the Universe to the Present*, NASA Mission Concept Study (2004).
11. http://universe.nasa.gov/program/bbo.html
12. N. Seto, S. Kawamura and T. Nakamura, *Phys. Rev. Lett.* **87** (2001) 221103.
13. W. Israel, *Phys. Rev.* **164** (1967) 1776.
14. W. Israel, *Commun. Math. Phys.* **8** (1967) 245.
15. B. Carter, *Phys. Rev. Lett.* **26** (1971) 331.
16. R. Price, *Phys. Rev. D* **5**, 2439 (1972).
17. D. C. Robinson, *Phys. Rev. Lett.* **34** (1975) 905.
18. L. Blanchet, *Living Rev. Relativ.* **5** (2002) 3; http://www.livingreviews.org/lrr-2002-3
19. M. Sasaki and H. Tagoshi, *Living Rev. Relativ.* **6** (2003) 6; http://www.livingreviews.org/lrr-2003-6
20. E. Poisson, *Phys. Rev. D* **52** (1995) 5719.
21. L. E. Simone *et al. Class. Quant. Grav.* **14** (1997) 237.
22. E. E. Flanagan and S. A. Hughes, *Phys. Rev. D* **57** (1998) 4535.
23. E. E. Flanagan and S. A. Hughes, *Phys. Rev. D* **57** (1998) 4566.
24. L. S. Finn and K. S. Thorne, *Phys. Rev. D* **62** (2000) 124021.
25. L. Barack and C. Cutler, *Phys. Rev. D* **69** (2004) 082005.
26. A. Buonanno, Y. Chen and T. Damour, gr-qc/0508067.
27. E. Berti, S. Iyer and C. M. Will, gr-qc/0607047
28. K. Glampedakis and S. Babak, *Class. Quant. Grav.* **23** (2006) 4167.
29. J. R. Gair and K. Glampedakis, *Phys. Rev. D* **73** (2006) 0604037.
30. S. Babak *et al.*, gr-qc/0607007.
31. R. van der Marel, astro-ph/0302101.
32. J. M. Miller, Present Evidence for Intermediate Mass Black Holes in ULXs and Future Prospects, *From X-ray Binaries to Quasars: Black Hole Accretion on All Mass Scales*, eds. T. J. Maccarone, R. P. Fender and L. C. Ho (Kluwer, Dordrecht, 2006).
33. H. Baumgardt, J. Makino and P. Hut, *Astrophys. J.* **620** (2005) 238.
34. R. P. van der Marel *et al. Astrophys. J.* **124** (2002) 3255.
35. K. Gephardt, R. M. Rich and L. Ho, *Astrophys. J.* **578** (2002) L41.
36. H. Baumgardt, J. Makino and T. Ebisuzuki, *Astrophys. J.* **613** (2004) 1143.
37. H. Baumgardt *et al.* astro-ph/0511752.

PART 4

PHYSICS BEYOND THE STANDARD MODEL

DETECTING STERILE DARK MATTER IN SPACE

ALEXANDER KUSENKO

Department of Physics and Astronomy,
University of California, Los Angeles, CA 90095-1547, USA

Space-based instruments provide new and, in some cases, unique opportunities to search for dark matter. In particular, if dark matter comprises sterile neutrinos, the X-ray detection of their decay line is the most promising strategy for discovery. Sterile neutrinos with masses in the keV range could solve several long-standing astrophysical puzzles, from supernova asymmetries and the pulsar kicks to star formation, reionization, and baryogenesis. The best current limits on sterile neutrinos come from Chandra and XMM-Newton. Future advances can be achieved with a high-resolution X-ray spectrometry in space.

Keywords: Dark matter; sterile neutrinos; X-ray astronomy.

1. Introduction

There is an overwhelming amount of evidence that most of the matter in the Universe is not made up of ordinary atoms but, rather, of new, yet-undiscovered particles.[1] The evidence for dark matter is based on several independent observations, including cosmic-microwave-background radiation, gravitational lensing, the galactic rotation curves, and the X-ray observations of clusters. None of the Standard Model particles can be dark matter. Hence, the identification of dark matter will be a discovery of new physics beyond the Standard Model.

To detect dark matter one must guess at its properties, which ultimately determine one's strategy for detection. One can base one's guesses on compelling theoretical ideas or on some observational clues.

One of the most popular theories for physics beyond the Standard Model is supersymmetry. A class of supersymmetric extensions of the Standard Model predict dark matter in the form of either the lightest supersymmetric particles[2] or SUSY Q balls.[3–7] Another theoretically appealing possibility is dark matter in the form of axions.[8–11] The axion is a very weakly interacting field which accompanies the Peccei–Quinn solution to the strong CP problem. There are several other dark matter candidates that are well motivated by theoretical reasoning. A comprehensive review of possibilities is not our purpose; rather, we will focus on the forms of dark matter which are well motivated and for which there are new opportunities in space research.

Right-handed or sterile neutrinos can be the dark matter.[12–15] The existence of such right-handed states is implied by the discovery of the active neutrino masses. Although it is not impossible to explain the neutrino masses otherwise, most models introduce gauge singlet fermions that give the neutrinos their masses via mixing. If one of these right-handed states has a mass in the ~ 1–$50\,\mathrm{keV}$ range, it can be the dark matter. Several indirect astrophysical clues support this hypothesis. Indeed, if the sterile neutrinos exist, they can explain the long-standing puzzle of pulsar velocities.[16] In addition, the X-rays produced in decays of the relic neutrinos could increase the ionization of the primordial gas and catalyze the formation of molecular hydrogen at redshift as high as 100. Since the molecular hydrogen is an important cooling agent, its increased abundance could cause the early and prompt start formation.[17,18] Sterile neutrinos can also help the formation of supermassive black holes in the early Universe.[21] For smaller masses, the sterile neutrinos have a large enough free-streaming length to rectify several reported inconsistencies between the predictions of cold dark matter on small scales and the observations. The consensus of these indirect observational hints helps make a stronger case for the sterile dark matter.

2. Sterile Neutrinos

The number of light "active" left-handed neutrinos — three — is well established from the LEP measurements of the Z-boson decay width. In the Standard Model, the three active neutrinos fit into the three generations of fermions. In its original form the Standard Model described massless neutrinos. The relatively recent but long-anticipated discovery of the neutrino masses has made a strong case for considering right-handed neutrinos, which are $SU(3)\times SU(2)\times U(1)$ singlets. The number of right-handed neutrinos may vary and need not be equal to three.[22,23] Depending on the structure of the neutrino mass matrix, one can end up with none, one, or several states that are light and (mostly) sterile, i.e. they interact only through their small mixing with the active neutrinos.

The sterile neutrino is not a new idea. The name "sterile" was coined by Bruno Pontecorvo in 1967.[24] Many seesaw models[25–29] assume that sterile neutrinos have very large masses, which makes them unobservable. However, one can consider a lighter sterile neutrino, which can be dark matter.[12] Emission of sterile neutrinos from a supernova could explain the pulsar kicks if the sterile neutrino mass was several keV.[30–32] More recently, a number of papers have focused on this range of masses because several indirect observational hints suggest the existence of a sterile neutrino with such a mass.

Unless some neutrino experiments are wrong, the present data on neutrino oscillations cannot be explained with only the active neutrinos. Neutrino oscillation experiments measure the differences between the squares of neutrino masses, and the results are: one mass squared difference is of the order of $10^{-5}\,\text{eV}^2$, the other one is $10^{-3}\,\text{eV}^2$, and the third is about $1\,\text{eV}^2$. Obviously, one needs more than three masses to get the three different mass splittings, which do not add up to zero. Since we know that there are only three active neutrinos, the fourth neutrino must be sterile. However, if the light sterile neutrinos exist, there is no compelling reason why their number should be limited to one.

The neutrino masses can be introduced into the Standard Model by means of the following addition to the Lagrangian:

$$\mathcal{L} = \mathcal{L}_{\text{SM}} + \bar{\nu}_{s,a}\left(i\partial_\mu\gamma^\mu\right)\nu_{s,a} - y_{\alpha a}H\bar{L}_\alpha\nu_{s,a} - \frac{M_{aa}}{2}\,\bar{\nu}^c_{s,a}\nu_{s,a} + \text{h.c.}, \tag{1}$$

where H is the Higgs boson and L_α $(\alpha = e, \mu, \tau)$ are the lepton doublets, while $\nu_{s,a}$ $(a = 1, \ldots, N)$ are the additional singlets. This model, dubbed νMSM,[15] provides a natural framework for considering sterile neutrinos. Of course, the gauge singlet fields may have some additional couplings omitted from Eq. (1). The neutrino mass matrix has the form

$$M = \begin{pmatrix} \tilde{m}_{3\times3} & D_{3\times N} \\ D^T_{N\times3} & M_{N\times N} \end{pmatrix}, \tag{2}$$

where the Dirac masses $D_{\alpha a} = y_{\alpha a}\langle H\rangle$ are the result of spontaneous symmetry breaking. For symmetry reasons one usually sets $\tilde{m}_{3\times3}$ to zero. As for the right-handed Majorana masses M, the scale of these masses can be either much greater or much smaller than the electroweak scale.

The seesaw mechanism[25–29] can explain the smallness of neutrino masses in the presence of the Yukawa couplings of order 1. For this purpose, one assumes that the Majorana masses are much larger than the electroweak scale, and the smaller eigenvalues of the mass matrix (2) are suppressed by the ratio of $\langle H\rangle$ to M.

However, the origin of the Yukawa couplings remains unknown, and, in the absence of the fundamental theory, there is no compelling reason to believe that these couplings must be of order 1. Indeed, the Yukawa couplings of most known fermions are much smaller than 1; for example, the Yukawa coupling of the electron is $\sim 10^{-6}$.

Thus, for all we know, the scale of the Majorana mass M in Eq. (2) can be much smaller than the electroweak scale. If $M \sim 1$ eV, the sterile neutrinos with the mass $m_s \sim 1$ eV can explain the LSND results.[33] If $M \sim 1$ keV, the sterile neutrinos with the corresponding mass could explain the pulsar kicks[30–32] and dark matter,[12] and they can also play a role in generating the matter–antimatter asymmetry of the Universe.[19,20]

3. Production of Sterile Neutrinos in the Early Universe

Sterile neutrinos can be produced in the early Universe from neutrino oscillations, as well as from other couplings, not included in Eq. (1). For example, dark matter in the form of sterile neutrinos can be produced by a direct coupling to the inflaton.[34]

At very high temperatures the active neutrinos have frequent interactions in plasma, which reduce the probability of their conversion into sterile neutrinos.[35,36] The mixing of sterile neutrinos with one of the active species in plasma can be represented by an effective, density- and temperature-dependent mixing angle[12–14]:

$$|\nu_1\rangle = \cos\theta_m\,|\nu_e\rangle - \sin\theta_m\,|\nu_s\rangle, \tag{3}$$

$$|\nu_2\rangle = \sin\theta_m\,|\nu_e\rangle + \cos\theta_m\,|\nu_s\rangle, \tag{4}$$

where

$$\sin^2 2\theta_m = \frac{(\Delta m^2/2p)^2 \sin^2 2\theta}{(\Delta m^2/2p)^2 \sin^2 2\theta + (\Delta m^2/2p\cos 2\theta - V_m - V_{_T})^2}. \tag{5}$$

Here V_m and V_T are the effective matter and temperature potentials. In the limit of small angles and small lepton asymmetry, the mixing angle can be approximated as

$$\sin 2\theta_m \approx \frac{\sin 2\theta}{1 + 0.27\zeta \left(\frac{T}{100\,\mathrm{MeV}}\right)^6 \left(\frac{\mathrm{keV}^2}{\Delta m^2}\right)}, \tag{6}$$

where $\zeta = 1.0$ for mixing with the electron neutrino and $\zeta = 0.30$ for ν_μ and ν_τ.

Obviously, thermal effects suppress the mixing significantly for temperatures $T > 150\,(\mathrm{m/keV})^{1/3}\,\mathrm{MeV}$. If the singlet neutrinos interact only through mixing, all the interaction rates are suppressed by the square of the mixing angle, $\sin^2\theta_m$. It is easy to see that these sterile neutrinos are *never* in thermal equilibrium in the early Universe. Thus, in contrast with the case of the active neutrinos, the relic population of sterile neutrinos is not a result of a freeze-out. One immediate consequence of this observation is that the Gershtein–Zeldovich bound[37] and the Lee–Weinberg bound[39] do not apply to sterile neutrinos. In general, the existing experimental constraints on sterile neutrinos[40] allow a wide range of parameters, especially for small mixing angles.

One can calculate the production of sterile neutrinos in plasma by solving the Boltzmann equation for the distribution function $f(p,t)$:

$$\left(\frac{\partial}{\partial t} - Hp\frac{\partial}{\partial p}\right) f_s(p,t) \equiv xH\partial_x f_s \tag{7}$$

$$= \Gamma_{(\nu_a\to\nu_s)}\left(f_a(p,t) - f_s(p,t)\right), \tag{8}$$

where H is the Hubble constant, $x = 1\,\mathrm{MeV}\ a(t)$ [$a(t)$ is the scale[41] factor] and Γ is the probability of conversion. The solution[12–14,41] is shown in Fig. 1 as "dark matter produced via mixing." One should keep in mind that this solution is subject to hadronic uncertainties.[42]

If the sterile neutrinos have additional interactions, not included in Eq. (1), the relic population of these particles can be produced via different mechanisms. One example is a direct coupling of sterile neutrinos to the inflaton.[34] In this case the production of sterile neutrinos may not be governed by the mixing angle, although the mixing angle still controls the decay rate and, therefore, some of the constraints discussed below depend on the mixing angles.

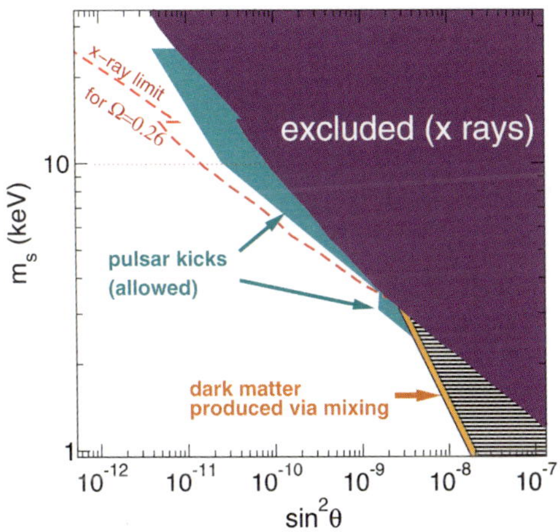

Fig. 1. The range of the sterile neutrino masses and mixing angles. The X-ray limits depend on the abundance of the relic sterile neutrinos, which in turn depends on their production mechanism. If the sterile neutrinos are produced only via their mixing with active neutrinos, they can be dark matter for masses below 3 keV, as shown in the figure. This range is in conflict with the 10 keV lower bound from the Lyman alpha forest shown by a dotted line. In sharp contrast, the observations of dwarf spheroid galaxies favor the masses of a few keV. If the sterile neutrinos are produced via some additional couplings, besides the mixing with the active neutrinos, and if the sterile neutrinos make up all the dark matter ($\Omega = 0.26$), the corresponding X-ray limit is shown as a dashed line. Also shown is the allowed range of parameters consistent with the pulsar kicks.

4. Constraints on Sterile Dark Matter

Although dark matter sterile neutrinos are stable on cosmological time scales, they nevertheless decay.[14,43–50] The dominant decay mode, into three light neutrinos, is "invisible" because the daughter neutrinos are beyond the detection capabilities of today's experiments. The most prominent "visible" mode is decay into one active neutrino and one photon, $\nu_s \to \nu_a \gamma$. Assuming the two-neutrino mixing for simplicity, one can express the inverse width of such a decay as[51]

$$\tau \equiv \Gamma^{-1}_{\nu_s \to \nu_a \gamma} = 1 \times 10^{26} \mathrm{s} \left(\frac{7\ \mathrm{keV}}{m_s} \right)^5 \left(\frac{1 \times 10^{-9}}{\sin^2 \theta} \right), \tag{9}$$

where m_s is the mass and θ is the mixing angle.

Since this is a two-body decay, the photon energy is half the mass of the sterile neutrino. The monochromatic line from dark matter decays can, in principle, be observed by X-ray telescopes. No such observation has been reported, and some important limits have been derived on the allowed masses and mixing angles. These constraints are based on different astrophysical objects, from Virgo and Coma clusters to Large Magellanic Clouds, to the Milky Way halo and its components.[43–50] There are different uncertainties in modeling the dark matter populations in these objects. Different groups have also used very different methods to derive these bounds: from a conservative assumption that the dark matter line should not exceed the signal, to more ambitious approaches that involved modeling the signal or merely fitting it with a smooth curve and requiring that the line-shaped addition not affect the quality of the fit. In any case, the limits apply to the flux of X-rays, which can be translated into the limits on the mass and mixing angle if the sterile neutrino abundance is known. As we discussed above, the production is possible via the mixing alone, but some additional couplings and other production mechanisms are by no means excluded.[34] Most published bounds[43–50] *assume* that sterile neutrinos make up all the dark matter, i.e. $\Omega_s = 0.26$. The limit based on this assumption is shown as a dashed (red) line in Fig. 1. However, it should not be used as the exclusion limit for sterile neutrinos in general, because it is possible that $\Omega_s < 0.26$, while the sterile neutrinos could still explain the pulsar velocities and they could play a role in the star formation.

A different kind of limit based on the same X-ray data[43–50] can be set without assuming that $\Omega_s = 0.26$. As long as there is a mixing due to the couplings in the Lagrangian (1), some sterile neutrinos are produced in the hot plasma, regardless of any additional couplings that may or may not be present. This amount corresponds to the lower bound on the sterile neutrino abundance, and the bound obtained this way is the most robust, model-independent limit. The corresponding exclusion region is shown in Fig. 1.

Additional constraints on dark matter come from the observations of the Lyman alpha forest,[52–54] which limit the sterile neutrino mass from below. Based on the high-redshift data from SDSS and some modeling of gas dynamics, one can set a limit as strong as 14 keV.[53] However, the high-redshift data may have systematic errors, and more conservative approaches, based on the relatively low-redshift data, have led to some less stringent bounds[52]. Recently Viel *et al.*[54] have reanalyzed the high-redshift data and arrived at the bound $m_s > 10$ keV. The mass bounds quoted depend on the production mechanism in the early Universe.

The Lyman alpha observations constrain the free-streaming lengths of dark matter particles, not their masses. For each cosmological production mechanism, the relation between the free-streaming length and the mass is different.[55] For example, the bound $m_s > 10$ keV[54] applies to the production model due to Dodelson and Widrow.[12] If the lepton asymmetry of the Universe (which is unknown *a priori*) is sufficiently large, then the sterile neutrinos can be produced through resonant Mikheev–Smirnov–Wolfenstein[56–58] (MSW) oscillations in the early Universe.[59] These neutrinos are nonthermal and colder because the adiabaticity condition selects the low-energy part of the neutrino spectrum. Even within a given cosmological scenario, there are uncertainties in the production rates of neutrinos for any given mass and mixing angle.[42] These uncertainties may further affect the interpretation of the Lyman alpha bounds in terms of the sterile neutrino mass.

It should also be mentioned that the Lyman alpha bounds appear to contradict the observations of dwarf spheroidal galaxies,[71,72] which suggest that dark matter is warm and which would favor the 1–5 keV mass range for sterile neutrinos. There are several inconsistencies between the predictions of N-body simulations of cold dark matter (CDM) and the observations.[60,60–69] Each of these problems may find a separate, independent solution. Perhaps a better understanding of CDM on small scales will resolve these discrepancies. It is true, however, that warm dark matter in the form of sterile neutrinos is free from all these small-scale problems altogether, while on large scales WDM fits the data as well as CDM.

If the sterile neutrinos make up only a part of dark matter, the Lyman alpha bounds do not apply. In this case, the sterile neutrinos may still be responsible for pulsar velocities, and they can play a role in star formation and reionization of the Universe. Also, if inflation ended with a low reheat temperature, the bounds are significantly weaker.[73]

5. Reionization and Star Formation

Sterile neutrinos decay in the early Universe, in particular during the "dark ages" following recombination. The ionizing photons are too few to affect the cosmic microwave background directly,[74] but they can have an important effect on star formation and reionization. Star formation requires cooling and collapse of gas clouds, which is impossible unless the fraction of molecular hydrogen is high enough.[75] It is accompanied by the reionization of gas in the Universe. The WMAP (three years) measurement[76] of the reionization redshift $z_r = 10.9^{+2.7}_{-2.3}$ has posed a new challenge to theories of star formation. On the one hand, stars have to form early enough to reionize gas at redshift 11. On the other hand, the spectra of bright distant quasars imply that reionization must be completed by redshift 6. Stars form in clouds of hydrogen, which collapse at different times, depending on their sizes: the small clouds collapse first, while the large ones collapse last. If the big clouds must collapse by redshift 6, then the small halos must undergo the collapse at an earlier time. It appears that the star formation in these small halos would have occurred at high redshift, when the gas density was very high, and it would have resulted in an unacceptable overproduction of the Thompson optical depth.[77] To be consistent with WMAP, the efficiency for the production of ionizing photons in minihalos must have been at least an order of magnitude lower than expected.[77] One solution is to suppress the star formation rate in small halos by some dynamical feedback mechanism. The suppression required is by at least an order of magnitude.

An alternative solution is to consider warm dark matter, in which case the small clouds are absent altogether. However, it has been argued that "generic" warm dark matter can delay the collapse of gas clouds.[78] This problem does not arise in the case of sterile neutrinos, because the X-ray photons from their slow decays could have increased the production of molecular hydrogen and could have precipitated rapid and prompt star formation at a high enough redshift.[17,18]

6. Pulsar Velocities

The space velocities of pulsars range from 250 km/s to 500 km/s.[79–93] Some 15% of pulsars[93] appear to have velocities greater than 1000 km/s, while the fastest pulsars have speeds as high as 1600 km/s. The origin of these velocities remains a puzzle.[16]

Since most of the supernova energy, as much as 99% of the total 10^{53} erg, is emitted in neutrinos, a few percent anisotropy in the distribution of these neutrinos would be sufficient to explain the pulsar kicks.

Neutrinos are always *produced* with an asymmetry, but they usually *escape* isotropically. The asymmetry in production comes from the asymmetry in the basic weak interactions in the presence of a strong magnetic field.[a] Indeed, if the electrons and other fermions are polarized by the magnetic field, the cross-section of the urca processes, such as $n+e^+ \rightleftharpoons p+\bar{\nu}_e$ and $p+e^- \rightleftharpoons n+\nu_e$, depends on the orientation of the neutrino momentum:

$$\sigma(\uparrow e^-, \uparrow \nu) \neq \sigma(\uparrow e^-, \downarrow \nu). \tag{10}$$

Depending on the fraction of the electrons in the lowest Landau level, this asymmetry can be as large as 30%, which is, seemingly, more than one needs to explain the pulsar kicks.[97] However, this asymmetry is completely washed out by scattering of neutrinos on their way out of the star.[98–100] This is intuitively clear because, as a result of scatterings, the neutrino momentum is transferred to and shared by the neutrons. In the approximate thermal equilibrium, no asymmetry in the production or scattering amplitudes can result in a macroscopic momentum anisotropy. This statement can be proved rigorously.[98–100]

However, if the neutron star cooling produced a particle whose interactions with nuclear matter were *even weaker* than those of ordinary neutrinos, such a particle could escape the star with an anisotropy equal to its production anisotropy. The sterile neutrinos, whose interactions are suppressed by $(\sin^2\theta_m)$, can play such a role.[30–32,101,102] The region of masses and mixing angles consistent with this explanation for the pulsar kicks is shown in Fig. 1. The neutrino-driven kicks have a number of ramifications: in particular, they can increase the energy of the shock and can generate asymmetric jets, the strongest of which is aligned with the direction of the pulsar motion.[103]

7. Conclusion

Several independent observational hints point to sterile neutrinos with masses in the keV range. Pulsar velocities can be explained by the emission of such sterile neutrinos from a supernova, because the sterile neutrino emission is anisotropic in the presence of the magnetic field. The X-ray photons from the decays of the sterile neutrinos can ionize the primordial gas and can cause an increase in the fraction of molecular hydrogen, which makes a prompt star formation possible at a relatively high redshift. The sterile neutrinos can be the dark matter. In addition, they could have played a role in generating the matter–antimatter asymmetry of the Universe.

[a]Here we disregard the neutrino magnetic moments, which are negligible in the Standard Model and its simplest extensions. Even for vanishing magnetic moments, neutrino oscillations are affected by the magnetic field through the polarization of the matter fermions.[94–96]

Future observations of X-ray telescopes may be able to discover the relic sterile neutrinos by detecting keV photons from their decays.

Acknowledgments

This work was supported in part by the US Department of Energy grant DE-FG03-91ER40662 and by the NASA ATP grants NAG 5-10842 and NAG 5-13399.

References

1. G. Bertone, D. Hooper and J. Silk, *Phys. Rep.* **405** (2005) 279.
2. G. Jungman, M. Kamionkowski and K. Griest, *Phys. Rep.* **267** (1996) 195.
3. A. Kusenko, *Phys. Lett. B* **405** (1997) 108.
4. A. Kusenko and M. E. Shaposhnikov, *Phys. Lett. B* **418** (1998) 46.
5. A. Kusenko *et al.*, *Phys. Rev. Lett.* **80** (1998) 3185.
6. K. Enqvist and A. Mazumdar, *Phys. Rep.* **380** (2003) 99.
7. M. Dine and A. Kusenko, *Rev. Mod. Phys.* **76** (2004) 1.
8. R. D. Peccei and H. R. Quinn, *Phys. Rev. Lett.* **38** (1977) 1440.
9. R. D. Peccei and H. R. Quinn, *Phys. Rev. D* **16** (1977) 1791.
10. S. Weinberg, *Phys. Rev. Lett.* **40** (1978) 223.
11. F. Wilczek, *Phys. Rev. Lett.* **40** (1978) 279.
12. S. Dodelson and L. M. Widrow, *Phys. Rev. Lett.* **72** (1994) 17.
13. K. Abazajian, G. M. Fuller and M. Patel, *Phys. Rev. D* **64** (2001) 023501.
14. A. D. Dolgov and S. H. Hansen, *Astropart. Phys.* **16** (2002) 339.
15. T. Asaka, S. Blanchet and M. Shaposhnikov, *Phys. Lett. B* **631** (2005) 151.
16. A. Kusenko, *Int. J. Mod. Phys. D* **13** (2004) 2065.
17. P. L. Biermann and A. Kusenko, *Phys. Rev. Lett.* **96** (2006) 091301.
18. J. Stasielak, P. L. Biermann and A. Kusenko, astro-ph/0606435.
19. E. K. Akhmedov, V. A. Rubakov and A. Y. Smirnov, *Phys. Rev. Lett.* **81** (1998) 1359.
20. T. Asaka and M. Shaposhnikov, *Phys. Lett. B* **620** (2005) 17.
21. F. Munyaneza and P. L. Biermann, astro-ph/0403511.
22. P. H. Frampton, S. L. Glashow and T. Yanagida, *Phys. Lett. B* **548** (2002) 119.
23. B. Kayser, *Nucl. Phys. Proc. Suppl.* **118** (2003) 425.
24. B. Pontecorvo, *J. Exp. Theor. Phys.* **53** (1967) 1717.
25. P. Minkowski, *Phys. Lett. B* **67** (1977) 421.
26. M. Gell-Mann, P. Ramond and R. Slansky, *Supergravity*, eds. P. van Nieuwenhuizen *et al.* (North-Holland, Amsterdam, 1980), p. 315.
27. T. Yanagida, in *Proc. Workshop on the Unified Theory and the Baryon Number in the Universe*, eds. O. Sawada and A. Sugamoto (KEK, Tsukuba, Japan, 1979), p. 95.
28. S. L. Glashow, The Future of Elementary Particle Physics, in *Proc. 1979 Cargèse Summer Institute on Quarks and Leptons*, eds. M. Lévy *et al.* (Plenum, New York, 1980), pp. 687.
29. R. N. Mohapatra and G. Senjanović, *Phys. Rev. Lett.* **44** (1980) 912.
30. A. Kusenko and G. Segrè, *Phys. Lett. B* **396** (1997) 197.
31. A. Kusenko and G. Segre, *Phys. Rev. D* **59** (1999) 061302.
32. G. M. Fuller *et al.*, *Phys. Rev. D* **68** (2003) 103002.
33. A. de Gouvea, *Phys. Rev. D* **72** (2005) 033005.
34. M. Shaposhnikov and I. Tkachev, hep-ph/0604236.
35. L. Stodolsky, *Phys. Rev. D* **36** (1987) 2273.

36. R. Barbieri and A. Dolgov, *Nucl. Phys. B* **349** (1991) 743.
37. S. S. Gershtein and Y. B. Zeldovich, *J. Exp. Theor. Phys. Lett.* **4** (1966) 120.
38. S. S. Gershtein and Y. B. Zeldovich, *Pisma Zh. Eksp. Teor. Fiz.* **4** (1966) 174.
39. B. W. Lee and S. Weinberg, *Phys. Rev. Lett.* **39** (1977) 165.
40. A. Kusenko, S. Pascoli and D. Semikoz, *J. High Energy Phys.* **0511** (2005) 028.
41. K. Abazajian, *Phys. Rev. D* **73** (2006) 063506.
42. T. Asaka, M. Laine and M. Shaposhnikov, *J. High Energy Phys.* **0606** (2006) 053.
43. K. Abazajian, G. M. Fuller and W. H. Tucker, *Astrophys. J.* **562** (2001) 593.
44. A. Boyarsky *et al.*, astro-ph/0512509.
45. A. Boyarsky *et al.*, *J. Exp. Theor. Phys. Lett.* **83** (2006) 133.
46. A. Boyarsky *et al.*, astro-ph/0603368.
47. A. Boyarsky *et al.*, astro-ph/0603660.
48. S. Riemer-Sorensen, S. H. Hansen and K. Pedersen, astro-ph/0603661.
49. K. Abazajian and S. M. Koushiappas, astro-ph/0605271.
50. C. R. Watson *et al.*, astro-ph/0605424.
51. P. B. Pal and L. Wolfenstein, *Phys. Rev. D* **25** (1982) 766.
52. M. Viel *et al.*, *Phys. Rev. D* **71** (2005) 063534.
53. U. Seljak *et al.*, astro-ph/0602430.
54. M. Viel *et al.*, astro-ph/0605706.
55. T. Asaka, A. Kusenko and M. Shaposhnikov, *Phys. Lett. B* **638** (2006) 401.
56. S. P. Mikheev and A. Yu. Smirnov, *Yad. Fiz.* **42** (1985) 1441.
57. S. P. Mikheev and A. Yu. Smirnov, *Sov. J. Nucl. Phys.* **42** (1985) 913.
58. L. Wolfenstein, *Phys. Rev. D* **17** (1978) 2369.
59. X. D. Shi and G. M. Fuller, *Phys. Rev. Lett.* **82** (1999) 2832.
60. G. Kauffmann, S. D. M. White and B. Guiderdoni, *Mon. Not. R. Astron. Soc.* **264** (1993) 201.
61. A. A. Klypin *et al.*, *Astrophys. J.* **522** (1999) 82.
62. B. Moore *et al.*, *Astrophys. J.* **524** (1999) L19.
63. B. Willman *et al.*, *Mon. Not. R. Astron. Soc.* **355** (2004) 159.
64. P. Bode, J. P. Ostriker and N. Turok, *Astrophys. J.* **556** (2001) 93.
65. P. J. E. Peebles, *Astrophys. J.* **557** (2001) 495.
66. J. J. Dalcanton and C. J. Hogan, *Astrophys. J.* **561** (2001) 35.
67. A. R. Zentner and J. S. Bullock, *Phys. Rev. D* **66** (2002) 043003.
68. F. Governato *et al.*, *Astrophys. J.* **607** (2004) 688.
69. G. Gentile *et al.*, *Mon. Not. R. Astron. Soc.* **351** (2004) 903.
70. J. Kormendy *et al.*, astro-ph/0601393.
71. M. I. Wilkinson *et al.*, astro-ph/0602186.
72. L. E. Strigari *et al.*, astro-ph/0603775.
73. G. Gelmini, S. Palomares-Ruiz and S. Pascoli, *Phys. Rev. Lett.* **93** (2004) 081302.
74. M. Mapelli, A. Ferrara and E. Pierpaoli, *Mon. Not. R. Astron. Soc.* **369** (2006) 1719.
75. M. Tegmark *et al.*, *Astrophys. J.* **474** (1997) 1.
76. D. N. Spergel *et al.*, astro-ph/0603449.
77. Z. Haiman and G. L. Bryan, astro-ph/0603541.
78. N. Yoshida *et al.*, *Astrophys. J. Lett.* **591** (2003) 1.
79. A. G. Lyne, B. Anderson and M. J. Salter, *Mon. Not. R. Astron. Soc.* **201** (1982) 503.
80. Bailes *et al.*, *Astrophys. J.* **343** (1989) L53.
81. Formalont *et al.*, *Mon. Not. R. Astron. Soc.* **258** (1992) 497.
82. P. A. Harrison *et al.*, *Mon. Not. R. Astron. Soc.* **261** (1993) 113.
83. A. G. Lyne and D. R. Lorimer, *Nature* **369** (1994) 127.
84. P. A. G. Scheuer, *Nature* **218** (1968) 920.

85. B. J. Rickett, *Mon. Not. R. Astron. Soc.* **150** (1970) 67.
86. J. A. Galt and A. G. Lyne, *Mon. Not. R. Astron. Soc.* **158** (1972) 281.
87. Slee *et al.*, *Mon. Not. R. Astron. Soc.* **167** (1974) 31.
88. A. G. Lyne and F. G. Smith, *Nature* **298** (1982) 825.
89. J. M. Cordes, *Astrophys. J.* **311** (1986) 183.
90. B. M. S. Hansen and E. S. Phinney, *Mon. Not. R. Astron. Soc.* **291** (1997) 569.
91. J. M. Cordes and D. F. Chernoff, *Astrophys. J.* **505** (1998) 315.
92. C. Fryer, A. Burrows and W. Benz, *Astrophys. J.* **496** (1998) 333.
93. Z. Arzoumanian, D. F. Chernoff and J. M. Cordes, *Astrophys. J.* **568** (2002) 289.
94. V. B. Semikoz, *Yad Fiz.* **46** (1987) 1592.
95. J. C. D'Olivo, J. F. Nieves and P. B. Pal, *Phys. Rev. D* **40** (1989) 3679.
96. J. C. D'Olivo and J. F. Nieves, *Phys. Rev. D* **56** (1997) 5898.
97. O. F. Dorofeev, V. N. Rodionov and I. M. Ternov, *Sov. Astron. Lett.* **11** (1985) 123.
98. A. Vilenkin, *Astrophys. J.* **451** (1995) 700.
99. A. Kusenko, G. Segrè and A. Vilenkin, *Phys. Lett. B* **437** (1998) 359.
100. P. Arras and D. Lai, astro-ph/9806285.
101. M. Barkovich *et al.*, *Phys. Rev. D* **66** (2002) 123005.
102. M. Barkovich, J. C. D'Olivo and R. Montemayor, *Phys. Rev. D* **70** (2004) 043005.
103. C. L. Fryer and A. Kusenko, *Astrophys. J. Suppl.* **163** (2006) 335.

ELECTRON ELECTRIC DIPOLE MOMENT EXPERIMENT WITH SLOW ATOMS*

HARVEY GOULD
Mail Stop 71-259, Lawrence Berkeley National Laboratory,
One Cyclotron Rd, Berkeley CA, 94720, USA
HAGould@lbl.gov

Discovering an electron electric dipole moment (e-EDM) would uncover new physics requiring an extension of the Standard Model. e-EDMs, large enough to be discovered by new experiments are now common predictions in extensions of the Standard Model, including extensions that describe baryogenesis, dark matter, and neutrino mass.

A cesium slow-atom e-EDM experiment (which is similar to an atomic clock) can improve the sensitivity to the e-EDM. And, as with an atomic clock, it could be more sensitive in microgravity than on Earth. As a first step an Earth-based demonstration Cs fountain e-EDM experiment has been carried out at LBNL.

Keywords: Electric dipole moment; Standard Model; laser cooling and trapping.

1. Introduction

The existence of an electron electric dipole moment (e-EDM) requires both parity (P) nonconservation and time-reversal (T) violation, where T violation is equivalent to charge conjugation (C)–parity violation (CP violation) provided that CPT is conserved. CP violation is presently observed in the decay of B and K^0 mesons, with the measurements of those decays being fully explained by the Cabbibo–Kobayashi–Maskawa (CKM) mechanism in the Standard Model.

The CKM mechanism, which arises in the quark sector, predicts only very small EDM's of nucleons and even smaller EDM's of leptons.[1–3] The predicted e-EDM is some ten orders of magnitude smaller that the present experimental sensitivity.[4–7] This makes e-EDM experiments particularly useful for searching for new sources of CP violation (that couple to leptons). There is nothing to subtract out and any observation of an e-EDM is *ipso facto* an observation of new physics.

*The experimental work reported here was done in collaboration with Jason Amini (LBNL) and Charles T. Munger Jr (SLAC).

The CKM mechanism also does not produce sufficient CP violation to account for the excess of matter over antimatter in the Universe. One or more new sources of CP violation are considered necessary for generating this observed asymmetry.[3,8]

Extensions of the Standard Model contain additional, non-CKM sources of CP violation that couple to leptons and that can give rise to a large e-EDM. Electron EDMs, large enough to be discovered by new experiments, are predicted[1–3,9] by Standard Model extensions such as supersymmetry,[10] multi-Higgs models, left–right symmetric models, lepton flavor-changing models and technicolor models.[11] Split supersymmetry[12–14] predicts an e-EDM within a factor of 100 of the present experimental limit. Merely improving the present e-EDM limit would place constraints on many Standard Model extensions and possibly on current models of neutrino physics.[15]

2. Electron EDM Experiments Using Neutral Atoms

e-EDM experiments measure any difference in interaction energy when an electric field is reversed relative to the electron spin.[a] An atom is used because it establishes an overall neutral system — handy when applying electric fields — and may provide an enhancement to the applied electric field. For convenience this enhancement is usually expressed as a ratio R of the atom EDM to any e-EDM.

Due to relativistic effects, R can reach large values in some high-atomic-number atoms such as cesium,[16,17] where in the ground state $R = 114 \pm 15$. The theory of the enhancement in alkali atoms and in thallium is very well established. And the discovery of an e-EDM (larger than the CKM prediction) does not depend upon a precise calculation of the enhancement factor because there is no observable Standard Model CKM effect to be subtracted out.

Cesium has one stable isotope, ^{133}Cs with $I = 7/2$, and its $6^2S_{1/2}$ ground state forms hyperfine levels with total angular momentum $F = I + J$ of $F = 4$ and $F = 3$, each of which has $2F + 1$ different m_F sublevels. An applied electric field lifts the degeneracy between levels of different $|m_F|$. The signature of an e-EDM is an interaction that is linear in applied electric field. To separate out the quadratic Stark shift due to the atom's polarizability, the polarity of the electric field is reversed.

Traditionally experiments search for the change in transition energy, upon reversal of the electric field, in a single photon $\Delta m_F = 1$ transition between adjacent m_F levels. However, in future experiments we plan to use a $\Delta m_F = 7$, seven-photon transition between the $m_F = \pm 4$ and $m_F = \mp 3$ sublevels.[18]

3. Atomic Beam Experiments

For nearly 40 years, atomic beams have been the most widely used technique for lowering the limit to the e-EDM (Fig. 1). They have few perturbing fields, there

[a]Alternatively, if the field is not fully aligned with the spin, a change in the rate of precession of the electron spin is observed.

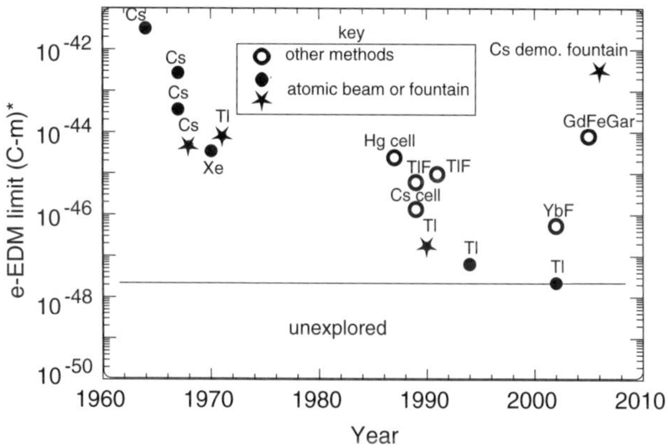

Fig. 1. Experimental upper limits to the e-EDM 1962–2006. Atomic beam experiments are shown as filled-in, black circles and other methods are shown as open circles. The atom, molecule, or solid used is also indicated. The author collaborated on atomic beam experiments (and a fountain experiment) marked with a filled-in star.

is ample flux, and the systematics are well understood even if not always easy to overcome. Improving the suppression of the systematic effect due to the motional magnetic field has been the most important factor in lowering the e-EDM limit in atomic beam experiments. However, after six orders of magnitude the technology is nearing its limits. Slow-atom experiments in a fountain or a slow-traveling beam in microgravity allows us to introduce new techniques for suppressing motional magnetic field effects. This is discussed in Sec. 4.

4. Suppressing Systematic Effects

Transitions sensitive to an e-EDM are also sensitive to magnetic fields and will exhibit a linear Zeeman effect. Therefore all e-EDM experiments must minimize systematic effects due to magnetic fields, of which the most important for atomic beam e-EDM experiments has been the motional magnetic field effect. The motional magnetic field $\mathbf{B}_{\text{mot}}$ seen in the rest frame of an atom, moving with velocity $\mathbf{v}$ transverse to a laboratory electric field $\mathbf{E}$, is given in lowest order (SI units) by

$$\mathbf{B}_{\text{mot}} = \mathbf{v} \times \frac{\mathbf{E}}{c^2}, \tag{1}$$

where $c = 3 \times 10^8$ m/s.

When a static magnetic field $\mathbf{B}_0$, such as may be used to lift the degeneracy between m_F sublevels, is also present, misalignment between $\mathbf{E}$ and $\mathbf{B}_0$ causes a component of $\mathbf{B}_{\text{mot}}$ to lie along $\mathbf{B}_0$. The total magnetic field then changes linearly with $\mathbf{E}$ and, through the atom's magnetic moment, mimics an EDM.[19]

A slow atom e-EDM experiment can use two effective methods to suppress motional magnetic field effects: atom-by-atom cancelation of the beam velocity by the rise and fall of the atoms under gravity in a fountain (or in a microgravity experiment by reflecting the atoms back with a small electric field gradient), and electric field quantization,[20] where no static magnetic field is needed because the electric field alone lifts the degeneracy of sublevels with different $|m_F|$. However, the tensor polarizabilities of the alkali atom ground state m_F sublevels are very small and electric field quantization has only been achieved with the combination of the strong electric fields and narrow instrumental line width available in our demonstration cesium fountain e-EDM experiment.[18]

In electric field quantization, energy shifts due to the motional magnetic field are absent to lowest order. The leading systematic term $W_{\text{sys}}(m_F)$ is given by

$$\frac{W_{\text{sys}}(m_F)}{h} = -2K_2(m_F)\frac{(g\mu)^3 B_{\perp\text{res}} B_{\text{mot}} B_{||}}{(\epsilon E^2)^2}. \tag{2}$$

$\epsilon = -3\alpha_T/56$, $\alpha_T = -3.5 \times 10^{-12}$ HzV^{-2}m^2 is the tensor polarizability,[21,22] $g\mu \approx 3.5 \times 10^9$ Hz/T, $B_{||}$ is the residual magnetic field component parallel to $\mathbf{E}$, B_{mot} is the motional magnetic field, $B_{\perp\text{res}}$ is the residual magnetic field parallel to B_{mot}, and

$$K_2(m_F) = \frac{81 m_F}{2(4m_F^2 - 1)^2}.$$

$W_{\text{sys}}(m_F)$ is odd in E (through B_{mot}) and odd in m_F (through K_2). However, it can be made very small when E and m_F are large and v, $B_{\perp\text{res}}$, and $B_{||}$ are small.

In an e-EDM experiment with $E = 13.5$ MV/m, where a rise and subsequent fall of atoms reduces the time-averaged average velocity to < 3 mm/s, and where residual magnetic fields are ≤ 20 pT, the systematic is 2.5×10^{-52} C-m — nearly four orders of magnitude below the present experimental sensitivity.[18]

5. Demonstration Experiment

To test the feasibility of using electric field quantization in a slow-atom e-EDM experiment with state preparation, analysis, and atom transport in field-free regions, we constructed a prototype e-EDM experiment. It is described in more detail in Ref. 18. Magnetic fields were reduced to 200 pT with a combination of static magnetic shielding, demagnetizing coils and, inside the shields, three sets of orthogonal magnetic field coils for nulling residual magnetic fields. These coils were also used for inducing transitions between m_F states.

The average beam velocity was set to about 3 m/s by increasing the launch velocity so that the upward-traveling atoms did not turn around inside the electric field plates, but instead exited and were analyzed and detected above the electric field plates. After launching from the fountain's magneto-optic trap, the packet of cesium atoms entered a magnetically shielded and nulled region where the states were prepared, transitions were induced by pulsed fields, the electric field was applied, and the final states were analyzed and detected. The apparatus is shown in Fig. 2.

Fig. 2. Photograph of the demonstration fountain e-EDM experiment. The large cylinder is the outer of four sets of static magnetic shields. Optical components for state preparation, analysis, and detection are seen on elevated platforms and vertical supports. A live image of the Cs trap (running when the photo was taken) is seen in the monitor at the left of the photo.

State preparation and analysis were done in the regions free of electric and magnetic fields ($B \leq 200$ pT in each orthogonal direction). The mixing of the m_F states by unnulled magnetic fields was still noticeable but small. Results are included in Fig. 1.

Acknowledgments

Support from the NASA Office of Biological and Physical Research and from a NIST Precision Measurements Grant is most gratefully acknowledged. The Lawrence Berkeley National Laboratory is operated for the US DOE under Contract No. DE-AC02-05CH11231.

References

1. W. Bernreuther and M. Suzuki, *Rev. Mod. Phys.* **63** (1991) 313.
2. W. Bernreuther and M. Suzuki, *Rev. Mod. Phys.* (errata) **66** (1992) 633.
3. M. Pospelov and A. Ritz, *Ann. Phys.* **318** (2005) 119.
4. B. C. Regan *et al.*, *Phys. Rev. Lett.* **88** (2002) 071805.
5. J. J. Hudson *et al.*, *Phys. Rev. Lett.* **89** (2002) 023003.
6. K. Abdullah *et al.*, *Phys. Rev. Lett.* **65** (1990) 2347.
7. S. A. Murthy *et al.*, *Phys. Rev. Lett.* **63** (1989) 965.
8. A. D. Sakharov, *Sm. Zh. Eksp. Teor. Fiz.* **5** (1967) 32 [*J. Exp. Theor. Phys. Lett.* **5** (1967) 24].
9. S. M. Barr, *Int. J. Mod. Phys. A* **8** (1993) 209.
10. A. Abel, S. Khalil and O. Lebedev, *Nucl. Phys. B* **606** (2001) 151.

11. T. Appleuist, M. Piai and R. Shrock, *Phys. Lett. B* **593** (2004) 175.
12. N. Arkani-Hamed *et al.*, *Nucl. Phys. B* **709** (2005) 3.
13. D. Chang, W.-F. Chang and W.-Y. Keung, *Phys. Rev. D* **71** (2005) 076006.
14. G. F. Giudice and A. Romanino, *Phys. Lett. B* **634** (2006) 307.
15. R. N. Mohaparta *et al.*, hep-ph/0510213.
16. P. G. H. Sandars, *Phys. Lett.* **22** (1966) 290.
17. W. R. Johnson *et al.*, *Phys. Rev. A* **34** (1986) 1043.
18. J. M. Amini, C. T. Munger Jr. and H. Gould, hep-ph/0602011.
19. P. G. H. Sandars and E. Lipworth, *Phys. Rev. Lett.* **13** (1964) 718.
20. M. A. Player and P. G. H. Sandars, *J. Phys. B* **3** (1970) 1620.
21. H. Gould, E. Lipworth and M. C. Weisskopf, *Phys. Rev.* **188** (1969) 24.
22. C. Ospelkaus, U. Rasbach and A. Weis, *Phys. Rev. A* **67** (2003) 011402R.

TESTING RELATIVITY AT HIGH ENERGIES USING SPACEBORNE DETECTORS

F. W. STECKER
NASA Goddard Space Flight Center,
Greenbelt, MD, USA

The Gamma-Ray Large Area Space Telescope (GLAST), to be launched in the fall of 2007, will measure the spectra of distant extragalactic sources of high energy γ-rays, particularly active galactic nuclei and γ-ray bursts. GLAST can look for energy-dependent γ-ray propagation effects from such sources as a signal of Lorentz invariance violation (LIV). These sources should also exhibit the high energy cutoffs predicted to be the result of intergalactic annihilation interactions with low energy photons having a flux level as determined by various astronomical observations. Such annihilations result in electron–positron pair production above a threshold energy given by $2m_e$ in the center-of-momentum frame of the system, assuming Lorentz invariance. If Lorentz invariance is violated, this threshold can be significantly raised, changing the predicted absorption turnover in the observed spectrum of the sources. Stecker and Glashow have shown that the existence of such absorption features in the spectra of extragalactic sources puts constraints on LIV. Such constraints have important implications for some quantum gravity and large extra dimension models. Future spaceborne detectors dedicated to measuring γ-ray polarization can look for birefringence effects as a possible signal of loop quantum gravity.

As shown by Coleman and Glashow, a much smaller amount of LIV has potential implications for possibly suppressing the "GZK cutoff" predicted to be caused by the interactions of cosmic rays having multijoule energies with photons of the 2.7 K cosmic background radiation in intergalactic space. Owing to the rarity of such ultrahigh energy cosmic rays, their spectra are best studied by a UV-sensitive satellite detector which looks down on a large volume of the Earth's atmosphere to study the nitrogen fluorescence tracks of giant air showers produced by these ultrahigh energy cosmic rays. We discuss here, in particular, a two-satellite mission called OWL, which would be suited for making such studies.

Keywords: Relativity; gamma rays; quantum gravity; cosmic rays; space telescopes.

1. Introduction

The theory of relativity is one of the fundamental pillars of modern physics. However, because of the problems associated with merging relativity and quantum theory, it has long been felt that relativity may have to be modified in some way. It has been suggested that relativity, i.e. Lorentz invariance (LI), may be only an approximate symmetry of nature.[1] There has been a particular interest in the possibility that a breakdown of relativity may be associated with the Planck scale of $M_{\rm QG} \sim 10^{19}$ GeV, where quantum effects are expected to become significant in gravitational theory. Although no true quantum theory of gravity exists, it was independently proposed that LI might be violated in such a theory with astrophysical consequences[2] being manifested at an energy scale $\ll M_{\rm QG}$. The subject of this paper is the potential use of observations of high energy phenomena from satellite detectors to search for the possible breakdown of LI.

2. A Lorentz Invariance Violation Formalism

A simple formulation for breaking LI by a small first order perturbation in the electromagnetic Lagrangian which leads to a renormalizable treatment has been given by Coleman and Glashow.[3] The small perturbative noninvariant terms are both rotationally and translationally invariant in a preferred reference frame, which one can assume to be the frame in which the cosmic background radiation is isotropic. These terms are also taken to be invariant under $\mathrm{SU}(3) \otimes \mathrm{SU}(2) \otimes \mathrm{U}(1)$ gauge transformations in the standard model.

With this form of LI violation (LIV), different particles can have differing maximum attainable velocities (MAV's), and these MAV's can be different from c. Using the formalism of Ref. 3, we denote the MAV of a particle of type i by c_i, a quantity which is not necessarily equal to $c \equiv 1$, the low energy *in vacua* velocity of light. We further define the difference $c_i - c_j \equiv \delta_{ij}$. These definitions will be used to discuss the physics implications of cosmic ray and cosmic γ-ray observations.[4–6]

In general, then, $c_e \neq c_\gamma$. The physical consequences of such a violation of LI depend on the sign of the difference between these two MAV's. Defining

$$c_e \equiv c_\gamma(1+\delta), \quad 0 < |\delta| \ll 1, \tag{1}$$

one can consider the two cases of positive and negative values of δ separately.[3,4]

Case I. If $c_e < c_\gamma$ ($\delta < 0$), the decay of a photon into an electron–positron pair is kinematically allowed for photons with energies exceeding

$$E_{\max} = m_e \sqrt{\frac{2}{|\delta|}}. \tag{2}$$

The decay would take place rapidly, so that photons with energies exceeding $E_{\max}$ could not be observed either in the laboratory or as cosmic rays. From the fact that photons have been observed with energies $E_\gamma \geq 50$ TeV from the Crab nebula, one deduces for this case that $E_{\max} \geq 50$ TeV, or that $-\delta < 2 \times 10^{-16}$.

Case II. For this possibility, where $c_e > c_\gamma$ ($\delta > 0$), electrons become superluminal if their energies exceed $E_{\max}/2$. Electrons traveling faster than light will emit light at all frequencies by a process of "vacuum Čerenkov radiation." This process occurs rapidly, so that superluminal electron energies quickly approach $E_{\max}/2$. However, because electrons have been seen in the cosmic radiation with energies up to ~ 2 TeV, it follows that $E_{\max} \geq 2$ TeV, which leads to an upper limit on δ for this case of 3×10^{-14}. Note that this limit is two orders of magnitude weaker than the limit obtained for Case I. However, this limit can be considerably improved by considering constraints obtained from studying the γ-ray spectra of active galaxies.[4]

3. Extragalactic Gamma Ray Constraints on LIV

A constraint on δ for $\delta > 0$ follows from a change in the threshold energy for the pair production process $\gamma + \gamma \rightarrow e^+ + e^-$. This arises from the fact that the square of the four-momentum is changed to give the threshold condition

$$2\epsilon E_\gamma(1 - \cos\theta) - 2E_\gamma^2\delta \geq 4m_e^2, \tag{3}$$

where ϵ is the energy of the low energy photon and θ is the angle between the two photons. The second term on the left-hand side comes from the fact that $c_\gamma = \partial E_\gamma/\partial p_\gamma$. It follows that the condition for a significant increase in the energy threshold for pair production is $E_\gamma\delta/2 \geq m_e^2/E_\gamma$ or, equivalently, $\delta \geq 2m_e^2/E_\gamma^2$. The observed γ-ray spectrum of the active galaxy Mkn 501 while flaring extended to $E_\gamma \geq 24$ TeV[7] and exhibited the high energy absorption expected from γ-ray annihilation by extragalactic pair production interactions with extragalactic infrared photons.[8,9] This has led Stecker and Glashow[4] to point out that the Mkn 501 spectrum presents evidence for pair production with no indication of LIV up to a photon energy of ~ 20 TeV and to thereby place a quantitative constraint on LIV given by $\delta < 2m_e^2/E_\gamma^2 \simeq 10^{-15}$, a factor of 30 better than that given in the previous section. GLAST will observe many more such active galaxies at different redshifts,

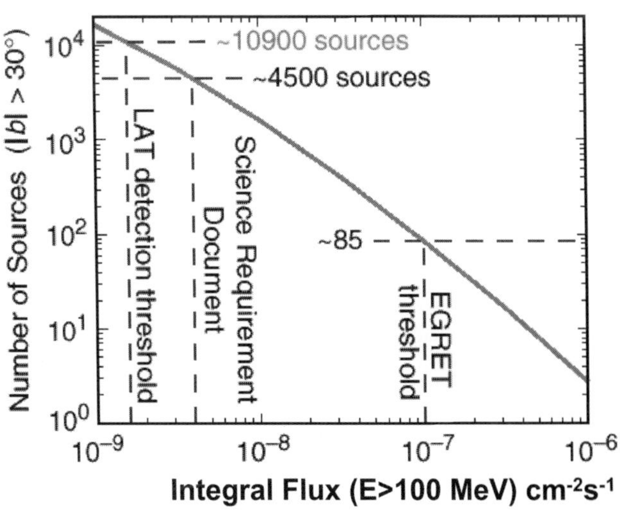

Fig. 1. The number of γ-ray-emitting active galaxies at high galactic latitudes (galactic latitude $|b| > 30°$) predicted to be seen by the GLAST LAT (Large Area Telescope) instrument. The approximate number of sources detected by the previous EGRET instrument on the Cosmic Gamma Ray Observatory is also shown. The curve shows the predicted integral source count versus threshold flux.[10]

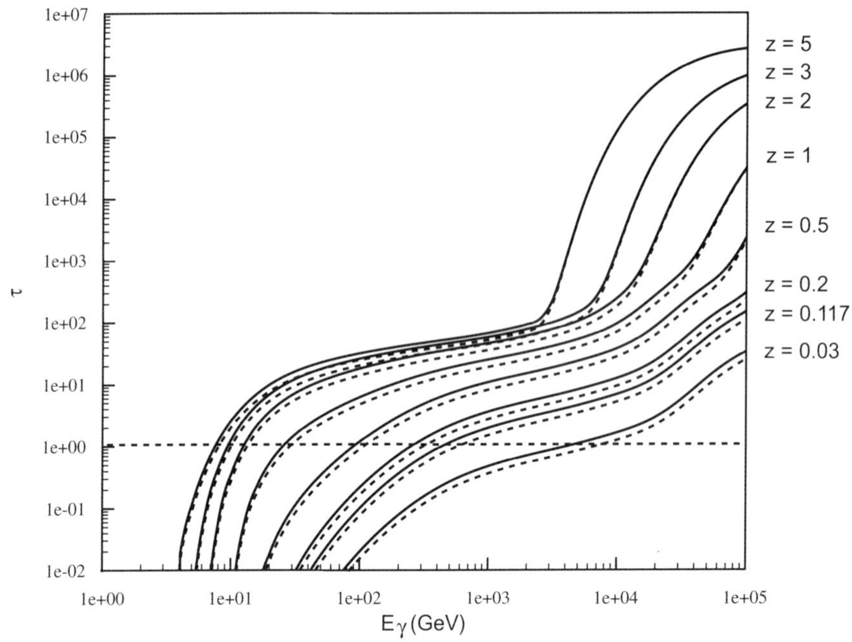

Fig. 2. The optical depth of the Universe to γ-rays from interactions with photons of the intergalactic background light and the 2.7 K cosmic background radiation for γ-rays having energies up to 100 TeV. This is given for a family of redshifts from 0.03 to 5, as indicated. The solid lines are for the fast evolution model; the dashed lines are for the baseline model.[11]

as shown in Fig. 1,[10] and thereby further test such constraints on LIV by looking for deviations from predicted absorption effects. Figure 2 shows the optical depth of the Universe to high energy γ-rays against pair production interactions for sources at various redshifts under the assumption that LI holds.[11]

4. Gamma Ray Constraints on Quantum Gravity and Extra Dimension Models

As previously mentioned, LIV has been proposed to be a consequence of quantum gravity physics at the Planck scale $M_{\rm Planck} = \sqrt{\hbar c/G} \simeq 1.22 \times 10^{19}$ GeV.[12,13] In models involving large extra dimensions, the energy scale at which gravity becomes strong can occur at a scale $M_{\rm QG} \ll M_{\rm Planck}$, even approaching a TeV.[14] In the most commonly considered case, the usual relativistic dispersion relations between the energy and momentum of the photon and the electron are modified[2,13] by a term of order $p^3/M_{\rm QG}$.[a]

[a]We note that there are variants of quantum gravity and large extra dimension models which do not violate LI and for which the constraints considered here do not apply. There are also variants for which there are no cubic terms in momentum but, rather, much smaller quartic terms of order $\sim p^4/M_{\rm QG}^2$.

Generalizing the LIV parameter δ to an energy-dependent form,

$$\delta \equiv \frac{\partial E_e}{\partial p_e} - \frac{\partial E_\gamma}{\partial p_\gamma} \simeq \frac{E_\gamma}{M_{\rm QG}} - \frac{m_e^2}{2E_e^2} - \frac{E_e}{M_{\rm QG}}, \tag{4}$$

the threshold condition from pair production implies that $M_{\rm QG} \geq E_\gamma^3/8m_e^2$. Since pair production occurs for energies of at least 20 TeV, we find a constraint on the quantum gravity scale[5] $M_{\rm QG} \geq 0.3 M_{\rm Planck}$. This constraint contradicts the predictions of some proposed quantum gravity models involving large extra dimensions and smaller effective Planck masses. In a variant model of Ref. 15, the photon dispersion relation is changed, but not that of the electrons. In this case, we find the even stronger constraint $M_{\rm QG} \geq 0.6 M_{\rm Planck}$. Future studies of the spectra of active galaxies can extend these constraints on quantum gravity models.

5. Energy-Dependent Time Variability of GRB Spectra and Tests of LIV

One possible manifestation of LIV, possibly from Planck scale physics produced by quantum gravity effects, is a change in the energy–momentum dispersion relation of a free particle or a photon which may be of first order in $E_\gamma/M_{\rm QG}$, where $M_{\rm QG}$ is the quantum gravity scale, usually assumed to be the Planck scale.[2,14] In a ΛCDM cosmology, where present observational data indicate that $\Omega_\Lambda \simeq 0.7$ and $\Omega_m \simeq 0.3$, the resulting difference in the propagation times of two photons having an energy difference ΔE_γ from a γ-ray burst (GRB) at a redshift z will be

$$\Delta t_{\rm LIV} = H_0^{-1} \frac{\Delta E_\gamma}{M_{\rm QG}} \int_0^z \frac{dz'}{\sqrt{\Omega_\Lambda + \Omega_m (1+z')^3}} \tag{5}$$

for a photon dispersion of the form $c_\gamma = c(1 \pm E_\gamma/M_{\rm QG})$, with c being the usual low energy velocity of light.[16,17] In other words, δ, as defined earlier, is given by $\pm E_\gamma/M_{\rm QG}$. Data on GRB021206 for $E_\gamma > 3\,{\rm MeV}$ imply a value for $M_{\rm QG} > 1.8 \times 10^{17}\,{\rm GeV}$.[18] Data from GRB051221A have given a constraint $M_{\rm QG} > 0.66 \times 10^{17}\,{\rm GeV}$.[19]

The dispersion effect will be smaller if the dispersion relation has a quadratic dependence on $E_\gamma/M_{\rm QG}$, as suggested by effective-field-theory considerations.[20,21] This will obviate the limits on $M_{\rm QG}$ given above. The possible effect of extra dimension models on γ-ray propagation has also been pointed out very recently.[22]

The GLAST satellite (see Fig. 3), with its γ-ray burst monitors (GBM's) covering an energy range from 10 keV to 25 MeV and its Large Area Telescope (LAT) covering an energy range from 20 MeV to > 300 GeV, can study both GRB's and flares from active galactic nuclei over a large range of both energy and distance. So our studies can be extended to GLAST observations of GRB's and blazar flares after the expected GLAST launch in the fall of 2007.

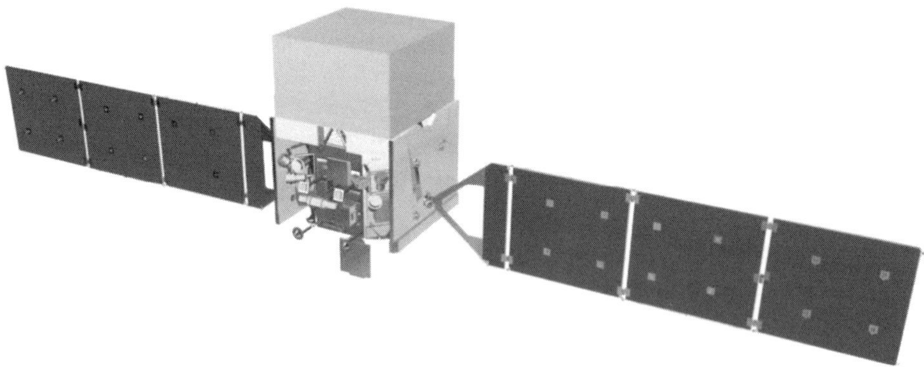

Fig. 3. Schematic of the GLAST satellite deployed in orbit. The LAT is in the top (yellow) area and the GBM's are located directly below.

6. Looking for Birefringence Effects from Quantum Gravity

A possible model for quantizing space–time which has been actively investigated is *loop quantum gravity* (see the review given in Ref. 23 and references therein). A signature of this model is that the quantum nature of space–time can produce an intrinsic birefringence effect. This is because electromagnetic waves of opposite circular polarizations will propagate with different velocities, which leads to a rotation of linear polarization direction through the angle

$$\theta(t) = [\omega_+(k) - \omega_-(k)]\, t/2 = \xi k^2 t/2M_{\text{Planck}} \tag{6}$$

for a plane wave with wave vector k.[24]

Some astrophysical sources emit highly polarized radiation. It can be seen from Eq. (6) that the rotation angle is reduced by the large value of the Planck mass. However, the small rotations given by Eq. (6) can add up over astronomical or cosmological distances to erase the polarization of the source emission. Therefore, if polarization is seen in a distant source, it puts constraints on the parameter ξ. Observations of polarized radiation from distant sources can therefore be used to place an upper bound on ξ.

Equation (6) indicates that the higher the wave number $|k|$ is, the stronger the rotation effect will be. Thus, the depolarizing effect of space–time-induced birefringence will be most pronounced in the γ-ray energy range. It can also be seen that this effect grows linearly with propagation time. The best secure bound on this effect, $|\xi| \lesssim 2 \times 10^{-4}$, was obtained using the observed 10% polarization of ultraviolet light from a distant galaxy.[25]

A few years ago, there was a report of strong linear γ-ray polarization from GRB021206 observed from the RHESSI satellite.[26] The survival of such polarization over cosmological distances would put a much stronger constraint on the value of the parameter ξ. The constraint arises from the fact that if the angle of polarization rotation (6) were to differ by more than $\pi/2$ over the 0.1–0.3 MeV energy range and by more than $3\pi/2$ over the 0.1–0.5 MeV energy range, the instantaneous polarization at the detector would fluctuate sufficiently for the net polarization of the signal to be suppressed well below the observed value. The difference in rotation angles for wave vectors k_1 and k_2 is

$$\Delta\theta = \xi(k_2^2 - k_1^2)d/2M_{\mathrm{Planck}}, \tag{7}$$

replacing the time t by the distance from the GRB to the detector, denoted by d.

While the distance to GRB021206 is unknown, it is well known that most cosmological bursts have redshifts in the range 1–2 corresponding to distances of greater than a Gpc. Using the distance distribution derived in Ref. 27 one can conservatively take the minimum distance to this burst as 0.5 Gpc, corresponding to a redshift of ~ 0.1. This yields the constraint

$$|\xi| < 5.0 \times 10^{-15}/d_{0.5}, \tag{8}$$

where $d_{0.5}$ is the distance to the burst in units of 0.5 Gpc.[21] However, the polarization measurement reported in Ref. 26 has been questioned in other analyses[28,29] and so remains controversial.

It should be noted that the RHESSI satellite detector was not designed specifically to measure γ-ray polarization. Detectors which are dedicated to polarization measurements in the X-ray and the γ-ray energy range and which can be flown in space to study the polarization from distant astronomical sources are now being designed.[30,31] We note that linear polarization in X-ray flares from GRB's has been predicted.[32]

A further discussion of astrophysical constraints on LIV may be found in Ref. 21.

7. LIV and the Ultrahigh Energy Cosmic Ray Spectrum

The flux of ultrahigh energy nucleons is expected to be attenuated by photomeson-producing interactions of these hadrons with the cosmic-microwave-background radiation (CBR). This predicted effect is now known as the "GZK effect."[33,34] The mean free path for this attenuation effect is less than 100 Mpc for cosmic ray nucleons of energy greater than 100 EeV.[35]

Coleman and Glashow[3] have shown that for interactions of protons with CBR photons of energy ϵ and temperature $T_{\mathrm{CBR}} = 2.73\,\mathrm{K}$, pion production is kinematically forbidden and thus *photomeson interactions are turned off* if

$$\delta_{p\pi} > 5 \times 10^{-24}(\epsilon/T_{\mathrm{CBR}})^2. \tag{9}$$

Thus, given even a very small amount of LIV, photomeson interactions of ultrahigh energy cosmic rays (UHECR's) with the CBR can be turned off. Such a violation of LI might be produced by Planck scale effects.[36,37]

Some "trans-GZK" hadronic showers with energies above the predicted "cutoff energy" (usually considered to be 100 EeV) have been observed by both scintillator and fluorescence detectors, particularly by the scintillator array AGASA group at

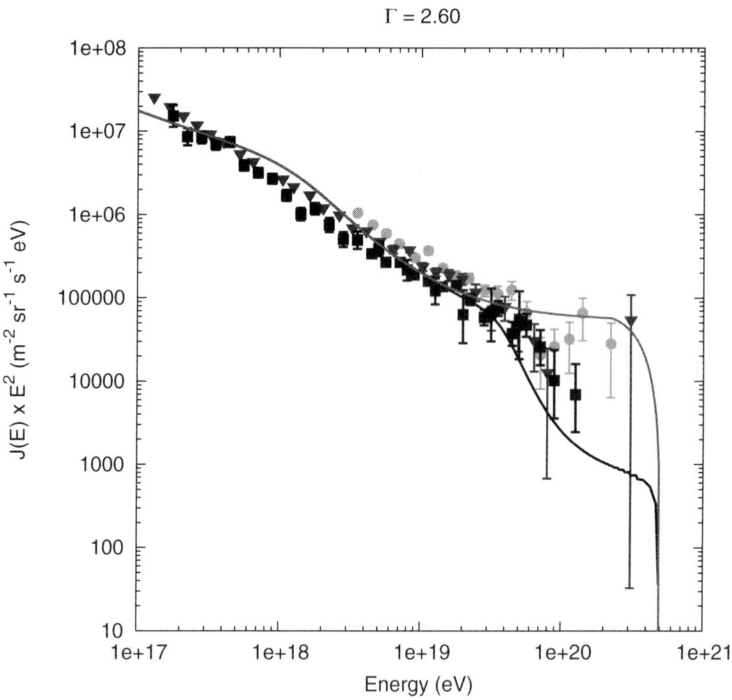

Fig. 4. Predicted spectra for an $E^{-2.6}$ source spectrum with redshift evolution and $E_{\max} = 500\,\mathrm{EeV}$, shown with pair production losses included and photomeson losses both included (black curve) and turned off [lighter (red) curve]. The curves are shown with UHECR spectral data from Fly's Eye (triangles), AGASA[38] (circles), and HiRes[39] monocular data (squares).[6]

Akeno, Japan,[38] possibly in contradiction to the expected attenuation effect. While there is less evidence for such interesting events from fluorescence detectors (see Fig. 4), we note that the Fly's Eye fluorescence detector reported the detection of a 320 EeV event,[40] an energy which is a factor of ~ 5 above the GZK cutoff energy. The subject of UHECR's having trans-GZK energies has not as yet been settled experimentally, even by the Pierre Auger ground-based detector array.[41]

If LIV is the explanation for a possibly missing GZK effect, indicated in the AGASA data but not in the HiRes data (see Fig. 4),[6] one can also look for the absence of a "pileup" spectral feature and for the absence of the neutrinos which should be produced by the GZK effect. The detection of ultrahigh energy nucleons and neutrinos at sufficiently high energies and with excellent event statistics can best be done from space. This possibility will be discussed in the next section.

8. The OWL Satellite Detectors

The OWL (Orbiting Wide-field Light-collectors) mission is designed to obtain data UHECR's and neutrinos in order to tackle the fundamental problems associated with their origin.[42] It aims to provide the event statistics and extended energy range which are crucial to addressing these issues. To accomplish this, OWL makes use of the Earth's atmosphere as a huge "calorimeter" to make stereoscopic measurements of the atmospheric UV fluorescence produced by air shower particles. This is the most accurate technique that has been developed for measuring the energy, arrival direction, and interaction characteristics of UHECR's.[43] To this end, OWL will consist of a pair of satellites placed in tandem in a low inclination, medium altitude orbit. The OWL telescopes will point down at the Earth and will together point at a section of the atmosphere about the size of the state of Texas ($\sim 6 \times 10^5\,\text{km}^2$), obtaining a much greater sensitivity than present ground-based detectors. The ability of OWL to detect cosmic rays, in units of km^2 sr, is called the *aperture*. The instantaneous aperture at the highest energies is $\sim 2 \times 10^6\ \text{km}^2\,\text{sr}$. The effective aperture, reduced by the effects of the Moon, man-made light, and clouds, will be conservatively $\sim 0.9 \times 10^5\,\text{km}^2$ sr. For each year of operation, OWL will have 90 times the aperture of the ground based HiRes detector and 13 times the aperture of the Pierre Auger detector array (130 times its most sensitive "hybrid" mode). The OWL detectors will observe the UV fluorescence light from the giant air showers produced by UHECR's on the dark side of the Earth. They will thus provide a stereoscopic picture of the temporal and spatial development of the showers.[b]

Following a stacked dual launch on a Delta rocket as shown in Fig. 5, the two satellites will fly in formation at an altitude of 1000 km and with a separation of 10–20 km for about 3 months to search for upward-going showers from ν_τ's propagating through the Earth. The spacecraft will then separate to 600 km for ~ 2.5 years to measure the high energy end of the UHECR spectrum. Following this period, the altitude is reduced to 600 km and the separation to 500 km in order to measure the cosmic ray flux closer to 10 EeV.

With the fluorescence technique, a fast, highly pixelized camera (or "eye") is used to resolve both the spatial and the temporal development of the shower. This detailed information provides a powerful tool for determining the nature of the primary particle. The UV emission, principally in the 300–400 nm range, is isotropic and the camera can view the shower from any direction except almost directly toward the camera. In the exceptional case, the camera may still be utilized as a Cherenkov detector. Thus, a single camera can view particles incident on the Earth from a hemisphere of the sky.

In monocular operation, precision measurements of the arrival times of UV photons from different parts of the shower track must be used to partly resolve spatial ambiguities. The angle of the shower relative to the viewing plane is resolvable using differential timing. Resolving distance, however, requires that the pixel crossing time be measured to an accuracy that is virtually impossible to achieve in a real instrument at orbit altitudes. Stereoscopic observation resolves both of these ambiguities. In stereo, fast timing provides supplementary information to reduce

[b] The technical details, as well as discussion of the science, including ultrahigh energy neutrino science with OWL, can be found at http://owl.gsfc.nasa.gov.

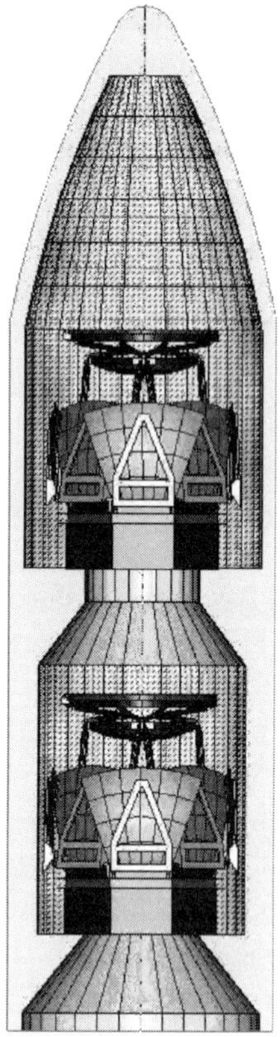

Fig. 5. Schematic of the stowed OWL satellites in the launch vehicle.

systematics and improve the resolution of the arrival direction of the UHECR's. By using stereo, differences in atmospheric absorption or scattering of the UV light can be determined. The results obtained by the HiRes collaboration viewing the same shower in both modes have clearly demonstrated the desirability of stereo viewing.

The light collector will use a Schmidt camera design as shown in Fig. 6. The Schmidt corrector has a spherical front surface and an aspheric back surface, while the primary mirror has a slight aspheric figure. The focal plane is a spherical surface tiled with flat detector elements. The corrector is slightly domed for strength. The primary mirror is made of a lightweight composite material with a central octagonal section and eight petals that fold upward for launch. The entire optical system is covered by an inflatable light and micrometeoroid shield and is closed out by a redundant shutter system. The shield will be made of a multilayer material with kevlar layers for strength. Figure 7 is a schematic of the OWL detectors in orbit looking down on the track of an extremely large air shower.

Monte Carlo simulations of the physics and response of orbiting instruments to the UV air fluorescence signals are crucial to the development of OWL. One such

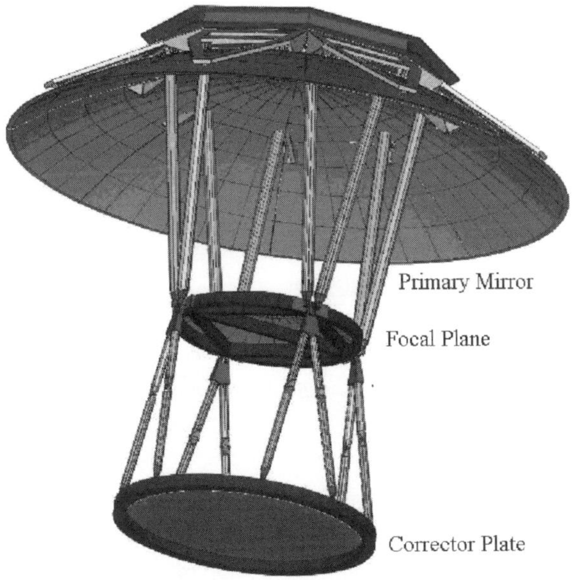

Fig. 6. Schematic of the Schmidt optics that form an OWL "eye" in the deployed configuration. The spacecraft bus, light shield, and shutter are not shown.

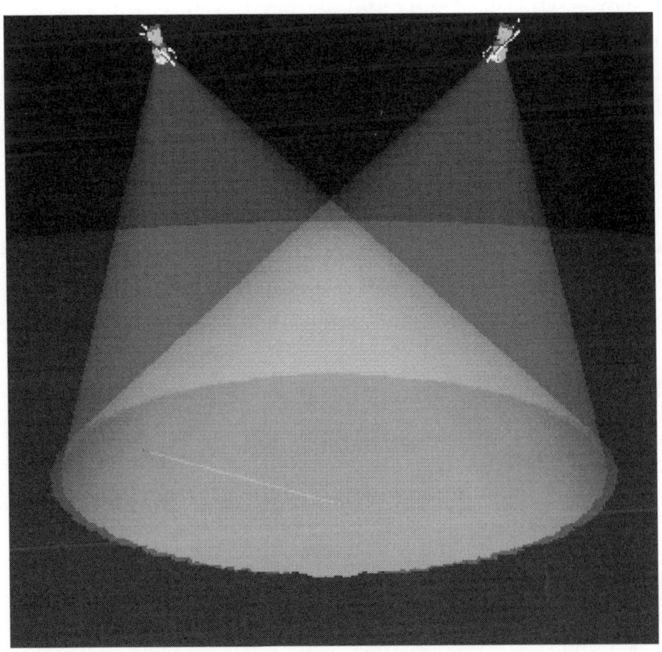

Fig. 7. OWL satellites observing the fluorescent track of a giant air shower. The shaded cones show the field of view for each satellite.

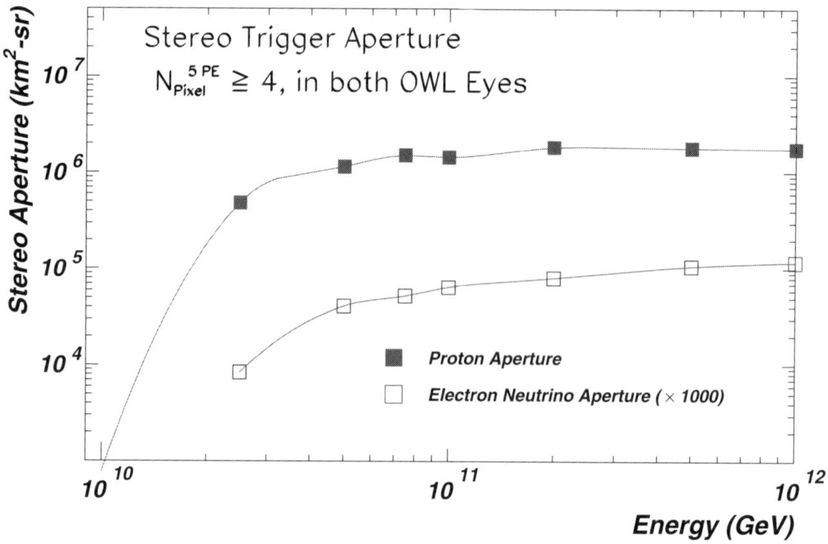

Fig. 8. Instantaneous aperture for proton-induced and deep ν_e-induced giant air showers as a function of energy.

Monte Carlo has been developed at the NASA Goddard Space Flight Center.[44] The simulation employs a hadronic event generator that includes effects due to fluctuation in the shower starting point and shower development, charged pion decay, neutral pion interactions, and the LPM (Landau–Migdal–Pomeranchuk) effect.

The number of events detected by OWL for a monoenergetic isotropic flux of protons and ν_e's with a standard model cross section is calculated by the Monte Carlo program, yielding the detection aperture as a function of energy, simulated trigger, and orbit parameters. Figure 8 shows the resultant proton and neutrino aperture for an altitude of 1000 km and a separation of 500 km. The asymptotic instantaneous proton aperture is $\sim 2 \times 10^6\,\mathrm{km}^2\,\mathrm{sr}$. The ν_e aperture determination includes the requirement that the observed starting point of the air shower, be in slant depth, $X_{\mathrm{start}} \geq 1500\,\mathrm{g\,cm}^{-2}$.

Acknowledgment

Part of this work was supported by NASA grant ATP03-0000-0057.

References

1. H. Sato and T. Tati, *Prog. Theor. Phys.* **47** (1972) 1788.
2. G. Amelino-Camelia *et al.*, *Nature* **393** (1998) 763.
3. S. Coleman and S. L. Glashow, *Phys. Rev. D* **59** (1999) 116008.
4. F. W. Stecker and S. L. Glashow, *Astropart. Phys.* **16** (2001) 97.
5. F. W. Stecker, *Astropart. Phys.* **20** (2003) 85.
6. F. W. Stecker and S. T. Scully, *Astropart. Phys.* **23** (2005) 203.
7. F. Aharonian *et al.*, *Astron. Astrophys.* **366** (2001) 62.

8. O. C. de Jager and F. W. Stecker, *Astrophys. J.* **566** (2002) 738.
9. A. Konopelko *et al.*, *Astrophys. J.* **597** (2003) 851.
10. F. W. Stecker and M. H. Salamon, *Astrophys. J.* **464** (1996) 600.
11. F. W. Stecker, M. A. Malkan and S. T. Scully, *Astrophys. J.* **648** (2006) 774 [astro-ph/0510449].
12. L. J. Garay, *Int. J. Mod. Phys. A* **10** (1995) 165.
13. J. Alfaro *et al.*, *Phys. Rev. D* **65** (2002) 103509.
14. J. Ellis *et al.*, *Phys. Rev. D* **63** (2001) 124025.
15. J. Ellis *et al.*, *Astropart. Phys.* **20** (2004) 669.
16. J. Ellis *et al.*, *Astron. Astrophys.* **402** (2003) 409.
17. J. Ellis *et al.*, *Astropart. Phys.* **25** (2006) 402.
18. S. E. Boggs *et al.*, *Astrophys. J.* **611** (2004) L77.
19. M. Rodriguez-Martinez, T. Piran and Y. Oren, *J. Cosmol. Astropart. Phys.* **5** (2006) 17.
20. C. Myers and M. Pospelov, *Phys. Rev. Lett.* **90** (2003) 211601.
21. T. Jacobson *et al.*, *Phys. Rev. Lett.* **93** (2004) 021101.
22. M. Gogberashvili, A. S. Sakharov and E. K. G. Sarkisyan, *Phys. Lett. B* **644** (2007) 79 [hep-ph/0605326].
23. A. Perez, in *Proc. 2nd Int. Conf. Fundamental Interactions* (2004), p. 1 [gr-qc/0409061].
24. R. Gambini and J. Pullin, *Phys. Rev. D* **59** (1999) 124021.
25. R. J. Gleiser and C. N. Kozameh, *Phys. Rev. D* **64** (2001) 083007.
26. W. Coburn and S. E. Boggs, *Nature* **423** (2003) 415.
27. T. Donaghy *et al.*, *AIP Conf. Proc.* **662** (2003) 450.
28. R. E. Rutledge and D. B. Fox, *Mon. Not. R. Astron. Soc.* **350** (2004) 1288.
29. C. Wigger *et al.*, *Astrophys. J.* **613** (2004) 1088.
30. T. Mizuno *et al.*, *Nucl. Instrum. Meth. A* **540** (2005) 158.
31. N. Produit *et al.*, *Nucl. Instrum. Meth. A* **550** (2005) 616.
32. Y. Z. Fan, B. Zhang and D. Proga, *Astrophys. J.* **635** (2005) L129.
33. K. Greisen, *Phys. Rev. Lett.* **16** (1966) 748.
34. G. T. Zatsepin and V. A. Kuz'min, *Zh. Eks. Teor. Fiz., Pis'ma Red.* **4** (1966) 144.
35. F. W. Stecker, *Phys. Rev. Lett.* **21** (1968) 1016.
36. R. Aloisio *et al.*, *Phys. Rev. D* **62** (2000) 053010.
37. J. Alfaro and G. Palma, *Phys. Rev. D* **67** (2003) 083003.
38. M. Takeda *et al.*, *Phys. Rev. Lett.* **81** (1998) 1163.
39. R. U. Abbasi *et al.*, *Phys. Rev. Lett.* **92** (2004) 151101.
40. D. J. Bird *et al.*, *Astrophys. J.* **441** (1995) 144.
41. Auger Collaboration (A. Zech), in *Proc. 41st Recontre de Moriond*, in press [astro-ph/0605344].
42. F. W. Stecker *et al.*, *Nucl. Phys. B* **136** (2004).
43. R. E. Streitmatter, in *Workshop on Observing Giant Cosmic Ray Airshowers from* $> 10^{20}$ *eV Particles from Space*, eds. J. F. Krizmanic, J. F. Ormes and R. E. Streitmatter; *AIP Conf. Pro.* **433** (1998) 95.
44. J. F. Krizmanic *et al.*, in *27th Int. Cosmic Ray Conf.* (Hamburg, 2001), p. 861.

NAMBU–GOLDSTONE MODES IN GRAVITATIONAL THEORIES WITH SPONTANEOUS LORENTZ BREAKING

ROBERT BLUHM
Department of Physics, Colby College,
Waterville, ME 04901, USA
rtbluhm@colby.edu

Spontaneous breaking of Lorentz symmetry has been suggested as a possible mechanism that might occur in the context of a fundamental Planck-scale theory, such as string theory or a quantum theory of gravity. However, if Lorentz symmetry is spontaneously broken, two sets of questions immediately arise: What is the fate of the Nambu–Goldstone (NG) modes, and can a Higgs mechanism occur? A brief summary of some recent work looking at these questions is presented here.

Keywords: Lorentz symmetry; Higgs mechanism; gravity.

1. Introduction

In gauge theory, spontaneous symmetry breaking has well-known consequences. The Goldstone theorem states that when a continuous global symmetry is spontaneously broken, massless Nambu–Goldstone (NG) modes appear.[1–3] On the other hand, if the symmetry is local, then a Higgs mechanism can occur in which the gauge fields acquire mass.[4–6]

In this work, those processes are examined for the case where the symmetry is Lorentz symmetry. In flat space–time, Lorentz symmetry is a global symmetry. Therefore, if the symmetry is spontaneously broken, it is expected that massless NG modes will appear. However, in curved space–time, in a gravitational theory, Lorentz symmetry is a local symmetry. It is in this context that the possibility of a Higgs mechanism arises.

The question of what the fate of the NG modes is when Lorentz symmetry is spontaneously broken was recently examined,[7] including the possibility of a Higgs mechanism. It is mainly the results of this work that are summarized here. However, the original motivation for considering the possibility that Lorentz symmetry might be spontaneously broken stems from the work in the late 1980s of Kostelecký and Samuel.[8–10] For example, they found that mechanisms occurring in the context of string field theory can lead to spontaneous Lorentz breaking. This prompted them to propose a vector model with spontaneous Lorentz violation, now known as the bumblebee model, which can be used to study the gravitational implications of spontaneous Lorentz violation. This model is summarized here as well, as are some of the original results of Kostelecký and Samuel concerning an alternative Higgs mechanism involving the metric. A number of additional studies concerning the bumblebee model have been carried out in recent years,[11–36] with many of them leading to new ideas concerning modified gravity and new phenomenological tests of relativity theory. However, space limitations do not permit a full review. Instead, the focus here will be on the NG modes and gravitational Higgs mechanism.

2. Spontaneous Lorentz Breaking

Lorentz symmetry is spontaneously broken when a local tensor field acquires a vacuum expectation value (vev),

$$\langle T_{abc} \rangle = t_{abc}. \tag{1}$$

The vacuum of the theory then has preferred space–time directions, which spontaneously breaks the symmetry.

In curved space–time, the Lorentz group acts locally at each space–time point. In addition to being locally Lorentz-invariant, a gravitational theory is also invariant under diffeomorphisms. There are thus two relevant symmetries, and it is important to consider them both. While Lorentz symmetry acts in local frames, and transforms tensor components with respect to a local basis, e.g. T_{abc} (the Latin indices denote components with respect to a local frame), diffeomorphisms act in the space–time manifold and transform components $T_{\lambda\mu\nu}$ defined (using Greek indices) with respect to the space–time coordinate system. These local and the space–time tensor components are linked by a vierbein. For example, the space–time metric and the local Minkowski metric are related by

$$g_{\mu\nu} = e_\mu{}^a e_\nu{}^b \eta_{ab}. \tag{2}$$

In a similar way, space–time tensor components are related to the components in a local frame using the vierbein:

$$T_{\lambda\mu\nu} = e_\lambda{}^a e_\mu{}^b e_\nu{}^c t_{abc}. \tag{3}$$

There are a number of reasons why it is natural to use a vierbein formalism to consider local Lorentz symmetry in the context of a gravitational theory. First, the introduction of vierbeins allows spinors to be incorporated into the theory. The vierbein formalism also naturally parallels gauge theory, with Lorentz symmetry acting as a local symmetry group. The spin connection $\omega_\mu{}^{ab}$ enters in covariant derivatives that act on local tensor components and plays the role of the gauge field for the Lorentz symmetry. In contrast, the metric excitations act as the gauge fields for the diffeomorphism symmetry. When one is working with a vierbein formalism, there are primarily two distinct geometries that must be distinguished. In a Riemannian geometry (with no torsion), the spin connection is nondynamical. It is purely an auxiliary field that does not propagate. However, in a Riemann–Cartan geometry (with nonzero torsion), the spin connection must be treated as independent degrees of freedom that in principle can propagate.[33,34]

3. Nambu–Goldstone Modes

Consider a theory with a tensor vev in a local Lorentz frame, $\langle T_{abc}\rangle = t_{abc}$, which spontaneously breaks Lorentz symmetry. The vacuum value of the vierbein is also a constant or fixed function; for example, for simplicity consider a background Minkowski space–time where

$$\langle e_\mu{}^a\rangle = \delta_\mu{}^a. \tag{4}$$

The space–time tensor therefore has a vev as well:

$$\langle T_{\lambda\mu\nu}\rangle = t_{\lambda\mu\nu}. \tag{5}$$

This means that diffeomorphisms are also spontaneously broken. Our first result is therefore that *spontaneous breaking of local Lorentz symmetry implies spontaneous breaking of diffeomorphisms.*

The spontaneous breaking of these symmetries implies that NG modes should appear (in the absence of a Higgs mechanism). This immediately raises the question of how many NG modes appear. In general, there can be up to as many NG modes as there are broken symmetries. Since the maximal symmetry-breaking case would yield six broken Lorentz generators and four broken diffeomorphisms, there can thus be up to ten NG modes. A natural follow-up question is to ask where these modes reside. In general, this depends on the choices of gauge. However, one natural choice is to put all the NG modes into the vierbein, as a simple counting argument shows is possible. The vierbein $e_\mu{}^a$ has 16 components. With no spontaneous Lorentz violation, the six Lorentz and four diffeomorphism degrees of freedom can be used to reduce the vierbein to six independent degrees of freedom. (Note that a general gravitational theory can have six propagating metric modes; however, general relativity is special in that there are only two.) In contrast, in a theory with spontaneous Lorentz breaking, where all ten space–time symmetries have been broken, up to all ten NG modes can potentially propagate. Thus, our second result is that *in a theory with spontaneous Lorentz breaking, up to ten NG modes can appear and all of them can naturally be incorporated as degrees of freedom in the vierbein.*

4. Bumblebee Model

The simplest case of a theory with spontaneous Lorentz breaking is the bumblebee model.[8–10] This is defined as theories in which a vector field B_μ acquires a vev:

$$\langle B_\mu \rangle = b_\mu. \tag{6}$$

The vev can be induced by a potential V in the Lagrangian that has a minimum for nonzero values of the vector field. A simple example of the bumblebee model has the form $\mathcal{L} = \mathcal{L}_\mathrm{G} + \mathcal{L}_\mathrm{B} + \mathcal{L}_\mathrm{M}$, where $\mathcal{L}_\mathrm{G}$ describes the pure-gravity sector, $\mathcal{L}_\mathrm{M}$ describes the matter sector, and (choosing a Maxwell form for the kinetic term)

$$\mathcal{L}_\mathrm{B} = \sqrt{-g}\left(-\frac{1}{4}B_{\mu\nu}B^{\mu\nu} - V(B_\mu) + B_\mu J^\mu\right) \tag{7}$$

describes the bumblebee field. Here, J^μ is a matter current, and the bumblebee field strength is $B_{\mu\nu} = D_\mu B_\nu - D_\nu B_\mu$, which in a Riemann space–time (with no torsion) reduces to $B_{\mu\nu} = \partial_\mu B_\nu - \partial_\nu B_\mu$. (For simplicity, we are neglecting additional possible interactions between the curvature tensor and B_μ.)

The potential V depends on B_μ and the metric $g_{\mu\nu}$. It is chosen so that its minimum occurs when B_μ and $g_{\mu\nu}$ acquire nonzero vev's. For a general class of theories, V is a function of $B_\mu B^\nu \pm b^2$, with $b^2 > 0$ equaling a constant, and where the minimum of the potential occurs when $B_\mu g^{\mu\nu} B_\nu \pm b^2 = 0$. The vacuum solutions for B_μ (which can be time-like or space-like, depending on the choice of sign) as well as for $g_{\mu\nu}$ must be nonzero and therefore spontaneously break both Lorentz and diffeomorphism symmetry.

Among the possible choices for the potential are a sigma-model potential $V = \lambda(B_\mu B^\nu \pm b^2)$, where λ is a Lagrange-multiplier field, and a squared potential $V = \frac{1}{2}\kappa(B_\mu B^\nu \pm b^2)^2$, where κ is a constant (of mass dimension zero). In the former case, only excitations that stay within the potential minimum (the NG modes) are allowed by the constraint imposed by λ. However, in the latter case, excitations out of the potential minimum are possible as well. In either of these models, three Lorentz symmetries and one diffeomorphism are broken, and therefore up to four NG modes can appear. However, the diffeomorphism NG mode does not propagate.[7] It drops out of all of the kinetic terms and is purely an auxiliary field. In contrast, the Lorentz NG modes do propagate. They are made up of a massless vector, with two independent transverse degrees of freedom (or polarizations). Indeed, they are found to propagate just like a photon.

5. Photons and Lorentz Violation

We find that the NG modes resulting from spontaneous local Lorentz violation can lead to an alternative explanation for the existence of massless photons [besides that of $U(1)$ gauge invariance]. Previous links between QED gauge fields, fermion composites and the NG modes had been uncovered in flat space–time (with global Lorentz symmetry).[35–39] Here, we propose a theory with just a vector field [but no $U(1)$ gauge symmetry] giving rise to photons in the context of a gravitational theory where local Lorentz symmetry is spontaneously broken.[7] Defining $B_\mu - b_\mu = A_\mu$, we find at lowest order that the Lorentz NG excitations propagate as transverse massless modes obeying an axial gauge condition, $b_\mu A^\mu = 0$. Hence, in summary, our third result is that *spontaneous local Lorentz violation may provide an alternative explanation for massless photons.* In the bumblebee model, the photon fields couple to the current J_μ as conventional photons, but also have additional Lorentz-violating background interactions like those appearing in the Standard Model extension (SME).[40–42] By studying these additional interactions, signatures can be searched for that might ultimately distinguish a photon theory based on local Lorentz breaking from conventional Einstein–Maxwell theory.

6. Higgs Mechanisms

Since there are two sets of broken symmetries (Lorentz and diffeomorphisms), there are potentially two associated Higgs mechanisms. However, in addition to the usual Higgs mechanism (in which a gauge-covariant-derivative term generates a mass term in the Lagrangian), it was shown[8–10] that an alternative Higgs mechanism can occur due to the gravitational couplings that appear in the potential V.

First, consider the case of diffeomorphisms. Here, it was shown that the usual Higgs mechanism involving the metric does not occur.[8–10] This is because the mass term that is generated by covariant derivatives involves the connection, which consists of derivatives of the metric and not the metric itself. As a result, no mass term for the metric is generated following the usual Higgs prescription. However, it was also shown that because of the form of the potential, e.g. $V = V(B_\mu g^{\mu\nu} B_\nu + b^2)$, quadratic terms for the metric can arise, resulting in an alternative form of the Higgs mechanism.[8–10] These can lead to mass terms that can potentially modify gravity in a way that avoids the van Dam, Veltmann and Zakharov discontinuity.[43,44] Summarizing the case of diffeomorphisms, *there is no conventional Higgs mechanism for the graviton; however, mass terms for the metric may arise from the potential V in an alternative mechanism.*

In contrast, for the case of Lorentz symmetry, it is found that a conventional Higgs mechanism can occur.[7] In this case the relevant gauge field (for the Lorentz symmetry) is the spin connection. This field appears directly in covariant derivatives acting on local tensor components, and for the case where the local tensors acquire a vev, quadratic mass terms for the spin connection can be generated following a similar prescription as in the usual Higgs mechanism. For example, in the bumblebee model, using a unitary gauge, the kinetic terms involving $B_{\mu\nu}$ generate quadratic mass terms for the spin connection ${\omega_\mu}^{ab}$. However, a viable Higgs mechanism involving the spin connection can occur only if the spin connection is a dynamical field. This then requires that there be nonzero torsion and that the geometry be Riemann–Cartan. Our final result is therefore that *a Higgs mechanism for the spin connection is possible, but only in a Riemann–Cartan geometry.*

Constructing a ghost-free model with a propagating spin connection is known to be a challenging problem.[45–47] Evidently, incorporating Lorentz violation leads to the appearance of additional mass terms, which in turn could create new possibilities for model building.

7. Summary and Conclusions

In theories with spontaneous Lorentz violation, up to ten NG modes can appear. They can all be incorporated naturally in the vierbein. For the vector bumblebee model, the Lorentz NG modes propagate like photons in an axial gauge. In principle, two Higgs mechanisms can occur: one associated with broken diffeomorphisms, the other with Lorentz symmetry. While a usual Higgs mechanism (for diffeomorphisms) does not occur involving the metric field, an alternative Higgs mechanism can lead to the appearance of quadratic metric terms in the Lagrangian. If in addition the geometry is Riemann–Cartan, then a Higgs mechanism (for the Lorentz symmetry) can occur in which the spin connection acquires a mass. Clearly, there are numerous phenomenological questions that arise in these processes, all of which can be pursued comprehensively using the SME.

Acknowledgment

This work was supported by NSF grant PHY-0554663.

References

1. Y. Nambu, *Phys. Rev. Lett.* **4** (1960) 380.
2. J. Goldstone, *Nuov. Cim.* **19** (1961) 154.
3. J. Goldstone, A. Salam and S. Weinberg, *Phys. Rev.* **127** (1962) 965.
4. F. Englert and R. Brout, *Phys. Rev. Lett.* **13** (1964) 321.
5. P. W. Higgs, *Phys. Rev. Lett.* **13** (1964) 508.
6. G. S. Guralnik, C. R. Hagen and T. W. B. Kibble, *Phys. Rev. Lett.* **13** (1964) 585.
7. R. Bluhm and V. A. Kostelecký, *Phys. Rev. D* **71** (2005) 0065008.
8. V. A. Kostelecký and S. Samuel, *Phys. Rev. D* **40** (1989) 1886.
9. V. A. Kostelecký and S. Samuel, *Phys. Rev. D* **39** (1989) 683.
10. V. A. Kostelecký and S. Samuel, *Phys. Rev. Lett.* **63** (1989) 224.
11. V. A. Kostelecký and R. Lehnert, *Phys. Rev. D* **63** (2001) 065008.
12. V. A. Kostelecký, *Phys. Rev. D* **69** (2004) 105009.
13. V. A. Kostelecký and R. Potting, *Gen. Relativ. Gravit.* **37** (2005) 1675.
14. B. Altschul and V. A. Kostelecký, *Phys. Lett. B* **628** (2005) 106.
15. V. A. Kostelecký and Q. G. Bailey, gr-qc/0603030.
16. T. Jacobson and D. Mattingly, *Phys. Rev. D* **64** (2001) 024028.
17. T. Jacobson and D. Mattingly, *Phys. Rev. D* **70** (2004) 024003.
18. C. Eling and T. Jacobson, *Phys. Rev. D* **69** (2004) 064005.
19. C. Eling, gr-qc/0507059.
20. P. Kraus and E. T. Tomboulis, *Phys. Rev. D* **66** (2002) 045015.
21. J. W. Moffat, *Int. J. Mod. Phys. D* **12** (2003) 1279.
22. B. M. Gripaios, *J. High Energy Phys.* **0410** (2004) 069.

23. M. L. Graesser, A. Jenkins and M. B. Wise, *Phys. Lett. B* **613** (2005) 5.
24. S. M. Carroll and E. A. Lim, *Phys. Rev. D* **70** (2004) 123525.
25. E. A. Lim, *Phys. Rev. D* **71** (2005) 063504.
26. O. Bertolami and J. Paramos, *Phys. Rev. D* **72** (2005) 044001.
27. J. W. Elliott, G. D. Moore and H. Stoica, *J. High Energy Phys.* **0508** (2005) 066.
28. J. L. Chkareuli *et al.*, hep-th/0412225.
29. A. T. Azatov and J. L. Chkareuli, hep-th/0511178.
30. M. V. Libanov and V. A. Rubakov, *J. High Energy Phys.* **0508** (2005) 001.
31. D. S. Gorbunov and S. M. Sibiryakov, *J. High Energy Phys.* **0509** (2005) 082.
32. H.-C. Cheng *et al.*, hep-th/0603010.
33. F. W. Hehl *et al.*, *Rev. Mod. Phys.* **48** (1976) 393.
34. I. L. Shapiro, *Phys. Rep.* **357** (2002) 113.
35. P. A. M. Dirac, *Proc. R. Soc. Lon. A* 209 (1951) 291.
36. W. Heisenberg, *Rev. Mod. Phys.* **29** (1957) 269.
37. P. G. O. Freund, Acta *Phys. Austriaca* **14** (1961) 445.
38. J. D. Bjorken, *Ann. Phys.* **24** (1963) 174.
39. Y. Nambu, *Prog. Theor. Phys. Suppl. Extra* (1968) 190.
40. V. A. Kostelecký and R. Potting, *Phys. Rev. D* **51** (1995) 3923.
41. D. Colladay and V. A. Kostelecký, *Phys. Rev. D* **55** (1997) 6760.
42. D. Colladay and V. A. Kostelecký, *Phys. Rev. D* **58** (1998) 116002.
43. For a review of the Standard model extension, see R. Bluhm, in *Special Relativity: Will It Survive the Next 101 Years?*, eds. J. Ehlers and C. Lämmerzahl (Springer, Berlin, 2006), hep-ph/0506054.
44. H. van Dam and M. Veltman, *Nucl. Phys. B* **22** (1970) 397.
45. V. I. Zakharov, *J. Exp. Theor. Phys. Lett.* **12** (1970) 312.
46. E. Sezgin and P. van Nieuwenhuizen, *Phys. Rev. D* **21** (1980) 3269.
47. K. Fukuma, *Prog. Theor. Phys.* **107** (2002) 191.

THE SEARCH FOR DARK MATTER FROM SPACE AND ON THE EARTH

DAVID B. CLINE

Astrophysics Division, Department of Physics & Astronomy,
University of California, Los Angeles, CA 90095, USA
dcline@physics.ucla.edu

We first show that the MOND concept is very unlikely: nonbaryonic dark matter exists. We then discuss dwarf special galaxies that in some cases appear to be nearly pure dark matter systems. The search for dark matter particles can be carried out from space (sterile neutrinos neutralino annihilation) or on the Earth (direct detection). We describe progress in these areas and focus on the progress of the ZEPLIN II detector now taking data.[1,2]

Keywords: MOND; dark matter; ZEPLIN II.

1. Evidence Concerning MOND

The search for dark matter is of great importance in the context of the general theory of relativity. A brief history of dark matter/dark energy is given in Table 1.[a]

There are suggested modifications to take into account the effects attributed to dark matter. The most noticeable is the MOND theory of Milgram.[3] Some say that this theory violates Einstein's strong equivalence principle — and therefore general relativity. Thus it is important to determine its correctness.

The MOND theory was invented to explain the star rotation curves for galaxies that are normally used to indicate the existence of cold dark matter. MOND assumes that gravity is modified beyond a gravitational acceleration of $a_0 \sim 10^{-8}\,\mathrm{cm/s^2}$.

For acceleration greater than this we get the normal law:

$$a_N = \frac{MG}{R^2}, \tag{1}$$

[a]The references for Secs. 1 and 2 can be found in Ref. 3.

Table 1. Brief history of the evidence for dark matter and dark energy.

~1933	F. Zwickey observes fast galaxies in comet clusters; suggests mixing mass in cluster is the cause.[4]
~1960's	Astronomers realize that galaxies have fast-moving stars in halo; suggest dark matter is the cause.
~1980	Suggestions for a MOND to explain radiation curves by modifying Newtonian gravity.[3]
~1998	Experiments on SN1A reported at dark matter meeting; indication of accelerating Universe; dark energy is suggested cause.
~2003	WMAP data strongly support both dark matter and dark energy[1] components of the Universe.
~2005	SDSS observes baryon acoustic oscillations; provides additional proof for dark matter (several independent measurements of $\Omega_{\rm DM}$ suggest a single origin: cold dark matter).[3]

for acceleration less than a_0 we get[3]

$$a = \sqrt{a_0 a_n} = \sqrt{a_0 MG}\frac{1}{R}. \tag{2}$$

The $1/R$ behavior then provides a good fit to the rotation curves of galaxies at large r. There are three more (at least) ways to test MOND[3]:

- Study of clusters in X-rays and weak lensing to determine whether dark matter is different from baryonic matter.
- Study of CMBR and Baryon oscillations that come from $Z \sim 1000$ — the surface of the last scatter — two independent ways to measure $\Omega_{\rm DM}$.
- Direct detection of dark matter particles (reviewed here).

2. Test of MOND with Gravity Lensing and X-ray Maps

The MOND theory describes galaxies as being made of only baryonic matter and with a modified Newtonian force law to explain high velocities in the halo. Any experiment that (1) finds evidence for dark matter with nongravitational probes or (2) finds dark matter displaced from baryonic matter will constrain MOND. Two such experiments have been published. We show in Figs. 1 and 2 the results of these experiments.

Figure 1 shows the predicted temperature profile for the X-rays from a cluster and the measured profile. In this version of MOND there is a poor fit to the data. Figure 2 shows the results of the study of interacting superclusters. The authors state that the location of the bulk of this mass (determined by weak lensing) is in a different place from the location of the visible (baryonic matter). Since in MOND there is only baryonic matter, the authors state that this indicates the existence of dark matter.

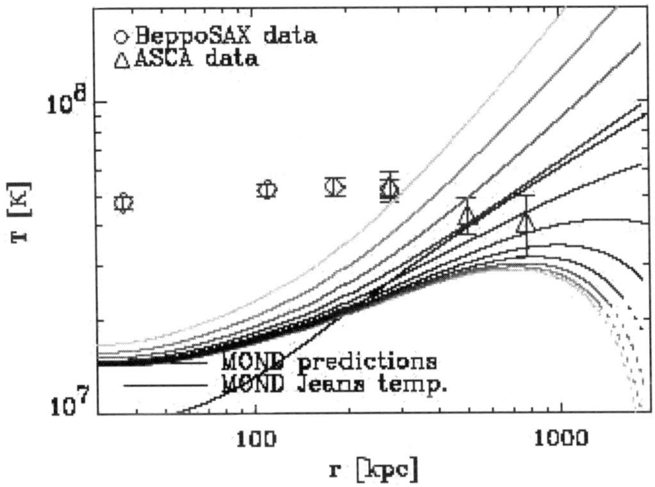

Fig. 1. Temperature profile from X-rays for a galactic cluster. See Ref. 3 for references.

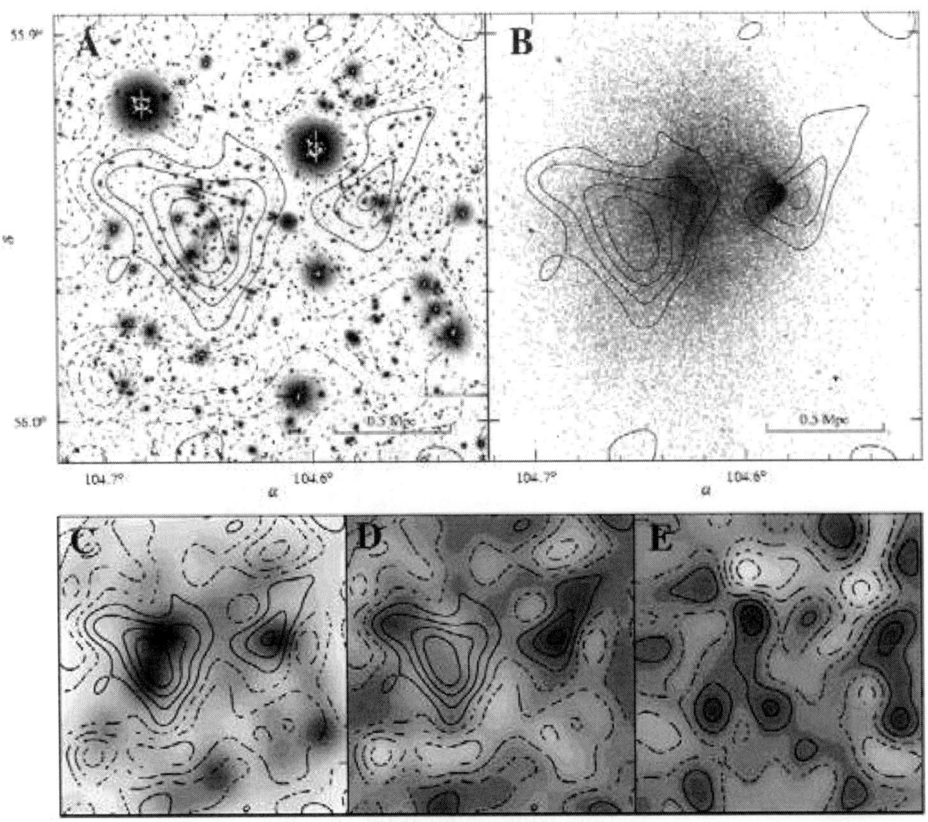

Fig. 2. Images of weak lensing and X-ray studies from Ref. 5.

3. Baryonic Acoustic Oscillations and Evidence for Dark Matter Compared to CMBR Evidence

Recently the SDSS collaboration has carried out an impressive study of galactic correlations showing a baryonic oscillation peak in the spectrum that derives its origin from the early Universe.[1] Table 2 lists some of the properties. The measurement of dark matter[1]

$$\Omega_{\rm DM} = 0.275 \pm 0.025 \tag{3}$$

for a flat, $\omega = -1$ Universe is in impressive agreement with the results for the CMBR from WMAP. Furthermore the authors state that this effect is a direct proof of the existence of dark matter. Figure 3 shows the baryonic oscillation structure.

It is hard to see how the MOND theory could reproduce such an effect including the precision value for $\Omega_{\rm DM}$ from these different measurements.

Table 2. Testing MOND with acoustic oscillations in the baryon density.

D. Eisenstein *et al.*, SDSS team (Astro-PH/0501171)
"The recent observation by the SDSS team of an Acoustic Peak in the LSS at a scale of about 100 MPC strongly supports the CDM origin of dark matter! One can show that this proves CDM must exist in the redshift ranges:
$0 < Z < 0.35$ $0.35 < Z > 1000$
for the growth structure.
It will be nearly impossible for MOND to produce such an effect."

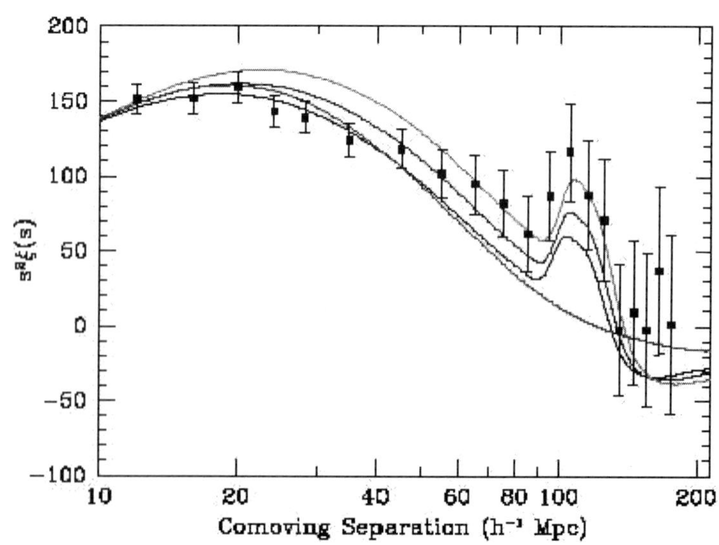

Fig. 3. Evidence for baryon acoustic oscillation signal (Ref. 6).

4. Dark Matter in the Milky Way: Halo Uncertainty and Streams

In order to detect dark matter particles we must have an understanding of the flux of particles through any given detector on the Earth. Therefore we must understand the halo of dark matter for our galaxy. In addition some models give clumps of dark matter, and others give causes of dark matter. These effects can increase or decrease the rate of interaction in an Earth-bound detector. At the recent Marina del Rey meeting we devoted an entire session to the knowledge of our halo.

The halo model is very important when one is attempting to compare different types of experiments — say, direct searches and annual variation searches.[5–8] There is no doubt that the ultimate test for the existence of dark matter will be the observation of an annual variation signal.[5–8] However, there is a strong debate among the experiments as to whether this annual variation signal should be carried out with discriminated events (reduced background) or with raw data (large background).

There are also models of dark matter caustics by P. Sikivie and colleagues that can give the opposite sign of the annual variation to that expected in the standard isothermal sphere model.[5–13] At the Marina del Rey meeting two notable contributions were given by Anne Green and Larry Krauss.

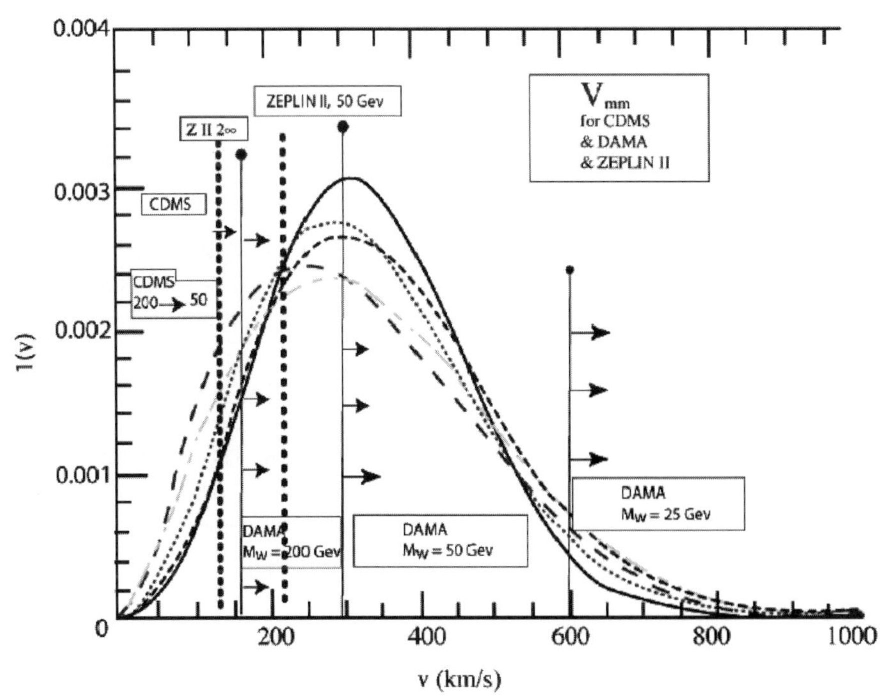

Fig. 4. Schematic of the halo velocity distribution with minimal velocities for CDMS, DAMA and ZEPLIN II (Modified from Ref. 6).

Table 3. Summary.

• A vast increase in precise stellar kinematic data allows more sophisticated derivation of mass profiles in the dSph.
• UMa — discovered in 2005 — extends to M/L $\sim$ 500.
• All are consistent with:
• Central mass cores, not cusps
• Central mass density $\leq 20\,\text{GeV/cc}$
• Dispersion $\sim 9\,\text{km/s}$
• Scale length $\sim$ few $\times$ 100 pc
• DM minimum mass $\sim 5 \times 10^7\,M_\odot$
• Somewhat preferring particle-mass $\ll$ GeV
• We have two new *dSph* under study (today), to extend the sample further, and see if these numbers are really meaningful.

5. Dwarf Spheroidal Galaxies as Pure Dark Matter Systems

At the Dark Matter 2006 meeting G. Gilmore gave an interesting talk about some implications of the study of *dSph* systems and indicated some possibly remarkable conclusions. Table 3 gives the conclusions. If these conclusions are correct they imply that dark matter is very low mass and could be warm dark matter. Sterile neutrinos of a-few-keV mass could be a condition. Under any circumstance the *dSph* systems consist of very little gas and are nearly pure dark matter systems. WIMP's could also be the dark matter in *dSph's* and in this case the GLAST detection could see a signal of gamma rays from the *dSph's*.[1]

6. The Search for Dark Matter from Space

The next generation of gamma ray detectors, GLAST, holds great promise for the detection of dark matter. There are two possible problems:

(1) The detector of high energy gammas by HESS and MAGIC from the galactic center with a simple power law spectrum seems not to be from dark matter but the source could constitute a new background for the detection of dark matter.[14]

(2) GLAST and other detectors rely on the concentration of dark matter into clumps. Recently V. Berezinsky *et al.*[15] have shown that such clumps are tidally destroyed in the galaxy environment, greatly reducing the effect for the dark matter search. So the prospects for dark matter detectors in space may be reduced by these two effects. The fact that *dSph's* offer nearly pure systems for dark matter could lead to the search for two forms of dark matter:

   (i) Sterile neutrinos by

$$\nu_s \to \nu + \gamma \tag{4}$$

   in the 1–10 keV mass (see the talk by A. Kusenko in Ref. 1) with detectors like Chandra.

(ii) Neutralinos from

$$x + x \to \gamma + x \tag{5}$$

with detectors like GLAST (see the GLAST talks in Ref. 1). For a summary of recent searches for sterile neutrinos see Ref. 16.

7. Methods for the Direct Search for Dark Matter Particles

The direct search for dark matter particles is among the hardest experiments ever undertaken in science.[20] Backgrounds exist for cosmic rays, natural radioactivity even at great depths underground. Early reviews can be found in Ref. 2. Therefore the next generation detector will almost certainly use a method to discriminate against the background as well as an active veto shield to reduce the neutron flux from cosmic-ray-induced events even at great depths underground.

The types of detectors can be generally classed as cryogenic, liquid xenon, neon or argon; and other methods, such as the bubble chamber or the nondiscriminating detector.[2]

To get some sense of the number of detectors and the time scale, we give a partial list in Table 5.

7.1. *Cryogenic detectors*

For more than 15 years several groups around the world have been studying the possibility of constructing a low temperature detector to measure the recoil energy of the nucleus having been hit by a WIMP. Since this energy is in the range of kiloelectron volts the detector must act as a bolometer to measure the "heat" produced by the recoil. Three groups have now made such detectors using this technique. They are CDMS, Edelweiss and Cresst.

All three groups have now reported limits in the search for dark matter particles. So far the nucleus of choice has been Ge or Si. However, the Cresst group has worked with CaW mixtures.

7.2. *Liquid noble gas detectors: xenon, argon and neon*

Another promising method to detect dark matter is to use the scintillation light produced in noble gas liquids. The process is very well known since excimer lasers use a similar concept. For example, the very first excimer laser was made in Russia in 1970 using liquid xenon. A key part of this method is to apply an electric field to the detector to drift out any electrons that are produced at the recoil vertex as a basis to discriminate against the background.[20,21]

This method was just invented by our group within the ICARUS collaboration and is the basis for the ZEPLIN II, III, IV and XENON, as well as the XMASS detector.[22] In Fig. 5 we show a schematic of the ZEPLIN II detector and the

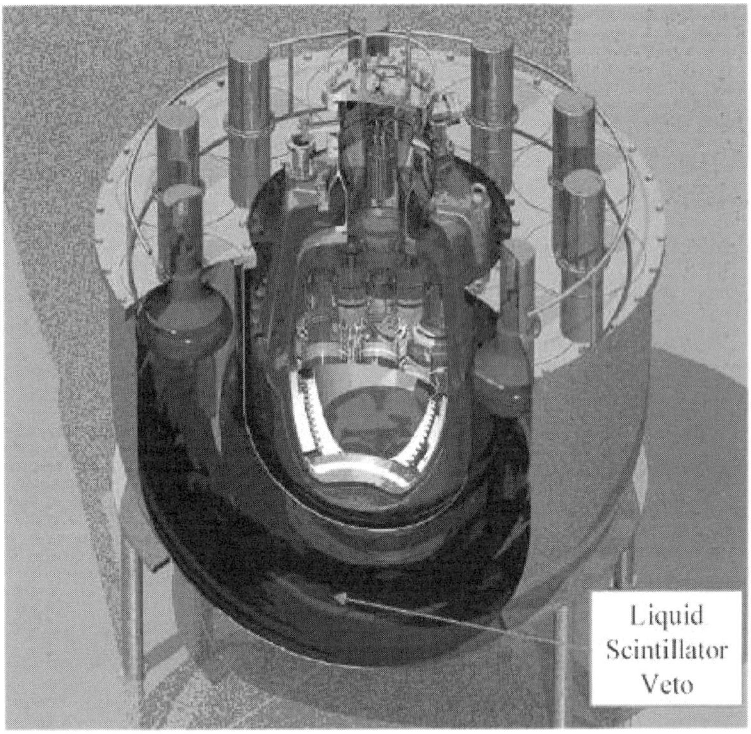

Fig. 5. Schematic of the ZEPLIN IV detector.

Table 4. Brief history of the ZEPLIN II dark matter detector.

~ 1980–90	ICARUS team at CERN studies properties of liquid argon.
~ 1990	DBC talk at Oxford meeting on LAr for dark matter detection.
~ 1992–93	Liquid xenon properties studied by UCLA/Torino group; LiXe dark matter detector invented.
~ 1992–97	Study 2KS LiXe detector at CERN; take detector to Mont Blanc Laboratory.
~ 1998	H. Wang thesis at UCLA.
~ 1995–98	1995–98 Form ZEPLIN collaboration: UCLA, Torino, UKDMC.
~ 2000	Publication of ZEPLIN concept by UCLA/Torino/CERN group of astroparticle physicists.
~ 2001–04	Construction of ZII at UCLA/TAMU/RAL.
~ 2004–05	Turn-on of detector and move to Boulby.
~ 2006	Start data-taking with ZII; goal 8000 kg day of data.

complete detector being tested at RAL. The XENON detector uses a similar design. Table 4 gives this history of the ZEPLIN detector.

More recently there have been studies[1] of the use of liquid argon (WARP) and liquid neon (Clean) as WIMP detectors. One virtue of the use of liquid xenon is the existence of different isotopes with different spins, thus testing the spin dependence of the WIMP interaction.

The ZEPLIN I team detector has reported a limit in the WIMP search using a partial discrimination method of pulse shape analysis. Of all the current detector concepts, the one most easily expanded to the one-ton scale seems to be liquid xenon. The US/UKDMC team is designing the ZEPLIN IV/MAX detector, which will have a mass in the range of one ton. Currently it is not clear if there will be a single one-ton detector or four 250 kg detectors. A schematic of the one-ton ZEPLIN IV/MAX detector is shown later in this paper.[23]

The goal of the one-ton detectors is to reach the cross-section level of absorb 10^{9}–10^{10} pb. Current calculations of the cross-section for SUSY WIMPS indicate that a discovery of dark matter is likely to be made in this cross-section range.[1]

8. Status of the Search for Dark Matter Particles

A serious search for dark matter particles started around 1995 with the use of NaI detectors at several locations. Table 5 gives the proposals for one-ton detectors in the future.

9. Future Detectors on the Ton Scale and Sensitivity

There were many new estimates for the SUSY DM cross-section range given at the Dark Matter 2004 symposium. One was published elsewhere by P. Nath and colleagues.[1] Note that in these types of calculations the most likely region of discovery is the 10^{-7}–10^{-8} pb cross-section, but that the signal could be as low as 10^{-9}–10^{-10} pb (Table 5).

While the next generation of detectors will likely reach 10^{-7} or even 10^{-8} pb (CDMS II, ZEPLIN II, Edelweiss II, etc.), there is no certainty that even 10^{-8} pb can be reached. For this case much larger detectors in the one-ton range will be needed. Even if a tentative signal is observed at 10^{-8}, a much larger detector will be needed to confirm this signal; see Fig. 6.

A new, third generation of detectors is being studied for this case. We consider the example of ZEPLIN IV/MAX here for such a detector.[23] In the case where a single one-ton detector is to be constructed, the detector will require some new concepts beyond that employed in the ZEPLIN II/III detectors. Of course the data

Table 5. One-ton dark matter detector proposals.

Detector	Material	Method	Proposal	Current prototype
GENIUS	Ge	Ultrapure detector in LNGS	1997	10 kg GENIUS
ZEPLIN IV (Max) (Boulby/SNOLAB)	Xe	2-phase discriminating detector	~1999	ZII/III Boulby
Super CDMS (SNOLAB)	Ge/Si	Ionization and phonons	~2001	CDMS II
XMass (Japan)	Xe	2-phase (?)	~2000	prototype
Xenon (LNGS)	Xe	2-phase detector	~2001	prototype
WARP (LNGS)	Ar	2-phase (possibly larger than 1 ton	~2003	prototype

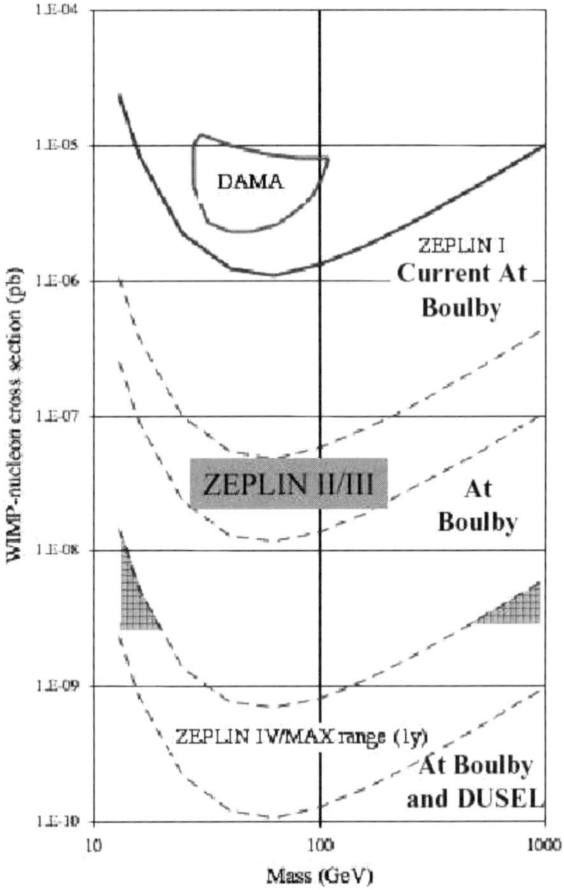

Fig. 6. Schematic of the possible future of the dark matter search (*R. Gaitskell*).

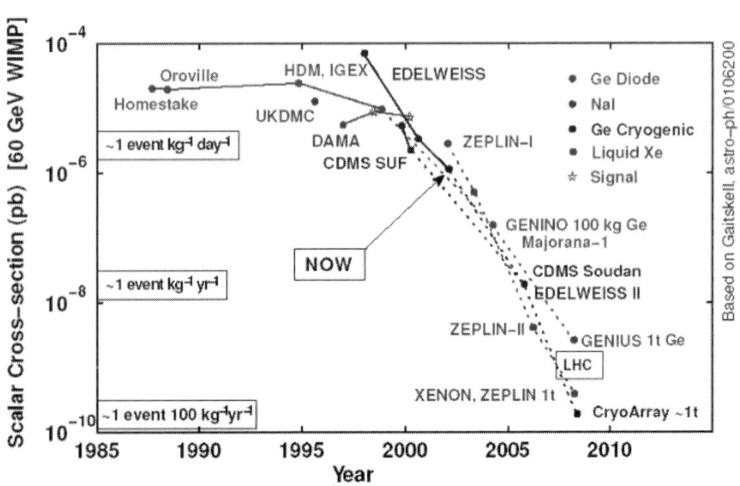

Fig. 7. Schematic of the possible future of the dark matter search (*R. Gaitskell*).

from these detectors will be crucial to the understanding of how such a detector will work underground. Future search sensitivity is shown in Fig. 7.

Acknowledgment

I wish to thank the organizers of this exciting meeting for a good scientific time.

References

1. See the talks in *7th UCLA Symposium on Sources and Detection of Dark Matter and Dark Energy in the Universe*, ed. Marina del Rey, Feb. 2006, at http://www.physics.ucla.edu/hep/dm06/dm06.htm.
2. D. Cline, The Dark Universe: The Search for Dark Matter and the Nature of Dark Energy, in *AIP Conference Proceedings Vol. 809, Advanced Summer School in Physics 2005: Frontiers in Contemporary Physics EAV05*, eds. Oscar Rosas-Ortiz, Mauricio Carbajal and Omar Miranda, Mexico City, July 2005, p. 11.
3. D. Cline, astro-ph/0510576.
4. F. Zwicky, *Helv. Phys. Act.* **6** (1993) 110.
5. P. Sivikie, *Phys. Lett. B* **432** (1998) 139.
6. K. Freese, J. A. Frieman and A. Gould, *Phys. Rev. D* **37** (1988) 3388.
7. G. Gelmini and P. Gondolo, *Phys. Rev. D* **64** (2001) 023504.
8. F. S. Long, P. Sikivie and S. Wick, Diurnal and Annual Modulation of Cold Dark Matter Signals, UFUFT-HET-04-6.
9. G. Fuller *et al.*, *Phys. Rev. D* **68** (2003) 103002.
10. A. Kusenko and Y. Segre, *Phys. Lett. B* **396** (1997) 197.
11. M. W. Goodwin and E. Witten, *Phys Rev D* **31** (1985) 3059.
12. A. Drukier, K. Freese and D. N. Spergel, *Phys. Rev. D* **187** (1986) 3495.
13. I. Wasserman, *Phys. Rev. D* **33** (1986) 2071.
14. G. Zaharijes and D. Hooper, *Phys. Rev. D* **73** (2006) 103501.
15. V. Berezinsky *et al.*, *Phys. Rev. D* **73** (2006) 063504.
16. S. Riemer-Sorenson *et al.*, *Astrophys. J.* **644** (2006) L33.
17. J. R. Primack, D. Seckel and B. Sadoulet, *Ann. Rev. Nucl. Part. Sci.* **38** (1988) 751.
18. P. F. Smith and J. D. Lewin, *Phys. Rep.* **187** (1990) 203.
19. G. Jungman, M. Kamionkowski and K. Griest, *Phys. Rep.* **267** (1996) 195.
20. D. Cline, *Sci. Am.* **288** (2003) 50.
21. D. Cline *et al.*, *Astropart. Phys.* **12** (1999) 373.
22. D. Cline *et al.*, Status of ZEPLIN II and ZEPLIN IV Study, in *Nucl. Phys. B, Proc. Supp., 5th Int. Symp. Search and Detection of Dark Matter and Dark Energy in the Universe*, ed. D. B. Cline (Elsevier, Amsterdam, 2003), p. 229.
23. D. Cline, ZEPLIN IV: A One-Ton WIMP Detector, paper given at the Dark 2002 meeting, Capetown, Jan. 2002, published in *Proc. Dark 2002* (Springer, Heidelberg, 2002), eds. H. V. Klapdor-Kleingrothaus and R. Viollier, p. 492.

NEW PHYSICS WITH 10^{20} eV NEUTRINOS AND ADVANTAGES OF SPACE-BASED OBSERVATION

THOMAS J. WEILER

Department of Physics and Astronomy,
Vanderbilt University,
Nashville Tennessee 37235, USA
tom.weiler@vanderbilt.edu

Nature accelerates cosmic particles to energies as high as 10^{20} eV. The rates for neutrino-initiated quasi-horizontal air showers (HASs) and upgoing air showers (UASs) have different dependences on $\sigma_{\nu N}$. Therefore, a measurement of HAS and UAS rates would allow an inference of $\sigma_{\nu N}$ at energies far beyond what is conceivable with terrestrial accelerators. At a minimum, such a measurement provides a microscope/telescope for QCD evolution. More ambitiously, such a measurement may reveal energy thresholds of completely new physics. The feasibility of this measurement is examined. Favorable conclusions result, especially for proposed space-based observatories. The latter benefit from a larger field of view and from a UAS rate enhanced by $\mathcal{O}(10)$ over oceans compared to over land.

Keywords: Neutrinos; space-based; extreme energy.

1. Introduction

Detection of *ultrahigh energy* ($E_\nu > 10^{18}\,\mathrm{eV} \equiv \mathrm{EeV}$) *neutrinos* is important for several reasons. *First* of all, neutrino primaries are not deflected by magnetic fields and so should point back to their cosmic sources. This contrasts with cosmic rays, which are charged and follow bent trajectories. *Secondly*, well above $E_{\mathrm{GZK}} \sim 5 \times 10^{19}\,\mathrm{eV}$, they may be the only propagating primaries. As such, they may be the only messengers revealing the ultimate energy reach of extreme cosmic accelerators, generally believed to be powered by black holes. Above E_{GZK}, the suppression of cosmic rays results from the resonant process $N + \gamma_{\mathrm{CMB}} \to \Delta \to N + \pi$; a handful of cosmic ray events have been detected with estimated energies exceeding 10^{20} eV. The observable neutrino spectrum could extend to much higher energies. *Thirdly*, in contrast to cosmic rays and photons, neutrinos are little affected by the ambient matter surrounding the central engines of Nature's extreme accelerators. Accordingly, neutrinos may carry information about the central engine itself, inaccessible with other primaries. In principle, neutrinos may be emitted from close to the black hole horizon, subject only to energy loss due to gravitational red-shifting. An analogy can be made with solar studies. Solar photons are emitted from the outer centimeter of the Sun's chromosphere, while solar neutrinos are emitted from the central core where fusion powers the Sun. *Fourthly*, neutrinos carry a quantum number that cosmic rays and photons do not have — "flavor." Neutrinos come in three flavors, ν_e, ν_μ, and ν_τ. One may think of this "extra" flavor degree of information as the neutrino's superb analog of polarization for the photon, or nucleon number A for the cosmic ray. Each of these attributes — flavor, polarization, and nucleon number — carries information about the nature and dynamics of the source, and about the environment and pathlength of the intergalactic journey. The flavor ratios of cosmic neutrinos are observable.[1,2] Several papers have recently analyzed the benefits that neutrino flavor identification offers for unraveling the dynamics of cosmic sources.[3,4] The *fifth* reason why ultrahigh energy neutrino primaries traveling over cosmic distances are interesting is that such travel allows studies of the fundamental properties of neutrinos themselves. For studying some properties of the neutrino, such as neutrino stability/lifetime,[5] and pseudo-Dirac mass patterns,[6] it is the cosmic distance that is essential; for other properties, it is the extreme energy that is essential. A clear example of the latter is any attempt to determine the neutrino cross-section at energies beyond the reach of our terrestrial accelerators.

This paper will summarize the potential for cosmic ray experiments designed to track ultrahigh energy air showers by monitoring their fluorescence yield, to infer the neutrino–nucleon cross-section $\sigma_{\nu N}$ at energies above 10^{19} eV. The idea here[7] is to measure both horizontal air shower (HAS) and upgoing air shower (UAS) events. Since HAS and UAS rates have very different dependences on $\sigma_{\nu N}^{\rm CC}$, one may infer $\sigma_{\nu N}^{\rm CC}$ from their measured ratio.

From the point of view of QCD, such a cross-section measurement would be an interesting microscope on the world of small-x parton evolution. The cross-section could be quite different from popular extrapolations. For example, parton saturation (overlap) effects can significantly reduce the total cross-section at these very high energies. On the other hand, if a new threshold is crossed between terrestrial neutrino energies $\sim$100 GeV and the extreme energies reached by cosmic rays, $\sim 10^{11}$ GeV, then the cross-section could much exceed the QCD extrapolations. The nine-orders-of-magnitude increase in the lab energy reach corresponds to a 4.5-orders-of-magnitude increase in the center-of-momentum (CoM) energy reach. Even the CoM energy at the e–p HERA collider is more than three orders of magnitude below the CoM energy of measured cosmic rays. Cosmic rays offer ample energy room for new physics beyond our Standard Model. Proposals for new physics thresholds in this energy region include low scale unification with gravity, nonperturbative electroweak instanton effects, compositeness models, a low energy unification scale in string-inspired models, and Kaluza–Klein modes from compactified extra dimensions. All of these models produce a strongly interacting neutrino cross-section above the new threshold. In addition, cosmic ray energies probe the ultraviolet completion region of extended electroweak models like the "little Higgs" model. Dispersion relations allow one to use low energy elastic scattering to place constraints on the high energy cross-section,[8] but the constraints are quite weak. For all we know, any of the above new physics may be present in existing cosmic ray data. The first available window on the presence of such new physics may be the cross-section measurement.

2. Event Rates

A horizontal shower, deeply initiated, is the classic signature for a neutrino primary. The weak nature of the neutrino cross-section means that horizontal events begin where the atmospheric target is most dense, low in the atmosphere. In contrast, the ultrahigh energy *pp* cross-section exceeds 100 mb, so the air-nucleon cross-section exceeds a barn! Even the vertical atmospheric column density provides hundreds of interaction lengths for a nucleon, and so the cosmic ray interacts high in the atmosphere. The weak nature of the neutrino cross-section also means that the event rate for neutrino-induced HAS is proportional to the neutrino–nucleon cross-section.

For a neutrino-induced UAS, the dependence on the neutrino cross-section is more complicated, and more interesting. The Earth is opaque to all known quanta except the neutrino, and in fact is opaque to neutrinos as well if their energies exceed about a PeV (10^{15} eV). However, "Earth-skimming" neutrinos, those with a short enough chord length through the Earth, will penetrate and exit, or penetrate and interact. In particular, there is much interest in the Earth-skimming process $\nu_\tau \to \tau$ in the shallow Earth, followed by τ decay in the atmosphere to produce an observable shower. In Ref. 7 it was shown that the rate for the Earth-skimming process $\nu_\tau \to \tau$ is *inversely* proportional to $\sigma_{\nu N}$. Two effects, one physical and one geometrical, bring this about. The first effect is that for the UAS events, the incident neutrino flux suffers absorption in the Earth, and so the UAS rate is proportional to the neutrino mfp in the Earth, λ_ν, which scales as the inverse of $\sigma_{\nu N}$. The second effect is that event rates of "surface" detectors, which include spaced-based observatories orbiting at a distance much larger than the $h = 8\,\text{km}$ scale height of the atmosphere, are proportional to the projection of the incident neutrino flux normal to the surface of view, i.e. to $\cos\theta_n$ (θ_n is the nadir angle), equal in the mean to $\lambda_\nu/2\,R_\oplus$ ($R_\oplus$ is the Earth's radius), which again scales as the inverse of $\sigma_{\nu N}$. Thus, compared to the HAS rate which scales as $\sigma_{\nu N}$, the UAS rate scales as the inverse of $\sigma_{\nu N}$.[a]

In Fig. 1 we show an interesting relation between the neutrino cross-section, the neutrino's mfp in the Earth, and roughly speaking, the maximum horizontal angle for which the neutrino may transit the Earth. In this figure, the Earth has

[a] A volume detector, such as IceCube in the South Polar cap, or KM3net in the Mediterranean Sea, does not involve projection of events onto a plane, and has a UAS rate which scales as $(\sigma_{\nu N}^{\rm CC})^0$, i.e. as a constant. Nuances on this theme have been explored in Ref. 9.

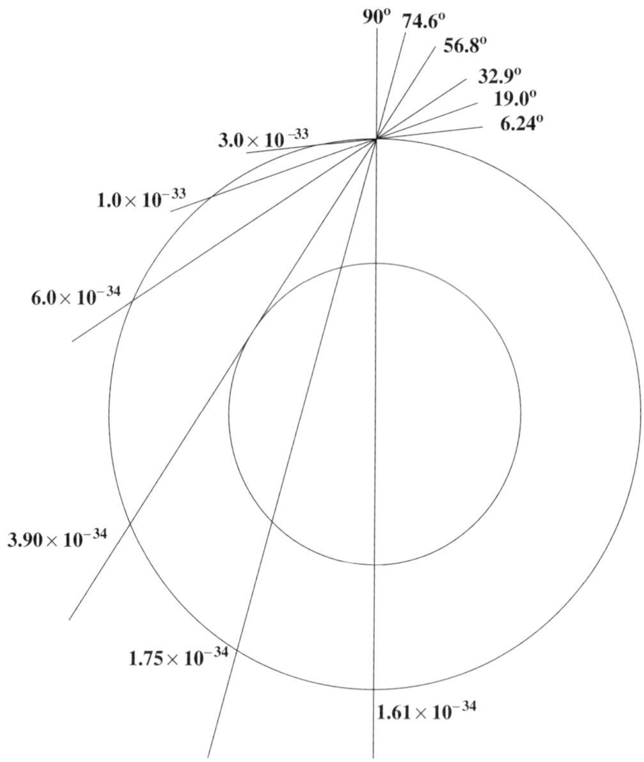

Fig. 1. Shown are neutrino trajectories for which the interaction mfp matches the chord length through the Earth. The various trajectories are parametrized values of the neutrino cross-section. Also shown is the trajectory's angle with respect to the horizontal.

been approximated according to the two-shell model. There is a central core with mean density $12\,\mathrm{g/cm^3}$ out to a radius of 3486 km, and a mantle with mean density $4.0\,\mathrm{g/cm^3}$ out to $R_\oplus = 6371\,\mathrm{km}$. The point of this figure is that, although the Earth is marginally transparent for neutrinos with the HERA cross-section of $2\times10^{-34}\,\mathrm{cm}^2$, the Earth quickly becomes opaque at a larger cross-section. The HERA accelerator presents the highest energy for which the neutrino cross-section has been measured. The HERA energy is $\sqrt{s} = 314\,\mathrm{GeV}$, which corresponds to an energy on a fixed target of $5.2\times10^{13}\,\mathrm{eV}$ (52 TeV). It is hard to imagine that $\sigma_{\nu N}^{\mathrm{CC}}$ at $10^{20}\,\mathrm{eV}$ would not have grown beyond the HERA value. A popular QCD extrapolation[10] of the known charged current neutrino cross-section to $E_\nu \sim 10^{20}\,\mathrm{eV}$ returns a value of $0.54\times10^{-31}\,\mathrm{cm}^2$. For such cross-section extrapolations to $\sim10^{20}\,\mathrm{eV}$, the concomitant horizontal angles are very small, and the trajectories are truly "Earth-skimming." The angle of the trajectory above the horizon ($\theta_{\mathrm{hor}} = \pi/2 - \theta_n$) is related to the chord length as $\sin\theta_{\mathrm{hor}} = L/2\,R_\oplus$. Setting the chord length equal to the neutrino mfp λ_ν, we get for the typical angle

$$\theta_{\mathrm{hor}} \simeq (2\,R_\oplus\,\sigma_{\nu N}\,\rho_{\mathrm{Earth}})^{-1} = 0.28^\circ/(\sigma_{31}\,\rho_{2.65})\,, \tag{1}$$

where σ_{31} is the cross-section in units of $10^{-31}\,\text{cm}^2$, and $\rho_{2.65}$ is the mean density (ρ) in units of the value for surface rock $\rho_{\text{sr}} = 2.65\,\text{g/cm}^3$.[b] The mean density of ocean water is $1.0\,\text{g/cm}^3$.

The inverse dependence of the UAS rate on $\sigma_{\nu N}$ is broken by the $\tau \rightarrow$ *shower* process in the atmosphere. As the cross-section decreases, the allowed chord length in the Earth increases, and the tau emerges with a larger angle from the Earth's tangent plane. This in turn provides a smaller path length in air in which the tau may decay and the resulting shower may evolve. This effect somewhat mitigates the inverse dependence of the UAS on $\sigma_{\nu N}$. Reference 7 provided an approximate calculation of the whole UAS process, including a simplified probability for the tau to decay in the atmosphere: $P_{\tau\text{DK}} = 1 - e^{-h/c\tau_\tau \cos\theta_n}$, where $h = 8\,\text{km}$ is the scale height of the exponentially declining atmosphere, and τ_τ is the tau lifetime in the atmosphere. The resulting approximate dependence of the HAS/UAS ratio on $\sigma_{\nu N}$ is shown in Fig. 2.

Reference 11 improved upon Ref. 7 in several ways. The energy dependences of the tau energy losses in Earth rock and, separately, in ocean water were included. For land-based detectors, only the propagation in Earth rock is relevant, whereas for space-based observation, both are relevant. It turns out, as we will see, that the event rate is predicted to be more than an order of magnitude larger over water than over land. Energy dependences of the tau lifetime in the atmosphere were included. On the issue of shower development, the dependence of atmospheric density on altitude was incorporated. Also imposed were visibility requirements for the resulting shower to satisfy experimental triggering. In the case of the upgoing showers, the pathlength of the predecayed tau may be so long that the Earth's curvature enters into the altitude dependence. The nonnegligible correction from curvature was included. Partial loss of visibility due to high cirrus or low cumulus

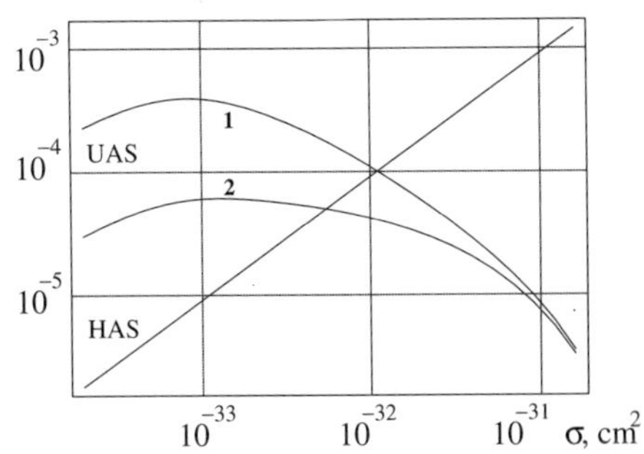

Fig. 2. The air shower probability per incident tau neutrino ($R_{\text{UAS}}/F_{\nu_\tau}\pi A$) as a function of the neutrino cross-section.[6] The incident neutrino energy is 10^{20} eV and the assumed energy threshold for detection of the UAS is $E_{\text{th}} = 10^{18}$ eV for curve 1 and 10^{19} eV for curve 2.

[b]Density, meaning number density, is usually expressed in units of g/cm^3, with the multiplicative factor of $N_{\text{A}} = 6.022 \times 10^{23}\,\text{g}^{-1}$ implicitly understood.

cloud layers was also calculated and included. For ground-based observation, it is mainly the low-lying cumulus clouds that limit visibility. For space-based observation, it is mainly the high cirrus clouds that limit visibility.[c] It is estimated that clouds will obscure the viewing area about 60–70% of the time, so the cloud study is highly relevant.

The event rate to flux ratio is known as the "instantaneous experimental acceptance" (or, sometimes, the "aperture"), with units of *area* × *solid angle*. Our results will be illustrated in a series of plots of acceptances, for ground-based and space-based experiments, for HAS and UAS events, versus the neutrino–nucleon cross-section. One has merely to multiply an experiment's acceptance by Nature's flux to arrive at an event rate for the experiment. Multiplying again by the experiment's run time (including the duty factor), one obtains the total number of events. Acceptance times run time is termed the experimental "exposure." Situations with and without cloud layers are analyzed, as are events over the solid Earth and over the ocean. Incident neutrino energies, energy thresholds for experimental detection of the air shower, and various shower trigger parameters are varied. Earth curvature effects are included in our UAS calculations. They typically reduce the event rate.

The results of Ref. 11 validate the qualitative conclusions of Ref. 7, but show quantitative differences. One common qualitative conclusion is that the HAS-to-UAS ratio is of the order of unity for cross-section values very near to the common extrapolation. This is fortunate, for it offers the best possibility that both HAS and UAS rates can be measured, and a true cross-section inferred.

We now discuss some of the physics details that enter into the rate calculations.

2.1. *Upgoing air-showers*

There are four probabilities which in convolution give the probability for an incident ν_τ to produce an observable UAS. The probabilities arise from:

(i) The ν_τ with energy E_ν must propagate without interaction to within a slant depth $w_{\rm max}$ of the Earth's surface.
(ii) Within $w_{\rm max}$ of the Earth's surface, the ν_τ must interact to produce a tau.
(iii) The produced tau, with initial energy (in the mean) $0.8\,E_\nu$, must emerge from the Earth after radiation losses with $E_\tau \geq E^\tau_{\rm th}$, where $E^\tau_{\rm th}$ is the minimum tau energy enabling a shower energy visible to the detector; this requirement determines the value of $w_{\rm max}$.
(iv) The emergent tau must decay sufficiently early in the atmosphere such that a visible shower develops in the remaining column density of atmosphere.

[c]In fact, low-lying cumulus clouds may aid in HAS identification for space-based observing. When the HAS hits the cloud layer, diffuse reflection of the forward Cerenkov cone can be seen as a one-time "Čerenkov flash." The time of the flash and the measured height of the cloud then provide the absolute (t, z) coordinates of the shower.

We have calculated these probabilities analytically, as a function of three variables, θ_n, and the production and decay sites of the tau. Then a numerical integration over these three variables was performed. In the next paragraphs we provide some relevant details that enter into these four probabilities.

First of all, the neutral current (NC) contribution to the neutrino mfp is assumed to be negligible, for three reasons. First, the NC cross-section is expected to be small compared to the charged current (CC) cross-section, as it is known to be at the lower energies of terrestrial accelerators.[d] Secondly, the NC interaction does not absorb the neutrino, but rather lowers the energy of the propagating neutrino by a small amount; in the SM, the energy loss is only $\langle y \rangle \sim 20\%$. Thirdly, the increase in complexity of our calculation, when the NC mfp is included, seems unwarranted. We also ignore multiple CC interactions due to the "tau regeneration" decay chain $\nu_\tau \to \tau \to \nu_\tau$, since the long decay length of the tau at $E_\tau > 10^{17}$ eV makes tau regeneration negligible.

It is useful to define the neutrino charged current (CC) interaction mean free path (mfp) as $\lambda_\nu = (\sigma_{\nu N}^{\rm CC} \rho)^{-1} = 63\,{\rm km}\,(\rho_{2.65}\,\sigma_{31})^{-1}$. The commonly used high energy neutrino–nucleon CC cross-section extrapolated from QCD[10] is $0.54\,\sigma_{31}$ $(E_\nu/10^{20}\,{\rm eV})^{0.363}$. As discussed in the introduction, chord lengths in the Earth matching the neutrino mfp lie on trajectories having a small angle with respect to horizontal given by Eq. (1).

The bound $w_{\rm max}(\theta_n, E_{\rm th}^\tau)$ on the depth of $w_{\rm int}$ integration is determined by the requirement that the tau emerge from the Earth with sufficient energy, $E_{\rm th}^\tau$, to produce air showers which trigger the detector apparatus. This constraint involves both β_{19} and $\rho_{\rm Earth}$, each of which depends on the Earth's composition, e.g. ocean versus land. The tau energy attenuation length is $\lambda_\tau = (\beta_\tau\,\rho)^{-1}$, where $\beta_\tau(E)$ is the coefficient in the tau energy loss equation $dE_\tau/dx = -\beta_\tau(E)\rho E_\tau$. For the energies of interest ($E_\nu > 10^{18}$ eV) tau energy losses are dominated by photonuclear processes. We find that recent calculations of $\beta_\tau(E)$ are well fitted in the energy region of interest by a simple power law, $\beta_\tau(E) = \beta_{19}\,(E_\tau/10^{19}\,{\rm eV})^\alpha$, with $\alpha = 0.2$, and the constant prefactor β_{19} scaling as $\langle A \rangle$ and equal to $1.0 \times 10^{-6}\,{\rm cm}^2/{\rm g}$ for surface rock and to $0.55 \times 10^{-6}\,{\rm cm}^2/{\rm g}$ for water. The tau energy attenuation length at $E_\tau = 10^{19}$ eV is $\lambda_\tau = 3.8$ km in surface rock, and 18 km in water. The tau decay mfp is much longer, $c\tau_\tau = 490\,(E_\tau/10^{19}\,{\rm eV})$ km, and the probability of decay within the Earth is negligible.

[d]It is an implicit assumption in this work that the ratio of the neutral to the CC cross-section is small, ~ 0.44, according to the Standard Model of particle physics. However, it is possible that above 10^{15} eV and below 10^{20} eV a threshold is passed at which the NC interaction becomes strong and the CC interaction does not. Such would be the case, for example, in models of low scale gravity unification. Crossing such a hypothetical threshold would change the physics in this paper dramatically. First of all, even though the NC interaction typically puts ~ 5 times less energy into the shower than does the ν_e CC interaction, with a much larger NC cross-section, even at fixed $E^{\rm sh}$ the NC events would dominate the ν_e CC events. Secondly, UAS acceptances would be reduced because the energy losses of neutrinos passing through the Earth would be larger.

The muon energy attenuation length is seven times smaller than that of the tau, and the electron energy attenuation length is many times smaller again. Because the energy attenuation length for a tau is an order of magnitude larger than that of a muon, UAS events are dominantly initiated by the CC interaction of tau neutrinos. Cosmic neutrinos are mostly expected to arise from the decay of pions and subsequently muons produced in astrophysical sources.[3] For this decay chain, the flavor mix at the source is devoid of ν_τ's, with $\nu_e : \nu_\mu : \nu_\tau = 1 : 2 : 0$. Fortunately, the (near) maximal mixing between ν_μ and ν_τ, and the (near) zero value of $(\sin\delta_{\rm CP}\ \sin\theta_{13})$, both inferred from terrestrial oscillation experiments, then leads after propagation for many oscillation lengths, to ν_μ–ν_τ equilibration and a flavor ratio at the Earth of $1 : 1 : 1$, i.e. "flavor democracy." Thus, a healthy ν_τ source for UAS events should exist.

The ratio of the tau energy attenuation length to the neutrino mfp $\lambda_\tau/\lambda_\nu = N_A\sigma_{\nu N}^{\rm CC}/\beta_\tau \sim (\sigma_{\nu N}^{\rm CC}/10^{-31}\,{\rm cm}^2)\times 0.06\,(0.11)$ for rock (water) is independent of ρ and only weakly dependent on tau energy. For $\sigma_{\nu N}^{\rm CC} \lesssim 2\times 10^{-30}\,{\rm cm}^2$, we expect most of the path length in the Earth (rock or water) to be neutrino; for $\sigma_{\nu N}^{\rm CC} \gtrsim 2\times 10^{-30}\,{\rm cm}^2$, we expect most of the path length to be tau.

After the tau emerges from the Earth, it must decay to produce an air shower. The tau has a 64% branching probability to decay to $\nu_\tau + hadrons$. For an unpolarized tau, $\sim 2/3$ of its energy goes into the hadronic shower. We define $E_{\rm th}^{\rm sh}$ to be the minimum energy trigger for the detector. Thus, we have the threshold relation $E_{\rm th}^\tau = \frac{3}{2}E_{\rm th}^{\rm sh}$. The tau also has 18% branching probabilities, each into $\nu+\bar\nu+e$ and $\nu+\bar\nu+\mu$. The electronic mode immediately creates an electromagnetic shower with $\sim 1/3$ of the tau energy, on average, and so $E_{\rm th}^\tau = 3E_{\rm th}^{\rm sh}$ for this mode. The muonic mode is ignorable, for the decay length of the muon exceeds the distance to the ground. In our calculation of the UAS acceptance, we weight each tau decay with 64% for the hadron mode, 18% for the electron mode, and 18% for the unobservable muon mode.

The tau must not only decay to produce an air shower, it must also decay relatively quickly to produce an air shower with time to evolve in brightness. The requirements for this are discussed in Sec. 2.3. We now turn to the derivation of the HAS event rate, must simpler than the convolution of probabilities just outlined for the UAS rate.

2.2. *Horizontal air showers*

Neutrino-induced HAS events come in several topologies.[1,2] All three neutrino flavors contribute equally to the NC events, but these transfer on average only 20% of the incident energy to the shower. Furthermore, the NC interaction rate is smaller, about 44% of the CC rate. Among the CC events, the leading muon and tau from incident ν_μ and ν_τ, respectively, are not visible in the air (unless the tau decays in a "double-bang" event). In the CC process, only 20% of the incident energy is transferred to the visible shower. For a ν_e-initiated CC event, the produced electron contributes electromagnetically to the shower, so the full incident energy converts to shower energy. In summary, about one event in four (the ν_e CC interaction) will transfer 100% of the incident energy to the shower, while three events in four will transfer $\sim 20\%$ of the energy. If the incident neutrino spectrum is falling as a power, then *at fixed energy* the ν_e CC events dominate the total rate. To be definite, we have assumed a ν_e CC interaction.[e] The horizontal air shower rate is simply

$$R_{\nu_e}(\text{HAS}) = F_{\nu_e} \int d\Omega \int \sigma_{\nu N}^{\text{CC}}\, \rho_{\text{atm}}(\mathbf{r}_{\text{int}}) d^3 r_{\text{int}}\,, \tag{2}$$

where $\mathbf{r}_{\text{int}}$ is the point of interaction. The atmospheric density function is $\rho_{\text{atm}}(z) = \rho_{\text{atm}}(0)\, e^{-z/h}$, where z is the altitude and $h = 8\,\text{km}$ is the atmospheric scale height. The absorption probability of the neutrino in the atmosphere is negligibly small. The natural scales of atmospheric column density are the vertical density

$$d_{\text{vert}} \equiv \int_0^\infty dz\, \rho_{\text{atm}}(z) = h\, \rho_{\text{atm}}(0) = 1030\,\text{g/cm}^2 \tag{3}$$

and the horizontal density

$$d_{\text{hor}} = \int_0^\infty dx\, \rho_{\text{atm}} \left(z = \sqrt{R_\oplus^2 + x^2} - R_\oplus \right) \approx \sqrt{\pi R_\oplus / 2h}\, d_{\text{vert}} = 36 d_{\text{vert}}\,. \tag{4}$$

In terms of the latter, the neutrino absorption probability in the atmosphere is

$$P(\nu - \text{air absorption}) = \sigma_{\nu N}^{\text{CC}}\, N_A\, d_{\text{hor}} \left(\frac{d}{d_{\text{hor}}} \right) = 2 \times 10^{-3}\, \sigma_{31} \left(\frac{d}{d_{\text{hor}}} \right), \tag{5}$$

where $d \le d_{\text{hor}}$ is the column density of the neutrino's trajectory in the atmosphere. Thus, for $\sigma_{\nu N}^{\text{CC}} \lesssim 10^{-29}\text{cm}^2$, atmospheric absorption is negligible even for horizontal neutrinos, and so the neutrino interaction rate scales linearly with $\sigma_{\nu N}^{\text{CC}}$.

Let us assume that the air shower must originate in the detector field of view (FOV) of area A. Then the straightforward integration of Eq. (2) gives

$$R_{\nu_e}(\text{HAS}) = 2\pi A F_{\nu_e} h \sigma_{\nu N}^{\text{CC}} \rho_{\text{atm}}(0)\,. \tag{6}$$

The value $h\sigma_{\nu N}^{\text{CC}} \rho_{\text{atm}}(0) = 0.62 \times 10^{-4}\, \sigma_{31}$ sets the scale for the interaction probability in the atmosphere per incident neutrino. The resulting value of the acceptance[f] is $\mathcal{A}cc \equiv R_{\nu_e}(\text{HAS})/F_{\nu_e} = 2\pi A h \sigma_{\nu N}^{\text{CC}} \rho_{\text{atm}}(0) = 3.9\sigma_{31}(A/10^4\,\text{km}^2)\,\text{km}^2\text{-sr}$. This value suggests that wide-angle, large-area detectors exceeding $10^4\,\text{km}^2$-sr, and cosmic neutrino fluxes exceeding $1/\text{km}^2$-sr yr, are needed for HAS event collection. Put

[e]The relevant HAS/UAS flux ratio is, therefore, the ratio of the ν_e flux to the ν_τ flux. The ν_e-to-ν_τ-flux ratio at $\sim 10^{20}$ eV is not known at present. Dynamics at the source, or new physics en route from the source, could alter the 1:1:1 flavor ratio expected from the flavor democracy theorem. *Caveat emptor*!

[f]The acceptance may also be written as $2\pi A(h/\lambda_\nu)$, where $\lambda_\nu{}^{-1} = \sigma_{\nu N}^{\text{CC}} \rho_{\text{atm}}(0)$ is the neutrino mfp. This expression is the $\lambda_\nu \gg h$ limit of $\mathcal{A}cc = 2\pi A(1 - e^{-h/\lambda_\nu})$. In this latter form, one sees the acceptance saturating its geometric value of $2\pi A$ in the strong cross-section limit.

another way, full sky coverage of an air mass of $\sim 10^5\,\text{km}^2 \times h\rho(0) \sim$ teraton is required.

2.3. *Constraints from development and identification of showers*

For the showers to be observable, some visibility conditions must be met. First of all, shower detection will require that within the FOV, the length of the shower track projected on a plane tangent to the Earth's surface (as would be seen from far above or far below) exceed some minimum length, l_{min}. We assign a relatively small value to l_{min} to maximize the observable event rate. As an example of a space-based observatory, the EUSO experiment (proposed for the ISS) maps a square kilometer of the Earth's surface onto a one pixel.[12] Thus, an l_{min} of 10 km corresponds to a signal in ten contiguous pixels. With ten contiguous pixels, the background is small and the angular reconstruction of the event direction is roughly 1/10 radian ($\sim 5°$). In addition, there are three "shower development" constraints. A minimum column density, d_{min}, beyond the point of shower initiation is required for the shower to develop in brightness. On the other hand, after a maximum column density, d_{max}, the shower particles are below threshold for further excitation of the N_2 molecules which provide the observable fluorescence signal. Visible showers terminate at d_{max}.

Finally, the fluorescent emission per unit length of the shower will decline exponentially with the air density at altitude. At $z = 2h$, the fluorescent emission is down to $e^{-2} = 14\%$ of that at sea level. At $z = 3h\,(4h)$, it is down to 5% (2%) of that at sea level. Atmospheric absorption of the emitted fluorescence also affects the signal. This absorption is thought to scale roughly as the atmospheric density, up to about 20 km.[12] Thus, it turns out that the fluorescence signal could roughly be taken as constant between zero and 20 km. Accordingly, we will take $z_{\text{thin}} = 3h = 24$ km as the "too thin" altitude beyond which the signal becomes imperceptible.

In summary, there are four constraints that render the shower observable. These are the l_{min}, d_{min}, d_{max}, and "too thin" (or z_{thin}) conditions. The d_{min} and d_{max} choices are inferred from the observed longitudinal development profiles of ultra-high energy cosmic ray showers (the famous Fly's Eye event at energy 3×10^{20} eV provides a splendid example[13]). Showers at 300–400 g/cm^2 of column density (also called "atmospheric depth" or "slant depth") comprise tens of billions of electrons, with a brightness roughly 10% of the shower maximum. The electrons in showers at $\gtrsim 1200\,\text{g/cm}^2$ are ranging out, reducing significantly the shower brightness. The values which we choose for the four shower development parameters are $z_{\text{thin}} = 3h$, $d_{\text{min}} = 400\,\text{g}\,\text{cm}^{-2}$, $d_{\text{max}} = 1200\,\text{g}\,\text{cm}^{-2}$, and $l_{\text{min}} = 10$ km or 5 km. We have also studied variations about these chosen values.

3. Determining the Neutrino Cross-Section at 10^{20} eV

In the figures to follow, the FOV and solid angle entering the acceptance calculations are taken to be those of the EUSO design report.[12] The area of this FOV is

$\pi \times (400/\sqrt{3})^2\,\text{km}^2$. The solid angle is 2π for either the HAS or the UAS events. The product of area and solid angle is then very nearly $10^6\,\text{km}^2$-sr.[g]

The cosmic neutrino flux is a matter of pure speculation at present. We will choose as our benchmark a neutrino flux which is ten times the integrated flux of cosmic rays at E_{GZK}, just below the GZK suppression. This benchmark (BM) value is

$$\frac{d\mathcal{F}_{\text{BM}}(E_\nu > E_{\text{GZK}})}{dA\,d\Omega\,dt} \equiv 10 \times \frac{d\mathcal{F}_{\text{CR}}(>E_{\text{GZK}})}{dA\,d\Omega\,dt} = \frac{1}{\text{km}^2\text{-sr yr}}\,. \tag{7}$$

The factor of ten is included to give a simple number for the benchmark flux. For this benchmark flux, an acceptance of $1\,\text{km}^2$-sr is required to yield one event per year. A popular alternative benchmark neutrino flux is that of Waxman and Bahcall,[14,15] who offered arguments relating the high energy neutrino flux to the observed high energy cosmic ray flux. They obtained

$$\frac{d\mathcal{F}_{\text{WB}}(E_\nu > E_*)}{dA\,d\Omega\,dt} = \frac{6 \times 10^{-2}\,(10^{20}\,\text{eV}/E_*)}{\text{km}^2\text{-sr yr}}\,. \tag{8}$$

A subsequent analysis of the dip structure in the cosmic ray spectrum around 10^{18}–10^{19} eV suggested an origin due to absorption of extragalactic $p+\gamma_{\text{CMB}} \to p+e^+e^-$. This in turn requires dominance of the extragalactic flux at a lower energy than previously assumed,[16,17] and a neutrino flux of extragalactic origin larger than the WB benchmark.[18] Proposed sources of a more exotic nature give still larger fluxes.

In Fig. 3 are plotted UAS (solid and dashed) and HAS (dotted) acceptances in our standard units (km^2-sr), versus fixed values of $\sigma_{\nu N}^{\text{CC}}$, for the ideal case of a cloudless sky. Five separate dependences are illustrated in this figure: UAS vs HAS; incident $E_\nu = 10^{20}$ eV (thin lines) vs 10^{21} eV (thick lines); over ocean (3.5 km uniform depth assumed) (solid lines) vs over land (dashed lines); shower threshold energy $E_{\text{th}}^{\text{sh}} = 10^{19}$ eV (upper panels) vs 5×10^{19} eV (lower panels); and minimum shower length $l_{\min} = 10$ km (left panels) vs 5 km (right panels). A sixth possible dependence is whether the shower is viewed from above by a space-based observatory, or from the below by a ground-based observatory. Within the approximations of this paper, there is no difference between the acceptances for ground-based and space-based detectors *in the cloudless case*, and only in the cloudless case.

[g] A simple estimate of the instantaneous EUSO acceptance for HAS cosmic ray events is readily obtained by multiplying this $A \times 2\pi$ value by $\frac{1}{2}$ to account for the mean projection of the FOV normal to the source. The result is a naïve HAS acceptance of $\sim 5 \times 10^5\,\text{km}^2$-sr for cosmic rays. For neutrinos, the detection efficiency is less than unity by a factor of $\sim 2h\rho(0)\sigma_{\nu N}^{\text{CC}} = 1.2 \times 10^{-4}\,\sigma_{31}$, leading to a naïve neutrino acceptance of $\sim 60\,\sigma_{31}\,\text{km}^2$-sr. The factor of 2 arises because the mean path length in the atmosphere of a neutrino is twice the vertical value. Put another way, the increased interaction probability for oblique trajectories compensates for the $\frac{1}{2}$ coming from projecting the FOV normal to the mean neutrino direction (cosines cancel). These simple HAS acceptances assume 100% detection efficiencies.

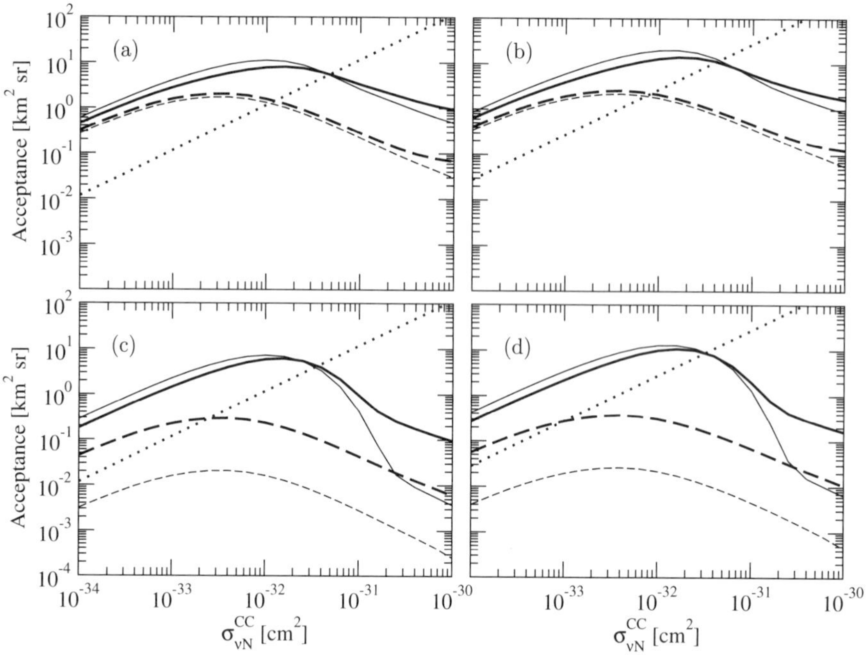

Fig. 3. Acceptances for space-based (or ground-based) detectors in the absence of clouds. Values of $d_{\min}$ and $d_{\max}$ are fixed at 400 and 1200 g/cm^2, respectively. The curves correspond to HASs (dotted line), which are independent of E_ν except through $\sigma_{\nu N}^{CC}(E_\nu)$; and UASs over ocean with $E_\nu = 10^{21}$ eV (thick solid line), ocean with $E_\nu = 10^{20}$ eV (thin solid line), land with $E_\nu = 10^{21}$ eV (thick dashed line), and land with $E_\nu = 10^{20}$ eV (thin dashed line). Panels are for (a) $E_{\rm th}^{\rm sh} = 10^{19}$ eV and $l_{\min} = 10$ km; (b) $E_{\rm th}^{\rm sh} = 10^{19}$ eV and $l_{\min} = 5$ km; (c) $E_{\rm th}^{\rm sh} = 5\times10^{19}$ eV and $l_{\min} = 10$ km; (d) $E_{\rm th}^{\rm sh} = 5 \times 10^{19}$ eV and $l_{\min} = 5$ km. For reference, a popular QCD extrapolation of the neutrino–nucleon cross-section[9] gives 0.54 and 1.2 times 10^{-31} cm^2 at $E_\nu = 10^{20}$ and 10^{21} eV, respectively; the known CC cross-section is 2×10^{-34} cm^2 at an equivalent fixed-target energy of 5×10^{13} eV, the highest energy for which measurement has been made (at HERA).

The HAS acceptance depends on neutrino energy only via $\sigma_{\nu N}^{CC}(E_\nu)$, and rises linearly with $\sigma_{\nu N}^{CC}$. The UAS acceptances have a complicated dependence on E_ν.

Several trends are evident in Fig. 3. The UAS acceptance (and thus also the rate) is typically an order of magnitude larger when neutrinos traverse ocean water, compared to a trajectory where they only cross rock. The value of this enhancement depends on the shower threshold energy $E_{\rm th}^{\rm sh}$ of the detector (upper versus lower panels) and on the cross-section in a nontrivial way. The sensitivity to $E_{\rm th}^{\rm sh}$ is partly due to the various energy transfers from the tau to the shower in the different tau-decay modes. We have remarked that for the hadronic/electronic/muonic decay modes, $\frac{2}{3}/\frac{1}{3}/0$ of the tau energy goes into the shower. The electronic mode is below the $E_{\rm th}^{\rm sh} = 5 \times 10^{19}$ eV threshold, and the hadronic mode is barely above. Both modes are above the $E_{\rm th}^{\rm sh} = 10^{19}$ eV threshold.

The benchmark flux gives a HAS rate exceeding 1/yr if $\sigma_{\nu N}^{CC}$ exceeds 10^{-32} cm^2; and a UAS rate exceeding 1/yr over water for the whole cross-section range with

$E_{\rm th}^{\rm sh} = 10^{19}$ eV, and over land if $\sigma_{\nu N}^{\rm CC} \lesssim 10^{-31}$ cm^2. When $E_{\rm th}^{\rm sh}$ is raised to 5×10^{19} eV, however, the UAS signal over land is seriously compromised, while UAS rates over the ocean are little changed.

We call attention to the fact that the shape of the UAS acceptance with respect to $\sigma_{\nu N}^{\rm CC}$ establishes the "can't-lose theorem,"[7] which states that although a large cross-section is desirable in order to enhance the HAS rate, a smaller cross-section still provides a robust event sample due to the contribution of the UAS. The latter is especially true over ocean. Importantly, the very different dependences on the cross-section of the HAS (linear) and UAS acceptances offer a practical method to measure $\sigma_{\nu N}^{\rm CC}$. One has simply to exploit the ratio of UAS-to-HAS event rates.

Finally, from the comparison of left ($l_{\rm min} = 10$ km) and right ($l_{\rm min} = 5$ km) panels, one infers a "factor of a few" sensitivity of acceptance to the experimental trigger for visible shower length. We have found that the sensitivity to the $l_{\rm min}$ trigger becomes extreme when a sky with clouds is considered. For example, with low-lying clouds such as cumulus at $z \sim 2$ km, the choice $l_{\rm min} = 10$ km (5 km) returns zero (nonzero) UAS acceptance for ground-based detectors while the space-based rates are virtually unaffected by low clouds. In Fig. 4, we continue the study of the dependence of space-based acceptances on cloud altitude, setting clouds at 4 km (thick curves) and at 12 km (thin curves). We model the cloud layer as infinitely thin, but with an infinite optical depth so that showers on the far side of the cloud layer are completely hidden. In Fig. 4 we also examine the suppressing effect of the Earth's curvature.

The symmetry between upward-looking ground-based detectors and downward-looking space-based detectors is broken by the cloud layer, so we must now show

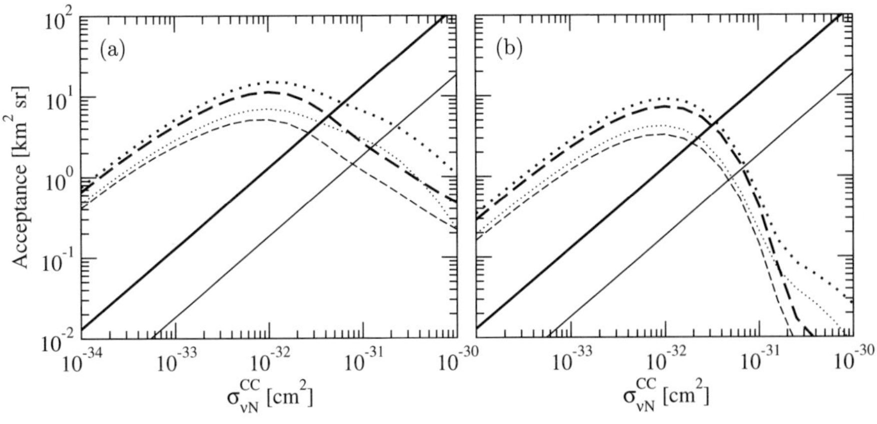

Fig. 4. Dependence of acceptance on cloud altitudes for space-based fluorescence detectors. Fixed values are $l_{\rm min} = 5$ km, and threshold energies of $E_{\rm th}^{\rm sh} = 10^{19}$ eV in the left panel and 5×10^{19} eV in the right panel. All curves representing UASs assume trajectories over water and an initial neutrino energy of 10^{20} eV; curves for HASs are valid for any energy exceeding $E_{\rm th}^{\rm sh}$. Solid lines show HASs, while dashed and dotted lines show UASs with and without Earth curvature effects, respectively. Thick lines are for a cloud layer at 4 km and thin lines for a cloud layer at 12 km.

acceptances separately for space-based (right panels) and ground-based (left panels) detectors. Results are to be compared with panels (b) and (d) of Fig. 3. We infer from comparing these figures that the effect on a space-based detector of higher cumulus clouds, and even higher cirrus clouds, is more dramatic for downgoing HASs than for upcoming UASs. The HAS acceptance is reduced by factors of ~ 1.5, 3, and 10 when the cloud layer lies at 2 km, 4 km, and 12 km, respectively. In contrast, the UAS acceptance is reduced by factors of ~ 1.5, 2, and 3 when the cloud layer lies at 2 km, 4 km, and 12 km, respectively. Since cloud layers commonly occur, they will compromise the acceptance of space-based detectors. The EUSO detector proposes to use lidar on an event-by-event basis to record the altitude of clouds.

Also shown in both panels of Fig. 4 are the UAS acceptances (dotted lines) for a flat Earth. One sees that correct inclusion of the Earth's curvature lowers the acceptance, since it puts the tau decay and the subsequent onset of shower evolution into the thinner air of higher altitudes. Curvature does little harm for smaller cross-sections, but reduces the acceptance for $\sigma_{\nu N}^{\rm CC} \gtrsim 0.5 \times 10^{-31}\,{\rm cm}^2$. The reduction of acceptance for larger cross-sections is understandable, because for larger $\sigma_{\nu N}^{\rm CC}$ the taus emerge from the Earth more horizontally, and hence travel a more lateral distance before they decay. The Earth "falls away" from the taus as (lateral displacement)$^2/2\,R_\oplus$. Beyond $\sim 10^{-31}\,{\rm cm}^2$, the reduction factor is about 2.5 for cloud layers at either 4 or 12 km (and quite different for E_ν near $E_{\rm th}^\tau$). It is dangerous to generalize that the Earth's curvature always causes event suppression, for one sees in the left panel of Fig. 4 that for high clouds, curvature effects may even *increase* the event rates.

4. Conclusions

We have calculated the acceptances of fluorescence detectors, both space-based and ground-based, for neutrino-initiated events, as a function of the unknown extreme-energy neutrino cross-section. For the downgoing HAS events, the dependence of acceptance on the cross-section is linear, but for upcoming UAS events the acceptance is quite complicated. It turns out to be somewhat flat and relatively large, which validates the "can't-lose theorem",[7] which says that if the HAS rate is suppressed by a small $\sigma_{\nu N}^{\rm CC}$, then the UAS rate compensates to establish a robust signal.

We have studied the dependence of acceptances on the incident neutrino energy, on the trigger energy $E_{\rm th}^{\rm sh}$ for the shower, on shower development parameters; and on the "environmental" conditions of cloud layers for HASs and UASs, and events over ocean versus over land for UASs. UASs typically originate at a considerable distance $[c\tau_\tau = 4900\,(E_\tau/10^{20}\,{\rm eV})\,{\rm km}]$ from the point on the Earth where the parent tau emerged. Therefore, due to the Earth's curvature, they originate at higher altitudes with thinner air. Thus, it is necessary to include the Earth's curvature in the calculation of UAS acceptances. We have done so. Clearly, lower shower trigger energies are better. This is especially true when clouds are present. We have quantified the sensitivity to $E_{\rm th}^{\rm sh}$ by comparing two realistic values, 10^{19} eV and 5×10^{19} eV, in the face of incident neutrino energies of 10^{20} eV and 10^{21} eV. Cloud layers may severely suppress acceptances. It turns out that[11] for UAS acceptances, if the shower trigger parameters can be chosen such that $(d_{\rm max} - d_{\rm min})/l_{\rm min} \geq \rho(0) = 129\,{\rm g\,cm^{-2}/km}$, then the rate suppression due to clouds is significantly minimized. We do not have the space to expand on this remark here.

Concerning UAS events over water versus over land, we find that acceptances over water are larger, typically by an order of magnitude. We have traced this enhancement to the increased path length in water of both neutrinos and taus, and to the increased path length in air for tau decay and increased column density in air for shower development, when a tau emerges with a small horizontal angle from the relatively shallow ocean. There is a further small enhancement from the fact that the atmospheric grammage over water integrates from sea level, whereas the grammage over land is often 15% less. It is difficult to imagine a ground-based detector over the ocean, so the "water advantage" clearly belongs to the orbiting space-based detectors.

We are led to three bottom-line conclusions:

(i) The "no-lose theorem" is valid, namely that acceptances are robust for the combined HAS plus UAS signal regardless of the cross-section value;
(ii) Inference of the neutrino cross-section at $\gtrsim 10^{20}$ eV from the ratio of UAS and HAS events appears feasible, assuming that a neutrino flux exists at these energies.
(iii) Space-based detectors enjoy advantages over ground-based detectors for enhancing the event rate. The advantages are a much higher UAS rate over water compared to land, and the obvious advantage that space-based FOVs greatly exceed ground-based FOVs.

Our hope is that space-based fluorescence detection becomes a reality, so that the advantages of point (iii) can be used to discover/explore extreme-energy cosmic neutrino physics. According to points (i) and (ii), part of the discovery/exploration can be the inference of the neutrino cross-section at $E_\nu \sim 10^{20}$ eV. The cross-section is sensitive to thresholds of new physics, inaccessible with terrestrial accelerators.

Acknowledgments

I wish to thank my excellent collaborators in the work presented here, namely Alex Kusenko and Sergio Palomares-Ruiz. I also acknowledge the omission of many references due to space limitations on this writeup. These references are listed in the extensive bibliography in Ref. 11. This work has been supported by NASA Grant ATP02-0000-0151 for EUSO studies, US Department of Energy Grant DE-FG05-85ER40226, and by a Vanderbilt University Discovery Award.

References

1. J. F. Beacom, N. F. Bell, D. Hooper, S. Pakvasa and T. J. Weiler, *Phys. Rev. D* **68** (2003) 093005; erratum, *ibid.* **72** (2005) 019901 [hep-ph/0307025].
2. Talks by T. DeYoung and D. Cowen at 2nd Workshop on TeV Particle Astrophysics (Madison, WI, USA; 28–31, 2006 Aug.).
3. L. A. Anchordoqui, H. Goldberg, F. Halzen and T. J. Weiler, *Phys. Lett. B* **621** (2005) 18 [hep-ph/0410003].
4. T. Kashti and E. Waxman, *Phys. Rev. Lett.* **95** (2005) 181101 [astro-ph/ 0507599].
5. J. F. Beacom, N. F. Bell, D. Hooper, S. Pakvasa and T. J. Weiler, *Phys. Rev. Lett.* **90** (2003) 181301 [hep-ph/0211305].
6. J. F. Beacom, N. F. Bell, D. Hooper, J. G. Learned, S. Pakvasa and T. J. Weiler, *Phys. Rev. Lett.* **92** (2004) 011101 [hep-ph/0307151].
7. A. Kusenko and T. J. Weiler, *Phys. Rev. Lett.* **88** (2002) 161101 [hep-ph/0106071].
8. H. Goldberg and T. J. Weiler, *Phys. Rev. D* **59** (1999) 113005 [hep-ph/9810533].
9. S. Hussain, D. Marfatia, D. W. McKay and D. Seckel, *Phys. Rev. Lett.* **97** (2006) 161101 [hep-ph/0606246].
10. R. Gandhi, C. Quigg, M. H. Reno and I. Sarcevic, *Phys. Rev. D* **58** (1998) 093009 [hep-ph/9807264].
11. S. Palomares-Ruiz, A. Irimia and T. J. Weiler, *Phys. Rev. D* **73** (2006) 083003 [astro-ph/0512231].
12. http://euso.riken.go.jp/ and http://aquila.lbl.gov/EUSO
13. Fly's Eye Collab. (D. J. Bird *et al.*), *Astrophys. J.* **441** (1995) 144.
14. E. Waxman and J. N. Bahcall, *Phys. Rev. D* **59** (1999) 023002 [hep-ph/9807282].
15. E. Waxman and J. N. Bahcall, *Phys. Rev. D* **64** (2001) 023002 [hep-ph/9902383].
16. V. Berezinsky, A. Z. Gazizov and S. I. Grigorieva, *Phys. Rev. D* **74** (2006) 043005 [hep-ph/0204357].
17. V. Berezinsky, A. Z. Gazizov and S. I. Grigorieva, *Phys. Lett. B* **612** (2005) 147 [astro-ph/0502550].
18. M. Ahlers, L. A. Anchordoqui, H. Goldberg, F. Halzen, A. Ringwald and T. J. Weiler, *Phys. Rev. D* **72** (2005) 023001 [astro-ph/0503229].

DETECTING LORENTZ INVARIANCE VIOLATIONS IN THE 10^{-20} RANGE

J. A. LIPA*, SUWEN WANG, J. NISSEN, M. KASEVICH
and J. MESTER

Physics Department, Stanford University,
Stanford, California 94305, USA
**jlipa@stanford.edu*

In recent years the possibility has been raised of Lorentz invariance violations arising from physics beyond the Standard Model. Some of these effects manifest themselves as small anisotropies in the velocity of light, c. By comparing the resonant frequencies of cavity modes with different spatial alignments, limits on the order $\delta c/c < 10^{-15}$ have been set and some further improvement can be expected. However, the largest Lorentz violations originating at the Planck scale are expected to manifest themselves as a fractional frequency variation at the 10^{-17} level in the absence of suppression factors. Space experiments have been proposed to approach the 10^{-18} level. Here we explore the possibilities for pushing further and show that it is possible in principle to reach well into the 10^{-20} range with existing technology. This could be done in a very quiet cryogenic environment, such as the drag-free orbiter being developed for the Satellite Test of the Equivalence Principle (STEP).

1. Introduction

Kostelecký and Mewes[1] have pointed out that in the Standard Model extension (SME), which describes general Lorentz violations, a number of terms exist which may lead to detectable effects in electromagnetic cavity experiments. They showed that in general the fractional beat frequency between two resonators with non-spherical mode symmetry is given by

$$\frac{\delta\nu}{\nu} = \text{Const} + A_S \sin(\omega t) + B_S \sin(2\omega t) + A_C \cos(\omega t) + B_C \cos(2\omega t), \qquad (1)$$

where ω is the rotation rate of the cavities relative to inertial space and the coefficients are linear combinations of the Lorentz-violating terms in the SME. The constant term and the other coefficients may contain slowly varying terms on the time scale of the precession rate of the rotation axis. A typical experiment that studies this beat note is then able to put bounds on the various parameters using standard statistical techniques. The expressions for the coefficients can be quite complicated: an example of the A_S term for the Superconducting Microwave Oscillator (SUMO) experiment[2] proposed for deployment on the International Space Station is as follows:

$$A_S = \frac{1}{4}\cos 2\zeta[\sin\alpha(\tilde{\kappa}_{e'})^{XZ} - \cos\alpha(\tilde{\kappa}_{e'})^{YZ}] + \frac{1}{8}\sin 2\zeta[(1+\sin^2\alpha)(\tilde{\kappa}_{e'})^{XX}$$
$$+ (1+\cos^2\alpha)(\tilde{\kappa}_{e'})^{YY} - \sin 2\alpha(\tilde{\kappa}_{e'})^{XY}] + \frac{1}{8}\beta_s[\sin\alpha((\tilde{\kappa}_{o'})^{XZ}$$
$$- (\tilde{\kappa}_{o'})^{ZX}) - \cos\alpha((\tilde{\kappa}_{o'})^{YZ} - (\tilde{\kappa}_{o'})^{YZ})] + \cdots 20 \text{ more terms.} \tag{2}$$

Here the $(\tilde{\kappa}_{e'})^{ij}$ and $(\tilde{\kappa}_{o'})^{ij}$ terms are matrix elements constructed from linear combinations of the coefficients of Lorentz violation, α is the azimuthal angle at which the orbital plane intersects the Earth's equatorial plane, ζ is the angle between the orbital plane of the spacecraft and the plane of the ecliptic, and β_s is the ratio of the spacecraft velocity to the speed of light. The coordinate system used in the expression is Sun-centric, with inertially aligned axes.[1] As pointed out by Kostelecký and Potting,[3] in some circumstances Lorentz violations originating at the Planck scale may manifest themselves as a fractional frequency variation at the 10^{-17} level in the absence of suppression factors. Thus measurements to the 10^{-20} range could result in some interesting new physics that goes beyond the Standard Model and would of course have implications for cosmology.

Since the development of SUMO a number of advances have occurred in the optics field and the tightest limits set on the coefficients of Lorentz violation are now obtained from optical etalons.[4] A space experiment called OPTIS, using optical etalons, has been proposed[5] to measure $\delta c/c \sim 3 \times 10^{-18}$, comparable with what could be achieved with SUMO in a free flier. Cost considerations tend to rule out dedicated cryogenic free fliers for Lorentz violation experiments, but a shared spacecraft is a viable alternative. It is therefore reasonable to ask what the basic limitations would be for an experiment in a spacecraft such as STEP[6] which is expected to have a very quiet, drag-free environment.

2. Experiment Considerations

The operation of optical etalons at cryogenic temperatures has been studied fairly extensively, motivated by the search for gravitational radiation.[7] Figure 1 shows a conceptual view of the optics required for the experiment. Liu *et al.*[8] showed that Allan deviations on the order of $\sigma_{\rm SN} \sim 4 \times 10^{-19}/\sqrt{\tau}$, where τ is the measurement time, could be achieved for a shot-noise-limited detection method with a sapphire etalon. These authors estimated that Brownian length fluctuations should set a limit of $\sigma_{\rm BM} \sim 3.7{\times}10^{-18}/\sqrt{\tau}$. However, they assumed a very conservative mechanical $Q = 10$ in their calculation, whereas at low temperatures $Q \sim 10^4$ is not uncommon. This would imply that $\sigma_{\rm BM} \sim 10^{-19}/\sqrt{\tau}$ is more realistic. Thus for $\tau > 100\,$s, it is

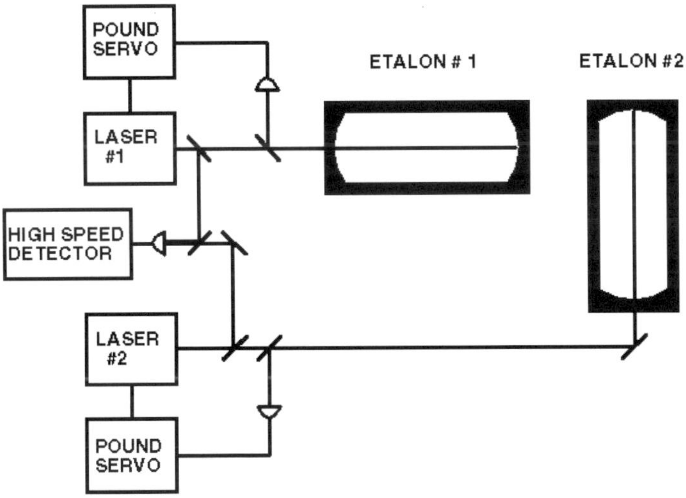

Fig. 1. Conceptual design of experiment optics. Etalons may be cut from a monolithic block.

easy to see that it is possible to detect a sinusoidal signal in the 10^{-20} range, given an etalon with sufficient mechanical stability.

A more serious issue common to the design of essentially all ultrastable oscillator experiments is thermal control. Typically a design with a frequency turning point as a function of temperature is preferred, but failing that, a very small expansion coefficient is essential. At low temperatures, many materials exhibit a Debye form of thermal expansion, with the coefficient given by $\alpha = A_E T^3$, where T is the temperature and A_E is inversely proportional to the cube of the Debye temperature. Thus we are led to consider materials with high Debye temperatures operating well below 100 K, where the Debye approximation is valid. It is well known that sapphire is an excellent material in this regard, with an exceptionally low value[7] of $A_E \sim 5.3 \times 10^{-13}$. For comparison,[7] niobium has $A_E \sim 4.3 \times 10^{-11}$. In space, cryogenic temperature control is typically limited by thermal gradient effects due to charged particle heating variations as a spacecraft traverses its orbit. Thus the thermal conductivity and energy absorption cross-section are key parameters and careful heat sinking to a controlled thermal node is an important design consideration. For a near-polar orbit a typical niobium cavity would exhibit a frequency fluctuation of $\sim 2.5 \times 10^{-15}$ at twice the orbital period due to cosmic ray flux variations, unless special precautions were taken. For a sapphire etalon, the corresponding fluctuation would be substantially less, $\sim 3.3 \times 10^{-20}$. Fortunately these effects can to a large extent be discriminated against at the spacecraft level by choosing an angular frequency, ω, that is asynchronous with, and higher than, the orbital angular frequency, ω_0. The thermal controller itself also gives rise to frequency fluctuations, primarily due to detector noise. With the use of bolometer-style thermometry, temperature variations should easily be controllable to < 10 nK without excessive demands on components. This implies a thermal signal limit of $\delta\nu/\nu \sim 3.4\times10^{-18}$ in niobium and $\sim 4.4\times10^{-20}$ in sapphire. Another two orders of magnitude of stability could be obtained by the use of very high-resolution magnetic thermometry based on paramagnetic materials.[9] At this level other factors would come into play, such as the stability of the frequency tracking and differencing schemes. The practical limit will most likely come from thermal effects in the electronics due to temperature variations at the roll rate. With a Sun-synchronous roll axis, the thermal environment of the electronics would be relatively benign, easing the intermediate frequency oscillator requirement. Also, the residual thermal fluctuations from Earth-shine will mostly be at $\omega+\omega_0$, again easing the requirements at roll frequency and its harmonics.

An issue yet to be fully addressed is the mechanical stability of an etalon which typically results in frequency drift. We note that sapphire has already been shown to be extremely promising in this regard, with an upper limit of $\delta\nu/\nu < 9\times10^{-20}$ per sec being observed,[4] with no special care being taken. Since sapphire is an anisotropic material, preserving the relative alignment of the crystal axes in the various components would help reduce stress from the thermal cycling. On the Gravity Probe B experiment, it was found that by cutting all fused quartz components from the same boule of material and preserving the relative locations of components cut from the boule, it was possible to reduce thermal stresses by about two orders of magnitude.[10] It would be a simple matter to extend this technique to include control of the relative rotation of the components.

We now briefly consider the possibility of operating sapphire etalons in a very quiet, drag-free cryogenic environment such as that provided by the STEP spacecraft. In Fig. 2 we show a conceptual view of the motion of a dual etalon system within the STEP spacecraft for a single orbit. For the gravity gradient contribution to the beat note, we estimate $\delta\nu/\nu \sim 10^{-17}$ at a frequency of $4(\omega-\omega_0)$, assuming an offset of 1 cm from the roll axis. Fortunately, this would be a very pure spectral line at a frequency that can easily be avoided for the Lorentz violation measurements. In this case the Lorentz violation signals would be at twice and four times the roll frequency rather than as indicated by Eq. (1). A judicious choice of roll frequency, asynchronous with orbital frequency, would greatly aid the data analysis. With the nominal rate of three rolls per orbit for the STEP spacecraft, the minimum value of $\sqrt{\tau}$ is $15.4\,\mathrm{s}^{1/2}$, reducing the laser-related noise sources to very low levels. Further noise reductions of an order of magnitude can be obtained by averaging over 100 roll periods, less than two days of mission time. This would result in all noise sources except for discrete spectral lines being reduced to the low 10^{-20} fractional frequency range. With a short signal averaging time compared to the Earth's orbital period, it is possible to take advantage of the slow rotation of the orbit plane during the six-month mission lifetime to probe those coefficients of Lorentz violation that were not accessed by the roll motion, as is done in some ground experiments.[4] This orbital plane rotation is depicted in Fig. 3, which shows the orbital motion of the etalons at different times of the year.

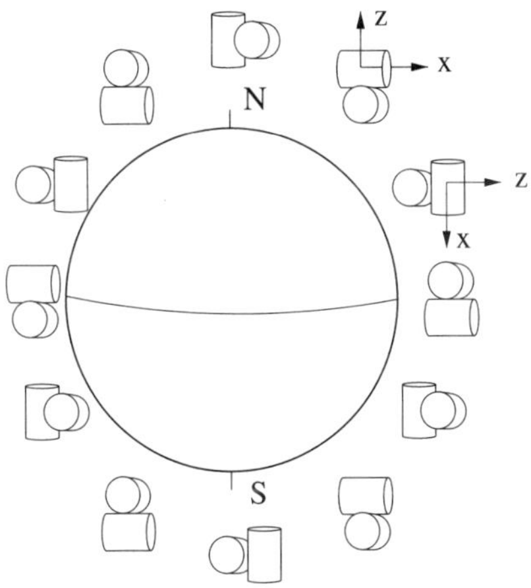

Fig. 2. The etalon orientation in 12 positions as the STEP spacecraft circles the Earth in a near-polar orbit making three rotations per orbit. The x axis lies along the optical axis of one of the etalons. The y axis (pointing out of the page) lies along the roll axis of the spacecraft as well as the optical axis of the other etalon.

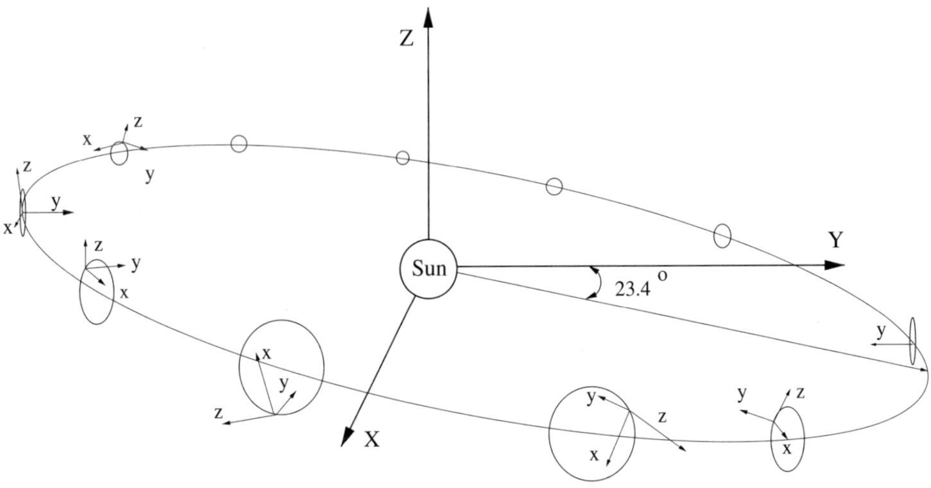

Fig. 3. The orientation of the orbit of the STEP spacecraft in several positions as the Earth revolves around the Sun. In the Sun-centered celestial equatorial frame, the Z axis points north along the axis of rotation of the Earth, the X axis points toward the vernal equinox and the Y axis completes the right-handed coordinate system. In this frame the plane of the Earth's orbit is tilted by 23.4°. The roll axis of the STEP spacecraft (y axis) always points toward the Z axis of the Sun-centered frame, while the x axis rotates in the Earth-orbit plane of the spacecraft.

The Lorentz violation tests considered here could easily be extended by adding a high quality atomic clock such as the rubidium clock being developed by Gibble.[5] If anisotropic atomic transitions were used, other sectors of the SME could be probed. Also, an intercomparison of the two types of clocks as a function of gravitational potential could give a much-improved bound on the differential redshift.

Acknowledgment

We wish to thank NASA for its support with grant #NAG3-2852.

References

1. V. A. Kostelecký and M. Mewes, *Phys. Rev. D* **66** (2002) 0056005.
2. J. A. Lipa *et al.*, *Adv. Space Res.* **35** (2005) 82.
3. V. A. Kostelecký and R. Potting, *Phys. Rev. D* **51** (1995) 3923.
4. H. Muller *et al.*, *Phys. Rev. Lett.* **91** (2003) 20401.
5. C. Lammerzahl *et al.*, *Class. Quant. Grav.* **18** (2001) 2499.
6. J. Mester *et al.*, *Class. Quant. Grav.* **18** (2001) 2475.
7. J. P. Richard, J. J. Hamilton and Y. Pang, *J. Low Temp. Phys.* **81** (1990) 189.
8. R. Liu, S. Schiller and R. L. Byer, Stanford internal report (1993).
9. X. Qin *et al.*, *Czech. J. Phys.* **46**, Suppl. S1 (1996) 2857.
10. S. Wang *et al.*, *Proc. 33rd COSPAR Meeting* (Warsaw, July 2000).

LIGHT SUPERCONDUCTING STRINGS IN THE GALAXY

FRANCESC FERRER and TANMAY VACHASPATI*

CERCA, Department of Physics,
Case Western Reserve University,
10900 Euclid Avenue, Cleveland,
OH 44106-7079, USA
**tanmay@monopole.phys.cwru.edu*

Observations of the Milky Way by the SPI/INTEGRAL satellite have confirmed the presence of a strong 511 keV gamma ray line emission from the bulge, which requires an intense source of positrons in the galactic center. These observations are hard to account for by conventional astrophysical scenarios, whereas other proposals, such as light DM, face stringent constraints from the diffuse gamma ray background. Here we suggest that light superconducting strings could be the source of the observed 511 keV emission. The associated particle physics, at the $\sim$ 1 TeV scale, is within the reach of planned accelerator experiments, while the distinguishing spatial distribution, proportional to the galactic magnetic field, could be mapped by SPI or by future, more sensitive satellite missions.

Keywords: Positrons; strings; galaxy.

1. Positron Sources in the Galaxy

The problem of the birth, propagation and annihilation of positrons in the Galaxy has been a major topic of astrophysical investigation, since the first detection[1] of the 511 keV gamma ray line signature. The SPI instrument on board ESA's INTEGRAL satellite has established the presence of a diffuse source of positrons in the galactic center (GC).[2–7] The observed photon flux of

$$9.9^{+4.7}_{-2.1} \times 10^{-4}\ \mathrm{cm}^{-2}\,\mathrm{s}^{-1} \tag{1}$$

with a line width of about 3 keV is in good agreement with previous measurements.[8]

For the spatial distribution of the 511 keV line component, the mapping results point to an intense bulge emission, better explained by an extended distribution than by a point source. Assuming a Gaussian spatial distribution for the flux, a full width at half maximum of 9° is indicated. The disk component has been either absent, or weakly detected in the initial results.

The origin of these galactic positrons remains a mystery. Several scenarios involving astrophysical sources have been proposed, including neutron stars or black holes, massive stars, supernovae, hypernovae, gamma ray bursts or cosmic rays.[9–13] However, the fraction of positrons produced in such processes is uncertain, and it is unclear that the positrons could fill the whole bulge.

Alternatively, mechanisms associated with the dark matter (DM) at the GC have been put forward. If DM is constituted by a light (MeV) scalar, its decay or annihilation could account for the observed signal.[14–18]

The positrons should be injected at nonrelativistic energies so that the associated bremsstrahlung emission does not violate the COMPTEL and EGRET measurements of diffuse radiation from the GC.[19–21] For the DM scenario, this implies that the DM particles should be lighter than $\sim$ 20–30 MeV. Moreover, inflight annihilation of positrons would also overproduce gamma rays unless the positrons are injected at energies below $\sim 3\,\mathrm{MeV}$,[22] thus disfavoring some of the scenarios in Refs. 9–18.

We will discuss here the possibility, proposed in Ref. 23, that a network of light superconducting strings[24] occurring in particle physics just beyond the standard model could be a source of the galactic positrons. This scenario predicts a characteristic positron distribution that could be used to distinguish this source from the other possibilities.

Assuming that a tangle of superconducting strings exists in the Milky Way, then the strings are frozen in the plasma as long as the radius of curvature is larger than a certain critical length scale. If the curvature radius is smaller, the string tension wins over the plasma forces and the string moves with respect to the magnetized plasma. During the string motion, the loop will cut across the Milky Way magnetic field, generating current as given by Faraday's law of induction.

The current is composed of zero modes of charged particles, including positrons, propagating along the string. The external magnetic field shifts the modes of the charge carriers into the bulk and modifies the dispersion relation dramatically so that their energy remains below the threshold for expulsion[25] (which for the positron zero modes is 511 keV). An additional perturbation, like inhomogeneities in the magnetic field, string motion and curvature, or scattering by counterpropagating particles,[26,27] ejects the zero modes at the threshold of 511 keV. The ejected positrons will annihilate with the ambient electrons, thus emitting 511 keV gamma rays.

2. Light Superconducting Strings in the Galaxy

The amount of positrons injected in the Milky Way will depend on how many strings are injecting positrons per unit volume and on the output rate of positrons per unit length of string. Let us first estimate the density of strings in the Galaxy.

The strings, being superconducting, can sustain currents which couple the dynamics of the string network to the Milky Way plasma. The density of strings will, thus, depend on the properties of both the string (like its tension, μ, radius of curvature, R, and the intensity of the current being carried, J) and the plasma (like its density, ρ).

The dynamics is determined by comparing the force due to string tension, F_s, with the plasma drag force, F_d. The analysis[28,29] shows that there is a critical radius of curvature,

$$R_c \sim \frac{\mu}{\sqrt{\rho}J}, \tag{2}$$

such that the plasma drag cannot check the force due to the string tension when $R < R_c$, and the strings move at relativistic speeds. String loops will then emit electromagnetic radiation and eventually dissipate.

On the other hand, less curved strings, i.e. for $R > R_c$, accelerate under their own tension until they reach a terminal velocity,

$$v_{\mathrm{term}} \sim \frac{\mu}{\sqrt{\rho}JR}. \tag{3}$$

In a turbulent plasma, such as in our Milky Way, there is another length scale of interest, called R_* ($R_* > R_c$), even when the string motion is overdamped. For $R > R_*$, the terminal speed of the strings is small compared to the turbulence speed of the plasma and the strings are carried along with the plasma. As the strings follow the plasma flow, they get more entangled due to turbulent eddies, and get more curved until the curvature radius drops below R_*. Then the string velocity is large compared to the plasma velocity, and hence the strings break away from the turbulent flow. Therefore, R_* is the smallest scale at which the string network follows the plasma flow. For $R_* > R > R_c$, the string motion is overdamped but independent of the turbulent flow. Hence, string curvature on these scales is not generated by the turbulence, and we can estimate the length density of strings in the plasma as $\rho_l \sim 1/R_*^2$. The scale R_* at which the terminal velocity (3) equals the turbulent velocity of the plasma, v_*, is given by[30]

$$R_* \sim l\left(\sqrt{\frac{\mu}{\rho}}\frac{1}{e\kappa v_l l}\right)^{4/5}, \quad v_* \sim v_l\left(\sqrt{\frac{\mu}{\rho}}\frac{1}{e\kappa v_l l}\right)^{1/5}, \tag{4}$$

where, for convenience, the dimensionless parameter κ has been introduced via $J \equiv \kappa e\sqrt{\mu}$.

3. Particle Emission by Superconducting Strings

When a string of typical length R_* moves at a velocity v_* with respect to the Milky Way plasma, it cuts across the galactic magnetic field lines and a current is generated on the strings, as described by Faraday's law of induction.

The same magnetic field that creates the current, changes the dispersion relation of the zero modes. In the absence of the external magnetic field, the zero modes behave as massless particles, with a linear dispersion relation. The current could increase indefinitely as the zero modes gain momentum. However, the presence of the external magnetic field changes the dispersion relation of the zero modes dramatically[25] and it can be approximated by

$$\omega_k = m_{e^+} \tanh\left(\frac{k}{k_*}\right), \tag{5}$$

where k_* is a parameter that depends on the magnetic field. Consequently, the current on the string saturates at $J_{\max} = em_{e^+}$, and the positron zero modes have energies approaching from below the threshold for expulsion, 511 keV.

The string will, in general, carry additional zero modes corresponding to other, heavier particles — say, heavy quarks. Then, the total current in the string, entering the network dynamics in Eq. (4), will be determined by the heavier particles. The positron current, though, will still be bounded to be below 511 keV.

A given charge carrier can, in principle, leave the string once it has enough energy. The escape is triggered by several factors, such as string motion and curvature or scattering by counterpropagating particles[26,27] (u quarks for electroweak strings[23]), but in any case the positrons will be emitted at their threshold energy, 511 keV.

As shown in Fig. 1, when a piece of string of length R_* cuts through a magnetic field B, it will produce electrons or positrons, with equal likelihood, at the rate

$$\frac{dN}{dt} \sim ev_* BR_*. \tag{6}$$

In a volume $V = 4\pi L^3/3$, there are of order L^3/R_*^3 such pieces of string and hence the rate of particle production in the entire volume is

$$\frac{dN_V}{dt} \sim ev_* B\frac{L^3}{R_*^2}. \tag{7}$$

The current in the positrons will grow at first, but then saturate at 511 keV. After that, further motion of the string across the galactic magnetic field will

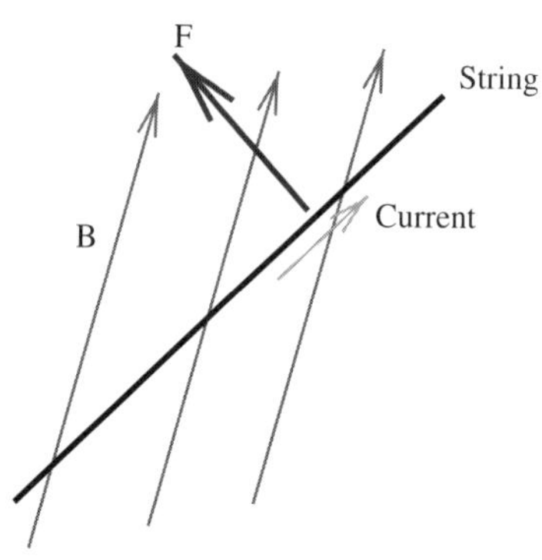

Fig. 1. The charge carriers run along the string, in the presence of a perpendicular external magnetic field.

generate positrons that leave the string. So N_V is also the number of positrons being produced in the volume V, which we denote by N_+. Inserting Eqs. (4) in (7) we get

$$\frac{dN_+}{dt} \sim 10^{42} B_3 \kappa^{7/5} V_1 \mu_1^{-7/10} \rho_{\rm GC}^{7/10} v_{l,6}^{12/5} l_{100}^{-3/5}\ {\rm s}^{-1}, \tag{8}$$

where we have introduced the parameters describing the plasma of the spherical region of radius 1 kpc around the GC, $\rho \sim 6 \times 10^{-24} \rho_{\rm GC}\,{\rm gm/cm}^3$, $B = B_3 10^{-3}\,{\rm G}$, $v_l = 10^6\ v_{l,6}\,{\rm cm/s}$ and $l = 100\ l_{100}$ pc. The string tension is given by $\mu = \mu_1$ $(1\ {\rm TeV})^2$.

Although the astrophysical parameters describing the GC are not known very accurately, assuming equipartition of plasma kinetic energy ($\sim \rho v_l^2$) and magnetic energy ($\sim B^2/8\pi$), with $l \sim \rho^{-1/3}$, we find that $v_{l,6} \sim 100$ and $l_{100} \sim 0.1$, which boosts the estimate in Eq. (8) by 10^5, yielding

$$\frac{dN_+}{dt} \lesssim 10^{47}\,{\rm s}^{-1}. \tag{9}$$

This should be compared with the actual positron production rate in the GC:

$$\frac{dN_+^{\rm obs}}{dt} \sim 1.2 \times 10^{43}\ {\rm s}^{-1}. \tag{10}$$

Comparing Eqs. (9) and (10) we conclude that light superconducting strings are possible sources of positrons that lead to the flux of 511 keV gamma rays observed by the INTEGRAL collaboration.

4. Observational Signatures

We see from Eq. (8) that a unique prediction of our scenario is that the gamma ray flux is proportional to the strength of the magnetic field in the Milky Way, with a milder dependence on the plasma density. In the disk, $B_3 \sim 10^{-3}$, and we estimate a photon flux of $\sim 10^{-6}\,{\rm cm}^{-2}\,{\rm s}^{-1}$ in a 16° field of view as in SPI. The target sensitivity of the SPI instrument, once sufficient exposure becomes available, is $2 \times 10^{-5}\ {\rm cm}^{-2}\,{\rm s}^{-1}$ at 511 keV.[31] This threshold is somewhat above what is needed to map the emission from the disk in our scenario.

That the flux should follow the magnetic field is in marked contrast with the MeV DM hypothesis. There, the flux follows $\rho_{\rm DM}^2$, and a signal from nearby DM-dominated regions, e.g. the Sagittarius dSph galaxy, is expected.[32] If superconducting strings source the observed 511 keV, however, at most a flux of $\sim 10^{-7}\ {\rm cm}^{-2}{\rm s}^{-1}$ in the direction of Sagittarius is expected,[23] some three orders of magnitude fainter than the MeV DM model prediction

The strings are expected to carry additional zero modes apart from the ones corresponding to $e^{\pm}$. We expect, thus, the presence of other currents, each saturated at the mass m_X of the particle in vacuum, which would potentially result in the ejection of these particles also at the threshold, although the presence of conserved charges might inhibit or delay some of these processes. For instance, since pion emission cannot deplete the baryonic current on the string, only at ~ 1 GeV energies can antiprotons be emitted, leading to another possible signature of galactic superconducting strings.[33]

Our scenario could also explain the excess of high-energy positrons in cosmic rays at energies around 10 GeV detected by the HEAT balloon experiment.[34] Since the positron current cannot build up beyond 511 keV in the presence of the external magnetic field, additional heavy-charged fermions would be responsible, after decaying or annihilating with ambient particles in the plasma, for the positrons in the 10 GeV energy range.[23]

5. What Can NASA Do to Check These Predictions?

The experimental results obtained by the SPI/INTEGRAL collaboration have confirmed the puzzle of the positron injection in the galactic bulge and sparked, in the few years since the publication of the first results, a plethora of possible explanations.

The distinguishing feature of our scenario is the spatial distribution tracking the magnetic field intensity. SPI will not be able to attain the sensitivity required to observe the emission from the galactic disk in our model. An improvement of roughly an order of magnitude in the sensitivity would suffice to pursue this task.[a] This order-of-magnitude improvement would also test predictions from other scenarios. For instance, the signal from nearby DM clumps expected in light DM scenarios could be unveiled, or else the models would be disproved (barring astrophysical uncertainties in the region of the clumps).

It is noteworthy that the analysis of complementary data coming from older satellites like COMPTEL and EGRET provides some of the most stringent constraints for all the scenarios.[19–22] In this respect, light superconducting strings remain a viable proposal, since the positrons are emitted at the threshold, well below the ~ 3 MeV bounds from inflight annihilation. It should be stressed that these bounds require knowledge of the diffuse gamma ray flux in the GC to a great accuracy. With the data at hand, extrapolations of data at different energies and from different regions are necessary, which add uncertainty to the bounds. The forthcoming GLAST satellite, partially funded by NASA, is better suited to energies above 1 GeV. The task remains to get more precise data at lower energies. SPI is already contributing to this effort, but a more precise experiment would make a difference.

References

1. W. N. Johnson III, F. R. Harnden Jr. and R. C. Haymes, *Astrophys. J.* **172** (1972) L1.
2. P. Jean *et al.*, *Astron. Astrophys.* **407** (2003) L55.

[a]Other sources, e.g. cosmic rays, could contribute to the emission from the disk at a level that could be mapped by SPI.

3. Knödlseder *et al.*, *Astron. Astrophys.* **411** (2003) L457.
4. G. Weidenspointner *et al.*, astro-ph/0601673.
5. P. Jean *et al.*, *Astron. Astrophys.* **445** (2006) 579.
6. J. Knodlseder *et al.*, *Astron. Astrophys.* **441** (2005) 513.
7. B. J. Teegarden *et al.*, *Astrophys. J.* **621** (2005) 296.
8. P. A. Milne *et al.*, *New Astron. Rev.* **46** (2002) 553, and references therein.
9. P. A. Milne, L. S. The and M. D. Leising, astro-ph/0104185.
10. M. Casse *et al.*, *Astrophys. J.* **602** (2004) L17.
11. G. Bertone *et al.*, *Phys. Lett. B* **636** (2006) 20.
12. N. Prantzos, *Astron. Astrophys.* **449** (2006) 869.
13. N. Guessoum, P. Jean and N. Prantzos, astro-ph/0607296.
14. C. Boehm *et al.*, *Phys. Rev. Lett.* **92** (2004) 101301.
15. D. Hooper and L. T. Wang, *Phys. Rev. D* **70** (2004) 063506.
16. D. H. Oaknin and A. R. Zhitnitsky, *Phys. Rev. Lett.* **94** (2005) 101301.
17. S. Kasuya and M. Kawasaki, *Phys. Rev. D* **73** (2006) 063007.
18. S. Kasuya and F. Takahashi, *Phys. Rev. D* **72** (2005) 085015.
19. J. F. Beacom, N. F. Bell and G. Bertone, *Phys. Rev. Lett.* **94** (2005) 171301.
20. P. Sizun, M. Casse and S. Schanne, astro-ph/0607374.
21. C. Boehm and P. Uwer, hep-ph/0606058.
22. J. F. Beacom and H. Yuksel, astro-ph/0512411.
23. F. Ferrer and T. Vachaspati, *Phys. Rev. Lett.* **95** (2005) 261302.
24. E. Witten, *Nucl. Phys. B* **249** (1985) 557.
25. F. Ferrer, H. Mathur, T. Vachaspati and G. D. Starkman, *Phys. Rev. D* **74** (2006) 025012.
26. S. M. Barr and A. M. Matheson, *Phys. Lett. B* **198** (1987) 146.
27. S. M. Barr and A. M. Matheson, *Phys. Rev. D* **39** (1989) 412.
28. E. M. Chudnovsky, G. B. Field, D. N. Spergel and A. Vilenkin, *Phys. Rev. D* **34** (1986) 944.
29. A. Vilenkin and E. P. S. Shellard, *Cosmic Strings and Other Topological Defects* (Cambridge University Press, 1994).
30. E. Chudnovsky and A. Vilenkin, *Phys. Rev. Lett.* **61** (1988) 1043.
31. http://smsc.cnes.fr/SPI.
32. D. Hooper *et al.*, *Phys. Rev. Lett.* **93** (2004) 161302.
33. G. D. Starkman and T. Vachaspati, *Phys. Rev. D* **53** (1996) 6711.
34. S. Coutu *et al.*, *Astropart. Phys.* **11** (1999) 429–435.
35. S. W. Barwick *et al.*, *Astrophys. J.* **482** (1997) L191.

ADVANCED HYBRID SQUID MULTIPLEXER CONCEPT FOR THE NEXT GENERATION OF ASTRONOMICAL INSTRUMENTS

I. HAHN*, P. DAY, B. BUMBLE and H. G. LEDUC

Jet Propulsion Lab, California Institute of Technology,
4800 Oak Grove Dr, Pasadena, CA 91109 8099, USA
**Inseob.Hahn@jpl.nasa.gov*

The Superconducting Quantum Interference Device (SQUID) has been used and proposed often to read out low-temperature detectors for astronomical instruments. A multiplexed SQUID readout for currently envisioned astronomical detector arrays, which will have tens of thousands of pixels, is still challenging with the present technology. We present a new, advanced multiplexing concept and its prototype development that will allow for the readout of 1,000–10,000 detectors with only three pairs of wires and a single microwave coaxial cable.

Keywords: SQUID; multiplexer; bolometer.

1. Introduction

A leading candidate detector array technology for the next generation of astronomical instruments for millimeter though far-infrared (FIR) wavelengths is the transition edge sensor (TES), which is being developed intensively at several laboratories, including JPL. The readout of TES bolometers is accomplished using SQUID amplifiers, for which multiplexing techniques have been developed that serve to reduce the number of wires needed between the cryogenic detector arrays and the warm electronics. In current SQUID multiplexers, the outputs of a small (8–32 elements) array of bolometers are encoded in either the time or frequency domain, and the combined set of signals is amplified by a relatively-high-bandwidth "series array" SQUID amplifier. The outputs of these series arrays are read out in a nonmultiplexed fashion, and hence the saving in wire count is a factor of 8–32, enabling arrays of thousands of detectors, but still requiring hundreds of wires. We propose to develop a new SQUID multiplexer that makes use of a technique recently demonstrated in our group for reading out a SQUID at very high frequency (~ 10 GHz). The "microwave SQUID" (MSQUID) has a greater bandwidth than the series array amplifier and its output is itself multiplexable. In our new multiplexer architecture, we use a microwave SQUID in place of the series array SQUID used in current designs. Each MSQUID can read out 100 detectors because of its greater bandwidth, and we can multiplex the output of at least 100 MSQUIDs. In this way we can achieve a significant increase in the multiplexing factor, potentially 10,000 detectors with a single set of wires.

To demonstrate the feasibility of the new concept, we are developing a prototype device. A series of microwave resonators with frequencies $\sim 10\,\text{GHz}$ are each loaded by a dc SQUID to a degree that depends on the flux state of the SQUID. By using resonators with high quality factors and slightly different resonance frequencies, many of these resonator-coupled SQUID's may be read out with a single excitation line and cryogenic amplifier. Recent noise measurements of the device demonstrated $\sim 5\mu\Phi_0/\sqrt{\text{Hz}}$ performance at 4.2 K. We also present a new technique for modulating the SQUID array in series that alleviates the need to individually flux-bias the SQUID's. The new MSQUID device has applications to the readout of detector arrays for astronomy and fundamental physics experiments in space.

2. Principles and Design

SQUID multiplexers have been demonstrated using both time and frequency division schemes.[1–2] Most recently, Irwin and Lehnert[3] have demonstrated a frequency division multiplex technique using an array of SQUID's operated at microwave frequency (600 MHz). In the microwave SQUID multiplexer, each SQUID is part of a resonant circuit with a unique resonance frequency. A comb of microwave frequencies is used to simultaneously excite all of the resonant circuits of the array. The quality factor, Q, of the resonance varies as the flux state of the dc SQUID changes. Typically SQUID readout electronics employ feedback to keep the flux in the SQUID loop at a sensitive part of the modulation function and to linearize the output. Despite the large bandwidth advantage of the new microwave technique, separate feedback lines for each SQUID of a large array would ultimately be impractical. To overcome this difficulty, we propose to operate the SQUID multiplexer system in non-flux-locked mode. Normally non-flux-locked operation is hampered by the periodic nature of the SQUID response function, which limits the dynamic range and leads to the possibility that stray magnetic fields bias the SQUID at a point of degraded sensitivity. We propose to circumvent this problem by applying a high frequency modulation to all of the SQUID's in series, which can eliminate the need for a multiplexed flux-biasing circuit. Figure 1 shows a schematic of the microwave SQUID multiplexer and a photo of a device that was recently designed and fabricated at the Microdevices Laboratory at JPL. This device contains four SQUID's with resonant circuits fed with a single microwave readout line. The SQUID's are ac-coupled to the meandering half-wave resonator circuits similar to the structures used to multiplex microwave kinetic inductance detectors.[4] Optimal values for the interdigital gap capacitors were determined to maximize the sensitivity of Q with respect to changes in the dynamic resistance of the SQUID. The dc current bias line

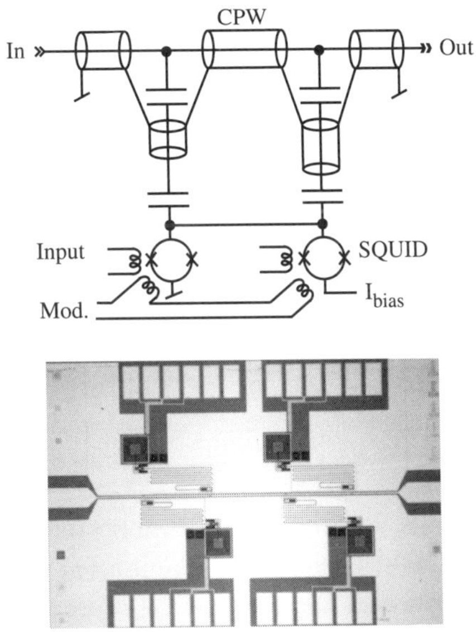

Fig. 1. Schematic of the microwave SQUID multiplexer and a photo of the prototype device. The device can be easily scalable for a large size array. Individual SQUID chips share common current bias and modulation. The SQUID multiplexer is operating in the open-loop mode such that the feedback line for individual SQUID's is not necessary. The size of the chip is 3 mm × 5 mm.

for the SQUID has an in-line inductive high frequency filter to minimize coupling between the bias line and the resonator. The coplanar wave guide (CPW) design is similar to the one used by P. Day *et al.*[4]

The main components of our multiplexer are (i) a set of tank circuits used for the ac bias of the bolometers,[2] (ii) the MSQUID, and (iii) a multiplexable microwave frequency backend electronics based on commercially available high-speed-digital-frequency generation/demodulation cards.[5] The last item is a critical component of the system, because it is important that the back-end electronics does not rely on individual components for each pixel of the detector array. In the digital back-end, the frequency comb used for the bolometer excitation is generated digitally. The signals from the MSQUID array, after down-conversion from the several-GHz carrier frequencies, are directly digitized with fast (100 MHz – 1 GHz) A/D converters, then demodulated using a fast FPGA.

The output of each SQUID is lock-in detected at both f and $2f$, giving outputs I and Q, where

$$I(\Phi) = \langle \sin \omega t \cdot S(\Phi + A \sin \omega t) \rangle, \tag{1}$$

$$Q(\Phi) = \langle \cos 2\omega t \cdot S(\Phi + A \sin \omega t) \rangle, \tag{2}$$

where $\langle\rangle$ indicates the time average, A is the amplitude of the flux modulation, ω is the angular frequency, S is the SQUID modulation function, and Φ represents the external magnetic flux. A phase angle can be defined by

$$\theta(\Phi) = \arctan\left(\frac{Q(\Phi)/Q_m}{I(\Phi)/I_m}\right), \tag{3}$$

where I_m and Q_m are the maximum values of I and Q. The demodulation and calculation of θ can be accomplished using a digital signal processor. For a non-sinusoidal SQUID modulation function, the function deviates slightly from linearity. The nonlinearity can be measured by sweeping the flux state of the SQUID with the input held at zero, then corrected for in the DSP software. A numerical simulation showed that the nonlinearity could be less than 1%. Signals greater than a single flux quantum can be measured by keeping track of the phase wrapping.[6]

3. Test Results

To demonstrate the concept, we have designed and fabricated a multiplexer chip that contains four SQUID's with resonant circuits fed with a single microwave readout line. Initial tests showed the four distinct resonance lines near 10 GHz.[6] The Q of the individual circuits was approximately 500. To demonstrate the flux sensitivity, we have measured the RF response of the CPW line at four different input bias levels. The single SQUID input flux sensitivity results are shown in Fig. 2. For this measurement, we have used a commercial network analyzer. Figure 3 shows

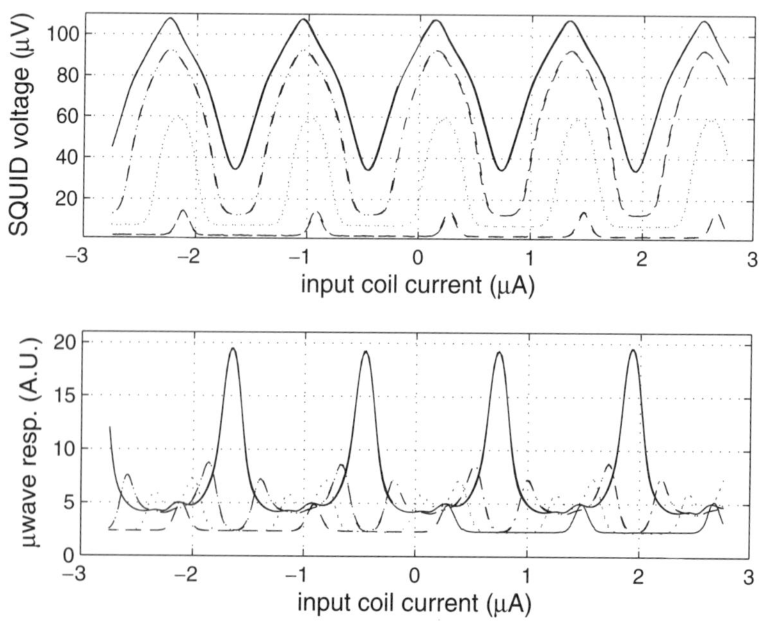

Fig. 2. Preliminary data on the input coil sensitivity of the first channel SQUID. The measurement showed dc (top) and rf (bottom) responses of the SQUID as a function of the input current at four different bias levels. The measurements were performed at 4.2 K.

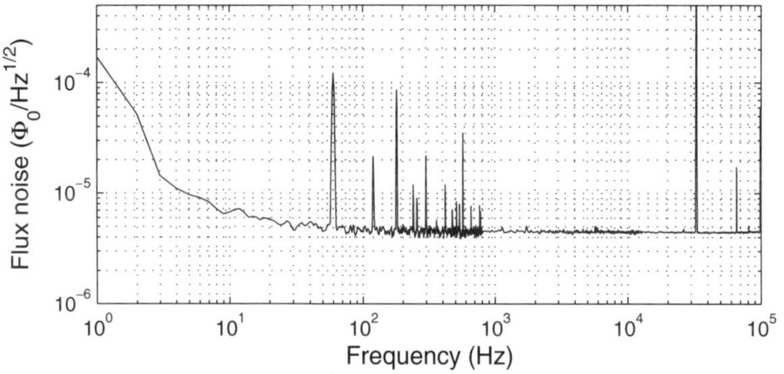

Fig. 3. Noise data of the first channel SQUID.

the noise measurements. Noise measurement at 4.2 K demonstrated $\sim 5\mu\Phi_0/\sqrt{\text{Hz}}$. The signal at 33 kHz is a test signal.

4. Summary

There are many applications of the new SQUID multiplexer readout scheme. The transition edge sensor (TES) and the magnetic microcalorimeter (MMC) detector have been advanced over many years. These detectors have been proposed for many missions, including SAFIR, CMB-Pol and Constellation-X, and have been used in dark-matter-searching experiments. The SQUID readout has been a leading technology for amplifying signals at low temperature close to the detectors without dissipating heat. As one requires more detectors at low temperature, it becomes very critical to minimize the number of leads to the first amplification stage at low temperature.

In this paper, we described a new SQUID multiplexing technique and a chip design utilizing X-band microwave frequencies. A new modulation scheme is also introduced to linearize the SQUID transfer function, enabling further minimization of required lead wires. By supplying ac bias signals for the entire array of detectors on a single set of wires, and similarly supplying the dc bias and modulation signals in series, we can potentially read out 1,000–10,000 detectors with only 3 pairs of wires and a single microwave coaxial cable.

Acknowledgments

This work was supported by the Jet Propulsion Laboratory, California Institute of Technology, under contract with the National Aeronautics and Space Administration.

References

1. J. A. Chervenak *et al.*, *Appl. Phys. Lett.* **74** (1999) 4043.
2. J. Yoon *et al.*, *Appl. Phys. Lett.* **78** (2001) 371.

3. K. D. Irwin and K. W. Lehnert, *Appl. Phys. Lett.* **85** (2004) 2107.
4. P. K. Day *et al.*, *Nature* **425** (2003) 817.
5. B. A. Mazin *et al.*, *Nucl. Instrum. Meth. Phys. Res. A* **559** (2006) 799.
6. I. Hahn *et al.*, *The 24th Int. Conference on Low Temperature Physics (LT24)* (Aug. 10–17, 2005, Orlando, Florida, USA), *Am. Inst. Phys. Conf. Proc.* (2006).

PART 5
ATOMS AND CLOCKS

NEW FORMS OF QUANTUM MATTER NEAR ABSOLUTE ZERO TEMPERATURE

WOLFGANG KETTERLE

Research Laboratory for Electronics,
MIT–Harvard Center for Ultracold Atoms

and

Department of Physics,
Massachusetts Institute of Technology,
Cambridge, MA 02139, USA
ketterle@mit.edu

In my talk at the workshop on fundamental physics in space I described the nanokelvin revolution which has taken place in atomic physics. Nanokelvin temperatures have given us access to new physical phenomena including Bose–Einstein condensation, quantum reflection, and fermionic superfluidity in a gas. They also enabled new techniques of preparing and manipulating cold atoms. At low temperatures, only very weak forces are needed to control the motion of atoms. This gave rise to the development of miniaturized setups including atom chips. In Earth-based experiments, gravitational forces are dominant unless they are compensated by optical and magnetic forces.

The following text describes the work which I used to illustrate the nanokelvin revolution in atomic physics. Strongest emphasis is given to superfluidity in fermionic atoms. This is a prime example of how ultracold atoms are used to create well-controlled strongly interacting systems and obtain new insight into many-body physics.

Keywords: Quantum matter; cold atoms; superfluidity; fermionic atoms.

1. The Role of Interactions in Quantum Reflection of Bose–Einstein Condensates

Quantum reflection is the phenomenon by which an atom is accelerated so abruptly by the Casimir–Polder potential that it reflects from the potential rather than being drawn into the surface. The usual model of quantum reflection treats the atom–surface interaction as a single atom in a potential. However, in a recent study of quantum reflection of Bose–Einstein condensates (BEC's), the reflection probability was limited to $\sim$15% at low velocity.[1] A theoretical paper simulating quantum reflection of BEC's could not explain the low reflectivity.[2]

In this work, we have studied how interatomic interactions affect quantum reflection of BEC's.[3] A silicon surface with a square array of pillars resulted in a higher

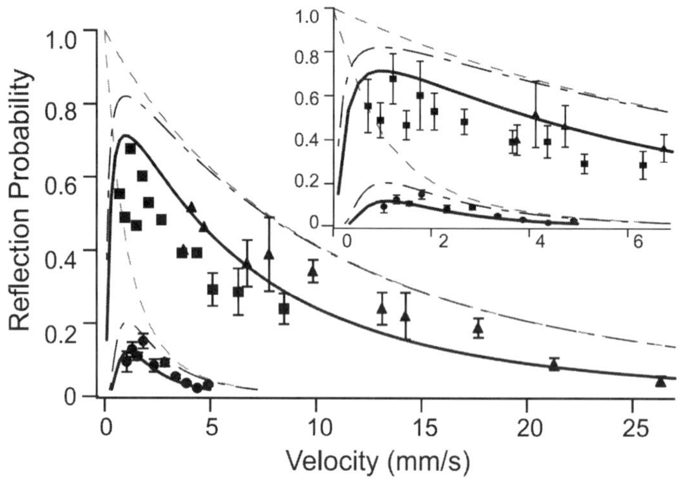

Fig. 1. Reflection probability vs incident velocity. Data are shown for a pillared (square) and solid (circle) Si surface. Single atom models give a monotonic rise to unity reflection. Our model, which includes interactions, shows saturation of reflection at low velocity, in qualitative agreement with our observations.

reflection probability than was previously observed with a solid silicon surface (see Fig. 1). For incident velocities greater than 2.5 mm/s, our observations agreed with single-particle theory. At velocities below 2.5 mm/s, the measured reflection probability saturated near 60% rather than increasing toward unity as predicted. We have extended the theory of quantum reflection to account for the mean field interactions of a condensate which suppress quantum reflection at low velocity. Our model predicts improvements for larger healing lengths and how the corresponding reduction in condensate density sets a limit on the incident flux of atoms.

2. Interference of Bose–Einstein Condensates Split with an Atom Chip

A major step toward compact matter wave sensors is an atom interferometer on an atom chip. We have used an atom chip to split a single BEC of sodium atoms into two spatially separated condensates[4] (see Fig. 2). Dynamical splitting was achieved by deforming the trap along the tightly confining direction into a purely magnetic double-well potential. We observed the matter wave interference pattern formed upon releasing the condensates from the microtraps. The intrinsic features of the quartic potential at the merge point, such as zero trap frequency and extremely high field sensitivity, caused random variations of the relative phase between the two split condensates. Moreover, the perturbation from the abrupt change of the trapping potential during the splitting was observed to induce vortices.

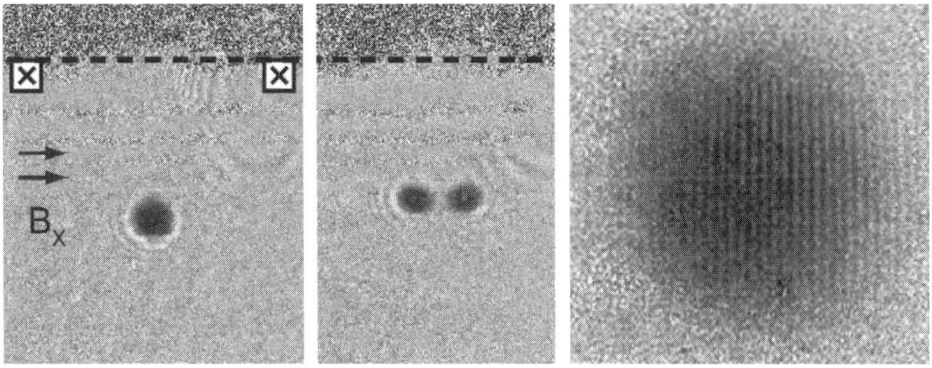

Fig. 2. Splitting of condensates. Condensates were (*left*) initially loaded and prepared in the bottom well and (*middle*) split into two parts by increasing the external magnetic field. For clarity, two condensates were split by 80 μm. The dashed line indicates the chip surface position. (*Right*): Two condensates were released from the magnetic double-well potential, and the matter wave interference pattern of two condensates formed after the time of flight.

3. Long Phase Coherence Time and Number Squeezing of Two Bose–Einstein Condensates on an Atom Chip

Precision measurements in atomic physics are usually done at low atomic densities to avoid collisional shifts and dephasing. This applies to both atomic clocks and atom interferometers. At high density, the atomic interaction energy results in so-called clock shifts and leads to phase diffusion in BEC's. Operating an atom interferometer at low density severely limits the flux and therefore the achievable signal-to-noise ratio.

Here we show that we can operate a BEC interferometer at high density, with mean field energies exceeding h 5 kHz.[5] Using a radio frequency (RF)–induced beam splitter, we demonstrate that condensates can be split reproducibly, so that even after 200 ms, or more than 1000 cycles of the mean field evolution, the two condensates still have a controlled phase. The observed coherence time of 200 ms is ten times longer than the phase diffusion time for a coherent state, i.e. a state with a perfectly defined relative phase at the time of splitting (see Fig. 3). Therefore, repulsive interactions during the beam splitting process have created a nonclassical squeezed state with relative number fluctuations ten times smaller than for a Poissonian distribution.

4. Vortices and Superfluidity in a Strongly Interacting Fermi Gas

Quantum-degenerate Fermi gases provide a remarkable opportunity to study strongly interacting fermions. In contrast to other Fermi systems, such as superconductors, neutron stars or the quark–gluon plasma, these gases have low densities and their interactions can be precisely controlled over an enormous range. By varying the pairing strength between two fermions near a Feshbach resonance,

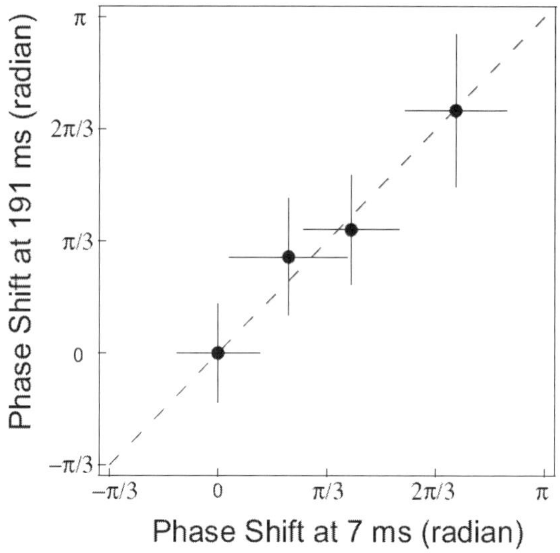

Fig. 3. Long phase coherence of two separated condensates. Various phase shifts were applied on the condensates 2 ms after splitting by pulsing on an additional magnetic field. The shifts of the relative phase were measured at 7 ms and 191 ms, showing strong correlation. The dotted line denotes the ideal case of perfect phase coherence.

one can explore the crossover from a BEC of molecules to a Bardeen–Cooper–Schrieffer (BCS) superfluid of loosely bound pairs whose size is comparable to, or even larger than, the interparticle spacing. The crossover realizes a novel form of high-T_C superfluidity and it may provide new insight into high-T_C superconductors. Earlier experiments with Fermi gases have revealed condensation of fermion pairs,[6–10] but not observed superfluidity. Our observation of vortex lattices directly displays superfluid flow in a strongly interacting, rotating Fermi gas.[11]

A strongly interacting cloud of fermions was created by laser cooling, sympathetic cooling with sodium in a magnetic trap, followed by evaporative cooling in an optical trap.[8,10] The trapped cloud was rotated about its long axis using a blue detuned laser beam (wavelength 532 nm). A two-axis acousto-optic deflector generated a two-beam pattern that was rotated symmetrically around the cloud at a variable angular frequency. Vortex lattices were generated both above and below the Feshbach resonance at 834 Gauss (see Fig. 4).

5. Fermionic Superfluidity with Imbalanced Spin Populations

We have established superfluidity in a two-state mixture of ultracold fermionic atoms with imbalanced spin populations[12] (see Fig. 5). This study relates to the long-standing debate about the nature of the superfluid state in Fermi systems. Indicators for superfluidity were condensates of fermion pairs, and vortices in rotating clouds. For strong interactions, near a Feshbach resonance, superfluidity was

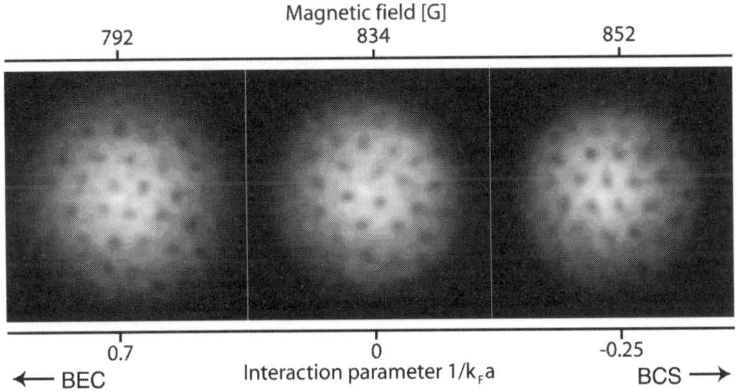

Fig. 4. Vortex lattices in the BEC–BCS crossover. After a vortex lattice was created at 812 G, the field was ramped in 100 ms to 792 G (BEC side), 833 G (resonance), and 853 G (BCS side), where the cloud was held for 50 ms. After 2 ms of ballistic expansion, the magnetic field was ramped to 735 G for imaging. The field of view of each image is 880 μm $\times$ 880 μm.

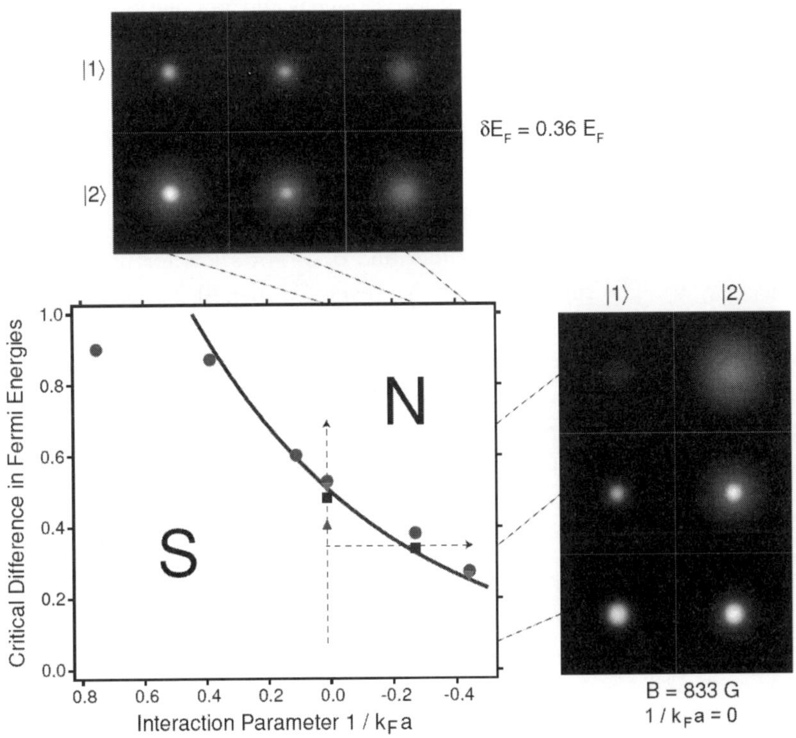

Fig. 5. Phase diagram for interacting Fermi systems with adjustable interactions (horizontal axis) and adjustable spin populations (vertical axis, expressed by the difference in Fermi energies between the two spin states) showing the normal (N) and superfluid (S) regimes. Representative density profiles illustrate the quantum phase transition for fixed interaction (top; induced by varying the population imbalanced) and for fixed population imbalance (right) along the dashed lines. The bright spot in the center of the profiles is the pair condensate which indicates superfluidity.

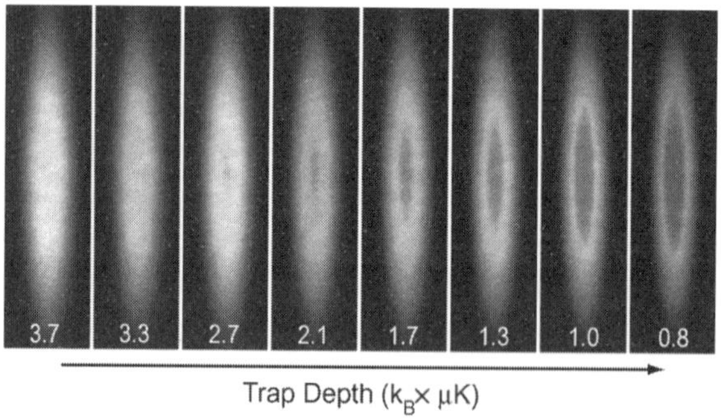

Fig. 6. Observation of phase separation in strongly interacting, unbalanced Fermi mixtures as the temperature was lowered. The images show the *in situ* optical density *difference* between the two spin species. The emergence of a central region of equal spin densities is directly seen as the growth of a central, "hollow" core, surrounded by a cloud at unequal densities.

observed for a broad range of population imbalances. We mapped out the superfluid regime as a function of interaction strength and population imbalance, and characterized the quantum phase transition to the normal state, known as the Pauli limit of superfluidity.

6. Observation of Phase Separation in a Strongly Interacting Imbalanced Fermi Gas

At zero temperature, a BCS-type superfluid does not allow for unequal spin densities (see Fig. 6). The superfluid gap Δ prevents unpaired fermions from entering the condensate. In a harmonic trap, this implies that an unbalanced Fermi mixture will phase-separate into a central superfluid core of equal densities surrounded by a normal state at unequal densities. To test this hypothesis, we developed a novel phase contrast imaging technique that allows us to directly measure the density *difference* of the spin mixture. This enabled us to observe the emergence of phase separation *in situ* (in the trap) as the Fermi mixture was cooled.[13] At our lowest temperatures, the presence of a condensate was correlated with equal densities for the two spin states.

7. Conclusions

The nanokelvin revolution is still in progress. For the future, we expect a rapid growth of studies of many-body physics using ultracold atoms. This includes mixtures of Bose and Fermi gases and spinor condensates, ultracold molecules, atoms in optical lattices, antiferromagnetism and other magnetic phases, fermionic superfluidity in a lattice, and the realization of various bosonic and fermionic Hubbard models in optical lattices.

Acknowledgments

This work was supported by the NSF, DARPA, ONR, and NASA.

References

1. T. A. Pasquini *et al.*, *Phys. Rev. Lett.* **93** (2004) 223201.
2. R. G. Scott *et al.*, *Phys. Rev. Lett.* **95** (2005) 073201.
3. T. A. Pasquini *et al.*, *Phys. Rev. Lett.* **97** (2006) 093201.
4. Y. Shin *et al.*, *Phys. Rev. A* **72** (2005) 021604(R).
5. G.-B. Jo *et al.*, cond-mat/0608585.
6. M. Greiner *et al.*, *Nature* **426** (2003) 537.
7. S. Jochim *et al.*, *Science* **302** (2003) 2101.
8. M. W. Zwierlein *et al.*, *Phys. Rev. Lett.* **91** (2003) 250401.
9. C. A. Regal, M. Greiner and D. S. Jin, *Phys. Rev. Lett.* **92** (2004) 040403.
10. M. W. Zwierlein *et al.*, *Phys. Rev. Lett.* **92** (2004) 120403.
11. M. W. Zwierlein *et al.*, *Nature* **435** (2005) 1047.
12. M. W. Zwierlein *et al.*, *Science* **311** (2006) 492.
13. Y. Shin *et al.*, *Phys. Rev. Lett.* **97** (2006) 030401.

ATOMIC QUANTUM SENSORS IN SPACE

T. VAN ZOEST, T. MÜLLER, T. WENDRICH, M. GILOWSKI,
E. M. RASEL and W. ERTMER

Institut für Quantenoptik,
Leibniz Universität Hannover, Welfengarten 1,
30167 Hannover, Germany
ertmer@iqo.uni-hannover.de

T. KÖNEMANN, C. LÄMMERZAHL and H. J. DITTUS

ZARM, Universität Bremen, Am Fallturm,
28359 Bremen, Germany

A. VOGEL, K. BONGS and K. SENGSTOCK

Institut für Laser-Physik, Universität Hamburg,
Luruper Chaussee 149, 22761 Hamburg, Germany

W. LEWOCZKO-ADAMCZYK and A. PETERS

Institut für Physik, Humboldt-Universität zu Berlin,
Hausvogteiplatz 5-7, 10117 Berlin, Germany

T. STEINMETZ and J. REICHEL

Laboratoire Kastler Brossel, ENS, 24 rue Lhomond,
75231 Paris, France

G. NANDI, W. SCHLEICH and R. WALSER

Abteilung Quantenphysik, Universität Ulm,
Alber-Einstein Allee 11, 89069 Ulm, Germany

In this article we present actual projects concerning high resolution measurements developed for future space missions based on ultracold atoms at the Institut für Quantenoptik (IQ) of the University of Hannover. This work involves the realization of a Bose–Einstein condensate in a microgravitational environment and of an inertial atomic quantum sensor.

Keywords: Cold atoms; quantum sensors; BEC.

1. Introduction

Microgravity is expected to be a decisive condition for the next leap in tests in fundamental physics of gravity, relativity, and theories beyond the standard model. Thanks to the recent progress in quantum engineering, fundamental tests can now be extended to the quantum domain. Promising techniques for fundamental tests in the quantum domain are matter-wave sensors based on cold atoms or atom lasers, which use atoms as unperturbed microscopic test bodies for measuring inertial forces or as frequency references. Microgravity is of high relevance to matter-wave interferometers and experiments with quantum matter, like Bose–Einstein condensates (BEC's)[1] or degenerate Fermi gases, as it permits the extension of an unperturbed free fall of these test particles (wave packets) in a low-noise environment. The HYPER project[2] was the first European initiative for hyperprecision atomic quantum sensors on board a satellite. Its scientific objective was the mapping of the relativistic Lense–Thirring effect close to the Earth using cold-atom interferometry.

In this paper, we present two projects, both dealing with the exploration of cold matter waves for space applications. First we present an experimental realization of a space atom laser which is undertaken at IQ within the QUANTUS project (QUANTen Gase *U*nter *S*chwerelosigkeit). QUANTUS was initiated by the IQ together with the ZARM in a DLR-funded cooperation with partners from the Institut für Laser-physik of the University of Hamburg, the Institut für Physik of the University of Berlin, the Laboratoire Kastler Brossel of the ENS and the Abteilung Quantenphysik of the University of Ulm. This project is a feasibility study of a compact, robust and mobile experiment for the creation of a BEC and interferometric diagnostics, which can withstand high accelerations of up to 50 g in a drop tower facility. The full experiment with all components (power supply, laser systems, etc.) has to be implemented in a drop capsule with an effective length of 215 cm and 60 cm diameter. The compact setup is based on an atom chip[4] and uses a robust DFB-diode laser system as a light source. In future, the apparatus will serve as an experimental platform for investigating various aspects of ultracold gases in microgravity, like adiabatic release, extended coherent evolution and features of atom lasers. The project is supported by the Deutsche Zentrum für Luft- und Raumfahrt (DLR, project number DLR 50 WM 0346).[5]

The second project we present is a compact differential interferometer named CASI (*C*old *A*tom *S*agnac *I*nterferometer). In this experiment we use ultracold ^{87}Rb atoms in a differential measurement scheme for extremely precise sensing of rotations and accelerations. As a long-term goal the combination of coherent atomic sources with high resolution atom interferometry could lead to even better accuracies, especially when profiting from low-noise environments like microgravity. This could give new insights into questions of fundamental physics, relativistic effects or gravitation. In this paper we present in detail these two projects developed for future space missions.

2. Quantus

The goal of QUANTUS is to continue the path toward lower energy scales by lifting Earth-bound laboratory restrictions and investigate BEC's in a microgravity environment. There are several reasons why weightlessness is important for fundamental research on cold quantum gases. First of all, in a microgravity environment it is possible to substantially lower the trapping potential adiabatically without the need for levitational fields to compensate for gravity. In this context, the preparation of atomic ensembles with temperatures in the fK regime seems to be possible, as the gravitional sag is limiting the achievable temperature in ground-based experiments.[6] Furthermore, the effect of ultraweak long-range forces becomes important in these condensates, which promises the discovery of new kinds of low-energy phase transitions. Another important point is that the time of free and unperturbed evolution can be significantly longer than in Earth-bound laboratories. This is crucial for the precision of atom interferometric metrology (see Sec. 3). Finally, we want to emphasize that weightlessness is a major advantage for research on mixtures of quantum gases, as atoms with different masses do not experience different potentials and can be perfectly overlapped in the trap.

For experiments at the drop tower facility at ZARM there are stringent requirements concerning important experimental parameters like weight and volume of the experiment or low power consumption which qualifies the project as a test-bed for future space missions. The specific experimental requirements are a cylindrical volume of 215 cm $\times$ 60 cm diameter which has to fit into a special drop capsule, a weight less than 230 kg, a power consumption less than 1120 W, and the experiment has to withstand a force of 50 g at the impact at the end of the free fall. The drop capsule holding the setup (see Fig. 1) reaches excellent acceleration suppression down to the microgravity level of 10^{-6} g, while it falls freely in the evacuated drop tower tube for 4.8 s. An enhancement of the time of the free fall can be accomplished by using a catapult in the drop tower which prolongates the free fall to about 9 s.

In order to keep the setup as simple as possible, we chose ^{87}Rb for the experiment, as it has a simple laser cooling scheme for which small-size and robust diode lasers are available. The atoms are released into the vacuum chamber using commercially available current-controlled dispensers. Additionally, the use of light-induced atom desorption (LIAD) as a switchable atomic source including a controllable background pressure has been tested.[7] The vacuum chamber is made from steel with a very low magnetic permeability, resulting in high mechanical stability and low field disturbances. The chamber is kept at low pressure (2×10^{-10} mbar) by a titanium sublimation pump and a modified ion getter pump (20 l/s).

The laser light for the manipulation of the atoms, like laser-cooling, optical pumping or detection, is created with three laser modules including the diode laser sources and one module for distributing the light for the different experimental steps. These modules fit into two 19-inch racks and provide the experiment with the needed light via optical fibers. They include all optical elements needed for the light

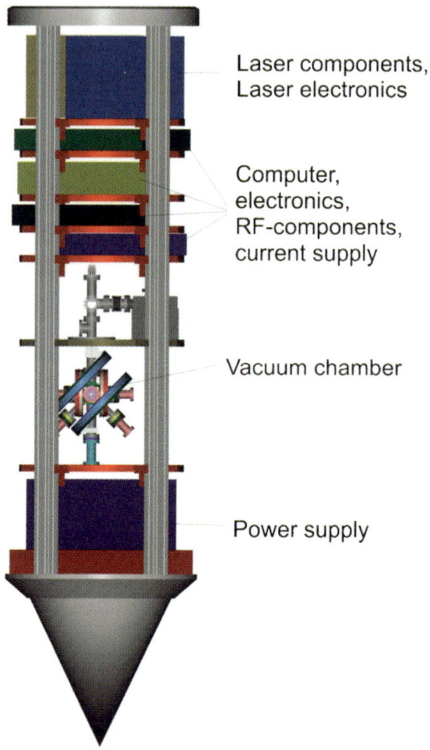

Fig. 1. Drop capsule with all the components.

manipulation, like acousto-optic modulators, Doppler-free saturation spectroscopy for the frequency stabilization of the lasers and a TA (tapered amplifier) diode chip for power amplification of the light. The lasers and their stability have already been tested under drop tower conditions, where the fullfillment of the requirments for a BEC experiment during the free flight could be demonstrated.[5] To assure the stability of the experimental conditions, the remaining laser optics at the fiber exits is rigidly fixed to the vacuum chamber.

In order to reach short evaporation times and low power consumption, we use a magnetic microtrap on a chip reaching strong magnetic confining potentials with moderate currents; see Fig. 2 for a schematic of the chip. The chip contains a U-shaped wire producing a quadrupole field for the operation of an on-chip magneto-optical trap (MOT)[3] and a Z-shaped wire for the creation of an Ioffe-type magnetic trap.[4] Additional magnetic fields needed for the creation of the desired magnetic potentials are created with external coils, which are fixed on the outside of the vacuum chamber.

The drop capsule also includes a computer system, which allows autonomous control of the experiment during the free fall. Additionally, the power supply for the complete experiment fits into the drop capsule.

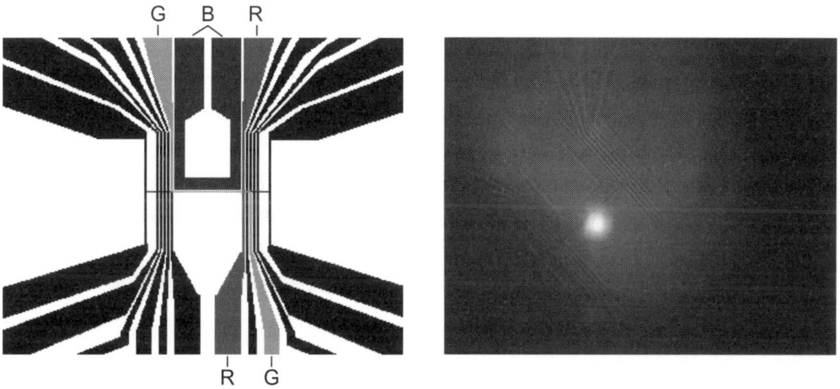

Fig. 2. *Left*: Schematic of the chip creating the magnetic fields for the magneto-optical trap (blue wire B) and an Ioffe-type magnetic trap (green wire G) with an additional dimple in the conservative potential (red wire R). All traps need another homogeneous magnetic field that is produced by an external coil. *Right*: Photograph of the chip with a cloud of trapped cold atoms.

Table 1. Experimental parameters for different trap types during the experiment.

Trap type	Number of atoms	Temperature (μK)
External MOT	1.3×10^7	230
U-MOT before shift	1.2×10^7	230
U-MOT after shift	1.0×10^7	230
Optical molasses	7×10^6	20
Initial magnetic trap	3–3.5×10^6	35–50

The experimental procedure to produce a degenerate Bose gas is as follows. In the first stage the atoms are trapped and precooled using laser cooling in a MOT with a quadrupole field made by external coils. Afterward, the atoms are loaded in the on-chip MOT. In this way, 1.2×10^7 atoms are loaded on the chip. After reaching the corresponding temperature limit in the μK range by applying a molasses cooling period, the sample is transferred to a conservative trapping potential (in our case realized by a magnetic field minimum). Afterward the phase-space density of the sample is enhanced by forced evaporative cooling. In the drop tower, there is the possibility of performing the phase transition to condensation either before or during the free fall of the ensemble to compare the advantages between these two experimental concepts. At the actual status of the experiment the ramps of the forced evaporation are optimized in ground-based measurements. The achieved experimental parameters for different trap types during the experiment are displayed in Table 1.

3. CASI

In recent years, atom interferometry has become an important technique for highly sensitive measurements of various kinds, and a series of experiments with impressive resolutions has been performed: a measurement of the fine-structure constant α based on the photon recoil,[8] a gravimeter[9] for the measurement of g and a double differential measurement to determine the gravitational constant,[10] G. Additionally, a high precision rotation measurement experiment has been performed,[11] which reaches a sensitivity comparable to state-of-the-art optical gyroscopes. The interferometric measurement of rotations is based on the Sagnac effect,[12] which indicates that between the two arms enclosing an area $\mathbf{A}$ of an interferometer, which is rotating with angular velocity $\mathbf{\Omega}$, a phase shift $\delta\phi = 4\pi\mathbf{A}\mathbf{\Omega}/\lambda c$ is induced. The use of matter wave interferometers is motivated by the fact that due to the sensitivity to the used wavelength an improvement of the order 10^{10} compared to light from the visible spectrum is in principle possible.

In our experiment CASI (Cold Atom Sagnac Interferometer),[2] we are setting up an experiment for high resolution inertial measurements similar to the one from Ref. 11. Besides a high accuracy, the additional goals of our experiment are a good, long-term stability for signal integration and a compact and transportable setup, which is needed for combined measurements with other state-of-the-art gyroscopes, for example the one from Ref. 13. Additionally, the miniaturization of the atom interferometer is essential for space-based experiments. In this context, we use cold and slow atoms in our interferometer, allowing large enclosed areas of similar size to thermal atomic beams and long interaction times while still retaining a compact experimental setup. For the realization of the interferometer, optimized vacuum concepts have been developed and the optical setup is completely based on fiber technology, similar to the QUANTUS experiment described before.

The basic schematic of our experiment is sketched in Fig. 3. We use a double interferometer setup for a differential measurement, to discriminate between rotations and accelerations.[11] To achieve this, we use two identical atomic sources, each emitting atoms on flat parabolic trajectories into the interferometer, but with opposite launch directions. The atomic sources consist of both a double magneto-optical trap and a two-dimensional MOT forming a brilliant (10^{10} at/s) and slow atomic beam, which is further manipulated in a following 3D-MOT/optical molasses. With

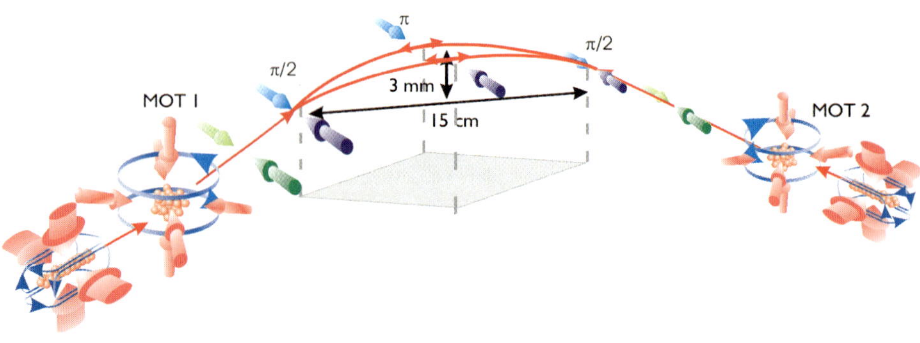

Fig. 3. Concept of the cold atom sagnac interferometer.

the laser cooling in a "moving" optical molasses, a well-definied velocity is imprinted on the atoms, what is important for precise control of the enclosed area **A** in the interferometer.

This flexible source concept is well suited for high resolution measurements: on one hand the achieved useful atom number will give rise to a good signal-to-noise ratio in the interferometer; on the other hand it allows for elaborated studies of different measurement concepts, e.g. pulsed vs continuous measurement.

In the interferometer, the magnetic insensitive hyperfine ground states $|F = 1, m_F = 0\rangle$ and $|F = 2, m_F = 0\rangle$ are used. A well-definied state preparation for the interferometer in one of these states is performed thanks to precisely controllable laser manipulation of the atoms. The interferometric beam splitting process relies on a velocity-sensitive two(optical)-photon Raman transition between the two quantum states mentioned before. To reduce measurement noise arising from phase variations of the beam splitters themselves, the phases imprinted on the atoms by the beam splitters have to be controlled to a high extent. By employing the Raman transition, only the frequency difference between the two optical fields (at $\sim$6.8 GHz) has to be phase-stabilized, which can be accomplished to a high degree (<1 mrad) thanks to well-established microwave techniques; see Fig. 4. By applying a Mach–Zehnder-like pulse sequence, which combines two 50/50 beam splitters ($\pi/2$) with one mirror pulse (π) in between, the inertial phase shifts experienced by the atoms are converted into a difference of the atom numbers of the two output ports of the interferometer. These two ports, the two atomic quantum

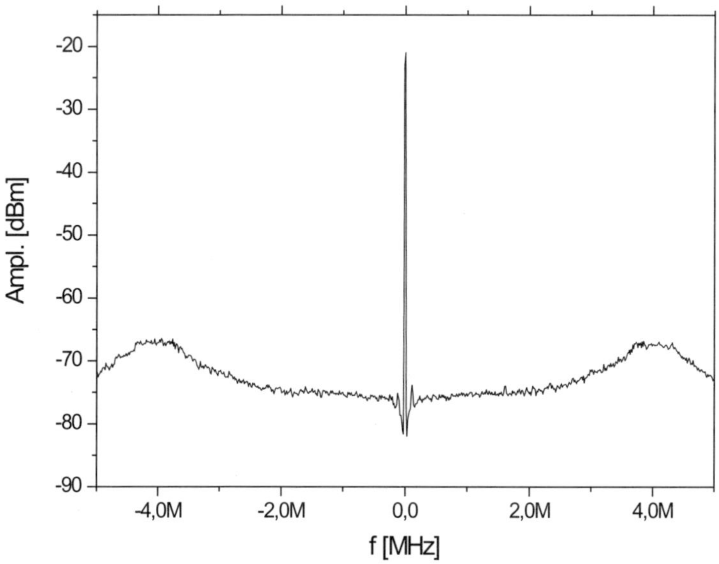

Fig. 4. Beat note of the two laser beams used for the Raman transition with a frequency difference of about 6.834 GHz. Both lasers are phase-stabilized to a high degree, resulting in a width of the beat note with an FWHM less than 1 Hz.

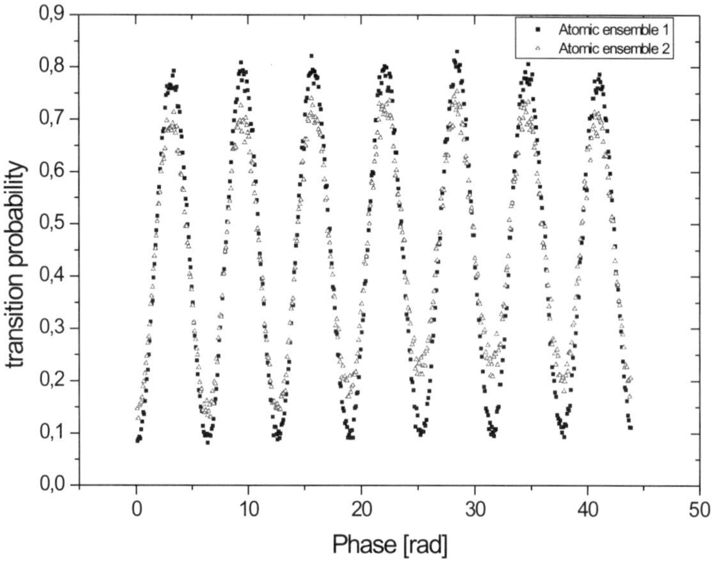

Fig. 5. Interference fringes in the Mach–Zehnder interferometer induced by scanning the phase of the last beam splitter pulse. The displayed measurement has been performed in a configuration with reduced sensitivity to inertial forces.

states, are each internal-state-selectively detected by laser excitation and fluorescence collection.

In the present phase of the project, we are evaluating the sensor in a mode of low resolutions for the inertial forces only: two synchronous atomic Mach–Zehnder interferometers are realized with a temporal sequence of light pulses ($\pi/2$–π–$\pi/2$ sequence) while the atoms cross a single interaction zone. An example of the realization of the Mach–Zehnder atom interferometer is displayed in Fig. 5. When fully extended to a spatially separated geometry, the estimated sensitivity of our device is $1 \cdot 10^{-9}$ rad/s $\mathrm{Hz}^{-1/2}$, when atomic shot noise is limited with 10^8 atoms/s.

4. Conclusion

In this paper we have show that atom optical experiments can give new insights into various exciting physical fields, where especially the microgravity environment or space would allow full realization of the great potential of these sensors.

Acknowledgments

We acknowledge financial funding from the following institutions: the QUANTUS project is supported by the Deutsche Zentrum für Luft- und Raumfahrt under contract number DLR 50 WM 0346; the CASI project is supported by the Deutsche Forschungsgemeinschaft as part of SFB407.

References

1. M. H. Anderson *et al.*, *Science* **269** (1995) 198.
2. C. Jentsch *et al.*, *Gen. Relativ. Gravit.* **36** (2004) 2197.
3. E. L. Raab *et al.*, *Phys. Rev. Lett.* **59** (1987) 2631.
4. W. Hänsel *et al.*, *Nature* **413** (2001) 498.
5. QUANTUS Collab. (A. Vogel *et al.*), *Appl. Phys. B* **84** (2006) 663.
6. A. E. Leanhardt *et al.*, *Science* **301** (2003) 1513.
7. C. Klempt *et al.*, *Phys. Rev. A* **73** (2006) 013410.
8. A. Wicht *et al.*, *Phys. Scripta T* **102** (2002) 82.
9. A. Peters, K. Y. Chung and S. Chu, *Metrologia* **38** (2001) 25.
10. M. Fattori *et al.*, *Phys. Lett. A* **318** (2003) 184.
11. T. L. Gustavsson, A. Landragin and M. A. Kasevich, *Class. Quant. Grav.* **17** (2000) 2385.
12. M. Sagnac, *Compt. Rend. des Sc. d. l'Acad. d. Sc.* **157** (1913) 1410.
13. F. Yver-Leduc *et al.*, *J. Opt. B* **5** (2003) S75.

COHERENT ATOM SOURCES FOR ATOM INTERFEROMETRY IN SPACE: THE ICE PROJECT

PHILIPPE BOUYER

Laboratoire Charles Fabry de l'Institut d'Optique, CNRS,
Université Paris-Sud, Campus Polytechnique,
RD127, 91127 Palaiseau Cedex, France
philippe.bouyer@institutoptique.fr
http://www.ice-space.fr
http://www.atomoptic.fr

Atomic quantum sensors are a major breakthrough in the technology of time and frequency standards as well as ultraprecise sensing and monitoring of accelerations and rotations. They apply a new kind of optics based on matter waves. Today, atomic clocks are the standard for time and frequency measurement at the highest precisions. Inertial and rotational sensors using atom interferometers have already shown similar potential for replacing state-of-the-art sensors in other fields. With Bose–Einstein condensates, also referred as atom lasers, the traditional experiments with atom interferometers can be greatly improved. Testing of fundamental principles, studies of atomic properties, applications as inertial sensors, and measurements of fundamental constants can benefit from the brightness (intensity and small momentum spread) of these coherent sources. In addition, the coherence properties of condensates may allow BEC-based atom interferometers to approach the Heisenberg detection limit. This corresponds to a measurement precision which scales like $1/N$ for N atoms and not like $1/\sqrt{N}$ as for independent measurements on N atoms. We present here the recent progress toward the acheivement of new coherent atomic sources, i.e. atom lasers, that will be use in space-based atom interferometers. We introduce new concepts of atom accelerometers and gyrometers that take advantage of the high collimation and coherence properties of atoms lasers and report on the development of a 0-g coherent atom interferometer (ICE) that will be used to test the ultimate performances of atom accelerometers in space.

Keywords: Atom interferometry; Bose–Einstein condensates; Atom laser.

1. Introduction

Inertial sensors are useful devices in both science and industry. Higher precision sensors could find scientific applications in the areas of general relativity,[1] geodesy and geology. There are also important applications of such devices in the field of navigation, surveying and analysis of earth structures. Matter-wave interferometry was envisaged for its potential to be an extremely sensitive probe for inertial forces.[2] First, neutron interferometers have been used to measure the acceleration due to gravity[3] and the rotation of the Earth[4] at the end of the 1970's. In 1991, atom interference techniques were used in proof-of-principle work to measure rotations[5] and accelerations.[6] In the following years, many theoretical and experimental works have been performed to investigate this new kind of inertial sensors.[7] Some of the recent works have shown very promising results leading to a sensitivity comparable to that of other kinds of sensors, as well as for rotation[8,9] and acceleration.[10–12]

Atom interferometry[2,5,7,13,14] is nowadays one of the most promising candidates for ultraprecise and ultra-accurate measurement of gravitoinertial forces[8–12,15–18] or for precision measurements of fundamental constants.[19] The realization of Bose–Einstein condensation (BEC) of a dilute gas of trapped atoms in a single quantum state[20–22] has produced the matter-wave analog of a laser in optics.[23–26] Like the revolution brought about by lasers in optical interferometry,[1,27,28] it is expected that the use of Bose–Einstein condensed atoms will take the science of atom optics, particularly atom interferometry, to an unprecedented level of accuracy.[29–31] In addition, BEC-based coherent atom interferometry will reach its full potential in space-based applications where microgravity will allow the atomic interferometers to reach their best performance.[32]

2. Inertial Sensors Based on Atom Interferometry: Basic Principle

Generally, atom interferometry is performed by applying successive *coherent* phase-locked beam-splitting processes separated by a time T to an ensemble of particles (see Fig. 1),[33,34] followed by detection of the particles in each of the two

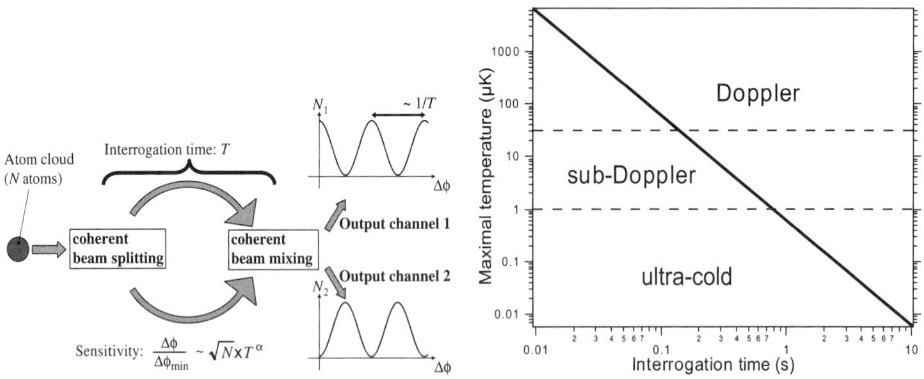

Fig. 1. *Left*: Principle of an atom interferometer. An initial atomic wave packet is split into two parts by the first beam-splitter. The wave packets then propagate freely along the two different paths for an "interrogation time" T, during which the two wave packets can accumulate different phases. A second pulse is then applied to the wave packets so that the number of atoms at each output is modulated with respect to this phase difference. *Right*: Maximum temperature of the atom source for a given interrogation time. The maximum interrogation time for a given initial temperature has been calculated for a detection area of $10\,\text{cm}^2$ and defined as the time at which half of the atoms are no longer detected. The dashed lines indicate the limits of Doppler and sub-Doppler cooling.

output channels. The interpretation in terms of matter waves follows from the analogy with optical interferometry. The incoming matter wave is separated into two different paths by the first beam-splitter. The accumulation of phases along the two paths leads to interference at the last beam-splitter, producing complementary probability amplitudes in the two output channels.[35–37] The detection probability in each channel is then a sine function of the accumulated phase difference, $\Delta\phi$.

Atomic clocks[38–40] can be considered one of the most advanced applications of atom interferometry.[41] In this "interferometer," the two different paths of Fig. 1 consist of the free evolution of atoms in different internal states with an energy separation $\hbar\Delta\omega$. An absolute standard of frequency is obtained by servo-locking a local oscillator to the output signal of the interferometer. The output signal of the clock then varies as $\cos(\delta\omega \times T)$, where $\delta\omega$ is the frequency difference between the transition frequency $\Delta\omega$ and the local oscillator frequency. Atom interferometers can also be used as a probe of gravitoinertial fields. In such applications, the beam-splitters usually consist of pulsed near-resonance light fields which interact with the atoms to create a coherent superposition of two different *external* degrees of freedom, by coherent transfer of momentum from the light field to the atoms.[2,33] Consequently, the two interferometer paths are separated in space, and a change in the gravitoinertial field in either path will result in a modification of the accumulated phase difference. Effects of acceleration and rotation can thus be measured with very high accuracy. To date, ground-based experiments using atomic gravimeters (measuring acceleration),[10–12] gravity gradiometers (measuring acceleration gradients)[15,16] and gyroscopes[8,9] have been realized and proved to be competitive with existing optical[42] or artifact-based devices.[43]

3. Ultracold Sources and Applications in Space

The ultimate phase sensitivity of an atom interferometer is, aside from technical difficulties, limited by the finite number of detected particles N and scales as $\Delta\phi_{\min} = 1/\sqrt{N}$ (quantum projection noise limit[44,45]). Of course, the relation between the relative phases accumulated along the two different paths and the actual physical property to be measured is a function of to the "interrogation" time T spent by the particles between the two beam-splitters. Thus, the ideal sensitivity of an atom interferometer is expected to scale[a] as $\sqrt{N}\,T^{\alpha}$ with $\alpha > 0$, and it is obviously of strong interest to increase these two factors. Using cold atomic sources helps this quest for higher performances in two ways. First, reducing the velocity dispersion of the atomic sample (a few millimeters per second) enables one to drastically reduce the longitudinal velocity of the atoms v_L (a few cm/s) and

[a]An atomic clock or an atomic gyrometer, for example, has a sensitivity proportional to T and an on-ground gravimeter has a sensitivity proportional to T^2 due to the quadratic nature of the free-fall trajectory in a constant gravitational field.

enhances in the same way the enclosed area and the sensitivity for a constant length. Second, the accuracy and the knowledge of the scaling factor depend directly on the initial velocity of the atoms and can be better controlled with cold atomic sources than with thermal beams, as has already been demonstrated with atomic clocks.[46]

Nevertheless, seeking to increase the sensitivity of on-ground atom interferometers by increasing the interrogation time T, one soon reaches a limit imposed by gravity. With the stringent requirements of an ultrahigh vacuum and a very-well-controlled environment, current state-of-the-art experimental apparatus does not allow more than a few meters of free fall, with corresponding interrogation times of the order of $T \sim 400\,\mathrm{ms}$. Space-based applications will enable much longer interrogation times to be used, thereby increasing dramatically the sensitivity and accuracy of atom interferometers.[32]

Even in space, atom interferometry with a *classical* atomic source will not outperform the highest-precision ground-based atom interferometers that use samples of cold atoms prepared with standard techniques of Doppler and sub-Doppler laser cooling.[47] Indeed, the temperature of such sub-Doppler laser-cooled atom clouds is typically $\sim 1\,\mu\mathrm{K}$ ($v_{\mathrm{rms}} \sim 1\,\mathrm{cm/s}$). In the absence of gravity, the time evolution of cold samples of atoms will be dominated by the effect of finite temperature: in free space, a cloud of atoms follows a ballistic expansion until the atoms reach the walls of the apparatus, where they are lost. Therefore the maximum interrogation time reasonably available for space-based atom interferometers will strongly depend on the initial temperature of the atomic source. As shown in Fig. 1, the 200 ms limit imposed by gravity for a 30 cm free fall is still compatible with typical sub-Doppler temperatures, whereas an interrogation time of several seconds is only accessible by using an "ultracold" source of atoms (far below the limit of laser cooling) with a temperature of the order of a few hundred nano kelvins.

4. Coherent Atom Sensors: BEC and Atom Lasers

Dense, *ultracold* samples of atoms are now routinely produced in laboratories all around the world. Using evaporative cooling techniques,[20–22] one can cool a cloud of a few 10^6 atoms to temperatures below 100 nK.[48] At a sufficiently low temperature and high density, a cloud of atoms undergoes a phase transition to quantum degeneracy. For a cloud of bosonic (integer spin) atoms, this is known as Bose–Einstein condensation, in which all the atoms accumulate in the same quantum state (the atom-optical analog of the laser effect in optics). A BEC exhibits long range correlation and can therefore be described as a coherent "matter wave": an ideal candidate for the future of atom interferometry in space. The extremely low temperature associated with a BEC results in a very slow ballistic expansion, which in turn leads to interrogation times of the order of several tens of seconds in a space-based atom interferometer. In addition, the use of such a coherent source for atom optics could give rise to novel types of atom interferometry.[29–31,35–37,49,50]

4.1. *Atom laser: a coherent source for future space applications*

The idea for an atom laser predates the demonstration of the exotic quantum phenomenon of BEC in dilute atomic gases. But it was only after the first such condensate was produced in 1995 that the pursuit of creating a laser-like source of atomic de Broglie waves became intense. This is illustrated in Fig. 2.

In a Bose–Einstein condensate all the atoms occupy the same quantum state and can be described by the same wave function. The condensate therefore has many unusual properties not found in other states of matter. In particular, a Bose condensate can be seen as a coherent source of matter waves. Indeed, in a (photonic) laser all the photons share the same wave function. This is possible because photons have an intrinsic angular momentum, or "spin," of the Planck constant h divided by 2π. Particles that have a spin that is an integer multiple of $\hbar = h/2\pi$ obey Bose–Einstein statistics. This means that more than one so-called boson can occupy the same quantum state. Particles with half-integer spin — such as electrons, neutrons and protons, which all have spin $\hbar/2$ — obey Fermi–Dirac statistics. Only one fermion can occupy a given quantum state.

A composite particle, such as an atom, is a boson if the sum of its protons, neutrons and electrons is an even number; the composite particle is a fermion if

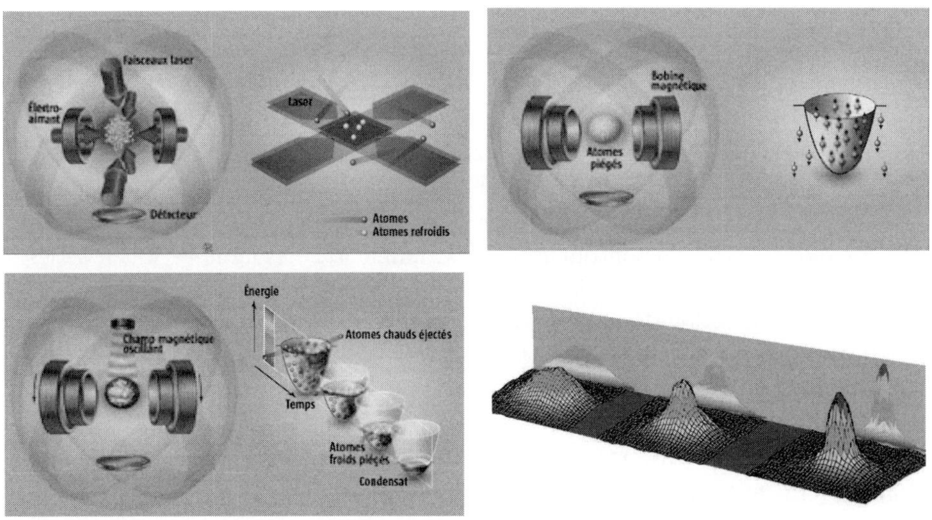

Fig. 2. Evaporative cooling toward Bose–Einstein codensation. Initially, atoms are trapped in optical molasses using radiative forces. Then, the atoms are transferred in a magnetic trap where they can stay trapped for hundreds of seconds. Since no damping exists in such traps (as opposed to radiative traps), an evaporative cooling technique is used to remove the hottest atoms. In this technique, the trap is capped at a chosen height (using RF induces spin flip) and the atoms with higher energy escape. By lowering the trap height, an ultracold high density sample of atoms is obtained. The bottom right picture shows the BEC transition where a tiny dense peak of atoms (a coherent matter wave) appears at the center of a Maxwell–Boltzman distribution (incoherent background).

this sum is an odd number. Rubidium-87 or Caesium-133 atoms, for example, are bosons, so a large number of them can be forced to occupy the same quantum state and therefore have the same wave function. To achieve this, a large number of atoms must be confined within a tiny trap and cooled to submillikelvin temperatures using a combination of optical and magnetic techniques (see for example Ref. 51). The Bose–Einstein condensates are produced in confining potentials such as magnetic or optical traps by exploiting either the atoms' magnetic moment or an electric dipole moment induced by lasers. In a magnetic trap, for instance, once the atoms have been cooled and trapped by lasers, the light is switched off and an inhomogeneous magnetic field provides a confining potential around the atoms. The trap is analogous to the optical cavity formed by the mirrors in a conventional laser. To make a laser we need to extract the coherent field from the optical cavity in a controlled way. This technique is known as "output coupling." In the case of a conventional laser the output coupler is a partially transmitting mirror. Output coupling for atoms can be achieved by transferring them from states that are confined to ones that are not, typically by changing an internal degree of freedom, such as the magnetic states of the atoms. The development of such atom lasers is providing atom sources that are as different from ordinary atomic beams as lasers are from classical light sources, and promises to outperform existing precision measurements in atom interferometry[29–31] or to study new tranport properties.[52–54]

The first demonstration of atomic output coupling from a Bose–Einstein condensate was performed with sodium atoms in a magnetic trap by W. Ketterle and coworkers at the Massachusetts Institute of Technology (MIT) in 1997. Only the atoms that had their magnetic moments pointing in the direction opposite to the magnetic field were trapped. The MIT researchers applied short radiofrequency pulses to "flip" the spins of some of the atoms and therefore release them from the trap [see Fig. 3(a)]. The extracted atoms then accelerated away from the trap under the force of gravity. The output from this rudimentary atom laser was a series of pulses that expanded as they fell due to repulsive interactions between the ejected atoms and those inside the trap. Later T. Hänsch and colleagues at the Max Planck Institute for Quantum Optics in Munich extracted a continuous atom beam that lasted for 0.1 s. The Munich team employed radiofrequency output coupling in an experimental setup that was similar to the one at MIT but used more stable magnetic fields [see Fig. 3(b)]. Except for a few cases,[55] the outcoupling methods do not allow one to choose either the direction or the wavelength of the atom laser beam. In addition, the intrinsic repulsion between the atom laser beam and the BEC has dramatic effects[56,57] and gravity plays a significant role,[58] such that the atom laser wavelength becomes rapidly small.

The way to overcome these limitations is either to apply coherent sources in space or to suspend the atom laser during its propagation. For the latter, many atomic wave guides have been developed for cold thermal beams or even for degenerate gases.[54] Nevertheless, as in optics, the transfer of cold atoms from magneto-optical traps into these small atom guides represents a critical step, and

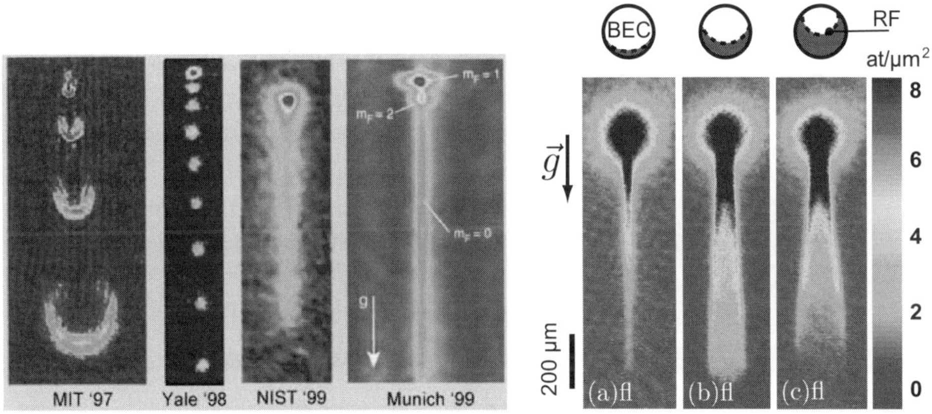

Fig. 3. Various types of atom lasers. (a) At MIT, intense RF pulses spin-flip the atoms from a trapped state to an untrapped state. They fall under gravity. (b) At Yale, the condensate is loaded in an optical lattice. The combination of tunnel effect and gravity produces coherent pulses of atoms. (c) At NIST, Raman pulses extract atoms pulses in a chosen direction. When the pulses overlap, a quasicontinuous atom laser is achieved. (d) In Munich, a weak RF coupler extracts a continuous atom wave from the condensate. *Right*: Absorption images of a nonideal atom laser, corresponding to density integration along the elongated axis x of the BEC. The figures correspond to different heights of RF outcoupler detunings with respect to the bottom of the BEC: (a) -0.37 μm, (b) $-2, 22$ μm, (c) -3.55 μm. The graph above shows the RF outcoupler (dashed line) and the BEC slice (red), which is crossed by the atom laser and results in the observation of caustics.

so far coupling attempts using either cold atomic beams[59,60] or cold have atomic clouds have led to relatively low coupling efficiency. To increase this efficiency, a solution consists in creating the atom laser directly into the guide,[61] leading eventually to a continuous guided atom laser analogous to the photonic fiber laser. This has recently been achieved in Orsay (LCFIO), where the BEC from which the atom laser is extracted from is *pigtailed* to the atom guide (see Fig. 4). In this setup, an atom laser is outcoupled from a hybrid optomagnetic trap to an optical guide. The propagation direction is fixed by the propagation direction of the dipole trap laser beam, and the velocity of the outcoupled atoms can be controlled by carefully adjusting the guide parameters. Using this scheme, an atomic de Broglie wavelength as high as 0.7 μm was observed.

4.2. *The prospects and limits of high density coherent samples*

The fact that ultracold bosons interact is a major drawback for precision measurements using atom interferometry. In the above experiment, interactions result in a systematic shift as well as a decrease in measurement precision. In principle, the systematic shifts can be calculated. However, the interaction parameter U is hard to measure and is generally not known to be better than $\sim 10^{-4}$. The atomic density is also subject to time fluctuations and is difficult to know to be better than $\sim 10^{-2}$, reducing the absolute accuracy. In addition, as shown in earlier experiments,[62–64]

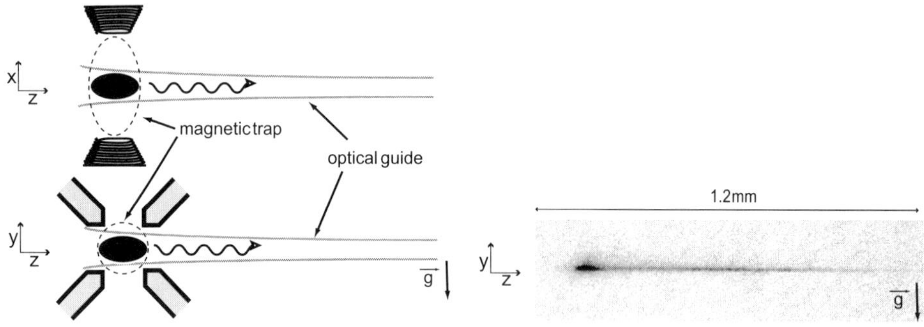

(a) (b)

Fig. 4. (a) Schematic view of the setup. The BEC is obtained in a crossed hybrid magnetic and optical trap. The optical trap is horizontal. Its focus is shifted in the longitudinal direction z so as to attract the atoms. (b) Experimental absorption image of a guided atom laser after 50 ms of outcoupling. The imaging is along the x axis.

interactions produce a loss of coherence of the atomic samples at ultralow, finite temperatures, limiting the maximum interrogation time of a coherent matter-wave atom interferometer. Finally, even at zero temperature, the mean-field energy due to interactions is converted into kinetic energy during free fall, giving rise to a faster ballistic expansion. This last effect will ultimately reduce interrogation times.

4.3. *The need of an ideal coherent atomic source*

From the observations of both MIT and Orsay, we conclude that one should ideally use an interaction-free, ultracold atomic source for ultimate-precision atom interferometry in space. Using bosons, one could think of two ways of decreasing interaction effects. Close to a Feshbach resonance,[65–69] one can control the interaction parameter U, which can be made equal to zero for a certain magnetic field. However, magnetic fields introduce further systematic shifts that are not controllable to within a reasonable accuracy. Alternatively, one could try to decrease the density of the sample of atoms, but the production of large-atom-number, ultralow-density Bose–Einstein condensate is a technical challenge not yet met.[70] A promising alternative solution is to use quantum-degenerate fermionic atomic sources.[71] The Pauli exclusion principle forbids symmetric two-body collision wave functions, so at zero temperature a sample of neutral atomic fermions has no interactions. An ultracold fermionic source may still allow very long interrogation times, even if limited by the excess energy of the Fermi pressure, and would therefore be an ideal candidate for atom interferometry in space with ultimate precision and accuracy.

5. ICE: Towards a Coherent Atom Sensor for Space Applications

The objective of ICE,[72] a CNES-funded project that shares the experience of various partners (SYRTE, ONERA and IOTA), is to produce an accelerometer for space with coherent atomic source. It uses a mixture of Bose–Einstein condensates with two species of atoms (Rb and K) to carry out a first comparison of accelerations measured by the two different types of atomic species (bosons and fermions). The central components of this project are the atomic-physics vacuum system, the optics, and their supports. The atomic manipulation starts with alkali-metal vapor dispensers for rubidium and potassium. A slow jet of atoms is sent from the collection chamber by a dual-species, two-dimensional, magneto-optical trap (2D-MOT) to the trapping chamber, for collection and cooling in a 3D-MOT. Atoms are then to be transferred to a conservative, far-off-resonance optical-dipole trap (FORT) for further cooling toward degeneracy. The sample is then ready for coherent manipulation in an atom interferometer. Raman two-photon transition will be used as atomic beam-splitters and mirrors. Three-pulse sequences ($\pi/2 - \pi - \pi/2$) will be used for accelerometry. All light for the experiment arrives by optical fibers, making the laser sources independent of the vacuum system. Transportable fibered laser sources for laser cooling and trapping have been fabricated with the required frequency stability. The techniques for mechanically stable power distribution by free-space fiber couplers function according to specifications. The vacuum chamber is compatible with the constraints of microgravity in an Airbus parabolic flight. Such a flight permits total interrogation times of up to 7 s, giving a potential sensitivity of better than 10^{-9} m/s^2 per shot, limited by phase noise on the frequency reference for the Raman transitions.

5.1. *Laser systems*

5.1.1. *Continuous-wave fiber-laser source at 780 nm for rubidium cooling*

An entirely pigtailed laser source is particularly appropriate in our case as it does not suffer from misalignments due to environmental vibrations. Moreover, telecommunications laser sources in the C-band (1530–1570 nm) have narrow line widths ranging from less than 1 MHz for laser diodes to a few kHz for erbium-doped fiber lasers. By second-harmonic generation (SHG) in a nonlinear crystal, these 1.56 μm sources can be converted to 780 nm sources.[73–75] Such devices avoid having to use extended cavities as their line widths are sufficiently narrow to satisfy the requirements of laser cooling.

The laser setup is sketched in Fig. 5. A 1560 nm erbium-doped fiber laser is amplified by a 500 mW polarization-maintaining (PM) erbium-doped fiber amplifier (EDFA). A 90/10 PM fiber-coupler directs 10% of the pump power to a pigtailed output. 90% of light is then sent into a periodically poled lithium-niobate wave guide (PPLN-WG). This crystal is pigtailed on both sides with 1560 nm single-mode fibers. The input fiber is installed in a polarization loop system in order to align the electric field with principal axes of the crystal. A fiber-coupler which is monomode at 780 nm filters pump light after the crystal and sends half of the 780 nm light into a saturated-absorption spectroscopy device for frequency servo-control. The other half is the frequency-stabilized pigtailed output. The whole device, including

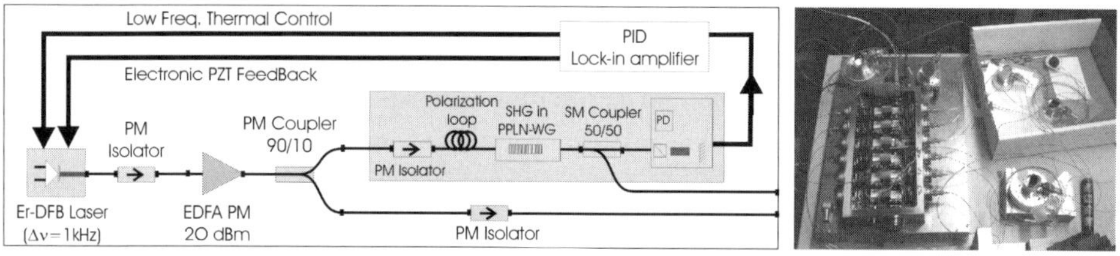

Fig. 5. *Left*: Transportable laser setup schematic. A double-loop feedback system is used for frequency control: the first 1000 returns a saturated absorption signal to the piezoelectric transducer; the second loop compensates for thermal for drifts of the fiber laser when the error signal of the first loop becomes large. *Right*: Fiber splitters developed at SYRTE.

the frequency control electronics, was implemented in a rack for ease of transport. Typical output from the first generation device was 500 μW of 780 nm light, with more than 86 dB attenuation of 1560 nm light after 3 m of monomode fiber.

5.1.2. *Fiber power splitters*

The optical bench and the vacuum chamber are not rigidly connected to each other, and laser light is transported to the vacuum chamber using optical fibers. Stability in trapping and coherent atom manipulation is assured by using only PM fibers. Six trapping and cooling laser beams are needed for the 3D-MOT and five for the 2D-MOT, with relative power stability better than a few percent. A fiber beam-splitters based on polarizing cubes and half-wave plates with one input fiber and the relevant number of output fibers. The stability of the beam splitters has been tested by measuring the ratio of output powers between different outputs as a function of time. Fluctuations are negligible on short time scales (less than 10^{-4} relative intensity over 1 s), and very small over typical periods of experimental operation (less than 1% over a day).

5.2. *Mechanical and vacuum systems*

The mechanical construction of the apparatus is critical to any free-fall experiment. Atomic-physics experiments require heavy vacuum systems and carefully aligned optics. The ICE design is based on a cuboidal frame of foam-damped hollow bars with one face being a vibration-damped optical breadboard; see Fig. 6. The outside dimensions are 1.2 m × 0.9 m × 0.9 m, and the total weight of the final system is estimated to be 400 kg (excluding power supplies, lasers, control electronics, air and water flow). The frame provides support for the vacuum system and optics, which are positioned independently of each other. The heavy parts of the vacuum system are rigged to the frame using steel chains and high-performance polymer slings under tension, adjusted using turnbuckles; most of the equipment is standard in recreational sailing or climbing. The hollow bars have precisely positioned grooves

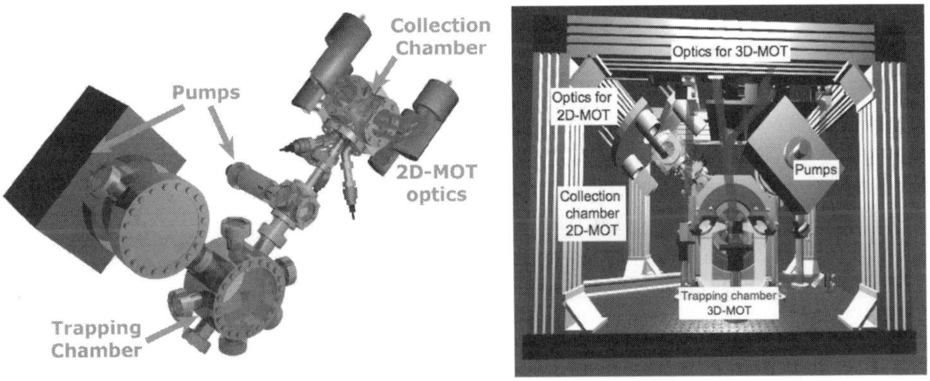

Fig. 6. *Left*: Artist's impression of the vacuum system. Atoms are transferred from the collection chamber, using a 2D-MOT, to the trapping chamber, where they are collected in a 3D-MOT. The trapping chamber has large optical accesses for the 3D-MOT, an optical-dipole trap (FORT), imaging, and interferometry. There is a getter pump between the two chambers to ensure a large pressure difference. The other pump is a combined ion pump–titanium sublimation pump. *Right*: ICE mechanical structure with optics and light paths represented.

which permit optical elements to be rigidly fixed (bolted and glued) almost anywhere in the volume within the frame. An adaptation for transportability will be to enclose the frame in a box, including acoustic and magnetic shielding, temperature control, air overpressure (dust exclusion), as well as ensuring safety in the presence of the high-power lasers.

The vacuum chamber has three main parts: the collection chamber (for the 2D-MOT), the trapping chamber (for the 3D-MOT and the FORT) and the pumps (combined ion pump–titanium sublimation pump) Between the collection and trapping chambers there is an orifice and a getter pump, allowing for a high differential pressure, permitting rapid collection by the 2D-MOT but low trap losses in the 3D-MOT and the FORT. The magnetic coils for the 2D-MOT are under vacuum, and consume just 5 W of electrical power.

To avoid heating due to vibrations in the FORT optics, or measurement uncertainties due to vibrations of the imaging system, the trapping chamber is as close to the breadboard as possible. For laboratory tests, the breadboard is lowest, and the 2D-MOT arrives at 45° to the vertical, leaving the vertical axis available for addition of interferometry for precise measurements, e.g. a standing light wave. Around the main chamber, large electromagnet coils in a Helmholtz configuration will be added, to produce homogeneous, stable fields of up to 0.12 T (1200 G), or gradients of up to 0.6 T/m (60 G/cm).

5.2.1. *2D-MOT*

The 2D-MOT is becoming a common source of cold atoms in two-chamber atomic-physics experiments, and is particularly efficient for mixtures[76] of ^{40}K and ^{87}Rb, if isotopically enriched dispensers are used. Briefly, a 2D-MOT has four sets of beams (two mutually orthogonal, counterpropagating pairs) transverse to the axis of the output jet of atoms, and a cylindrical-quadrupole magnetic field generated by elongated electromagnet pairs (one pair, or two orthogonal pairs). Atoms are cooled transverse to the axis, as well as collimated. Implicitly, only slow atoms spend enough time in the 2D-MOT to be collimated, so the output jet is longitudinally slow. The number of atoms in the jet can be increased by the addition of the push beam, running parallel to the jet: a 2D-MOT$^+$. Typically the output jet has a mean velocity below 30 m/s, with up to 10^{10} atoms per second of ^{87}Rb and 10^8 atoms per second of ^{40}K.

The ICE design uses 40 mW per species for each of the four transverse beams, each divided into two zones of about 20 mm using nonpolarizing beam-splitter cubes, corresponding to about three times the saturation intensity for the trapping transitions. The push beam uses 10 mW of power, and is about 6 mm in diameter. Each beam comes from an individual PM optical fiber, with the light at 766.5 nm and 780 nm being superimposed on entry to the fibers.

5.2.2. *3D-MOT and optical-dipole trap*

The atomic jet from the 2D-MOT is captured by the 3D-MOT in the trapping chamber (see Fig 7). The 3D-MOT uses one PM fiber input per species. Beams are superimposed and split into six arms (on a small optical breadboard fixed near one face of the frame) for the three orthogonal, counterpropagating beam pairs. Once enough atoms are collected in the 3D-MOT, the 2D-MOT is to be turned off, and the 3D-MOT optimized for transfer to the FORT, which consists of two

Fig. 7. *Left*: Artist's impression of the 3D-MOT (dark, red beams and the electromagnets) and the far-off-resonance optical-dipole trap (pale, yellow beams). *Right*: Photograph of the vacuum chamber, the support structure and the optics for magneto-optical traps. The main chamber has two very large viewports as well as seven side windows (and one entry for the atoms from the 2D-MOT). Thus there is very good optical access for the 3D-MOT, the FORT, imaging and interferometry. To preserve this optical access, the magnetic coils are outside of the chamber, although this markedly increases their weight and power consumption.

nearly orthogonal (70°) beams making a crossed dipole trap using 50 W of light at 1565 nm. Rapid control over intensity is acheived using an electro-optical modulator, and beam size using a mechanical zoom, after the design of Kinoshita *et al.*[77] Optimization of transfer from the 3D-MOT to the FORT, and the subsequent evaporative cooling, can be enhanced with strong, homogeneous, magnetic fields that can be used to control interspecies interactions via Feshbach resonances,[65–69] to expedite sympathetic cooling of ^{40}K by ^{87}Rb. With the expected loading of the 3D-MOT during less than 5 s, then cooling to degeneracy in the optical-dipole trap in around 3–10 s, ICE will be able to prepare a sample for interferometry in less than the free-fall time of a parabolic flight (around 20 s).

Acknowledgments

The ICE collaboration is funded by CNES. The ICE team members are P. Boyer, R. Nyman, G. Varoquaux, J.-F. Clement and J.-P. Brantut from IOTA, A. Landragin, F. Pereira and C. Borde from SYRTE, A. Bresson, Y. Bidel, F. Deysac and P. Touboul from ONERA, and L. Mondin and M. Rouze from CNES. Further support comes from the European Union STREP consortium FINAQS.

References

1. W. W. Chow *et al.*, *Rev. Mod. Phys.* **72** (1985) 61.
2. J. F. Clauser, *Physica B* **151** (1988) 262.
3. S. A. Werner, J.-L. Staudenmann and R. Colella, *Phys. Rev. Lett.* **42** (1979) 1103.
4. R. Colella, A. W. Overhauser and S. A. Werner, *Phys. Rev. Lett.* **34** (1975) 1472.
5. F. Riehle *et al.*, *Phys. Rev. Lett.* **67** (1991) 177.
6. M. Kasevich and S. Chu, *Appl. Phys. B* **54** (1992) 321.
7. P. R. Berman (ed.), *Atom Interferometry* (Academic, 1997).
8. T. L. Gustavson, P. Bouyer and M. A. Kasevich, *Phys. Rev. Lett.* **78** (1997) 2046.
9. T. L. Gustavson *et al.*, *Class. Quant. Grav.* **17** (2000) 1.
10. A. Peters *et al.*, *Phil. Trans. Roy. Soc. Lond. A* **355** (1997) 2223
11. A. Peters, K. Y. Chung and S. Chu, *Metrologia* **38** (2001) 25.
12. A. Peters, K. Chung and S. Chu, *Metrologia* **38** (2001) 25.
13. M. Kasevich and S. Chu, *Phys. Rev. Lett.* **67** (1991) 181.
14. D. W. Keith *et al.*, *Phys. Rev. Lett.* **66** (1991) 2693.
15. M. J. Snadden *et al.*, *Phys. Rev. Lett.* **81** (1998) 971.
16. J. M. McGuirk *et al.*, *Phys. Rev. A* **65** (2002) 033608.
17. Ch. J. Bordé, in *Advances in the Interplay Between Quantum and Gravity Physics*, eds. P. G. Bergmann and V. de Sabbada (Kluwer, 2002).
18. M. Fattori *et al.*, *Phys. Lett. A* **318** (2003) 184.
19. A. Wicht *et al.*, *Proc. 6th Symposium on Frequency Standards and Metrology*, ed. Patrick Gill (World Scientific, 2001) p. 193.
20. M. H. Anderson *et al.*, *Science* **269** (1995) 198.
21. K. B. Davis *et al.*, *Phys. Rev. Lett.* **75** (1995) 3969.
22. C. C. Bradley, C. A. Sackett and R. G. Hulet, *Phys. Rev. Lett.* **75** (1995) 1687.
23. M.-O. Mewes *et al.*, *Phys. Rev. Lett.* **78** (1997) 582.
24. B. P. Anderson and M. A. Kasevich, *Science* **282** (1998) 1686.
25. E. W. Hagley *et al.*, *Science* **283** (1999) 1706.

26. I. Bloch, T. W. Hänsch and T. Esslinger, *Phys. Rev. Lett.* **82** (1999) 3008.
27. G. E. Stedman *et al.*, *Phys. Rev. A* **51** (1995) 4944.
28. G. E. Stedman, *Rep. Prog. Phys.* **60** (1997) 615.
29. P. Bouyer and M. Kasevich, *Phys. Rev. A* **56** (1997) R1083.
30. S. Gupta *et al.*, *Phys. Rev. Lett.* **89** (2002) 140401.
31. Y. Le Coq *et al.*, *Appl. Phys. B* **84** (2006) 627.
32. C. Jentsch *et al.*, *Gen. Relativ. Gravit.* **36** (2004) 2197.
33. C. Bordé, *Phys. Lett. A* **140** (1989) 10.
34. D. M. Giltner, R. W. McGowan and S. A. Lee, *Phys. Rev. Lett.* **75** (1995) 2638.
35. C. Cohen-Tannoudji, *Cours au Collège de France* (1992–1993).
36. Ch. Antoine and C. J. Bordé, *Phys. Lett. A* **306** (2003) 277.
37. P. Storey and C. Cohen-Tannoudji, *J. Phys. II France* **4** (1994) 1999.
38. M. Kasevich *et al.*, *Phys. Rev. Lett.* **63** (1989) 612.
39. A. Clairon *et al.*, *Europhys. Lett.* **16** (1991) 165.
40. Y. Sortais, S. Bize and M. Abgrall, *Phys. Scripta* **2001** (2001) 50.
41. Ch. J. Bordé, *Metrologia* **39** (2002) 435.
42. K. U. Schreiber *et al.*, *J. Geophys. Res.* **109** (2004) B06405.
43. T. Niebauer *et al.*, *Metrologia*, **32** (1995) 159.
44. G. Santarelli *et al.*, *Phys. Rev. Lett.* **82** (1999) 4619.
45. D. Wineland *et al.*, *Phys. Rev. A* **46** (1992) 6797.
46. H. Marion *et al.*, *Phys. Rev. Lett.* **90** (2003) 150801.
47. H. J. Metcalf and P. Van Der Straten, *Laser Cooling and Trapping* (Springer-Verlag, Berlin, 1999).
48. W. Ketterle, ScientificAmerican.com, "Ask the Experts," Jan. 19, 2004. *Sci. Am.*, May 2004, p. 120.
49. J. Stenger *et al.*, *Phys. Rev. Lett.* **82** (1999) 4569.
50. Y. Shin *et al.*, *Phys. Rev. Lett.* **92** (2004) 050405.
51. *"When Atoms Behave as Waves: Bose–Einstein Condensation and the Atom Laser,"* in *Les Prix Nobel* 2001 (The Nobel Foundation, Stockholm, 2002), pp. 118–154. Reprinted in *ChemPhysChem* **3** (2002) 736; *Rev. Mod. Phys.* **74** (2002) 1131.
52. W. Hänsel *et al.*, *Phys. Rev. Lett.* **86** (2001) 608.
53. T. Paul *et al.*, *Phys. Rev. A* **72** (2005) 063621.
54. D. Clément *et al.*, *Phys. Rev. Lett.* **95** (2005) 170409.
55. N. P. Robins *et al.*, *Phys. Rev. A* **72** (2005) R031606.
56. Y. Le Coq *et al.*, *Phys. Rev. Lett.* **87** (2001) 17.
57. J.-F. Riou *et al.*, *Phys. Rev. Lett.* **96** (2006) 070404.
58. F. Gerbier, P. Bouyer and A. Aspect. *Phys. Rev. Lett.* **86** (2001) 4729.
59. Z. T. Lu *et al.*, *Phys. Rev. Lett.* **77** (1996) 3331.
60. D. Müller *et al.*, *Phys. Rev. A* **61** (2000) 033411.
61. T. Lahaye *et al.*, *Phys. Rev. Lett.* **93** (2004) 093003.
62. S. Richard *et al.*, cond-mat/0303137.
63. S. Richard *et al.*, *Phys. Rev. Lett.* **91** (2003) 010405.
64. D. Hellweg *et al.*, *Phys. Rev. Lett.* **91** (2003) 010406.
65. W. C. Stwalley, *Phys. Rev. Lett.* **37** (1976) 1628.
66. E. Tiesinga *et al.*, *Phys. Rev. A* **46** (1992) R1167.
67. P. Fedichev *et al.*, *Phys. Rev. Lett.* **77** (1996) 2913.
68. M. Theis *et al.*, *Phys. Rev. Lett.* **93** (2004) 123001.
69. J. L. Roberts *et al.*, *Phys. Rev. Lett.* **86** (2001) 4211.
70. A. E. Leanhardt *et al.*, *Science* **301** (2003) 1513.
71. G. Roati *et al.*, *Phys. Rev. Lett.* **92** (2004) 230402.

72. R. Nyman *et al.*, *Appl. Phys. B.* **84** (2006) 673. See also http://ice-space.fr.
73. V. Mahal *ei al.*, *Opt. Lett.* **21** (1996) 1217.
74. R. J. Thompson *et al.*, *Opt. Exp.* **11** (2003) 1709.
75. J. Dingjan *et al.*, *App. Phys. B* **82** (2006) 47.
76. C. Ospelkaus *et al.*, *Phys. Rev. Lett.* **96** (2006) 020401.
77. T. Kinoshita, T. R. Wenger and D. S. Weiss. *Phys. Rev. A* **71** (2005) R01162.

RUBIDIUM BOSE–EINSTEIN CONDENSATE UNDER MICROGRAVITY

A. PETERS and W. LEWOCZKO-ADAMCZYK*
Institut für Physik, Humboldt-Universität zu Berlin,
Hausvogteiplatz 5-7, 10117 Berlin, Germany
**lewoczko@physik.hu-berlin.de*

T. VAN ZOEST, E. RASEL and W. ERTMER
Institut für Quantenoptik, Universität Hannover,
Welfengarten 1, 30167, Hannover, Germany

A. VOGEL, S. WILDFANG, G. JOHANNSEN,
K. BONGS and K. SENGSTOCK
Institut für Laser-Physik, Universität Hamburg,
Luruper Chaussee 149, 22761 Hamburg, Germany

T. STEIMNETZ and J. REICHEL
Laboratorie Kastler Brossel de l'ENS,
24 rue Lhomond, 75231 Paris Cedex 05, France

T. KÖNEMANN, W. BRINKMANN, C. LÄMMERZAHL
and H. J. DITTUS
ZARM Universität Bremen,
Am Fallturm, 28359 Bremen, Germany

G. NANDI, W. P. SCHLEICH and R. WALSER
Abteilung Quantenphysik, Universität Ulm,
Albert-Einstein-Allee 11, 89069 Ulm, Germany

Weightlessness promises to substantially extend the science of quantum gases toward presently inaccessible regimes of low temperatures, macroscopic dimensions of coherent matter waves, and enhanced duration of unperturbed evolution. With the long-term goal of studying cold quantum gases on a space platform, we currently focus on the implementation of an ^{87}Rb Bose–Einstein condensate (BEC) experiment under microgravity conditions at the ZARM[a] drop tower in Bremen (Germany). Special challenges in the construction of the experimental setup are posed by a low volume of the drop capsule

[a]Center of Applied Space Technology and Microgravity.

($< 1\,\mathrm{m}^3$) as well as critical vibrations during capsule release and peak decelerations of up to 50 g during recapture at the bottom of the tower. All mechanical and electronic components have thus been designed with stringent demands on miniaturization, mechanical stability and reliability. Additionally, the system provides extensive remote control capabilities as it is not manually accessible in the tower two hours before and during the drop. We present the robust system and show results from first tests at the drop tower.

Keywords: Bose–Einstein condensate; microgravity; chip trap.

1. Introduction

For a few decades, the trapping, cooling, and manipulation of neutral atoms[1] has been an especially active field in modern physics. Continuous improvement toward achievement of ever-lower temperatures temporarily culminated in the experimental realization of Bose–Einstein condensation in 1995.[2,3] Since that time, laboratory experiments have been able to almost routinely produce Bose–Einstein condensation (BEC's) and many studies have extensively investigated the properties of this new state of matter. Fundamental insights into matter wave interference, superfluidity and vortex lattices, solitons and four-wave mixing in matter waves, atom lasers, quantum phase transitions, and controlled cold molecule production, to name only a few (see e.g. Ref. 4 for a review), have been provided. At the same time, further efforts to reduce the energy of trapped atoms resulted in reaching a low-temperature record of 500 pK.[5] The corresponding average energy per atom at this temperature equals the gravitational potential energy of a single Rb atom at a height of 5 nm, much smaller than a typical physical dimension of a condensate. Therefore, in the low-temperature regime, Earth gravity presents a major perturbation to the system.

Weightlessness offers significant potential to extend the physics of degenerate quantum gases in new directions. First of all, in the absence of disturbing gravitational force, it is possible to adiabatically "open" the trapping potential without the need for any levitational fields compensating gravity. Reducing the confinement of the trap enlarges the physical extension of the ground state. Resulting ultralarge condensates, possibly extending over 10 mm, hold promise for investigating the matter waves more precisely, since optical detection and manipulation can be performed with higher relative spatial resolution. On the other hand, lowered ground state energy and expected corresponding temperature in the femtokelvin range represent an improvement of up to three orders of magnitude as compared to present experiments. Another important point is a significantly extended time of free and unperturbed evolution of a condensate released from the trap, which is crucial for precision interferometric measurements with coherent dilute matter waves.[6] Last but not least, microgravity is a suitable environment for investigating mixtures of cold gases, since atoms with different masses do not experience different gravitational forces and can be equally well held by the trapping potential.

In this paper, we briefly present the current status of our collaborative effort to realize a ^{87}Rb BEC under microgravity. More details, including a theoretical description of a freely falling BEC, can be found in Ref. 7.

2. Microgravity at the Drop Tower

Our long-term goal is to establish an experimental platform in space to allow the investigation of ultracold quantum matter in free fall for an unlimited time. At the present, we focus on the implementation of an ^{87}Rb BEC experiment at the ZARM drop tower in Bremen. The drop tower environment is similar to that of a space platform in several aspects. First, it offers an excellent acceleration suppression down to the microgravity level of 10^{-6} g, which is four orders of magnitude better than during a parabolic flight of a zero-g aircraft. Second, strict requirements concerning the limited volume, low power consumption, and the impact deceleration of the drop capsule have to be fulfilled. Thus, special technical challenges in the construction of the experimental setup make it to a large extent different from common Earth-bound setups. To go into detail, the following points have been crucial for the design of all mechanical, optical, and electronic components:

- *Miniaturization.* A common laboratory BEC experiment usually fills the area of the whole optical laboratory. In contrast to this, our setup has to fit into the drop capsule (shown in Fig. 1), the volume of which is less than 1 cubic meter.
- *Low mass.* The drop capsule can carry a maximum mass of 230 kg. In all space missions weight is an important factor in terms of launch costs.
- *Low power consumption.* Electrical power for the experiments in the drop tower is supplied from batteries placed in the bottom of the drop capsule. They can provide a total energy amount of as much as 0.56 kWh, with a maximal momentary power of about 3 kW. Obviously, there are even stronger energy constraints on the space platform.

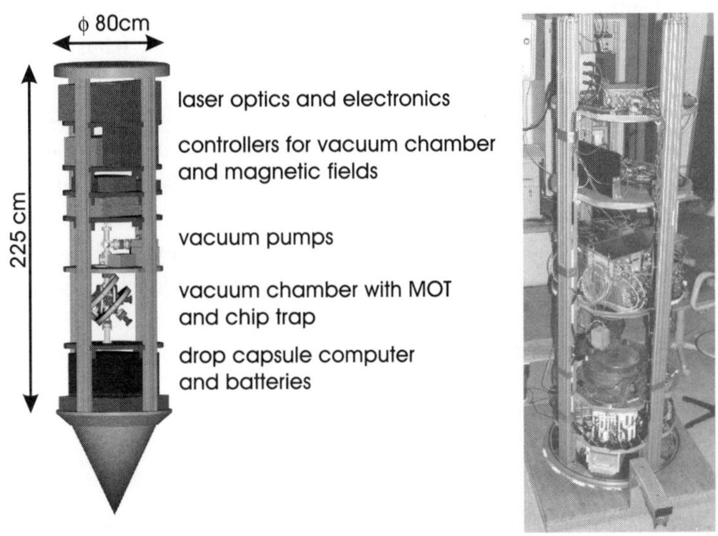

Fig. 1. Scheme of the drop capsule (*left*) and the capsule with the laser system only (*right*).

- *High mechanical stability.* Residual vibrations of the platforms in the drop capsule at the moment of capsule release can be critical to laser frequency and light power stability. Moreover, all the components used have to withstand the peak deceleration of about 50 g at the end of the drop. By the maximal drop rate of three times per day, there is only less than an hour for likely corrections and readjustment of the setup between the flights. Therefore, it is desired that the experiment is not permanently misaligned after each drop. Fulfilling this requirement is also important because of the presence of violent shocks and vibrations during the launch phase in future space missions.
- *Fast BEC preparation.* In order to dump out the effect of the above-mentioned vibrations on atoms, we plan to keep the atoms during and after release in the nonconservative magneto-optical trap (MOT). Thus, all following cooling phases, in particular the evaporation in the magnetic trap, have to be faster than the total drop time of about 4.5 s minus the time needed to investigate the condensate in weightlessness.
- *Remote control capability.* While the drop tube is evacuated, the experiment is for two hours accessible only via remote control. In particular, one has to be able to lock the lasers to an adequate atomic line without manual access to them.

In the following section we describe the experimental setup in detail, emphasizing how it addresses those issues. We also present data from the first drop tower tests of the laser system.

3. Experimental Setup

As already mentioned, BEC's can nowadays be almost routinely produced in many optical laboratories and a number of different experimental techniques are being extensively discussed in the literature.[1,8] The main requirements are, however, similar. The most important are good thermal decoupling from the environment using contact-free storage in ultrahigh-vacuum chambers (typically 10^{-10} mbar) and a sophisticated two-stage trapping and cooling process. The latter begins with precooling using Doppler and sub-Doppler laser cooling in a magneto-optical trap down to the temperatures in the μK range. Subsequently, the cold atomic sample is transferred to a conservative trapping potential (magnetic or optical) and cooled further by evaporative cooling.[9,10] In order to keep the setup as simple as possible, we trap ^{87}Rb atoms, which, like other alkalis, have a simple laser cooling scheme for which laser diodes are commercially available. Furthermore, the relatively high ratio between elastic and inelastic collision rates in ^{87}Rb is advantageous for evaporative cooling.

3.1. *Atomic chip trap*

In order to meet the need for short evaporation times and low power consumption, we use a magnetic microtrap on a chip.[11–13] In a chip trap, the required magnetic

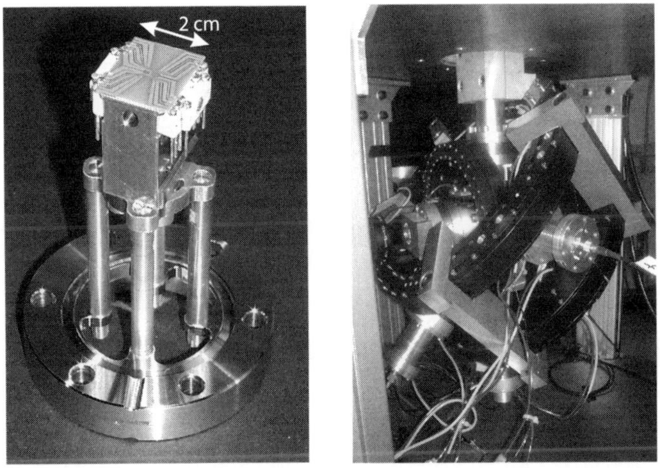

Fig. 2. Vacuum flange with the atom chip (*left*) and the vacuum chamber with MOT (*right*).

field gradients are produced by currents flowing through a microstrip line (Fig. 2). An atom chip requires only a power of 10 W (as compared to about 1 kW in conventional, coil-based magnetic traps) plus an additional 350 W for external bias field coils and external MOT coils. Nevertheless, the magnetic field gradients and therefore the confinement in the magnetic chip trap are stronger by one order of magnitude compared with magnetic traps realized with external coils only. This promises short evaporation times of the order of 1 s, well within the time of free fall in the drop tower.

3.2. *Vacuum chamber*

The atom chip is mounted in a nonmagnetic stainless steel vacuum chamber, which is kept at a low pressure of 10^{-10} mbar by a titanium sublimation pump and an ion-getter pump. It is important that there are no moving elements inside either of the pumps, which could be damaged during deceleration of the capsule. Moreover, they could disturb the quality of the microgravity. The laser light needed to prepare, cool, and detect the atomic ensemble is transmitted to the vacuum chamber via polarization-maintaining optical fibers and is expanded to a diameter of 20 mm with a telescope arrangement. All optics are rigidly attached to the steel body of the vacuum chamber, giving maximum stability and minimizing possible sources of misalignment.

3.3. *Laser system*

Full control over both the frequencies of the lasers and that of the light power at the moment of capsule release and during the flight are critical parameters that assure a satisfactory number of trapped atoms at the temperature low enough to efficiently load the magnetic trap. For the sake of mechanical stability, we have intentionally excluded the use of (common in laser trapping) extended cavity diode lasers (ECDL's). Instead, we drive our laser system with distributed feedback (DFB) laser diodes, which have an intrinsic grating in the active semiconductor area. Commercially available Eagleyard DFB diodes (EYP-DFB-0780-00080-1500-TOC03-0000) mounted in a TO3 housing with an internal Peltier element fit well our requirements regarding compactness. Moreover, the diode's emission line width lies just in the order of magnitude of the ^{87}Rb D2 transition natural line width ($\sim 6\,\mathrm{MHz}$). Out of several diodes, we selected the one with the smallest line width (2 MHz) to use it as a cooling laser. The line width was measured by recording a beat note of two independent, freely running diode lasers. We stabilize the temperature of the diodes, and the only variable parameter influencing the wavelength is a current. DFB diodes have an extremely wide mode-hop-free operation range of more than 100 GHz, which greatly facilitates their use. Information on rubidium spectroscopy with DFB diode lasers can be found in Ref. 14.

At present, the laser system consists of two atomic transition stabilized master lasers, a MOPA (master oscillator power amplifier), and a power distribution and control module with acousto-optic modulators (AOM's). The first master laser, which provides the light to repumping transition in the cooling cycle, is based on a Doppler-free dichroic atomic vapor laser lock (DFDL).[15] The second one is locked utilizing modulation transfer spectroscopy (MTS)[16,17] and is used as a reliable frequency reference for all other required transitions. The MOPA is frequency offset-locked to the MTS master laser.

All mounts for optics with the beam height of 20 mm were designed by us, with a special emphasis on mechanical stability. Both master modules and the MOPA are placed in stable and robust housing ($210 \times 190 \times 60\,\mathrm{mm}$) made of stress-free aluminium. The three lasers together with the AOM-intensity-control module (see Fig. 3) are integrated into an assembly with dimensions chosen to be compatible with a standard 19" electronic rack. The laser system including electronics has a total mass of about 45 kg and fits the area of one platform (80 cm diameter) in the drop capsule.

4. Drop Tower Test of the Laser System

Stability tests of the laser system have been carried out at the drop tower in Bremen. Figure 4(a) shows the error signals of both DFDL and MTS master lasers and of the offset lock during capsule release. All three lasers remain locked and are not sensitive to residual capsule vibrations. Figure 4(b) shows the stability of the fiber coupling during the flight. No significant change in the intensity of light coming out of the fiber during the capsule release has been observed. The test results indicate that the laser system in the present form can be used in the drop BEC experiment.

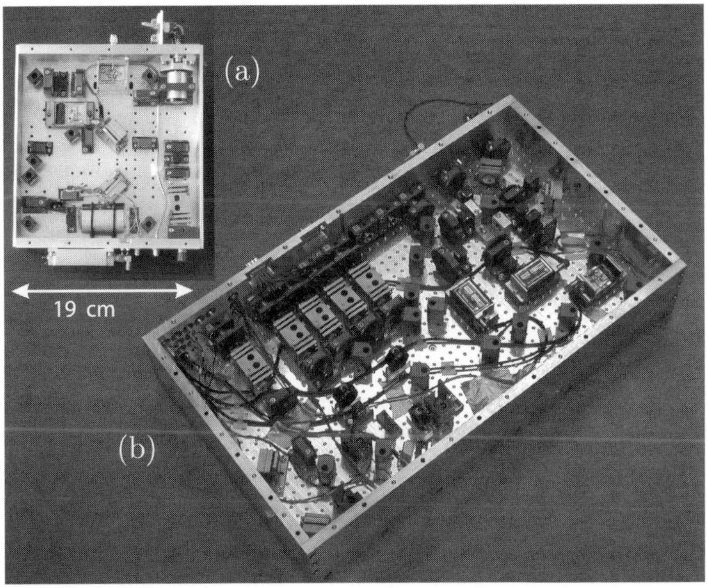

Fig. 3. (a) MTS master laser and (b) AOM-power-control module.

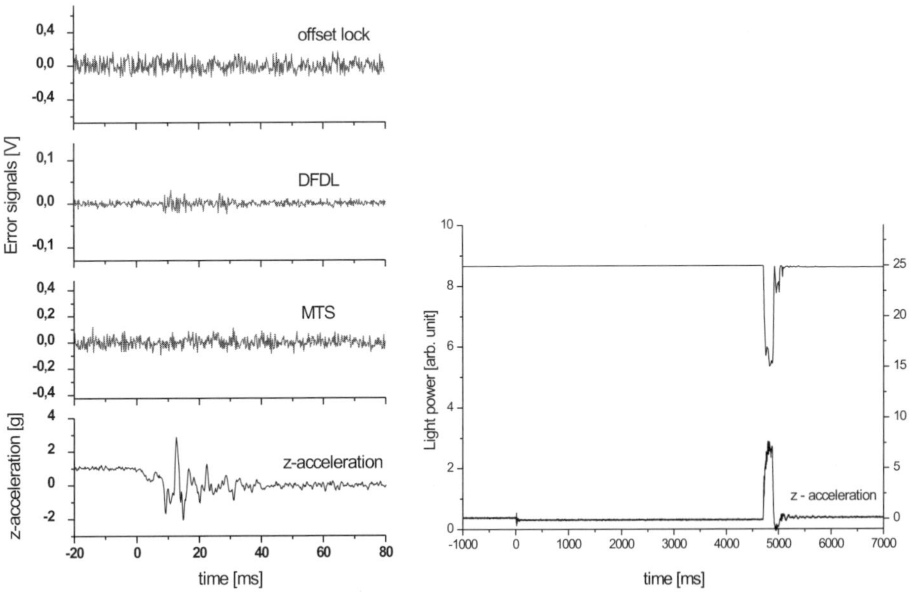

(a) (b)

Fig. 4. (a) Error signals of the three laser locks during the capsule release. No change of the laser frequency has been observed. (b) Intensity of the light coming out of the fiber during the flight. At the moment of capsule release the intensity remains stable; it drops down by about 40% during recapture to finally reach its beginning value. The latter indicates no need to readjust the fiber coupling after each drop.

5. Current Status

At the moment of writing this paper we are able to capture 1.2×10^7 atoms in the external MOT and transfer nearly 85% of them to the chip trap. The atoms are further cooled in optical molasses to a temperature of about $40\,\mu$K and subsequently transferred to the magnetic trap on the chip. The corresponding phase space density is about 10^{-6}. Optimizing transfer to the magnetic trap and preparing the final evaporative cooling are being done.

6. Summary

To sum up, we have presented a compact and robust experimental setup for producing a Bose–Einstein condensate under microgravity during a free fall in the drop tower. At the moment of writing this paper we optimize the parameters of the magnetic trap and prepare the last stage on the way to the BEC — the evaporative cooling. First drop experiments with the condensate are planned for the end of the year.

Acknowledgment

The project is funded under grant number 50 WM 0346 by the German Federal Ministry of Economics and Technology (BMWi) via the German Space Agency (DLR).

References

1. H. J. Metcalf and P. van der Straten, *Laser Cooling and Trapping* (Springer-Verlag, New York, 1999).
2. M. H. Anderson *et al.*, *Science* **269** (1995) 198.
3. K. B. Davis *et al.*, *Phys. Rev. Lett.* **75** (1995) 3969.
4. K. Bongs and K. Sengstock, *Rep. Prog. Phys.* **67** (2004) 907.
5. A. E. Leanhardt, *Science* **301** (2003) 1513.
6. S. Bize and P. Laurent, *C. R. Phys.* **5** (2004) 829.
7. A. Vogel *et al.*, *Appl. Phys. B* **84** (2006) 663.
8. W. Ketterle, D. S. Durfee and D. M. Stamper-Kurn, cond-mat/9904034.
9. K. B. Davis *et al.*, *Phys. Rev. Lett.* **74** (1995) 5202.
10. W. Ketterle and N. J. Van Druten, *Adv. Atom Mol. Opt. Phys.* **37** (1996) 181.
11. W. Hänsel *et al.*, *Nature* **413** (2001) 498.
12. J. Reichel, W. Hänsel and T. W. Hänsch, *Phys. Rev. Lett.* **83** (1999) 3398.
13. H. Ott *et al.*, *Phys. Rev. Lett.* **87** (2001) 230401.
14. S. Kraft *et al.*, *Laser Phys. Lett.* **2** (2005) 71.
15. G. Wasik *et al.*, *Appl. Phys. B* **75** (2002) 613.
16. J. H. Shirley, *Opt. Lett.* **7** (1982) 537.
17. J. Zhang *et al.*, *Opt. Express* **11** (2003) 1338.

TIME, CLOCKS AND FUNDAMENTAL PHYSICS

CLAUS LÄMMERZAHL and HANSJÖRG DITTUS
ZARM, University of Bremen,
Am Fallturm, 28359 Bremen, Germany
laemmerzahl@zarm.uni-bremen.de
dittus@zarm.uni-bremen.de

Time is the most basic notion in physics. Correspondingly, clocks are the most basic tool for the exploration of physical laws. We show that most of the fundamental physical principles and laws valid in today's description of physical phenomena are related to clocks. Clocks are an almost universal tool for exploring the fundamental structure of theories related to relativity. We describe this structure and give examples where violations of standard physics are predicted and, thus, may be important in the search for a theory of quantum gravity. After stressing the importance for future precise clock experiments to be performed in space, we refer to the OPTIS mission, to which another article in this issue is devoted. It is also outlined that clocks are not only important for fundamental tests but at the same time are also indispensable for practical purposes like navigation, Earth sciences, metrology, etc.

Keywords: Special relativity; general relativity; experimental tests.

1. Introduction

Time is the primary and most basic notion of human existence and of physics. Accordingly, clocks — based on astrophysical or Earth-bound phenomena — have very early been considered as a tool for giving phenomena a causal order. Today precise time-keeping — now based on atomic clocks — is the most fundamental and important activity of all bureaus of standards around the world. Without precise time-keeping it is not possible to uncover the nature of physical phenomena and to set up physical laws. In fact, we will see that clocks are the most basic tools in order to set up the equations describing gravitation phenomena and in general the laws of physics. Only if clocks behave in a certain way do we arrive at the physical laws we know to be valid today. As a consequence, clocks are a fundamental tool for confirming and testing the present description of physical phenomena encoded in the Standard Model and in Einstein's special and general theory of relativity (SR and GR).

In this article we emphasize how clocks and their particular behavior determine, and are determined by, the structure of SR and GR. Furthermore, clocks may be of great importance in the search for a new — still unknown — theory combining GR and quantum theory, since it has been shown that these two theories in their present forms show a fundamental incompatibility. Since this new theory is different from the standard theory of gravity and quantum mechanics, one or more of the basic principles underlying these theories should be violated. This is strongly suggested by the low energy limit of string theory or loop quantum gravity, for example. And since most of these principles are related to the behavior of clocks, clocks will play a central and crucial role in the search for effects related to the theory of quantum gravity.

2. The Situation of Standard Physics

For an introduction to the subject and the related important questions, we briefly describe the present structure of GR. We also outline the deep connection of clocks with the structure of SR and GR, which is also basic for metrology, and show that the present system of physical units will completely break down if SR and/or GR turn out to be violated or, equivalently, clocks behave differently.

2.1. *The structure of the general theory of relativity*

The structure of GR is shown in the scheme in Fig. 1. This scheme has two aspects[1]:

(i) The Einstein equivalence principle (EEP) primarily fixes the equations of motion for the matter fields, i.e. the Maxwell equations, the Dirac equation, etc.
(ii) Through the characterization of the still-possible structure of couplings in these field equations, the EEP implies that gravity has to be described by a pseudo-Riemannian space–time metric.

The structure of the equations of motion for matter is determined by the EEP (see Fig. 1), which consists of:

The universality of free fall (UFF). This principle states that all kinds of structureless matter fall in the same way in the gravitational field. In order to test that, one needs, in principle, to compare in a huge number of experiments the free fall of all sorts of available materials. However, in the frame of elementary particle theory this just means that all elementary particles fall in the gravitational field in the same way. This is an amazing fact: Why should all particles behave in the same way? From what do all the particles know how the other particles behave? This fact is characteristic of the gravitational interaction. No other interaction shares this property.

The universality of the gravitational redshift (UGR). This principle states that all kinds of clocks based on nongravitational physics (pendula or sand clocks are not

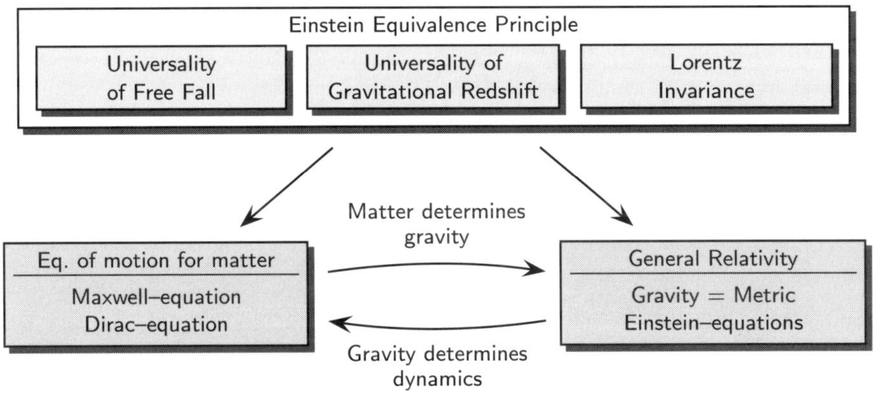

Fig. 1. The scheme of general relativity: the EEP determines the structure of the equations of motion of matter and, consequently, of the gravitational field.

allowed) behave in the same way when transported together through a position-dependent gravitational field. This again means that all particles and all (nongravitational) interactions (also represented by particles) couple in the same way to the gravitational field. Again, one has to test this for all kinds of clocks and analyzes the result in terms of the coupling of elementary particles to gravity.

The local validity of Lorentz invariance (LLI). The third underlying principle is the local validity of Lorentz invariance. This means that the outcome of all local small-scale experiments is independent of the orientation and the state of motion of the freely falling laboratory: by means of experiments it should not be possible to single out a particular reference system. In particular, this means that the velocity of light is constant and all limiting velocities of elementary particles are again given by the velocity of light. This again is an amazing fact: From what do all the particles know the properties of the other particles? Since LLI is a property which applies to all physics, one has to perform tests with all physical systems. The Michelson–Morley, Kennedy–Thorndike, and Ives–Stilwell tests are the best-known tests of this type with photons which have to completed by tests with electrons, protons, etc. which are given by, for example, the Hughes–Drever experiments.

One can show[2] that from these three principles the gravitational field has to be described by a space–time metric; see Fig. 1. Some more assumptions are needed in order to see that the metric has to obey Einstein's field equations.

The EEP thus determines the structure of the equations of motion of matter, i.e. of the

- light rays and point particles,[3,4]
- Maxwell equations,[5,6]
- Dirac equation[7] (which in the nonrelativistic limit leads to the Schrödinger equation),
- structure of the Standard Model,

and by that also fixes the metrical structure of gravity. Only if the equations of motion of these matter field possess a certain structure are they compatible with the metrical structure of the gravitational field.

2.2. *The consequences*

All aspects of GR are experimentally well tested and confirmed. The tests split into tests of the EEP and tests of the predictions of GR. No single test contradicts its foundation or its predictions. These predictions are[2]:

- Solar system effects
  - — perihelion shift,
  - — gravitational redshift,
  - — deflection of light,
  - — gravitational time delay,
  - — Lense–Thirring effect,
  - — Schiff effect.
- GR in the strong field regime has been proven to be valid to very high accuracy by means of the observation of binary systems.
- The existence of gravitational waves as predicted by GR has been indirectly proven by the energy loss experienced by binary systems.
- Cosmology.

3. Tests of SR and GR

From the EEP it is evident that clocks provide an almost universal tool for testing the foundations of physics: besides the tests of the UFF and of the universality of the maximum velocity of particles[8,9] (including photons), all other tests can be regarded as clock tests. The tests to be carried out are clock comparison tests. In one class of clock tests the rates of neighboring clocks, hypothetically depending on their position, their velocity and their orientation, are compared; see Fig. 2. In the other class of tests, clocks at different positions with different velocities are compared. For a review of the SR test, see Refs. 9 and 10.

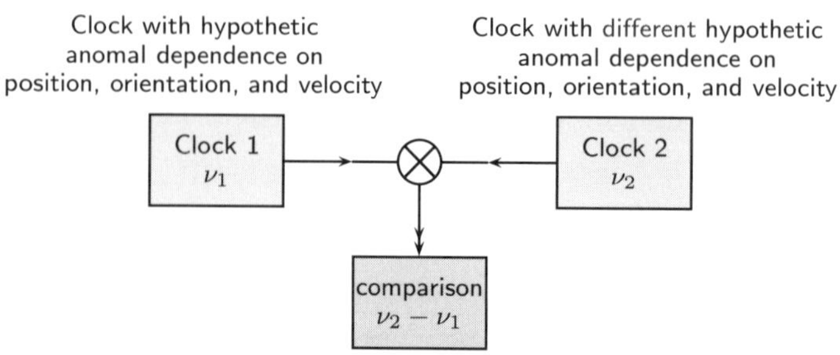

Fig. 2. Comparison of clocks of different nature and in different states of motion yields a complete test of special relativity and also tests UGR.

The first of the three classical tests of SR, i.e. the test of the isotropy of the speed of light (Michelson–Morley tests), compares the velocity of light in two orthogonally oriented optical resonators which constitute light clocks. The relative difference of the speed of light in different directions is now smaller than[11–14] $\Delta_\vartheta c/c \leq 10^{-16}$ and is approaching the 10^{-17} level.

The difference of the speed of light in differently moving inertial systems is now smaller than[15] $\Delta_v c/c \leq 10^{-16}$. In these Kennedy–Thorndike type experiments one compares an optical resonator with an atomic clock during the change of the state of motion, which in most cases is given by the motion of the Earth around its own axis and around the Sun.

With the time dilation factor γ in the parametrization $\gamma(v) = 1+\left(\frac{1}{2}+\alpha\right)\frac{v^2}{c^2}+\cdots$ (for SR we have $\alpha = 0$) the most recent experiment[16] gave $|\alpha| \leq 2.2 \times 10^{-7}$. Time dilation has also been verified by the decay of moving elementary particles; see Ref. 17 for a recent version of these experiments where time dilation has been verified at the 10^{-3} level. Though these experiments are not as precise as the spectroscopic ones, they prove the time dilation for a different physical process and, thus, its universality. Another class of time dilation experiments are rotor experiments.[18–20] The achieved accuracy was $|\alpha| \leq 10^{-5}$.

4. Clocks

There are many clocks which rely on completely different physical mechanisms. This is related to the dependence of the energy transitions or frequencies on different combinations of physical constants. These constants are the mass of the electron and the proton, m_e and m_p, the fine structure constant $\alpha = e^2/\hbar c$, and the coupling constant for the weak interaction G_F. Accordingly, we have the following scaling properties of transition energies in units of the Rydberg energy:

Hyperfine energies: $g(m_e/m_p)\alpha^2 f(\alpha)$, where g and f are some functions;
Vibrational energies in molecules: $\sqrt{m_e/m_p}$;
Rotational energies in molecules: m_e/m_p;
Fine-structure energies: α^2;
Electronic energies: $f(\alpha)$;
Cavity frequency: Since the length of a cavity is given by the atomic Bohr radius, which is proportional to the fine-structure constant, the frequency of a standing wave in a cavity is proportional to $1/\alpha$.
Weak interaction splitting/Zeeman frequency: G_{F}.

Since these expressions are derived from the Dirac equation coupled to the electromagnetic potential, it is clear that they will change if the equation of motion for the Dirac and/or the Maxwell field will be modified.

Further clocks are given by:

Decay of particles: The lifetime of decaying particles also defines a time unit. This is much less precise than atomic clocks but is based on a completely different physical process than atomic clocks.

The Earth's rotation: For long time the rotation of the Earth gave the time standard. This has been replaced by atomic clocks.
Pulsars: Pulsars define a very precise time standard,[21] which is, however, a bit less precise than atomic clocks.

The essential point is that different clocks depend on different physical laws. Therefore, clocks tests are basically tests of the properties of the underlying physical laws in gravitational fields and in different states of motion. That also means that one has to compare *all* types of clocks in order to explore the structure of all physical laws.

5. Implications for Metrology

Modern metrology consists of the definition, preparation and reproduction of physical units. The most important of these units is the second; see Fig. 3. Since now SR and GR are the physics of space and time and since time and distances today are measured with high precision clocks, modern metrology can be regarded as a practical realization of SR and GR. In fact, one needs SR and GR in order to set up the International Atomic Time (TAI), and thus today's definition of the second. To establish the TAI one needs to compare clocks at different positions on the surface of the Earth. Each of these positions possesses another velocity and gravitational potential in the Earth-centered nonrotating reference system. For a precise comparison of the clocks at these positions one has to consider the time dilation, the Sagnac effect and the gravitational redshift.

Furthermore, the meter is today defined as the distance light travels within a certain fraction of a second and, thus, is reduced to the measurement of time; see Fig. 3. The uniqueness of this definition requires the validity of SR and GR. If the

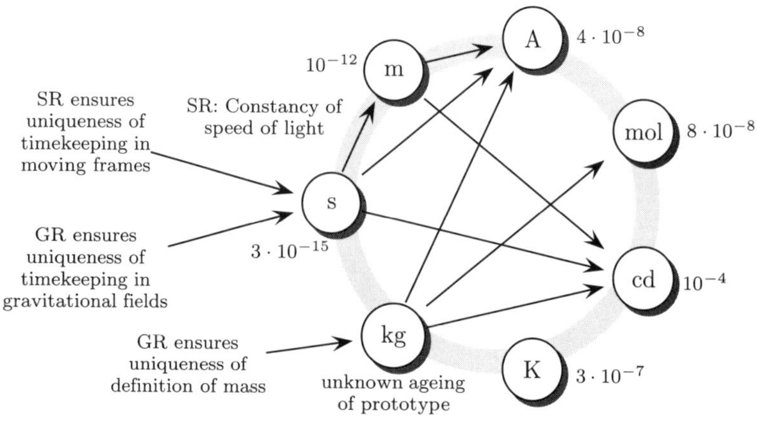

Fig. 3. The SI units (s = second, m = meter, A = ampere, mol = mole, cd = candela, K = kelvin, kg = kilogram) and their interdependences.[22] The numbers indicate the stability of the corresponding unit. The uniqueness of the transport of the definition of the second and of the meter depends on the validity of the EEP.

velocity of light depends on the direction or on the velocity of the laboratory, for example, then we obtain for each direction and for each laboratory moving with different speed, a different meter standard.

In a bit more complicated way, a future definition of the kilogram by means of, for example, a Watt balance is also related to the definition of time. The Watt balance depends on the quantum Hall and the Josephson effect, which rely on the validity of the Dirac and Maxwell equations, which are also at the basis of the TAI.

In general, the validity of the EEP is a prerequisite for modern metrology, whose task is to replace all units by universally reproducible quantum realizations. This is possible only if one assumes a theoretical description of these effects based on the ordinary Schrödinger and Maxwell equations; see Fig. 4.

6. Practical Use of High Precision Clocks

The physics of clocks and the validity of SR and GR are not only important for fundamental questions and modern metrology but are a necessary tool for practical purposes. One of these practical purposes is the Global Positioning System (GPS), which will fail by more than 10 km per day if one does not take into account SR and GR.[23,24] High precision positioning is also used for high precision navigation of spacecraft, since for that one needs very precisely the location of the ground stations used for tracking and ranging the spacecraft. This is important, for example, for the analysis of the Pioneer 10 and 11 data.[25]

Furthermore, the precise measurement of the rotation of the Earth yields a lot of useful information on the state of the Earth,[26] i.e. on Earth tides, climate changes (the El Niño and El Niña phenomenan, for example), ocean warming, etc.

7. Are There Problems in This Scheme?

Besides the fact that, as we will see below, GR and/or quantum theory have to be modified to a theory of quantum gravity which leads to tiny deviations from

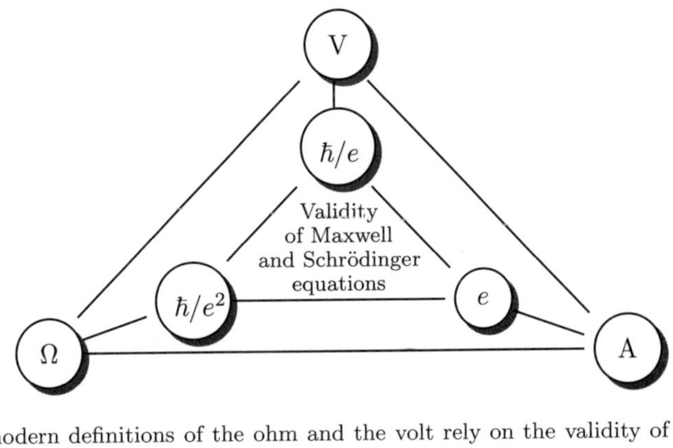

Fig. 4. The modern definitions of the ohm and the volt rely on the validity of basic equations. Any modifications of these equations will influence these definitions.

standard physics, there are already a few phenomena for which there is a lack of understanding within the standard theory of GR. These phenomena are:

- dark matter,
- dark energy,
- the Pioneer anomaly,
- the flyby anomaly,[27–29]
- the quadrupole/octupole anomaly,[30]
- a reported tiny increase of the astronomical unit.[31,32]

The question now is whether this might already be regarded as a signature for "new physics." And have these effects something to do with clocks, or can these phenomena be explored better by using more precise clocks? For example, the Pioneer anomaly and an increase of the astronomical unit can be simulated by a drift of the clocks on the Earth. In order to exclude that, one should perform a high precision comparison of clocks with astrophysical phenomena like pulsars and binary systems. For a further exploration of the Pioneer or flyby anomaly by future space missions, clocks may be of help in performing measurements independent of ranging and Doppler tracking and, thus, may reveal additional information about the gravitational field.

8. The Need and Search for Quantum Gravity

The present status of the theoretical description of the physical world is given by four universal theories and four interactions. The universal theories — theories which apply to all kinds of matter and phenomena — are quantum theory, SR, GR and statistics (or condensed matter). On the other side, we have the four interactions — electromagnetism, gravity, weak interaction and strong interaction. It is one of the big wishes that there might be a true unification of all these interactions. This is not a logical necessity but it might be very useful for understanding the physical world.

Frame theories	Interactions
Quantum theory Special relativity General relativity Statistical mechanics	Electromagnetism Gravity Weak interaction Strong interaction
Problem:	Wish:
Incompatibility of quantum theory and general relativity	Unification of all interactions

On the other side, however, there is a big problem, namely the incompatibility of quantum mechanics and GR. Since both theories have to be applied to all phenomena, this incompatibility necessarily has to be resolved. The theory which leads to a consistent coexistence between some kind of quantum mechanics and some kind of a theory of gravity is called "quantum gravity." Therefore, the most fundamental quest of modern physics is the search for a theory of quantum gravity.

Further reasons for the need to have a quantized version of GR are[33]:

- The problem of time: While time in quantum mechanics is an externally given parameter, in GR it is a dynamical variable which is influenced by the gravitational field and, thus, by the matter content in the Universe.
- According to the discussion of Bohr and Rosenfeld, it has been shown that if matter is quantized (and this is without any doubt), then the interactions between the matter, e.g. the electromagnetic field, have to be quantized, too. Therefore, gravity also has to be quantized.
- The role of singularities (particularly black holes): In classical GR, singularity theorems state that under very general assumptions singularities will occur where all known physics will break down. Quantization of the gravitational field may circumvent the breakdown of physics in such singularities and, in particular, in the early Universe and in black holes.

If we accept that there has to be some underlying theory for quantum gravity which is different from the standard theories, then at some stage deviations from the experimental outcomes as predicted by standard theories should be expected. These deviations come from tiny modifications of the standard equations like the Dirac and Maxwell equations and are related to tiny violations of the EEP. Other searches look for a fundamental decoherence or a modification of the dispersion relation. Possible effects which are expected to follow from a theory of quantum gravity are:

- birefringence[34–36]
- anisotropic speed of light,[34,35]
- anisotropy in quantum fields,[37]
- violation of UFF,[38–41]
- violation of UGR,[38–41]
- anomalous dispersion,[42]
- modification of Newton potential at short,[43] and long distances,[44,45]
- modification of Coulomb potential,[46,47]
- decoherence, fundamental noise,[48]
- nonlocalities,
- charge nonconservation.[5]

For an estimate of the magnitude of observable effects from quantum gravity, one first derives an effective low energy limit. For string theory one arrives in particular at a dilaton scenario[40,41] which explicitly predicts a violation of the UGR. Further predictions of a violation of the UGR are given in Refs. 49–51. Most of these "predictions" are in the range to be testable within the next few years with improved clocks.[1]

9. Space Conditions

In many cases the experimental conditions on the Earth are too restrictive for obtaining better results. Here we list the cases where it is of great advantage to perform experiments is space. We give particular space conditions and mention the corresponding tests and give examples for already performed or planned space tests.

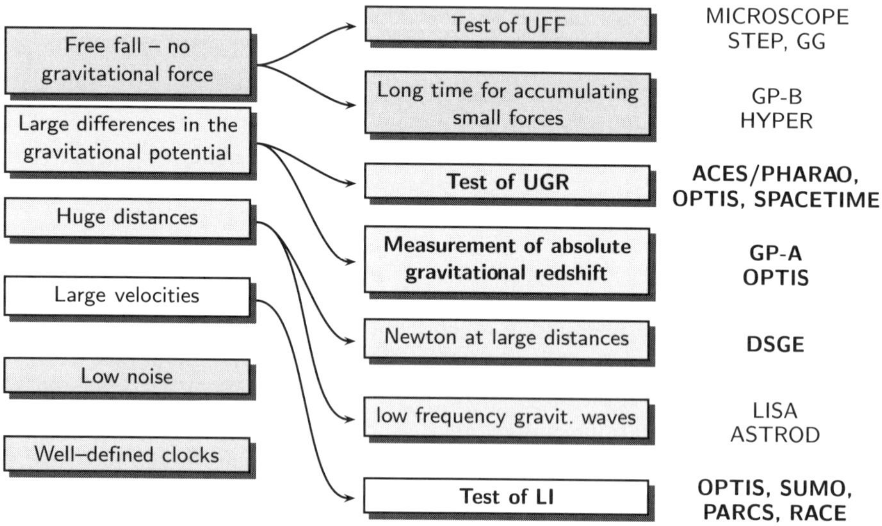

Clock tests (boldface) will benefit most from the large differences in the gravitational potential obtainable in space and from large velocity changes [ACES = Atomic Ensembe in Space[52,53]; PHARAO = Projet d'Horloge Atomique par Refroidissement d'Atomes en Orbite[52,53]; GP-A = Gravity Probe A[54]; GP-B = Gravity Probe B[55]; OPTIS = Optical Tests of the Isotropy of Space[53]; STEP = Satellite Test of the Equivalence Principle[56]; MICROSCOPE = MICRO-Satellite à trainée Compensée pour l'Observation du Principle d'Equivalence[57]; GG = Galileo Galilei[58]; HYPER = HYPER precision atomic interferometry[59]; SPACETIME[60]; DSGE = Deep Space Gravity Explorer[61] (= Pioneer anomaly mission), for the clock projects PARCS, RACE, and SUMO; see Ref. 53.]

The low noise environment is not specific for particular missions; it is a benefit for all missions. For a technical application of new high precision clocks, the environment on the Earth is too ill-defined (the height of the clock changes due to movements of the surface of the Earth, etc.) so that one has to go to space; only there can the reading of clocks be interpreted in a well-defined manner.

10. The OPTIS Mission

The OPTIS mission is a satellite performing all possible clock tests described above in space. It uses a variety of clocks and laser ranging and tracking facilities. This mission makes advantage of the space conditions of large differences in the velocity and the gravitational potential. The proposed OPTIS mission aims at an improvement of the complete test of the LLI and of the UGR by three orders of magnitude compared to the present ground experiments. Therefore, together with the test of the UFF by MICROSCOPE and STEP, we will have a complete test of the foundations of GR.

For more information about this mission, refer to the corresponding article in this issue, Ref. 62.

Acknowledgments

We would like to thank the German Aerospace Agency (DLR) and the German Research Foundation (DFG) for financial support.

References

1. C. Lämmerzahl, *Appl. Phys. B* **84** (2006) 551.
2. C. M. Will, *Theory and Experiment in Gravitational Physics*, revised edn. (Cambridge University Press, Cambridge, 1993).
3. J. Ehlers, F. A. E. Pirani and A. Schild, The Geometry of Free Fall and Light Propagation, in *General Relativity: Papers in Honour of J. L. Synge*, ed. L. O'Raifeartaigh (Clarendon Press, Oxford, 1972), p. 63.
4. J. Ehlers, Survey of general relativity theory, in *Relativity, Astrophysics and Cosmology*, ed. W. Israel (Reidel, Dordrecht, 1973), p. 1.
5. C. Lämmerzahl, A. Macias and H. Müller, *Phys. Rev. D* **71** (2005) 025007.
6. W.-T. Ni, *Phys. Rev. Lett.* **38** (1977) 301.
7. J. Audretsch and C. Lämmerzahl, A new constructive axiomatic scheme for the geometry of space–time, in *Semantical Aspects of Space–Time Geometry*, eds. U. Majer and H.-J. Schmidt (BI Verlag, Mannheim, 1993), p. 21.
8. C. Lämmerzahl, *Ann. Phys. (Leipzig)* **14** (2005) 71.
9. G. Amelino-Camelia *et al.*, The search for quantum gravity signals, in *Gravitation and Cosmology*, eds. A. Macias, C. Lämmerzahl and D. Nunez (*AIP Conf. Proc. 758*, Melville, New York, 2005), p. 30.
10. D. Mattingly, *Living Rev. Relativ.* **8** (2005) 5; http://www.livingreviews.org/lrr–2005–5, 2005.
11. P. Antonini *et al.*, *Phys. Rev. A* **71** (2005) 050101.
12. P. Stanwix *et al.*, *Phys. Rev. Lett.* **95** (2005) 040404.
13. S. Herrmann and A. Peters, Test of Lorentz invariance using a continuously rotating optical resonator, in *Special Relativity: Will It Survive the Next 100 Years?*, eds. J. Ehlers and C. Lämmerzahl (Springer-Verlag, Berlin, 2006), Lecture Notes in Physics, Vol. 702, p. 385.
14. S. Herrmann *et al.*, *Phys. Rev. Lett.* **95** (2005) 150401.
15. P. Wolf *et al.*, *Phys. Rev. D* **70** (2004) 051902.
16. S. Saathoff *et al.*, *Phys. Rev. Lett.* **91** (2003) 190403.

17. J. Bailey *et al.*, *Nature* **268** (1977) 301.
18. M. Ruderfer, *Phys. Rev. Lett.* **5** (1960) 191.
19. D. C. Champeney, G. R. Isaak and A. M. Khan, *Phys. Lett.* **7** (1963) 241.
20. D. C. Champeney, G. R. Isaak and A. M. Khan, *Proc. Phys. Soc.* **85** (1965) 583.
21. N. Wex, Pulsar timing — Strong gravity clock experiments, in *Gyros, Clocks, and Interferometers: Testing Relativistic Gravity in Space*, eds. C. Lämmerzahl, C. W. F. Everitt and F. W. Hehl (Springer-Verlag, Berlin, 2001) p. 381.
22. T. Quinn, *Metrologia* **31** (1995) 515.
23. N. Ashby, *Phys. Today*, May 2002, p. 42.
24. N. Ashby, *Living Rev. Relativ.* **6** (2003) 1; http://www.livingreviews.org/lrr–2003–1.
25. J. D. Anderson *et al.*, *Phys. Rev. D* **65** (2002) 082004.
26. C. Lämmerzahl, *Ann. Phys.* (*Leipzig*) **15** (2006) 5.
27. P. G. Antreasian and J. R. Guinn, Investigations into the unexpected delta-v increase during the earth gravity assist of GALILEO and NEAR, in *Astrodynamics Specialist Conference and Exhibition* (American Institute of Aeronautics and Astronautics, Washington, 1998), pp. 98–4287.
28. J. D. Anderson and J. G. Williams, *Class. Quant. Grav.* **18** (2001) 2447.
29. C. Lämmerzahl, O. Preuss and H. Dittus, Is the phyiscs in the solar system really understood? in *Lasers, Clocks, and Drag-Free: Exploration of Relativistic Gravity in Space*, eds. H. Dittus, C. Lämmerzahl and S. G. Turyshev (Springer-Verlag, Berlin, 2007), p. 75.
30. D. J. Schwarz *et al.*, *Phys. Rev. Lett.* **93** (2004) 221301.
31. G. A. Krasinsky and V. A. Brumberg, *Celest. Mech. Dyn. Astron.* **90** (2004) 267.
32. E. M. Standish, The astronomical unit now, in *Transits of Venus: New Views of the Solar System and Galaxy: Proceedings IAU Colloquium No. 196*, ed. D. W. Kurtz (Cambridge University Press, 2005), p. 163.
33. C. Kiefer, *Quantum Gravity* (Oxford University Press, 2004).
34. R. Gambini and J. Pullin, *Phys. Rev. D* **59** (1999) 124021.
35. J. Alfaro, H. A. Morales-Tecotl and L. F. Urrutia, *Phys. Rev. D* **65** (2002) 103509.
36. A. Kostelecky and M. Mewes, *Phys. Rev. Lett.* **87** (2001) 251304.
37. J. Alfaro, H. A. Morales-Tecotl and L. F. Urrutia, *Phys. Rev. D* **66** (2002) 124006.
38. T. Damour and A. M. Polyakov, *Nucl. Phys. B* **423** (1994) 532.
39. T. Damour and A. M. Polyakov, *Gen. Relativ. Gravit.* **12** (1996) 1171.
40. T. Damour, F. Piazza and G. Veneziano, *Phys. Rev. Lett.* **89** (2002) 081601.
41. T. Damour, F. Piazza and G. Veneziano, *Phys. Rev. D* **66** (2002) 046007.
42. G. Amelino-Camelia and C. Lämmerzahl, *Class. Quant. Grav.* **21** (2004) 899.
43. I. Antoniadis, Physics with large extra dimensions and non-Newtonian gravity at sub-mm, distances, in *Quantum Gravity — From Theory to Experimental Search*, eds. D. Giulini, C. Kiefer and D. Lämmerzahl (Springer-Verlag, Berlin, 2003), p. 337.
44. D. Dvali, G. Gabadadze and M. Porrati, *Phys. Lett. B* **485** (2000) 208.
45. M.-T. Jaekel and S. Reynaud, *Class. Quant. Grav.* **22** (2005) to appear.
46. A. Kostelecky and M. Mewes, *Phys. Rev. D* **66** (2002) 056005.
47. H. Müller *et al.*, *Phys. Rev. D* **67** (2003) 056006.
48. J. Ellis *et al.*, *Nucl. Phys. B* **241** (1984) 381.
49. H. B. Sandvik, J. D. Barrow and J. Magueijo, *Phys. Rev. Lett.* **88** (2002) 031302.
50. C. Wetterich, *Phys. Lett. B* **561** (2003) 10.
51. C. Wetterich, *Astropart. Phys.* **10** (2003) 2.
52. C. Salomon *et al.*, *C. R. Acad. Sci. Paris* **4** (2004) 1313.
53. C. Lämmerzahl *et al.*, *Gen. Relativ. Gravit.* **36** (2004) 615.
54. R. F. C. Vessot *et al.*, *Phys. Rev. Lett.* **45** (1980) 2081.

55. C. W. F. Everitt *et al.* of the Gravity Probe B team, Gravity Probe B: Countdown to launch, in *Gyros, Clocks, and Interferometers: Testing Relativistic Gravity in Space*, eds. C. Lämmerzahl, C. W. F. Everitt and F. W. Hehl (Springer-Verlag, Berlin, 2001), p. 52.
56. N. Lockerbie *et al.*, STEP: A status report, in *Gyros, Clocks, and Interferometers: Testing Relativistic Gravity in Space*, eds. C. Lämmerzahl, C. W. F. Everitt and F. W. Hehl (Springer-Verlag, Berlin, 2001), p. 213.
57. P. Touboul, *Comptes Rendus de l'Aced. Sci. Série IV: Physique Astrophysique* **2** (2001) 1271.
58. http://tycho.dm.unipi.it/nobili
59. C. Jentsch *et al.*, *Gen. Relativ. Gravit.* **36** (2004) 2197.
60. L. Maleki and J. Prestage, SpaceTime mission: Clock test of relativity at four solar radii, in *Gyros, Clocks, and Interferometers: Testing Relativistic Gravity in Space*, eds. C. Lämmerzahl, C. W. F. Everitt and F. W. Hehl (Springer-Verlag, Berlin, 2001), p. 369.
61. H. Dittus and Pioneer Explorer Collaboration, A mission to explore the Pioneer anomaly, in *Trends in Space Science and Cosmic Vision 2020*, eds. Á. Giménez *et al.* (ESA, Noordwijk, 2005), p. 3 [gr-qc/0506139].
62. H. Dittus and C. Lämmerzahl, *Int. J. Mod. Phys. D* **16** (2007) 2499–2510.

PROBING RELATIVITY USING SPACE-BASED EXPERIMENTS

NEIL RUSSELL
Physics Department, Northern Michigan University,
Marquette, MI 49855, USA
nrussell@nmu.edu

An overview of space tests searching for small deviations from special relativity arising at the Planck scale is given. Potential high-sensitivity space-based experiments include ones with atomic clocks, masers, and electromagnetic cavities. We show that a significant portion of the coefficient space in the Standard Model extension, a framework that covers the full spectrum of possible effects, can be accessed using space tests. Some remarks on Lorentz violation in the gravitational sector are also given.

Keywords: Lorentz violation; Standard Model extension.

1. Introduction

The Standard Model coupled to general relativity is considered to be the best existing physical theory of nature. It is thought to be the effective low-energy limit of an underlying fundamental theory that unifies the gravity and matter sectors at the Planck scale. This underlying theory may well include Lorentz violation, detectable in experiments with appropriate types and levels of sensitivity. These nonstandard effects can be described for practical purposes using effective field theory.[1] If one takes the general-relativity-coupled Standard Model and adds appropriate terms that involve operators for Lorentz violation, the result is the Standard Model extension (SME), which has provided a framework for Lorentz testing for more than a decade. Fundamental theories describing such violation could involve string theory[2–5] and spontaneous symmetry breaking.[6–10]

The Minkowski-space–time limit of the SME[11,12] has been examined in several dozen experiments. Theoretical aspects of the photon physics of the SME have been looked at,[13–17] one of them being radiative corrections.[18–24] Experimental investigations of electromagnetism of the SME have considered microwave and optical cavities,[25–36] Doppler-shift experiments,[37] and Cerenkov radiation.[38–47] Work aimed at investigating the electron physics in the SME includes studies and measurements of electron coefficients made using Penning traps,[48–52] torsion pendula,[53–55] and scattering physics.[56–58] Lorentz-violating effects involving protons and neutrons have been investigated,[59–62] and a variety of tests involving clock-comparison experiments have been performed.[63–72] Similar tests with antihydrogen[73,74] may be possible in the near future. This article will focus on the possibility of testing Lorentz symmetry in the space environment.[75,76] The SME physics associated with muons has been studied and experiments performed.[77–79] Similarly, SME physics has been researched for neutrinos,[80–86] the Higgs,[87,88] and baryogenesis and nucleosynthesis.[89–92] Several studies have been done with neutral mesons.[93–106] The relationship of the SME to noncommutative geometry has been uncovered,[107–112] and numerous other aspects have been investigated.[113–119] Several general reviews of the SME are available.[120–125]

In this article, we first consider the use of clock-comparison experiments for measuring the coefficients for Lorentz violation in the Minkowski limit of the SME. The clocks referred to here take several forms, including atomic clocks, masers, and electromagnetic cavity oscillators. We also discuss the pure-gravity sector of the SME.[126–129]

2. Fermions and the Minkowski Limit of the SME

The Lagrangian describing a spin-$\frac{1}{2}$ Dirac fermion ψ of mass m in the presence of Lorentz violation is:[11,12]

$$\mathcal{L} = \frac{1}{2} i \overline{\psi} \Gamma_\nu \overleftrightarrow{\partial^\nu} \psi - \overline{\psi} M \psi, \tag{1}$$

where

$$M := m + a_\mu \gamma^\mu + b_\mu \gamma_5 \gamma^\mu + \frac{1}{2} H_{\mu\nu} \sigma^{\mu\nu}, \tag{2}$$

$$\Gamma_\nu := \gamma_\nu + c_{\mu\nu} \gamma^\mu + d_{\mu\nu} \gamma_5 \gamma^\mu + e_\nu + i f_\nu \gamma_5 + \frac{1}{2} g_{\lambda\mu\nu} \sigma^{\lambda\mu}. \tag{3}$$

In this expression, the conventional Lorentz-symmetric case is contained in the first terms in the expressions for M and Γ_ν above. The other terms contain conventional Dirac matrices $\{1, \gamma_5, \gamma^\mu, \gamma_5\gamma^\mu, \sigma^{\mu\nu}\}$ and coefficients for Lorentz violation a_μ, b_μ, $c_{\mu\nu}$, $d_{\mu\nu}$, e_μ, f_μ, $g_{\lambda\mu\nu}$, $H_{\mu\nu}$. Various possible mechanisms can be envisaged for an underlying theory giving rise to such coefficients, including spontaneous symmetry breaking.[6–10] The coefficients $c_{\mu\nu}$ and $d_{\mu\nu}$ are traceless, while $H_{\mu\nu}$ is antisymmetric and $g_{\lambda\mu\nu}$ is antisymmetric in its first two indices. The terms in the equation for M have dimensions of mass, and those in the equation for Γ_ν are dimensionless.

The coefficients for Lorentz violation appearing in the Lagrangian can be thought of as fixed geometrical background objects in space–time. Thus, any experiment rotating in space could detect time-dependent projections of these geometric quantities. Boost-dependent effects could in principle be detected by considering two identical experiments with differing velocity vectors. The general approach to finding Lorentz violation is therefore to compare identical experiments with differing rotations and boosts. Equivalently, one can seek time dependence in a single experiment as it rotates in space. The SME shows that Lorentz symmetry is violated under these "particle transformations" where the entire experimental configuration moves relative to another one, or to itself. Perturbation theory can be used to calculate the effects of the coefficients for Lorentz violation, since they are known to be small. In the case of atomic clocks, various simplifying assumptions are made, and it is then possible to calculate the energy-level corrections that affect the frequency of the clock.

In contrast, "observer transformations" preserve Lorentz symmetry. This is important since it guarantees agreement between different observers: different experimenters observing one experimental system from differently boosted or rotated inertial reference frames will find that the components of the parameters a_μ, b_μ, $c_{\mu\nu}$, $d_{\mu\nu}$, e_μ, f_μ, $g_{\lambda\mu\nu}$, $H_{\mu\nu}$ transform like conventional tensors under Lorentz transformations.

3. Clock-Comparison Experiments

The basic function of an atomic clock is to produce a stable frequency based on an atomic energy-level transition. In common configurations, there is a quantization axis defined by a magnetic field. If the third coordinate axis of the laboratory reference frame is defined to run parallel to this axis, then the output frequency is a function of this magnetic-field component: $f(B_3)$.

One of the key issues relating to stability is the reduction of this dependence on the magnetic field, which will drift over time. In the general SME formalism, the output frequency is of the form

$$\omega = f(B_3) + \delta\omega, \tag{4}$$

where the small correction $\delta\omega$ carries all the Lorentz-violating contributions to the output frequency. It can contain terms that are orientation-dependent, an example being the dot product of the spatial part of b^μ and $\mathbf{B}$. In addition, $\delta\omega$ can depend on the boost velocity of the clock. These effects arise because of the motion of the laboratory reference frame. Therefore, searching for Lorentz violation requires a detailed knowledge of the motion of the laboratory relative to a reference frame that is known to be inertial over the life of the experiment.

A number of simplifying assumptions must be made to perform calculations of the effects of Lorentz violation on atomic clocks, due to the complexity of these atomic systems. The Hamiltonian is split into a conventional part describing the atom within the chosen nuclear model, and a perturbative part containing the coefficients for Lorentz violation. The perturbative Hamiltonian has separate terms for each proton, electron, and neutron, indexed by the letter w:

$$h' = \sum_w \sum_{N=1}^{N_w} \delta h_{w,N}. \tag{5}$$

Here, the atom or ion w has N_w particles of type w, and $\delta h_{w,N}$ is the Lorentz-violating correction for the Nth particle of type w. Each of the three particle species in the atom has a set of Lorentz-violation parameters, so a superscript w must be placed on each of the parameters a_μ, b_μ, $c_{\mu\nu}$, $d_{\mu\nu}$, e_μ, f_μ, $g_{\lambda\mu\nu}$, $H_{\mu\nu}$.

The perturbations in the energy levels due to Lorentz violation are calculated by finding the expectation value of the Hamiltonian h'. Usually, the total angular momentum $\mathbf{F}$ of the atom or ion, and its projection along the quantization axis, are conserved to a good approximation. It is therefore possible to label the unperturbed quantum states for the atomic-clock atoms by the quantum numbers $|F, m_F\rangle$. The energy-level shift for the state $|F, m_F\rangle$, calculated in the laboratory frame with the 3-coordinate taken as the quantization axis, is

$$\begin{aligned}\delta E(F, m_F) &= \langle F, m_F|h'|F, m_F\rangle \\ &= \widehat{m}_F \sum_w (\beta_w \tilde{b}_3^w + \delta_w \tilde{d}_3^w + \kappa_w \tilde{g}_d^w) + \widetilde{m}_F \sum_w (\gamma_w \tilde{c}_q^w + \lambda_w \tilde{g}_q^w)\,. \quad (6)\end{aligned}$$

In this expression, the quantities β_w, δ_w, κ_w, γ_w, λ_w are expectation values of combinations of spin and momentum operators calculated using the extremal states $|F, m_F = F\rangle$. The quantities $\widehat{m}_F$ and $\widetilde{m}_F$ are particular ratios of Clebsch–Gordan coefficients. The calculations of quantities appearing in (6) are only approximate, depending on the nuclear model used. The quantities with tildes are specific combinations of Lorentz-violation parameters that are the only possible parameter combinations to which clock-comparison experiments are sensitive. There are five such combinations; for example, $\tilde{b}_3^w$ is given by

$$\tilde{b}_3^w := b_3^w - m_w d_{30}^w + m_w g_{120}^w - H_{12}^w. \quad (7)$$

More details of these quantities can be found in Refs. 59 and 60. Once the energy-level shifts are known, the effects of Lorentz violation on the clock frequency for the transition $(F, m_F) \to (F', m'_F)$ can be found from the difference

$$\delta\omega = \delta E(F, m_F) - \delta E(F', m'_F). \quad (8)$$

4. The Standard Inertial Reference Frame

The SME framework shows that Lorentz violation could occur in nature in the form of a variety of background *observer* Lorentz tensors. Experimental Lorentz tests aim to measure these tensor components using a reference frame that is unavoidably noninertial.

Since different experiments use different laboratory coordinates, it is important to standardize the reference frame in which the measurements are reported so that comparisons can be made. The conventional reference frame that has been used has the origin at the center of the Sun, and the Z axis running northward parallel to the Earth's rotation axis. The X axis points toward the vernal equinox on the celestial sphere, and the orthogonal right-handed system is completed by the appropriate choice of Y axis. The standard time coordinate, denoted by T, is measured by a clock at the center of the Sun with the time origin at the vernal equinox in the year

2000. This frame is approximately inertial over periods of thousands of years. We note that the choice of an Earth-centered frame would be approximately inertial only for periods on the order of 10 days.

Any given experiment is conducted in a laboratory with spatial coordinates (x_1, x_2, x_3), where the third coordinate is defined to be the quantization axis. This frame may be on the Earth or in space, and is not generally inertial. To obtain the experimental result in the standard frame, details of the laboratory trajectory in space–time need to be included in the analysis.

To illustrate the approach for the case of a satellite, we consider the superposition of two circular motions, one being that of the Earth around the Sun, and the other being that of the satellite around the Earth. The plane of the Earth's circular motion is inclined at angle $\eta \approx 23°$ from the equatorial plane, and its path intersects the positive X axis at the vernal equinox. To specify the satellite orbital plane, we use the angle of inclination ζ relative to the Z axis, and the right ascension α of the point on its orbit where it intersects the equatorial plane in the northward direction. The experimental laboratory can itself be oriented in various ways within the satellite, so, for definiteness, we choose the x_1 axis to point toward the center of the Earth, and the quantization axis x_3 to point along the velocity vector of the satellite relative to the Earth. Using appropriate rotation and boost matrices, results in the laboratory frame can be transformed into the inertial frame. As an example, the expression $\tilde{b}_3$ has the form

$$\begin{aligned}\tilde{b}_3 = \cos\omega_s T_s\{&[-\tilde{b}_X \sin\alpha\cos\zeta + \tilde{b}_Y \cos\alpha\cos\zeta + \tilde{b}_Z \sin\zeta] \\ &+ \beta_\oplus[\text{seasonal terms}\ldots]\} + \sin\omega_s T_s\{[-\tilde{b}_X\cos\alpha - \tilde{b}_Y \sin\alpha] \\ &+ \beta_\oplus[\text{seasonal terms}\ldots]\} + \cos 2\omega_s T_s\{\beta_s[\text{constant terms}\ldots]\} \\ &+ \sin 2\omega_s T_s\{\beta_s[\text{constant terms}\ldots]\} + \beta_s[\text{constant terms}\ldots]. \end{aligned} \tag{9}$$

The satellite orbital frequency ω_s appears as expected, and we note that signals appear also at twice this frequency. The much lower orbital frequency of the Earth, $\Omega_\oplus$, appears in the seasonal terms.

Since the transformations between the laboratory and inertial reference frames include a boost, the appearance of the speed $\beta_\oplus \approx 10^{-4}$ of the Earth relative to the Sun is as might be expected. Also, the speed of the satellite relative to the Earth is expected; in the case of the International Space Station, this speed is $\beta_s \approx 10^{-5}$. For further details of $\tilde{b}_3$, see Refs. 75 and 76. The energy shifts appearing in Eq. (6) depend also on the particular atoms of the clock and the choice of transition used. Transitions that change the component of the angular momentum along the quantization axis are often the most favorable for Lorentz tests.[73,74]

5. Microwave and Optical Cavities

Experiments involving optical and microwave cavities have attracted much recent interest. Typically, the centimeter dimensions of microwave cavities are approximately equal to the wavelength of the radiation, while optical cavities have much higher frequencies and correspondingly smaller wavelengths.

To perform calculations for these systems, we need to consider the pure-photon sector of the SME. Considering only renormalizable terms, which are constructed from operators that have mass dimension 4 or less, the Lagrangian is[11,12]

$$\mathcal{L} = -\frac{1}{4}F_{\mu\nu}F^{\mu\nu} + \frac{1}{2}(k_{AF})^{\kappa}\epsilon_{\kappa\lambda\mu\nu}A^{\lambda}F^{\mu\nu} - \frac{1}{4}(k_F)_{\kappa\lambda\mu\nu}F^{\kappa\lambda}F^{\mu\nu}, \tag{10}$$

with $F_{\mu\nu} \equiv \partial_\mu A_\nu - \partial_\nu A_\mu$. Conventional photon physics is contained in the first term, while the second and third terms describe Lorentz-violating interactions. They contain the coefficients for Lorentz violation $(k_{AF})^{\kappa}$ and $(k_F)_{\kappa\lambda\mu\nu}$, which are CPT odd and CPT even respectively. The coefficient $(k_F)_{\kappa\lambda\mu\nu}$ has the symmetries of the Riemann tensor and a zero double trace, giving it a total of 19 independent components.

The second term has been studied extensively,[18–24] and since it has been exceptionally tightly constrained by polarization data from distant astronomical sources,[16,17] $(k_{AF})^{\kappa}$ is set to zero. A number of the components of the third term have also been tightly constrained by optical data from distant cosmological sources.[13,14] The remaining components, nine in total, have been the focus of a variety of experimental tests with cavities in the last few years.

In cavity oscillators, the resonant-frequency shift $\delta\nu/\nu$ is the important experimental quantity. For a cavity with resonant angular frequency ω_0, we take $\mathbf{E}_0$, $\mathbf{B}_0$, $\mathbf{D}_0$, and $\mathbf{H}_0$ to be the conventional fields. These can be perturbed by nonzero k_F coefficients, giving perturbed fields $\mathbf{E}$, $\mathbf{B}$, $\mathbf{D}$, and $\mathbf{H}$ with resonant frequency changed by $\delta\nu = \delta\omega/2\pi$. The Lorentz-violating Maxwell equations derived from (10) lead to the following expression for the fractional resonant-frequency shift,

$$\begin{aligned}\frac{\delta\nu}{\nu} &= -\left(\int_V d^3x\,(\mathbf{E}_0^*\cdot\mathbf{D} + \mathbf{H}_0^*\cdot\mathbf{B})\right)^{-1} \\ &\quad\times \int_V d^3x(\mathbf{E}_0^*\cdot\mathbf{D} - \mathbf{D}_0^*\cdot\mathbf{E} - \mathbf{B}_0^*\cdot\mathbf{H} + \mathbf{H}_0^*\cdot\mathbf{B} \\ &\quad - i\omega_0^{-1}\boldsymbol{\nabla}\cdot(\mathbf{H}_0^*\times\mathbf{E} - \mathbf{E}_0^*\times\mathbf{H})),\end{aligned} \tag{11}$$

with integrals taken over the volume V of the cavity. Various assumptions regarding boundary conditions and cavity losses are made, and since Lorentz violation can be expected to be small, only leading-order effects on $\delta\nu/\nu$ need to be considered. For further details see Ref. 15. In the following, we consider optical cavities first and then microwave cavities.

Optical tests of Lorentz invariance started with the classic tests involving the speed of light which were done by Michelson and Morley to search for spatial anisotropy. Later tests by Kennedy and Thorndyke sought dependence of the speed of light on the speed of the laboratory. The SME provides a general framework for both these types of experiments, and indeed for experiments from all areas of physics. In the following, modern versions of these experiments are considered.

An optical cavity can be analyzed by considering it to be two parallel reflecting surfaces with plane waves traveling between them. With this approach and some simplifying assumptions, Eq. (11) gives the following expression for the fractional frequency shift $\delta\nu/\nu$ due to Lorentz-violating effects:

$$\frac{\delta\nu}{\nu} = -\frac{1}{2|\mathbf{E}_0|^2}\left[\mathbf{E}_0^* \cdot (\kappa_{DE})_{\text{lab}} \cdot \mathbf{E}_0/\epsilon - (\hat{N} \times \mathbf{E}_0^*) \cdot (\kappa_{HB})_{\text{lab}} \cdot (\hat{N} \times \mathbf{E}_0)\right]. \quad (12)$$

Here, $\hat{N}$ is the unit vector along the cavity axis, $\mathbf{E}_0$ specifies the polarization, and ϵ is the transverse relative permittivity. The coefficients for Lorentz violation $(\kappa_{DE})_{\text{lab}}$ and $(\kappa_{HB})_{\text{lab}}$ are particular laboratory-frame linear combinations of $k_F^{\mu\nu\lambda\tau}$, as is more fully explained in Ref. 15.

Thus, in the presence of Lorentz violation, the fractional frequency shift of an optical-cavity oscillator depends on the cavity orientation and on the polarization direction of the light. In any laboratory, Earth-based or space-based, laser light incident on a cavity will have the fractional frequency shift

$$\begin{aligned}\frac{\delta\nu}{\nu} &= -\frac{1}{4}[2(\kappa_{DE})^{33}_{\text{lab}}/\epsilon - (\kappa_{HB})^{11}_{\text{lab}} - (\kappa_{HB})^{22}_{\text{lab}}] - \frac{1}{2}(\kappa_{HB})^{12}_{\text{lab}} \sin 2\theta \\ &\quad - \frac{1}{4}[(\kappa_{HB})^{11}_{\text{lab}} - (\kappa_{HB})^{22}_{\text{lab}}] \cos 2\theta, \end{aligned} \quad (13)$$

where 1,2, and 3 are the three orthogonal spatial directions. In general, the laboratory frame is noninertial, since it is either on the surface of the Earth or on a spacecraft. To analyze the results in a standardized form as per the conventions of other SME works, the fractional frequency shift must be expressed in the Sun-centered celestial equatorial basis through the use of a suitable coordinate transformation.

Microwave-frequency cavity oscillators are also capable of excellent Lorentz-symmetry tests. Superconducting cavity-stabilized oscillators have been considered for use as clocks on the International Space Station. Niobium superconducting cavities have achieved Q factors of 10^{11} or better, and frequency stabilities of 3×10^{-16} have been attained.

The fractional resonant-frequency shift $\delta\nu/\nu$ for a superconducting microwave cavity of any geometry can be found from Eq. (11). For a cylindrical cavity of circular cross section that is evacuated and operated in the fundamental TM_{010} mode, the fractional frequency shift takes the form

$$\begin{aligned}\left.\frac{\delta\nu}{\nu}\right|_{TM_{010}} &= -\frac{1}{4}\hat{N}^j\hat{N}^k R^{jJ} R^{kK}[3(\tilde{\kappa}_{e+})^{JK} + (\tilde{\kappa}_{e-})^{JK}] \\ &\quad - \frac{1}{2}(\delta^{jk} + \hat{N}^j\hat{N}^k) R^{jJ} R^{kK} \epsilon^{JPQ}\beta^Q \\ &\quad \times [3(\tilde{\kappa}_{o-})^{KP} + (\tilde{\kappa}_{o+})^{KP}] - \tilde{\kappa}_{\text{tr}}. \end{aligned} \quad (14)$$

By convention, lower-case Roman letters denote the laboratory frame and upper-case ones denote the inertial frame; R^{jJ} is the spatial rotation from the

Sun-centered frame to the laboratory frame, $\hat{N}$ is a unit vector parallel to the symmetry axis of the cavity, and β^Q are the components of the boost vector in the inertial reference frame. The various $\tilde{\kappa}$ quantities are coefficients for Lorentz violation constructed from particular linear combinations of the $(k_F)_{\kappa\lambda\mu\nu}$ quantities. The above equation applies for experiments in any laboratory. For a specific experiment, a variety of parameters are substituted to include information such as the geometry of the cavities used, the trajectory of the laboratory, and the materials in the system. The analysis proceeds in a manner similar to that for clock-comparison experiments, using transformations to produce the results in the standard reference frame. One of the advantageous configurations involves cavities oriented perpendicular to each other. For more details on these issues, see Ref. 15.

6. The Gravitational Sector

The gravitational sector of the SME consists of a framework for addressing Lorentz and CPT violation in Riemann–Cartan space–times.[126] The usual Riemann and Minkowski geometries are recovered as limiting cases. A special case of interest is the quantum-electrodynamics extension in a Riemann–Cartan background. The coefficients for Lorentz violation $(a_\mu, b_\mu, \ldots)$ typically vary with position, and the fermion and photon sectors couple to these coefficients. To obtain the full SME, one needs to consider the actions for the conventional Standard Model, the Lorentz- and CPT-violating terms built from conventional Standard Model fields, and the pure-gravity sector. Each of these is modified by gravitational couplings appropriate for a background Riemann–Cartan space–time. This approach allows identification of the dominant terms in the effective low-energy action for the gravitational sector, thereby completing the formulation of the leading-order terms in the SME with gravity. In principle, Lorentz-symmetry breaking can occur in one of two ways: either it must be explicit in the action, or it must occur in the solutions obtained from an action that is itself Lorentz-symmetric. In other words, Lorentz violation must be either explicit or spontaneous. One of the profound and surprising results obtained in Ref. 126 is the incompatibility of explicit breaking with generic Riemann–Cartan geometries. No such difficulties occur in the case of spontaneous Lorentz violation.

The Nambu–Goldstone theorem basically states that a massless particle arises whenever a continuous global symmetry of the action is not a symmetry of the vacuum. This means, then, that spontaneous Lorentz violation may give rise to various massless particles, or Nambu–Goldstone modes. In nature, known massless particles include the photon and the graviton, both of which have two independent polarizations. It is natural to ask whether the massless modes following from spontaneous Lorentz violation are compatible with the photon and the graviton. This issue is addressed in Ref. 127. A variety of interesting issues arise. One is that spontaneous violation of Lorentz-symmetry is always associated with spontaneous violation of diffeomorphism symmetry and vice versa. Another aspect is the ten possible Nambu–Goldstone modes associated with the six generators for Lorentz transformations and the four generators for diffeomorphisms. The fate of these modes has been found to depend on the space–time geometry and the dynamics of the tensor field triggering the spontaneous Lorentz violation.

Explicit models involving tensor fields that take on vacuum values spontaneously can be used to study generic features of the Nambu–Goldstone modes in the case of Minkowski, Riemann, and Riemann–Cartan space–times. One example is the so-called Bumblebee model,[127] involving a vector field B_μ. This idea of using a potential to spontaneously break Lorentz-symmetry, thus enforcing a nonzero vacuum value for B_μ, was introduced by Kostelecký and Samuel.[9,10]

Remarkably, in Minkowski and Riemann space–times, the bumblebee model generates a photon as a Nambu–Goldstone boson for spontaneous Lorentz violation. In principle, such theories can be experimentally tested, because there are unconventional Lorentz-violating and Lorentz-preserving couplings that could be observed in sensitive experiments. It has also been shown in Ref. 128 that in Riemann–Cartan space–time, the Nambu–Goldstone modes could be absorbed into the torsion component of the gravitational field through a Higgs mechanism for the spin connection.

Another study of the Nambu–Goldstone modes arising from spontaneous Lorentz violation has been done for the case of a two-index symmetric "cardinal" field $C_{\mu\nu}$ taking on a vacuum value.[128] As in the case of the bumblebee field, the result found is intriguing: two massless modes are generated that correspond to the two polarizations of the graviton. At low energy and temperature, conventional gravity is recovered, and small experimentally testable differences from conventional gravity. These testable findings, that the photon and the graviton could be evidence of spontaneous local Lorentz and diffeomorphism violation, hold the potential for a profound impact on conventional physics.

An extensive study of the pure-gravity sector of the minimal SME in the limit of Riemann space–time has been performed.[129] Under simple assumptions, there are 20 Lorentz-violating coefficient fields that take on vacuum values and lead to modified Einstein field equations in the limit of small fluctuations about the Minkowski vacuum. The equations can be solved to obtain the post-Newtonian metric. This work[129] includes a detailed theoretical investigation of experimental tests and some estimated bounds for experiments involving lunar and satellite laser ranging, laboratory experiments with gravimeters and torsion pendula, measurements of the spin precession of orbiting gyroscopes, timing studies of signals from binary pulsars, and the classic tests involving the perihelion precession and the time delay of light. The sensitivities in these experiments range from parts in 10^4 to parts in 10^{15}.

References

1. V. A. Kostelecký and R. Potting, *Phys. Rev. D* **51** (1995) 3923 [hep-ph/9501341].
2. V. A. Kostelecký and S. Samuel, *Phys. Rev. Lett.* **66** (1991) 1811.
3. V. A. Kostelecký and R. Potting, *Phys. Lett. B* **381** (1996) 89 [hep-th/9605088].
4. V. A. Kostelecký and R. Potting, *Phys. Rev. D* **63** (2001) 046007 [hep-th/0008252].

5. V. A. Kostelecký, M. Perry and R. Potting, *Phys. Rev. Lett.* **84** (2000) 4541 [hep-th/9912243].
6. V. A. Kostelecký and S. Samuel, *Phys. Rev. D* **39** (1989) 683.
7. V. A. Kostelecký and R. Potting, *Nucl. Phys. B* **359** (1991) 545.
8. B. Altschul and V. A. Kostelecký, *Phys. Lett. B* **628** (2005) 106 [hep-th/0509068].
9. V. A. Kostelecký and S. Samuel, *Phys. Rev. Lett.* **63** (1989) 224.
10. V. A. Kostelecký and S. Samuel, *Phys. Rev. D* **40** (1989) 1886.
11. D. Colladay and V. A. Kostelecký, *Phys. Rev. D* **55** (1997) 6760 [hep-ph/9703464].
12. D. Colladay and V. A. Kostelecký, *Phys. Rev. D* **58** (1998) 116002 [hep-ph/9809521].
13. V. A. Kostelecký and M. Mewes, *Phys. Rev. Lett.* **87** (2001) 251304 [hep-ph/0111026].
14. V. A. Kostelecký and M. Mewes, hep-ph/0607084.
15. V. A. Kostelecký and M. Mewes, *Phys. Rev. D* **66** (2002) 056005 [hep-ph/0205211].
16. S. M. Carroll, G. B. Field and R. Jackiw, *Phys. Rev. D* **41** (1990) 1231.
17. M. P. Haugan and T. F. Kauffmann, *Phys. Rev. D* **52** (1995) 3168 [gr-qc/9504032].
18. R. Jackiw and V. A. Kostelecký, *Phys. Rev. Lett.* **82** (1999) 3572 [hep-ph/9901358].
19. M. Perez-Victoria, *J. High Energy Phys.* **104** (2001) 32 [hep-th/0102021].
20. V. A. Kostelecký, C. D. Lane and A. G. M. Pickering, *Phys. Rev. D* **65** (2002) 056006 [hep-th/0111123].
21. B. Altschul, *Phys. Rev. D* **69** (2004) 125009 [hep-th/0311200].
22. B. Altschul, *Phys. Rev. D* **70** (2004) 101701 [hep-th/0407172].
23. H. Belich *et al.*, *Eur. Phys. J. C* **42** (2005) 127 [hep-th/0411151].
24. T. Mariz *et al.*, *J. High Energy Phys.* **510** (2005) 19 [hep-th/0509008].
25. P. Antonini *et al.*, *Phys. Rev. A* **72** (2005) 066102 [physics/0602115].
26. M. E. Tobar, P. Wolf and P. L. Stanwix, *Phys. Rev. A* **72** (2005) 066101 [physics/0601186].
27. S. Herrmann *et al.*, *Phys. Rev. Lett.* **95** (2005) 150401 [physics/0508097].
28. M. E. Tobar *et al.*, hep-ph/0506200.
29. P. L. Stanwix *et al.*, *Phys. Rev. Lett.* **95** (2005) 040404 [hep-ph/0506074].
30. P. Antonini *et al.*, *Phys. Rev. A* **71** (2005) 050101 [gr-qc/0504109].
31. M. E. Tobar *et al.*, *Phys. Rev. D* **71** (2005) 025004 [hep-ph/0408006].
32. P. Wolf *et al.*, *Phys. Rev. D* **70** (2004) 051902 [hep-ph/0407232].
33. P. Wolf *et al.*, *Gen. Relativ. Gravit.* **36** (2004) 2352 [gr-qc/0401017].
34. H. Muller *et al.*, *Phys. Rev. D* **68** (2003) 116006 [hep-ph/0401016].
35. H. Müller *et al.*, *Phys. Rev. Lett.* **91** (2003) 020401.
36. J. A. Lipa *et al.*, *Phys. Rev. Lett.* **90** (2003) 060403 [physics/0302093].
37. J. P. Cotter and B. Varcoe, physics/0603111.
38. C. Adam and F. R. Klinkhamer, *Nucl. Phys. B* **657** (2003) 214 [hep-th/0212028].
39. T. Jacobson, S. Liberati and D. Mattingly, *Phys. Rev. D* **67** (2003) 124011 [hep-ph/0209264].
40. V. A. Kostelecký and A. G. M. Pickering, *Phys. Rev. Lett.* **91** (2003) 031801 [hep-ph/0212382].
41. R. Lehnert and R. Potting, *Phys. Rev. Lett.* **93** (2004) 110402 [hep-ph/0406128].
42. R. Lehnert and R. Potting, *Phys. Rev. D* **70** (2004) 125010; erratum, *ibid. D* **70** (2004) 129906 [hep-ph/0408285].
43. T. A. Jacobson *et al.*, *Phys. Rev. Lett.* **93** (2004) 021101 [astro-ph/0309681].
44. B. Altschul, *Phys. Rev. D* **72** (2005) 085003 [hep-th/0507258].
45. F. R. Klinkhamer and C. Rupp, *Phys. Rev. D* **72** (2005) 017901 [hep-ph/0506071].
46. C. Kaufhold and F. R. Klinkhamer, *Nucl. Phys.* **734** (2006) 1.
47. B. Altschul, *Phys. Rev. D* **70** (2004) 056005 [hep-ph/0405084].

48. R. Bluhm *et al.*, *Phys. Rev. Lett.* **79** (1997) 1432 [hep-ph/9707364].
49. R. Bluhm *et al.*, *Phys. Rev. D* **57** (1998) 3932 [hep-ph/9809543].
50. H. Dehmelt *et al.*, *Phys. Rev. Lett.* **83** (1999) 4694 [hep-ph/9906262].
51. R. K. Mittleman *et al.*, *Phys. Rev. Lett.* **83** (1999) 2116.
52. G. Gabrielse *et al.*, *Phys. Rev. Lett.* **82** (1999) 3198.
53. R. Bluhm and V. A. Kostelecký, *Phys. Rev. Lett.* **84** (2000) 1381 [hep-ph/9912542].
54. B. R. Heckel, in *CPT and Lorentz Symmetry III*, ed. V. A. Kostelecký (World Scientific, Singapore, 2004), p. 133.
55. L.-S. Hou, W.-T. Ni and Y.-C. M. Li, *Phys. Rev. Lett.* **90** (2003) 201101.
56. D. Colladay and V. A. Kostelecký, *Phys. Lett. B* **511** (2001) 209 [hep-ph/0104300].
57. B. Altschul, *Phys. Rev. D* **70** (2004) 056005 [hep-ph/0405084].
58. B. Altschul, *Phys. Rev. Lett.* **96** (2006) 201101 [hep-ph/0603138].
59. V. A. Kostelecký and C. D. Lane, *Phys. Rev. D* **60** (1999) 116010 [hep-ph/9908504].
60. V. A. Kostelecký and C. D. Lane, *J. Math. Phys.* **40** (1999) 6245 [hep-ph/9909542].
61. C. D. Lane, *Phys. Rev. D* **72** (2005) 016005 [hep-ph/0505130].
62. D. Colladay and P. McDonald, hep-ph/0602071.
63. P. Wolf *et al.*, *Phys. Rev. Lett.* **96** (2006) 060801 [hep-ph/0601024].
64. P. Wolf *et al.*, hep-ph/0509329.
65. P. Wolf *et al.*, physics/0506168.
66. F. Cane *et al.*, *Phys. Rev. Lett.* **93** (2004) 230801 [physics/0309070].
67. D. F. Phillips *et al.*, *Phys. Rev. D* **63** (2001) 111101 [physics/0008230].
68. M. A. Humphrey *et al.*, *Phys. Rev. A* **68** (2003) 063807 [physics/0103068].
69. M. A. Humphrey, D. F. Phillips and R. L. Walsworth, *Phys. Rev. A* **62** (2000) 063405.
70. D. Bear *et al.*, *Phys. Rev. Lett.* **85** (2000) 5038; erratum, *ibid.* **89** (2002) 209902 [physics/0007049].
71. R. L. Walsworth *et al.*, *AIP Conf. Proc.* **539** (2000) 119 [physics/0007063].
72. L. Hunter, in *CPT and Lorentz Symmetry*, ed. V. Kostelecký (World Scientific, Singapore, 1999), p. 180.
73. R. Bluhm *et al.*, *Phys. Rev. Lett.* **82** (1999) 2254 [hep-ph/9810269].
74. G. M. Shore, *Nucl. Phys. B* **717** (2005) 86 [hep-th/0409125].
75. R. Bluhm *et al. Phys. Rev. Lett.* **88** (2002) 090801 [hep-ph/0111141].
76. R. Bluhm *et al. Phys. Rev. D* **68** (2003) 125008 [hep-ph/0306190].
77. R. Bluhm, V. A. Kostelecký and C. D. Lane, *Phys. Rev. Lett.* **84** (2000) 1098 [hep-ph/9912451].
78. V. W. Hughes *et al.*, *Phys. Rev. Lett.* **87** (2001) 111804 [hep-ex/0106103].
79. Muon g-2 Collab. (M. Deile *et al.*), hep-ex/0110044.
80. V. A. Kostelecký and M. Mewes, *Phys. Rev. D* **69** (2004) 016005 [hep-ph/0309025].
81. V. A. Kostelecký and M. Mewes, *Phys. Rev. D* **70** (2004) 031902 [hep-ph/0308300].
82. V. A. Kostelecký and M. Mewes, *Phys. Rev. D* **70** (2004) 076002 [hep-ph/0406255].
83. T. Katori *et al.*, hep-ph/0606154.
84. LSND Collab. (L. B. Auerbach *et al.*), *Phys. Rev. D* **72** (2005) 076004 [hep-ex/0506067].
85. SuperKamiokande Collab. (M. D. Messier *et al.*), in *CPT and Lorentz Symmetry III*, ed. V. A. Kostelecký (World Scientific, Singapore, 2005).
86. MINOS Collab. (B. J. Rebel and S. F. Mufson), in *CPT and Lorentz Symmetry III*, ed. V. A. Kostelecký (World Scientific, Singapore, 2005).
87. D. L. Anderson, M. Sher and I. Turan, *Phys. Rev. D* **70** (2004) 016001 [hep-ph/0403116].
88. E. O. Iltan, *Mod. Phys. Lett. A* **19** (2004) 327 [hep-ph/0309154].

89. O. Bertolami *et al.*, *Phys. Lett. B* **395** (1997) 178 [hep-ph/9612437].
90. G. Lambiase, *Phys. Rev. D* **72** (2005) 087702 [astro-ph/0510386].
91. J. M. Carmona *et al.*, *Mod. Phys. Lett. A* **21** (2006) 883 [hep-th/0410143].
92. S. M. Carroll and J. Shu, *Phys. Rev. D* **73** (2006) 103515 [hep-ph/0510081].
93. KTeV Collab. (H. Nguyen), in *CPT and Lorentz Symmetry II*, ed. V. A. Kostelecký (World Scientific, Singapore, 2002) [hep-ex/0112046].
94. OPAL Collab. (K. Ackerstaff *et al.*), *Z. Phys. C* **76** (1997) 401 [hep-ex/9707009].
95. DELPHI Collab. (M. Feindt, C. Kreuter and O. Podobrin), DELPHI 97–98, CONF 80 (1997).
96. Belle Collab. (K. Abe *et al.*), *Phys. Rev. Lett.* **86** (2001) 3228 [hep-ex/0011090].
97. BABAR Collab. (B. Aubert *et al.*), hep-ex/0607103.
98. FOCUS Collab. (J. M. Link *et al.*), *Phys. Lett. B* **556** (2003) 7 [hep-ex/0208034].
99. V. A. Kostelecký, *Phys. Rev. Lett.* **80** (1998) 1818 [hep-ph/9809572].
100. V. A. Kostelecký, *Phys. Rev. D* **61** (2000) 016002 [hep-ph/9909554].
101. V. A. Kostelecký, *Phys. Rev. D* **64** (2001) 076001 [hep-ph/0104120].
102. D. Colladay and V. A. Kostelecký, *Phys. Lett. B* **344** (1995) 259 [hep-ph/9501372].
103. D. Colladay and V. A. Kostelecký, *Phys. Rev. D* **52** (1995) 6224 [hep-ph/9510365].
104. V. A. Kostelecký and R. Van Kooten, *Phys. Rev. D* **54** (1996) 5585 [hep-ph/9607449].
105. V. A. Kostelecký and A. Roberts, *Phys. Rev. D* **63** (2001) 096002 [hep-ph/0012381].
106. N. Isgur *et al.*, *Phys. Lett. B* **515** (2001) 333 [hep-ph/0106353].
107. I. Mocioiu, M. Pospelov and R. Roiban, *Phys. Lett. B* **489** (2000) 390 [hep-ph/0005191].
108. S. M. Carroll *et al.*, *Phys. Rev. Lett.* **87** (2001) 141601 [hep-th/0105082].
109. Z. Guralnik *et al.*, *Phys. Lett. B* **517** (2001) 450 [hep-th/0106044].
110. C. E. Carlson, C. D. Carone and R. F. Lebed, *Phys. Lett. B* **518** (2001) 201 [hep-ph/0107291].
111. A. Anisimov *et al.*, *Phys. Rev. D* **65** (2002) 085032 [hep-ph/0106356].
112. A. Das *et al.*, *Phys. Rev. D* **72** (2005) 107702 [hep-th/0510002].
113. H. Muller *et al.*, *Phys. Rev. D* **70** (2004) 076004.
114. H. Muller, *Phys. Rev. D* **71** (2005) 045004 [hep-ph/0412385].
115. Q. G. Bailey and V. A. Kostelecký, *Phys. Rev. D* **70** (2004) 076006 [hep-ph/0407252].
116. V. A. Kostelecký and R. Lehnert, *Phys. Rev. D* **63** (2001) 065008 [hep-th/0012060].
117. M. S. Berger and V. A. Kostelecký, *Phys. Rev. D* **65** (2002) 091701 [hep-th/0112243].
118. V. A. Kostelecký, R. Lehnert and M. J. Perry, *Phys. Rev. D* **68** (2003) 123511 [astro-ph/0212003].
119. B. Altschul, hep-th/0602235.
120. V. A. Kostelecký (ed.), *CPT and Lorentz Symmetry* (World Scientific, Singapore, 1999); *CPT and Lorentz Symmetry II* (World Scientific, Singapore, 2002); *CPT and Lorentz Symmetry III* (World Scientific, Singapore, 2005).
121. R. Bluhm, hep-ph/0506054.
122. D. Mattingly, *Living Rev. Relativ.* **8** (2005) 5 [gr-qc/0502097].
123. G. Amelino-Camelia *et al.*, *AIP Conf. Proc.* **758** (2005) 30 [gr-qc/0501053].
124. H. Vucetich, gr-qc/0502093.
125. N. Russell, *Phys. Scripta* **72** (2005) C38 [hep-ph/0501127].
126. V. A. Kostelecký, *Phys. Rev. D* **69** (2004) 105009 [hep-th/0312310].
127. R. Bluhm and V. A. Kostelecký, *Phys. Rev. D* **71** (2005) 065008 [hep-th/0412320].
128. V. A. Kostelecký and R. Potting, *Gen. Relativ. Gravit.* **37** (2005) 1675 [gr-qc/0510124].
129. Q. G. Bailey and V. A. Kostelecký, *Phys. Rev. D* **74** (2006) 045001 [gr-qc/0603030].

PRECISION MEASUREMENT BASED ON ULTRACOLD ATOMS AND COLD MOLECULES

JUN YE*, SEBASTIAN BLATT, MARTIN M. BOYD, SETH M. FOREMAN, ERIC R. HUDSON, TETSUYA IDO, BENJAMIN LEV, ANDREW D. LUDLOW, BRIAN C. SAWYER, BENJAMIN STUHL and TANYA ZELINSKY

JILA, National Institute of Standards and Technology, and University of Colorado

Department of Physics, University of Colorado, Boulder, CO 80309-0440, USA

**Ye@jila.colorado.edu*

http://jilawww.colorado.edu/YeLabs

Ultracold atoms and molecules provide ideal stages for precision tests of fundamental physics. With microkelvin neutral strontium atoms confined in an optical lattice, we have achieved a fractional resolution of 4×10^{-15} on the ${}^1S_0$–${}^3P_0$ doubly forbidden ^{87}Sr clock transition at 698 nm. Measurements of the clock line shifts as a function of experimental parameters indicate systematic errors below the 10^{-15} level. The ultrahigh spectral resolution permits resolving the nuclear spin states of the clock transition at small magnetic fields, leading to measurements of the ${}^3P_0$ magnetic moment and metastable lifetime. In addition, photoassociation spectroscopy performed on the narrow ${}^1S_0$–${}^3P_1$ transition of ^{88}Sr shows promise for efficient optical tuning of the ground state scattering length and production of ultracold ground state molecules. Lattice-confined Sr_2 molecules are suitable for constraining the time variation of the proton–electron mass ratio. In a separate experiment, cold, stable, ground state polar molecules are produced from Stark decelerators. These cold samples have enabled an order-of-magnitude improvement in the measurement precision of ground state, Λ doublet microwave transitions in the OH molecule. Comparing the laboratory results to those from OH megamasers in interstellar space will allow a sensitivity of 10^{-6} for measuring the potential time variation of the fundamental fine structure constant $\Delta\alpha/\alpha$ over 10^{10} years. These results have also led to improved understandings of the molecular structure. The study of the low magnetic field behavior of OH in its ${}^2\Pi_{3/2}$ ro-vibronic ground state precisely determines a differential Landé g factor between opposite parity components of the Λ doublet.

Keywords: Coherent interactions; optical lattice; frequency comb; optical atomic clocks; cold molecules; precision measurement.

1. Introduction

The unique atomic structure of alkaline earth atoms such as strontium permits studies of narrow line physics based on the forbidden ${}^1S_0$–${}^3P_0$ and ${}^1S_0$–${}^3P_1$

transitions. The millihertz-wide ${}^1S_0$–${}^3P_0$ ^{87}Sr line at 698 nm is especially attractive for an optical atomic clock. Using optically cooled ^{87}Sr atoms in a zero-Stark-shift, one-dimensional optical lattice and an ultrastable probe laser with sub-Hz line width, we have achieved repeatable Fourier-limited line widths of below 2 Hz. This represents a fractional resolution of $\sim 4 \times 10^{-15}$. We have characterized the systematic uncertainty of the clock at $< 1 \times 10^{-15}$.

The Hz-level line widths allowed us to resolve all hyperfine components of the clock transition (the nuclear spin is $I = 9/2$ for ^{87}Sr), and measure the differential ground-excited g factor that arises from hyperfine mixing of ${}^3P_0$ with ${}^3P_1$ and ${}^1P_1$. This measurement yielded an experimental determination of the ${}^3P_0$ lifetime.

We have also carried out narrow line photoassociation studies with ^{88}Sr near the ${}^1S_0$–${}^3P_1$ dissociation limit. The 15 kHz natural width of the molecular line allowed observation of nine least-bound molecular states. The line shapes were sensitive to thermal effects even at 2 μK ultracold temperatures and to zero-point shifts by the optical lattice confinement. The combination of a narrow width of the least-bound state and its strong coupling to the scattering state should allow efficient tuning of the ground state scattering length with the optical Feshbach resonance technique. We also predict that the deepest-bound level we observed decays to a single ground electronic molecular state with 90% efficiency. This is promising for producing ultracold molecules through photoassociation, which is the subject of our ongoing and future research.

There has been substantial progress recently[1–4] in the control of molecular degrees of freedom, with the goal of preparing molecules in a single quantum state for both internal and external degrees of freedom. These molecules provide a new paradigm for precision measurement. For example, when both the electronic and vibrational transitions are probed precisely, one would be comparing clocks built from two fundamentally different interactions — one of the origin of quantum electrodynamics (α) and the other of strong interaction (electron–proton mass ratio β). Molecular systems will therefore provide unique tests of possible time-variations of fundamental constants.

Stark deceleration currently provides relatively large numbers of polar molecules, but at temperatures limited to a few mK. In our laboratory we have demonstrated deceleration of both hydroxyl radicals (OH) and formaldehyde (H_2CO) molecules to near rest.[5] We demonstrate acceleration/deceleration of a supersonic beam of OH to a mean speed adjustable between 550 m/s and rest, with a translational temperature tunable from 1 mK to 1 K. These velocity-manipulated stable "bunches" contain 10^4–10^6 molecules at a density of 10^5–$10^7\,\mathrm{cm}^{-3}$ in the beam.

These slow, cold molecular packets are ideal for high resolution microwave spectroscopy using Rabi or Ramsey interrogation techniques. The entire manifold of the astrophysically important $J = 3/2$ Λ doublet,[6,7] including both the main lines ($\Delta F = 0$) and the magnetically sensitive satellite lines ($\Delta F = \pm 1$), is measured with a tenfold accuracy improvement. These measurements highlight the ability of cold molecules not only to enhance our understanding of unexplored regimes of molecular coupling, but also to contribute to searches of non–Standard Model physics, for example the variation of fundamental constants such as α and β, which may be measured by comparing Earth-bound OH with that found in OH megamasers.[8,9] These distant sources are spatially well-defined, and by combining our recent measurements with astrophysical studies of comparable resolution, we will be able to constrain — with spatial dependence — fine-structure variation below 1 ppm for $\Delta\alpha/\alpha$ over 10^{10} years.[10]

The precise knowledge gained from these experiments on the polar molecule hyperfine Zeeman behavior allows us to refine the theory of angular momentum couplings in the molecule. The low-B field behavior revealed in OH hyperfine structure may also occur in other molecules and certainly in those with $^2\Pi$ structure. Methods we have explored are crucial if B field effects are to be enhanced for molecular cooling schemes in a magnetic trap. Additionally, because fluctuating B fields are a nuisance for long-term qubit coherence and precision measurements, we need to understand accurately the B field effects in molecule-based clocks, qubits,[11,12] and measurements of the electron electric dipole moment in molecules.[13]

2. Optical Atomic Clock

Optical clocks based on neutral atoms tightly confined in optical lattices have recently begun to show promise as future time/frequency standards.[14–17] These optical lattice clocks enjoy a relatively high signal-to-noise ratio from the large numbers of atoms, while at the same time allowing Doppler-free interrogation of the clock transitions for long probing times, a feature typically associated with single trapped ions. Currently, detailed evaluations of systematic effects for the $^1S_0$–$^3P_0$ sub-Hz line in ^{87}Sr are being performed in several independent systems.[14,17,18] Here we present our recent progress in evaluation of systematic effects below the 10^{-15} level. These measurements have greatly benefited from an ultranarrow spectral line width corresponding to a line quality factor $Q \sim 2.4 \times 10^{14}$.[19–21]

2.1. *Experimental technique*

Neutral strontium atoms are loaded into a dual-stage magneto-optical trap, where they are first cooled to mK temperatures using the strong (32 MHz) $^1S_0$–$^1P_1$ line and then to μK temperatures using the weak (7 kHz) $^1S_0$–$^3P_1$ intercombination line.[22] Approximately 10^4 atoms at $\sim$2 μK are loaded into a one-dimensional, $\sim$300 mW standing wave (optical lattice). The lattice wavelength of 813 nm is chosen to zero the net Stark shift of the clock transition,[23] thus also eliminating line broadening due to the trapping potential inhomogeneity. The atoms are confined in the Lamb–Dicke regime, such that the recoil frequency (5 kHz) is much smaller than the axial trapping frequency (50 kHz). As long as the probe is carefully aligned along the lattice axis, spectroscopy is both Doppler-free and recoil-free. In the transverse direction, the lattice provides a trapping frequency of about 150 Hz, which is smaller than the recoil frequency, but still much larger than the clock transition line width. Atoms can be held in the perturbation-free lattice for times exceeding 1 s, which is important for Hz-level spectroscopy.

We probe the extremely narrow natural line width (1 mHz) of the clock transition in ^{87}Sr with a cavity-stabilized diode laser operating at 698 nm. The high finesse cavity is mounted in a vertical orientation to reduce sensitivity to vibrations.[24] To characterize the probe laser, we compare it to a second stable laser locked to an identical cavity. This comparison shows laser line widths below 0.2 Hz for a 3 s integration time (resolution-limited) and ~2.1 Hz for a 30 s integration time (limited by nonlinear laser drift).

For absolute frequency measurements of the clock transition, we frequency-count the probe laser against a hydrogen maser microwave signal that is calibrated by the NIST primary Cs fountain clock. A self-referenced octave spanning frequency comb[25] is locked to the probe laser, while its repetition rate is counted against the maser. The instability of this frequency counting signal is $2.5 \times 10^{-13}\tau^{-1/2}$, where τ is the integration time. This is the primary limitation on our statistics.

2.2. *Systematic effects*

When the Zeeman sublevels of the ground and excited clock states are degenerate (nuclear spin $I = 9/2$ for $^1S_0$, $^3P_0$), line widths of < 5 Hz ($Q \sim 10^{14}$) are achieved. This spectral resolution should allow us to push the measurements of systematic effects below the 10^{-15} level of uncertainty, limited by our microwave frequency reference.

The systematic effect benefiting most directly from the high line Q is the frequency shift associated with an ambient magnetic field. The differential magnetic moment between the ground and excited states leads to a first-order Zeeman shift of the clock transition. This can lead to shifts or broadening from stray magnetic fields, depending on the population distribution among the magnetic sublevels. By varying the strength of an applied magnetic field in three orthogonal directions and measuring the spectral line width as a function of field strength, the uncertainty of the residual magnetic field has been reduced to < 5 mG for each axis. The resulting net uncertainty for magnetically induced frequency shifts is now < 0.2 Hz ($< 5 \times 10^{-16}$). Understanding and controlling the magnetic shifts is essential for the ^{87}Sr optical clock since the accuracy of all recent measurements has been limited by the sensitivity to magnetic fields.[14,15,17]

Reduction of other systematic uncertainties (due to lattice intensity, probe intensity, and atom density) is straightforward with our high spectral resolution. Recent results in JILA indicate an overall systematic uncertainty below 1×10^{-15} for the Sr lattice clock. Although we have stabilized the microwave phase of the fiber that transmits the maser reference from NIST to JILA to take full advantage of the accuracy of the microwave reference, the averaging times necessary for achieving 10^{-15} uncertainties for all systematic effects are still quite long. Our approach to studying most systematic effects is thus to make frequency measurements under several values of the same systematic parameter within a time interval sufficiently short that the mode frequency of the ultrastable optical cavity used as a reference does not drift over the desired level of uncertainty. In addition, we have locked the probe laser to the clock transition with the goal of directly comparing the Sr clock to the Hg^+ and Al^+ optical atomic clocks at NIST.

2.3. *High spectral resolution*

The recent improvements to the probe laser stabilization have allowed us to probe the clock transition at an unprecedented level of spectral resolution. With the nuclear spin degeneracy removed by a small magnetic field, individual transition components allow exploration of the ultimate limit of our resolution by eliminating any broadening due to residual magnetic fields or light shifts. Figure 1 shows sample spectra of the ${}^1S_0(m_F = 5/2)$–${}^3P_0(m_F = 5/2)$ transition, where m_F is the nuclear spin projection onto the lattice polarization axis. The line widths are probe time limited to ~1.8 Hz, representing a line Q of $\sim 2.4 \times 10^{14}$. This Q value can be reproduced reliably, with some scatter of the measured line widths in the ~ 1–3 Hz range.

Besides the single-pulse spectroscopy of the clock line, two-pulse optical Ramsey experiments were performed on an isolated Zeeman component. When a system is limited by the atom or trap lifetime, the Ramsey technique can yield higher spectral resolution at the expense of signal contrast. An additional motivation for Ramsey spectroscopy in the Lamb–Dicke regime is the ability to use long interrogation pulses, which results in a drastically narrowed Rabi pedestal compared to free-space atoms. The reduced number of Ramsey fringes facilitates the identification of the central fringe.

Figure 2(a) shows a sample Ramsey spectrum, where the preparation and probe pulses are 20 ms and the free evolution time is 25 ms, yielding a pattern with a

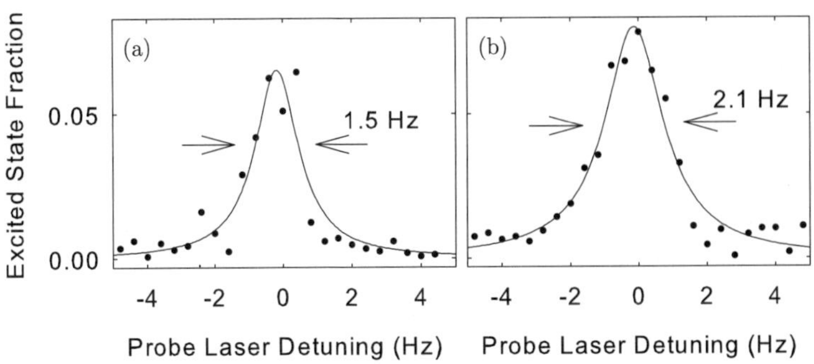

Fig. 1. Typical spectra of the ${}^1S_1$–${}^3P_0$ clock transition, exhibiting a line quality factor $Q \sim 2.4 \times 10^{14}$. The line widths are (a) 1.5(2) Hz and (b) 2.1(2) Hz, in good agreement with the probe time limit of 1.8 Hz. One such trace takes approximately 30 s to collect (since the atom trap must be reloaded for each data point), and involves no averaging or normalization.

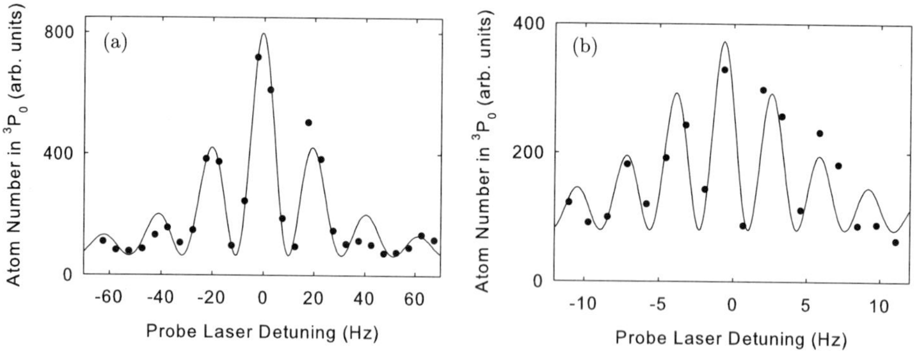

Fig. 2. Ramsey spectra of the $^1S_1$–$^3P_0$ clock transition. (a) The preparation and probe pulses are 20 ms; the evolution time is 25 ms. The fringe width is 10.4(2) Hz. (b) The preparation and probe pulses are 80 ms; the evolution time is 200 ms. The fringe width is 1.7(1) Hz.

fringe width of 10.4(2) Hz, as expected. Figure 2(b) shows the same transition with the preparation and probe pulses of 80 ms and the evolution time of 200 ms. Here the width of the central fringe is reduced to 1.7(1) Hz. Both spectra exhibit no degradation of the fringe contrast. However, the quality of the spectra deteriorated at longer evolution times. Our inability to increase the resolution as compared to single-pulse Rabi spectroscopy suggests that the line width is not limited by the atom or trap lifetime, but rather by phase decoherence between the light and atoms, most likely due to nonlinear laser frequency fluctuations during the scan. This is supported by Rabi spectroscopy where the the laser stability appeared to limit the line width repeatability at the probe time limit near 0.9 Hz.

2.4. *Magnetic moment and lifetime measurements on the clock transition*

This high spectral resolution allowed us to perform NMR-type experiments in the optical domain. We applied a small magnetic bias field and made direct observations of the magnetic sublevels associated with the nuclear spin. The magnetic moments of $^1S_0$ and $^3P_0$ differ because of hyperfine-induced state mixing of $^3P_0$ with $^3P_1$ and $^1P_1$. The differential Landé g factor, Δg, leads to an $\sim$110 Hz/(G m_F) linear Zeeman shift of the clock line. This Zeeman shift measurement is also a direct determination of the perturbed wave function of $^3P_0$, and hence of its metastable ($\sim$100 s) lifetime. Our approach uses only a small magnetic field, while traditional NMR experiments performed on either $^1S_0$ or $^3P_0$ would need large magnetic fields to induce splitting in the radiofrequency range. As larger fields can give rise to additional field-induced state mixing between $^3P_0$, $^3P_1$, and $^1P_1$ (we estimate that a field as weak as 16 G causes a 1% change in Δg), the use of a small field permits an accurate, unperturbed measurement of mixing effects. Additionally, this measurement on the resolved transitions is helpful for the optical clock accuracy evaluation, as one can look for changes in the splitting as a function of the lattice polarization to measure the effect of polarization-dependent light shifts.

The differential nuclear magnetic moment can be determined by setting the probe light polarization parallel to the lattice polarization axis and magnetic field direction, and fitting the resulting Zeeman splitting of the π transitions. However, such a measurement requires an independent calibration of the magnetic field in the trap region, and is sensitive to the linear drift of the probe laser during the scan, which manifests as a fictitious magnetic field. In a more powerful measurement scheme, we polarize the probe laser perpendicular to the quantum axis to excite σ^+ and σ^- transitions, and we obtain spectra as in Fig. 3. Each spectrum yields information about the Δg value from the splitting between the neighboring σ^+ or σ^- lines and about the magnetic field from the splitting between the two manifolds, since the ground state magnetic moment is well known.[26] Thus, the nuclear spin is used as a magnetometer. Since the field calibration and the Δg measurements are done simultaneously, this approach is immune to any linear laser drift or magnetic field variation. This method also eliminates a Δg sign ambiguity that is present in the π transition approach.

We found that the linear Zeeman shift of the clock transition is $-108.8(4)$ Hz/(G m_F), corresponding to a $^3P_0$ magnetic moment of $-1.735(6)\mu_N$, where μ_N is the nuclear magneton. To check for systematic errors related to light shifts, we varied the intensities of the lattice laser (by $> 50\%$) and the probe laser (by a factor of 10) and observed no statistically significant change in the measured splitting. We have also searched for possible m_F-dependent systematics by comparing the splitting frequency of different pairs of sublevels. The magnetic field was varied to

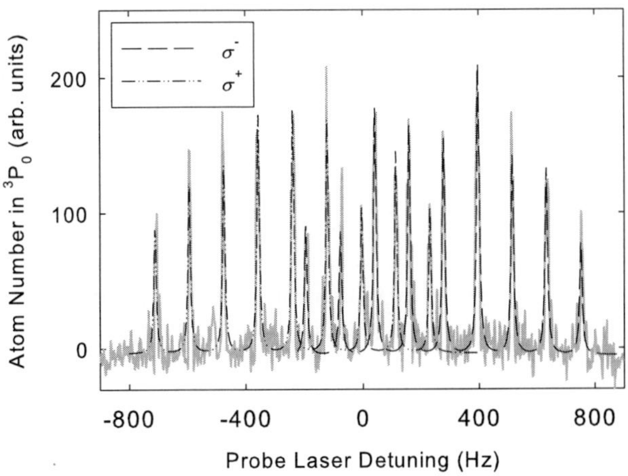

Fig. 3. Clock transition with resolved nuclear spin sublevels. The probe light is polarized perpendicular to the magnetic field direction and lattice polarization, in order to probe σ transitions. The σ^+ and σ^- components are seen as two combs with nine individual lines (nuclear spin $I = 9/2$). The spacing between these lines is determined by the differential g factor of the two clock states, Δg.

verify the field independence of the measurement. A combination of our Δg measurements with atomic theory of hyperfine mixing[27,28] can predict the metastable lifetime of $^3P_0$. Depending on the specifics of the chosen model, the predicted lifetimes range between 100 and 180 s. Although our experimental measurement error is small, it results in a relatively large model-dependent error associated with the lifetime measurement, which gives the value of 140(40) s. Still, it is a very useful confirmation of recently calculated values,[29,30] since a direct and accurate measurement of the natural lifetime is difficult due to limitations from the trap lifetime and black body quenching.

3. Narrow Line Photoassociation

While we worked with the doubly forbidden transition of ^{87}Sr in the optical clock experiments, we used ^{88}Sr for cold collision studies. The even isotope is the most abundant (83%), and has a zero nuclear spin. The large abundance increases the sample density in the trap for more efficient collisions, and the absence of the nuclear spin greatly simplifies the molecular potentials, thus facilitating comparisons of experiment and theory. With ultracold ^{88}Sr in a zero-Stark-shift optical lattice (914 nm wavelength), we performed narrow line photoassociation (PA) spectroscopy near the $^1S_0$–$^3P_1$ intercombination transition.[31] Nine least-bound vibrational molecular levels associated with the long-range 0_u and 1_u excited molecular potentials were measured and identified. The measured PA resonance strengths showed that optical tuning of the ground state scattering length should be possible without significant atom loss. The calculated decay strengths of the photoassociated molecules to the ground electronic state indicate great promise for ultracold stable molecule production.

In contrast to prior PA work that utilizes strongly allowed transitions with typical line widths in the MHz range, here the spin-forbidden atomic $^1S_0$–$^3P_1$ line has a natural width of ~7 kHz. This narrow width allows us to measure the least-bound vibrational levels that would otherwise be obscured by a broad atomic line, and to observe characteristic thermal line shapes even at μK atom temperatures. It also permits examination of the unique crossover regime between the van der Waals and resonant dipole–dipole interactions that occurs very close to the dissociation limit. This access to the van der Waals interactions ensures large bound–bound Franck–Condon factors, and may lead to more efficient creation of cold ground state Sr_2 molecules with two-color PA than what is possible using broad transitions.

Figure 4(a) illustrates the relevant potential energy curves for the Sr_2 dimer as a function of interatomic separation R. The PA laser induces allowed transitions from the separated $^1S_0$ atom continuum at the temperature $T \sim 2\,\mu$K to the bound vibrational levels of the excited potentials V_{0u} and V_{1u}, corresponding to the total atomic angular momentum projections onto the internuclear axis of 0 and 1, respectively. The long-range potentials are determined by the C_6/R^6 (van der Waals) and C_3/R^3 (resonant dipole–dipole) interactions. The values of the C_3

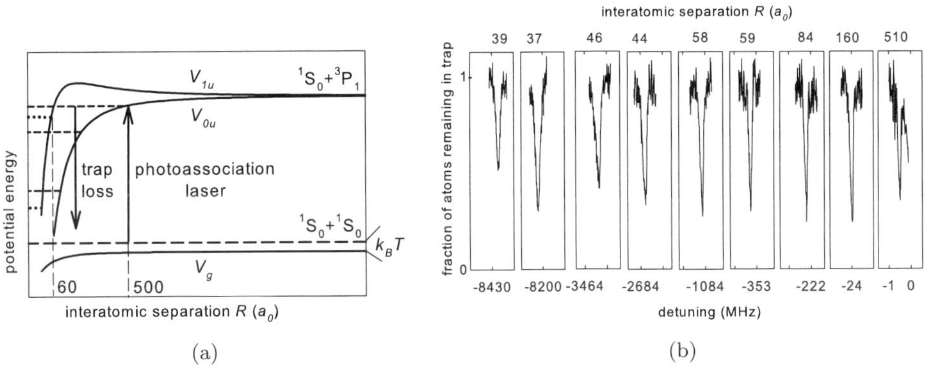

Fig. 4. (a) Schematic diagram of the long-range Sr_2 molecular potentials. The ground state has *gerade* symmetry and its energy is given by the potential V_g, while the excited state *ungerade* potentials that support transitions to the ground state are V_{0u} and V_{1u}. All vibrational states of 0_u and 1_u (dashed and dotted lines, respectively) are separated by more than the natural line width, permitting high-resolution PA spectroscopy very close to the dissociation limit when the atoms are sufficiently cold. (b) The spectrum of the long-range Sr_2 molecule near the $^1S_0$–$^3P_1$ dissociation limit. The horizontal scale is marked on the rightmost panel and is the same for each of the nine blocks; different PA laser intensities were used for each line due to largely varying transition strengths. The top labels indicate the interatomic separations that correspond to the classical outer turning points of each resonance.

and C_6 coefficients are adjusted in the multichannel theoretical model[32] so that bound states exist at the experimentally determined resonance energies. The C_3 coefficient can be expressed in terms of the atomic lifetime τ as $C_3 = 3\hbar c^3/4\tau\omega^3$, where $\hbar\omega$ is the atomic transition energy and c is the speed of light. Our data and theoretical model yielded a C_3 coefficient that corresponds to the $^3P_1$ atomic lifetime of 21.5(2) μs.

To trace out the molecular line spectra, a frequency-stabilized 689 nm diode laser is tuned near the $^1S_0$–$^3P_1$ intercombination line. The laser frequency is stepped, and after 320 ms of PA at a fixed frequency the atoms are released from the optical lattice trap and counted by a strongly resonant light pulse. At a PA resonance, the atom number drops as excited molecules form and subsequently decay to ground state molecules in high vibrational states or hot atoms that cannot remain trapped. Figure 4(b) shows the nine observed PA line spectra near the dissociation limit. The individual lines were fit to convolutions of a Lorentzian profile with an initial thermal distribution, and distinct thermal tails were observed even at the ultracold $\sim$1–3 μK temperatures. In addition, the strong axial lattice confinement alters the collisional dynamics, and the two-dimensional effects factor into the line shapes as a larger density of states at small thermal energies, and as a redshift by the lattice zero-point confinement frequency.

Intercombination transitions of alkaline earths such as Sr are particularly good candidates for optical control of the ground state scattering length, $a + a_{\rm opt}$, because there is a possibility of large gains in $a_{\rm opt}$ with small atom losses. These optical

Feshbach resonances are of great interest for Sr, since magnetic Feshbach resonances are absent for the $^1S_0$ ground state, and the background scattering length is too small to allow evaporative cooling.[33] Using the -0.4 MHz PA line [see Fig. 4(b)] should allow tuning of the ground state scattering length[34] by $\pm 300\ a_0$ (a_0 is the Bohr radius), where the PA laser with intensity $I = 10$ W/cm^2 is far-detuned by $\delta = \pm 160$ MHz from the molecular resonance. In contrast, optical tuning of the scattering length in alkali ^{87}Rb [35] achieved tuning of $\pm 90\ a_0$ at much larger PA laser intensities of 500 W/cm^2. In addition, the Sr system at the given parameter values will have a loss rate of $\sim 2 \times 10^{-14}$ cm^3/s,[34] while the loss rate in the ^{87}Rb experiment was 2×10^{-10} cm^3/s. The overall efficiency gain of over five orders of magnitude is possible for Sr because the narrow intercombination transition allows access to the least-bound molecular state, and the PA line strength exponentially increases with decreasing detuning from the atomic resonance.[34] The above scattering length tunings and atom loss values accessible with the -0.4 MHz resonance result in the elastic and inelastic collision rates of $\sim$600/s and $\sim$0.1/s, respectively.

Narrow line PA is also suitable for efficient production of ultracold molecules in the ground electronic state. The bound–bound Franck–Condon factor calculations show that about 90% of the molecules photoassociated into the -8.4 GHz bound state decay to a single ground state vibrational level (distributed between only two rotational sublevels). Because the Sr excited state molecular potential is strongly influenced by van der Waals (C_6) interactions, the wave function overlap with the ground molecular bound states is large, making it possible to use low-power lasers in a Raman configuration to coherently transfer the molecules into the absolute ground ro-vibrational molecular state. Other exciting prospects include using the Raman scheme to perform precision measurements of the molecular ground state vibrational spacings with the goal of constraining temporal variations of the proton–electron mass ratio, and combining strontium with other alkaline earth atoms to create strongly interacting ultracold polar molecules in an optical lattice.

4. Precision Measurement Based on Cold OH Molecules

Comparison of different atomic clock systems[36] has provided tight constraints on the time variation of various fundamental constants, including α, during the modern epoch. However, observation of absorption lines in distant quasars[37,38] provides conflicting results about possible α variation over cosmological time. Recently, there has been much interest in using OH megamasers in interstellar space to constrain the evolution of fundamental constants[8–10] with several key advantages. Most importantly, the multiple lines (which have different dependence on the fundamental constants) arising from one of these localized sources differentiate the relative Doppler shift from true variation in the transition frequency. Furthermore, if it is assumed that the only fundamental constant to vary is α, it can be shown that the sum (Δ_Λ) and difference (Γ^f and Γ^e) of the $\Delta F = 0$ (F is total angular momentum) transition frequencies in the ground Λ doublet of OH depend on α as $\alpha^{0.4}$ and α^4, respectively (see Fig. 5).[8] Thus, by comparing these quantities as measured from OH megamasers to laboratory values, it is possible to remove the Doppler shift systematic and constrain α over cosmological time. Furthermore, because of the unique properties of the Λ doublet, the $\Delta F = 0$ transitions are extremely insensitive to magnetic fields, while the $\Delta F = \pm 1$ satellite transitions can be used to calibrate the B field. However, as pointed out by Darling,[8] for the current limits on $\Delta\alpha/\alpha$ the change in the relevant measurable quantities is on the order of 100 Hz, which, prior to this work, was the accuracy of the best laboratory-based measurement.[39] Moreover, an astrophysical measurement of OH megamasers, scheduled for later this year, expects a resolution better than 100 Hz,[40] and thus better laboratory measurements of the OH Λ doublet microwave transitions will soon be needed to allow for tighter constraints on $\Delta\alpha/\alpha$.

Experimentally, we focus on the electric-dipole-allowed, π ($\Delta m_F = 0$) and σ ($\Delta m_F = \pm 1$) transitions between the Λ double parity states, as shown by arrows in Fig. 5(a). We write these states in the $|F, m_F, p\rangle$ basis and distinguish π transitions from states by substituting Δp for the parity value, p. The hyperfine energies of the parity states are written as $\Gamma^{f,e} = \bar{\Gamma} \pm \xi/2$, where the average splitting, $\bar{\Gamma}$, and difference, ξ, are approximately 54.1 and 1.96 MHz, respectively.[39] Similarly, the g factors are written as $g^{f,e} = \bar{g} \pm \delta g/2$, where δg is 0.3% of $\bar{g}$.[41]

We utilize the Stark decelerator and microwave spectroscopy[6,7] to obtain high spectral resolutions. In brief, the Stark decelerator provides OH primarily in the strongest, weak electric (E) field–seeking states, $|F, m_F, p\rangle = |2, \neq 0, f\rangle$. Typical operation produces a 10^6 cm^{-3} dense packet with mean velocity controllable from 410 m/s down to nearly rest and a minimum temperature of 5 mK.[3] We choose to run the decelerator at 200 m/s as a compromise between slowing efficiency, microwave power, and interrogation time. We perform Rabi or Ramsey spectroscopy by interrogating the OH packet with one or more microwave pulses referenced to the Cs standard.[6,7] Laser-induced fluorescence provides state detection by exciting

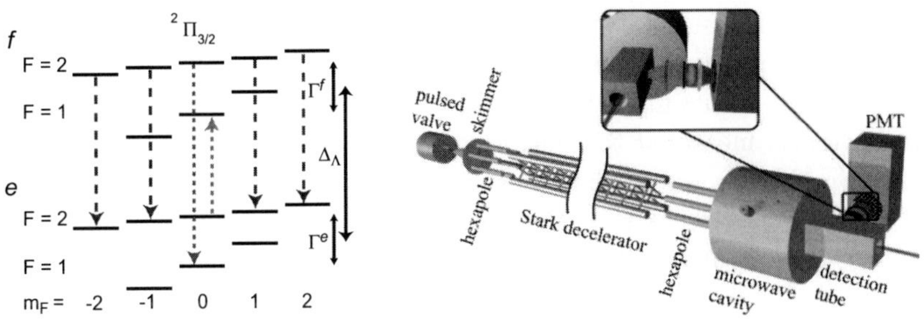

(a) (b)

Fig. 5. (a) OH ground Λ doublet state. The arrows represent the transitions induced by the applied microwave pulses. Long-dashed arrows depict main-line π transitions ($\Delta F = 0$) while short-dashed arrows show measured satellite lines ($\Delta F = \pm 1$). (b) Schematic of the experiment (the inset depicts the detection region).

the f manifold — equally driving the two hyperfine states — with 282 nm light and registering a fraction of the subsequent 313 nm emission with a photomultiplier tube. Main and satellite transitions originating from $|2, m_F, f\rangle$ are driven with a single microwave pulse and detected by registering a reduction in photon counts [see Fig 5(a)].

The 10-cm-long, TM_{010} microwave cavity is tuned near the $|2, m_F, \Delta p\rangle$ main line. A solenoid is wound around the cavity and encased in a μ-metal magnetic shield. The latter reduces the ambient field to $\leq 6\,\mathrm{mG}$, while the former can apply a B field > 10 G along the resonant TM_{010} axis, which itself is aligned along the OH beam path. At the $|2, m_F, \Delta p\rangle$ frequency, the cavity E and B fields are nearly collinear — determined by the absence of observed σ transitions — and the E field magnitude is constant over 80% of the cavity. In contrast, the satellite line frequencies are detuned far enough from cavity resonance to distort the associated intracavity E field magnitude and direction, and we observe both π ($\mathbf{E} \parallel \mathbf{B}$) and σ ($\mathbf{E} \perp \mathbf{B}$) transitions on these lines.

We calibrate the intracavity B field by tracking the $|2, m_F, f\rangle \rightarrow |1, m'_F, e\rangle$ satellite splitting as a function of the B field in the $< 1\,\mathrm{G}$ regime, where the Zeeman theory is well understood. Concurrent observation of the π and σ lines provides cross-checks on the B field calibration. Their relative peak heights and frequency shifts versus cavity position also provide a tool for mapping E field direction and power inhomogeneities. Under a residual B field of $< 6\,\mathrm{mG}$, the transition frequencies of the two main lines ($|2, m_f, f\rangle \rightarrow |2, m_f, e\rangle$ and $|1, m_f, f\rangle \rightarrow |1, m_f, e\rangle$) are measured to be $(1,667,358,996 \pm 4)$ Hz and $(1,665,401,803 \pm 12)$ Hz, respectively. These results are limited only by statistical uncertainties as other systematic shifts, such as from collisions, stray electric field, doppler shift, and black body radiation, are collectively $< 3\,\mathrm{Hz}$.[6] When a bias B field is applied to test the hyperfine structure theory, four peaks become visible for the $|2, m_f, f\rangle \rightarrow |2, m_f, e\rangle$ transition since for π transitions $m_F = m'_F = 0$ is selection-rule-forbidden when $\Delta F = 0$. As expected, the $m_F = m'_F = \pm 2$ lines are linearly shifted under an increasing B field while we detect curvature in the $m_F = m'_F = \pm 1$ lines at fields greater than $\sim$1 G. The measured $\bar{\delta g}_F = 1.267(5) \times 10^{-3}$ is 1σ, consistent with Radford measurements[41] in the J basis at B = 0.6–0.9 T of $\delta g_J = 1.29(2) \times 10^{-3}$. Higher precision measurements in both regimes may detect a difference.

We have also measured the magnetically sensitive satellite line transition frequencies. Specifically, we trace the $\nu_{21} = |2, 0, f\rangle \rightarrow |1, 0, e\rangle$ and $\nu_{12} = |2, 0, e\rangle \rightarrow |1, 0, f\rangle$ frequencies versus the B field. These Zeeman transitions are first-order-insensitive to the B field, which suppresses spurious frequency shift contributions from calibration and ambient field uncertainties. Nearly overlapping π and σ transitions preclude an accurate measurement at a zero applied field, and instead we must magnetically split the lines. The theory fit is used to project the data down to a zero applied B field. The uncertainty due to the ambient field on the line frequencies is estimated at $< 0.6\,\mathrm{Hz}$. Thus, $\nu_{21} = 1,720,529,887(10)\,\mathrm{Hz}$. For the ν_{12} line, we use the distorted E field at the cavity entrance to drive $|2, \pm 1, f\rangle \rightarrow |2, 0, e\rangle$

σ transitions. A second microwave pulse then drives the π transition to $|1, 0, f\rangle$. We measure $\nu_{12} = 1{,}612{,}230{,}825(15)\,\text{Hz}$. These values provide a tenfold improvement over the previous measurement accuracy.[39]

Variation of α may be constrained by taking the sum and difference of either the main or the satellite lines. The difference is sensitive to α while the sum measures the redshift systematic. The satellite lines offer a 55-fold increase in resolution over the main lines due to the $2\bar{\Gamma}$ versus ξ separation.[8] Now that the complete Λ doublet is measured with high precision, one can simultaneously constrain variation of α and β.[9] All four lines must be detected from the same source, and the closure criterion, i.e. the zero difference of the averages of the main and satellite line frequencies, provides a critical systematics check. We obtain an Earth-bound closure of $44 \pm 21\,\text{Hz}$. Imminent astrophysical measurements of the satellite lines below the 100 Hz level[40] highlight the prescience of our study for observing the variation of fundamental constants.

Acknowledgments

We thank S. Diddams, T. Parker, M. Notcutt, J. L. Hall, H. Lewandowski, and S. Jefferts for technical help and discussions. We also acknowledge collaboration with R. Ciuryło, P. Naidon and P. Julienne on the photoassociation theory. The Sr work is funded by ONR, NASA, NIST, and NSF. The OH work is funded by DOE, NIST, and NSF.

References

1. J. Weinstein *et al.*, *Nature* **395** (1998) 148.
2. H. L. Bethlem *et al.*, *Phys. Rev. Lett.* **83** (1999) 1558.
3. J. Bochinski *et al.*, *Phys. Rev. Lett.* **91** (2003) 243001.
4. J. Sage *et al.*, *Phys. Rev. Lett.* **94** (2005) 203001.
5. E. R. Hudson *et al.*, *Phys. Rev. A* **73** (2006) 063404.
6. E. R. Hudson *et al.*, *Phys. Rev. Lett.* **96** (2006) 143004.
7. B. Lev *et al. Phys. Rev. A* **74** (2006) 061402.
8. J. Darling, *Phys. Rev. Lett.* **91** (2003) 011301.
9. J. Chengalur and N. Kanekar, *Phys. Rev. Lett.* **91** (2003) 241302.
10. N. Kanekar *et al.*, *Phys. Rev. Lett.* **95** (2005) 261301.
11. D. DeMille, *Phys. Rev. Lett.* **88** (2002) 067901.
12. A. Micheli, G. Brennen and P. Zoller, *Nature Phys.* **2** (2006) 341.
13. M. Kozlov and L. Labzowsky, *J. Phys. B* **28** (1995) 1933.
14. A. Ludlow *et al.*, *Phys. Rev. Lett.* **96** (2006) 033003.
15. M. Takamoto *et al.*, *Nature* **435** (2005) 321.
16. Z. Barber *et al.*, *Phys. Rev. Lett.* **96** (2006) 083002.
17. A. Brusch *et al.*, *Phys. Rev. Lett.* **96** (2006) 103003.
18. M. Takamoto *et al.*, *J. Phys. Soc. Jpn.* **75** (2006) 104302.
19. M. M. Boyd *et al.*, *Science* **314** (2006) 1430.
20. M. M. Boyd *et al.*, *Phys. Rev. Lett.* **98** (2007) 083602.
21. M. M. Boyd *et al.*, *Phys. Rev. A* **76** (2007) 022510.
22. T. Loftus *et al.*, *Phys. Rev. A* **70** (2004) 063413.

23. T. Ido and H. Katori, *Phys. Rev. Lett.* **91** (2003) 053001.
24. M. Notcutt *et al.*, *Opt. Lett.* **30** (2005) 1815.
25. T. Fortier *et al.*, *Opt. Lett.* **28** (2003) 2198.
26. L. Olschewski, *Z. Phys.* **249** (1972) 205.
27. H. Kluge and H. Sauter, *Z. Phys.* **270** (1974) 295.
28. B. Lahaye and J. Margerie, *J. Phys.* **36** (1975) 943.
29. S. Porsev and A. Derevianko, *Phys. Rev. A* **69** (2004) 042506.
30. R. Santra *et al.*, *Phys. Rev. A* **69** (2004) 042510.
31. T. Zelevinsky *et al.*, *Phys. Rev. Lett.* **96** (2006) 203201.
32. R. Ciuryło *et al.*, *Phys. Rev. A* **70** (2004) 062710.
33. P. Mickelson *et al.*, *Phys. Rev. Lett.* **95** (2005) 223002.
34. R. Ciuryło *et al.*, *Phys. Rev. A* **71** (2005) 030701(R).
35. M. Theis *et al.*, *Phys. Rev. Lett.* **93** (2004) 123001.
36. E. Peik *et al.*, *Phys. Rev. Lett.* **93** (2004) 170801.
37. J. K. Webb *et al.*, *Phys. Rev. Lett.* **87** (2001) 091301.
38. R. Quast, D. Reimers and S. Levshakov, *Astron. Astrophys.* **415** (2004) L7.
39. J. ter Meulen and A. Dymanus, *Astrophys. J.* **172** (1972) L21.
40. N. Kanekar, private communication (2006).
41. H. E. Radford, *Phys. Rev.* **122** (1961) 114.

ATOMIC CLOCKS AND PRECISION MEASUREMENTS

KURT GIBBLE

Department of Physics, The Pennsylvania State University,
104 Davey Laboratory 232, University Park,
Pennsylvania 16802, USA
kgibble@psu.edu

We present a review of our clock science conducted under the NASA Microgravity Fundamental Physics program. Our work has led to the development of rubidium atomic clocks, designs for ground- and space-based clocks that juggle atoms to achieve ultrahigh stability and accuracy, improved microwave cavities for atomic clocks, and elucidation of new systematic errors such as the atomic recoil from microwave photons. High stability clocks can be used for precise tests of fundamental physics and accurate deep-space navigation.

Keywords: Atomic clocks; photon recoil; photon momentum.

1. Introduction

The NASA Microgravity Fundamental Physics program has enabled many contributions to basic science and the development of technologies for ground- and space-based experiments. As part of this program, we have made a number of advances related to atomic clocks and precision measurements. In this paper we review some of those developments toward better atomic clocks. One interesting systematic for atomic fountain clocks is a potential frequency shift due to the momentum imparted to the atoms from the microwave photons.[1]

2. Improving Atomic Clocks

Our initial goal in this program was to find a solution to the frequency shift caused by ultracold collisions, which is the dominant problem for the design and operation of laser-cooled cesium clocks.[2,3] We built a laser-cooled fountain clock based on ^{87}Rb and showed that the ultracold collision shift is 50 times smaller than that for cesium clocks (see Fig. 1).[4] As a result, in 2004, the International Consultative Committee on Time and Frequency recommended that ^{87}Rb be adopted as a secondary definition of the SI second.[5] Because of the small collision shift, at least four groups are operating or constructing ^{87}Rb fountain clocks around the world.

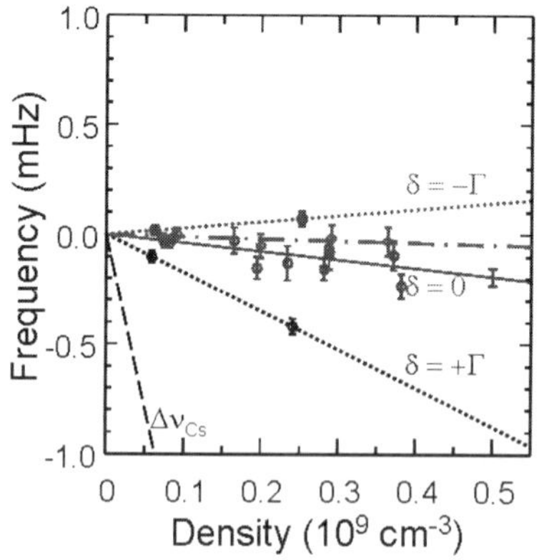

Fig. 1. Frequency shifts due to ultracold collisions for ^{87}Rb (solid) and Cs (dashed). The ^{87}Rb frequency shift is 50 time smaller than that for Cs. When the microwave cavity is detuned by one line width (dotted lines), the cavity pulls the transition frequency proportional to the number of atoms in the cavity so that the frequency shift can be canceled (dashed–dotted).

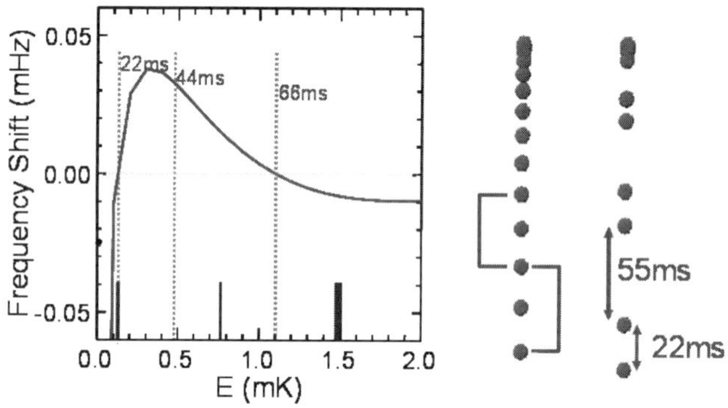

Fig. 2. *s*-wave juggling frequency shift for ^{87}Rb. For a constant juggling delay of 22 ms, the collision of every second ball at 44 ms gives a large frequency shift. A juggling pattern with alternate delays of 22 and 55 ms gives a juggling frequency shift of 0.

One, the US Naval Observatory, is building six ^{87}Rb fountain clocks[6] and these will soon be the master system clocks for the Global Positioning System.

The Rubidium Atomic Clock Experiment (RACE)[7] aimed to achieve extremely high short-term stability and accuracy on the International Space Station. It was based on juggling ^{87}Rb atoms in microgravity. Juggling[8] many balls of atoms at once, i.e. launching atoms with time delays much shorter than the interrogation time as in Fig. 2, allows many more atoms to pass through the clock in a given time. It dramatically reduces the averaging time that is required to achieve a given accuracy.[9] Our work for RACE has also directly led to a better understanding of systematic errors due to the microwave cavities, and improved cavities for use in both space- and ground-based clocks.[10,11] Clocks in space, like RACE, can be used for precise test of fundamental physics[7] and precise interplanetary navigation.[12]

The phenomenal accuracy of clocks makes them sensitive to some surprising effects. Cesium clocks have progressed to accuracies near 5×10^{-16}.[3] To achieve this accuracy, we need to understand the kinetic energy transferred to an atom when it absorbs a microwave photon.[13,14] The simplest picture suggests that a traveling-wave microwave photon has a momentum of $\hbar k$ and so the atom's kinetic energy after it absorbs a photon is $(\hbar k)^2/2m$. This extra energy has to come from the photon's energy and therefore the transition energy is shifted by this amount, which is fractionally 1.5×10^{-16} for the cesium transition. However, in a standing wave, there are multiphoton processes. When an atom is localized near the antinode of the standing wave, the momentum transfer from the left- and right-going photons destructively interferes so that the atom is transferred to the excited state with no photon recoil.[1] For small displacements from the antinode, the multiphoton processes act as positive and negative lenses on the atomic wave functions. The frequency shifts that result are fortunately often slightly smaller than the usual recoil shift above, for the transverse wave vector of the standing wave.[1] This effect is also present in precision measurements with atom interferometers, where the finite laser beams have transverse wave vectors in the microwave regime.[15]

Acknowledgments

We gratefully acknowledge financial support from the NASA Microgravity program, the Office of Naval Research, and Penn State University.

References

1. K. Gibble, *Phys. Rev. Lett.* **97** (2006) 073002.
2. K. Gibble and S. Chu, *Phys. Rev. Lett.* **70** (1993) 1771.
3. R. Wynands and S. Weyers, *Metrologia* **42** (2005) S64.
4. C. Fertig and K. Gibble, *Phys. Rev. Lett.* **85** (2000) 1622.
5. http://www.bipm.fr/wg/CCL/CCL-CCTF/Allowed/2005/CCTF04-rec1.pdf
6. S. Peil *et al.*, *Proc. 2006 IEEE Freq. Contr. Symp.*, Exposition, pp. 304–306.
7. C. Lammerzahl *et al.*, *Gen. Relativ. Gravit.* **36** (2004) 615.
8. R. Legere and K. Gibble, *Phys. Rev. Lett.* **81** (1998) 5780.
9. C. Fertig, J. I. Rees and K. Gibble, *Proc. 2001 IEEE Freq. Contr. Symp.* (2001), p. 18.
10. R. Li and K. Gibble, *Metrologia* **41** (2004) 376.
11. R. Li and K. Gibble, *Proc. 2005 IEEE Int. Freq. Contr. Symp. and Exposition* (2005), p. 99.
12. K. Gibble, *Proc. 2004 NASA/JPL Workshop* (Solvang, CA).
13. J. L. Hall, C. J. Borde and K. Uehara, *Phys. Rev. Lett.* **37** (1976) 1339.
14. C. Vian *et al.*, *IEEE Trans. Instrum. Meas.* **54** (2005) 833.
15. A. Wicht *et al.*, *Phys. Scr.* **T102** (2002) 82.

THE CLOCK MISSION OPTIS

HANSJÖRG DITTUS* and CLAUS LÄMMERZAHL†
ZARM, University of Bremen,
Am Fallturm, 28359 Bremen, Germany
**dittus@zarm.uni-bremen.de*
†laemmerzahl@zarm.uni-bremen.de

Clocks are an almost universal tool for exploring the fundamental structure of theories related to relativity. For future clock experiments, it is important for them to be performed in space. One mission which has the capability to perform and improve all relativity tests based on clocks by several orders of magnitude is OPTIS. These tests consist of (i) tests of the isotropy of light propagation (from which information about the matter sector which the optical resonators are made of can also be drawn), (ii) tests of the constancy of the speed of light, (iii) tests of the universality of the gravitational redshift by comparing clocks based on light propagation, like light clocks and various atomic clocks, (iv) time dilation based on the Doppler effect, (v) measuring the absolute gravitational redshift, (vi) measuring the perihelion advance of the satellite's orbit by using very precise tracking techniques, (vii) measuring the Lense–Thirring effect, and (viii) testing Newton's gravitational potential law on the scale of Earth-bound satellites. The corresponding tests are not only important for fundamental physics but also indispensable for practical purposes like navigation, Earth sciences, metrology, etc.

Keywords: Test of special relativity; test of general relativity; clock comparison experiment.

1. Introduction

A wide range of the physical laws can be read off from the rates of clocks. In fact, special relativity (SR), for example, can be completely based on particular properties or behavior of clocks when moving together or separately through space. Also, general relativity (GR) can, except for the universality of free fall, be based on the behavior of clocks; see e.g. Refs. 1 and 2 in this issue for more details. Furthermore, today it is a general agreement that high precision clocks have to be taken to space since the conditions on Earth are poorly defined, in order to obtain a well-defined interpretation of the time which the clock is showing. Therefore it is highly natural to propose a clock space mission with a variety of high precision clocks by which all the relevant clock tests for SR and GR can be carried out. Such a mission, called OPTIS, will be described below. Other missions using clocks are ACES/PHARAO,[3,4] SUMO,[4] PARCS,[4] RACE,[4] and SPACETIME.[5] Clocks also may be of great use in the exploration of the gravitational field of the Earth and of the solar system, like in a future deep space Gravity Explorer mission.[6] Furthermore, the use for practical purposes as well as the space conditions and the variety of clocks available have been described in Ref. 2.

The OPTIS mission is a satellite equipped with a variety of clocks and laser ranging and tracking facilities for performing improved tests of the foundations as well as predictions of SR and GR. This mission makes advantage of the space conditions of large differences in the velocity and the gravitational potential. Here we report on recent progress made in the studies of the behavior of the resonator in the field of a gravity gradient.

It has been outlined[2,7] that Einstein's GR is mainly based on the Einstein equivalence principle, i.e. on the UFF, the UGR and on LLI. The proposed OPTIS mission aims at an improvement of the complete test of LLI and of UGR by three orders of magnitude compared to the present ground experiments. Therefore, together with the test of UFF by MICROSCOPE and STEP, we will have a complete test of the foundations of GR.

2. Science Objectives of OPTIS

The mission OPTIS consists of a collection of clocks in a highly elliptic Earth-bound orbit. Laser tracking devices and a laser link to the Earth complete the scientific hardware components. As a consequence, many issues related to clocks and to the orbit can be measured with high precision. Therefore, OPTIS aims at improving tests of the *foundations of SR and GR* by up to three orders of magnitude. The scientific basis for this has been outlined in detail in Ref. 4. The science objectives are listed in Table 1.

Except for the universality of free fall, which will be tested by MICROSCOPE[8] and STEP[9], OPTIS represents a complete test of the foundations of metric theories of gravity; see Fig. 1. Furthermore, relativistic orbital effects as predicted by Einstein's theory of gravity will be tested.

Table 1. The scientific objectives of OPTIS.

Test	Method	Present Accuracy	OPTIS Accuracy
1. Isotropy of speed of light	Cavity–cavity comparison	10^{-16}	10^{-19}
2. Constancy of speed of light	Cavity–clock comparison	10^{-16}	10^{-19}
3. Time dilation — Doppler effect	Laser link	$2 \cdot 10^{-7}$	10^{-9}
4. Universality of gravit. redshift I	Cavity–clock comparison	$1.7 \cdot 10^{-2}$	10^{-4}
5. Universality of gravit. redshift II	Clock–clock comparison	$2.5 \cdot 10^{-5}$	10^{-7}
6. Absolute gravitational redshift	Time transfer	$1.4 \cdot 10^{-4}$	10^{-8}
7. Lense–Thirring effect	Laser tracking	10^{-1}	10^{-3}
8. Einstein perigee advance	Laser tracking	$3 \cdot 10^{-3}$	$6 \cdot 10^{-4}$
9. Test of Newton potential	Laser tracking	10^{-5}	10^{-12}

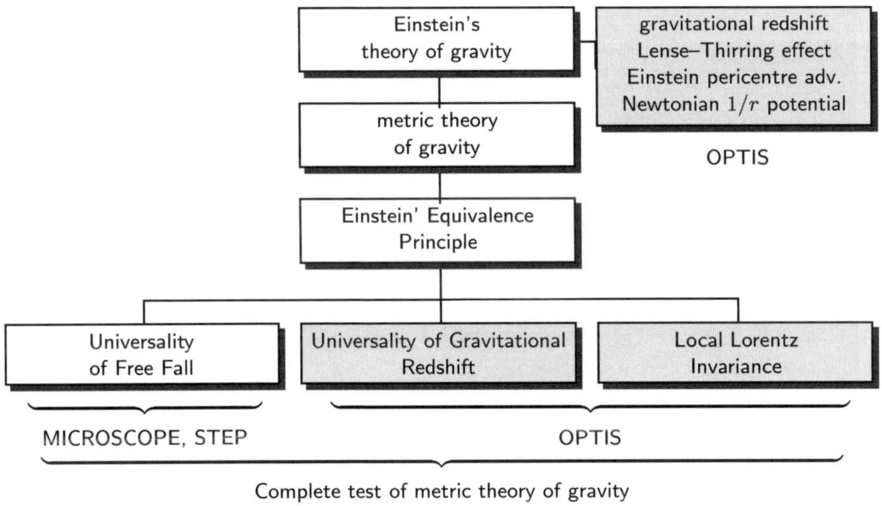

Fig. 1. Together with MICROSCOPE or STEP, OPTIS will provide a full test of the metric structure of the gravitational field. Furthermore, effects characteristic of the Einstein field equations, namely the Lense–Thirring effect, the Einstein pericenter advance and the validity of the Newtonian $1/r$ potential, will be tested by OPTIS.

3. Mission Design

In order to have a good test of the universality of the gravitational redshift, clocks should move through a large gravitational potential difference. Therefore, a highly elliptic orbit is preferable. For tests of SR, a large variation of the velocity is also needed. This again can be obtained from a highly elliptic orbit. From tracking and ranging, highly precise orbital data can be obtained which are of use for the exploration of the structure of the gravitational field of the Earth, the Sun and the Moon.

The mission scenario is shown in Fig. 2; see also Ref. 4.

4. Mission Technology

For this mission technologies are required which have been used recently to carry out the most precise tests of special relativity. The precision of these tests can be further increased under space conditions thanks to a more quiet environment, larger changes in the orbital velocity, and larger differences of the gravitational potential. Furthermore, very precise laser tracking and linking of satellites is a well-established technique and will provide, in combination with the active drag-free control system, very accurate orbit data.

The core technologies for OPTIS are:

- optical cavities,
- highly stabilized lasers,
- gravitational reference sensors,

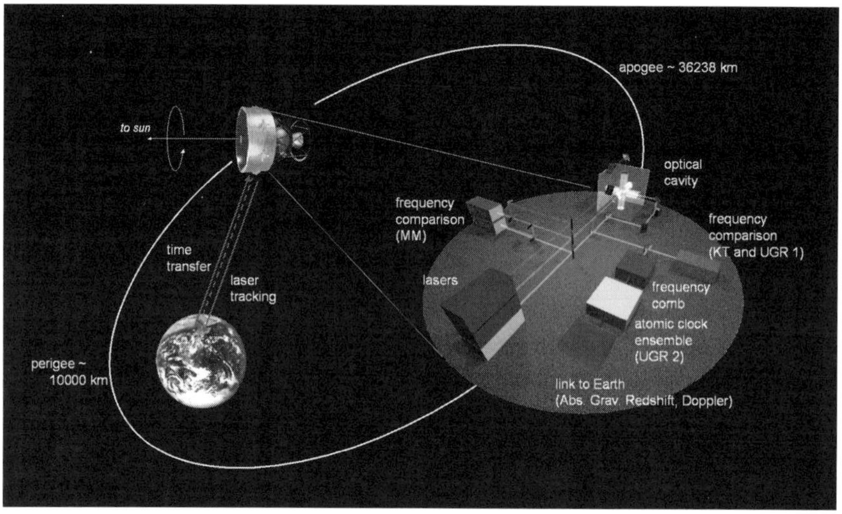

Fig. 2. Mission scenario (apogee and perigee heights measured from the Earth's surface).

- drag-free control,
- H maser,
- ion clocks,
- optical clocks,
- frequency combs,
- laser tracking systems,
- laser links.

These technologies are also key technologies for other future missions.

5. Science Requirements

In order to meet the planned accuracy of tests and measurements, the scientific payload, the tracking and ranging systems as well as the drag-free Attitude and Orbit Control System (AOCS) have to fulfill the following requirements:

***Residual accelerations*:** Accelerations will distort the resonator and thus lead to unwanted signals. In order to keep the accuracy stated in the science goals, residual, unmodelizable accelerations must be below 3×10^{-11} m/s^2. This can be achieved with available drag-free techniques.

***Residual rotations*:** Unknown rotations also will distort the resonator and, thus, have to be constrained to within $4 \cdot 10^{-5}$ Hz.

***Temperature stability*:** Changes in the temperature of the optical resonator will lead to reduced accuracy. Suppression of these fluctuations to the aimed accuracy requires a temperature stabilization at a level of about 10^{-6} μK. This is the most challenging technical task of the mission.

***Laser lock stability*:** Drifts of the laser-to-cavity lock and mechanical misalignments of the laser beam into the resonator must be reduced to appropriate levels.

***Tracking accuracy*:** Today, Earth-bound satellites can be laser-tracked with an accuracy of a few centimeters. This is sufficient for the science goals stated above. Current development in laser tracking may lead to an accuracy of a few millimeters in the near future. This would lead to further improvements of tests 7–9. Improvements in laser tracking accuracy are independent of the OPTIS mission and can be applied even during the operation of the mission.

6. Payload and Mission Relevant Technology: Optical Resonators

Optical resonators are one of the essential parts of the experimental payload. Locking of lasers to the resonators will give highly stabilized frequencies which carry, via $\nu = nc/2L$, (L is the length of the cavity), the information about the velocity of light c propagation along the cavity axis. The length of the cavity has to be very stable, since otherwise this could mask or simulate the searched effect. For cryogenic resonators the stability is[10] $\delta L/L \sim 10^{-16}$.

There are many influences on the cavities: temperature, temperature gradients, inertial forces like residual acceleration and centrifugal forces, gravitationally induced nongeodesic forces and torques, and the gravitational tidal force (gravity gradient). In order to avoid accelerations induced by the gravity gradient on an extended body, a cubic symmetry of the cavity is preferable. The gravity gradient will unavoidably distort the cavity. Therefore a stiff material is preferred. The high-dimensional stability requires materials with small thermal expansion coefficients. For OPTIS the options are (i) ULE glass with a thermal expansion coefficient of $\beta = 10^{-9}/\mathrm{K}$ at room temperature, or (ii) silicon with a vanishing thermal expansion coefficient at 140 K and a second order thermal expansion coefficient of $2 \cdot 10^{-9}/\mathrm{K}^2$.

Since the tidal gravitational force which acts through every extended body cannot be eliminated by choosing an appropriate frame, it will induce distortions on the resonator which have to be calculated and subtracted from the signal. We give a rough estimate of the expected effect of the tidal gravitational force on a freely moving cube of length L. If the position of the cube is at a distance R from the center of the Earth, then the difference of the Earth's acceleration on the top and the bottom of the cube is $\Delta a = (\partial^2 U/\partial r^2)L$, where $U = GM_\oplus/R$ is the Earth's Newtonian potential. For $R = 10^4$ km and a typical resonator length of $L = 5$ cm, we have $\Delta a \approx GM_\oplus L/R^3 \approx 2 \cdot 10^{-8}\ \mathrm{m/s}^2$. In a rough estimate we assume this Δa to act on the top surface of the cube. From Hook's law $F/A = E\Delta L/L$, giving the change of the length ΔL of the cube due to a force F acting on the area A, and taking $F = m\Delta a = \rho L^3 \Delta a$, we get $\frac{\Delta L}{L} = \frac{\rho L}{E}\Delta a \approx 10^{-17}$ for an elasticity modulus of $E = 90$ GPa and a density of 2350 kg/m^3, typical for Zerodur. Therefore the tidal gravitational force will lead to systematic deformations which are two orders of magnitude larger than the expected accuracy. Hence we have to subtract this effect

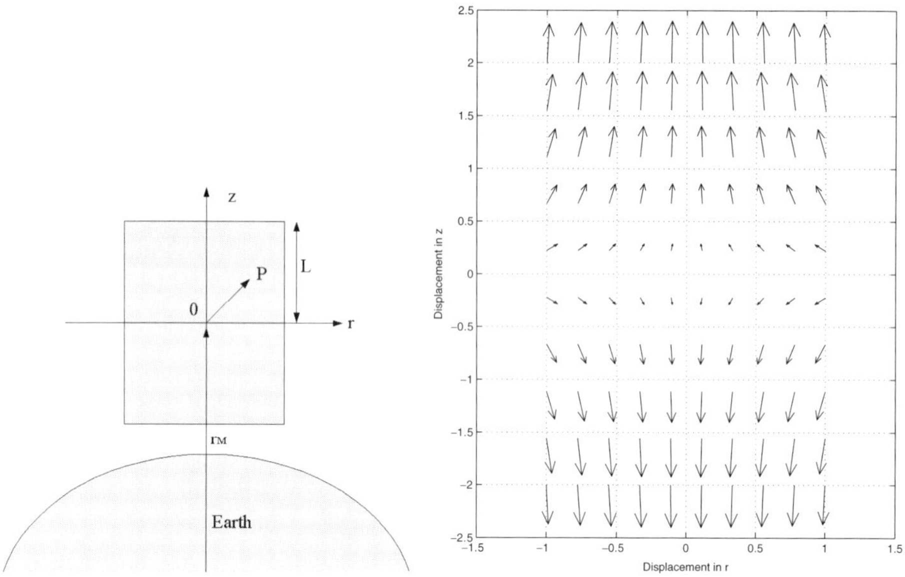

Fig. 3. *Left*: Simplified model of an optical resonator on a geodetic Earth orbit. *Right*: Displacement field of a cylinder under the influence of a spherical tidal gravitational force field. The displacements are plotted over the body coordinates r and z. The cylinder boundaries are at $z = \pm L = \pm 2$ and $r = R = 1$.

from the signal. This effect has to be calculated from the equations of elasticity in the force field given by the gravity gradient.

For the actual resonator with drillings etc., the effect has to be determined numerically.[11] Since the deformations are of the order 10^{-17}, we checked the numerical accuracy with an analytical solution for a cylindrically symmetric resonator in a tidal gravitational field found for the first time in Ref. 12. The analytical solution for the displacement is shown in Fig. 3. As expected, the cylinder in squeezed along the horizontal direction and stretched along the vertical direction. The numerical simulation in Fig. 4 verifies the analytic solution.

Figure 5 shows the displacements of the mirrors Δx, Δy, and Δz for the two opposing mirrors on each resonator axis as well as the relative displacements dx, dy, and dz on these axes. The data show one complete rotation of the resonator around its body z axis when the resonator is at point $\phi_y = \pi/2$ on its way around the Earth, which means when the gravity gradient force is acting in the body x direction.

The important information we need is the relative displacement between two opposing mirrors. Figure 6 shows contour plots of the displacements dx, dy, and dz along the x axis. The displacements along the y and z axes are similar. In all plots the data for $\phi_y = \phi_z = 0$, corresponding to a certain initial state of deformation, have already been subtracted. That means that the numbers on the contour lines give the relative displacements of the mirrors with respect to this starting state.

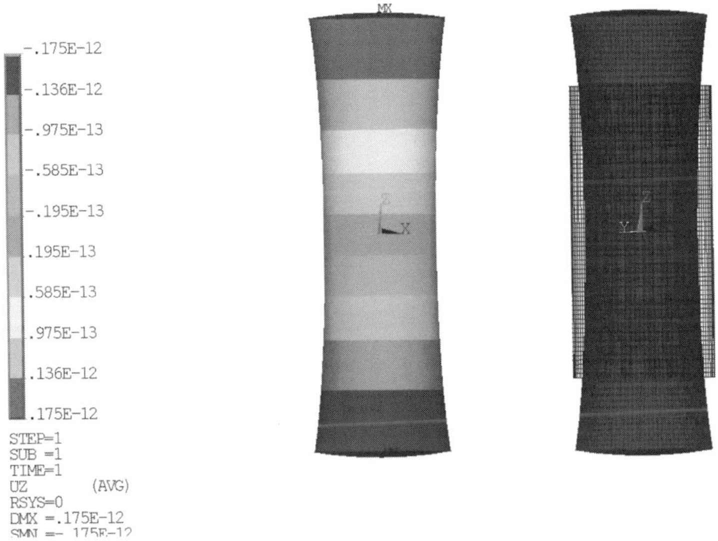

Fig. 4. Finite element solution: deformation of the cylinder under the influence of a spherical tidal gravitational force field. The deformation is scaled by a factor of $6 \cdot 10^{13}$. *Right*: Deformed cylinder shape and original finite element mesh. *Left*: The scale shows the z displacements.

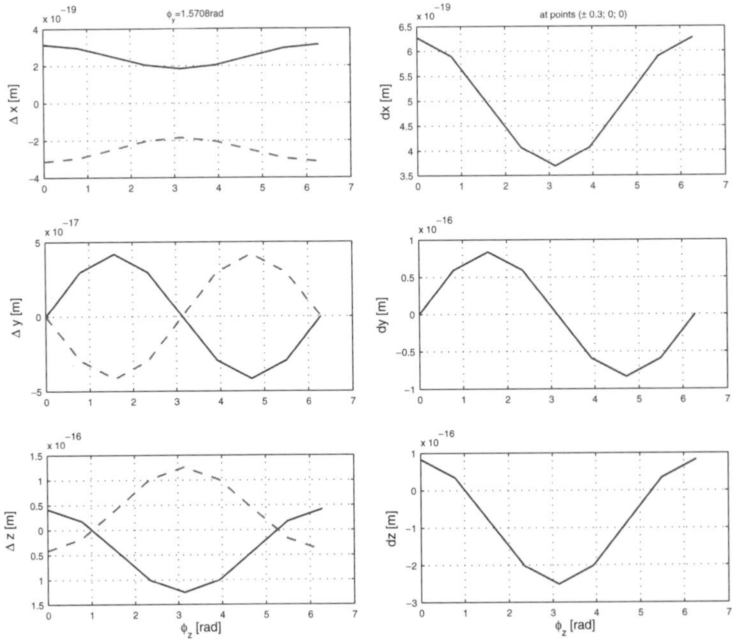

(a) Mirrors on the x axis.

Fig. 5. OPTIS resonator with a gravity gradient at point $\phi_y = \pi/2$ on its orbit around the Earth. *Left side of each subfigure*: Displacements at the mirror midpoints: dotted line = mirror on the negative original coordinate point; full line = mirror on the positive original coordinate point. *Right side*: Relative displacements between two opposing mirrors.

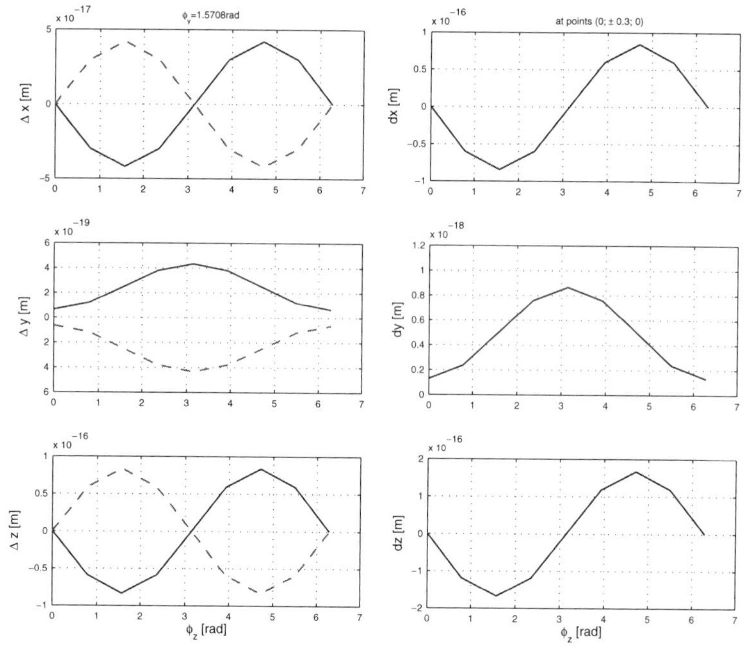

(b) Mirrors on the y axis.

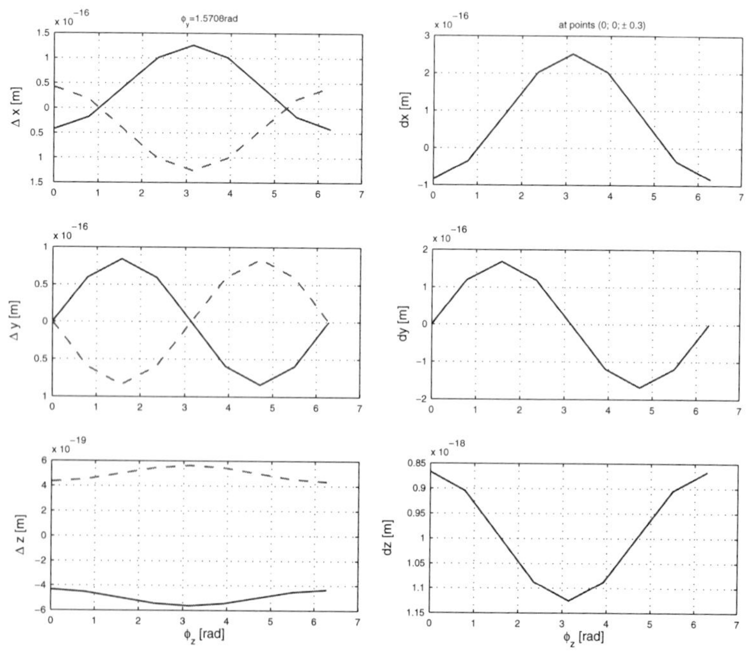

(c) Mirrors on the z axis.

Fig. 5. (*Continued*)

7. OPTIS Resonator Under a Thermal Gradient

To analyze the influence of thermal gradients on the deformation of the OPTIS resonator, a thermal analysis within the FEM program ANSYS was carried out. Figure 7 shows the deformation of the resonator assuming a thermal gradient of

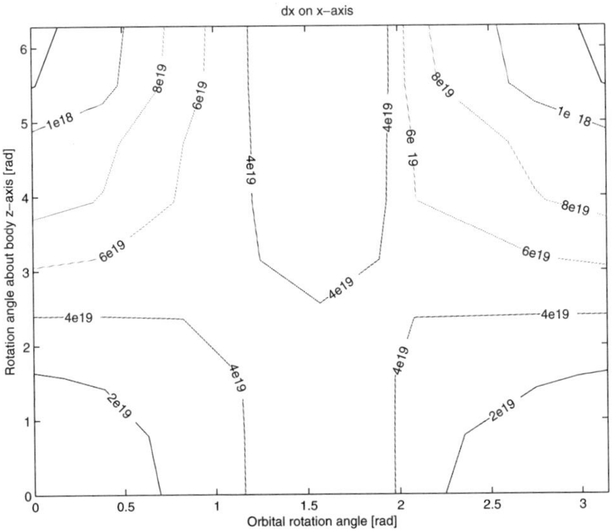

(a) Contour plots of the relative mirror displacements dx on the x axis.

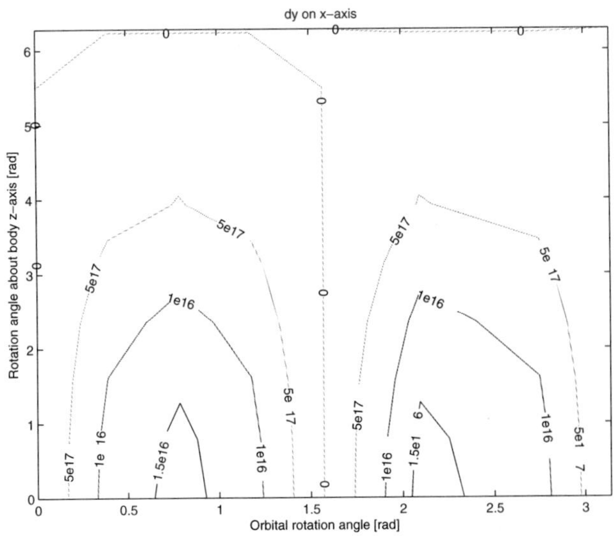

(b) Contour plots of the relative mirror displacements dy on the x axis.

Fig. 6. OPTIS resonator with a gravity gradient. The data correspond to an orbital rotation around the inertial y axis for half an orbit with angle ϕ_y and a rotation of the resonator around its body z axis with angle ϕ_z.

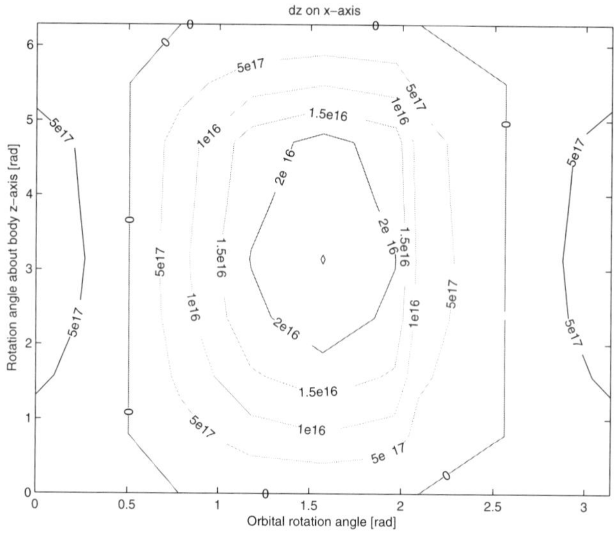

(c) Contour plots of the relative mirror displacements dz on the x axis.

Fig. 6. (*Continued*)

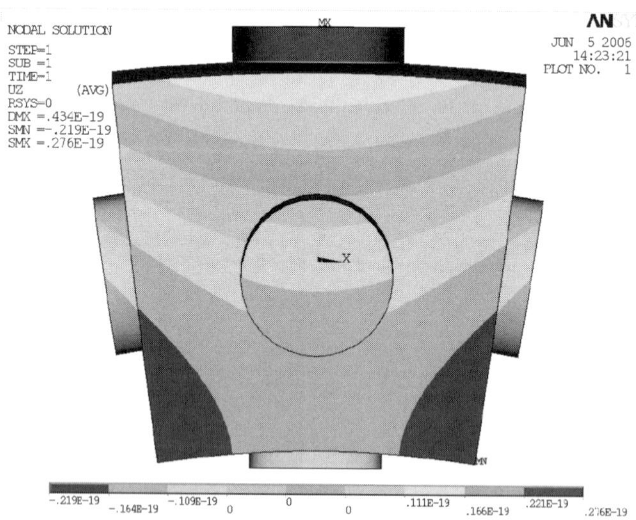

Fig. 7. OPTIS resonator under the influence of a linear temperature gradient field of 10^{-9} K in the z axis direction. The color bar gives the values of the z displacements in meters. The deformation is scaled by a factor of $3 \cdot 10^{17}$.

10^{-9} K between the upper and the lower surface in the z direction. We used the material properties of Corning ULE glass, which has so far the lowest thermal expansion coefficient of $\alpha \approx 10^{-9}$ K^{-1}.

The relative deformations between the mirrors on the x, y and z axes resulting from this temperature gradient are $dx = dy = dz = 2.99 \cdot 10^{-20}$ m. This is still within the science requirement $dL/L \leq 10^{-18}$, which means for a resonator of $L = 6$ cm that $dL \leq 6 \cdot 10^{-20}$ m.

Thus we can conclude that for the OPTIS resonator made of ULE the temperature gradient must be smaller than 10^{-7} K/m in order to fulfill the science requirements. Furthermore, this means that the temperature stability between the endpoints of the resonator must be better than 10^{-9} K on a time scale (L^2/χ_T) that is determined by the temperature conductivity χ_T of the material.

The first step in providing an appropriate thermal stabilization is the analysis of the thermal conditions given by the orbit and the corresponding thermal properties inside the satellite in the case where we provide the satellite with superinsolation layers. The temperature outside the satellite varies between –110 K and –50 K. If we assume four superinsolation layers made of coated Capton film, each being 20 mm thick leads to a temperature variation inside the satellite of around 20 mK. The required thermal stability and thermal gradient then have to be provided by an active shielding.

8. Further Components

Due to lack of space we skip the discussion of further technological elements needed for the OPTIS missions, like lasers, laser locking, atomic clocks, optical clocks, frequency comb, gravity reference sensors, satellite laser ranging, time transfer by laser link, microthrusters, orbit selection, and drag-free and attitude control system, and refer to the review Ref. 13, where these components have been described extensively.

Acknowledgments

We would like to thank I. Ciufolini, L. Iorio, H. Müller, A. Peters, E. Samain, S. Scheithauer, S. Schiller, and A. Wicht for fruitful collaboration and discussions, and the German Aerospace Agency (DLR) and the German Research Foundation (DFG) for financial support.

References

1. H. Dittus, C. Lämmerzahl and S. G. Turyshev eds., *Lasers, Clocks, and Drag-Free: Exploration of Relativistic Gravity in Space*, Astrophysics and Space Science Library, Vol. 349 (Springer-Verlag, Berlin, 2006), in press.
2. C. Lämmerzahl and H. Dittus, *Int. J. Mod. Phys. D*, this issue.
3. C. Salomon *et al.*, *C. R. Acad. Sci. Paris* **4** (2004) 1313.
4. C. Lämmerzahl *et al.*, *Gen. Relativ. Gravit.* **36** (2004) 615.
5. L. Maleki and J. Prestage, SpaceTime mission: Clock test of relativity at four solar radii, in *Gyros, Clocks, and Interferometers: Testing Relativistic Gravity in Space*, eds. C. Lämmerzahl, C. W. F. Everitt and F. W. Hehl (Springer-Verlag, Berlin, 2001), p. 369.
6. H. Dittus and the Pioneer Explorer Collaboration, A mission to explore the pioneer anomaly, in *Trends in Space Science and Cosmic Vision 2020*, eds. Á. Giménez *et al.* (ESA, Noordwijk, 2005, p. 3) [gr-qc/0506139].

7. C. Lämmerzahl, "The Search for Quantum Gravity Effects I," *Appl. Phys. B*: in press.
8. P. Touboul, *Comptes Rendus de l'Aced. Sci. Série IV: Phys. Astrophys.* **2** (2001) 1271.
9. N. Lockerbie *et al.*, STEP: A status report, in *Gyros, Clocks, and Interferometers: Testing Relativistic Gravity in Space*, eds. C. Lämmerzahl, C. W. F. Everitt and F. W. Hehl (Springer-Verlag, Berlin, 2001), p. 213.
10. S. Herrmann *et al.*, *Phys. Rev. Lett.* **95** (2005) 150401.
11. S. Scheithauer, "Analyse der elastischen Verformungen von optischen Resonatoren in erdgebundenen und Weltraumnexperimenten," PhD thesis (ZARM, University of Bremen, 2006).
12. S. Scheithauer and C. Lämmerzahl, physics/0606250.
13. C. Lämmerzahl *et al.*, *Gen. Relativ. Gravit.* **36** (2004) 2373.

ATOMIC CLOCK ENSEMBLE IN SPACE: AN UPDATE

C. SALOMON

Laboratoire Kastler Brossel, ENS,
24, rue Lhomond, 75005 Paris, France
salomon@lkb.ens.fr

L. CACCIAPUOTI

European Space Agency,
Research and Scientific Support Department,
ESTEC, Keplerlaan 1–PO Box 299,
2200 AG Noordwijk ZH, The Netherlands

N. DIMARCQ

SYRTE-CNRS UMR8630, Observatoire de Paris,
61, avenue de l'Observatoire 75014 Paris, France

Atomic Clock Ensemble in Space (ACES) is a mission in fundamental physics that will operate a new generation of atomic clocks in the microgravity environment of the International Space Station. Fractional frequency stability and accuracy of a few parts in 10^{16} will be achieved. The on-board time base, distributed on the Earth via a microwave link, will be used to perform space-to-ground as well as ground-to-ground comparisons of atomic frequency standards. Based on these comparisons, ACES will perform fundamental physics tests (Einstein's theories of special and general relativity, the search for drift of fundamental constants, the Standard Model extension and tests of string theories) and develop applications in time and frequency metrology, time scales, geodesy, global positioning and navigation. After an overview of the mission concept and its scientific objectives, the present status of ACES instruments and subsystems will be discussed.

Keywords: Cold atoms in space; ACES mission; precision clocks.

1. The ACES Mission Concept

Atomic Clock Ensemble in Space (ACES)[1,2] is a mission in fundamental physics whose main objective is to demonstrate and use the performances of a new generation of atomic clocks in the microgravity environment of the International Space Station (ISS). Scheduled for launch in the 2011 time frame, ACES will be accommodated on board the ISS, on the External Payload Facility of the Columbus module. The station is orbiting at a mean elevation of 400 km with a 90 min period and an inclination angle of 51.6°.

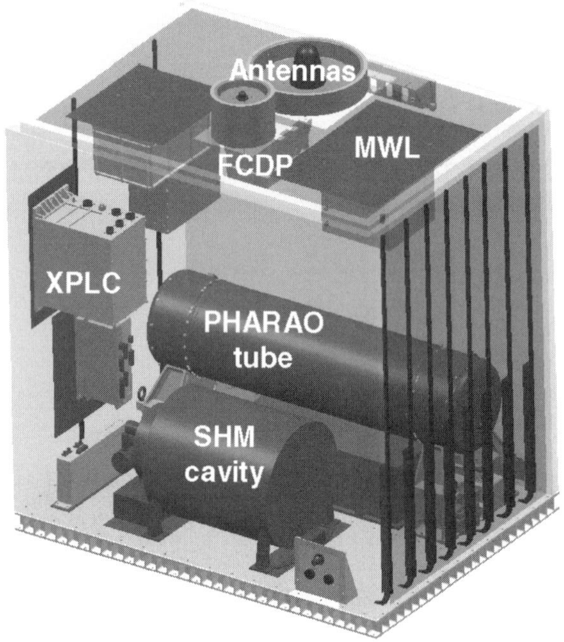

Fig. 1. ACES payload. The two atomic clocks (PHARAO and SHM), the frequency comparison and distribution package (FCDP), and the microwave link (MWL) fit into a thermally regulated payload with a total mass of 227 kg and a power consumption of 450 W. Antennas point toward the Earth.

ACES is composed of a flight payload and ground terminals. The payload occupies a volume of $1\,\mathrm{m}^3$ and involves both state-of-the-art instruments and subsystems (see Fig. 1). The heart of the payload is represented by an atomic clock based on laser-cooled cesium atoms. The performances of the cesium frequency standard PHARAO (Projet d'Horloge Atomique par Refroidissement d'Atomes en Orbit) developed by CNES, France, are combined with the characteristics of a space hydrogen maser (SHM) developed by the Neuchatel Observatory, Switzerland. The ACES clock signal will therefore merge together the good medium-term frequency stability of the hydrogen maser and the long-term stability and accuracy of a primary frequency standard based on cold atoms. The on-board clock-to-clock comparison (PHARAO–SHM) and the distribution of the clock signal are ensured by the Frequency Comparison and Distribution Package (FCDP), while all data handling processes are controlled by the eXternal PayLoad Computer (XPLC).

One of the main objectives of the ACES mission consists in maintaining a stable and accurate on-board time scale that can be used for space-to-ground as well as ground-to-ground comparisons of frequency standards. Stable and accurate time and frequency transfer is achieved by using a microwave link (MWL), important not only for characterizing the ACES clock ensemble with respect to ground clocks, but also for performing general relativity tests of high scientific relevance.

The planned mission duration is 18 months. During the first 6 months, the performances of the SHM and PHARAO will be established. Thanks to the microgravity environment, the line width of the atomic resonance will be varied by two orders of magnitude (from 11 Hz to 110 mHz). Performances in the 10^{-16} regime for both frequency stability and accuracy are expected. In the second part of the mission, the on-board clocks will be compared to a number of ground-based clocks operating in both the microwave and the optical domain. The recent development of optical frequency combs[3,4] and of optical clocks has made it easy to connect the optical and microwave domains and to enable ACES to perform worldwide advanced clock comparisons of different types.

2. ACES Scientific Objectives

ACES will demonstrate the high performances of a new generation of atomic clocks for space applications. Based on the direct comparison of the ACES clocks and ground-based clocks, accurate tests of Einstein's theory of general relativity and of the Standard Model extension will be performed.

Atomic fountain clocks have already demonstrated their excellent performances in terms of both frequency stability and accuracy.[5] A stability of 1.4×10^{-16} at 50,000 s and an accuracy of 7×10^{-16} have been measured. These clocks will reach excellent performances in space where the long interrogation times due to the reduced gravity conditions allow the detection of very narrow atomic resonances. The PHARAO clock, based on samples of laser-cooled cesium atoms, will reach a fractional frequency instability below $1 \times 10^{-13}\tau^{-1/2}$, where τ is the integration time in seconds, and an inaccuracy at the 1×10^{-16} level. These performances will be established thanks to the excellent stability of the SHM, able to ensure a mid-term frequency instability of 1.5×10^{-15} for 10,000 s of integration time.

The ACES clock signal will be sent to the Earth via MWL and used to demonstrate time transfer capability with rms time deviation better than 0.3 ps at 300 s, 6 ps at 1 day, and 23 ps at 10 days of integration time. MWL will be a key element of the overall mission, important not only for comparing the space clocks to ground-based clocks, but also for performing distant comparisons of clocks arbitrarily distributed around the Earth, both in common view and in noncommon view. When the two ground clocks are close enough, they can be simultaneously compared to the same on-board signal. In this case, the comparison budget depends mainly on the short-term stability of the link. ACES will allow common view comparisons at an uncertainty level below 300 fs per ISS pass (~300 s). If the two ground clocks are geographically distant, they can be compared to the same on-board time at different epochs, separated by a well-defined time delay, T. During this dead time, the ACES clock signal will act as a flywheel. The time stability of the clock-to-clock comparison will be limited by the stability of the ACES time scale over the time interval T. ACES will allow noncommon view comparisons of ground clocks at an uncertainty level of 2 ps for $T = 1000$ s, 5 ps for $T = 10,000$ s, and 20 ps for

$T = 1$ day. These stability levels will outperform by at least one order of magnitude present time and frequency transfer systems based on GPS satellites or current two-way satellite time and frequency transfer systems.[6,7]

The comparison of primary frequency standards at the 10^{-16} level will give a relevant contribution to the definition and the characterization of a worldwide atomic time scale and, for the first time, will demonstrate the possibility of synchronizing distant atomic clocks at the 100 ps uncertainty level.

These performances are at the basis of accurate tests of Einstein's theory of general relativity. During the mission lifetime, ACES will measure the gravitational redshift at an unprecedented level of accuracy, search for time variations of fundamental constants, test isotropy and constancy of the speed of light, and implement a new type of relativistic geodesy based on the redshift.

2.1. *Measurement of the gravitational redshift*

As a direct consequence of Einstein's equivalence principle (EEP), identical clocks placed at different positions in stationary gravitational fields experience a gravitational redshift that, in the frame of the post-Newtonian approximation (PPN), can be expressed as a function of the Newtonian potential at the clock positions:

$$\frac{\nu(x_1)}{\nu(x_0)} = 1 - \frac{U(x_1) - U(x_0)}{c^2}. \tag{1}$$

In 1960, Pound and Rebka made a pioneer laboratory measurement of the gravitational redshift using the Mössbauer effect.[8] The result confirmed the prediction of general relativity within 1%. To date, the most precise measurement of the gravitational redshift goes back to the 1979 Gravity Probe A experiment (GPA).[9] GPA was based on the direct comparison of two H masers, one on the ground and the other in a spacecraft launched nearly vertically upward to a height of about 20,000 km. The experiment demonstrated an agreement between the observed frequency shift and the theoretical prediction at the 70×10^{-6} level.

Knowing precisely the orbital parameters of the space station and ground stations (position and velocity), the frequency difference between the ground clocks and the PHARAO clock will be measured and compared to theory. A single pass over a given ground station will be sufficient to achieve a relative accuracy of 50×10^{-6} on Einstein's gravitational redshift. This value is already below the result obtained in 1980 by the GPA experiment. With the full accuracy of ground and space clocks at the 10^{-16} level or better, Einstein's effect can be tested with a relative uncertainty of 2×10^{-6}, yielding a factor-of-35 improvement with respect to the previous experiment.

2.2. *Time variations of fundamental constants*

In general relativity, as in other metric theories of gravitation, a time variation of nongravitational physical constants is forbidden. This is a direct consequence of the principle of local position invariance (LPI). However several modern theories predict the existence of new interactions which violate EEP. Damour and Polyakov,[10] as well as several other authors, predict time variation of fundamental constants and in particular of the fine structure constant. High-accuracy atomic clocks are unique instruments to test these theories and to detect LPI violations.[11–14] Such measurements interestingly complement tests of the local Lorentz invariance (LLI) and of the universality of the free fall. All together these tests experimentally establish the validity of EEP. Nearly all unification theories (particularly string theories) violate EEP at some level, strongly motivating experimental searches for such deviations.

A recent determination of the fine structure constant[15] gives

$$\alpha = \frac{e^2}{4\pi\epsilon_0\hbar c} = 1/137.035999710(96), \tag{2}$$

with 0.7 ppb uncertainty. α characterizes the strength of the electromagnetic interaction in atoms and molecules. In atoms and molecules, any transition between two energy levels can be expressed in terms of α, m_e/m_p and the ratio of a magnetic moment to the proton magnetic moment g_p. Within QCD these constants are related to three (and only three) dimensionless constants, α, $m_q/\Lambda_{\rm QCD}$, and $m_e/\Lambda_{\rm QCD}$, derived from the quark mass m_q, the mass scale of quantum chromodynamics $\Lambda_{\rm QCD}$, and the electron mass m_e.[16,17]

Time variations of fundamental constants can be measured by comparing clocks based on different transitions or different atomic species. When distant clocks are compared, the precision of such tests relies on the frequency stability and accuracy of the contributing clocks, on the noise of the time transfer link, and on the uncertainties of relativistic corrections. Using clocks with an accuracy of 10^{-16} or better and the ACES clock-to-clock comparison scheme will allow one to measure any frequency drift due to variations of fundamental constants with a resolution below 10^{-16} per year, improving on current experimental results.[12,18] Further, by combining the results of several optical and microwave clock comparisons, it will be possible to deduce stringent limits not only to the time variations of α, but also of m_e/m_p, $m_q/\Lambda_{\rm QCD}$, and $m_e/\Lambda_{\rm QCD}$.

2.3. *Isotropy and constancy of the light velocity*

The foundations of special relativity (SR) lie on the hypothesis of LLI. According to this principle, the outcome of any local test experiment is independent of the velocity of the freely falling apparatus. More generally, LLI can be seen as one of the cornerstones of the EEP. Over the last century a large number of tests have provided demonstration of the validity of SR at different accuracy levels.

Direct measurements of the speed of light can be performed by continuously comparing a space clock to a ground clock during each passage of the spacecraft on the ground clock location. The difference of the measured reception and emission times provides the one-way travel time of the signal plus an unknown but constant offset depending on the fact that the ground and the space clocks are not synchronized by slow clock transport. In the case of an LLI violation, the difference of the up and down travel times will be sensitive to a nonzero value of $\delta c/c$ along a preferred direction and to variations of c depending on the relative velocity of the clocks. LLI tests based on this technique have already been performed in 1997 by comparing clocks on board GPS satellites to a ground-based hydrogen maser.[19]

With an overall time instability over one ISS pass as low as 1 ps, the expected sensitivity of ACES to $\delta c/c$ measurements would be in the low 10^{-10} region, representing an improvement of a factor 10 over previous measurements.

2.4. *Relativistic geodesy*

Another promising application of ACES deals with geodesy. A new kind of "relativistic geodesy" based on Einstein's gravitational redshift will be demonstrated. Near the Earth surface the Einstein effect amounts to a relative frequency shift of 10^{-16} per meter of elevation. Using ground optical clocks with 10^{-17} accuracy and ACES MWL, the Earth potential difference between two arbitrarily distant clocks will be determined with an equivalent height difference of about 20 cm. Such a "geodesy" would nicely complement the current space geodetic missions CHAMP, GRACE and the coming mission GOCE. As the clock frequency is sensitive to the gravitational potential rather than the local gravity acceleration, the information provided by clocks is of different nature. Optical clocks with 7×10^{-17} accuracy have been demonstrated at NIST, USA, in 2006[21] and the prospects for reaching 10^{-18} with trapped ion clocks and lattice neutral atom clocks[22] are excellent. With an improved time transfer link along the lines of the fiber link developed in Ref. 23, geodesy at the 1 cm level over one day of averaging appears feasible in future missions with optical clocks.

More generally, because of the time-dependent fluctuations of the local Earth-gravity field, clocks with 10^{-18} stability and accuracy will lose their universality as time standards precisely because of the geodetic effects described above.[24] It is therefore tempting to transfer these ultraprecise time instruments into orbiting satellites where the local Earth potential fluctuations are averaged out. With a set of four satellites and links between them it is possible to define an ultrastable space–time reference frame with a variety of applications in navigation, positioning, astrometry, and fundamental physics.

2.5. *Time scales and navigation*

ACES will improve the accuracy of the atomic time and will contribute to the definition of global time scales (TAI, UTC, GPS, and GALILEO). The third generation of navigation systems will benefit from technology developments related to the ACES mission: better clocks and high-performance time and frequency transfer systems will soon be available. New concepts for global positioning systems based on a reduced set of ultrastable space clocks in orbit associated with simple transponding satellites could be studied.

3. ACES Instruments and Subsystems

3.1. *The cesium clock PHARAO*

PHARAO is a cesium clock based on laser-cooled atoms developed by SYRTE, LKB, and CNES. Its operation is very similar to that of ground-based atomic fountains. Atoms launched in free flight cross two microwave cavities tuned to the transition between the two hyperfine levels of the cesium ground state. The interrogation method, based on two separate oscillating fields (Ramsey scheme), allows the detection of an atomic line whose typical width is inversely proportional to the transit time between the two microwave cavities. The resonant microwave field at 9.192631770 GHz (SI definition of the second) is synthesized starting from a quartz oscillator and stabilized to the clock line using the error signal generated by the cesium resonator. In this way, the intrinsic qualities of the cesium hyperfine transition, in terms of both accuracy and frequency stability, are transferred to the macroscopic oscillator. In a microgravity environment, the velocity of atoms along the ballistic trajectories is constant and can be changed continuously over almost two orders of magnitude (5–500 cm/s). Therefore, very long interaction times (up to a few seconds) are possible, while keeping reasonable the size of the instrument.

PHARAO will provide a clock signal with fractional frequency instability below $1 \times 10^{-13}\tau^{-1/2}$ (Fig. 2) and inaccuracy near 10^{-16}. Further, the error signal generated by the cesium resonator will be sent to XPLC, processed, and used to correct

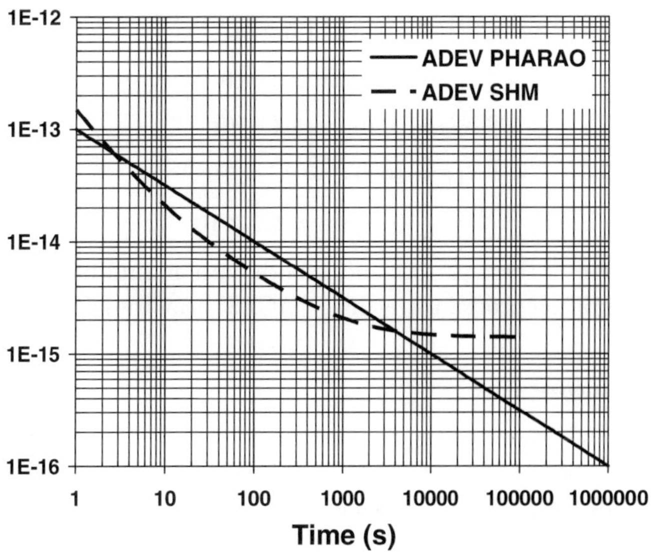

Fig. 2. Expected fractional frequency instability of PHARAO and SHM. The ACES clock signal will combine the short- and medium-term frequency stability of SHM with the long-term stability and accuracy of a primary frequency standard based on laser-cooled cesium atoms.

SHM for long-term frequency drifts. PHARAO will also provide all the frequency correction parameters necessary for evaluating the clock accuracy. According to ACES mission objectives, PHARAO performances will be verified through preliminary tests on the ground and a full in-flight validation.

3.2. *The space hydrogen maser*

Because of their simplicity and reliability, H masers are used in a variety of ground applications. Passive and active masers are expected to be key instruments in future space missions, satellite positioning systems, and high resolution VLBI (very-long-baseline interferometry) experiments.

H masers are based on the hyperfine transition of atomic hydrogen at 1.420405751 GHz. H_2 molecules are dissociated in a plasma discharge and the resulting beam of H atoms is state-selected and sent in a storage bulb. The bulb is surrounded by a sapphire-loaded microwave cavity that, tuned on the resonance frequency, induces the maser action.

The ACES mission will be a test-bed for the space qualification of the active hydrogen maser SHM, developed by the Neuchâtel Observatory. The on-board frequency comparison between SHM and PHARAO will be a key element for the evaluation of the accuracy and the short/medium-term stability of the cesium clock. Further, it will allow one to identify the optimal operating conditions for PHARAO and to choose the right compromise between frequency accuracy and stability. SHM will provide a clock signal at 100 MHz with the following fractional frequency instability (Fig. 2):

$$\begin{aligned}
\sigma_y^{\mathrm{SHM}}(\tau = 1\,\mathrm{s}) &= 1.5 \times 10^{-13}, \\
\sigma_y^{\mathrm{SHM}}(\tau = 10\,\mathrm{s}) &= 2.1 \times 10^{-14}, \\
\sigma_y^{\mathrm{SHM}}(\tau = 100\,\mathrm{s}) &= 5.1 \times 10^{-15}, \\
\sigma_y^{\mathrm{SHM}}(\tau = 1000\,\mathrm{s}) &= 2.1 \times 10^{-15}, \\
\sigma_y^{\mathrm{SHM}}(\tau = 10,000\,\mathrm{s}) &= 1.5 \times 10^{-15}.
\end{aligned} \tag{3}$$

The demonstration of these performances, both with ground-based tests and with an on-flight calibration procedure, is one of ACES' primary mission objectives.

3.3. *The microwave link*

Ensuring stable and accurate time and frequency transfer from space to the ground MWL is a key element of the ACES mission. MWL is under ESA's responsibility and was developed by EADS/KT/Timetech. The proposed MWL concept is an upgraded version of the Vessot two-way technique used for the GP-A experiment in 1976.[9] The system operates continuously with a carrier frequency in the Ku band (near 15 GHz). The high carrier frequency of the up- and down-link allows for a noticeable reduction of the ionospheric delay. A third frequency in the S band is used to determine the ionosphere total electron content (TEC). A PN code phase measurement removes the phase ambiguity between successive comparison sessions separated by large dead times. The system is designed for multiple access capabilities, allowing up to four simultaneous ground users, distinguished by the different PN codes and Doppler shifts.

Due to the ISS orbit, time and frequency transfer will be possible only for continuous periods of short duration ($\sim 300\,\mathrm{s}$). This condition defines the MWL performances in the short term. Depending on its latitude, a given ground terminal has typically 5 ISS paths per day. The noise introduced by the link must be minimized on the ISS single pass duration. As measurements rely on phase comparisons, white phase noise is assumed to limit the performances of MWL for integration times $10\,\mathrm{s} \leq \tau \leq 300\,\mathrm{s}$. Therefore, short-term frequency instability has to satisfy the requirement

$$\sigma_x^{\mathrm{MWL}}(10\,\mathrm{s} \leq \tau \leq 300\,\mathrm{s}) \leq 4.1 \times 10^{-12}\tau^{-1/2}, \tag{4}$$

corresponding to a time deviation

$$\sigma_x^{\mathrm{MWL}}(\tau = 300\,\mathrm{s}) \leq 0.24\,\mathrm{ps}. \tag{5}$$

Clock-to-clock comparisons in the low 10^{-16} level require excellent long-term stability for the time and frequency transfer link (up to 10 days). Assuming that MWL exhibits white frequency noise behavior for durations longer than 1000 s,

$$\sigma_x^{\mathrm{MWL}}(\tau > 1000\,\mathrm{s}) \leq 1.7 \times 10^{-14}\tau^{1/2}, \tag{6}$$

corresponding to

$$\begin{aligned} \sigma_x^{\mathrm{MWL}}(\tau = 1\,\mathrm{day}) &\leq 5.1\,\mathrm{ps}, \\ \sigma_x^{\mathrm{MWL}}(\tau = 10\,\mathrm{day}) &\leq 16.2\,\mathrm{ps}. \end{aligned} \tag{7}$$

The direct comparison between ACES and a ground clock will be performed both on board the ISS and on the ground. Data will allow one to calculate the cumulative stability of the two clocks, including the corrections for all propagation error terms and of course relativistic effects.

4. ACES Status

The ACES mission is in the C/D phase. All instruments and subsystems are at an advanced stage of development, with engineering models delivered or in final assembly. In September 2006, ACES started the Preliminary Design Review (PDR), which will consolidate the overall mission concept, the status of instruments and subsystems, and will decide on the development of the flight models.

4.1. *PHARAO status*

The cesium clock PHARAO is composed of four main subsystems: the cesium tube (Fig. 3), the optical bench, the microwave source, and the computer control.

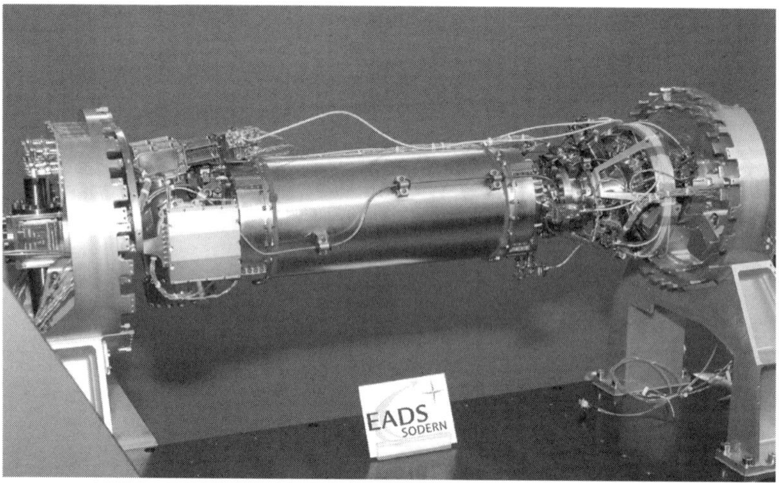

Fig. 3. PHARAO cesium tube without the two external μ-metal magnetic shields. The total length is 990 mm and the mass 44 kg.

Contracted to different manufacturers, the engineering models (EM's) of PHARAO subsystems have been completed, successfully tested, and delivered. The EM PHARAO clock, fully assembled on the CNES premises in Toulouse, is presently under test. Its design and recent performance tests are described in Ref. 20. Cesium atoms have been loaded in the optical molasses, cooled down, and detected. Samples of cold atoms, launched along the PHARAO tube, have been probed on the clock transition by using the PHARAO microwave frequency source.

Microwave resonance signals (Ramsey fringes) have been successfully recorded, demonstrating the correct interfacing of PHARAO subsystems and the correct operation of the clock. The frequency scan near 9, 192, 631, 770 Hz reveals the Ramsey fringes shown in Fig. 4. On the ground, for an upward launch velocity of 3.42 m/s, the atoms spend 100 ms between the two Ramsey interactions and the fringe width is 5 Hz, more than ten times wider than during operation in microgravity. The signal-to-noise ratio is several hundreds per cycle. The central pattern contrast is only 75% as the atoms are decelerated by gravity and therefore do not undergo the same Rabi excitation in the two interaction sections. Next, a performance verification campaign will evaluate the clock stability and accuracy, and fabrication of the PHARAO flight model will start in 2007.

4.2. *SHM status*

SHM is composed of an electronic package (EP) and a physics package (PP). The heart of the physics package is represented by a sapphire-loaded microwave cavity responsible for stimulating maser action in the hydrogen atoms contained in a storage bulb. The main elements of the EP are the RF unit, the power supply unit, and the SHM controller.

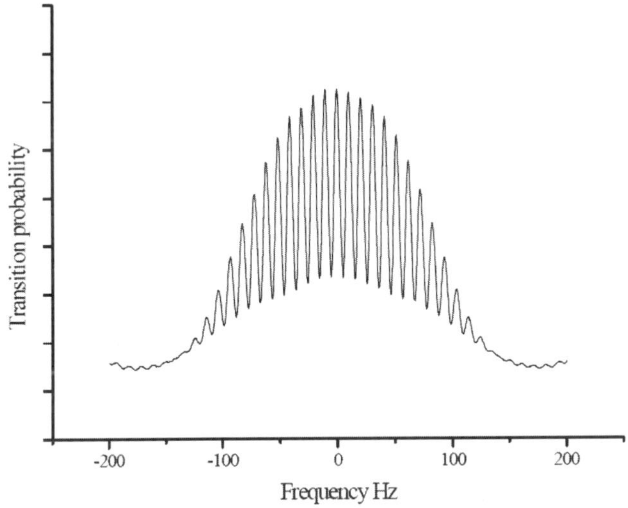

Fig. 4. Cesium clock resonance in the PHARAO instrument. Transition probability as a function of the microwave field detuning around 9, 192, 631, 770 Hz. For an upward launch velocity of 3.4 m/s, the width of the central resonance is 5 Hz. Although the atomic preparation was not used for this measurement, the signal-to-noise ratio reaches several hundreds per point.

The key elements of the SHM PP engineering model have been manufactured and assembled (Fig. 5). At present, SHM PP is undergoing functional and performance tests in combination with a development model of the maser electronics manufactured at the Neuchâtel Observatory. The measurements performed have already demonstrated the correct operation of the automatic cavity tuning (ACT) system.

Fig. 5. Physics package of the active maser SHM.

ACT stabilizes the resonance of the maser cavity against temperature variations. First Allan deviation measurements show that the EM of SHM PP, in combination with the development model of the maser electronics, performs correctly, meeting the specifications on both short and long integration times. The EM of SHM EP is presently under development. Once delivered SHM will be fully assembled and end-to-end tests will evaluate the clock performances.

4.3. *Subsystems' status*

The MWL development model has been manufactured and successfully tested. In particular, specific tests related to the operation of the delay locked loop have been concluded, demonstrating the adequacy of the present design. The MWL EM will be completed by 2007, when functional and performance tests will take place.

Another important element of the ACES payload is FCDP. This subsystem has three main functions: it distributes the clock signal to MWL, compares the PHARAO and SHM clock signals, and phase-lock them. The FCDP EM has been completed and tested. In July 2006 a performance test involving FCDP, the PHARAO microwave source, a ground maser, and a cryogenic sapphire oscillator was carried out on the CNES premises in Toulouse. The tests have shown the successful operation of the servo loop that locks the PHARAO 100 MHz signal on to the SHM 100 MHz signal. Results will be published elsewhere.[26]

Acknowledgments

The authors express their warm thanks to all the members of the ACES project team, the PHARAO team at SYRTE and CNES, and the SHM team for their valuable contribution's to the mission development. In particular, A. Clairon, P. Laurent, P. Lemonde, G. Santarelli, D. Svehla, P. Wolf, and L. Duchayne are gratefully acknowledged. This work is supported by CNES and ESA. C. Salomon and N. Dimarcq are members of IFRAF (Institut Francilien de Recherche sur les Atomes Froids), supported by Region Ile de France. Laboratoire Kastler Brossel is a unit associated with CNRS UMR 8552 and with University Pierre and Marie Curie. SYRTE is a unit associated with CNRS UMR 8630.

References

1. C. Salomon *et al.*, *C. R. Acad. Sci. Paris* **t.2 Série 4** (2001) 1313.
2. L. Cacciapuoti *et al.*, in *Proc. 1st ESA International Workshop on Optical Clocks*, 8–10 June 2005, Noordwijk, The Netherlands (ESA Publication, 2005), p. 45.
3. R. Holzwarth *et al.*, *Phys. Rev. Lett.* **85** (2000) 2265.
4. S. Diddams *et al.*, *Phys. Rev. Lett.* **84** (2000) 5102.
5. S. Bize *et al.*, *C. R. Phys.* **5** (2004) 829.
6. T. E. Parker and D. Matsakis, *GPS World, Nov* (2004), 32.
7. A. Bauch *et al.*, *Metrologia* **43** (2006) 109120.
8. R. V. Pound and G. A. Rebka, *Phys. Rev. Lett.* **4** (1960) 337.

9. R. F. C. Vessot *et al.*, *Phys. Rev. Lett.* **45** (1980) 2081.
10. T. Damour and A. Polyakov, *Nucl. Phys. B* **423** (1994) 532.
11. J. D. Prestage, R. L. Tjoelker and L. Maleki, *Phys. Rev. Lett.* **74** (1995) 3511.
12. H. Marion *et al.*, *Phys. Rev. Lett.* **90** (2003) 150801.
13. S. Bize *et al.*, *J. Phys. B* **38** (2005) S449.
14. E. Peik *et al.*, *Phys. Rev. Lett.* **93** (2004) 170801.
15. G. Gabrielse *et al.*, *Phys. Rev. Lett.* **97** (2006) 030802.
16. V. V. Flambaum, physics/0302015.
17. V. V. Flambaum *et al.*, *Phys. Rev. D* **69** (2004) 115006.
18. S. Bize *et al.*, *Phys. Rev. Lett.* **90** (2003) 150802.
19. P. Wolf and G. Petit, *Phys. Rev. A* **56** (1997) 4405.
20. P. Laurent *et al.*, *Appl. Phys. B* **84** (2006) 683.
21. W. H. Oskay *et al. Phys. Rev. Lett.* **97** (2006) 020801.
22. M. Takamoto *et al.*, *Nature* **435** (2005) 321.
23. C. Daussy *et al.*, *Phys. Rev. Lett.* **94** (2005) 203904.
24. D. Kleppner, *Phys. Today, March* (2006) 10.
25. S. Schiller *et al.*, *Proc. Space Parts Symposium* (Beijing) to appear in *Nucl. Phys. B*.
26. G. Santarelli *et al.*, to be published.

SpaceTime: PROBING FOR 21ST CENTURY PHYSICS WITH CLOCKS NEAR THE SUN

LUTE MALEKI* and JOHN PRESTAGE†

Quantum Sciences and Technology Group,
Jet Propulsion Laboratory,
California Institute of Technology, Pasadena,
CA 91109, USA
**Lute.Maleki@jpl.nasa.gov*
†John.D.Prestage@jpl.nasa.gov

We will describe a space mission study based on three high-precision atomic clocks, flying to within six solar radii of the Sun, for a test of the possible variation of the fine structure constant, α. The three clocks are based on transitions in three different atomic species. Measurement of the drift in ratios between the frequencies generated by each clock will probe for the variation of α. Since the response of each atomic species to a change in α has a specific signature, this measurement will provide sensitive and unambiguous results. The sensitivity of this experiment to a changing α is comparable to the sensitivity of recent tests based on observational astronomy, exceeding the geophysical bounds on α variations. Thus, the experiment will provide a compelling reaffirmation or refutation of astronomical observations, and represents an important test of the models aimed at bridging physics of the quantum to the cosmos.

Keywords: General relativity; fundamental constants; SpaceTime mission.

1. Motivation

The notion of a fundamental model of the physical world was radically changed in the early part of the 20th century by the discovery of quantum mechanics and relativity. In its early days, quantum mechanics appeared as a mysterious framework that defied the conventional, classical model and demanded a fundamental shift in how the world at the subatomic scale was to be viewed. Despite the debate about its reality, the physics community accepted this model, though at times quite reluctantly, because it so accurately accounted for all phenomena at the microscopic scale. In a parallel manner, the view of space and time provided by relativity also demanded that the conventional notions be abandoned in favor of a counterintuitive model that nevertheless did not fail any tests of the experimental scrutiny. By the middle of the 20th century, the world of physics appeared to be closing on a comprehensive picture that could account for all observed phenomena, and a unified theory of all forces seemed to be around the corner.

This neat picture of physics began to lose its luster as the sophisticated theory of quantum filed theory that had been evolved from the original quantum theory, and so completely described the nature of fields and particles, refused to find a merger with general relativity. The latter had accurately and consistently provided a picture of gravitational interactions that included both the detailed dynamics of the planetary motion and exotic objects such as black holes. Meanwhile, cosmology, supported by new and powerful observatories in space, emerged as the area of science that provided data with the potential to clarify those features of fundamental physical models that stubbornly defied a resolution.

In the last 15 years of the century, however, the world of physics learned that it was sorely lacking models that reconcile what was observed in the realm of cosmology with the existing understanding. Observational data pointed to exotic notions that for lack of any understanding were termed "dark matter" and "dark energy." These exotic notions were required so that physics could make sense of the observed dynamics of the Universe, at both its early and later stages. Other unconventional models, such as inflation, were called in to account for data that otherwise could not be explained. We have learned that our Universe is expanding, yet flat, and could even be but a piece of a larger structure that would for ever remain out of our observational reach. In short, the picture of the physical reality that was neatly emerging in the first 75 years of the 20th century was found to be almost completely lacking to explain the observed data.

Today, there is a widespread belief in physics that the current models are incomplete, but a more complete model will indeed emerge before the current century reaches its half-way point. This is because candidate models such as M theory and non-commutative quantum theories hold clues that lead us to believe they can resolve the unresolved issues. In particular, they appear to be bringing physics a step closer to a unification of quantum and gravity, the two jewels of the early years of modern physics. These new candidate models also include steps toward resolving and clarifying the compelling data associated with cosmology that are being received by space probes.

A major feature that all candidate models seem to include is the existence of scalar fields, which require that we abandon the notion of constancy of the "constants" of nature. In particular, an examination of whether the nondimensional constants, such as the fine structure constant α, change, or have changed, in space and time could provide important clues as to the viability of the proposed new theories. Clues to a changing α have been hinted at by observational astronomy, but are not widely accepted by the physics community. A clear and complete test of α variation is needed to help a new physics emerge to better fit what we observe. Such a test could also bridge physics at the Planck scale with physics at the scale of the current Universe, from the quantum to the cosmos.

Once at hand, such a model could impact our notion of reality, and lead to the same kind of advances as those brought about with the 20th century's quantum physics and relativity, which led to a wide range of technological capabilities, from the computer to navigation via global positioning satellites.

SpaceTime is a mission concept aimed at testing the constancy of α at a level that can clearly inform the models being devised. This mission concept will be described in the following sections.

2. Background

There have been important developments on both the theoretical and observational fronts that have fueled considerable interest in the search for a variation of the fine structure constant. On the observational side, Webb *et al.*[1] have found evidence for a cosmological variation of the fine structure constant through an analysis of the absorption lines in galactic halos from quasar-emitted light. Their results indicate that the fractional change in α, averaged over the redshift in the range of $0.2 \leq z \leq 3.7$, is $(-0.57 \pm 0.10) \times 10^{-5}$. On the theoretical side, many of the outstanding issues confronting fundamental physics, such as the failure to include gravity in the standard model, and puzzles of cosmology, such as inflation and the apparent accelerated rate of the expansion of the Universe, appear to imply the existence of massless or nearly massless scalar fields. These fields appear as dilaton or moduli in the M theory, supporting the unification of gravity with other forces, as well as suggesting a possible breakdown of the equivalence principle (EP). They also appear as quintessence in models of cosmology aimed at resolving fine-tuning and other outstanding problems, including those mentioned above.[2] The scalar fields in these models imply a spatiotemporal variation of constants of nature, such as the fine structure and other field coupling constants.

Despite these important developments, at this writing there is no clear consensus amongst researchers regarding the validity of the theoretical predictions, and the observational conclusions are not regarded as inconvertible. The question of whether, how, and why the fine structure constant varies remains an open one.

It is clear then that a controlled experiment with sufficient measurement sensitivity beyond the current capabilities will be enormously important in clarifying some of the questions associated with α variations. SpaceTime is a space mission study aimed at providing such an experiment. It is based on flying an instrument based on three clocks that run on ground state hyperfine transitions of three different singly ionized atoms to within six solar radii of the Sun. The "triclock" instrument of SpaceTime is capable of testing a variation of α with four-orders-of-magnitude more sensitivity, as compared with the results of quasar observations. As discussed below, the choice of the atomic clocks as the instrument was made to ensure that the results would be conclusive and free of many questions that have confronted previous investigations searching for a varying α.

At this point it is worthwhile to consider some of the consequences of a varying fine structure constant. The fine structure constant has been a point of fascination with physicists since it was introduced, and named, by Sommerfeld in 1916 as a useful constant in spectroscopy; it is a measure of the doublet structure of hydrogen and other atoms with a single valence electron. Sommerfeld also considered α as an indication of an intimate relation between charge and quantum. In the years following Sommerfeld's introduction of α, various physicists, starting with Eddington, have considered the relation between α and other constants of nature. This interest was also fueled by suggestive numerology that relates specific functions of π to the value of α.

The conjecture of varying fundamental constants has also a relatively long history and dates back to Dirac's "large number hypothesis," which was based on the notion that there exists an underlying relationship between constants of nature, as manifested by large numbers, on the order of 10^{39}, that could be obtained by arranging them in various combinations.[3] Other *ad hoc* conjectures similarly have pointed to possible variation of constants, especially the gravitational constant G, through which a variation of α may also arise. These models, nevertheless, were all generally qualitative, and more importantly, lacked any observational support. The picture has changed in the last few years. Since a change in α implies a changing e, the charge of the electron, or c, the speed of light, or Planck's constant, h, through $\alpha = 2\pi e^2/ch$, several models based on variations of any of these dimensional constants have been devised.[4–8] There is, however, a good bit of controversy regarding the validity of these models, and whether their predictions do or do not support[9,10] a violation of the EP, as well.

Atomic clocks have traditionally been used to test the prediction of general relativity. The first such test was performed in 1976 by NASA's Gravity Probe A, where the rate of a hydrogen maser clock on a rocket in a suborbital trajectory was compared to that of a similar clock on the Earth's surface.[11] This measurement verified the exact prediction of a clock shift by general relativity to a part in 10^4, a precision that still stands unchallenged today. In a recent investigation it was shown that it is also possible to search for a variation in α by comparing the rate of drift of two clocks based on hydrogen and mercury ions.[12] This is because the energy of the hyperfine transition in atoms, which forms the basis of microwave clocks, has an αZ dependence, where Z is the atomic number. This first laboratory attempt to search for a varying α set a limit of $\sim 4 \times 10^{-14}$ per year for its temporal variation. This approach has recently been extended to the comparison of a rubidium and a cesium fountain clock, both based on microwave transitions,[13] as well as the comparison of a cesium fountain with an optical mercury ion clock, where an optical transition in the ion was used.[14] These more recent experiments set the limit for a varying α to be less than about 10^{-15}/yr. This is a less stringent limit than that obtained with an analysis of the neutron capture rate applied to a natural thermonuclear reaction that occurred 1.5 billion years ago in an Oklo mine in Africa,[15] which places the limit on α variation to be less than 5×10^{-17}/yr. SpaceTime's instrument is designed to provide sensitivity to a variation in α at the level of 10^{-20}/yr by searching for any spatial dependence of α.

For alkali atoms, an expression for the hyperfine interval may be obtained, as follows:

$$A_s = \frac{8}{3}\alpha^2 g_l Z \frac{z^2}{n^{*3}} \left(1 - \frac{d\Delta_n}{dn}\right) F(\alpha Z)(1-\delta)(1-\epsilon)\frac{m_e}{m_p} R_\infty. \tag{1}$$

Here, z is the net charge of the ion without the valence electron, and n^* is the effective quantum number with $\Delta n = nn^*$, δ, and ϵ are related to the corrections for finite size of the nucleus. Thus the sensitivity of different clocks, based on atoms of different Z, to a change in the fine structure constant displays specific signatures. In particular, the Casimir correction factor, $F(\alpha Z)$ (for the relativistic wave equation of the electron), leads to the differential sensitivity in the alkali microwave hyperfine clock transition frequencies f.

For alkali atoms, an expression for the hyperfine interval may be obtained, as follows:

$$f = \alpha^4 \frac{m_e}{m_p} \frac{m_e c^2}{\hbar} F(\alpha Z). \tag{2}$$

It is clear from this equation that different atomic systems with different Z display different frequency dependencies on a variation of α through the αZ-dependent terms. A direct test for a time variation of α can then be devised through a comparison of two clocks, based on two atomic species with different atomic number, Z.

This is a key feature of the SpaceTime instrument which, in conjunction with the individual sensitivity of each atomic species to an α variation, can produce clean and unambiguous results.

Since the changing α in all model predictions is mediated by the coupling of a scalar field to matter, the fall in the $1/R$ potential near the Sun will allow a direct test of general relativity, where only the tensor field is allowed, and where the constants are not allowed any variation. This is an important point to consider in clock tests, and other tests searching for an α variation based on a signature of the failure of the EP. Since the EP is currently tested at about the 10^{-12} level[16] with no violations found, any test searching for α variations must have a sensitivity higher than 10^{-12} to EP violation to produce a new result. The expected sensitivity of the differential redshifts as measured by the three clocks that are within six solar radii is at the level of 10^{-13} of the EP, or about six orders of magnitude larger than the GP-A experiment. Thus results of SpaceTime will improve the current state of art in EP violation by an order of magnitude, as well as improving on the results of Webb *et al.* by four orders of magnitude, beyond the capability of all existing and future Earth-bound clock experiments.

To improve the measurement sensitivity, our instrument consists of three clocks based on three different atomic species that can be intercompared for individual signatures. To reduce the influence of systematic errors that can mimic our signal, the three clocks share the same environment. To improve the source of the signal, the triclock instrument flies to within six solar radii of the largest body of matter in the solar system, the Sun. Thus the entire experiment is designed to provide a clean and unambiguous result, based on a technology that is proven, and has an outstanding chance for success. Finally, the spinning spacecraft, moving at 300 km/s, or 1/1000 of the speed of light, at its closest approach will test another important question with fundamental underpinning: Is Lorentz symmetry robust, or does it fail at some limit? This question is important since string theory, and theories that extend beyond the Standard Model,[17] result in physics without Lorentz and other global symmetries such as CPT.

Beyond this, as mentioned above, a consequence of a changing α is that either c, the speed of light, or e, the charge of the electron, or h, Planck's constant, must change. Theories based on a changing velocity of light have received considerable attention since they solve the outstanding problems in cosmology: the horizon, flatness, cosmological constant, entropy, and homogeneity problems.[5] They nonetheless violate Lorentz invariance. SpaceTime will provide a tenfold sensitivity for a test of Lorentz invariance, as compared to an Earth-bound test, due to the order-of-magnitude smaller orbital speed of the Earth.[18]

3. The Instrument

In the strongly time-dilated space–time curvature at six solar radii (4.2 Gm), time runs slower than on the Earth by about one half microsecond per second. Three atomic clocks based on hyperfine transitions of Hg^+ ($Z = 80$), Cd^+ ($Z = 48$), and Yb^+ ($Z = 70$) are different in their electromagnetic composition (given by the Casimir factor) and will be simultaneously monitored during a solar flyby to determine whether these different clocks will measure the same time interval near the Sun. The atomic clock hardware for the SpaceTime mission is a modification of the linear ion trap frequency standard (LITS) currently being deployed in the Deep Space Network stations worldwide. A laboratory prototype has shown ultrastable operation in a package far smaller than other clock technologies and represents the state of the art for atomic clocks.

Atomic clocks based on hyperfine transitions and ion traps are the most suitable technology for space applications. This is because of the inherent simplicity of this approach, which does not rely on resonant cavities. In lamp-based trapped ion clocks, as in the SpaceTime instrument, the risks associated with the use of lasers in space are eliminated. Ions confined in electromagnetic traps are significantly shielded from environmental perturbations such as collisions with the walls or with each other. The relatively large hyperfine splitting of singly ionized systems also reduces their sensitivity to ambient magnetic fields, as compared with atoms with smaller hyperfine frequencies.

The classical ion trap, consisting of a three-electrode structure made with hyperbolic electrodes, confines charge particles of a particular charge to mass ratios based on the applied dc and rf potentials. In this geometry, ions are confined in a spherical region as a result of the applied pondermotive forces. A geometry based on linear electrodes, first introduced at JPL for clock applications, improves the clock stability by providing a geometry whereby the temperature (kinetic energy) of the ions resulting from the micromotion in the trap is reduced.[19] This configuration was further refined at JPL[20] to put the ability to move the charged particles from one region of space to another, to separate the ion preparation region from the region where the microwave field produced by a local oscillator (LO) interacts with the clock transition of ions. By separating these regions it is possible to significantly reduce the requirement that magnetic shielding protect the ions undergoing interaction with the microwave field. Higher pole traps are also employed in order to further reduce ion density space-charge-related ion-heating. This is key to the reduction of the size and weight of the clock, parameters that are particularly important for space instruments.

The instrument for this mission is composed of three ion trap clocks in a package where much of the hardware is common to all of the clocks. Because some of the clock systematic frequency perturbations will be common to all three clocks and will have a characteristic signature that can be identified and removed from the difference of the clock frequencies, relative stability's to 10^{-16} in the intercomparison can be reached. The LO will simultaneously interrogate each of the three clock transitions, thereby removing LO noise in the intercomparison, and greatly improving short-term clock noise so that 10^{-16} resolution in the difference in clock rates can be obtained within the 15-hour close encounter. Because ion-trap-based clocks are relatively immune to temperature and magnetic field changes, a simple, robust electronics package is sufficient for ultrastable operation.

The basic architecture of the triclock instrument is three LITE (linear ion trap extended) units, each operating with a single element Hg^+, Cd^+ or Yb^+, and will be packaged into one housing with many shared components for mass reduction. Each separate clock is based upon a linear multipole trap.[20] For optical state selection, ions are trapped around the rf quadrupole electric field node along the center line, where they are prevented from escaping by dc fields applied at each end. By applying dc positive bias to all trap rods in one region along the length of the trap, ions can be excluded from that region and transported into another section where the rods are at the dc ground. Ions can thus be moved from one end of the trap to the other. This allows the optical state selection and interrogation to be carried out in an unshielded region while the much more critical clock hyperfine resonance is probed in a small, well-shielded region, away from magnetic optical components and openings in the shields for light entry and exit. The ion-number (space-charge) induced frequency pulling is reduced by more than 20 times in the multipole arrangement as compared to the linear quadrupole.[20–23]

The three traps will be operated with a common rf voltage source so that related trapping forces confine the three different ion species. In this way small variations in the trapping strength will affect each ion cloud in a characteristic manner that can be readily identified. Another unique feature of this clock comparison is the use of the ultrastable LO. Space-qualified quartz oscillators achieve short-term stabilities of 10^{-13} over tens-of-seconds averaging intervals. This will limit a conventional high performance atomic clock to about 10^{-13} at 1 s averaging time, falling from there as $\tau^{-1/2}$, where τ is the averaging interval in seconds. For the clock comparison at the near-solar flyby, the largest change in gravitational potential occurs over a 15-hour period, i.e. 54,000 s. This LO-limited performance gives 4×10^{-16} at 15 hours and falls short of the design goal. We have demonstrated atomic clock performance at $2-3 \times 10^{-14}$ at 1 s but LO noise degrades the performance for a single operating atomic clock. For a comparison between two or more clocks, however, a single LO can be used to interrogate all clock transitions simultaneously, and the LO noise will be common. This common noise in individual atomic line-center measurements will not be present in the differences of these and, we can recover the $2-3 \times 10^{-14}/\sqrt{\tau}$ and reach the 10^{-16} stability level in 15 hours averaging.

The triclock measurement offers a suppression of other common mode frequency shifts of the three atomic transitions. The suppression of systematic frequency pulling can also be applied to variations of the solar magnetic field along the spacecraft (S/C) trajectory. This approach will save mass and power in magnetic shielding. A set of four layers of magnetic shields will enclose the clock resonance tube. An additional layer will house the final package. Since the unshielded Hg^+ atom sensitivity is about 2×10^{-13}/mG (at an operating point of 50 mG), 20×10^{-13}/mG for Yb^+, and 15×10^{-13}/mG for Cd^+, a shielding factor of 10^7 is required to reduce a 1 G solar field variation during the S/C flyby to below one-part-in-10^{16} relative clock stability. A 1 G field variation might be expected during the solar flyby. This level of shielding is very difficult to achieve within the mass and power budget.

The differential response of the three clocks to a common field variation has a characteristic signature that will identify this systematic shift and will enable its removal in postanalysis. The magnetic sensitivity of the three hyperfine levels is well understood in the atomic physics of the clock transitions. The change of the clock frequency as the operating field changes by δH_0 is given by $\delta_y \equiv \delta f/f_0 = 2\beta H_0 \delta H_0$, where the constant β describes the field sensitivity of each of the three clock transitions. The atoms with a smaller hyperfine splitting, f_0, shift more. Note that this behavior is very different from the sensitivity to a change in α as given in Ref. 12. In that paper it is shown that the atoms with larger atomic number Z shift more with a change in α than the low Z atoms.

The two simultaneous equations for the variation of the difference frequencies are

$$\begin{aligned}\delta y_{AB} &= (L(Z_A) - L(Z_B))\frac{\delta\alpha}{\alpha} + \left(1 - \frac{\beta_B f_A}{\beta_A f_B}\right)\frac{2\beta_A}{f_A}\frac{H_0 \delta H}{S},\\ \delta y_{AC} &= (L(Z_A) - L(Z_C))\frac{\delta\alpha}{\alpha} + \left(1 - \frac{\beta_C f_A}{\beta_A f_C}\right)\frac{2\beta_A}{f_A}\frac{H_0 \delta H}{S}.\end{aligned} \tag{3}$$

We have taken the variation of the clock transitions with operating field, H_0, to be given by $f = f_0 + \beta H_0^2$, and the shielding factor for external fields to be S, i.e. $\delta H_0 = \delta H/S$. δH is the variation of the solar magnetic field along the S/C trajectory. The α sensitivities, $L(Z)$, are found in Fig. 1 of Ref. 12.

For the hyperfine clock transitions in Hg, Cd, and Yb, these equations can be inverted to solve for $\delta\alpha/\alpha$ and $(2\beta_A/\nu_A)H_0\delta H/S$ along the trajectory of the near-Sun flyby. Thus, even with imperfect magnetic shielding and the accompanying clock frequency pulling, an unambiguous variation of α could be extracted.

3.1. *Temperature-induced frequency shifts*

Ambient temperature changes of the clocks can cause spurious frequency pulling δy_{AB} and δy_{AC} and must be completely removed to the 10^{-16} level. Unlike magnetic sensitivities, which can to a large extent be understood as incomplete shielding of the atomic transition, temperature-induced frequency shifts are more difficult to predict from first principles. The only definitive measurement of temperature sensitivity must be carried out with a fully assembled and operating system. The differential sensitivity coefficients to be used in separating any observed effect from temperature-induced δy_{AB} and δy_{AC} must be generated *in situ*. Once these sensitivities are measured, we can use the two return data channels to distinguish temperature effects from any observed violations.

Some temperature effects have very clear signatures, completely distinguishable from any α variation along the S/C trajectory. For example, ion temperature variations will lead to clock frequency changes via second-order Doppler shifts, by an amount proportional to $-kT/mc^2$, where T is the ion temperature and m is the ion mass. Any temperature change, δT, common to all three ionic species will shift the three clock frequencies by an amount inversely proportional to their mass. This will allow this systematic frequency offset to be removed as in the magnetic case above. For these shifts,

$$\begin{aligned}\delta y_{AB} &= (L(Z_A) - L(Z_B))\frac{\delta\alpha}{\alpha} - \left(1 - \frac{m_A}{m_B}\right)\frac{k\delta T}{m_A c^2},\\ \delta y_{AC} &= (L(Z_A) - L(Z_C))\frac{\delta\alpha}{\alpha} - \left(1 - \frac{m_A}{m_C}\right)\frac{k\delta T}{m_A c^2},\end{aligned} \tag{4}$$

showing that these temperature variations can be separated from the variations that come from a nonzero $\delta\alpha/\alpha$ along the solar flyby trajectory. We have assumed no mass-dependent heating, δT, which will almost certainly be present. However, a prelaunch ground measurement will be carried out to catalog differential frequency shifts versus rf trap level, buffer gas pressure, etc.

3.2. *Mission design*

The only economical technique to get sufficient change in velocity to fly near the Sun is to go via Jupiter. This is because the angular momentum associated with the orbiting Earth must be lost so the spacecraft will fall to the Sun in a reasonable length of time. Thus, SpaceTime will launch in a direct transfer orbit to Jupiter and then a fast trajectory to the Sun. A kick stage is integrated with the S/C on a "spin table" that spins the entire integrated package during the launch. The spinning S/C does not have to be despun following injection, as with a typical three-axis stabilized S/C. This eliminates the mass and reliability penalties of a despin hardware.

Figure 1 illustrates the entire interplanetary trajectory to the Sun, including the first leg after injection. The time tics are 50-day intervals. Approaching Jupiter, a precision orbit determination is completed using only radio tracking data, and a precise final aiming maneuver is completed. The gravity assist flyby is used to: (1) reduce (almost canceling) the trajectory angular momentum, allowing the S/C to fall into a $6R_S$ perihelion, (2) rotate the plane of the heliocentric orbit to a final inclination of 90.0° and (3) establish the time of perihelion to produce a quadrature trajectory geometry (Sun–S/C–Earth angle = 90.0°) at perihelion. This latter condition is fundamental to the S/C architecture, which always has the shield pointed at the Sun and the high gain antenna (HGA) pointed at the Earth.

Following the Jupiter flyby, the S/C is on its final trajectory toward the perihelion. The perihelion flyby trajectory is shown in the figure E-2 (F/O E-1) from P-24 to P+24 h. This is the prime data acquisition period for the mission. The view in the figure E-2 is from the Earth, illustrating the effects of the quadrature trajectory geometry by the schematic drawings of the S/C. The S/C is a spinning drum with the direction of its spin axis toward the Earth (out of the page). The thermal shield for the S/C, as the S/C spins, maintains its orientation toward the Sun at all times, protecting the sensitive elements from the extreme thermal environment. This is a passive attitude control technique that simplifies the control of the S/C and allows a very robust design in this otherwise hostile environment.

It is interesting to point out that the most challenging aspect of the mission, affecting orbital trajectory and the number of passes (single) by the Sun, is the power requirements. Because of the extreme heat encountered near the Sun, solar panels, even those designed for high temperature, cannot be used. Instead, a bank

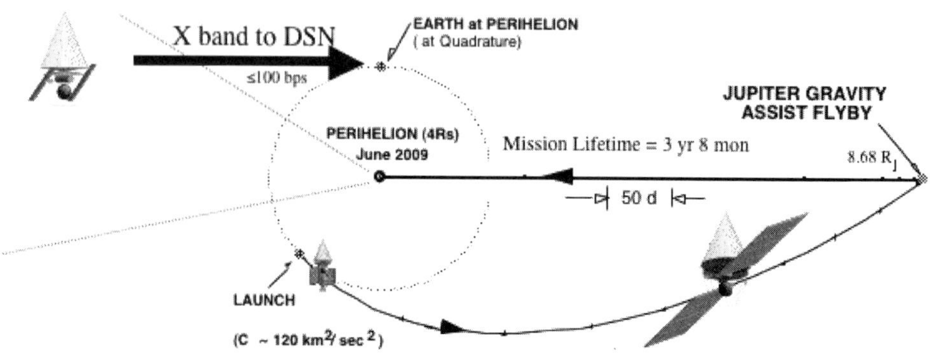

Fig. 1. Trajectory of the SpaceTime mission's spacecraft.

of batteries must provide the needed power to the spacecraft systems, and the instrument. The mass associated with the batteries ultimately limits the choices of a trajectory with a given launch vehicle, as well as the size of the S/C and associated systems. This ironic limitation (shortage of power while so near the Sun) is the major design issue that affects virtually all aspects of the mission.

4. Conclusion

We have briefly discussed a mission design study based on the intercomparison of the oscillation frequencies of three atomic clocks based on three different species of singly ionized atoms. By flying this instrument to within six solar radii of the Sun it is possible to search for a variation of the fine structure constant to a level that is not accessible to Earth-based instruments. At this point two other points regarding this approach are worth noting. First, one may question the choice of atomic clocks, as opposed to other instruments. As briefly mentioned above, the details of theories that predict a temporal or spatial variation in fine structure constant, such as M theory or theories based on varying c or e, are rather tentative. Experimental tests of these theories based on a search for varying α must then produce direct and unambiguous results to be most valuable. The three-clock comparison discussed here is indeed such an approach. As discussed above, each atomic clock will drift in a specific manner with varying α and intercomparison of these variations ensures that an observed signal produces a clear result. Secondly, the technology of atomic clocks is well developed, and a space test based on clocks has an inherently large probability of success.

Acknowledgments

This work was carried out at the Jet Propulsion Laboratory, California Institute of Technology, under a contract with the National Aeronautics and Space Administration. The SpaceTime science team in this study consisted of: Eric Adelberger, University of Washington, USA; John Armstrong, JPL, USA; Thibault Damour, Institut des Hautes Etudes Scientifiques, France; Kenneth Johnston, US Naval Observatory, USA; Alan Kostelecky, Indiana University, USA; Claus Lämmerzhal, Heinrich-Heine-Universitaet Duesseldorf, Germany; Lute Maleki (PI), JPL, USA; Kenneth Nordtvedt, Montana State University, USA; John Prestage, JPL, USA.

References

1. J. K. Webb *et al.*, *Phys. Rev. Lett.* **87** (2001) 091301.
2. T. Damour and A. M. Polyakov, *Nucl. Phys. B* **423** (1994) 532.
3. P. A. M. Dirac, *Nature* **139** (1937) 323.
4. J. D. Bekenstein, *Phys. Rev. D* **25** (1982) 1527.
5. J. D. Barrow and J. Magueijo, *Phys. Lett. B* **443** (1998) 104 [astro-ph/9811072].
6. H. Sandvik, J. D. Barrow and J. Magueijo, *Phys. Rev. Lett.* **88** (2002) 031302 [astro-ph/0107512].

7. J. W. Moffat, astro-ph/0109350.
8. D. Youm, hep-th/0108237.
9. J. D. Bekenstein, *Phys. Rev. D* **66** (2002) 123514.
10. T. Damour, *Astro. Phys. Space Sci.* **283** (2003) 445.
11. R. Vessot *et al.*, *Phys. Rev. Lett.* **45** (1980) 2081.
12. J. D. Prestage, R. L. Tjoelker and L. Maleki, *Phys. Rev. Lett.* **74** (1995) 3511.
13. S. Bize *et al.*, *Phys. Rev. Lett.* **90** (2003) 150802.
14. M. Fischer *et al.*, *Phys. Rev. Lett.* **92** (2004) 230802.
15. T. Damour and F. Dyson, *Nucl. Phys. B* **480** (1996) 37.
16. E. G. Adelberger, *Class. Quant. Grav.* **18** (2001) 2397.
17. V. A. Kostelecky, *CPT and Lorentz Symmetry* (World Scientific, Singapore, 1999).
18. R. Bluhm *et al.*, *Phys. Rev. Lett.* **88** (2001) 090801.
19. J. D. Prestage, G. J. Dick and L. Maleki, *J. Appl. Phys.* **66** (1989) 1013.
20. J. D. Prestage, R. L. Tjoelker and L. Maleki, *Top. Appl. Phys.* **79** (2001) 195.
21. J. D. Prestage, R. L. Tjoelker and L. Maleki, *Proc. 2000 IEEE Frequency Control Symp.* (2000), p. 459.
22. J. D. Prestage *et al.*, *Proc. 2002 IEEE Frequency Control Symp.* (2002), p. 706.
23. R. L. Tjoelker *et al.*, *Proc. 2003 IEEE Frequency Control Symp.* (2003), p. 1066.

OPTICAL CLOCKS AND FREQUENCY METROLOGY FOR SPACE

HUGH KLEIN
National Physical Laboratory (NPL),
Teddington, Middlesex TW11 0LW, UK
hugh.klein@npl.co.uk

Optical frequency standards and femtosecond comb measurement capabilities now rival and in some cases exceed those of microwave devices, with further improvements anticipated. Opportunities are emerging for the application of highly stable and accurate optical frequency devices to fundamental physics space science activities, and the European Space Agency (ESA) has recently commissioned studies on different aspects of optical clocks in space. This paper highlights some examples, including the difficulty of comparing very accurate terrestrial clocks at different locations due to fluctuations of the geoid; by locating a primary frequency standard in space, one could avoid geoid-related gravitational redshifts.

Keywords: Optical clocks; space experiments; femtosecond comb.

1. Introduction

Increasingly stringent demands on atomic timekeeping, driven by applications such as global navigation satellite systems (GNSSs), communications, and very-long-baseline interferometry (VLBI) radio astronomy, have motivated the development of improved time and frequency standards. There are many scientific applications of such devices in space.[1] The experiments that will form the next generation of space-based time and frequency programs will use optical frequency devices. Such "space science" research will aim to answer the third question addressed by the ESA 2015-25 COSMIC VISION: What are the fundamental laws of the Universe? In Europe, a road map covering the space application of time and frequency metrology is being constructed under the auspices of iMERA.[2] Over the few last decades several exciting new fundamental space-based experiments have emerged which will need very good optical clocks and frequency standards.[3,4] An example of the trend towards worldwide collaboration in this field is the Laser Interferometer Space Antenna (LISA). Aiming to detect gravitational waves using a space-based interferometer, with arm lengths of a million kilometers, LISA has very demanding requirements for laser stability. More recent mission proposals (e.g.

Ref. 5) suggesting very stringent multiple tests of both relativistic theory (special and general) and the stability of fundamental constants[6,7] will require lasers that are both ultra-narrow (linewidth below 1 Hz), and very stable.

The accuracy and stability of any frequency standard is closely related to its quality factor Q, the ratio of its resonant frequency to the width of the resonant feature.[8] A clock based on a weak optical transition between a long-lived metastable state and a lower ground state can have an at least 10^5 times higher Q than microwave standards.[9] The best caesium primary standards have a Q of about 10^{10}. To achieve an uncertainty of better than one part in 10^{15}, the centre of the resonant feature therefore has to be found to better than one part in 10^5. This is achieved using the signal from many atoms. In contrast, various atoms and ions possess metastable states and hence "clock" transitions with natural linewidths of below 1 Hz, corresponding to a Q of around 10^{15} or higher. Thus improved precision can be obtained using only a single ion. Ion-based clocks have the advantage that a single ion can be confined in a nearly perturbation-free environment.[10] Atom-based clocks may display better short-term stability by using a large ensemble of atoms.[9]

The use of optical frequency standards as optical clocks became possible as a result of the development, by Ted Hänsch, Jan Hall and others, of wide span optical frequency combs of accurately known and equally spaced frequencies. Based on femtosecond lasers, combs enable the stability of an optical frequency standard to be transferred to the microwave region for comparison with the caesium primary frequency standard, and made possible the first demonstration of a trapped-ion optical clock at the National Institute of Standards and Technology (NIST).[11]

2. Space Applications of Optical Clocks and Frequency Metrology

In the late 1990s the European Space Agency (ESA) recognised the potential for various time and frequency applications in space arising from a technological shift from the microwave to the optical frequencies which was already under way on the ground. Optical frequency standards based on ultra-narrow transitions in trapped ions and atoms already offer better stability than terrestrial microwave fountains and maser devices. The new goal is to extend this capability to space-based applications. An ESA study[12] and presentation at the 13th EFTF and IFCS[13] highlighted three application areas: communications, navigation and space science. The last area is of most interest to the "Quantum to Cosmos" workshop. Following the 1st ESA International Workshop on Optical Clocks,[3] a series of ESA-commissioned studies are concentrating on different aspects of optical clocks for space. For example, an ESA study on optical synthesizers for spaceborne optical frequency metrology led by NPL in Teddington began in February 2006, with partners in Munich, Neuchätel and Düsseldorf.

The feasibility of developing fully automated trapped-ion and atom-based optical frequency standards and clocks with compact laser sources is under consideration. For example, at NPL, a trapped $^{88}Sr^+$ ion optical frequency standard, based on the narrow $5s\,^2S_{1/2}-4d\,^2D_{5/2}$ quadrupole "optical clock" transition at 674 nm, has been developed. It has a natural linewidth of 0.4 Hz due to the 0.4 s lifetime of the $4d\,^2D_{5/2}$ state. Very accurate (Hz level) optical frequency measurements of such a trapped-ion standard were made[14]; recently, Hz-level probe laser linewidths have been achieved.

There are a number of opportunities for conducting fundamental physics space science activities using leading edge optical frequency standards and femtosecond comb measurement capabilities, which now rival and in some cases exceed the capability of microwave frequency devices (see Fig. 1). Today the best primary caesium standards can realize the SI second with an accuracy of better than one part in 10^{15}. However, it is getting increasingly difficult to improve on this value. Standards based on optical transitions are anticipated to achieve two or more orders of magnitude greater accuracy and stability. It is widely expected that the development of such optical clocks will lead to the redefinition of the SI second in terms of an optical transition in due course. As precision improves, it becomes difficult to compare frequency standards on the ground at different locations because of the gravitational red shift. Comparisons at the part-in-10^{18} level require knowledge of altitude to 1 cm. To avoid local fluctuations of the geoid, which change local time, it has been suggested that it would be sensible "to locate the primary frequency standard in space".[15] The challenge for atomic physics posed by this suggestion, together with the other emerging applications of optical clocks in space, is one to which I am sure the international time and frequency community will rise.

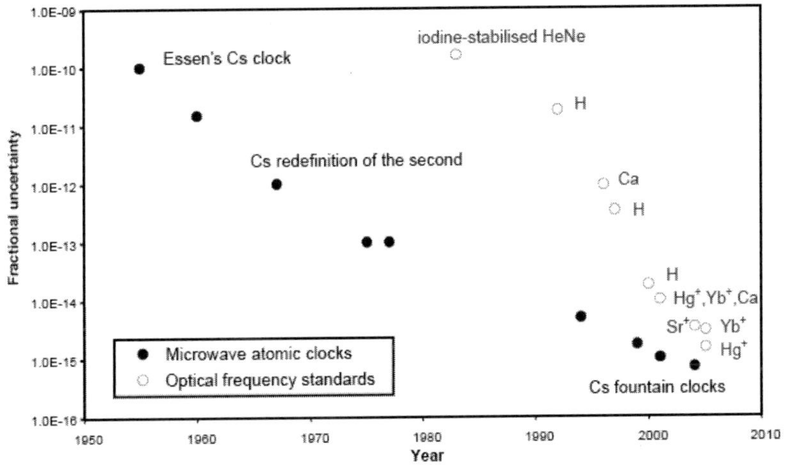

Fig. 1. A comparison of the accuracy of microwave-based atomic clocks and optical frequency standards, showing the rapid improvement achieved with femtosecond combs in the last decade (prepared by HS Margolis).

Acknowledgments

Valuable discussions with Lute Maleki and colleagues at JPL together with ESA staff and study partners are acknowledged. This work was partially funded by ESA and the UK NMSD Quantum Metrology and Length programmes.

References

1. L. Maleki (ed.), *Proc. Workshop on the Scientific Application of Clocks in Space* (JPL Publication 97–15, 1997).
2. http://www.euromet.org/projects/imera/
3. *Proc. 1st ESA International Workshop on Optical Clocks* (ESA-ESTEC, Noordwijk, The Netherlands; 8–10 Jun. 2005).
4. S. G. Karshenboim and E. Peik, *Lect. Notes Phys.* **648** (2005) 297.
5. S. Schiller *et al.*, Precision Tests of General Relativity and of the Equivalence Principle Using Ultrastable Optical Clocks: A Mission Proposal, in *Proc. 39th ESLAB Symposium* (19–21 Apr. 2005), eds. F. Favata and A. Gimenez (Apr. 2005).
6. S. N. Lea, *Rep. Prog. Phys.* **70** (2006) 1473.
7. J. F. Flowers, H. A. Klein and H. S. Margolis, *Contemp. Phys.* **45** (2004) 123.
8. F. Riehle, *Frequency Standards* (WILEY-VCH, 2004) and references therein.
9. *Metrologia* Vol. 42, Special Issue: 50 Years of Atomic Time-Keeping, 1955 to 2005 (Jun. 2005), with particular reference to P. Gill, *Metrologia* 42 (2005) S125.
10. S. N. Lea and H. A. Klein, Trapped ion optical clocks, in *Yearbook of Science and Technology* (McGraw-Hill, 2006).
11. S. A. Diddams *et al.*, *Science* **293** (2001) 825.
12. H. A. Klein and D. J. E. Knight, Ultra-Stable Optical Clocks, Frequency Sources and Standards for Space Applications, in *Final Report to ESA under contract 13290/98/NL/MV* (1999).
13. C. C. Hodge *et al.*, Ultra-stable Optical Frequencies for Space, *Proc. 13th EFTF and IFCS* (Besancon, Apr. 1999).
14. H. S. Margolis *et al.*, *Science* **306** (2004) 1355.
15. D. Kleppner, *Phys. Today* **59** (2006) 10.

ON ARTIFICIAL BLACK HOLES

ULF LEONHARDT* and THOMAS G. PHILBIN

School of Physics and Astronomy, University of St Andrews,
North Haugh, St Andrews KY16 9SS, Scotland
**ulf@st-andrews.ac.uk*

We explain some of the main motivations for creating laboratory analogs of horizons (artificial black holes). We present a concise derivation of the Hawking effect, the quantum radiation of black holes, using a simple analog model.

Keywords: Quantum black holes; analog models of gravity.

1. The Case for Artificial Black Holes

Experiments in space may serve to test fundamental physics, but, conversely, laboratory experiments on Earth may test elusive aspects of astrophysics. Probably the best example is the quantum physics of the black hole. Classical black holes are strictly black: nothing, not even light, can escape. Yet, according to quantum physics, the black hole spontaneously emits particles.[1–4] These particles are virtual particles of the quantum vacuum that materialize at the horizon of the black hole: the vacuum resembles a sea of particle and antiparticle pairs that are continuously created and annihilated. When a particle has the misfortune of being created at one side of the horizon while leaving its partner on the other side, the pair can no longer annihilate each other and is forced to materialize. This explanation is of course only a cartoon picture and not a quantitative physical model: it explains why particle creation is possible at horizons, but it is not suitable for making a quantitative prediction of the particle flux. Calculations show[1–4] that the black hole emits black-body radiation, radiation in thermal equilibrium, with the temperature

$$K_B T = \frac{\hbar\alpha}{2\pi}, \quad \alpha = \frac{c^3}{4GM} = \frac{c}{2R}, \tag{1}$$

where G is the gravitational constant, K_B denotes Boltzmann's constant, c is the speed of light in vacuum, M denotes the mass of the black hole and R the Schwarzschild radius. Both the spectrum and the quantum statistics of the radiation are consistent with the thermal prediction (1) first made by Hawking.[1,2] This is a remarkable and mysterious result: it is simple and universal; it connects the physics of the very small, quantum mechanics characterized by $\hbar$, to the physics of the very large, general relativity with G and thermodynamics with K_B. It implies that the black hole has an entropy,[5] a celebrated relationship that has been frequently used as a benchmark for potential quantum theories of gravity, such as superstring theory[6] and loop quantum gravity.[7,8]

The problem is that it appears to be extremely difficult to observe Hawking's effect in astrophysics, as one sees from a simple estimation of the temperature (1). The characteristic frequency ω of the radiation is $c/4\pi R$, therefore the characteristic wavelength $8\pi^2 R$ is about two orders of magnitude larger than the Schwarzschild radius. A solar-mass black hole with $R \approx 3 \times 10^3\,\text{m}$ emits an electromagnetic Planck spectrum peaked at wavelengths of hundreds of kilometers. It is very cold indeed, with a Hawking temperature of about $6 \times 10^{-8}\,\text{K}$, much below the cosmic microwave background. Smaller black holes radiate stronger, for a simple reason: in our cartoon picture, the particles are separated by tidal forces. Smaller black holes need to reach the central singularity in a smaller distance and therefore have higher gravity gradients at their horizons.

In the laboratory we can make small things. However, with present technology, we cannot create real black holes. What we can make are artificial black holes,[9,10] objects that behave like black holes, analog models of black holes. Most proposals of artificial black holes are based on the simple idea[10–12] illustrated in Fig. 1. Imagine a moving medium — say, a river — with variable speed. This river represents gravity. Consider waves propagating in the moving medium. The waves are dragged by the medium; copropagating waves move faster and counterpropagating waves are slowed down. Suppose that at some point the river exceeds the speed of the waves. Clearly, beyond this point no counterpropagating wave would be able to propagate upstream anymore. The point where the moving medium matches the wave velocity is a horizon. In the first practical demonstration of a horizon,[13] water waves in a narrow channel were sent against the current. The simple analogy between gravity and

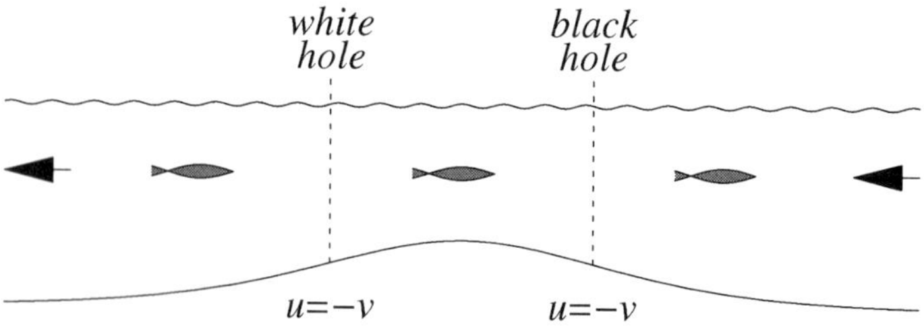

Fig. 1. Black holes are space–time rivers. Imagine a river populated by fish. Suppose that the fish swim with speed v and that the river flows with variable velocity u. Where the river flows faster than the fish they are trapped. The places where u matches $-v$ are horizons. No fish can enter a white-hole horizon and no fish can leave a black-hole horizon. This aquatic analog of the event horizon is surprisingly accurate if the fish are replaced by waves in moving media.

moving media is surprisingly accurate and universal; for example, the propagation of sound in moving fluids is exactly equivalent to the propagation of scalar waves in general relativity, as has first been noted by Moncrief[14] and shortly later by Unruh.[11] Most other candidates of artificial black holes are based on variations of this idea.

Of course, this analogy is limited. Waves in moving media behave like waves in gravitational fields, in curved space–time, but the mechanism for creating this effective space–time geometry is different from gravity. In place of mass in general relativity, moving media generate effective geometries. Artificial black holes capture the kinematic aspects of black holes, not the dynamic ones. On the other hand, concerning the kinematic aspects, analog models of gravity demonstrate not only the classical physics of black holes, showing for example the trapping of waves, but also the quantum physics, because the interaction between the moving medium and the wave is linear, regardless of how small the wave amplitudes are, even down to the quantum scale.

However, quantum effects such as Hawking radiation are subtle. In moving media, the Hawking temperature should be related to the velocity gradient at the horizon. The gradient of a velocity has the dimension of a frequency. Consequently, one might guess that the characteristic frequency α in Hawking's formula (1) corresponds to the velocity gradient at the horizon. We show in the next section that this is correct, using a simple model. The observation of the Hawking effect presumes that the moving fluid is colder than the Hawking radiation, so that the radiation is detectable, and usually requires cooling to ultralow temperatures. Bose–Einstein condensates or quantum liquids like superfluid He-3[10] could still behave as fluids even at the extremely low temperatures needed, whereas the water-wave analog of the black hole[13] is likely to remain a demonstration of the classical physics of the black hole. Optical black holes, where media move faster than the speed of light in the medium, the speed of light in vacuum c divided by the refractive index n, are excellent candidates for observing the quantum effects of black holes, because velocity gradients within a few wavelengths of light are conceivable. Such gradients would correspond to Hawking temperatures in the order of 10^3K. We found a trick to create such a superluminally moving medium in the laboratory[15–17] and are setting up an experiment.

What would we learn from artificial black holes? In general relativity, the physics at horizons appears to tend to extremes. Although the gravitational field is regular near the horizon (unless expressed in irregular coordinates such as Schwarzschild coordinates), waves freeze here, because they are trapped. In the vicinity of a horizon, waves are forced to oscillate with ever-decreasing wavelengths the closer they are to the horizon; see Fig. 2. The wavelength decreases below all scales, including the Planck scale, where one commonly is in doubt whether the presently known physics is applicable anymore. This issue is known as the trans-Planckian problem.[18,19] Some unknown mechanism must regularize the horizon of the astrophysical black hole.

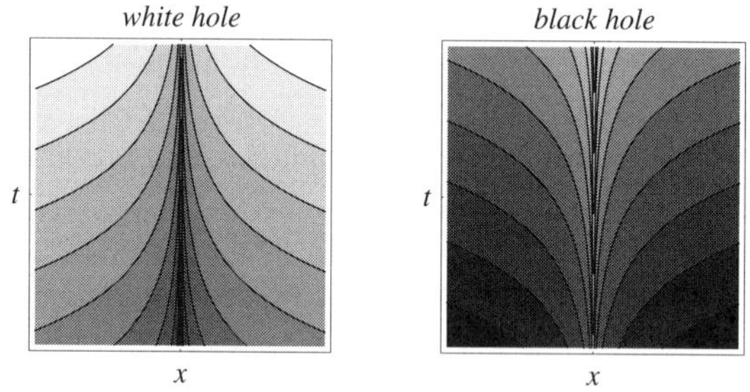

Fig. 2. Space–time diagrams of waves near white-hole horizons (*left*) and black-hole horizons (*right*). Waves cannot enter white holes and cannot leave black holes. Near the horizons waves freeze, oscillating with decreasing wavelengths and developing the logarithmic phase singularity (17), shown in the figures as contour plots of the phase. Rays, the lines of equal phase, are trapped at the white-hole horizon and, in the right figure, take an exponentially long time to get away from the black hole.

Artificial black holes have natural and known mechanisms to regularize the extremes of waves at horizons. Understanding them in detail may illuminate some of the trans-Planckian physics of gravity. The simplest mechanism is dispersion: decreasing wavelengths in the laboratory frame correspond, via the Doppler effect, to increasing local frequencies in the medium. In our optical model, the refractive index will change with frequency due to optical dispersion[20] and the waves are able to escape from the horizon. Consider the ray tracing at the black-hole horizon backward in time shown in Fig. 3, remembering that backward in time the flow changes direction. For normal dispersion[20] the refractive index increases with frequency, and the speed of light decreases such that, backward in time, the light wave is carried away from the horizon. This means that, forward in time, the light comes from the subluminal region in front of the horizon. In the case of anomalous dispersion, where the refractive index decreases with frequency, the frequency-shifted light stems from the region of superluminal flow behind the horizon.

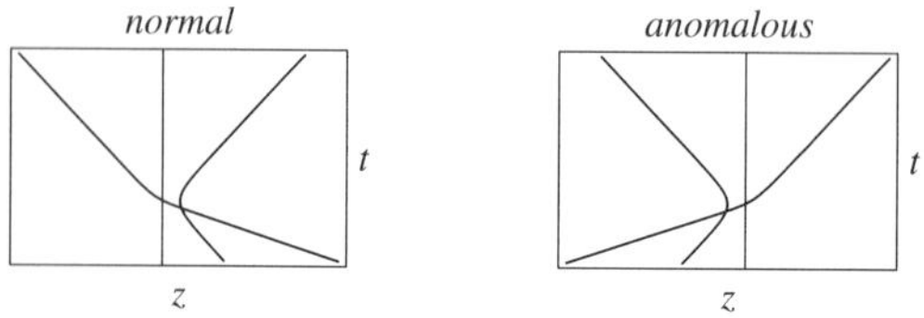

Fig. 3. Rays at realistic black-hole horizons with dispersion. Traced backward in time, such that the flow changes direction, light rays are blueshifted near the horizon. In the case of normal dispersion (*left*) the speed of light is reduced and the rays are dragged with the moving medium to the right. For anomalous dispersion, the speed of light increases with increasing frequency and the rays are dragged to the left. Seen forward in time, the rays stem from the subluminal region for normal dispersion and from the superluminal region for anomalous dispersion.

Strictly speaking, genuine horizons do not exist due to dispersion. However, the Hawking effect is remarkably robust against moderate dispersion.[21–23] Realistic experiments, however, tend to operate in a strongly dispersive regime. Here several natural questions arise. The horizon is the place where the speed of the medium is equal to the speed of the wave. Which wave speed matters — the group or the phase velocity or something in between? Both the group and the phase velocity depend on frequency. Which frequency matters? What is indispensable for the Hawking effect and what is not? What really is a horizon? Real experiments pose questions that otherwise one would have the liberty to avoid. They pose a challenge and they will give real data to support our understanding. It is highly likely that analyzing laboratory analogs of black holes will shed new light on one of the most fascinating effects in physics — the quantum creation at horizons.

2. A Simple Example

In this section we develop an elementary theory of the Hawking effect at horizons based on a simple analog model. Consider light in a one-dimensional moving medium with spatially variable speed $u(x)$ and uniform refractive index n. For definiteness and without loss of generality, assume that the medium moves from the right to the left, i.e. u is negative. We ignore optical dispersion,[20] i.e. we assume that n does not depend on frequency and we use geometrical optics[20] to describe the light propagation. Furthermore, we assume that the refractive index is very large such that light propagates at nonrelativistic speed, which is only a technical point to simplify the analysis.

2.1. *Light rays*

Geometrical optics[20] presumes that $u(x)$ varies over longer scales than the wavelength of light λ. Think of the medium as consisting of droplets that are larger than λ but smaller than the variation scale of $u(x)$. In each droplet the light propagates as a wave with the dispersion relation

$$k'^2 = n^2(x)\,\frac{\omega'^2}{c^2}\,, \tag{2}$$

as seen in a locally comoving frame, where ω' denotes the comoving frequency and k' the wave number $2\pi/\lambda'$. The frequency ω' is positive, but the dispersion relation has solutions with two different signs for k'. The minus sign corresponds to light propagating in negative direction, i.e. to copropagating light, whereas positive wave numbers k' describe counterpropagating light. Only counterpropagating waves experience significant effects at the horizon, whereas copropagating light is simply swept away. Consequently, in the following, we focus on counterpropagating light with the dispersion relation

$$k' = n(x)\,\frac{\omega'}{c}\,. \tag{3}$$

Frequencies and wave numbers are derivatives of the phase φ with respect to the coordinates,

$$\omega' = -\frac{\partial \varphi}{\partial t'}, \quad k' = \frac{\partial \varphi}{\partial x'}, \quad \omega = -\frac{\partial \varphi}{\partial t}, \quad k = \frac{\partial \varphi}{\partial x}. \tag{4}$$

The locally comoving coordinates are related to the laboratory coordinates by the Gallilei transformation

$$t' = t\,, \quad x' = x - ut. \tag{5}$$

Since

$$\frac{\partial}{\partial t} = \frac{\partial}{\partial t'} - u\frac{\partial}{\partial x'}, \quad \frac{\partial}{\partial x} = \frac{\partial}{\partial x'}, \tag{6}$$

we obtain the relations

$$\omega = \omega' + uk', \quad k = k', \tag{7}$$

which describe the Doppler effect: using the dispersion relation in the comoving frame (3) we find for the frequency in the laboratory frame the Doppler formula

$$\omega = \left(1 + u\frac{n}{c}\right)\omega'. \tag{8}$$

Our starting point, the dispersion relation (3) in frames comoving with the medium, is valid only in such frames, in each of the small droplets of the medium. In order to describe how light propagates through the medium, we express the dispersion relation in the laboratory frame. We obtain from Eqs. (3) and (7) the relation

$$\omega = \frac{ck}{n} + uk. \tag{9}$$

We interpret ω as the Hamiltonian of light rays and obtain from one of Hamilton's equations,

$$\frac{dx}{dt} = \frac{\partial \omega}{\partial k} = \frac{c}{n} + u(x), \tag{10}$$

the velocity-addition theorem of light and medium. Equation (10) represents an ordinary differential equation for the light rays with the implicit solution

$$t - t_0 = \int \frac{dx}{c/n + u(x)}, \tag{11}$$

where the time t_0 distinguishes the various ray trajectories in a space–time diagram. Close to a horizon, assumed, without loss of generality, at $x = 0$, we linearize the velocity of the medium as

$$u \sim -\frac{c}{n} + \alpha x\,. \tag{12}$$

The constant α describes the velocity gradient at the horizon. At black-hole horizons the (negative) speed turns from superluminal to subluminal: α is positive. A white-hole horizon would correspond to the opposite: a medium, which changes from subluminal to superluminal speed, where α is negative. In this case, no wave would enter the white hole, as no wave escapes from the region behind the black-hole horizon. Close to the horizon, in the linear velocity profile (12), we obtain the explicit solution

$$x(t) \sim x_0 e^{\alpha(t-t_0)}, \tag{13}$$

where the integration constant x_0 describes the initial position of the light ray. In the case of black holes, waves originally at positive x_0 take an exponential time to get away from the horizon and waves at negative x_0 drift back equally slowly. We obtain from the other Hamilton equation

$$\frac{dk}{dt} = -\frac{\partial \omega}{\partial x} = -k\frac{\partial u}{\partial x} \sim -\alpha k \tag{14}$$

with the solution

$$k(t) \sim k_0 e^{-\alpha(t-t_0)}, \tag{15}$$

which describes the exponential redshift at black-hole horizons or the exponential blueshift at white holes. We gain another important insight from our simple geometrical optics in moving media: in the superluminal region rays oscillate with negative frequencies ω in the laboratory frame. This is a consequence of the Doppler effect: the frequency ω' in the medium is always positive but, in the laboratory frame, the Doppler-shifted frequency (8) is negative for superluminal media with $|u| > c/n$. Therefore we should chose a positive frequency ω for light rays on the subluminal side and a negative ω for rays on the superluminal side.

2.2. *Light waves*

Waves $A(t,x)$ combine various rays in their phase φ, because any constant phase value φ_0 propagates with the corresponding ray trajectory for $\varphi_0 = -\omega t_0$. In this way, we obtain

$$A(t,x) = A_0 e^{i\varphi}, \quad \varphi = -\omega t + \omega \int \frac{dx}{c/n + u(x)}. \tag{16}$$

Alternatively, we can deduce Eq. (16) as being the solution to the Hamilton–Jacobi equation, the dispersion relation (9) with the relations (4). How do waves behave near the horizon? We see that the phase φ develops a logarithmic singularity at the horizon

$$\varphi \sim -\omega t + \frac{\omega}{\alpha} \ln x. \tag{17}$$

Consequently, the wave number is proportional to the inverse of the distance from the horizon, in agreement with the previous results (13) and (15) on the exponential shift of light rays. The horizon clearly separates rays, but waves are global objects and might be connected, exhibiting perhaps an exponentially small evanescent tail. Consider black holes with $\alpha > 0$. Imagine that we analytically continue the phase φ on the complex z plane around the horizon at the origin. The two sides of the horizon are the positive and the negative real axis x. The logarithm is a multi-valued function; we could choose to connect the two sides on the upper or the lower half plane. Suppose that we chose the connection on the upper half plane. On the left side of the horizon, φ gains the constant $i\pi\omega/\alpha$ from the integration of $(\omega/\alpha)/z$ along a half circle around the origin. For positive ω, waves A_+ are suppressed by the factor $\exp(-\pi\omega/\alpha)$ on the left side of the horizon, which is consistent with the idea that light mainly propagating on the subluminal side has positive frequencies in the laboratory frame. Conversely, for negative ω, waves A_- are suppressed on the superluminal side. Therefore, the analytic continuation of the phase on the upper half plane is consistent with our physical picture. We get a more precise insight from considering the spatial Fourier transformation

$$\widetilde{A}(t,k) = \int A(t,x)\,\exp(-ikx)\,dx\,. \tag{18}$$

When $A(t,z)$ is analytic on the upper half z plane, the Fourier integral vanishes for negative k. Consequently, we describe light propagating with purely positive wave numbers. On the other hand, we could form superpositions of waves that are strictly localized on either side of the horizon as follows. These are superpositions of $A_\pm$ and $A_\mp^*$, the complex conjugate of $A_\mp$ with frequency $-\omega$, such that $A_\pm$ and $A_\mp^*$ oscillate with the same ω. If we form, for any constant Z,

$$A'_\pm = Z^{1/2}(A_\pm - e^{-\pi\omega/\alpha}A_\mp^*), \tag{19}$$

the wave A'_+ vanishes on the negative real axis, in the superluminal region; whereas A'_- vanishes on the subluminal side. Consequently, the waves $A'_\pm$ describe light that is strictly separated by the horizon.

2.3. *Light quanta*

The horizon truly separates physical space into two disconnected regions, even for waves. We could regard the analytically continued waves $A_\pm$ as artifacts. On the other hand, the horizon was not eternal, but formed at some moment in time. At that stage, light waves were connected until the very moment when the horizon was formed. The outgoing waves, separated by the horizon, are superpositions of the $A'_\pm$. It is tempting to interpret the $A_\pm$ as the constituents of the incident light, because, as we know, they correspond to forward-propagating light with positive wave number k. We call $A_\pm$ the ingoing modes and $A'_\pm$ the outgoing modes. In quantum optics, light is regarded as a superposition of modes[24]:

$$\hat{A}(t,x) = \sum_\nu (A_\nu(t,x)\hat{a}_\nu + A_\nu^*(t,x)\hat{a}_\nu^\dagger). \tag{20}$$

A mode $A_\nu(t,x)$ describes the electromagnetic wave of a single light quantum, a photon. The mode operators $\hat{a}_\nu$ describe the amplitude of light, depending on how many photons are excited and which quantum state they form. The operators $\hat{a}_\nu$ are subject to the Bose communtation relations[24]

$$[\hat{a}_\nu, \hat{a}_{\nu'}^\dagger] = \delta_{\nu\nu'}\,, \quad [\hat{a}_\nu, \hat{a}_{\nu'}] = 0\,. \tag{21}$$

The $\hat{a}_\nu$ are annihilation operators and the $\hat{a}_{\nu'}^\dagger$ creation operators for the quanta of independent light modes. We could choose arbitrary sets of mode functions, provided that they obey the classical wave equation of light in moving media,[24] as long as the A_ν establish a complete set and the corresponding $\hat{a}_\nu$ satisfy the commutation relations (21). For example, we could use the incoming or the outgoing modes for various frequencies. Since the total field operator $\hat{A}(t,x)$ is independent of the choice of modes, the mode operators must compensate for the transformation of one set of modes into another, i.e. in our case for each frequency component

$$\sum_{\pm}(A_\pm \hat{a}_\pm + A_\pm^* \hat{a}_\pm^\dagger) = \sum_{\pm}(A'_\pm \hat{a}'_\pm + A'^*_\pm \hat{a}'^\dagger_\pm), \tag{22}$$

which implies that

$$\hat{a}_\pm = Z^{1/2}(\hat{a}'_\pm - e^{-\pi\omega/\alpha}\hat{a}'^\dagger_\mp). \tag{23}$$

The electromagnetic wave of a single photon cannot have an arbitrary amplitude and therefore Z should be fixed. Indeed, we obtain from the Bose commutation relations (21) of both the outgoing and the incoming mode operators

$$1 = [\hat{a}_\pm, \hat{a}_\pm^\dagger] = Z(1 - e^{-2\pi\omega/\alpha}) \Rightarrow Z = \frac{1}{1 - e^{-2\pi\omega/\alpha}}. \tag{24}$$

Suppose that the incoming light is in the vacuum state $|0,0\rangle$, i.e. no light at all is incident. To find out whether and how many quanta are spontaneously emitted by the horizon, we express the incoming vacuum in terms of the outgoing modes. We denote the outgoing photon-number eigenstates, the outgoing Fock states,[24] by $|n_+, n_-\rangle'$ with the integers $n_\pm$. The vacuum state $|0,0\rangle$ is the zero eigenstate of $\hat{a}_\pm$. Using the standard relations for the annihilation and creation operators

$$\hat{a}'_\pm|n\rangle' = \sqrt{n}\,|n-1\rangle'\,, \quad \hat{a}'^\dagger_\pm|n\rangle' = \sqrt{n+1}\,|n+1\rangle'\,, \tag{25}$$

one verifies that $\hat{a}_\pm|0\rangle$ vanishes for the state

$$|0,0\rangle = Z^{-1/2}\sum_{n=0}^{\infty} e^{-n\pi\omega/\alpha}|n,n\rangle'. \tag{26}$$

This is a most remarkable result.[1,2] First, it shows that the horizon spontaneously generates radiation from the incident quantum vacuum. Second, the emitted radiation consists of correlated photon pairs, and each photon on one side is correlated to a partner photon on the other side, because they are always produced in pairs. The total quantum state turns out to be an Einstein–Podolski–Rosen state,[24] the strongest entangled state for a given energy. Third, light on either side of the horizon consists of an ensemble of photon-number eigenstates with probability $Z^{-1}e^{-2n\pi\omega/\alpha}$. This is a Boltzmann distribution of $|n\rangle'$ photons with energies

$n\hbar\omega$ and temperature $K_B T = \hbar\alpha/2\pi$. Consequently, the horizon emits a Planck spectrum of black-body radiation with the Hawking temperature (1). Fourth, this Planck spectrum is consistent with Bekenstein's black-hole thermodynamics[6]: black holes seem to have an entropy and a temperature.

Acknowledgments

This work is supported by the Leverhulme Trust and EPSRC. We are indebted to Stephen Hill, Frieder König, Chris Kuklewicz and Renaud Parentani for our discussions about black holes.

References

1. S. M. Hawking, *Nature* **248** (1974) 30.
2. S. M. Hawking, *Commun. Math. Phys.* **43** (1975) 199.
3. N. D. Birrell and P. C. W. Davies, *Quantum Fields in Curved Space* (Cambridge University Press, 1984).
4. R. Brout *et al.*, *Phys. Rep.* **260** (1995) 329.
5. J. D. Bekenstein, *Phys. Rev. D* **7** (1973) 2333. see also gr-qc/0009019 for a review.
6. M. B. Green, J. H. Schwarz and E. Witten, *Superstring Theory* (Cambridge University Press, 1987).
7. C. Rovelli, *Living Rev. Relat.* **1** (1998) 1 [gr-qc/9710008].
8. D. Oriti, *Rep. Prog. Phys.* **64** (2001) 1489.
9. M. Novello, M. Visser and G. E. Volovik (eds.), *Artificial Black Holes* (World Scientific, Singapore, 2002).
10. G. E. Volovik, *The Universe in a Helium Droplet* (Clarendon, Oxford, 2003).
11. W. G. Unruh, *Phys. Rev. Lett.* **46** (1981) 1351.
12. M. Visser, *Class. Quant. Grav.* **15** (1998) 1767.
13. G. Rousseaux *et al.*, gr-qc/0711.4767
14. V. Moncrief, *Astrophys. J.* **235** (1980) 1038.
15. U. Leonhardt and F. König, *Invention Disclosure, Research and Enterprise Services, University of St Andrews*, 27 Jan. 2005.
16. T. G. Philbin *et al.*, gr-qc/0711.4796
17. T. G. Philbin *et al.*, gr-qc/0711.4797
18. G. 't Hooft, *Nucl. Phys. B* **256** (1985) 727.
19. T. Jacobson, *Phys. Rev. D* **44** (1991) 1731.
20. M. Born and E. Wolf, *Principles of Optics* (Cambridge University Press, 1999).
21. W. G. Unruh, *Phys. Rev. D* **51** (1995) 2827.
22. R. Brout *et al.*, *Phys. Rev. D* **52** (1995) 4559.
23. R. Balbinot *et al.*, *Riv. Nuovo Cimento* **28** (2005) 1.
24. U. Leonhardt, *Rep. Prog. Phys.* **66** (2003) 1207.

PART 6

COSMOLOGY AND DARK ENERGY

DARK ENERGY TASK FORCE: FINDINGS AND RECOMMENDATIONS

ROBERT N. CAHN

Lawrence Berkeley National Laboratory, 1 Cyclotron Road, Berkeley, CA 94720, USA
rncahn@lbl.gov

NASA, NSF, and the Department of Energy established the Dark Energy Task Force to examine the opportunities for exploration of dark energy and to report on the various techniques that might be employed to further our understanding of this remarkable phenomenon. The Task Force examined dozens of white papers outlining the plans of teams to measure dark energy's properties. I present here the findings and recommendations of the Task Force. Our methodology is explained in another talk at this conference.

Keywords: Dark energy.

1. Introduction

Dark energy appears to be the dominant component of the physical Universe, yet there is no persuasive theoretical explanation. The acceleration of the Universe is, along with dark matter, the observed phenomenon which most directly demonstrates that our fundamental theories of particles and gravity are either incorrect or incomplete. Most experts believe that nothing short of a revolution in our understanding of fundamental physics will be required to achieve a full understanding of the cosmic acceleration. For these reasons, the nature of dark energy ranks among the very most compelling of all outstanding problems in physical science. These circumstances demand an ambitious observational program to determine the dark energy properties as well as possible.

Thus begins the report of the Dark Energy Task Force.

2. Dark Energy Primer

The Friedmann–Lemaître equations are the basis for cosmology:

$$H^2 \equiv \left(\frac{\dot{a}}{a}\right)^2 = \frac{8\pi G_N \rho}{3} - \frac{k}{a^2} + \frac{\Lambda}{3}, \tag{1}$$

$$\frac{\ddot{a}}{a} = \frac{\Lambda}{3} - \frac{4\pi G_N}{3}(\rho + 3p), \tag{2}$$

where ρ and p represent the energy density and momentum density of a uniform isotropic fluid. We can combine these to obtain a very useful relation:

$$\dot{\rho} = -3H(\rho + p). \tag{3}$$

If there are several "fluids," each satisfies this separately. We define the equation of state by

$$w(a) = \frac{\rho(a)}{p(a)}. \tag{4}$$

From the first Friedmann–Lemaître equation we see that the accelerating Universe requires either $\Lambda > 0$ or $w < -1/3$. For ordinary nonrelativistic matter, $w \approx (v/c)^2$, while for relativistic matter, $w = 1/3$. Since the energy density associated with Λ is time-independent, for it $w = -1$. If w is constant,

$$\frac{\dot{\rho}}{\rho} = -3\frac{\dot{a}}{a}(1+w). \tag{5}$$

$$\rho = \rho_0 a^{-3(1+w)}. \tag{6}$$

More generally,

$$\frac{\dot{\rho}}{\rho} = -3\frac{\dot{a}}{a}[1+w(a)], \tag{7}$$

$$\rho(a) = \rho_0 \exp\left(3\int_a^1 da \frac{1+w(a)}{a}\right). \tag{8}$$

Remember that $a < 1$ for times prior to the present. As an example, suppose that $w(a) = w_0 + (1-a)w_a$. Then

$$\rho_{\rm DE} = \rho_{\rm DE0} a^{-3(1+w_0+w_a)} e^{-3(1-a)w_a}. \tag{9}$$

Now return to the Friedman–Lemaître equation:

$$H(a)^2 = \frac{8\pi G_N}{3}\left[\frac{\rho_{m0}}{a^3} + \frac{\rho_{r0}}{a^4} + \rho_{\rm DE0}\Delta(a) - \frac{k}{a^2}\right]. \tag{10}$$

We define

$$\Omega_m = \frac{8\pi G_N}{3H_0^2}\rho_{m0}, \quad \Omega_r = \frac{8\pi G_N}{3H_0^2}\rho_{r0}, \tag{11}$$

$$\Omega_{\rm DE} = \frac{8\pi G_N}{3H_0^2}\rho_{\rm DE0}, \quad \Omega_k = -\frac{k}{H_0^2}. \tag{12}$$

The Hubble constant today is $H_0 = 72 \pm 8$ (km/s/Mpc) $\equiv h \times 100$ (km/s/Mpc). A convenient mnemonic is

$$\frac{c}{H_0} = 3\,\text{Gpc}. \tag{13}$$

We now have

$$\frac{H^2}{H_0^2} = \frac{\Omega_m}{a^3} + \frac{\Omega_r}{a^4} + \frac{\Omega_k}{a^2} + \Omega_{\rm DE}\Delta(a). \tag{14}$$

Usually we can neglect Ω_r. Doing that and evaluating at the present,

$$1 = \Omega_m + \Omega_{\rm DE} + \Omega_k. \tag{15}$$

Note that Ω_k is not really an energy density. The notation just makes things look nice.

3. Observable Quantities

The distance element is given in terms of the metric $g_{\mu\nu}$:

$$ds^2 = g_{\mu\nu}dx^\mu dx^\nu. \tag{16}$$

The most general form that is isotropic and homogenous can be written as

$$ds^2 = dt^2 - a(t)^2\left\{\frac{dr^2}{1-kr^2} + r^2(d\theta^2 + \sin^2\theta d\phi^2)\right\}. \tag{17}$$

This is the Robertson–Walker metric.

While many authors use dimensions such that k is $-1, 0$, or 1, I find this not helpful. My r has dimensions of length and k has dimensions of (length)$^{-2}$. The scale parameter, a, is dimensionless and is unity today. It is conventional to write the Robertson–Walker metric above in a second way:

$$ds^2 = dt^2 - a(t)^2\{d\chi^2 + S_k(\chi)^2(d\theta^2 + \sin^2\theta d\phi^2)\}, \tag{18}$$

so that

$$\frac{dr}{\sqrt{1-kr^2}} = d\chi, \quad \chi(r) = \frac{1}{\sqrt{k}}\sin^{-1}\sqrt{k}r, \quad k > 0, \tag{19}$$

$$\chi(r) = r, \quad k = 0, \tag{20}$$

$$\chi(r) = \frac{1}{\sqrt{-k}}\sinh^{-1}\sqrt{-k}r, \quad k < 0. \tag{21}$$

Independent of the sign of k we can expand to get

$$r + \frac{1}{6}kr^3 + \cdots = \chi \tag{22}$$

for small curvature. It is conventional to define $S_k(\chi) = r$, and for small kr^2

$$S_k(\chi) = \chi - \frac{1}{6}k\chi^3 \cdots. \tag{23}$$

If light travels radially,

$$d\chi = \frac{cdt}{a} = \frac{da}{da}\frac{cdt}{a} = \frac{cda}{a^2H(a)} = \frac{dz}{H(z)}. \tag{24}$$

Thus, measuring z gives us χ, *if we know the cosmology*. The observations of supernovae, weak lensing, baryon acoustic oscillations, and clusters give us information on the "distances"

$$D_L = (1+z)\left(\chi - \frac{1}{6}k\chi^3\cdots\right), \tag{25}$$

$$D_{co} = \left(\chi - \frac{1}{6}k\chi^3\cdots\right), \tag{26}$$

$$D_A = (1+z)^{-1}\left(\chi - \frac{1}{6}k\chi^3\cdots\right). \tag{27}$$

These measurements give us access to $\Omega_{\rm DE}, \Omega_m \cdot w$, etc.

4. Growth

Once fluctuations in density, $g = \delta\rho/\rho$, are present, gravity will magnify them. The expansion of the Universe, however, damps the effect.

$$\frac{d^2g}{dt^2} + 2H\frac{dg}{dt} = 4\pi G\rho_m g = \frac{3\Omega_m H_0^2}{2a^3}g. \tag{28}$$

If GR is correct, there is a 1–1 map between $D(z)$ and $g(z)$. If GR is incorrect, observed quantities may fail to obey this relation. The growth factor is determined by measuring the density fluctuations in nearby dark matter (!), in comparison with those seen at $z = 1088$ by WMAP.

5. Dark Energy Techniques

5.1. *Supernovae*

Type Ia supernovae provided the first astonishing results indicating that the expansion of the Universe was occurring ever more rapidly. To the extent that SN Ia are uniform in their intrinsic brightness, they provide direct access to the luminosity distance through the relation

$$\mu = 5\log_{10} d_L + \text{const}, \tag{29}$$

where the constant depends on H_0. In practice, SN Ia are not perfect "standard candles," but rather their peak brightness is correlated with particular features of the supernova. The DETF report recommends detailed study of 500 nearby supernovae, both to increase our understanding of SN Ia and to provide a low-z anchor for the Hubble diagram, which ultimately provides information on the cosmological constants.

5.2. *Baryon acoustic oscillations*

Acoustic waves propagate in the baryon–photon plasma starting at the end of inflation. When plasma combines to neutral hydrogen, sound propagation ends. The total travel distance, the sound horizon $r_s \approx 140\,\text{Mpc}$, is imprinted on the matter density pattern. Identify the angular scale subtending r_s, then use $\theta_s = r_s/D(z)$.

WMAP and Planck determine r_s and the distance to $z = 1088$. Surveys of galaxies (as signposts for dark matter) recover $D(z), H(z)$ at $0 < z < 5$. By observing the oscillations in the transverse direction, we obtain $D(z)$, while the radial oscillations give $H(z)$. Photometric surveys generally do not have enough z resolution to obtain useful information on $H(z)$. This means that spectroscopic surveys have a big advantage.

5.3. *Galaxy clusters*

Galaxy clusters are the largest structures in the Universe to undergo gravitational collapse. They provide markers for locations with density contrast above a critical value. Theory predicts the mass function $dN/dMdV$. We observe $dN/dzd\Omega$.

$$\frac{dV}{d\Omega dz} \propto \frac{D^2(z)}{H(z)}. \tag{30}$$

The distribution is very sensitive to M and to $g(z)$. It is also very sensitive to misestimation of mass, which is not directly observed.

5.4. *Weak lensing*

Mass concentrations in the Universe deflect photons from distant sources. Displacement of background images is unobservable, but their distortion (shear) is measurable. The extent of distortion depends upon the size of mass concentrations and relative distances. The depth information comes from red shifts. Obtaining 10^8 red shifts from optical spectroscopy is infeasible, so "photometric" red shifts are used instead.

6. Stages in Dark Energy Research

We distinguish four stages of investigation of dark energy:

(i) What is known now (12/31/05).
(ii) Anticipated state upon completion of ongoing projects.
(iii) Near-term, medium-cost, currently proposed projects.
(iv) Large-Survey Telescope (LST) and/or Square Kilometer Array (SKA), and/or Joint Dark Energy (Space) Mission (JDEM).

7. DETF Figure of Merit

To quantify progress in understanding energy, we focus on the equation of state, $w(a)$, and take as a simple, but by no means unique, parametrization $w(a) = w_0 + (1 - a)w_a$. While this has no special theoretical justification, in the absence of compelling physical models, we opt for simplicity. A projection of the achievements of a proposed experiment can be turned into an error ellipse in the $w_0 - w_a$ plane. We take as our figure of merit the reciprocal of the area that encompasses the 95% CL limit.

8. Findings of the DETF

(i) Four observational techniques dominate the white papers received by the Task Force. In alphabetical order:

(a) Baryon acoustic oscillations (BAO's) are observed in large-scale surveys of the spatial distribution of galaxies. The BAO technique is sensitive to dark energy through its effect on the angular-diameter distance vs red shift relation and through its effect on the time evolution of the expansion rate.

(b) Galaxy cluster (CL) surveys measure the spatial density and distribution of galaxy clusters. The CL technique is sensitive to dark energy through its effect on a combination of the angular-diameter distance vs red shift relation, the time evolution of the expansion rate, and the growth rate of structure.

(c) Supernova (SN) surveys use Type Ia supernovae as standard candles to determine the luminosity distance vs red shift relation. The SN technique is sensitive to dark energy through its effect on this relation.

(d) Weak lensing (WL) surveys measure the distortion of background images due to the bending of light as it passes by galaxies or clusters of galaxies. The WL technique is sensitive to dark energy through its effect on the angular-diameter distance vs red shift relation and the growth rate of structure.

Other techniques discussed in white papers, such as using γ-ray bursts or gravitational waves from coalescing binaries as standard candles, merit further investigation. At this time, they have not yet been practically implemented, so it is difficult to predict how they might be part of a dark energy program. We do note that if dark energy dominance is a recent cosmological phenomenon, very-high-red shift ($z \gg 1$) probes will be of limited utility.

(ii) Different techniques have different strengths and weaknesses and are sensitive in different ways to the dark energy properties and to other cosmological parameters.

(iii) Each of the four techniques can be pursued by multiple observational approaches, e.g. radio, visible, near-infrared (NIR), and/or X-ray observations, and a single experiment can study dark energy with multiple techniques. Individual missions need not cover multiple techniques; combinations of projects can achieve the same overall goals.

(iv) The techniques are at different levels of maturity:

(a) The BAO technique has only recently been established. It is less affected by astrophysical uncertainties than other techniques.

(b) The CL technique has the statistical potential to exceed the BAO and SN techniques but at present has the largest systematic errors. Its eventual accuracy is currently very difficult to predict and its ultimate utility as a dark energy technique can only be determined through the development of techniques that control systematics due to nonlinear astrophysical processes.

(c) The SN technique is at present the most powerful and best-proven technique for studying dark energy. If red shifts are determined by multiband photometry, the power of the supernova technique depends critically on the accuracy achieved for photo-z's. (Multiband photometry measures the intensity of the object in several colors. A red shift determined by multiband photometry is called a photometric red shift, or a photo-z.) If spectroscopically measured red shifts are used, the power of the experiment as reflected in the DETF figure of merit is much better known, with the outcome depending on the uncertainties in supernova evolution and in the astronomical flux calibration.

(d) The WL technique is also an emerging technique. Its eventual accuracy will also be limited by systematic errors that are difficult to predict. If the systematic errors are at or below the level asserted by the proponents, it is likely to be the most powerful individual Stage IV technique and also the most powerful component in a multitechnique program.

(v) A program that includes multiple techniques at Stage IV can provide an order-of-magnitude increase in the DETF figure of merit. This would be a major advance in our understanding of dark energy. A program that includes multiple techniques at Stage III can provide a factor-of-3 increase in the DETF figure of merit. This would be a valuable advance in our understanding of dark energy. In the absence of a persuasive theoretical explanation for dark energy, we must be guided by ever-more-precise observations.

(vi) We find that no single observational technique is sufficiently powerful and well established that it is guaranteed to achieve by itself an order-of-magnitude increase in the DETF figure of merit. Combinations of the principal techniques have substantially more statistical power, much greater ability to discriminate among dark energy models, and more robustness to systematic errors than any single technique. The case for multiple techniques is supported as well by the critical need for confirmation of results from any single method. (The results for various model combinations can be found at the end of Section IX.)

(vii) Results on structure growth, obtainable from weak lensing or cluster observations, provide additional information not obtainable from other techniques. In particular, they allow for a consistency test of the basic paradigm: spatially constant dark energy plus general relativity.

(viii) In our modeling we assume constraints on H0 from current data and constraints on other cosmological parameters expected to come from further measurement of CMB temperature and polarization anisotropies.

(a) These data, though insensitive to $w(a)$ on their own, contribute to our knowledge of $w(a)$ when combined with any of the dark energy techniques we have considered.

(b) Increased precision in a particular cosmological parameter may improve dark energy constraints from a single technique. Increased precision is valuable for the important task of comparing dark energy results from different techniques.

(ix) Increased precision in cosmological parameters tends not to improve significantly the overall DETF figure of merit obtained from a multitechnique program. Indeed, a multitechnique program would itself provide powerful new constraints on cosmological parameters within the context of our parametric dark energy model.

(x) Setting the spatial curvature of the Universe to zero greatly strengthens the dark energy constraints from supernovae, but has a modest impact on the other techniques once a dark energy parameterization is selected. When techniques are combined, setting the spatial curvature of the Universe to zero makes little difference to constraints on parametrized dark energy, because the curvature is one of the parameters well determined by a multitechnique approach.

(xi) Optical, NIR, and X-ray experiments with very large numbers of astronomical targets will rely on photometrically determined red shifts. The ultimate accuracy that can be attained for photo-z's is likely to determine the power of such measurements. [Radio HI (neutral hydrogen) surveys produce precise red shifts as part of the survey.]

(xii) Our inability to forecast systematic error levels reliably is the biggest impediment to judging the future capabilities of the techniques. Assessments of effectiveness could be made more reliably with:

(a) For BAO theoretical investigations of how far into the nonlinear regime the data can be modeled with sufficient reliability and further understanding of galaxy bias on the galaxy power spectrum.

(b) For CL combined lensing, Sunyaev–Zeldovich, and X-ray observations of large numbers of galaxy clusters to constrain the relationship between galaxy cluster mass and observables.

(c) For SN detailed spectroscopic and photometric observations of about 500 nearby supernovae to study the variety of peak explosion magnitudes and any associated observational signatures of effects of evolution, metallicity, or reddening, as well as improvements in the system of photometric calibrations.

(d) For WL spectroscopic observations and multiband imaging of tens to hundreds of thousands of galaxies out to high red shifts and faint magnitudes in order to calibrate the photometric red shift technique and understand its limitations. It is also necessary to establish how well corrections can be made for the intrinsic shapes and alignments of galaxies, the effects of optics, (from the ground) the atmosphere, and the anisotropies in the point-spread function.

(xiii) Six types of Stage III projects have been considered. They include:

(a) A BAO survey on a 4-m-class telescope using photo-z's.
(b) A BAO survey on an 8-m-class telescope employing spectroscopy.
(c) A CL survey on a 4-m-class telescope obtaining optical photo-z's for clusters detected in ground-based SZ surveys.
(d) A SN survey on a 4-m-class telescope using photo-z's.
(e) A SN survey on a 4-m-class telescope employing spectroscopy from an 8-m-class telescope.
(f) A WL survey on a 4-m-class telescope using photo-z's. These projects are typically projected by proponents to cost in the range of tens of millions of dollars. (Cost projections were not independently checked by the DETF.)

(xiv) Our findings regarding Stage III projects are:

(a) Only an incremental increase in knowledge of dark energy parameters is likely to result from a Stage III BAO project using photo-z's. The primary benefit from a Stage III BAO photo-z project would be in exploring systematic photo-z uncertainties.
(b) A modest increase in knowledge of dark energy parameters is likely to result from a Stage III SN project using photo-z's. Such a survey would be valuable if it were to establish the viability of photometric determination of supernova red shifts, types, and evolutionary effects.
(c) A modest increase in knowledge of dark energy parameters is likely to result from any single Stage III CL, WL, spectroscopic BAO, or spectroscopic SN survey.
(d) The SN, CL, or WL techniques could, individually, produce factor-of-2 improvements in the DETF figure of merit, if the systematic errors are close to what the proponents claim.
(e) If executed in combination, Stage III projects would increase the DETF figure of merit by a factor in the range of approximately 3–5, with the large degree of uncertainty due to uncertain forecasts of systematic errors.

(xv) Four types of Stage IV projects have been considered:

(a) An optical Large Survey Telescope (LST), using one or more of the four techniques.
(b) An optical/NIR Joint Dark Energy Mission (JDEM) satellite, using one or more of the four techniques.
(c) An X-ray JDEM satellite, which would study dark energy by the cluster technique.
(d) A radio Square Kilometer Array, which could probe dark energy by WL and/or BAO techniques through a hemisphere-scale survey of 21 cm and continuum emission. The very large range of frequencies currently

demanded by the SKA specifications would likely require more than one type of antenna element. Our analysis is relevant to a lower frequency system, specifically to frequencies below 1.5 GHz. Each of these projects is projected by proponents to cost in the $0.3–1 billion range, but dark energy is not the only (in some cases not even the primary) science that would be done by these projects. (Cost projections were not independently checked by the DETF.) According to the white papers received by the Task Force, the technical capabilities needed to execute LST and JDEM are largely in hand. (The Task Force is not constituted to undertake a study of the technical issues.)

(xvi) Each of the Stage IV projects considered (LST, JDEM and SKA) offers compelling potential for advancing our knowledge of dark energy as part of a multitechnique program.

(xvii) The Stage IV experiments have different risk profiles:

(a) The SKA would likely have very low systematic errors, but it needs technical advances to reduce its costs and risk. Particularly important is the development of wide-field imaging techniques that will enable large surveys. The effectiveness of an SKA survey for dark energy would also depend on the number of galaxies it could detect, which is uncertain.
(b) An optical/NIR JDEM can mitigate systematic errors because it would likely obtain a wider spectrum of diagnostic data for SN, CL, and WL than possible from the ground, and it has no systematics associated with atmospheric influence, though it would incur the usual risks and costs of a space-based mission.
(c) LST would have higher systematic-error risk than an optical/NIR JDEM, but could in many respects match the power of JDEM if systematic errors, especially if those due to photo-z measurements, are small. An LST Stage IV program could be effective only if photo-z uncertainties on very large samples of galaxies could be made smaller than what has been achieved to date.

(xviii) A mix of techniques is essential for a fully effective Stage IV program. The technique mix may comprise elements of a ground-based program, or elements of a space-based program, or a combination of elements from ground- and space-based programs. No unique mix of techniques is optimal (aside from doing them all), but the absence of weak lensing would be the most damaging provided this technique proves as effective as projections suggest.

9. Recommendations of the DETF

(i) We strongly recommend that there be an aggressive program to explore dark energy as fully as possible, since it challenges our understanding of fundamental physical laws and the nature of the cosmos.

(ii) We recommend that the dark energy program have multiple techniques at every stage, at least one of which is a probe sensitive to the growth of cosmological structure in the form of galaxies and clusters of galaxies.

(iii) We recommend that the dark energy program include a combination of techniques from one or more Stage III projects designed to achieve, in combination, at least a factor-of-3 gain over Stage II in the DETF figure of merit, based on critical appraisals of likely statistical and systematic uncertainties.

(iv) We recommend that the dark energy program include a combination of techniques from one or more Stage IV projects designed to achieve, in combination, at least a factor-of-10 gain over Stage II in the DETF figure of merit, based on critical appraisals of likely statistical and systematic uncertainties. Because JDEM, LST, and SKA all offer promising avenues to greatly improved understanding of dark energy, we recommend continued research-and-development investments to optimize the programs and to address remaining technical questions and systematic-error risks.

(v) We recommend that high priority for near-term funding be given as well to projects that will improve our understanding of the dominant systematic effects in dark energy measurements and, wherever possible, reduce them, even if they do not immediately increase the DETF figure of merit.

(vi) We recommend that the community and the funding agencies develop a coherent program of experiments designed to meet the goals and criteria set out in these recommendations.

10. Membership of the Dark Energy Task Force

Andy Albrecht (UC Davis)
Gary Bernstein (Penn)
Bob Cahn (LBNL)
Wendy Freedman (Carnegie Inst.)
Jackie Hewitt (MIT)
Wayne Hu (Chicago)
John Huth (Harvard)
Mark Kamionkowski (Caltech)
Rocky Kolb (Fermilab, Chair)
Lloyd Knox (UC Davis)
John Mather (GSFC)
Suzanne Staggs (Princeton)
Nick Suntzeff (Texas A&M)

Agency Representatives: Kathy Turner (DOE), Dana Lehr (NSF), Michael Salamon (NASA).

Acknowledgments

I wish to thank all the members of the DETF and especially Gary Bernstein, whose presentations to various panels I have used extensively in this report.

CMB POLARIZATION: THE NEXT DECADE

B. WINSTEIN

The University of Chicago, 5640 S. Ellis Avenue,
Chicago, Illinois 60637, USA
bruce@cfcp.uchicago.edu

I review the exciting science that awaits cosmologists in precision measurements of the cosmic microwave background radiation, particularly its polarization. The conclusions of the Interagency Taskforce ("Weiss Panel") will also be presented. I conclude with an update based primarily on the new WMAP results from their three-year analysis.

Keywords: Cosmology; microwave radiation; CMB polarization.

1. The Science

Figure 1 shows the expected power spectra for both temperature and polarization anisotropies. As is well known, the polarization pattern on the sky can be split into two types, called "E" and "B." Figure 2 shows a picture of these patterns. Most of the interesting science is in the power spectra for the B polarization modes on the sky.

2. Sources of B Modes

The largest B mode signal is expected from "lensing" of the dominant E modes, as these patterns are warped by gravitational lensing in the passage through collapsing structures on the way to us. From Fig. 1, we see that these modes are a factor of about 300 below the EE spectrum and have their largest effect in the region around $l \approx 1000$. One piece of fundamental physics in this power spectrum is in its overall size: for a 1 ev change in the mean neutrino mass, this level changes by about a factor of 2. More precise studies of the lensing power spectrum can give unique information on both dark matter and dark energy.

An even more fundamental source for B modes comes from metric perturbations in the very early universe. These produce gravitational waves in the matter distribution which, after photons scatter, are manifest in B modes. For a level of T/S (the primordial ratio of tensor to scalar perturbations) of 0.001, corresponding in the

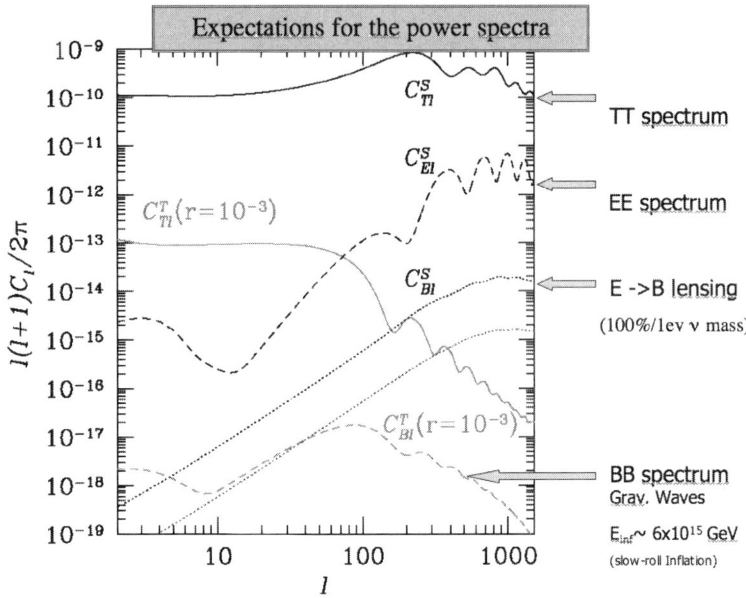

Fig. 1. The expected level of CMB power spectra (adapted from Ref. 1). Two B mode spectra are shown: the larger one from the lensing of E modes[2] by collapsing structures, and the smaller one from gravity waves produced during an inflationary era near the GUT scale. These two spectra are the targets for a host of new and planned CMB experiments.

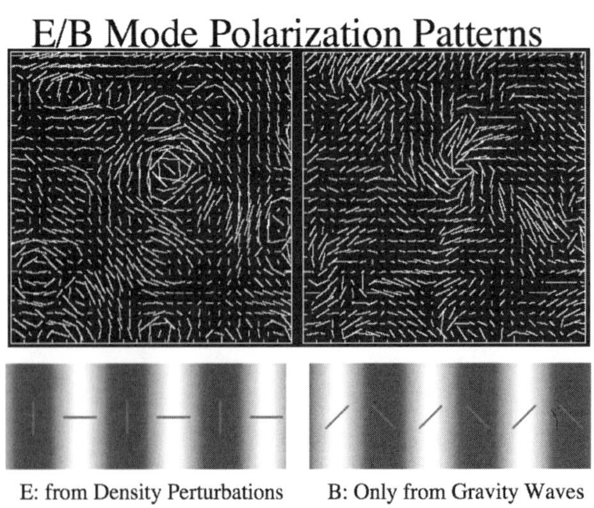

Fig. 2. Patterns for E modes (upper left) and B modes (upper right). The latter are produced from gravitational lensing or gravitational radiation. The lower figures show single modes of each. B modes have the electric field oriented 45° to the direction of the temperature anisotropy.

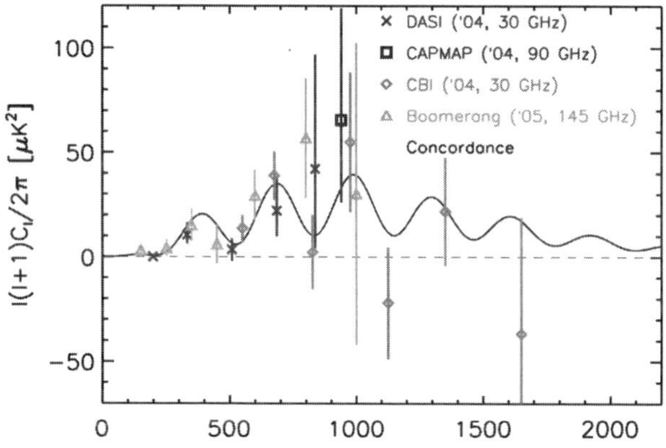

Fig. 3. The current status of all experiments reporting EE polarization detections.

simplest models of inflation to an energy scale near 10^{16} GeV (the Grand Unification Scale of particle physics), one has a possibly detectable level of B mode power.

From recent information on reionization from WMAP,[3] we now know that there are two plasmas from which photons can scatter: the "surface of last scattering" at $l \approx 1000$ gives a B mode spectrum around $l \approx 100$ while the reionized plasma shows power at the very smallest l values.

3. The EE Power Spectrum Today

The current status of EE polarization measurements is shown in Fig. 3. Four experiments (DASI,[4] Capmap,[5] CBI,[6] and Boomerang[7]) have reported positive results (WMAP is the fifth; see below) consistent with the concordance model. So far, this is a good confirmation of the model but polarization has yet to add more constraints, and sensitivities are not great enough to begin to explore the B mode physics mentioned above. A healthy program is in place which should significantly improve the scientific reach.

4. Future Experiments

Ground- and balloon-based experiments are already underway, being built, or being proposed — experiments with far greater sensitivities, enough to be able to explore B mode science. We will discuss them and then a future satellite experiment to follow up on WMAP and Planck.

Figure 4 shows the target power spectra together with the reach of two possible "ultimate" ground-based experiments. Most likely a ground-based experiment will be limited in reach to $l > 20$ but such experiments should have the sensitivity to explore values of $r = T/S < 0.01$ using the signal from the surface of last scattering. The satellite experiment will be able to detect gravity waves at this level from *both* the surface of last scattering and the reionized plasma at the lowest l values.

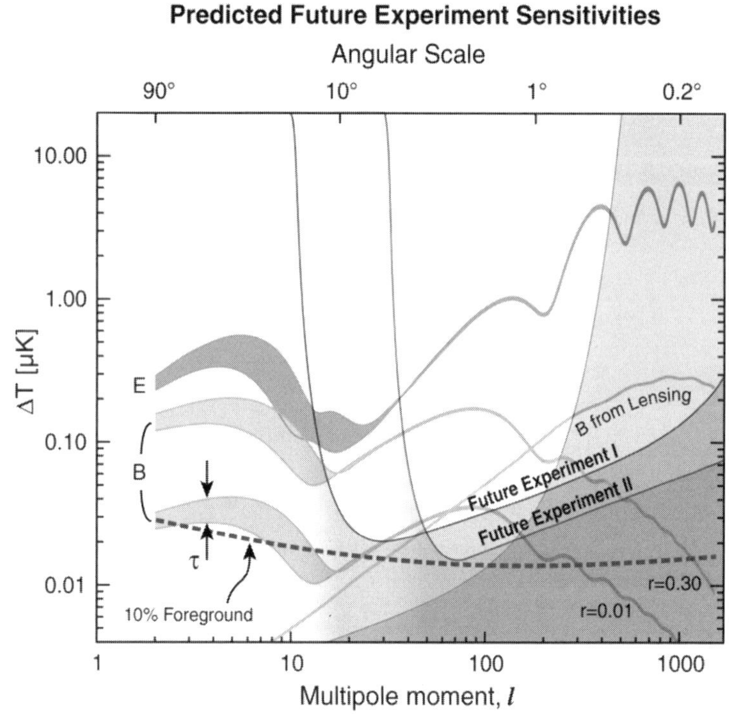

Fig. 4. The reach of two possible ground-based "future experiments" and of a satellite experiment (shaded gray in the upper right part of the figure). The regions above the respective curves show where the polarization anisotropy can be measured with good signal-to-noise. The two ground-based experiments differ in their beam sizes and number of detectors; both should have sensitivity to a gravity wave signal with $r > 0.01$. The dashed red line is the expected detection floor from contamination from galactic foregrounds. (Figure from the Interagency Task Force report on the CMB.[8])

4.1. *Technology*

We give just two examples of the technologies being explored — one an incoherent detector and the other coherent. These are expected to have similar sensitivities from the ground and different systematics. Figure 5 shows an antenna-coupled bolometer. Figure 6 shows "polarimeters on a chip" developed at JPL for the QUIET[9] experiment.

5. Ground-Based Experiments Go Deep

Before we consider the next satellite experiment, we comment that while ground-based experiments are not able to cover the full sky, they do concentrate their sensitivity (often as much as or greater than that for a satellite) into a smaller region of the sky and are therefore more sensitive to small-angular-scale science and to potential new physics requiring very deep coverage. This is illustrated in Fig. 7, which shows that the polarization signal can be seen on small scales well above the noise.

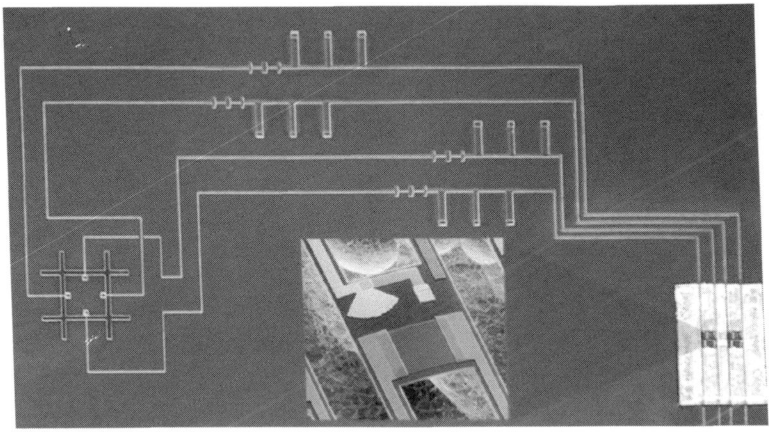

Fig. 5. A planar-antenna-coupled bolometer. A dual-polarization antenna is on the left. Each double-slot dipole coherently adds the signal from two slot dipoles to form a relatively symmetric antenna pattern. The 1 mm slots in this chip are lithographed in a superconducting Nb ground plane. Microstrip transmission lines and transmission line filters are used. The filter combination (top of photograph) includes a low pass filter (left) and a band pass filter (right). The design bandpass is centered at 220 GHz. The transmission lines terminate in the matched loads on the leg-isolated TES bolometers at the lower right. (Photograph courtesy of Adrian Lee; from the Interagency Task Force Report.[8])

Fig. 6. Left: Photograph of a prototype HEMT polarimeter with cover on and input wave guides shown. Center: The complete 90 GHz Q/U module shown with cover off. Right: A 40 GHz Q/U module shown with lid off. Coherent detectors can manipulate the amplified signals before detection, permitting the *simultaneous* detection of Stokes Q and U. (Photographs courtesy of Todd Gaier.)

6. The Interagency Task Force

The report of this panel, chaired by Rai Weiss, has been public for some time.[8] It was charged with defining a path between now and the next satellite experiment. Here we will summarize its main conclusions, although the reader is urged to look directly at the report. The recommendations were put into two groups: a science group and a technology group.

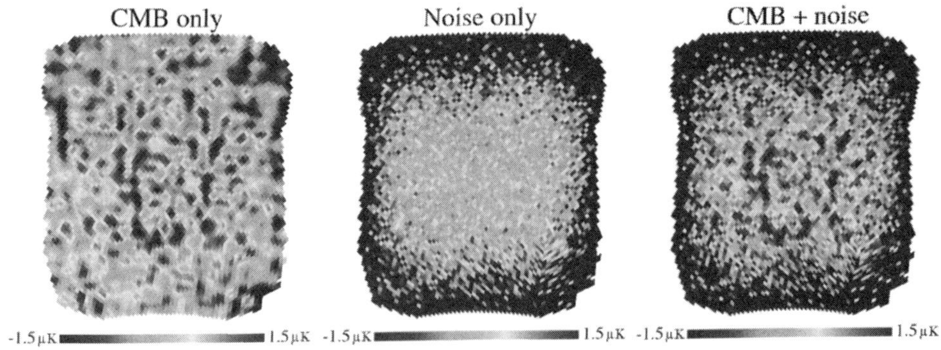

Fig. 7. Simulation for the QUIET experiment.[9] The patch size is 400 square degrees; the beam FWHM is 0.15°. Plotted is the Stokes Q parameter, which is clearly distinguishable above the noise. (Figure courtesy of H. K. Eriksen.)

6.1. *Science recommendations*

- *Recommendation S1.* Finding: A unique CMB polarization signal on large angular scales directly tests inflation and probes its energy scale.

  Recommendation: As our highest priority, we recommend a phased program to measure the large-scale CMB polarization signal expected from inflation. The primary emphasis is to test whether GUT-scale inflation occurred by measuring the signal imprinted by gravitational waves to a sensitivity limited only by our ability to remove the astrophysical foregrounds.

  The phased program begins with a strong ground- and balloon-based program that will make polarization measurements on small and medium angular scales, and culminates in a space mission for larger angular scales specifically optimized, for the first time, to measure CMB polarization to a sensitivity limited only by our ability to remove the astrophysical foreground emission. We estimate that limits at or below $r = 0.01$ can be set on the amplitude of primordial gravitational waves; to reach this level a sensitivity at least 10 times that of Planck will be required. The new mission is known as "CMBPOL" and is a candidate Beyond Einstein Inflation Probe.

- *Recommendation S2.* Finding: The CMB temperature anisotropy on small angular scales contains a wealth of additional information about inflation and the evolution of cosmic structure.

  Recommendation: We also recommend a program to measure the temperature and polarization anisotropy on small angular scales, including the signals induced by gravitational lensing and by the Sunyaev–Zeldovich effect.

- *Recommendation S3.* Finding: Foreground signals, particularly emission from our galaxy, will limit measurements of polarized fluctuations in the CMB.

  Recommendation: We recommend a systematic program to study polarized astrophysical foreground signals, especially from our galaxy.

6.2. *Technology recommendations*

- *Technology Recommendation T1.* We recommend technology development leading to receivers that contain a thousand or more polarization sensitive detectors, and adequate support for the facilities that produce these detectors.

  To meet the timeline outlined in this report there is a need to fund the development of polarization-sensitive detectors at a level of US$7 million per year for the next 5–6 years. This would roughly restore the pre-2003 level of funding for the field, which has been especially hard-hit by the shift in NASA's priorities toward exploration. It is important to keep open a variety of approaches until a clear technological winner has emerged. Nevertheless, highest priority needs to be given to the development of bolometer-based polarization-sensitive receivers.
- *Technology Recommendation T2.* We recommend a strategy that supports alternative technical approaches to detectors and instruments.

  Advances in CMB science have been based on a variety of technologies. Though we expect that bolometers will be the clear choice for CMBPOL, it is premature to shut down the development of alternatives. We recommend the continued development of HEMT-based detectors, as they might lead to an alternative space mission and will certainly be used in ground-based measurements. These relatively inexpensive enhancements would lower risk by keeping a wider set of technology channels open until an accepted best method has emerged.
- *Technology Recommendation T3.* We recommend funding for development of technology and for planning for a satellite mission to be launched in 2018.

  We recommend funding for both development of technology and planning for a satellite mission to be launched in 2018. Background(CMB)-noise-limited receivers with thousands of elements and sub-Kelvin cryogenics, required for these detectors, are part of the technical development required for the satellite mission. Another need is for modeling the mission based on improved knowledge of foreground emissions, to decide on the optimal spatial scale and frequency bands to separate the B mode signals from the polarized foreground emission and to control systematic effects. As detailed in Ref. 10, preparation for a 2011 AO and a 2018 launch requires adequate funding, starting at US$1 million in 2007 and rising to US$5 million per year in 2011, for systems planning and technology development and assessment leading to CMBPOL.
- *Technology Recommendation T4.* We recommend strong support for CMB modeling, data analysis and theory.

7. New Developments

The major new development in the last few months has been the release of the WMAP three-year data results.[3] These are shown in Fig. 8. WMAP is now reporting measurements of EE polarization in a few l bands, their lowest-l point now providing the strongest evidence for an epoch of reionization. Unlike the experiments that

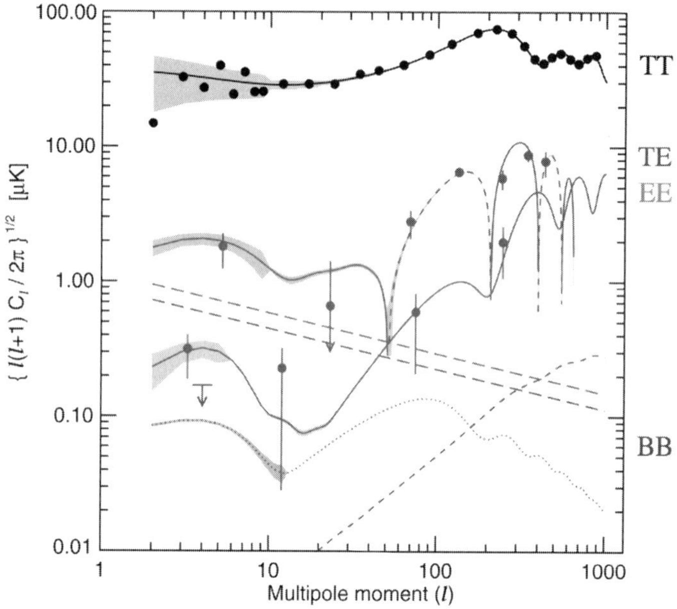

Fig. 8. WMAP three-year power spectra.

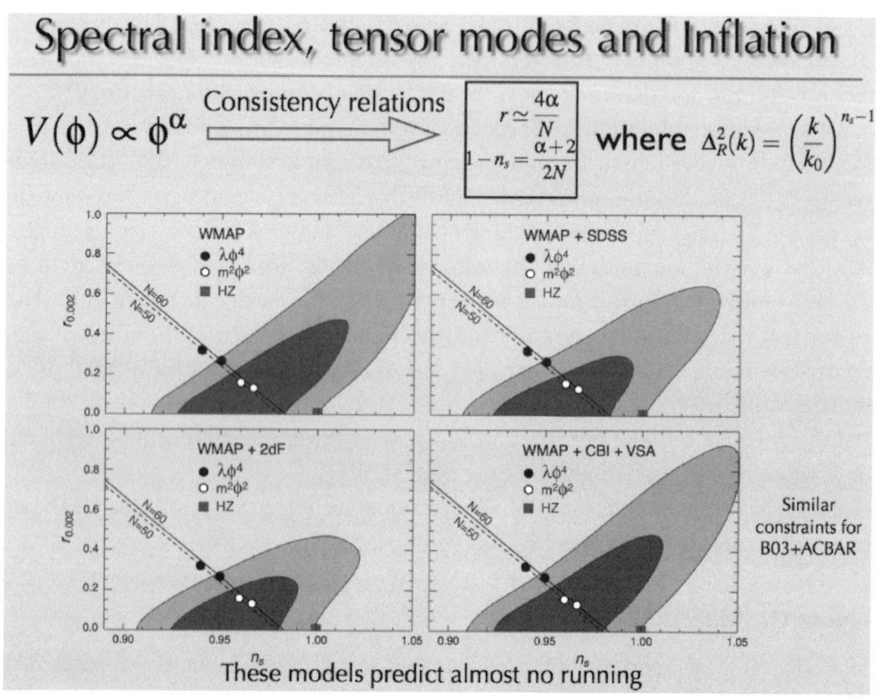

Fig. 9. Constraining the parameters of inflation. Shown are the allowed regions (at one and two standard deviations) in the space of the spectral index and tensor power. With WMAP and other data sets, these constraints are beginning to rule out the simplest models of inflation.

had already published polarization detections, WMAP was faced with significant foreground polarizations — one of the most valuable parts of the WMAP new release is the information they provide on polarized foregrounds. While the levels were not such a surprise, dealing with such an ill-characterized non-Gaussian field in the real world is messier than in simulations.

The WMAP data are providing increasing evidence for a spectral tilt to the primordial fluctuations[10] (see Fig. 9). While this finding is being debated, we can await data from the Planck satellite which should be able to confirm and extend this roughly 3σ finding.

8. Conclusions

With improving technologies, CMB experiments have come a long way in the past decade. Very important features of nature have been uncovered and there is every reason to believe that this path will continue. Foregrounds are now (finally) a serious obstacle but the importance and fundamental nature of the science assures us that this field will stay focussed and do the very best that is possible. We look forward to experiments that are in preparation or taking data and to some even more sensitive new initiatives. In particular, a new satellite experiment will be necessary to detect gravity waves from the reionized plasma.

References

1. L. Knox and Y.-S. Song, *Phys. Rev. Lett.* **89** (2002) 011303.
2. W. Hu and T. Okamoto, *Astrophys. J.* **574** (2002) 566.
3. L. Page *et al.*, astro-ph/0603450.
4. E. M. Leitch *et al.*, *Astrophys. J.* **624** (2005) 10.
5. D. Barkats *et al.*, *Astrophys. J.* **619** (2005) L127.
6. A. C. S. Readhead *et al.*, *Science* **306** (2004) 836.
7. T. E. Montroy *et al.*, astro-ph/0507514.
8. J. Bock *et al.*, astro-ph/0604101.
9. QUIET Experiment, quiet.uchicago.edu
10. D. N. Spergel *et al.*, astro-ph/0603449.

NATURAL INFLATION: STATUS AFTER WMAP THREE-YEAR DATA

KATHERINE FREESE
Michigan Center for Theoretical Physics,
Department of Physics, University of Michigan,
Ann Arbor, MI 48109, USA
ktfreese@umich.edu

WILLIAM H. KINNEY
Department of Physics, University at Buffalo,
SUNY, Buffalo, NY 14260, USA
whkinney@buffalo.edu

CHRISTOPHER SAVAGE
Michigan Center for Theoretical Physics,
Department of Physics, University of Michigan,
Ann Arbor, MI 48109, USA
cmsavage@umich.edu

Inflationary cosmology, a period of accelerated expansion in the early Universe, is being tested by cosmic microwave-background measurements. Generic predictions of inflation have been shown to be correct, and in addition individual models are being tested. The model of natural inflation is examined in light of recent three-year data from the Wilkinson Microwave Anisotropy Probe and shown to provide a good fit. The inflaton potential is naturally flat due to shift symmetries, and in the simplest version is $V(\phi) = \Lambda^4[1 \pm \cos(N\phi/f)]$. The model agrees with WMAP3 measurements as long as $f > 0.7\, m_{\rm Pl}$ (where $m_{\rm Pl} = 1.22 \times 10^{19}$ GeV) and $\Lambda \sim m_{\rm GUT}$. The running of the scalar spectral index is shown to be small — an order of magnitude below the sensitivity of WMAP3. The location of the field in the potential when perturbations on observable scales are produced is examined; for $f > 5\, m_{\rm Pl}$, the relevant part of the potential is indistinguishable from a quadratic, yet has the advantage that the required flatness is well motivated. Depending on the value of f, the model falls into the large field ($f \geq 1.5\, m_{\rm Pl}$) or small field ($f < 1.5\, m_{\rm Pl}$) classification scheme that has been applied to inflation models. Natural inflation provides a good fit to WMAP3 data.

Keywords: Inflation; cosmic microwave background; WMAP.

PACS Number(s): 98.80.Bp, 98.80.Cq

1. Introduction

Over the past five years, cosmic microwave background (CMB) data have taught us a tremendous amount about the global properties of the Universe. We have learned about the geometry of the Universe, precise measurements of the age of the Universe, the overall content of the Universe, and other properties. A "standard cosmology" with precision measurements is emerging. However, the standard Hot Big Bang has inconsistencies, many of which can be resolved by an accelerated period of expansion known as inflation.

Inflation was proposed[1] to solve several cosmological puzzles: the homogeneity, isotropy, and flatness of the Universe, as well as the lack of relic monopoles. While inflation results in an approximately homogeneous universe, inflation models also predict small inhomogeneities. Observations of inhomogeneities via the cosmic microwave background (CMB) anisotropies and structure formation are now providing tests of inflation models.

The release of three years of data from the Wilkinson Microwave Anisotropy Probe (WMAP3)[2] satellite has generated a great deal of excitement in the inflationary community. First, generic predictions of inflation match the observations: the Universe has a critical density ($\Omega = 1$), the density perturbation spectrum is nearly scale-invariant, and superhorizon fluctuations are evident. Second, current data are beginning to differentiate between inflationary models and already rule some of them out. For example, quartic potentials and generic hybrid models do not provide a good match to the data.[2–4] We illustrate here that the model known as natural inflation is an excellent match to current data.

Inflation models predict two types of perturbations, scalar and tensor, which result in density and gravitational wave fluctuations, respectively. Each is typically characterized by a fluctuation amplitude ($P_{\mathcal{R}}^{1/2}$ for scalar and $P_T^{1/2}$ for tensor, with the latter usually given in terms of the ratio $r \equiv P_T/P_{\mathcal{R}}$) and a spectral index ($n_s$ for scalar and n_{T} for tensor) describing the scale dependence of the fluctuation amplitude. As only two of these four degrees of freedom are independent parameters (as discussed below), theoretical predictions as well as data are presented in the r–n_s plane.

Most inflation models suffer from a potential drawback: to match various observational constraints, notably CMB anisotropy measurements as well as the requirement of sufficient inflation, the height of the inflaton potential must be of a much smaller scale than the width, by many orders of magnitude (i.e. the potential must be very flat). This requirement of two very different mass scales is what is known as the "fine-tuning" problem in inflation, since very precise couplings are required in the theory to prevent radiative corrections from bringing the two mass scales back to the same level. The natural inflation model (NI) uses shift symmetries to generate a flat potential, protected from radiative corrections, in a natural way.[5] In this regard, NI is one of the best-motivated inflation models.

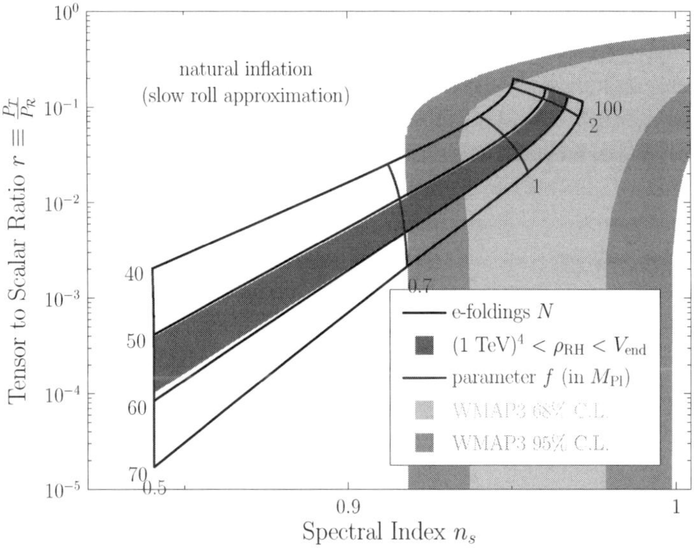

Fig. 1. Natural inflation predictions and WMAP3 constraints in the r–n_s plane. Solid/blue lines running from approximately the lower left to the upper right are predictions for constant N and varying f, where N is the number of e-foldings prior to the end of inflation at which current horizon size modes were generated, and f is the width of the potential. The remaining (dashed/red) lines are for constant f and varying N. The light blue band corresponds to the values of N for standard postinflation cosmology with (1 GeV)$^4 < \rho_{RH} < V_{end}$. Filled (nearly vertical) regions are the parameter spaces allowed by WMAP3 at 68% and 95% CL's. Natural inflation is consistent with the WMAP3 data for $f \gtrsim 0.7\, m_{Pl}$ and essentially all likely values of N.

One of our major results is shown in Fig. 1. The predictions of NI are plotted in the r–n_s plane for various parameters: the width f of the potential and the number of e-foldings N before the end of inflation at which present day horizon size fluctuations were produced. N depends upon the postinflationary Universe and is ~50–60. Also shown in the figure are the observational constraints from WMAP's recent three-year data, which provide some of the tightest constraints on inflationary models to date.[2] The primary result is that NI, for $f \gtrsim 0.7\, m_{Pl}$, is consistent with current observational constraints.[a]

We emphasize two further results as well. First, the running of the spectral index in natural inflation, i.e. the dependence of n_s on scale, is shown to be small: an order of magnitude smaller than the sensitivity of WMAP3. Second, we find how far down the potential the field is at the time structure is produced, and find that for $f > 5\, m_{Pl}$ the relevant part of the potential is indistinguishable from a quadratic potential. Still, the naturalness motivation for NI renders it a superior model to a quadratic potential, as the latter typically lacks an explanation for its flatness.

[a]We take $m_{Pl} = 1.22 \times 10^{19}$ GeV. Our result expands upon a previous analysis of NI[6] that was based upon WMAP's first year data.[7]

2. The Model of Natural Inflation

Motivation: To satisfy a combination of constraints on inflationary models (sufficient inflation and CMB measurements), the potential for the inflaton field must be very flat. For models with a single, slowly rolling field, it has been shown that the ratio of the height to the (width)^4 of the potential must satisfy[8]

$$\chi \equiv \frac{\Delta V}{(\Delta\phi)^4} \le \mathcal{O}(10^{-6} - 10^{-8}), \tag{1}$$

where ΔV is the change in the potential $V(\phi)$ and $\Delta\phi$ is the change in the field ϕ during the slowly rolling portion of the inflationary epoch. The small ratio of mass scales required by Eq. (1) is known as the "fine-tuning" problem in inflation.

Three approaches have been taken to this required flat potential characterized by a small ratio of mass scales. First, some simply say that there are many as-yet-unexplained hierarchies in physics, and inflation requires another one. The hope is that all these hierarchies will someday be explained. Second, models have been attempted where the smallness of χ is stabilized by supersymmetry. However, the required mass hierarchy, while stable, is itself unexplained. In addition, existing models have limitations.

Hence, in 1990, a third approach was proposed — natural inflation,[5] in which the inflaton potential is flat due to shift symmetries. Nambu–Goldstone bosons (NGB's) arise whenever a global symmetry is spontaneously broken. Their potential is exactly flat due to a shift symmetry under $\phi \to \phi +$ constant. As long as the shift symmetry is exact, the inflaton cannot roll and drive inflation, and hence there must be additional explicit symmetry breaking. Then these particles become pseudo-Nambu–Goldstone bosons (PNGB's), with "nearly" flat potentials, exactly as required by inflation. The small ratio of mass scales required by Eq. (1) can easily be accommodated. For example, in the case of the QCD axion, this ratio is of order 10^{-64}. While inflation clearly, requires different mass scales than the axion, the point is that the physics of PNGBs can easily accommodate the required small numbers.

The NI model was first proposed in Ref. 5. Then, in 1993, a second paper followed which provides a much more detailed study.[9] Many types of candidates have subsequently been explored for natural inflation; see e.g. Refs. 10–16. We focus here on the original version of NI, in which there is a single rolling field.

Potential: The PNGB potential resulting from explicit breaking of a shift symmetry in single field models is

$$V(\phi) = \Lambda^4 \left[1 \pm \cos\left(\frac{M\phi}{f}\right)\right]. \tag{2}$$

We will take the positive sign in Eq. (2) and $M = 1$, so the potential, of height $2\Lambda^4$, has a unique minimum at $\phi = \pi f$ (the periodicity of ϕ is $2\pi f$).

For appropriately chosen values of the mass scales, e.g. $f \sim m_{\rm Pl}$ and $\Lambda \sim m_{\rm GUT} \sim 10^{15}$ GeV, the PNGB field ϕ can drive inflation. This choice of parameters indeed produces the small ratio of scale required by Eq. (1), with $\chi \sim (\Lambda/f)^4 \sim 10^{-13}$. While $f \sim m_{\rm Pl}$ seems to be a reasonable scale for the potential width, there is no reason to believe that f cannot be much larger than $m_{\rm Pl}$. In fact, Kim, Nilles and Peloso,[17] as well as the idea of N-flation[18] showed that an *effective* potential of $f \gg m_{\rm Pl}$ can be generated from two or more axions, each with sub-Planckian scales. We shall thus include the possibility of $f \gg m_{\rm Pl}$ is our analysis and show that these parameters can fit the data.

Evolution of the inflaton field: The evolution of the inflaton field is described by

$$\ddot{\phi} + 3H\dot{\phi} + \Gamma\dot{\phi} + V'(\phi) = 0\,, \tag{3}$$

where Γ is the decay width of the inflaton. A sufficient condition for inflation is the slow roll (SR) condition $\ddot{\phi} \ll 3H\dot{\phi}$. The expansion of the scale factor a, with $H = \dot{a}/a$, is determined by the scalar-field-dominated Friedmann equation,

$$H^2 = \frac{8\pi}{3m_{\rm Pl}^2}V(\phi). \tag{4}$$

The SR condition implies that two conditions are met:

$$\epsilon(\phi) \approx \frac{m_{\rm Pl}^2}{16\pi}\left[\frac{V'(\phi)}{V(\phi)}\right]^2 = \frac{1}{16\pi}\left(\frac{m_{\rm Pl}}{f}\right)^2\left[\frac{\sin(\phi/f)}{1+\cos(\phi/f)}\right]^2 \ll 1, \tag{5}$$

$$\eta(\phi) \approx \frac{m_{\rm Pl}^2}{8\pi}\left[\frac{V''(\phi)}{V(\phi)} - \frac{1}{2}\left(\frac{V'(\phi)}{V(\phi)}\right)^2\right] = -\frac{1}{16\pi}\left(\frac{m_{\rm Pl}}{f}\right)^2 \ll 1. \tag{6}$$

Inflation ends when the field ϕ reaches a value $\phi_{\rm e}$ such that $\epsilon(\phi) < 1$ is violated, or

$$\cos\left(\frac{\phi_{\rm e}}{f}\right) = \frac{1-16\pi(f/m_{\rm Pl})^2}{1+16\pi(f/m_{\rm Pl})^2}\,. \tag{7}$$

More accurate results can be attained by numerically solving the equation of motion, (3), together with the Friedmann equations. Such calculations have been performed in Ref. 9, where it was shown the SR analysis is accurate to within a few percent for the $f \gtrsim 0.5\, m_{\rm Pl}$ parameter space we will be examining. Thus, we are justified in using the SR approximation in our calculations.

To test inflationary theories, present day observations must be related to the evolution of the inflaton field during the inflationary epoch. A comoving scale k today can be related back to a point during inflation by finding the value of N_k, the number of e-foldings before the end of inflation, at which structures on scale k were produced.[19]

Under a standard postinflation cosmology, once inflation ends, the Universe undergoes a period of reheating. Reheating can be instantaneous or last for a prolonged period of matter-dominated expansion. Then reheating ends at $T < T_{\rm RH}$, and the Universe enters its usual radiation-dominated and subsequent matter-dominated history. Instantaneous reheating ($\rho_{\rm RH} = \rho_e$) gives the minimum number of e-folds as one looks backward to the time of perturbation production, while a prolonged period of reheating gives a larger number of e-folds.

Henceforth we will use N to refer to the number of e-foldings prior to the end of inflation that correspond to the current horizon scale. Under the standard cosmology, the current horizon scale corresponds to $N \sim$50–60 (smaller N corresponds to smaller $\rho_{\rm RH}$), with a slight dependence on f. However, if one were to consider nonstandard cosmologies,[20] the range of possible N would be broader. Hence we will show results for the range $40 \leq N \leq 70$.

3. Perturbations

As the inflaton rolls down the potential, quantum fluctuations are generated which later give rise to galaxy formation and leave their imprint on the cosmic microwave background (CMB). We will examine the scalar (density) and tensor (gravitational wave) purturbations predicted by NI and compare them with the WMAP three year (WMAP3) data.[2]

3.1. *Scalar (density) fluctuations*

The perturbation amplitude for the density fluctuations (scalar modes) produced during inflation is given by[21,22]

$$P_{\mathcal{R}}^{1/2}(k) = \frac{H^2}{2\pi\dot{\phi}_k}\,. \tag{8}$$

Here, $P_{\mathcal{R}}^{1/2}(k)$ denotes the perturbation amplitude when a given wavelength re-enters the Hubble radius, and the right hand side of Eq. (8) is to be evaluated when the same comoving wavelength $(2\pi/k)$ crosses outside the horizon during inflation.

Normalizing to the COBE[23] or WMAP[2] anisotropy measurements gives $P_{\mathcal{R}}^{1/2} \sim 10^{-5}$. This normalization can be used to approximately fix the height of the potential (2) to be $\Lambda \sim 10^{15}$–10^{16} GeV for $f \sim m_{\rm Pl}$, yielding an inflaton mass $m_\phi = \Lambda/f^2 \sim 10^{11}$–$10^{13}$ GeV. Thus, a potential height Λ of the GUT scale and a potential width f of the Planck scale are required in NI in order to produce the fluctuations responsible for large scale structure. For $f \gg m_{\rm Pl}$, the potential height scales as $\Lambda \sim (10^{-3} m_{\rm Pl})\sqrt{f/m_{\rm Pl}}$.

The fluctuation amplitudes are, in general, scale-dependent. The spectrum of fluctuations is characterized by the spectral index n_s:

$$n_s - 1 \equiv \frac{d P_{\mathcal{R}}}{d\ln k} \approx -\frac{1}{8\pi}\left(\frac{m_{\rm Pl}}{f}\right)^2 \frac{3-\cos(\phi/f)}{1+\cos(\phi/f)}\,. \tag{9}$$

The spectral index for natural inflation is shown in Fig. 2. For small f, n_s is essentially independent of N, while for $f \gtrsim 2m_{\rm Pl}$, n_s has essentially no f dependence. Analytical estimates can be obtained in these two regimes:

$$n_s \approx \begin{cases} 1 - \dfrac{m_{\rm Pl}^2}{8\pi f^2}\,, & \text{for } f \lesssim \dfrac{3}{4}\, m_{\rm Pl}, \\[2ex] 1 - \dfrac{2}{N}\,, & \text{for } f \gtrsim 2\, m_{\rm Pl}\,. \end{cases} \tag{10}$$

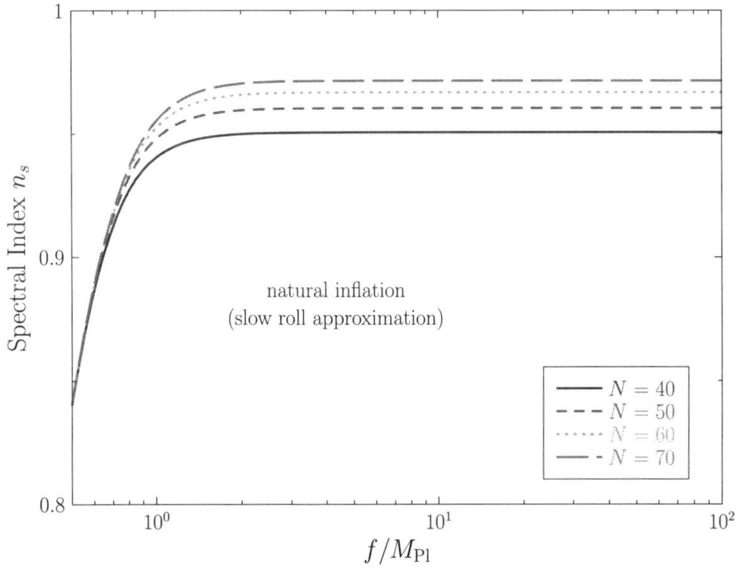

Fig. 2. The spectral index n_s is shown as a function of the potential width f for various numbers of e-foldingss N before the end of inflation.

The WMAP three-year data yield $n_s = 0.951^{+0.015}_{-0.019}$ ($n_s = 0.987^{+0.019}_{-0.037}$ when tensor modes are included in the fits) on the $k = 0.002\ \text{Mpc}^{-1}$ scale.[b] The WMAP3 results lead to the constraint on the width of the natural inflation potential, $f \gtrsim 0.7\, m_{\rm Pl}$ at 95% CL.

3.2. *Tensor (gravitational wave) fluctuations*

In addition to scalar (density) perturbations, inflation produces tensor (gravitational wave) perturbations with amplitude

$$P_T^{1/2}(k) = \frac{4H}{\sqrt{\pi} m_{\rm Pl}}\,. \tag{11}$$

Here, we examine the tensor mode predictions of natural inflation and compare them with WMAP data. Conventionally, the tensor amplitude is given in terms of the tensor/scalar ratio

$$r \equiv \frac{P_T}{P_{\mathcal{R}}} = 16\epsilon\,, \tag{12}$$

which is shown in Fig. 3 for NI. For small f, r rapidly becomes negligible, while $f \to \frac{8}{N}$ for $f \gg m_{\rm Pl}$. In all cases, $r \lesssim 0.2$, well below the WMAP limit of $r < 0.55$ (95% CL, no running).

[b]As discussed in Sec. 4, the running of the spectral index in NI is so small that the amplitude on the scale of the WMAP3 measurements is virtually identical to the amplitude on the horizon scale.

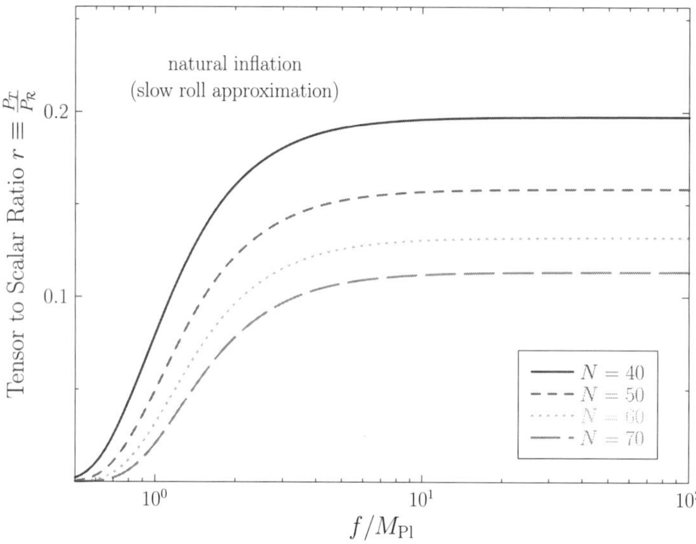

Fig. 3. The tensor to scalar ratio $r \equiv \frac{P_T}{P_\mathcal{R}}$ is shown as a function of the potential width f for various numbers of e-foldingss N before the end of inflation.

As mentioned in the introduction, in principle there are four parameters describing scalar and tensor fluctuations: the amplitude and spectra of both components, with the latter characterized by the spectral indices n_s and $n_{\rm T}$ (we are ignoring any running here). The amplitude of the scalar perturbations is normalized by the height of the potential (the energy density Λ^4). The tensor spectral index $n_{\rm T}$ is not an independent parameter since it is related to the tensor/scalar ratio r by the inflationary consistency condition $r = -8n_{\rm T}$. The remaining free parameters are the spectral index n_s of the scalar density fluctuations, and the tensor amplitude (given by r).

Hence, a useful parameter space for plotting the model predictions versus observational constraints is on the r–n_s plane.[24,25] Natural inflation generically predicts a tensor amplitude well below the detection sensitivity of current measurements such as WMAP. However, the situation will improve markedly in future experiments with greater sensitivity, such as QUIET[26] and PLANCK,[27] as well as proposed experiments, such as CMBPOL.[28]

In Fig. 1, we show the predictions of natural inflation for various choices of the number of e-folds N and the mass scale f, together with the WMAP3 observational constraints. For a given N, a fixed point is reached for $f \gg m_{\rm Pl}$, i.e. r and n_s become essentially independent of f for any $f \gtrsim 10\, m_{\rm Pl}$. This is apparent from the $f = 10\, m_{\rm Pl}$ and $f = 100\, m_{\rm Pl}$ lines in the figure, which are both shown, but are indistinguishable. As seen in the figure, $f \lesssim 0.7\, m_{\rm Pl}$ is excluded. However, $f \gtrsim 0.8\, m_{\rm Pl}$ falls well into the WMAP3-allowed region and is thus consistent with the WMAP3 data.

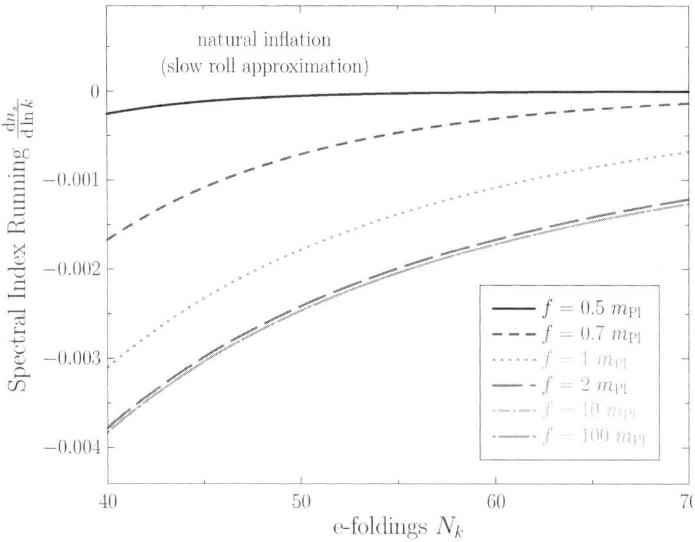

Fig. 4. The spectral index running $\frac{dn_s}{d\ln k}$ is shown as a function of the number of e-foldings N_k before the end of inflation for several values of the potential width f (note that larger N_k corresponds to smaller values of k). Results are shown using the slow-roll approximation, which is numerically inaccurate for this parameter, and may be off by up to a factor of 2–3.

4. Running of the Spectral Index

In general, n_s is not constant: its variation can be characterized by its running, $\frac{dn_s}{d\ln k}$. In this section, we will use the SR approximation, which is numerically inaccurate for this parameter, and may lead to inaccuracies of a factor of 2–3. However, our basic result, that the predicted running is small, is unaffected. As shown in Fig. 4, natural inflation predicts a small, $\mathcal{O}(10^{-3})$, negative spectral index running. This is negligibly small for WMAP sensitivities and this model is essentially indistinguishable from zero running in the WMAP analysis. While WMAP data prefer a nonzero, negative running of $\mathcal{O}(10^{-1})$ when running is included in the analysis, zero running is not excluded at 95% CL. Small scale CMB experiments such as CBI,[29] ACBAR,[30] and VSA[31] will provide more stringent tests of the running and hence of specific inflation models. If these experiments definitively detect a strong running (i.e. excluding a zero/trivial running), natural inflation in the form discussed here will be ruled out.

5. Inflaton Potential and Inflationary Model Space

In this section, we will examine the evolution of the inflaton field ϕ along the potential. We will show that the location on the potential at which the final ~60 e-foldings of inflation occur depends on the width f of the potential. We will also show that natural inflation can fall into either the "large field" or "small field" categorization defined by Ref. 24, depending again on the value of f.

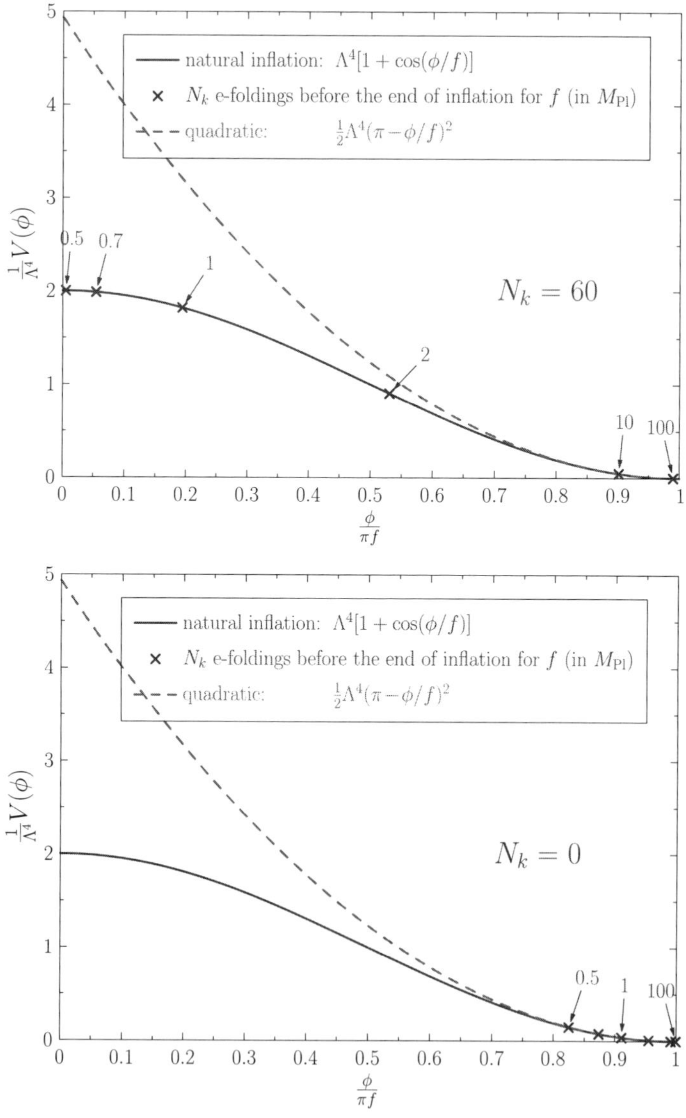

Fig. 5. The natural inflation potential is shown, along with a quadratic expansion around the potential minimum. Also shown are the positions on the potential at 60 e-foldings prior to the end of inflation (top panel) and at the end of inflation (bottom panel) for potential widths $f = (0.5, 0.7, 1, 2, 10, 100)\, m_{Pl}$. For $f \gtrsim 5\, m_{Pl}$, the relevant portion of the potential is essentially quadratic during the last 60 e-foldings of inflation.

The natural inflation potential is shown in Fig. 5. For comparison, a quadratic expansion about the minimum at $\phi = \pi f$ is also shown. Inflation occurs when the field slowly rolls down the potential and ends at the point where the field begins to move rapidly (technically, when $\epsilon \geq 1$). In the bottom panel of the figure, we show the location along the potential where inflation ends ($N_k = 0$) for various values of the potential width f. We see that inflation ends near the bottom of the potential for $f > 0.5\,m_{\rm Pl}$. In the top panel, the location along the potential is shown at $N_k = 60$ e-foldings prior to the end of inflation, the approximate time when fluctuations were produced that correspond to the current horizon — the largest scales observable in the CMB. The start of the observable portion of rolling is spread widely over the potential. For $f \lesssim 1\,m_{\rm Pl}$, current horizon modes were produced while the field was near the top of the potential. Conversely, for $f \gtrsim 3\,m_{\rm Pl}$, those modes were produced near the bottom of the potential. For $f \geq 5\,m_{\rm Pl}$, the observationally relevant portion of the potential is essentially a ϕ^2 potential; note, however, that in natural inflation this effectively power law potential is produced via a natural mechanism.

Due to the variety of inflation models, there have been attempts to classify models into a few groups. Reference 24 has proposed a scheme with three categories: small field, large field, and hybrid inflation models, which are easily distinguishable in the SR approximation by the parameters ϵ and η. To first order in SR, the categories have distinct regions in the r–n_s plane, as shown in Fig. 6. Also shown in the figure are the predictions for natural inflation; parameters are labeled as in Fig. 1 (which showed the same predictions, albeit with a logarithmic rather than linear scale). From Fig. 6, it can be seen that natural inflation does not fall into a single category, but may be either a small field or large a field, depending on the potential width f. This should not be surprising from the preceding

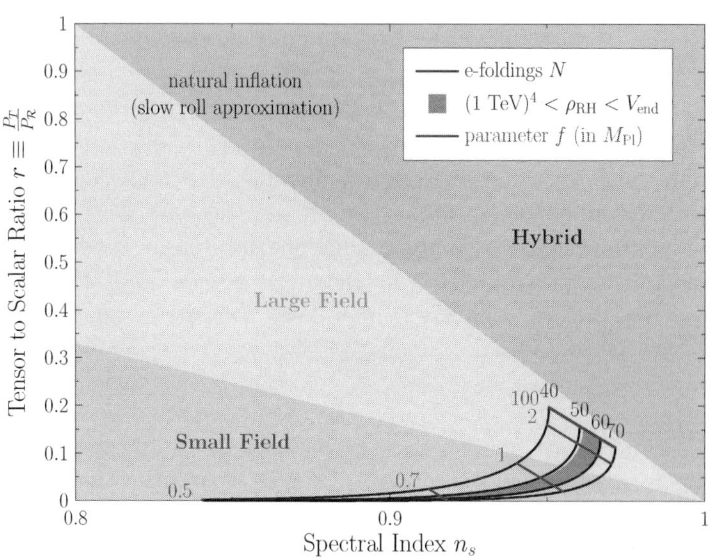

Fig. 6. Natural inflation predictions in the r–n_s plane (parameters and regions labeled as in Fig. 1), as well as the regions classifying small field, large field, and hybrid inflation models. Natural inflation falls into different classes, depending on the potential width f: for $f \lesssim 1.5\,m_{\rm Pl}$, natural inflation can be classified as a small field model, while for $f \gtrsim 1.5\,m_{\rm Pl}$, natural inflation can be classified as a large field model.

discussion of the potential. For $f \lesssim 1.5\, m_{\rm Pl}$, ϕ is on the upper part of the potential, where $V''(\phi) < 0$, at $N_k = 60$, and thus falls into the small field regime. For $f \gtrsim 1.5\, m_{\rm Pl}$, ϕ is lower down the potential, where $V''(\phi) > 0$, at $N_k = 60$, and falls into the large field regime along with power law [$V(\phi) \sim \phi^p$ for $p > 1$] models. The WMAP3 constraints shown in Fig. 1 and discussed in Sec. 3, requiring $f \gtrsim 0.7\, m_{\rm Pl}$, still allow natural inflation to fall into either of the small and large field categories.

6. Conclusion

Remarkable advances in cosmology have taken place in the past decade thanks to cosmic-microwave-background experiments. The release of the three-year data set by the Wilkinson Microwave Anisotropy Probe is leading to exciting times for inflationary cosmology. Not only are generic predictions of inflation confirmed (though there are still outstanding theoretical issues), but indeed individual inflation models are beginning to be tested.

Currently the natural inflation model, which is extremely well motivated on theoretical grounds of naturalness, is a good fit to existing data. For potential width $f > 0.7\, m_{\rm Pl}$ and height $\Lambda \sim m_{\rm GUT}$ the model is in good agreement with WMAP3 data. Natural inflation predicts very little running, an order of magnitude lower than the sensitivity of WMAP. The location of the field in the potential while perturbations on observable scales are produced was shown to depend on the width f. Even for values $f > 5\, m_{\rm Pl}$, where the relevant parts of the potential are indistinguishable from the quadratic, natural inflation provides a framework free of fine-tuning for the required potential.

There has been some confusion in the literature as to whether natural inflation should be characterized as a "small-scale" or a "large-scale" model. In Fig. 6 we demonstrated that either categorization is possible, depending on the value of f, and that both are in agreement with data.

Natural inflation makes definite predictions for tensor modes, as shown in Fig. 1. Polarization measurements in the next decade have the capability of testing these predictions and of nailing down the right type of inflationary potentials.

Acknowledgments

C. S. and K. F. acknowledge the support of the DOE and the Michigan Center for Theoretical Physics via the University of Michigan. K. F. thanks R. Easther and L. Verde for useful discussions.

References

1. A. H. Guth, *Phys. Rev. D* **23** (1981) 347.
2. D. N. Spergel *et al.*, astro-ph/0603449.

3. W. H. Kinney *et al.*, *Phys. Rev. D* **74** (2006) 023502 [astro-ph/0605338].
4. L. Alabidi and D. H. Lyth, astro-ph/0603539.
5. K. Freese, J. A. Frieman and A. V. Olinto, *Phys. Rev. Lett.* **65** (1990) 3233.
6. K. Freese and W. H. Kinney, *Phys. Rev. D* **70** (2004) 083512 [hep-ph/0404012].
7. WMAP Collab. (D. N. Spergel *et al.*), *Astrophys. J. Suppl.* **148** (2003) 175 [astro-ph/0302209].
8. F. C. Adams, K. Freese and A. H. Guth, *Phys. Rev. D* **43** (1991) 965.
9. F. C. Adams *et al.*, *Phys. Rev. D* **47** (1993) 426 [hep-ph/9207245].
10. M. Kawasaki, M. Yamaguchi and T. Yanagida, *Phys. Rev. Lett.* **85** (2000) 3572 [hep-ph/0004243].
11. N. Arkani-Hamed *et al.*, *Phys. Rev. Lett.* **90** (2003) 221302 [hep-th/0301218].
12. N. Arkani-Hamed *et al.*, *J. Cosmol. Astropart. Phys.* **0307** (2003) 003 [hep-th/0302034].
13. D. E. Kaplan and N. J. Weiner, *J. Cosmol. Astropart. Phys.* **0402** (2004) 005 [hep-ph/0302014].
14. H. Firouzjahi and S. H. H. Tye, *Phys. Lett. B* **584** (2004) 147 [hep-th/0312020].
15. J. P. Hsu and R. Kallosh, *J. High Energy Phys.* **0404** (2004) 042 [hep-th/0402047].
16. K. Freese, *Phys. Rev. D* **50** (1994) 7731 [astro-ph/9405045].
17. J. E. Kim, H. P. Nilles and M. Peloso, *J. Cosmol. Astropart. Phys.* **0501** (2005) 005 [hep-ph/0409138].
18. S. Dimopoulos *et al.*, hep-th/0507205.
19. J. E. Lidsey *et al.*, *Rev. Mod. Phys.* **69** (1997) 373 [astro-ph/9508078].
20. A.-R. Liddle and S. M. Leach, *Phys. Rev. D* **68** (2003) 103503 [astro-ph/0305263].
21. V. F. Mukhanov, H. A. Feldman and R. H. Brandenberger, *Phys. Rep.* **215** (1992) 203.
22. E. D. Stewart and D. H. Lyth, *Phys. Lett. B* **302** (1993) 171 [gr-qc/9302019].
23. G. F. Smoot *et al.*, *Astrophys. J.* **396** (1992) L1.
24. S. Dodelson, W. H. Kinney and E. W. Kolb, *Phys. Rev. D* **56** (1997) 3207 [astro-ph/9702166].
25. W. H. Kinney, *Phys. Rev. D* **58** (1998) 123506 [astro-ph/9806259].
26. B. Winstein, *2nd Irvine Cosmology Conference* (2006).
27. Planck Collaboration, astro-ph/0604069.
28. J. Bock *et al.*, astro-ph/0604101.
29. B. S. Mason *et al.*, *Astrophys. J.* **591** (2003) 540 [astro-ph/0205384].
30. ACBAR Collab. (C. l. Kuo *et al.*), *Astrophys. J.* **600** (2004) 32 [astro-ph/0212289].
31. C. Dickinson *et al.*, *Mon. Not. R. Astron. Soc.* **353** (2004) 732 [astro-ph/0402498].

IS DARK ENERGY ABNORMALLY WEIGHTING?

JEAN-MICHEL ALIMI* and ANDRÉ FÜZFA†

Laboratory Universe and Theories, CNRS UMR-8102, Observatoire de Paris, Université Denis-Diderot, Paris 7, 92190 Meudon, France

**jean-michel.alimi@obspm.fr*

†andre.fuzfa@obspm.fr

We investigate the possibility that dark energy does not couple to gravitation in the same way as ordinary matter, yielding a violation of the weak and strong equivalence principles on cosmological scales. We build a transient mechanism in which gravitation is pushed away from general relativity (GR) by a Born–Infeld (BI) gauge interaction acting as an "abnormally weighting (dark) energy" (AWE). This mechanism accounts for the Hubble diagram of far-away supernovae by cosmic acceleration and time variation of the gravitational constant while accounting naturally for the present tests on GR.

Keywords: Cosmology; accelerating universe; dark energy; Born–Infeld; equivalence principle.

1. Motivations

In recent years, there have been increasing evidences in favor of an unexpected energy component — often called dark energy — which dominates the present universe and affects the recent cosmic expansion. The first evidence came from the Hubble diagrams of type Ia supernovae (see Ref. 1 and references therein) and was thereafter confirmed by the measurements of cosmic microwave background (CMB) anisotropies, the large-scale distribution of galaxies, and indirectly by the observed properties of galaxy clusters and peculiar velocities. If the existence of dark energy seems unavoidable in the light of these various observational evidences, a definitive interpretation of dark energy has still to be provided.

We propose a completely new interpretation of dark energy that does not require violation of the energy condition. We assume that "dark energy" violates the weak equivalence principle on large scales, i.e. it does not couple to gravitation as ordinary matter and weights abnormally.

2. Cosmomolgy with AWE

We consider that the energy content of the universe is divided into three parts: a gravitational sector described by pure spin 2 (graviton) and spin 0 (dilaton) degrees of freedom, a matter sector containing the usual fluids of cosmology (baryons, photons, dark matter, neutrinos, etc.) and an "abnormally weighting energy" (AWE) sector, here composed of an $SU(2)$-valued gauge interaction ruled by BI type gauge dynamics.[2,3]

$$S = \frac{1}{2\kappa}\int \sqrt{-g}d^4x\{R - 2g^{\mu\nu}\partial_\mu\varphi\partial_\nu\varphi\} + S_{\rm M}[\psi_{\rm M}, A_{\rm M}^2(\varphi)g_{\mu\nu}] + S_{\rm BI}[A_\mu, A_{\rm BI}^2(\varphi)g_{\mu\nu}]. \quad (1)$$

If the consideration of a scalar sector of gravitation breaks down the strong equivalence principle (SEP), the nonuniversality of the coupling to the metric $g_{\mu\nu}$ $[A_{\rm BI}(\varphi) \neq A_M(\varphi)]$ violates the weak equivalence principle (WEP). The action (1) is written in the so-called "Einstein frame," where the metric components are measured by using purely gravitational rods and clocks. We define the "Dicke–Jordan" observable frame by the conformal transformation $\tilde{g}_{\mu\nu} = A_{\rm M}^2(\varphi)g_{\mu\nu}$ using the coupling function to ordinary matter. In this frame, the metric $\tilde{g}_{\mu\nu}$ couples universally to ordinary matter and is measured by clocks and rods made of ordinary matter (and not built upon the new gauge interaction we introduced as the AWE sector).

The BI character of the $SU(2)$-valued gauge interaction playing the role of AWE is given by the Lagrangian $\mathcal{L}_{\rm BI} = \epsilon_c(\mathcal{R} - 1)$, where

$$\mathcal{R} = \sqrt{1 + A_{\rm BI}^{-4}(\varphi)/(2\epsilon_c)F_{\mu\nu}F^{\mu\nu} - A_{\rm BI}^{-8}(\varphi)/(16\epsilon_c^2)(F_{\mu\nu}\tilde{F}^{\mu\nu})^2}$$

(see Ref. 3 and references therein). In this expression, ϵ_c is the BI critical energy and $A_{\rm BI}(\varphi)$ is the dilaton coupling function to the gauge field. The BI critical energy ϵ_c defines the scale above which nonlinear effects due to the term in $(F_{\mu\nu}\tilde{F}^{\mu\nu})^2$ become important in gauge dynamics, therefore breaking the scale invariance of the gauge fields. $SU(2)$-valued gauge fields ruled by the previous Lagrangian obey the following equation of state in a cosmological context[2,4]:

$$\omega_{\rm BI} = \frac{P_{\rm BI}}{\rho_{\rm BI}} = \frac{1}{3}\left(\frac{\epsilon_c - A_{\rm BI}^{-4}(\varphi)\rho_{\rm BI}}{\epsilon_c - A_{\rm BI}^{-4}(\varphi)\rho_{\rm BI}}\right), \quad (2)$$

where $P_{\rm BI}$ is the gauge pressure in the Einstein frame and the gauge energy density is

$$\rho_{\rm BI} = \epsilon_c A_{\rm BI}^4(\varphi)\left(\sqrt{1 + \mathcal{C}/(A_{\rm BI}^4(\varphi)a^4)} - 1\right), \quad (3)$$

where C is some integration constant.

When the condition $A_{\rm BI}^{-4} \gg \epsilon_c$ occurs, $\omega_{\rm BI} \simeq -1/3$ (Nambu–Goto string gas) and the related gauge field energy density scales as $(A_{\rm BI}(\varphi)a)^{-2}$, while the field strength is frozen. In the low-energy regime $A_{\rm BI}^{-4} \ll \epsilon_c$, the fluid becomes relativistic, $\omega_{\rm BI} \simeq 1/3$ (radiations) and the density scales as $(A_{\rm BI}(\varphi)a)^{-4}$. The transition between these two regimes occurs smoothly.

Let us now write down the FLRW homogeneous and isotropic cosmological equation induced by the action (1). The Friedmann equation (or Hamiltonian constraint) is

$$\left(\frac{\dot{a}}{a}\right)^2 = \frac{\dot{\varphi}^2}{3} + \frac{\kappa}{3}[\rho_{\mathrm{BI}} + \rho_{\mathrm{M}}], \tag{4}$$

and the acceleration equation written down in the Einstein frame is

$$\frac{\ddot{a}}{a} = -\frac{2}{3}\dot{\varphi}^2 - \frac{\kappa}{6}[(\rho_{\mathrm{BI}} + 3P_{\mathrm{BI}}) + (\rho_{\mathrm{M}} + 3P_{\mathrm{M}})]. \tag{5}$$

There cannot be any cosmic acceleration in terms of the metric $g_{\mu\nu}$ (the dilaton φ has been considered massless), because the highest value of $\ddot{a}$ that can be achieved in this frame is identically zero in the limit of the pure Einstein–Born–Infeld system at high energies. The BI gauge interaction therefore never violates the strong energy condition. The dark energy effects will occur only in the observable frame $\tilde{g}_{\mu\nu}$.

The scalar gravitational dynamics is given by the Klein–Gordon equation:

$$\ddot{\varphi} + 3\frac{\dot{a}}{a}\dot{\varphi} + \frac{\kappa}{2}\alpha_{\mathrm{BI}}(\varphi)(\rho_{\mathrm{BI}} - 3P_{\mathrm{BI}}) + \frac{\kappa}{2}\alpha_{\mathrm{M}}(\varphi)(\rho_{M} - 3P_{\mathrm{M}}) = 0, \tag{6}$$

where $\alpha_i(\varphi) = d\ln A_i(\varphi)/d\varphi$.

The violation of the WEP by the different coupling of the BI gauge interaction to the dilaton ($\alpha_{\mathrm{BI}} \neq \alpha_{\mathrm{M}}$) implies that the history of the universe can be seen as a competition between ordinary matter and AWE, particularly if the former attracts the field toward values corresponding to GR (here φ and its derivative vanish) while the latter repulses away from it. Furthermore, as the dark energy sector is constituted by a BI gauge interaction, this competition is temporary because of the equation of state [see Eq. (2)]. In the strong-field limit ($\omega_{\mathrm{BI}} \simeq -1/3$), the negative pressure first allows a late domination of AWE. Then, the phase transition to low energies and radiation behavior ensure a subdominance of AWE and a transient character of the dark energy mechanism. Furthermore, the radiative phase of the BI gauge interaction ($\omega_{\mathrm{BI}} \simeq 1/3$) decouples it from the scalar sector [see Eq. (6)] and once in this regime we retrieve the usual tensor–scalar theory of gravitation. A well-known and remarkable feature of ST theory is that they present a natural attraction toward GR during the matter-dominated era, which is ensured when the coupling function $\alpha_{\mathrm{M}}(\varphi)$ has for example a global minimum (see Ref. 5 and references therein). In order to introduce a competition between attraction by ordinary matter and repulsion by AWE in Eq. (6), it suffices to assume the usual coupling functions $A_{\mathrm{BI}}(\varphi) = \exp(k_{\mathrm{BI}}\varphi)$ and $A_{\mathrm{M}}(\varphi) = \exp\left(k_m \frac{\varphi^2}{2}\right)$.

The deviation from GR might occur when the dilaton φ is pushed away from the minimum of the matter coupling function $A_{\mathrm{M}}(\varphi)$, which actually corresponds to GR, by the dominance of the BI interaction in dilaton dynamics. This deviation from GR is achieved in Eq. (6) when $\alpha_{\mathrm{BI}}(\varphi)(\rho_{\mathrm{BI}} - 3P_{\mathrm{BI}}) \gg \alpha_{\mathrm{M}}(\varphi)\rho_{\mathrm{M}}$, where we have assumed that the dominant matter is dust ($P_{\mathrm{M}} \simeq 0$). The deviation from GR during the process is ensured by a constant drag term, $\alpha_{\mathrm{BI}} = k_{\mathrm{BI}}$, while the late convergence to GR at the end of the mechanism is ensured by the efficiency of the attraction mechanism associated with the coupling function $\alpha_{\rm M} = k_{\rm M}\varphi$. The convergence occurs when the force term due to ordinary matter in Eq. (6) dominates, which happens either when matter dominates or when the BI gauge interaction behaves like radiation (Yang–Mills dynamics, $\omega_{\rm BI} = 1/3$).

3. Observational Confirmations

Let us illustrate now the plausibility of this mechanism, where dark AWE never violates the strong energy condition $P < -\rho/3$ in the Einstein frame, by trying to reproduce a Hubble diagram built upon recent available data on far-away type Ia supernovae (SNLS data) (see Fig. 1). Within the framework of tensor–scalar gravity, the dimmed magnitude of such objects could be explained both by an acceleration of cosmic expansion and a time variation of the gravitational constant (see Figs. 2 and 3). In this sense, type Ia supernovae are no more standard candles for cosmology. There has been proposed in the literature the following toy model for the modulus distance versus red shift relation of type Ia supernovae:

$$\mu(\tilde{z}) = m - M = 5\log_{10} d_L(\tilde{z}) + \frac{15}{4}\log_{10}\frac{G_{\rm eff}(\tilde{z})}{G_0}. \qquad (7)$$

In the definition of the modulus distance $G_{\rm eff}$ is the effective gravitational "constant" at the epoch z and G_0 is the (bare) value of this constant today, where gravitation is well described by GR. In tensor–scalar theory, the effective

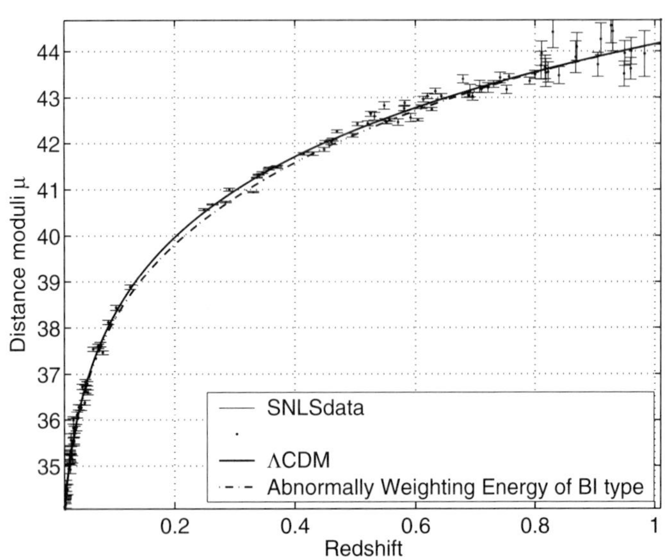

Fig. 1. Hubble diagram of the SNLS first year data set with the best fit ΛCDM flat model [solid line, $\Omega_m(a_0) = 0.26$, $\bar{\chi}^2/\text{dof} = 1.03$] and the AWE model (dashed–dotted line, $\bar{\chi}^2/\text{dof} = 1.09$) ($H_0 = 70\,\text{km/s/Mpc}$).

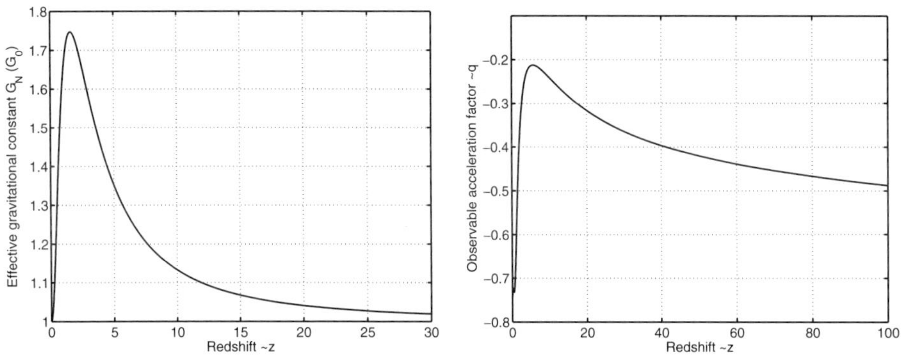

Fig. 2. *Left*: Cosmological evolution of G_N (in units of bare G). *Right*: Evolution of the acceleration factor $\tilde{q} = 1 + (d\tilde{H}/d\tilde{t})/\tilde{H}^2$ with the scale factor $\tilde{a}$ in the observable frame.

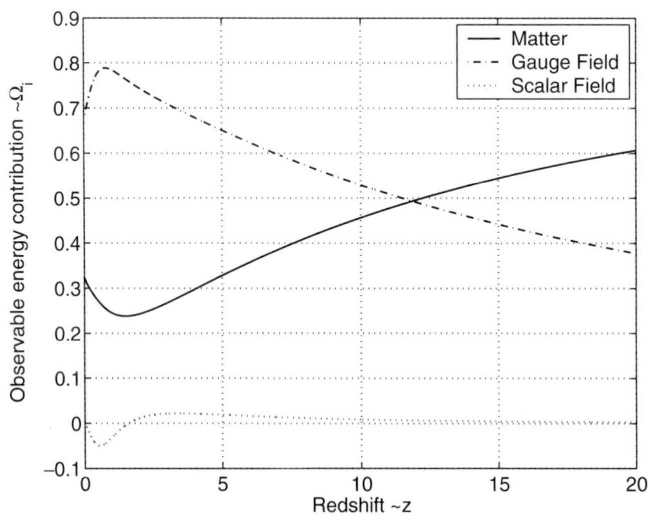

Fig. 3. Evolution of different energy contribution Ω_i in the observable frame.

gravitational constant for experiments with compact objects is given by $G_N = G_0 A_m^2(\varphi)(1 + \alpha_m^2(\varphi))$. In addition to accounting for modulus distance data, any dark energy mechanism based on the tensor–scalar theory of gravitation should be in agreement with the present tests of GR. The usual tests include post-Newtonian dynamics of compact bodies like spacecraft, planets and binary systems, which are obviously made of ordinary matter. The constraints on post-Newtonian parameters are given by

$$|\gamma - 1| = 2\frac{\alpha_m^2(\varphi)}{1 + \alpha_m^2(\varphi)} < 2 \times 10^{-5}, \tag{8}$$

$$|\beta - 1| = \left|\frac{d\alpha_m}{d\varphi}\frac{\alpha_m^2(\varphi)}{2(1 + \alpha_m^2(\varphi))^2}\right| < 6 \times 10^{-4}. \tag{9}$$

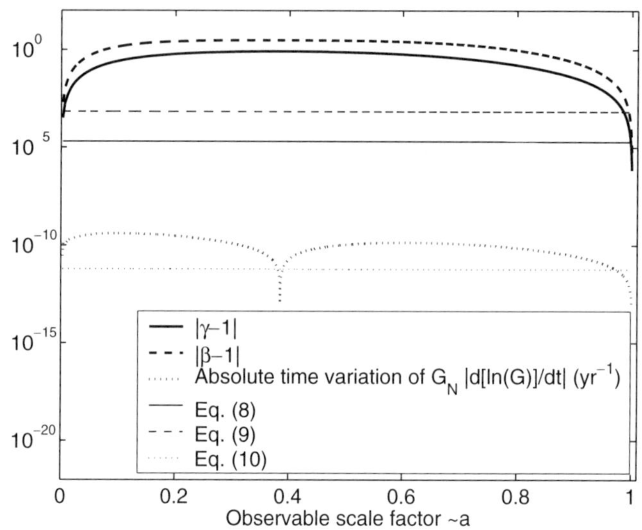

Fig. 4. Evolution of post-Newtonian parameters with the scale factor $|\gamma - 1|$ (solid line), $|\beta - 1|$ (dashed line), $|\dot{G}/G|$ (dots). Current observable constraints are indicated by the horizontal lines.

Another constraint is the time variation of the gravitational constant:

$$\left|\frac{\dot{G}}{G}\right| = \left|2\dot{\varphi}\alpha_m(\varphi)\frac{1+\dfrac{d\alpha_m}{d\varphi}}{1+\alpha_m^2(\varphi)}\right| < 6\times 10^{-12}\,\mathrm{yr}^{-1}. \tag{10}$$

To these constraints on the violation of the strong equivalence principle, one should add the constraints on the weak equivalence principle, which is tested at the 10^{-12} level by the universality of free fall of inertial masses with different compositions. Although the BI gauge interaction acting as AWE violates this universality of free fall, we might consider that this effect is extremely weak (and not observed in practice) provided the dark energy density (AWE sector) at our scale is of the order of its cosmological value. This is true if the BI AWE does not cluster too much at our scale, an assumption that should be verified in forthcoming works. Therefore, we will only consider all the previous constraints (see Fig. 4) while discarding the effects on the universality of free fall for the moment.

References

1. P. Astier *et al.*, *Astron. Astrophys.* **447** (2006) 31.
2. A. Füzfa and J.-M. Alimi, *Phys. Rev. D* **73** (2006) 023520.
3. V. V. Dyadichev *et al.*, *Phys. Rev. D* **65** (2002) 084007.
4. A. Füzfa, and J.-M. Alimi, *Phys. Rev. Lett.* **97** (2006) 061301 [astro-ph/0604517].
5. A. Serna, J.-M. Alimi and A. Navarro, *Class. Quant. Grav.* **19** (2002) 857.

COHERENT ACCELERATION OF MATERIAL WAVE PACKETS

FARHAN SAIF

Department of Electronics, Quaid-i-Azam University, Islamabad 45320, Pakistan
saif@fulbrightweb.org

PIERRE MEYSTRE

Department of Physics, University of Arizona, Tucson, AZ 85721, USA

We study the quantum dynamics of a material wave packet bouncing off a modulated atomic mirror in the presence of a gravitational field. We find the occurrence of coherent accelerated dynamics for atoms. The acceleration takes place for certain initial phase space data and within specific windows of modulation strengths. The realization of the proposed acceleration scheme is within the range of present day experimental possibilities.

Keywords: Matter waves; acceleration; coherence.

1. Introduction

Accelerating particles using oscillating potentials is an area of extensive research,[1,2] first triggered by the ideas of Fermi on the origin of cosmic rays. In his seminal paper "On the origin of cosmic rays," he stated that "cosmic rays are originated and accelerated primarily in the interstellar space of the galaxy by *collisions* against moving magnetic fields."[3] This understanding led to the development of two major models: The Fermi–Ulam accelerator, which deals with the bouncing of a particle off an oscillating surface in the presence of another fixed surface parallel to it; and the Fermi–Pustyl'nikov accelerator, where the particle bounces off an oscillating surface in the presence of gravity. In the case of the Fermi–Ulam accelerator,[4–6] it was shown that the energy of the particle remains bounded and the unlimited acceleration proposed by Fermi is absent.[7–11] In the Fermi–Pustyl'nikov accelerator, by contrast, there exists a set of initial data within specific domains of phase space that result in trajectories speeding up to infinity.

In recent years the acceleration of laser-cooled atoms has become a topic of great interest for applications such as atom interferometry and the development of matter-wave-based inertial sensors. Possible schemes of matter-wave acceleration have been proposed and studied. For example, a Bose–Einstein condensate in a frequency-chirped optical lattice[12] and an atom in an amplitude-modulated optical lattice in the presence of a gravitational field display acceleration.[13,14] The δ-kicked accelerator in the latter case operates for certain sets of initial data that originate in stable islands of phase space.

Here, we discuss an experimentally realizable technique to accelerate a material wave packet in a coherent fashion. It consists of an atom optics version of the Fermi–Pustyl'nikov accelerator[15,16] where a cloud of ultracold atoms falling in a gravitational field bounces off a spatially modulated atomic mirror. This scheme is different from previous accelerator schemes in the following ways: (i) the regions of phase space that support acceleration are located in the mixed phase space rather than in the islands of stability (or nonlinear resonances); (ii) the acceleration of the wave packet is coherent; (iii) it occurs only for certain windows of oscillation strengths.

2. The Model

We consider a cloud of laser-cooled atoms that move along the vertical $\tilde{z}$ direction under the influence of gravity and bounce back off an atomic mirror.[17] This mirror is formed by a laser beam incident on a glass prism and undergoing total internal reflection, thereby creating an optical evanescent wave of intensity $I(\tilde{z}) = I_0 \exp(-2k\tilde{z})$ and characteristic decay length k^{-1} outside of the prism.

The laser intensity is modulated by an acousto-optic modulator as[18] $I(\tilde{z}, \tilde{t}) = I_0 \exp(-2k\tilde{z} + \epsilon \sin \omega\tilde{t})$, where ω is the frequency and ϵ the amplitude of modulation. The laser frequency is tuned far from any atomic transition, so that there is no significant upper-state atomic population. The excited atomic level(s) can then be adiabatically eliminated, and the atoms behave for all practical purposes as scalar particles of mass m whose center-of-mass motion is governed by the one-dimensional Hamiltonian

$$\tilde{H} = \frac{\tilde{p}^2}{2m} + mg\tilde{z} + \frac{\hbar\Omega_{\text{eff}}}{4} e^{-2k\tilde{z}+\epsilon \sin \omega\tilde{t}}, \tag{1}$$

where $\tilde{p}$ is the atomic momentum along $\tilde{z}$ and g is the acceleration of gravity.

We proceed by introducing the dimensionless position and momentum coordinates $z \equiv \tilde{z}\omega^2/g$ and $p \equiv \tilde{p}\omega/mg$, the scaled time $t \equiv \omega\tilde{t}$, the dimensionless intensity $V_0 \equiv \hbar\omega^2\Omega_{\text{eff}}/4\,mg^2$, the steepness $\kappa \equiv 2kg/\omega^2$, and the modulation strength $\lambda \equiv \omega^2\epsilon/2\,kg$ of the evanescent wave field.

When extended to an ensemble of noninteracting particles, the classical dynamics obeys the condition of incompressibility of the flow,[4,5] and the phase space distribution function $P(z, p, t)$ satisfies the Liouville equation. In the absence of mirror modulation, the atomic dynamics is integrable. For very weak modulations the incommensurate motion almost follows the integrable evolution and remains rigorously stable, as prescribed by the KAM theorem. As the modulation increases, though, the classical system becomes chaotic.

In the quantum regime, the atomic evolution obeys the corresponding Schrödinger equation. The commutation relation, $[z, p] = i(\omega^3/mg^2)\hbar \equiv i\bar{k}$, naturally leads to the introduction of the dimensionless Planck constant, $\bar{k} \equiv \hbar\omega^3/mg^2$. It can easily be varied by changing for instance ω, thereby permitting one to study the transition from the semiclassical to the purely quantum dynamics of the atoms.

3. Accelerated Dynamics

The classical version of the system is characterized by the existence of a set of initial conditions resulting in trajectories that accelerate without bound.[19,20] More precisely, the classical evolution of the Fermi accelerator displays the onset of global diffusion above a critical modulation strength $\lambda_l = 0.24$,[6] while the quantum evolution remains localized until a larger value λ_u of the modulation.[15,16,21,22] Above that point both the classical and the quantum dynamics are diffusive. However, for specific sets of initial conditions that lie within phase space disks of radius ρ, accelerating modes appear for values of the modulation strength λ within the windows[19,20]

$$s\pi \leq \lambda < \sqrt{1 + (s\pi)^2}, \tag{2}$$

where s can take integer and half-integer values for the sinusoidal modulation of the reflecting surface considered here.

We found numerically that for a modulation strength outside the windows of Eq. (2) the dynamics is dominantly diffusive. However, as the fundamental requirement for the acceleration is met by choosing a modulation strength within the windows, the ensemble displays a nondispersive and coherent acceleration; see Fig. 1. A small diffusive background results from a small part of the initial distribution which is residing outside the area of the phase space supporting acceleration. This coherent acceleration restricts the momentum space variance Δp, which then remains very small, indicating the absence of diffusive dynamics.[23]

In the quantum case the Heisenberg uncertainty principle imposes a limit on the smallest size of the initial wave packet. Thus, in order to form an initial wave packet that resides entirely within regions of phase space leading to coherent dispersionless acceleration, an appropriate value of the effective Planck constant must be chosen — for example, by controlling the frequency ω.[24] For a broad wave packet, the coherent acceleration manifests itself as regular spikes in the marginal probability distributions $P(p, t) = \int dx P(x, p, t)$ and $P(x, t) = \int dp P(x, p, t)$.

This is illustrated in Fig. 2, which shows the marginal probability distribution $P(p, t)$ for (a) $\lambda = 1.7$ and (b) $\lambda = 2.4$, in both the classical and the quantum domains. In this example, the initial area of the particle phase space distribution is taken to be large compared to the size of the phase space regions leading to purely unbounded dispersionless acceleration. The sharp spikes in $P(p, t)$ appear when the modulation strength satisfies the condition of Eq. (2), and gradually disappear as it exits these windows. These spikes are therefore a signature of the coherent accelerated dynamics. In contrast, the portions of the initial probability

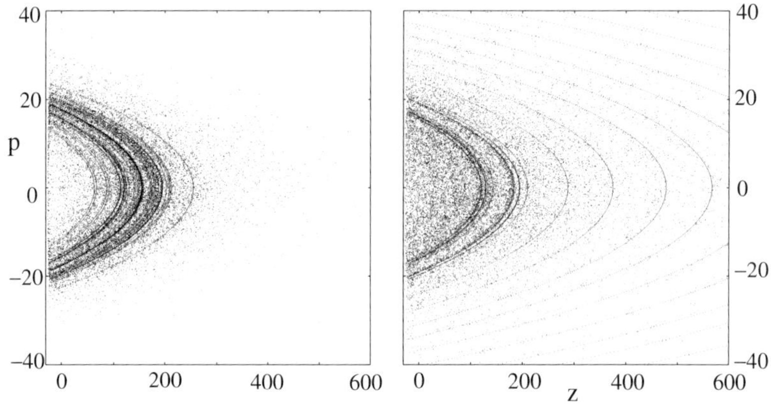

Fig. 1. Phase space evolution of a classical ensemble of particles initially in a narrowly peaked Gaussian distribution originating from the area of phase space that supports accelerated trajectories. The initial distribution, centered at $\bar{z} = 0$ and $\bar{p} = 2\pi^2$ with $\Delta p(0) = \Delta z(0) = 0.1$, is propagated for $\lambda = 1$ (*left*) and $\lambda = 1.7$ (*right*) for time $t = 1000$. The numerical calculations correspond to cesium atoms of mass $m = 2.2 \times 10^{-25}$ kg bouncing off an atomic mirror with an intensity modulation of $\epsilon = 0.55$. The modulation frequencies extend to the megahertz range, and $\kappa^{-1} = 0.55\,\mu$m.

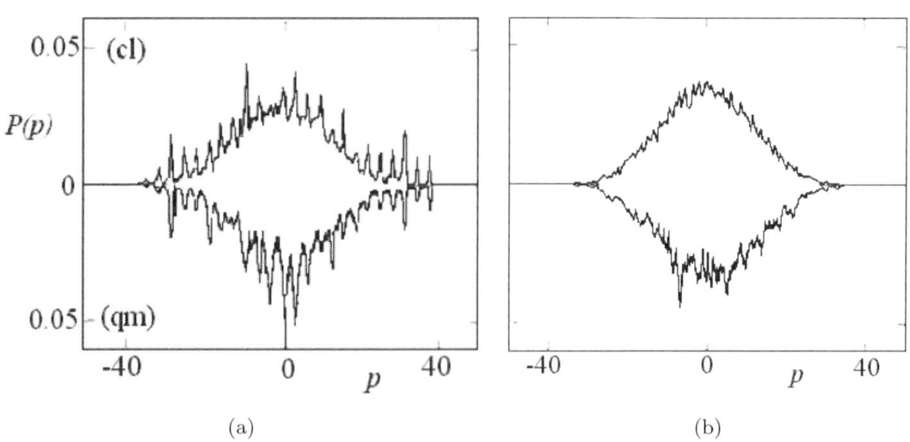

Fig. 2. Mirror images of the classical and quantum-mechanical momentum distributions, $P(p)$, plotted for (a) $\lambda = 1.7$ within the acceleration window, and (b) $\lambda = 2.4$ outside the acceleration window. The spikes in the momentum distribution for $\lambda = 1.7$ are a signature of coherent accelerated dynamics. The initial width of the momentum distribution is $\Delta p = 0.5$ and the probability distributions are recorded after a scaled propagation time $t = 500$. The initial probability distributions have variance $\Delta p = 0.5$, which fulfills the minimum uncertainty relation.

distribution originating from the regions of the phase space that do not support accelerated dynamics undergo diffusive dynamics.

From the numerical results of Fig. 2, we conjecture that the spikes are well described by a sequence of Gaussian distributions separated by a distance π, in

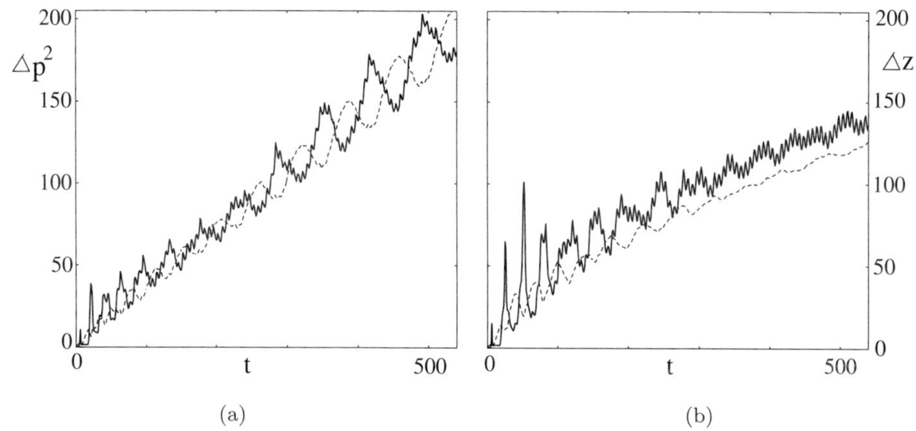

Fig. 3. (a) Square of the momentum variance (dark lines) and the coordinate space variance (gray lines) as a function of time for $\lambda = 1.7$. The coherent acceleration results in a breathing of the atomic wave packet, as evidenced by the out-of-phase oscillations of the variances. (b) Dynamics for $\lambda = 2.4$, a modulation strength that does not result in coherent acceleration. Note the absence of breathing in that case. Same parameters as in Fig. 2.

both momentum space and coordinate space. We can therefore express the complete time-evolved wave packet composed of a series of sharply peaked Gaussian distributions superposed to a broad background due to diffusive dynamics, such that

$$P(p) = \mathcal{N} e^{-p^2/4\Delta p^2} \sum_{n=-\infty}^{\infty} e^{-(p-n\pi)^2/4\epsilon^2}, \tag{3}$$

where $\epsilon \ll \Delta p$ and $\mathcal{N}$ is a normalization constant.

Further insight into the quantum acceleration of the atomic wave packet is obtained by studying its temporal evolution. We find that within the window of acceleration the atomic wave packet displays a linear growth in the square of the momentum variance and in the coordinate space variance. Figure 3 illustrates that for modulation strengths within the acceleration window, the growth in the square of the momentum variance displays oscillations of increasing periodicity whereas the variance in coordinate space follows with a phase difference of 180°. The out-of-phase oscillatory evolutions of Δp^2 and Δz indicate a breathing of the wave packet and is a signature of the coherence in accelerated dynamics. As a final point we note that outside of the acceleration window the linear growth in the square of the momentum variance, a consequence of normal diffusion, translates into a t^α law, with $\alpha < 1$, which is a consequence of anomalous diffusion.

4. Summary

We have investigated the classical and quantum evolution of atoms in a Fermi accelerator beyond the regime of dynamical localization where diffusive behavior occurs in both the classical and quantum domains. We have identified the conditions leading to the coherent acceleration of the atoms, and found signatures of the behavior both for an ensemble of classical particles and for a quantum wave packet. A quantum wave packet with a broad initial variance, restricted by the Heisenberg uncertainty principle, results in the coherent acceleration occurring on top of a diffusive background.

Acknowledgment

This work is supported in part by the US Office of Naval Research, the National Science Foundation, the US Army Research Office, the Joint Services Optics Program, the National Aeronautics and Space Administration, and the J. William Fulbright Foundation.

References

1. F. Saif, *Phys. Rep.* **419** (2005) 207.
2. F. Saif, *Phys. Rep.* **425** (2006) 369.
3. E. Fermi, *Phys. Rev.* **75** (1949) 1169.
4. A. J. Lichtenberg and M. A. Lieberman, *Regular and Stochastic Motion* (Springer, Berlin, 1983).
5. A. J. Lichtenberg and M. A. Lieberman, *Regular and Chaotic Dynamics* (Springer, New York, 1992).
6. L. E. Reichl, *The Transition to Chaos in Conservative Classical Systems: Quantum Manifestations* (Springer-Verlag, Berlin, 1992).
7. G. M. Zaslavskii and B. Chirikov, *Dokl. Akad. Nauk SSSR* **159** (1964) 306.
8. G. M. Zaslavskii and B. Chirikov, *Sov. Phys. Dokl.* **9** (1965) 989.
9. L. D. Pustilnikov, *Teor. Mat. Fiz.* **57** (1983) 128.
10. L. D. Pustilnikov, *Dokl. Akad. Nauk SSSR* **292** (1987) 549.
11. L. D. Pustilnikov, *Sov. Math. Dokl.* **35** (1987) 88.
12. S. Pötting *et al.*, *Phys. Rev. A* **64** (2001) 023604.
13. A. Buchleitner *et al.*, 0501146.
14. Z.-Y. Ma *et al.*, *Phys. Rev. A* **73** (2006) 013401 and references therein.
15. F. Saif *et al.*, *Phys. Rev. A* **58** (1998) 4779.
16. F. Saif, *Phys. Lett. A* **274** (2000) 98.
17. C. G. Aminoff *et al.*, *Phys. Rev. Lett.* **71** (1993) 3083.
18. A. Steane *et al.*, *Phys. Rev. Lett.* **74** (1995) 4972.
19. L. D. Pustyl'nikov, *Trudy Moskov. Mat. Obšč. Tom* **34**(2) (1977) 1.
20. L. D. Pustyl'nikov, *Trans. Moscow Math. Soc.* **2** (1978) 1.
21. F. Benvenuto *et al.*, *Z. Phys. B* **84** (1991) 159.
22. C. R. de Oliveira, I. Guarneri and G. Casati, *Europhys. Lett.* **27** (1994) 187.
23. F. Saif and I. Rehman, *Phys. Rev. A* **75** (2007) 043610.
24. F. Saif and P. Meystre, in preparation.

GRAVITOELECTROMAGNETISM AND DARK ENERGY IN SUPERCONDUCTORS

CLOVIS JACINTO DE MATOS

ESA-HQ, European Space Agency,
8-10 rue Mario Nikis, Paris, 75015, France
Clovis.de.Matos@esa.int

A gravitomagnetic analog of the London moment in superconductors could explain the anomalous Cooper pair mass excess reported by Janet Tate. Ultimately the gravitomagnetic London moment is attributed to the breaking of the principle of general covariance in superconductors. This naturally implies nonconservation of classical energy–momentum. A possible relation with the manifestation of dark energy in superconductors is questioned.

Keywords: Principle of general covariance; gravitomagnetism; superconductivity; dark energy.

1. Introduction

In 1989 Tate et *al.*[1,2] reported a Cooper pair mass excess in niobium of 84 parts per million greater than twice the free electron mass (m_e), whereas a theoretical calculation based on the theory of relativity predicts a value that is 8 parts per million less than $2m_e$. This disagreement between theory and experiment has not been resolved so far.[3–6] Conjecturing an additional gravitomagnetic term in the Cooper pair's canonical momentum accounts for Tate's observations. This naturally leads to the conjecture that Tate's excess of mass is not real, but instead a rotating superconductor would simply exhibit a gravitomagnetic analog of the well-known magnetic London moment. However, the magnitude of this conjectured gravitomagnetic field would be 10 orders of magnitude higher than Earth's natural gravitomagnetic field (about 10^{-14} rad/s).

The electromagnetic properties of superconductors could be explained through the breaking of gauge symmetry and consequently through a massive photon in the superconductive material. A similar mechanism for gravitation involving the breaking of the principle of general covariance (PGC) in superconductors would lead to a set of Proca type equations for gravitoelectromagnetism, with an associated massive spin 1 boson to convey the gravitoelectromagnetic interaction. Requiring that the PGC be recovered from this set of equations in the case of normal matter, we find back for the case of physical systems made simultaneously of coherent and normal matter, like superconductors, the anomalously high gravitomagnetic London moment conjectured from Tate's experiment. Ultimately it appears that Tate's measurements can be expressed in terms of the ratio between the Cooper pair mass density, ρ^*_m, and the superconductor's bulk density, ρ_m.

$$\frac{m^* - m}{m} \simeq \frac{\rho^*_m}{\rho_m}, \tag{1}$$

where m^* and m are respectively the experimentally measured and the theoretically predicted Cooper pair mass.

The breaking of the PGC implies that energy–momentum would not be conserved in superconductors. Could this be related with some still unknown properties of dark energy? The investigation of the physical nature of dark energy in quantum materials is a fascinating possibility, which Beck and Mackey[7] are already exploring for the case of Josephson junctions.

2. Gravitomagnetic London Moment

Tate *et al.* used a SQUID to measure the magnetic field generated by the rotation ($\omega = 2\pi\nu\,[\mathrm{Rad/s}]$) of a thin (on the order of the London penetration depth) niobium superconductive ring, also called the London moment. Following the Ginzburg–Landau theory of superconductivity, the total magnetic flux (including the Cooper pair's current density) cancels at regular frequency intervals $\Delta\nu\,[\mathrm{s}^{-1}]$.

$$\frac{\hbar}{m^*} = 2S\Delta\nu, \tag{2}$$

where S is the area bounded by the niobium ring. Based on Eq. (2) Tate estimated the Cooper pair's mass, $m^* = 1.82203 \times 10^{-30}\,[\mathrm{Kg}]$, which is higher than the theoretically expected $m = 1.82186 \times 10^{-30}\,[\mathrm{Kg}]$, including relativistic corrections.[8]

In order to assess Tate's experiment, following DeWitt and Ross, a gravitomagnetic term is added to the Copper pair's canonical momentum,[9,10] $\mathbf{p}_s$.

$$\mathbf{p}_s = m\mathbf{v}_s + e\mathbf{A} + m\mathbf{A}_g, \tag{3}$$

where e is the Cooper pair's electric charge, $\mathbf{v}_s$ is the Cooper pair's velocity, and $\mathbf{A}$ and $\mathbf{A}_g$ are respectively the magnetic and gravitomagnetic vector potentials.

Using again Ginzburg–Landau theory, we find that the zero-flux condition, Eq. (2), depends now also on a gravitomagnetic component.[11,12]

$$\frac{\hbar}{m} = 2S\left(\Delta\nu + \frac{\Delta B_g}{4\pi}\right). \tag{4}$$

Putting Eq. (2) into Eq. (4), we find that the gravitomagnetic variation needed to account for Tate's anomalous excess of mass[13] is $\Delta B_g = 1.65 \times 10^{-5}\,[\mathrm{rad/s}]$.

Subtracting the classical Cooper pair's canonical momentum, $\mathbf{p}^*_s = m^*\mathbf{v}_s + e\mathbf{A}$, from the generalized one, Eq. (3), and taking the rotational, we obtain the gravitomagnetic analog of the London moment in superconductors, which we call the "gravitomagnetic London moment."[14]

$$\mathbf{B}_g = \frac{m^* - m}{m} 2\boldsymbol{\omega} = \frac{\Delta m}{m} 2\boldsymbol{\omega}. \tag{5}$$

Therefore we conclude that the Cooper pair's mass does not increase, but instead when a superconductor is set rotating it generates simultaneously a (homogeneous) magnetic field (London moment) with a (homogeneous) gravitomagnetic field (gravitomagnetic London moment). If the latest phenomenon is neglected it is naturally interpreted, as Tate did, as being an anomalous excess of mass of the Cooper pairs.

The gravitomagnetic field generated by the rotating niobium ring in Tate's experiment would then be

$$\mathbf{B}_g = 1.84 \times 10^{-4} \boldsymbol{\omega}, \tag{6}$$

which is very large even for small angular velocities compared to classical astronomical sources, but cannot be ruled out based on the experimental results achieved so far.[15] However, how can we explain this conjecture? We will see that the answer passes through the investigation of the validity of the principle of general covariance for superconductors.

3. Spontaneous Breaking of Gauge Invariance in Superconductors

The properties of superconductors (zero resistivity, Meissner effect, London moment, flux quantization, Josephson effect, etc.) can be understood likewise through spontaneous breaking of electromagnetic gauge invariance when the material is in the superconductive phase.[16,17] In field theory, this symmetry breaking leads to massive photons via the Higgs mechanism. In this case the Maxwell equations transform to the so-called Maxwell–Proca equations, which are given by

$$\nabla \mathbf{E} = \frac{\rho^*}{\epsilon_0} - \frac{1}{\lambda_\gamma^2}\phi, \tag{7}$$

$$\nabla \mathbf{B} = 0, \tag{8}$$

$$\nabla \times \mathbf{E} = -\dot{\mathbf{B}}, \tag{9}$$

$$\nabla \times \mathbf{B} = \mu_0 \rho^* \mathbf{v}_s + \frac{1}{c^2}\dot{\mathbf{E}} - \frac{1}{\lambda_\gamma^2}\mathbf{A}, \tag{10}$$

where $\mathbf{E}$ is the electric field, $\mathbf{B}$ is the magnetic field, ϵ_0 is the vacuum electric permittivity, $\mu_0 = 1/\epsilon_0 c^2$ is the vacuum magnetic permeability, ϕ is the scalar electric potential, $\mathbf{A}$ is the magnetic vector potential, ρ^* is the Cooper pair's fluid electric density, $\mathbf{v}_S$ is the cooper pair's velocity, and $\lambda_\gamma = \hbar/m_\gamma c$ is the photon's Compton wavelength, which is equal to the London penetration depth $\lambda_L = \sqrt{\frac{m}{\mu_0 \rho^* e}}$.

Taking the rotational of Eq. (10) and neglecting the term coming from the displacement current, we get the following equation for the magnetic field:

$$\nabla^2 \mathbf{B} + \frac{1}{\lambda_\gamma^2}\mathbf{B} = \frac{1}{\lambda_L^2}\frac{m}{e}2\boldsymbol{\omega}. \tag{11}$$

Solving Eq. (11) for a one-dimensional case, we obtain the Meissner effect and the London moment.

$$B = B_0 e^{-x/\lambda_\gamma} + 2\omega\frac{m}{e}\left(\frac{\lambda_\gamma}{\lambda_L}\right)^2. \tag{12}$$

Following the argument from Becker *et al.*[18] and London,[19] the London moment is developed by a net current that is lagging behind the positive lattice matrix. The Cooper pair current density direction sign has to show in the opposite direction to the angular velocity of the superconducting bulk. This is important as the London moment in all measurements, due to the negative charge of the Cooper pair, shows in the same direction as the angular velocity. Having $\lambda_\gamma = \lambda_L$ we finally get

$$B = B_0 e^{-x/\lambda_L} - 2\omega\frac{m}{e}. \tag{13}$$

4. Spontaneous Breaking of the Principle of General Covariance in Superconductors

General relativity is founded on the *principle of equivalence*, which rests on the equality between the inertial and the gravitational mass of any physical system, and formulates that *at every space-time point in an arbitrary gravitational field it is possible to choose a "locally inertial coordinate system" such that, within a sufficiently small region of the point in question, the laws of nature take the same form as in unaccelerated Cartesian coordinate systems in the absence of gravity.* In other words, the inertial frames, i.e. the "freely falling coordinate systems," are indeed determined by the local gravitational field, which arises from all the matter in the Universe, far and near. However, once in an inertial frame, the laws of motion are completely unaffected by the presence of nearby masses, either gravitationally or in any other way.

Following Steven Weinberg, the principle of general covariance (PGC) is an alternative version of the principle of equivalence,[20] which is very appropriate for investigating the field equations for electromagnetism and gravitation. It states that *a physical equation holds in a general gravitational field, if two conditions are met*:

(i) *The equation holds in the absence of gravitation; that is, it agrees with the laws of special relativity when the metric tensor* $g_{\alpha\beta}$ *equals the Minkowsky tensor* $\eta_{\alpha\beta}$ *and when the affine connection* $\Gamma^\alpha_{\beta\gamma}$ *vanishes.*
(ii) *The equation is generally covariant; that is, it preserves its form under a general coordinate transformation* $x \to x'$.

It should be stressed that general covariance by itself is empty of physical content. The significance of the PGC lies in its statement about the effects of gravitation, that a physical equation by virtue of its general covariance will be true in a gravitational field if it is true in the absence of gravitation. The PGC is not an invariance principle, like the principle of Galilean or special relativity, but is instead a statement about the effects of gravitation, and about nothing else. In particular, general covariance does not imply Lorentz invariance. Any physical principle such as the PGC, which takes the form of an invariance principle but whose content is actually limited by the interaction of one particular field, is called a dynamic symmetry. As discussed above, local gauge invariance, which governs the electromagnetic interaction, is another important dynamical symmetry. We can actually say that the PGC in general relativity is the analog of the principle of gauge invariance in electrodynamics.

The Maxwell–Proca equations for electromagnetism Eqs. (7)–(10), which apply in a superconductor, are not gauge-invariant, just as they are not generally covariant. If we assume that the PGC is spontaneously broken in a superconductor, like gauge invariance is, the weak field approximation of Einstein field equations would lead to the following set of Proca equations for gravitoelectromagnetism, which contains a spin 1 massive boson, called a graviphoton, to convey the gravitoelectromagnetic interaction[21,22]:

$$\nabla \mathbf{g} = -\frac{\rho_m^*}{\epsilon_{0g}} - \frac{1}{\lambda_g^2}\phi_g, \tag{14}$$

$$\nabla \mathbf{B}_g = 0, \tag{15}$$

$$\nabla \times \mathbf{g} = -\dot{\mathbf{B}}_g, \tag{16}$$

$$\nabla \times \mathbf{B}_g = -\mu_{0g}\rho_m^*\mathbf{v}_s + \frac{1}{c^2}\dot{\mathbf{g}} - \frac{1}{\lambda_g^2}\mathbf{A}_g, \tag{17}$$

where $\mathbf{g}$ is the gravitational field, $\mathbf{B}_g$ is the gravitomagnetic field, $\epsilon_{0g} = 1/4\pi G$ is the vacuum gravitational permittivity, $\mu_{0g} = 4\pi G/c^2$ is the vacuum gravitomagnetic permeability, ϕ_g is the scalar gravitational potential, $\mathbf{A}_g$ is the gravitomagnetic vector potential, ρ_m^* is the Cooper pair's mass density, $\mathbf{v}_S$ is the cooper pair's velocity, and $\lambda_g = \hbar/m_g c$ is the Compton wavelength of the graviphoton.

Taking the gradient of Eq. (14), and the rotational of Eq. (17), and solving the resulting differential equations for the one-dimensional case, we find respectively the form of the principle of equivalence and of the gravitomagnetic Larmor theorem[23] in superconductive cavities.

$$\mathbf{g} = -\mathbf{a}\mu_{0g}\rho_m^*\lambda_g^2, \tag{18}$$

$$\mathbf{B}_g = 2\boldsymbol{\omega}\mu_{0g}\rho_m^*\lambda_g^2, \tag{19}$$

where for Eq. (19) we had to introduce Becker's argument that the Cooper pairs are lagging behind the lattice so that the current is flowing in the opposite direction of ω.

In order to find a phenomenological law for the graviphoton wavelength,[24] we require that from Eqs. (18) and (19) the PGC be restored, in the case of normal matter. In that case the mass density reduces to the materials bulk density, ρ_m, and no condensate phase is present within the material.

$$\mathbf{g} = -\mathbf{a}\mu_{0g}\rho_m\lambda_g^2, \tag{20}$$

$$\mathbf{B}_g = 2\boldsymbol{\omega}\mu_{0g}\rho_m\lambda_g^2. \tag{21}$$

Since normal matter complies with the PGC, Eqs. (20) and (21) must reduce to

$$\mathbf{g} = -\mathbf{a}, \tag{22}$$

$$\mathbf{B}_g = 2\boldsymbol{\omega}. \tag{23}$$

Comparing Eqs. (22) and (23) with Eqs. (20) and (21), we find that the graviphoton Compton wavelength must be inversely proportional to the local density of the bulk material mass:

$$\frac{1}{\lambda_g^2} = \mu_{0g}\rho_m. \tag{24}$$

Putting Eq. (24) into Eqs. (18) and (19), we get

$$\mathbf{g} = -\mathbf{a}\frac{\rho_m^*}{\rho_m}, \tag{25}$$

$$\mathbf{B}_g = 2\boldsymbol{\omega}\frac{\rho_m^*}{\rho_m}, \tag{26}$$

which clearly indicate a breaking of general covariance. Notice that in the case of Bose–Einstein condensates (BEC's) we have only one single condensate phase in our material ($\rho_m^* = \rho_m$), implying that the PGC is not violated in BEC's.

Comparing Eq. (26) with Eqs. (5) and (6), we can explain the additional gravitomagnetic term in the Cooper pairs canonical momentum, in the function of the mass density ratio of the coherent and the normal phase.

$$9.2 \times 10^{-5} = \frac{m^* - m}{m} \simeq \frac{\rho_m^*}{\rho_m} = 3.9 \times 10^{-6}. \tag{27}$$

The numerical values in Eq. (27) correspond to the case of niobium in Tate's experimental conditions.

5. Dark Energy in Superconductors?

It is well known that breaking of general covariance leads to nonconservation of energy–momentum (in the covariant sense).[20] Would that mean that in a superconductor we could observe some manifestation of dark energy?

Presently the physical nature of dark energy is unknown. What is clear is that various astronomical observations (supernovae, CMB fluctuations, large-scale structure) provide rather convincing evidence that around 73% of the energy contents of the Universe is a rather homogeneous form of energy, the so-called dark energy.

A large number of theoretical models exist for dark energy, but an entirely convincing theoretical breakthrough has not yet been achieved. Popular models are based on quintessence fields, phantom fields, quintom fields, Born–Infeld quantum condensates, the Chaplygin gas, fields with nonstandard kinetic terms, and possible links between the cosmological constant and the graviton mass, to name just a few (see for example Refs. 25 and 26 for reviews). All of these approaches contain "new physics" in one way or another, though at different levels. However, it is clear that the number of possible dark energy models that are based on new physics is infinite. Only experiments will ultimately be able to confirm or refute the various theoretical constructs.[27]

Beck is currently exploring the possibility that vacuum fluctuations, allowed by the uncertainty relation, in the Josephson junction create dark energy. This is *a priori* the simplest explanation for dark energy. Assuming that the total vacuum energy density associated with zero-point fluctuations cannot exceed the presently measured dark energy density of the Universe, Beck predicts an upper cutoff frequency of $\nu_c = (1.69 \pm 0.05) \times 10^{12}$ Hz for the measured frequency spectrum of the zero-point fluctuations in the Josephson junction. The largest frequencies that have been reached in the experiments are of the same order of magnitude as ν_c and provide a lower bound on the dark energy density of the Universe. If this is confirmed by future experiments, how would that be related with the photon and graviphoton mass in superconductors? Where does the mass of these particles come from in the superconductive material? Finally, can we break general covariance in a superconductor without violating energy–momentum conservation, if a superconductor contains a new form of energy?

6. Conclusion

The close analogy between the principle of general covariance and gauge invariance allows us to investigate the gravitoelectromagnetic properties of quantum materials in the framework of massive gravitoelectromagnetic Proca equations. We find that the breaking of the PGC in superconductors leads to a gravitomagnetic London moment and an associated additional gravitomagnetic term in the Cooper pair's canonical momentum, which can explain the anomalous excess of mass of Cooper pairs reported by Tate.

The breaking of the PGC in superconductors implies the nonequivalence between a rigid reference frame made with superconductive walls (superconductive cavity), being uniformly accelerated in a gravitational field free region,

$$\mathbf{g} = -\mathbf{a}\left(1 + \frac{\rho_m^*}{\rho_m}\right), \quad \mathbf{B}_g = 2\boldsymbol{\omega}\left(1 + \frac{\rho_m^*}{\rho_m}\right), \tag{28}$$

and a classical rigid reference frame (made with normal matter) in a similar situation,

$$\mathbf{g} = -\mathbf{a}, \quad \mathbf{B}_g = 2\boldsymbol{\omega}. \tag{29}$$

However, breaking the PGC leads to a violation of the law of conservation of energy–momentum. It is not clear yet if this would be a sign for some manifestation of dark energy in superconductive materials. However, it is worth further investigation.

Acknowledgments

I am grateful to Profs. Orfeu Bertolami, John Moffat, Francis Everitt and Alan Kostelecky for fruitful discussions and pertinent comments. I would like also to thank Prof. Raymond Chiao for his encouragement and for being a rich source of inspiration for the present work. My profound gratitude also goes to Dr Slava Turyshev and to all the organizers of the "From Quantum to Cosmos" conference for the fantastic forum of debates which they succeeded in establishing during the conference. Many thanks also to Dr. Martin Tajmar for many stimulating discussions.

References

1. J. Tate *et al.*, *Phys. Rev. Lett.* **62** (1989) 845.
2. J. Tate *et al.*, *Phys. Rev. B* **42** (1990) 7885.
3. M. Liu, *Phys. Rev. Lett.* **81** (1998) 3223.
4. Y. Jiang and M. Liu, *Phys. Rev. B* **63** (2001) 184506.
5. H. Capellmann, *Euro. Phys. J. B* **25** (2002) 25.
6. K. Capelle and E. K. U. Gross, *Phys. Rev. B* **59** (1999) 7140.
7. C. Beck and M. C. Mackey, *Phys. Lett. B* **605** (2005) 295.
8. B. Cabrera, H. Gutfreund and W. A. Little, *Phys. Rev. B* **25** (1982) 6644.
9. B. S. DeWitt, *Phys. Rev. Lett.* **16** (1966) 1092.
10. D. K. Ross, *J. Phys. A* **16** (1983) 1092.
11. C. Kiefer and C. Weber, *Ann. Phys.* **14** (2005) 253.
12. O. Bertolami and M. Tajmar, Gravity control and possible influence on space propulsion, in *ESA Contract Report CR(P)4365* (2002).
13. M. Tajmar and C. J. de Matos, *Physica C* **385** (2003) 551.
14. M. Tajmar and C. J. de Matos, *Physica C* **420** (2005) 56.
15. M. Tajmar *et al.*, Search for frame-dragging-like signals close to spinning superconductors, in *Time & Matter Conf. Proc.*, eds. D. Zavrtanik and S. Stanic (World Scientific, 2007) [gr-qc/0707.3806].
16. L. H. Ryder, *Quantum Field Theory*, 2nd edn. (Cambridge University Press, 1996), p. 296.
17. S. Weinberg, Modern Applications, The Quantum Theory of Fields, Vol. 2 (Cambridge University Press, 1996), p. 332.
18. R. Becker, G. Heller and F. Sauter, *Z. Phys.* **85** (1933) 772.
19. F. London, *Superfluids* (John Wiley and Sons, New York, 1950).
20. S. Weinberg, *Gravitation and Cosmology Principles and Applications of the General Theory of Relativity* (John Wiley and Sons, New York, 1972), pp. 91, 111, 361.
21. J. Argyris, *Aust. J. Phys.* **50** (1997) 879.
22. C. J. de Matos and M. Tajmar, *Physica C* **432** (2005) 167.
23. B. Mashhoon, *Phys. Lett. A* **173** (1993) 347.
24. M. Tajmar and C. J. de Matos, gr-qc/0603032.
25. P. J. E. Peebles and B. Ratra, *Rev. Mod. Phys.* **75** (2003) 559.
26. T. Padmanabhan, *Phys. Rep.* **380** (2005) 235.
27. C. Beck, *J. Phys. Conf. Ser.* **31** (2006) 123.

AUTHOR INDEX

SUBJECT INDEX